System Dynamics

Fourth Edition

Katsuhiko Ogata
University of Minnesota

Upper Saddle River, NJ 07458

Library of Congress Cataloging-in-Publication Data on file.

Vice President and Editorial Director, ECS: *Marcia J. Horton*
Acquisitions Editor: *Laura Fischer*
Vice President and Director of Production and Manufacturing, ESM: *David W. Riccardi*
Executive Managing Editor: *Vince O'Brien*
Managing Editor: *David A. George*
Production Editor: *Scott Disanno*
Director of Creative Services: *Paul Belfanti*
Creative Director: *Jayne Conte*
Art Editor: *Greg Dulles*
Manufacturing Manager: *Trudy Pisciotti*
Manufacturing Buyer: *Lisa McDowell*
Marketing Manager: *Holly Stark*

 © 2004, 1998, 1992, 1978 Pearson Education, Inc.
Pearson Prentice Hall
Pearson Education, Inc.
Upper Saddle River, New Jersey 07458

Printed in the United States of America

20 19 18 17 16 15 14 13

ISBN 0-13-142462-9

Pearson Education Ltd., *London*
Pearson Education Australia Pty. Ltd., *Sydney*
Pearson Education Singapore, Pte. Ltd.
Pearson Education North Asia Ltd., *Hong Kong*
Pearson Education Canada Inc., *Toronto*
Pearson Educatión de Mexico, S.A. de C.V.
Pearson Education—Japan, *Tokyo*
Pearson Education Malaysia, Pte. Ltd.
Pearson Education, Inc., *Upper Saddle River, New Jersey*

Contents

4 TRANSFER-FUNCTION APPROACH TO MODELING DYNAMIC SYSTEMS   106

5 STATE-SPACE APPROACH TO MODELING DYNAMIC SYSTEMS   169

6 ELECTRICAL SYSTEMS AND ELECTROMECHANICAL SYSTEMS   251

7 FLUID SYSTEMS AND THERMAL SYSTEMS   323

Preface

A course in system dynamics that deals with mathematical modeling and response analyses of dynamic systems is required in most mechanical and other engineering curricula. This book is written as a textbook for such a course. It is written at the junior level and presents a comprehensive treatment of modeling and analyses of dynamic systems and an introduction to control systems.

Prerequisites for studying this book are first courses in linear algebra, introductory differential equations, introductory vector-matrix analysis, mechanics, circuit analysis, and thermodynamics. Thermodynamics may be studied simultaneously.

Main revisions made in this edition are to shift the state space approach to modeling dynamic systems to Chapter 5, right next to the transfer function approach to modeling dynamic systems, and to add numerous examples for modeling and response analyses of dynamic systems. All plottings of response curves are done with MATLAB. Detailed MATLAB programs are provided for MATLAB works presented in this book.

This text is organized into 11 chapters and four appendixes. Chapter 1 presents an introduction to system dynamics. Chapter 2 deals with Laplace transforms of commonly encountered time functions and some theorems on Laplace transform that are useful in analyzing dynamic systems. Chapter 3 discusses details of mechanical elements and simple mechanical systems. This chapter includes introductory discussions of work, energy, and power.

Chapter 4 discusses the transfer function approach to modeling dynamic systems. Transient responses of various mechanical systems are studied and MATLAB is used to obtain response curves. Chapter 5 presents state space modeling of dynamic systems. Numerous examples are considered. Responses of systems in the state space form are discussed in detail and response curves are obtained with MATLAB.

Chapter 6 treats electrical systems and electromechanical systems. Here we included mechanical–electrical analogies and operational amplifier systems. Chapter 7 deals with mathematical modeling of fluid systems (such as liquid-level systems, pneumatic systems, and hydraulic systems) and thermal systems. A linearization technique for nonlinear systems is presented in this chapter.

Chapter 8 deals with the time-domain analysis of dynamic systems. Transient-response analysis of first-order systems, second-order systems, and higher order systems is discussed in detail. This chapter includes analytical solutions of state-space equations. Chapter 9 treats the frequency-domain analysis of dynamic systems. We first present the sinusoidal transfer function, followed by vibration analysis of mechanical systems and discussions on dynamic vibration absorbers. Then we discuss modes of vibration in two or more degrees-of-freedom systems.

Chapter 10 presents the analysis and design of control systems in the time domain. After giving introductory materials on control systems, this chapter discusses transient-response analysis of control systems, followed by stability analysis, root-locus analysis, and design of control systems. Finally, we conclude this chapter by giving tuning rules for PID controllers. Chapter 11 treats the analysis and design of control systems in the frequency domain. Bode diagrams, Nyquist plots, and the Nyquist stability criterion are discussed in detail. Several design problems using Bode diagrams are treated in detail. MATLAB is used to obtain Bode diagrams and Nyquist plots.

Appendix A summarizes systems of units used in engineering analyses. Appendix B provides useful conversion tables. Appendix C reviews briefly a basic vector-matrix algebra. Appendix D gives introductory materials on MATLAB. If the reader has no prior experience with MATLAB, it is recommended that he/she study Appendix D before attempting to write MATLAB programs.

Throughout the book, examples are presented at strategic points so that the reader will have a better understanding of the subject matter discussed. In addition, a number of solved problems (A problems) are provided at the end of each chapter, except Chapter 1. These problems constitute an integral part of the text. It is suggested that the reader study all these problems carefully to obtain a deeper understanding of the topics discussed. Many unsolved problems (B problems) are also provided for use as homework or quiz problems. An instructor using this text for his/her system dynamics course may obtain a complete solutions manual for B problems from the publisher.

Most of the materials presented in this book have been class tested in courses in the field of system dynamics and control systems in the Department of Mechanical Engineering, University of Minnesota over many years.

If this book is used as a text for a quarter-length course (with approximately 30 lecture hours and 18 recitation hours), Chapters 1 through 7 may be covered. After studying these chapters, the student should be able to derive mathematical models for many dynamic systems with reasonable simplicity in the forms of transfer function or state-space equation. Also, he/she will be able to obtain computer solutions of system responses with MATLAB. If the book is used as a text for a semester-length course (with approximately 40 lecture hours and 26 recitation hours), then the first nine chapters may be covered or, alternatively, the first seven chapters plus Chapters 10 and 11 may be covered. If the course devotes 50 to 60 hours to lectures, then the entire book may be covered in a semester.

Finally, I wish to acknowledge deep appreciation to the following professors who reviewed the third edition of this book prior to the preparation of this new edition: R. Gordon Kirk (Virginia Institute of Technology), Perry Y. Li (University of Minnesota), Sherif Noah (Texas A & M University), Mark L. Psiaki (Cornell University), and William Singhose (Georgia Institute of Technology). Their candid, insightful, and constructive comments are reflected in this new edition.

KATSUHIKO OGATA

Introduction to System Dynamics

1-1 INTRODUCTION

System dynamics deals with the mathematical modeling of dynamic systems and response analyses of such systems with a view toward understanding the dynamic nature of each system and improving the system's performance. Response analyses are frequently made through computer simulations of dynamic systems.

Because many physical systems involve various types of components, a wide variety of different types of dynamic systems will be examined in this book. The analysis and design methods presented can be applied to mechanical, electrical, pneumatic, and hydraulic systems, as well as nonengineering systems, such as economic systems and biological systems. It is important that the mechanical engineering student be able to determine dynamic responses of such systems.

We shall begin this chapter by defining several terms that must be understood in discussing system dynamics.

Systems. A *system* is a combination of components acting together to perform a specific objective. A *component* is a single functioning unit of a system. By no means limited to the realm of the physical phenomena, the concept of a system can be extended to abstract dynamic phenomena, such as those encountered in economics, transportation, population growth, and biology.

1

A system is called *dynamic* if its present output depends on past input; if its current output depends only on current input, the system is known as *static*. The output of a static system remains constant if the input does not change. The output changes only when the input changes. In a dynamic system, the output changes with time if the system is not in a state of equilibrium. In this book, we are concerned mostly with dynamic systems.

Mathematical models.   Any attempt to design a system must begin with a prediction of its performance before the system itself can be designed in detail or actually built. Such prediction is based on a mathematical description of the system's dynamic characteristics. This mathematical description is called a *mathematical model*. For many physical systems, useful mathematical models are described in terms of differential equations.

Linear and nonlinear differential equations.   Linear differential equations may be classified as linear, time-invariant differential equations and linear, time-varying differential equations.

A *linear, time-invariant differential equation* is an equation in which a dependent variable and its derivatives appear as linear combinations. An example of such an equation is

$$\frac{d^2x}{dt^2} + 5\frac{dx}{dt} + 10x = 0$$

Since the coefficients of all terms are constant, a linear, time-invariant differential equation is also called a *linear, constant-coefficient differential equation*.

In the case of a *linear, time-varying differential equation*, the dependent variable and its derivatives appear as linear combinations, but a coefficient or coefficients of terms may involve the independent variable. An example of this type of differential equation is

$$\frac{d^2x}{dt^2} + (1 - \cos 2t)x = 0$$

It is important to remember that, in order to be linear, the equation must contain no powers or other functions or products of the dependent variables or its derivatives.

A differential equation is called *nonlinear* if it is not linear. Two examples of nonlinear differential equations are

$$\frac{d^2x}{dt^2} + (x^2 - 1)\frac{dx}{dt} + x = 0$$

and

$$\frac{d^2x}{dt^2} + \frac{dx}{dt} + x + x^3 = \sin \omega t$$

Linear systems and nonlinear systems.   For linear systems, the equations that constitute the model are linear. In this book, we shall deal mostly with linear systems that can be represented by linear, time-invariant ordinary differential equations.

The most important property of linear systems is that the *principle of superposition* is applicable. This principle states that the response produced by simultaneous applications of two different forcing functions or inputs is the sum of two individual responses. Consequently, for linear systems, the response to several inputs can be calculated by dealing with one input at a time and then adding the results. As a result of superposition, complicated solutions to linear differential equations can be derived as a sum of simple solutions.

In an experimental investigation of a dynamic system, if cause and effect are proportional, thereby implying that the principle of superposition holds, the system can be considered linear.

Although physical relationships are often represented by linear equations, in many instances the actual relationships may not be quite linear. In fact, a careful study of physical systems reveals that so-called linear systems are actually linear only within limited operating ranges. For instance, many hydraulic systems and pneumatic systems involve nonlinear relationships among their variables, but they are frequently represented by linear equations within limited operating ranges.

For nonlinear systems, the most important characteristic is that the principle of superposition is *not* applicable. In general, procedures for finding the solutions of problems involving such systems are extremely complicated. Because of the mathematical difficulty involved, it is frequently necessary to linearize a nonlinear system near the operating condition. Once a nonlinear system is approximated by a linear mathematical model, a number of linear techniques may be used for analysis and design purposes.

Continuous-time systems and discrete-time systems.   Continuous-time systems are systems in which the signals involved are continuous in time. These systems may be described by differential equations.

Discrete-time systems are systems in which one or more variables can change only at discrete instants of time. (These instants may specify the times at which some physical measurement is performed or the times at which the memory of a digital computer is read out.) Discrete-time systems that involve digital signals and, possibly, continuous-time signals as well may be described by difference equations after the appropriate discretization of the continuous-time signals.

The materials presented in this text apply to continuous-time systems; discrete-time systems are not discussed.

1–2 MATHEMATICAL MODELING OF DYNAMIC SYSTEMS

Mathematical modeling.   Mathematical modeling involves descriptions of important system characteristics by sets of equations. By applying physical laws to a specific system, it may be possible to develop a mathematical model that describes the dynamics of the system. Such a model may include unknown parameters, which must then be evaluated through actual tests. Sometimes, however, the physical laws governing the behavior of a system are not completely defined, and formulating a mathematical model may be impossible. If so, an experimental modeling process can be used. In this process, the system is subjected to a set of known inputs, and its outputs are measured. Then a mathematical model is derived from the input–output relationships obtained.

Simplicity of mathematical model versus accuracy of results of analysis. In attempting to build a mathematical model, a compromise must be made between the simplicity of the model and the accuracy of the results of the analysis. It is important to note that the results obtained from the analysis are valid only to the extent that the model approximates a given physical system.

In determining a reasonably simplified model, we must decide which physical variables and relationships are negligible and which are crucial to the accuracy of the model. To obtain a model in the form of linear differential equations, any distributed parameters and nonlinearities that may be present in the physical system must be ignored. If the effects that these ignored properties have on the response are small, then the results of the analysis of a mathematical model and the results of the experimental study of the physical system will be in good agreement. Whether any particular features are important may be obvious in some cases, but may, in other instances, require physical insight and intuition. Experience is an important factor in this connection.

Usually, in solving a new problem, it is desirable first to build a simplified model to obtain a general idea about the solution. Afterward, a more detailed mathematical model can be built and used for a more complete analysis.

Remarks on mathematical models. The engineer must always keep in mind that the model he or she is analyzing is an approximate mathematical description of the physical system; it is not the physical system itself. In reality, no mathematical model can represent any physical component or system precisely. Approximations and assumptions are always involved. Such approximations and assumptions restrict the range of validity of the mathematical model. (The degree of approximation can be determined only by experiments.) So, in making a prediction about a system's performance, any approximations and assumptions involved in the model must be kept in mind.

Mathematical modeling procedure. The procedure for obtaining a mathematical model for a system can be summarized as follows:

1. Draw a schematic diagram of the system, and define variables.
2. Using physical laws, write equations for each component, combine them according to the system diagram, and obtain a mathematical model.
3. To verify the validity of the model, its predicted performance, obtained by solving the equations of the model, is compared with experimental results. (The question of the validity of any mathematical model can be answered only by experiment.) If the experimental results deviate from the prediction

to a great extent, the model must be modified. A new model is then derived and a new prediction compared with experimental results. The process is repeated until satisfactory agreement is obtained between the predictions and the experimental results.

1–3 ANALYSIS AND DESIGN OF DYNAMIC SYSTEMS

This section briefly explains what is involved in the analysis and design of dynamic systems.

Analysis.   *System analysis* means the investigation, under specified conditions, of the performance of a system whose mathematical model is known.

The first step in analyzing a dynamic system is to derive its mathematical model. Since any system is made up of components, analysis must start by developing a mathematical model for each component and combining all the models in order to build a model of the complete system. Once the latter model is obtained, the analysis may be formulated in such a way that system parameters in the model are varied to produce a number of solutions. The engineer then compares these solutions and interprets and applies the results of his or her analysis to the basic task.

It should always be remembered that deriving a reasonable model for the complete system is the most important part of the entire analysis. Once such a model is available, various analytical and computer techniques can be used to analyze it. The manner in which analysis is carried out is independent of the type of physical system involved—mechanical, electrical, hydraulic, and so on.

Design.   *System design* refers to the process of finding a system that accomplishes a given task. In general, the design procedure is not straightforward and will require trial and error.

Synthesis.   By *synthesis*, we mean the use of an explicit procedure to find a system that will perform in a specified way. Here the desired system characteristics are postulated at the outset, and then various mathematical techniques are used to synthesize a system having those characteristics. Generally, such a procedure is completely mathematical from the start to the end of the design process.

Basic approach to system design.   The basic approach to the design of any dynamic system necessarily involves trial-and-error procedures. Theoretically, a synthesis of linear systems is possible, and the engineer can systematically determine the components necessary to realize the system's objective. In practice, however, the system may be subject to many constraints or may be nonlinear; in such cases, no synthesis methods are currently applicable. Moreover, the features of the components may not be precisely known. Thus, trial-and-error techniques are almost always needed.

Design procedures.   Frequently, the design of a system proceeds as follows: The engineer begins the design procedure knowing the specifications to be met and the dynamics of the components, the latter of which involve design parameters. The specification may be given in terms of both precise numerical values and vague qualitative descriptions. (Engineering specifications normally include statements on such factors as cost, reliability, space, weight, and ease of maintenance.) It is important to note that the specifications may be changed as the design progresses, for detailed analysis may reveal that certain requirements are impossible to meet. Next, the engineer will apply any applicable synthesis techniques, as well as other methods, to build a mathematical model of the system.

Once the design problem is formulated in terms of a model, the engineer carries out a mathematical design that yields a solution to the mathematical version of the design problem. With the mathematical design completed, the engineer simulates the model on a computer to test the effects of various inputs and disturbances on the behavior of the resulting system. If the initial system configuration is not satisfactory, the system must be redesigned and the corresponding analysis completed. This process of design and analysis is repeated until a satisfactory system is found. Then a prototype physical system can be constructed.

Note that the process of constructing a prototype is the reverse of mathematical modeling. The prototype is a physical system that represents the mathematical model with reasonable accuracy. Once the prototype has been built, the engineer tests it to see whether it is satisfactory. If it is, the design of the prototype is complete. If not, the prototype must be modified and retested. The process continues until a satisfactory prototype is obtained.

1-4 SUMMARY

From the point of view of analysis, a successful engineer must be able to obtain a mathematical model of a given system and predict its performance. (The validity of a prediction depends to a great extent on the validity of the mathematical model used in making the prediction.) From the design standpoint, the engineer must be able to carry out a thorough performance analysis of the system before a prototype is constructed.

The objective of this book is to enable the reader (1) to build mathematical models that closely represent behaviors of physical systems and (2) to develop system responses to various inputs so that he or she can effectively analyze and design dynamic systems.

Outline of the text.   Chapter 1 has presented an introduction to system dynamics. Chapter 2 treats Laplace transforms. We begin with Laplace transformation of simple time functions and then discuss inverse Laplace transformation. Several useful theorems are derived. Chapter 3 deals with basic accounts of mechanical systems. Chapter 4 presents the transfer-function approach to modeling dynamic systems. The chapter discusses various types of mechanical systems. Chapter 5 examines the state-space approach to modeling dynamic systems. Various types of mechanical systems are considered. Chapter 6 treats electrical systems and electromechanical systems, including operational-amplifier systems. Chapter 7 deals with fluid systems, such as liquid-level systems, pneumatic systems, and hydraulic systems, as well as thermal systems. A linearization technique for nonlinear systems is explored.

Chapter 8 presents time-domain analyses of dynamic systems—specifically, transient-response analyses of dynamic systems. The chapter also presents the analytical solution of the state equation. Chapter 9 treats frequency-domain analyses of dynamic systems. Among the topics discussed are vibrations of rotating mechanical systems and vibration isolation problems. Also discussed are vibrations in multi-degrees-of-freedom systems and modes of vibrations.

Chapter 10 presents the basic theory of control systems, including transient-response analysis, stability analysis, and root-locus analysis and design. Also discussed are tuning rules for PID controllers. Chapter 11 deals with the analysis and design of control systems in the frequency domain. The chapter begins with Bode diagrams and then presents the Nyquist stability criterion, followed by detailed design procedures for lead, lag, and lag–lead compensators.

Appendix A treats systems of units, Appendix B summarizes conversion tables, and Appendix C gives a brief summary of vector–matrix algebra. Appendix D presents introductory materials for MATLAB.

Throughout the book, MATLAB is used for the solution of most computational problems. Readers who have no previous knowledge of MATLAB may read Appendix D before solving any MATLAB problems presented in this text.

The Laplace Transform

2–1 INTRODUCTION

The Laplace transform is one of the most important mathematical tools available for modeling and analyzing linear systems. Since the Laplace transform method must be studied in any system dynamics course, we present the subject at the beginning of this text so that the student can use the method throughout his or her study of system dynamics.

The remaining sections of this chapter are outlined as follows: Section 2–2 reviews complex numbers, complex variables, and complex functions. Section 2–3 defines the Laplace transformation and gives Laplace transforms of several common functions of time. Also examined are some of the most important Laplace transform theorems that apply to linear systems analysis. Section 2–4 deals with the inverse Laplace transformation. Finally, Section 2–5 presents the Laplace transform approach to the solution of the linear, time-invariant differential equation.

2–2 COMPLEX NUMBERS, COMPLEX VARIABLES, AND COMPLEX FUNCTIONS

This section reviews complex numbers, complex algebra, complex variables, and complex functions. Since most of the material covered is generally included in the basic mathematics courses required of engineering students, the section can be omitted entirely or used simply for personal reference.

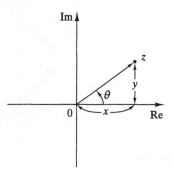

Figure 2–1 Complex plane representation of a complex number z.

Complex numbers.   Using the notation $j = \sqrt{-1}$, we can express all numbers in engineering calculations as

$$z = x + jy$$

where z is called a *complex number* and x and jy are its *real* and *imaginary parts,* respectively. Note that both x and y are real and that j is the only imaginary quantity in the expression. The complex plane representation of z is shown in Figure 2–1. (Note also that the real axis and the imaginary axis define the complex plane and that the combination of a real number and an imaginary number defines a point in that plane.) A complex number z can be considered a point in the complex plane or a directed line segment to the point; both interpretations are useful.

The magnitude, or absolute value, of z is defined as the length of the directed line segment shown in Figure 2–1. The angle of z is the angle that the directed line segment makes with the positive real axis. A counterclockwise rotation is defined as the positive direction for the measurement of angles. Mathematically,

$$\text{magnitude of } z = |z| = \sqrt{x^2 + y^2}, \qquad \text{angle of } z = \theta = \tan^{-1}\frac{y}{x}$$

A complex number can be written in rectangular form or in polar form as follows:

$$\left. \begin{aligned} z &= x + jy \\ z &= |z|(\cos\theta + j\sin\theta) \end{aligned} \right\} \text{rectangular forms}$$

$$\left. \begin{aligned} z &= |z|\ \underline{/\theta} \\ z &= |z|\ e^{j\theta} \end{aligned} \right\} \text{polar forms}$$

In converting complex numbers to polar form from rectangular, we use

$$|z| = \sqrt{x^2 + y^2}, \qquad \theta = \tan^{-1}\frac{y}{x}$$

To convert complex numbers to rectangular form from polar, we employ

$$x = |z| \cos\theta, \qquad y = |z| \sin\theta$$

Complex conjugate.   The *complex conjugate* of $z = x + jy$ is defined as

$$\overline{z} = x - jy$$

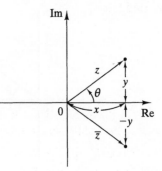

Figure 2–2 Complex number z and its complex conjugate $\bar{z}$.

The complex conjugate of z thus has the same real part as z and an imaginary part that is the negative of the imaginary part of z. Figure 2–2 shows both z and $\bar{z}$. Note that

$$z = x + jy = |z| \underline{/\theta} = |z| (\cos \theta + j \sin \theta)$$
$$\bar{z} = x - jy = |z| \underline{/-\theta} = |z| (\cos \theta - j \sin \theta)$$

Euler's theorem. The power series expansions of $\cos \theta$ and $\sin \theta$ are, respectively,

$$\cos \theta = 1 - \frac{\theta^2}{2!} + \frac{\theta^4}{4!} - \frac{\theta^6}{6!} + \cdots$$

and

$$\sin \theta = \theta - \frac{\theta^3}{3!} + \frac{\theta^5}{5!} - \frac{\theta^7}{7!} + \cdots$$

Thus,

$$\cos \theta + j \sin \theta = 1 + (j\theta) + \frac{(j\theta)^2}{2!} + \frac{(j\theta)^3}{3!} + \frac{(j\theta)^4}{4!} + \cdots$$

Since

$$e^x = 1 + x + \frac{x^2}{2!} + \frac{x^3}{3!} + \cdots$$

it follows that

$$\cos \theta + j \sin \theta = e^{j\theta}$$

This is known as *Euler's theorem.*

Using Euler's theorem, we can express the sine and cosine in complex form. Noting that $e^{-j\theta}$ is the complex conjugate of $e^{j\theta}$ and that

$$e^{j\theta} = \cos \theta + j \sin \theta$$
$$e^{-j\theta} = \cos \theta - j \sin \theta$$

we find that

$$\cos \theta = \frac{e^{j\theta} + e^{-j\theta}}{2}$$

$$\sin \theta = \frac{e^{j\theta} - e^{-j\theta}}{2j}$$

Complex algebra.   If the complex numbers are written in a suitable form, operations like addition, subtraction, multiplication, and division can be performed easily.

Equality of complex numbers.   Two complex numbers z and w are said to be equal if and only if their real parts are equal and their imaginary parts are equal. So if two complex numbers are written

$$z = x + jy, \qquad w = u + jv$$

then $z = w$ if and only if $x = u$ and $y = v$.

Addition.   Two complex numbers in rectangular form can be added by adding the real parts and the imaginary parts separately:

$$z + w = (x + jy) + (u + jv) = (x + u) + j(y + v)$$

Subtraction.   Subtracting one complex number from another can be considered as adding the negative of the former:

$$z - w = (x + jy) - (u + jv) = (x - u) + j(y - v)$$

Note that addition and subtraction can be done easily on the rectangular plane.

Multiplication.   If a complex number is multiplied by a real number, the result is a complex number whose real and imaginary parts are multiplied by that real number:

$$az = a(x + jy) = ax + jay \qquad (a = \text{real number})$$

If two complex numbers appear in rectangular form and we want the product in rectangular form, multiplication is accomplished by using the fact that $j^2 = -1$. Thus, if two complex numbers are written

$$z = x + jy, \qquad w = u + jv$$

then

$$zw = (x + jy)(u + jv) = xu + jyu + jxv + j^2yv$$
$$= (xu - yv) + j(xv + yu)$$

In polar form, multiplication of two complex numbers can be done easily. The magnitude of the product is the product of the two magnitudes, and the angle of the product is the sum of the two angles. So if two complex numbers are written

$$z = |z| \, \underline{/\theta}, \qquad w = |w| \, \underline{/\phi}$$

then

$$zw = |z||w| \, \underline{/\theta + \phi}$$

Multiplication by j.   It is important to note that multiplication by j is equivalent to counterclockwise rotation by 90°. For example, if

$$z = x + jy$$

then

$$jz = j(x + jy) = jx + j^2y = -y + jx$$

or, noting that $j = 1\ \underline{/90°}$, if

$$z = |z|\ \underline{/\theta}$$

then

$$jz = 1\ \underline{/90°}\ |z|\ \underline{/\theta} = |z|\ \underline{/\theta + 90°}$$

Figure 2–3 illustrates the multiplication of a complex number z by j.

Division.   If a complex number $z = |z|\ \underline{/\theta}$ is divided by another complex number $w = |w|\ \underline{/\phi}$, then

$$\frac{z}{w} = \frac{|z|\ \underline{/\theta}}{|w|\ \underline{/\phi}} = \frac{|z|}{|w|}\ \underline{/\theta - \phi}$$

That is, the result consists of the quotient of the magnitudes and the difference of the angles.

Division in rectangular form is inconvenient, but can be done by multiplying the denominator and numerator by the complex conjugate of the denominator. This procedure converts the denominator to a real number and thus simplifies division. For instance,

$$\frac{z}{w} = \frac{x + jy}{u + jv} = \frac{(x + jy)(u - jv)}{(u + jv)(u - jv)} = \frac{(xu + yv) + j(yu - xv)}{u^2 + v^2}$$

$$= \frac{xu + yv}{u^2 + v^2} + j\frac{yu - xv}{u^2 + v^2}$$

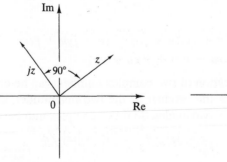

Figure 2–3   Multiplication of a complex number z by j.

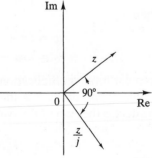

Figure 2–4   Division of a complex number z by j.

Division by j.   Note that division by j is equivalent to clockwise rotation by 90°. For example, if $z = x + jy$, then

$$\frac{z}{j} = \frac{x + jy}{j} = \frac{(x + jy)j}{jj} = \frac{jx - y}{-1} = y - jx$$

or

$$\frac{z}{j} = \frac{|z| \angle\theta}{1 \angle 90°} = |z| \angle \theta - 90°$$

Figure 2–4 illustrates the division of a complex number z by j.

Powers and roots.   Multiplying z by itself n times, we obtain

$$z^n = (|z| \angle\theta)^n = |z|^n \angle n\theta$$

Extracting the nth root of a complex number is equivalent to raising the number to the $1/n$th power:

$$z^{1/n} = (|z| \angle\theta)^{1/n} = |z|^{1/n} \left| \frac{\theta}{n} \right.$$

For instance,

$$(8.66 - j5)^3 = (10 \angle -30°)^3 = 1000 \angle -90° = 0 - j\,1000 = -j\,1000$$
$$(2.12 - j2.12)^{1/2} = (9 \angle -45°)^{1/2} = 3 \angle -22.5°$$

Comments.   It is important to note that

$$|zw| = |z||w|$$

and

$$|z + w| \neq |z| + |w|$$

Complex variable.   A complex number has a real part and an imaginary part, both of which are constant. If the real part or the imaginary part (or both) are variables, the complex number is called a *complex variable*. In the Laplace transformation, we use the notation s to denote a complex variable; that is,

$$s = \sigma + j\omega$$

where σ is the real part and $j\omega$ is the imaginary part. (Note that both σ and ω are real.)

Complex function.   A complex function $F(s)$, a function of s, has a real part and an imaginary part, or

$$F(s) = F_x + jF_y$$

where F_x and F_y are real quantities. The magnitude of $F(s)$ is $\sqrt{F_x^2 + F_y^2}$, and the angle θ of $F(s)$ is $\tan^{-1}(F_y/F_x)$. The angle is measured counterclockwise from the positive real axis. The complex conjugate of $F(s)$ is $\overline{F}(s) = F_x - jF_y$.

Complex functions commonly encountered in linear systems analysis are single-valued functions of s and are uniquely determined for a given value of s. Typically, such functions have the form

$$F(s) = \frac{K(s + z_1)(s + z_2) \cdots (s + z_m)}{(s + p_1)(s + p_2) \cdots (s + p_n)}$$

Points at which $F(s)$ equals zero are called *zeros*. That is, $s = -z_1, s = -z_2, \ldots,$ $s = -z_m$ are zeros of $F(s)$. [Note that $F(s)$ may have additional zeros at infinity; see the illustrative example that follows.] Points at which $F(s)$ equals infinity are called *poles*. That is, $s = -p_1, s = -p_2, \ldots, s = -p_n$ are poles of $F(s)$. If the denominator of $F(s)$ involves k-multiple factors $(s + p)^k$, then $s = -p$ is called a *multiple pole* of order k or *repeated pole* of order k. If $k = 1$, the pole is called a *simple pole*.

As an illustrative example, consider the complex function

$$G(s) = \frac{K(s + 2)(s + 10)}{s(s + 1)(s + 5)(s + 15)^2}$$

$G(s)$ has zeros at $s = -2$ and $s = -10$, simple poles at $s = 0, s = -1$, and $s = -5$, and a double pole (multiple pole of order 2) at $s = -15$. Note that $G(s)$ becomes zero at $s = \infty$. Since, for large values of s,

$$G(s) \doteq \frac{K}{s^3}$$

it follows that $G(s)$ possesses a triple zero (multiple zero of order 3) at $s = \infty$. If points at infinity are included, $G(s)$ has the same number of poles as zeros. To summarize, $G(s)$ has five zeros ($s = -2, s = -10, s = \infty, s = \infty, s = \infty$) and five poles ($s = 0, s = -1, s = -5, s = -15, s = -15$).

2–3 LAPLACE TRANSFORMATION

The Laplace transform method is an operational method that can be used advantageously in solving linear, time-invariant differential equations. Its main advantage is that differentiation of the time function corresponds to multiplication of the transform by a complex variable s, and thus the differential equations in time become algebraic equations in s. The solution of the differential equation can then be found by using a Laplace transform table or the partial-fraction expansion technique. Another advantage of the Laplace transform method is that, in solving the differential equation, the initial conditions are automatically taken care of, and both the particular solution and the complementary solution can be obtained simultaneously.

Laplace transformation.   Let us define

$f(t)$ = a time function such that $f(t) = 0$ for $t < 0$

s = a complex variable

$\mathcal{L}$ = an operational symbol indicating that the quantity upon which it operates is to be transformed

by the Laplace integral $\int_0^\infty e^{-st} \, dt$

$F(s)$ = Laplace transform of $f(t)$

Then the Laplace transform of $f(t)$ is given by

$$\mathscr{L}[f(t)] = F(s) = \int_0^\infty e^{-st}\, dt[f(t)] = \int_0^\infty f(t)e^{-st}\, dt$$

The reverse process of finding the time function $f(t)$ from the Laplace transform $F(s)$ is called *inverse Laplace transformation*. The notation for inverse Laplace trans-formation is $\mathscr{L}^{-1}$. Thus,

$$\mathscr{L}^{-1}[F(s)] = f(t)$$

Existence of Laplace transform.   The Laplace transform of a function $f(t)$ exists if the Laplace integral converges. The integral will converge if $f(t)$ is piecewise continuous in every finite interval in the range $t > 0$ and if $f(t)$ is of exponential order as t approaches infinity. A function $f(t)$ is said to be of exponential order if a real, positive constant σ exists such that the function

$$e^{-\sigma t}|f(t)|$$

approaches zero as t approaches infinity. If the limit of the function $e^{-\sigma t}|f(t)|$ approaches zero for σ greater than σ_c and the limit approaches infinity for σ less than σ_c, the value σ_c is called the *abscissa of convergence*.

It can be seen that, for such functions as t, sin ωt, and t sin ωt, the abscissa of convergence is equal to zero. For functions like e^{-ct}, te^{-ct}, and e^{-ct} sin ωt, the abscis-sa of convergence is equal to $-c$. In the case of functions that increase faster than the exponential function, it is impossible to find suitable values of the abscissa of convergence. Consequently, such functions as e^{t^2} and te^{t^2} do not possess Laplace transforms.

Nevertheless, it should be noted that, although e^{t^2} for $0 \le t \le \infty$ does not possess a Laplace transform, the time function defined by

$$f(t) = e^{t^2} \qquad \text{for } 0 \le t \le T < \infty$$
$$= 0 \qquad \text{for } t < 0, T < t$$

does, since $f(t) = e^{t^2}$ for only a limited time interval $0 \le t \le T$ and not for $0 \le t \le \infty$. Such a signal can be physically generated. Note that the signals that can be physically generated always have corresponding Laplace transforms.

If functions $f_1(t)$ and $f_2(t)$ are both Laplace transformable, then the Laplace transform of $f_1(t) + f_2(t)$ is given by

$$\mathscr{L}[f_1(t) + f_2(t)] = \mathscr{L}[f_1(t)] + \mathscr{L}[f_2(t)]$$

Exponential function.   Consider the exponential function

$$f(t) = 0 \qquad \text{for } t < 0$$
$$= Ae^{-\alpha t} \qquad \text{for } t \ge 0$$

where A and α are constants. The Laplace transform of this exponential function can be obtained as follows:

$$\mathscr{L}[Ae^{-\alpha t}] = \int_0^\infty Ae^{-\alpha t}e^{-st}\, dt = A\int_0^\infty e^{-(\alpha+s)t}\, dt = \frac{A}{s+\alpha}$$

In performing this integration, we assume that the real part of s is greater than $-\alpha$ (the abscissa of convergence), so that the integral converges. The Laplace transform $F(s)$ of any Laplace transformable function $f(t)$ obtained in this way is valid throughout the entire s plane, except at the poles of $F(s)$. (Although we do not present a proof of this statement, it can be proved by use of the theory of complex variables.)

Step function.   Consider the step function

$$f(t) = 0 \qquad \text{for } t < 0$$
$$ = A \qquad \text{for } t > 0$$

where A is a constant. Note that this is a special case of the exponential function $Ae^{-\alpha t}$, where $\alpha = 0$. The step function is undefined at $t = 0$. Its Laplace transform is given by

$$\mathscr{L}[A] = \int_0^\infty Ae^{-st}\, dt = \frac{A}{s}$$

The step function whose height is unity is called a *unit-step function*. The unit-step function that occurs at $t = t_0$ is frequently written $1(t - t_0)$, a notation that will be used in this book. The preceding step function whose height is A can thus be written $A1(t)$.

The Laplace transform of the unit-step function that is defined by

$$1(t) = 0 \qquad \text{for } t < 0$$
$$ = 1 \qquad \text{for } t > 0$$

is $1/s$, or

$$\mathscr{L}[1(t)] = \frac{1}{s}$$

Physically, a step function occurring at $t = t_0$ corresponds to a constant signal suddenly applied to the system at time t equals t_0.

Ramp function.   Consider the ramp function

$$f(t) = 0 \qquad \text{for } t < 0$$
$$ = At \qquad \text{for } t \geq 0$$

where A is a constant. The Laplace transform of this ramp function is

$$\mathscr{L}[At] = A \int_0^\infty te^{-st}\, dt$$

To evaluate the integral, we use the formula for integration by parts:

$$\int_a^b u\, dv = uv \Big|_a^b - \int_a^b v\, du$$

In this case, $u = t$ and $dv = e^{-st}\,dt$. [Note that $v = e^{-st}/(-s)$.] Hence,

$$\mathcal{L}[At] = A \int_0^\infty t e^{-st}\,dt = A\left(t \frac{e^{-st}}{-s}\Big|_0^\infty - \int_0^\infty \frac{e^{-st}}{-s}\,dt \right)$$

$$= \frac{A}{s} \int_0^\infty e^{-st}\,dt = \frac{A}{s^2}$$

Sinusoidal function. The Laplace transform of the sinusoidal function

$$f(t) = 0 \qquad\qquad \text{for } t < 0$$
$$= A \sin \omega t \qquad \text{for } t \geq 0$$

where A and ω are constants, is obtained as follows: Noting that

$$e^{j\omega t} = \cos \omega t + j \sin \omega t$$

and

$$e^{-j\omega t} = \cos \omega t - j \sin \omega t$$

we can write

$$\sin \omega t = \frac{1}{2j}(e^{j\omega t} - e^{-j\omega t})$$

Hence,

$$\mathcal{L}[A \sin \omega t] = \frac{A}{2j} \int_0^\infty (e^{j\omega t} - e^{-j\omega t})e^{-st}\,dt$$

$$= \frac{A}{2j} \frac{1}{s - j\omega} - \frac{A}{2j} \frac{1}{s + j\omega} = \frac{A\omega}{s^2 + \omega^2}$$

Similarly, the Laplace transform of $A \cos \omega t$ can be derived as follows:

$$\mathcal{L}[A \cos \omega t] = \frac{As}{s^2 + \omega^2}$$

Comments. The Laplace transform of any Laplace transformable function $f(t)$ can be found by multiplying $f(t)$ by e^{-st} and then integrating the product from $t = 0$ to $t = \infty$. Once we know the method of obtaining the Laplace transform, however, it is not necessary to derive the Laplace transform of $f(t)$ each time. Laplace transform tables can conveniently be used to find the transform of a given function $f(t)$. Table 2–1 shows Laplace transforms of time functions that will frequently appear in linear systems analysis. In Table 2–2, the properties of Laplace transforms are given.

Translated function. Let us obtain the Laplace transform of the translated function $f(t - \alpha)1(t - \alpha)$, where $\alpha \geq 0$. This function is zero for $t < \alpha$. The functions $f(t)1(t)$ and $f(t - \alpha)1(t - \alpha)$ are shown in Figure 2–5.

By definition, the Laplace transform of $f(t - \alpha)1(t - \alpha)$ is

$$\mathcal{L}[f(t - \alpha)1(t - \alpha)] = \int_0^\infty f(t - \alpha)1(t - \alpha)e^{-st}\,dt$$

TABLE 2–1 Laplace Transform Pairs

	$f(t)$	$F(s)$
1	Unit impulse $\delta(t)$	1
2	Unit step $1(t)$	$\dfrac{1}{s}$
3	t	$\dfrac{1}{s^2}$
4	$\dfrac{t^{n-1}}{(n-1)!}$ $(n = 1, 2, 3, \ldots)$	$\dfrac{1}{s^n}$
5	t^n $(n = 1, 2, 3, \ldots)$	$\dfrac{n!}{s^{n+1}}$
6	e^{-at}	$\dfrac{1}{s+a}$
7	te^{-at}	$\dfrac{1}{(s+a)^2}$
8	$\dfrac{1}{(n-1)!}t^{n-1}e^{-at}$ $(n = 1, 2, 3, \ldots)$	$\dfrac{1}{(s+a)^n}$
9	$t^n e^{-at}$ $(n = 1, 2, 3, \ldots)$	$\dfrac{n!}{(s+a)^{n+1}}$
10	$\sin \omega t$	$\dfrac{\omega}{s^2 + \omega^2}$
11	$\cos \omega t$	$\dfrac{s}{s^2 + \omega^2}$
12	$\sinh \omega t$	$\dfrac{\omega}{s^2 - \omega^2}$
13	$\cosh \omega t$	$\dfrac{s}{s^2 - \omega^2}$
14	$\dfrac{1}{a}(1 - e^{-at})$	$\dfrac{1}{s(s+a)}$
15	$\dfrac{1}{b-a}(e^{-at} - e^{-bt})$	$\dfrac{1}{(s+a)(s+b)}$
16	$\dfrac{1}{b-a}(be^{-bt} - ae^{-at})$	$\dfrac{s}{(s+a)(s+b)}$
17	$\dfrac{1}{ab}\left[1 + \dfrac{1}{a-b}(be^{-at} - ae^{-bt})\right]$	$\dfrac{1}{s(s+a)(s+b)}$

TABLE 2–1 (*continued*)

	$f(t)$	$F(s)$
18	$\dfrac{1}{a^2}(1 - e^{-at} - ate^{-at})$	$\dfrac{1}{s(s + a)^2}$
19	$\dfrac{1}{a^2}(at - 1 + e^{-at})$	$\dfrac{1}{s^2(s + a)}$
20	$e^{-at}\sin \omega t$	$\dfrac{\omega}{(s + a)^2 + \omega^2}$
21	$e^{-at}\cos \omega t$	$\dfrac{s + a}{(s + a)^2 + \omega^2}$
22	$\dfrac{\omega_n}{\sqrt{1 - \zeta^2}}e^{-\zeta\omega_n t}\sin \omega_n\sqrt{1 - \zeta^2}\,t$	$\dfrac{\omega_n^2}{s^2 + 2\zeta\omega_n s + \omega_n^2}$
23	$-\dfrac{1}{\sqrt{1 - \zeta^2}}e^{-\zeta\omega_n t}\sin(\omega_n\sqrt{1 - \zeta^2}\,t - \phi)$ $\phi = \tan^{-1}\dfrac{\sqrt{1 - \zeta^2}}{\zeta}$	$\dfrac{s}{s^2 + 2\zeta\omega_n s + \omega_n^2}$
24	$1 - \dfrac{1}{\sqrt{1 - \zeta^2}}e^{-\zeta\omega_n t}\sin(\omega_n\sqrt{1 - \zeta^2}\,t + \phi)$ $\phi = \tan^{-1}\dfrac{\sqrt{1 - \zeta^2}}{\zeta}$	$\dfrac{\omega_n^2}{s(s^2 + 2\zeta\omega_n s + \omega_n^2)}$
25	$1 - \cos \omega t$	$\dfrac{\omega^2}{s(s^2 + \omega^2)}$
26	$\omega t - \sin \omega t$	$\dfrac{\omega^3}{s^2(s^2 + \omega^2)}$
27	$\sin \omega t - \omega t \cos \omega t$	$\dfrac{2\omega^3}{(s^2 + \omega^2)^2}$
28	$\dfrac{1}{2\omega}t \sin \omega t$	$\dfrac{s}{(s^2 + \omega^2)^2}$
29	$t \cos \omega t$	$\dfrac{s^2 - \omega^2}{(s^2 + \omega^2)^2}$
30	$\dfrac{1}{\omega_2^2 - \omega_1^2}(\cos \omega_1 t - \cos \omega_2 t)\quad (\omega_1^2 \neq \omega_2^2)$	$\dfrac{s}{(s^2 + \omega_1^2)(s^2 + \omega_2^2)}$
31	$\dfrac{1}{2\omega}(\sin \omega t + \omega t \cos \omega t)$	$\dfrac{s^2}{(s^2 + \omega^2)^2}$

TABLE 2–2   Properties of Laplace Transforms

1	$\mathcal{L}[Af(t)] = AF(s)$
2	$\mathcal{L}[f_1(t) \pm f_2(t)] = F_1(s) \pm F_2(s)$
3	$\mathcal{L}_\pm\left[\dfrac{d}{dt}f(t)\right] = sF(s) - f(0\pm)$
4	$\mathcal{L}_\pm\left[\dfrac{d^2}{dt^2}f(t)\right] = s^2F(s) - sf(0\pm) - \dot{f}(0\pm)$
5	$\mathcal{L}_\pm\left[\dfrac{d^n}{dt^n}f(t)\right] = s^nF(s) - \displaystyle\sum_{k=1}^{n} s^{n-k}\overset{(k-1)}{f}(0\pm)$ where $\overset{(k-1)}{f}(t) = \dfrac{d^{k-1}}{dt^{k-1}}f(t)$
6	$\mathcal{L}_\pm\left[\displaystyle\int f(t)\,dt\right] = \dfrac{F(s)}{s} + \dfrac{[\int f(t)\,dt]_{t=0\pm}}{s}$
7	$\mathcal{L}_\pm\left[\displaystyle\iint f(t)\,dt\,dt\right] = \dfrac{F(s)}{s^2} + \dfrac{[\int f(t)\,dt]_{t=0\pm}}{s^2} + \dfrac{[\iint f(t)\,dt\,dt]_{t=0\pm}}{s}$
8	$\mathcal{L}_\pm\left[\displaystyle\int \cdots \int f(t)(dt)^n\right] = \dfrac{F(s)}{s^n} + \displaystyle\sum_{k=1}^{n} \dfrac{1}{s^{n-k+1}}\left[\int \cdots \int f(t)(dt)^k\right]_{t=0\pm}$
9	$\mathcal{L}\left[\displaystyle\int_0^t f(t)\,dt\right] = \dfrac{F(s)}{s}$
10	$\displaystyle\int_0^\infty f(t)\,dt = \lim_{s\to 0} F(s)$ \quad if $\displaystyle\int_0^\infty f(t)\,dt$ exists
11	$\mathcal{L}[e^{-at}f(t)] = F(s + a)$
12	$\mathcal{L}[f(t - \alpha)1(t - \alpha)] = e^{-\alpha s}F(s)$ \quad $\alpha \geq 0$
13	$\mathcal{L}[tf(t)] = -\dfrac{dF(s)}{ds}$
14	$\mathcal{L}[t^2f(t)] = \dfrac{d^2}{ds^2}F(s)$
15	$\mathcal{L}[t^nf(t)] = (-1)^n\dfrac{d^n}{ds^n}F(s)$ \quad $n = 1, 2, 3, \ldots$
16	$\mathcal{L}\left[\dfrac{1}{t}f(t)\right] = \displaystyle\int_s^\infty F(s)\,ds$ \quad if $\displaystyle\lim_{t\to 0}\dfrac{1}{t}f(t)$ exists
17	$\mathcal{L}\left[f\left(\dfrac{t}{a}\right)\right] = aF(as)$

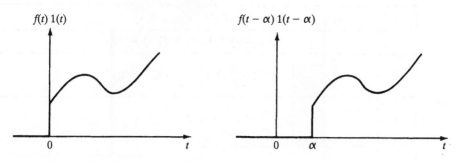

Figure 2–5   Function $f(t)1(t)$ and translated function $f(t - \alpha)1(t - \alpha)$.

By changing the independent variable from t to τ, where $\tau = t - \alpha$, we obtain

$$\int_0^\infty f(t - \alpha)1(t - \alpha)e^{-st}\, dt = \int_{-\alpha}^\infty f(\tau)1(\tau)e^{-s(\tau+\alpha)}\, d\tau$$

Noting that $f(\tau)1(\tau) = 0$ for $\tau < 0$, we can change the lower limit of integration from $-\alpha$ to 0. Thus,

$$\int_{-\alpha}^\infty f(\tau)1(\tau)e^{-s(\tau+\alpha)}\, d\tau = \int_0^\infty f(\tau)1(\tau)e^{-s(\tau+\alpha)}\, d\tau$$

$$= \int_0^\infty f(\tau)e^{-s\tau}e^{-\alpha s}\, d\tau$$

$$= e^{-\alpha s}\int_0^\infty f(\tau)e^{-s\tau}\, d\tau = e^{-\alpha s}F(s)$$

where

$$F(s) = \mathcal{L}[f(t)] = \int_0^\infty f(t)e^{-st}\, dt$$

Hence,

$$\mathcal{L}[f(t - \alpha)1(t - \alpha)] = e^{-\alpha s}F(s) \qquad \alpha \geq 0$$

This last equation states that the translation of the time function $f(t)1(t)$ by α (where $\alpha \geq 0$) corresponds to the multiplication of the transform $F(s)$ by $e^{-\alpha s}$.

Pulse function.   Consider the pulse function shown in Figure 2–6, namely,

$$f(t) = \frac{A}{t_0} \qquad \text{for } 0 < t < t_0$$
$$= 0 \qquad \text{for } t < 0, t_0 < t$$

where A and t_0 are constants.

The pulse function here may be considered a step function of height A/t_0 that begins at $t = 0$ and that is superimposed by a negative step function of height A/t_0

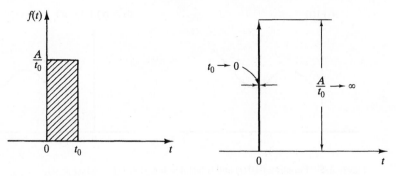

Figure 2–6   Pulse function.   **Figure 2–7**   Impulse function.

beginning at $t = t_0$; that is,

$$f(t) = \frac{A}{t_0}1(t) - \frac{A}{t_0}1(t - t_0)$$

Then the Laplace transform of $f(t)$ is obtained as

$$\mathcal{L}[f(t)] = \mathcal{L}\left[\frac{A}{t_0}1(t)\right] - \mathcal{L}\left[\frac{A}{t_0}1(t - t_0)\right]$$

$$= \frac{A}{t_0 s} - \frac{A}{t_0 s}e^{-st_0}$$

$$= \frac{A}{t_0 s}(1 - e^{-st_0}) \tag{2-1}$$

Impulse function.   The impulse function is a special limiting case of the pulse function. Consider the impulse function

$$f(t) = \lim_{t_0 \to 0}\frac{A}{t_0} \qquad \text{for } 0 < t < t_0$$

$$= 0 \qquad \text{for } t < 0, t_0 < t$$

Figure 2–7 depicts the impulse function defined here. It is a limiting case of the pulse function shown in Figure 2–6 as t_0 approaches zero. Since the height of the impulse function is A/t_0 and the duration is t_0, the area under the impulse is equal to A. As the duration t_0 approaches zero, the height A/t_0 approaches infinity, but the area under the impulse remains equal to A. Note that the magnitude of the impulse is measured by its area.

From Equation (2–1), the Laplace transform of this impulse function is shown to be

$$\mathcal{L}[f(t)] = \lim_{t_0 \to 0}\left[\frac{A}{t_0 s}(1 - e^{-st_0})\right]$$

$$= \lim_{t_0 \to 0}\frac{\dfrac{d}{dt_0}[A(1 - e^{-st_0})]}{\dfrac{d}{dt_0}(t_0 s)} = \frac{As}{s} = A$$

Thus, the Laplace transform of the impulse function is equal to the area under the impulse.

The impulse function whose area is equal to unity is called the *unit-impulse function* or the *Dirac delta function.* The unit-impulse function occurring at $t = t_0$ is usually denoted by $\delta(t - t_0)$, which satisfies the following conditions:

$$\delta(t - t_0) = 0 \qquad \text{for } t \neq t_0$$
$$\delta(t - t_0) = \infty \qquad \text{for } t = t_0$$
$$\int_{-\infty}^{\infty} \delta(t - t_0) \, dt = 1$$

An impulse that has an infinite magnitude and zero duration is mathematical fiction and does not occur in physical systems. If, however, the magnitude of a pulse input to a system is very large and its duration very short compared with the system time constants, then we can approximate the pulse input by an impulse function. For instance, if a force or torque input $f(t)$ is applied to a system for a very short time duration $0 < t < t_0$, where the magnitude of $f(t)$ is sufficiently large so that $\int_0^{t_0} f(t) \, dt$ is not negligible, then this input can be considered an impulse input. (Note that, when we describe the impulse input, the area or magnitude of the impulse is most important, but the exact shape of the impulse is usually immaterial.) The impulse input supplies energy to the system in an infinitesimal time.

The concept of the impulse function is highly useful in differentiating discontinuous-time functions. The unit-impulse function $\delta(t - t_0)$ can be considered the derivative of the unit-step function $1(t - t_0)$ at the point of discontinuity $t = t_0$, or

$$\delta(t - t_0) = \frac{d}{dt} 1(t - t_0)$$

Conversely, if the unit-impulse function $\delta(t - t_0)$ is integrated, the result is the unit-step function $1(t - t_0)$. With the concept of the impulse function, we can differentiate a function containing discontinuities, giving impulses, the magnitudes of which are equal to the magnitude of each corresponding discontinuity.

Multiplication of $f(t)$ by $e^{-\alpha t}$. If $f(t)$ is Laplace transformable and its Laplace transform is $F(s)$, then the Laplace transform of $e^{-\alpha t} f(t)$ is obtained as

$$\mathcal{L}[e^{-\alpha t} f(t)] = \int_0^{\infty} e^{-\alpha t} f(t) e^{-st} \, dt = F(s + \alpha) \qquad (2\text{-}2)$$

We see that the multiplication of $f(t)$ by $e^{-\alpha t}$ has the effect of replacing s by $(s + \alpha)$ in the Laplace transform. Conversely, changing s to $(s + \alpha)$ is equivalent to multiplying $f(t)$ by $e^{-\alpha t}$. (Note that α may be real or complex.)

The relationship given by Equation (2-2) is useful in finding the Laplace transforms of such functions as $e^{-\alpha t} \sin \omega t$ and $e^{-\alpha t} \cos \omega t$. For instance, since

$$\mathcal{L}[\sin \omega t] = \frac{\omega}{s^2 + \omega^2} = F(s) \qquad \text{and} \qquad \mathcal{L}[\cos \omega t] = \frac{s}{s^2 + \omega^2} = G(s)$$

it follows from Equation (2–2) that the Laplace transforms of $e^{-\alpha t} \sin \omega t$ and $e^{-\alpha t} \cos \omega t$ are given, respectively, by

$$\mathcal{L}[e^{-\alpha t} \sin \omega t] = F(s + \alpha) = \frac{\omega}{(s + \alpha)^2 + \omega^2}$$

and

$$\mathcal{L}[e^{-\alpha t} \cos \omega t] = G(s + \alpha) = \frac{s + \alpha}{(s + \alpha)^2 + \omega^2}$$

Comments on the lower limit of the Laplace integral.   In some cases, $f(t)$ possesses an impulse function at $t = 0$. Then the lower limit of the Laplace integral must be clearly specified as to whether it is $0-$ or $0+$, since the Laplace transforms of $f(t)$ differ for these two lower limits. If such a distinction of the lower limit of the Laplace integral is necessary, we use the notations

$$\mathcal{L}_+[f(t)] = \int_{0+}^{\infty} f(t)e^{-st}\, dt$$

and

$$\mathcal{L}_-[f(t)] = \int_{0-}^{\infty} f(t)e^{-st}\, dt = \mathcal{L}_+[f(t)] + \int_{0-}^{0+} f(t)e^{-st}\, dt$$

If $f(t)$ involves an impulse function at $t = 0$, then

$$\mathcal{L}_+[f(t)] \neq \mathcal{L}_-[f(t)]$$

since

$$\int_{0-}^{0+} f(t)e^{-st}\, dt \neq 0$$

for such a case. Obviously, if $f(t)$ does not possess an impulse function at $t = 0$ (i.e., if the function to be transformed is finite between $t = 0-$ and $t = 0+$), then

$$\mathcal{L}_+[f(t)] = \mathcal{L}_-[f(t)]$$

Differentiation theorem.   The Laplace transform of the derivative of a function $f(t)$ is given by

$$\mathcal{L}\left[\frac{d}{dt}f(t)\right] = sF(s) - f(0) \tag{2–3}$$

where $f(0)$ is the initial value of $f(t)$, evaluated at $t = 0$. Equation (2–3) is called the differentiation theorem.

For a given function $f(t)$, the values of $f(0+)$ and $f(0-)$ may be the same or different, as illustrated in Figure 2–8. The distinction between $f(0+)$ and $f(0-)$ is important when $f(t)$ has a discontinuity at $t = 0$, because, in such a case, $df(t)/dt$ will

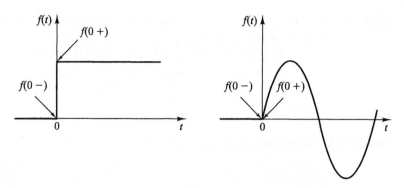

Figure 2–8 Step function and sine function indicating initial values at $t = 0-$ and $t = 0+$.

involve an impulse function at $t = 0$. If $f(0+) \neq f(0-)$, Equation (2–3) must be modified to

$$\mathscr{L}_+\left[\frac{d}{dt}f(t)\right] = sF(s) - f(0+)$$

$$\mathscr{L}_-\left[\frac{d}{dt}f(t)\right] = sF(s) - f(0-)$$

To prove the differentiation theorem, we proceed as follows: Integrating the Laplace integral by parts gives

$$\int_0^\infty f(t)e^{-st}\,dt = f(t)\frac{e^{-st}}{-s}\bigg|_0^\infty - \int_0^\infty \left[\frac{d}{dt}f(t)\right]\frac{e^{-st}}{-s}\,dt$$

Hence,

$$F(s) = \frac{f(0)}{s} + \frac{1}{s}\mathscr{L}\left[\frac{d}{dt}f(t)\right]$$

It follows that

$$\mathscr{L}\left[\frac{d}{dt}f(t)\right] = sF(s) - f(0)$$

Similarly, for the second derivative of $f(t)$, we obtain the relationship

$$\mathscr{L}\left[\frac{d^2}{dt^2}f(t)\right] = s^2F(s) - sf(0) - \dot{f}(0)$$

where $\dot{f}(0)$ is the value of $df(t)/dt$ evaluated at $t = 0$. To derive this equation, define

$$\frac{d}{dt}f(t) = g(t)$$

Then

$$\mathcal{L}\left[\frac{d^2}{dt^2}f(t)\right] = \mathcal{L}\left[\frac{d}{dt}g(t)\right] = s\mathcal{L}[g(t)] - g(0)$$

$$= s\mathcal{L}\left[\frac{d}{dt}f(t)\right] - \dot{f}(0)$$

$$= s^2F(s) - sf(0) - \dot{f}(0)$$

Similarly, for the nth derivative of $f(t)$, we obtain

$$\mathcal{L}\left[\frac{d^n}{dt^n}f(t)\right] = s^nF(s) - s^{n-1}f(0) - s^{n-2}\dot{f}(0) - \cdots - \overset{(n-1)}{f}(0)$$

where $f(0), \dot{f}(0), \ldots, \overset{(n-1)}{f}(0)$ represent the values of $f(t), df(t)/dt, \ldots, d^{n-1}f(t)/dt^{n-1}$, respectively, evaluated at $t = 0$. If the distinction between $\mathcal{L}_+$ and $\mathcal{L}_-$ is necessary, we substitute $t = 0+$ or $t = 0-$ into $f(t), df(t)/dt, \ldots, d^{n-1}f(t)/dt^{n-1}$, depending on whether we take $\mathcal{L}_+$ or $\mathcal{L}_-$.

Note that, for Laplace transforms of derivatives of $f(t)$ to exist, $d^nf(t)/dt^n$ ($n = 1, 2, 3, \ldots$) must be Laplace transformable.

Note also that, if all the initial values of $f(t)$ and its derivatives are equal to zero, then the Laplace transform of the nth derivative of $f(t)$ is given by $s^nF(s)$.

Final-value theorem. The final-value theorem relates the steady-state behavior of $f(t)$ to the behavior of $sF(s)$ in the neighborhood of $s = 0$. The theorem, however, applies if and only if $\lim_{t\to\infty} f(t)$ exists [which means that $f(t)$ settles down to a definite value as $t \to \infty$]. If all poles of $sF(s)$ lie in the left half s plane, then $\lim_{t\to\infty} f(t)$ exists, but if $sF(s)$ has poles on the imaginary axis or in the right half s plane, $f(t)$ will contain oscillating or exponentially increasing time functions, respectively, and $\lim_{t\to\infty} f(t)$ will not exist. The final-value theorem does not apply to such cases. For instance, if $f(t)$ is a sinusoidal function $\sin \omega t$, then $sF(s)$ has poles at $s = \pm j\omega$, and $\lim_{t\to\infty} f(t)$ does not exist. Therefore, the theorem is not applicable to such a function.

The final-value theorem may be stated as follows: If $f(t)$ and $df(t)/dt$ are Laplace transformable, if $F(s)$ is the Laplace transform of $f(t)$, and if $\lim_{t\to\infty} f(t)$ exists, then

$$\lim_{t\to\infty} f(t) = \lim_{s\to 0} sF(s)$$

To prove the theorem, we let s approach zero in the equation for the Laplace transform of the derivative of $f(t)$, or

$$\lim_{s\to 0} \int_0^\infty \left[\frac{d}{dt}f(t)\right]e^{-st}\,dt = \lim_{s\to 0} [sF(s) - f(0)]$$

Since $\lim_{s\to 0} e^{-st} = 1$, if $\lim_{t\to\infty} f(t)$ exists, then we obtain

$$\int_0^\infty \left[\frac{d}{dt}f(t)\right]dt = f(t)\Big|_0^\infty = f(\infty) - f(0)$$

$$= \lim_{s\to 0} sF(s) - f(0)$$

from which it follows that

$$f(\infty) = \lim_{t\to\infty} f(t) = \lim_{s\to 0} sF(s)$$

Initial-value theorem.   The initial-value theorem is the counterpart of the final-value theorem. Using the initial-value theorem, we are able to find the value of $f(t)$ at $t = 0+$ directly from the Laplace transform of $f(t)$. The theorem does not give the value of $f(t)$ at exactly $t = 0$, but rather gives it at a time slightly greater than zero.

The initial-value theorem may be stated as follows: If $f(t)$ and $df(t)/dt$ are both Laplace transformable and if $\lim_{s\to\infty} sF(s)$ exists, then

$$f(0+) = \lim_{s\to\infty} sF(s)$$

To prove this theorem, we use the equation for the $\mathcal{L}_+$ transform of $df(t)/dt$:

$$\mathcal{L}_+\left[\frac{d}{dt}f(t)\right] = sF(s) - f(0+)$$

For the time interval $0+ \le t \le \infty$, as s approaches infinity, e^{-st} approaches zero. (Note that we must use $\mathcal{L}_+$ rather than $\mathcal{L}_-$ for this condition.) Hence,

$$\lim_{s\to\infty} \int_{0+}^{\infty} \left[\frac{d}{dt}f(t)\right] e^{-st}\, dt = \lim_{s\to\infty} [sF(s) - f(0+)] = 0$$

or

$$f(0+) = \lim_{s\to\infty} sF(s)$$

In applying the initial-value theorem, we are not limited as to the locations of the poles of $sF(s)$. Thus, the theorem is valid for the sinusoidal function.

Note that the initial-value theorem and the final-value theorem provide a convenient check on the solution, since they enable us to predict the system behavior in the time domain without actually transforming functions in s back to time functions.

Integration theorem.   If $f(t)$ is of exponential order, then the Laplace transform of $\int f(t)\, dt$ exists and is given by

$$\mathcal{L}\left[\int f(t)\, dt\right] = \frac{F(s)}{s} + \frac{f^{-1}(0)}{s} \tag{2–4}$$

where $F(s) = \mathcal{L}[f(t)]$ and $f^{-1}(0) = \int f(t)\, dt$, evaluated at $t = 0$. Equation (2–4) is called the integration theorem.

The integration theorem can be proven as follows: Integration by parts yields

$$\mathcal{L}\left[\int f(t)\, dt\right] = \int_0^{\infty} \left[\int f(t)\, dt\right] e^{-st}\, dt$$

$$= \left[\int f(t)\, dt\right] \frac{e^{-st}}{-s}\bigg|_0^{\infty} - \int_0^{\infty} f(t) \frac{e^{-st}}{-s}\, dt$$

$$= \frac{1}{s} \int f(t)\, dt \bigg|_{t=0} + \frac{1}{s} \int_0^\infty f(t) e^{-st}\, dt$$

$$= \frac{f^{-1}(0)}{s} + \frac{F(s)}{s}$$

and the theorem is proven.

Note that, if $f(t)$ involves an impulse function at $t = 0$, then $f^{-1}(0+) \neq f^{-1}(0-)$. So if $f(t)$ involves an impulse function at $t = 0$, we must modify Equation (2–4) as follows:

$$\mathscr{L}_+\left[\int\int f(t)\, dt\right] = \frac{F(s)}{s} + \frac{f^{-1}(0+)}{s}$$

$$\mathscr{L}_-\left[\int\int f(t)\, dt\right] = \frac{F(s)}{s} + \frac{f^{-1}(0-)}{s}$$

We see that integration in the time domain is converted into division in the s domain. If the initial value of the integral is zero, the Laplace transform of the integral of $f(t)$ is given by $F(s)/s$.

The integration theorem can be modified slightly to deal with the definite integral of $f(t)$. If $f(t)$ is of exponential order, the Laplace transform of the definite integral $\int_0^t f(t)\, dt$ can be given by

$$\mathscr{L}\left[\int_0^t f(t)\, dt\right] = \frac{F(s)}{s} \tag{2–5}$$

To prove Equation (2–5), first note that

$$\int_0^t f(t)\, dt = \int f(t)\, dt - f^{-1}(0)$$

where $f^{-1}(0)$ is equal to $\int f(t)\, dt$, evaluated at $t = 0$, and is a constant. Hence,

$$\mathscr{L}\left[\int_0^t f(t)\, dt\right] = \mathscr{L}\left[\int f(t)\, dt - f^{-1}(0)\right]$$

$$= \mathscr{L}\left[\int f(t)\, dt\right] - \mathscr{L}[f^{-1}(0)]$$

Referring to Equation (2–4) and noting that $f^{-1}(0)$ is a constant, so that

$$\mathscr{L}[f^{-1}(0)] = \frac{f^{-1}(0)}{s}$$

we obtain

$$\mathscr{L}\left[\int_0^t f(t)\, dt\right] = \frac{F(s)}{s} + \frac{f^{-1}(0)}{s} - \frac{f^{-1}(0)}{s} = \frac{F(s)}{s}$$

Note that, if $f(t)$ involves an impulse function at $t = 0$, then $\int_{0+}^{t} f(t)\, dt \neq \int_{0-}^{t} f(t)\, dt$, and the following distinction must be observed:

$$\mathcal{L}_{+}\left[\int_{0+}^{t} f(t)\, dt\right] = \frac{\mathcal{L}_{+}[f(t)]}{s}$$

$$\mathcal{L}_{-}\left[\int_{0-}^{t} f(t)\, dt\right] = \frac{\mathcal{L}_{-}[f(t)]}{s}$$

2–4 INVERSE LAPLACE TRANSFORMATION

The inverse Laplace transformation refers to the process of finding the time function $f(t)$ from the corresponding Laplace transform $F(s)$. Several methods are available for finding inverse Laplace transforms. The simplest of these methods are (1) to use tables of Laplace transforms to find the time function $f(t)$ corresponding to a given Laplace transform $F(s)$ and (2) to use the partial-fraction expansion method. In this section, we present the latter technique. [Note that MATLAB is quite useful in obtaining the partial-fraction expansion of the ratio of two polynomials, $B(s)/A(s)$. We shall discuss the MATLAB approach to the partial-fraction expansion in Chapter 4.]

Partial-fraction expansion method for finding inverse Laplace transforms.
If $F(s)$, the Laplace transform of $f(t)$, is broken up into components, or

$$F(s) = F_1(s) + F_2(s) + \cdots + F_n(s)$$

and if the inverse Laplace transforms of $F_1(s), F_2(s), \ldots, F_n(s)$ are readily available, then

$$\mathcal{L}^{-1}[F(s)] = \mathcal{L}^{-1}[F_1(s)] + \mathcal{L}^{-1}[F_2(s)] + \cdots + \mathcal{L}^{-1}[F_n(s)]$$
$$= f_1(t) + f_2(t) + \cdots + f_n(t)$$

where $f_1(t), f_2(t), \ldots, f_n(t)$ are the inverse Laplace transforms of $F_1(s), F_2(s), \cdots, F_n(s)$, respectively. The inverse Laplace transform of $F(s)$ thus obtained is unique, except possibly at points where the time function is discontinuous. Whenever the time function is continuous, the time function $f(t)$ and its Laplace transform $F(s)$ have a one-to-one correspondence.

For problems in systems analysis, $F(s)$ frequently occurs in the form

$$F(s) = \frac{B(s)}{A(s)}$$

where $A(s)$ and $B(s)$ are polynomials in s and the degree of $B(s)$ is not higher than that of $A(s)$.

The advantage of the partial-fraction expansion approach is that the individual terms of $F(s)$ resulting from the expansion into partial-fraction form are very simple functions of s; consequently, it is not necessary to refer to a Laplace transform table if we memorize several simple Laplace transform pairs. Note, however, that in applying the partial-fraction expansion technique in the search for the inverse Laplace transform of $F(s) = B(s)/A(s)$, the roots of the denominator poly-nomial $A(s)$ must be known in advance. That is, this method does not apply until the denominator polynomial has been factored.

Consider $F(s)$ written in the factored form

$$F(s) = \frac{B(s)}{A(s)} = \frac{K(s + z_1)(s + z_2) \cdots (s + z_m)}{(s + p_1)(s + p_2) \cdots (s + p_n)}$$

where $p_1, p_2, \ldots, p_n$ and $z_1, z_2, \ldots, z_m$ are either real or complex quantities, but for each complex p_i or z_i, there will occur the complex conjugate of p_i or z_i, respective-ly. Here, the highest power of s in $A(s)$ is assumed to be higher than that in $B(s)$.

In the expansion of $B(s)/A(s)$ into partial-fraction form, it is important that the highest power of s in $A(s)$ be greater than the highest power of s in $B(s)$ because if that is not the case, then the numerator $B(s)$ must be divided by the denominator $A(s)$ in order to produce a polynomial in s plus a remainder (a ratio of polynomials in s whose numerator is of lower degree than the denominator). (For details, see **Example 2–2.**)

Partial-fraction expansion when F(s) involves distinct poles only. In this case, $F(s)$ can always be expanded into a sum of simple partial fractions; that is,

$$F(s) = \frac{B(s)}{A(s)} = \frac{a_1}{s + p_1} + \frac{a_2}{s + p_2} + \cdots + \frac{a_n}{s + p_n} \tag{2–6}$$

where $a_k(k = 1, 2, \ldots, n)$ are constants. The coefficient a_k is called the *residue* at the pole at $s = -p_k$. The value of a_k can be found by multiplying both sides of Equation (2–6) by $(s + p_k)$ and letting $s = -p_k$, giving

$$\left[(s + p_k) \frac{B(s)}{A(s)} \right]_{s=-p_k} = \left[\frac{a_1}{s + p_1}(s + p_k) + \frac{a_2}{s + p_2}(s + p_k) + \cdots \right.$$

$$\left. + \frac{a_k}{s + p_k}(s + p_k) + \cdots + \frac{a_n}{s + p_n}(s + p_k) \right]_{s=-p_k}$$

$$= a_k$$

We see that all the expanded terms drop out, with the exception of a_k. Thus, the residue a_k is found from

$$a_k = \left[(s + p_k) \frac{B(s)}{A(s)} \right]_{s=-p_k} \tag{2–7}$$

Note that since $f(t)$ is a real function of time, if p_1 and p_2 are complex conjugates, then the residues a_1 and a_2 are also complex conjugates. Only one of the conjugates, a_1 or a_2, need be evaluated, because the other is known automatically.

Since

$$\mathcal{L}^{-1} \left[\frac{a_k}{s + p_k} \right] = a_k e^{-p_k t}$$

$f(t)$ is obtained as

$$f(t) = \mathcal{L}^{-1}[F(s)] = a_1 e^{-p_1 t} + a_2 e^{-p_2 t} + \cdots + a_n e^{-p_n t} \qquad t \geq 0$$

Example 2–1

Find the inverse Laplace transform of

$$F(s) = \frac{s + 3}{(s + 1)(s + 2)}$$

The partial-fraction expansion of $F(s)$ is

$$F(s) = \frac{s + 3}{(s + 1)(s + 2)} = \frac{a_1}{s + 1} + \frac{a_2}{s + 2}$$

where a_1 and a_2 are found by using Equation (2–7):

$$a_1 = \left[(s + 1) \frac{s + 3}{(s + 1)(s + 2)} \right]_{s=-1} = \left[\frac{s + 3}{s + 2} \right]_{s=-1} = 2$$

$$a_2 = \left[(s + 2) \frac{s + 3}{(s + 1)(s + 2)} \right]_{s=-2} = \left[\frac{s + 3}{s + 1} \right]_{s=-2} = -1$$

Thus,

$$f(t) = \mathcal{L}^{-1}[F(s)]$$

$$= \mathcal{L}^{-1}\left[\frac{2}{s + 1} \right] + \mathcal{L}^{-1}\left[\frac{-1}{s + 2} \right]$$

$$= 2e^{-t} - e^{-2t} \qquad t \geq 0$$

Example 2–2

Obtain the inverse Laplace transform of

$$G(s) = \frac{s^3 + 5s^2 + 9s + 7}{(s + 1)(s + 2)}$$

Here, since the degree of the numerator polynomial is higher than that of the denominator polynomial, we must divide the numerator by the denominator:

$$G(s) = s + 2 + \frac{s + 3}{(s + 1)(s + 2)}$$

Note that the Laplace transform of the unit-impulse function $\delta(t)$ is unity and that the Laplace transform of $d\delta(t)/dt$ is s. The third term on the right-hand side of this last equation is $F(s)$ in **Example 2–1**. So the inverse Laplace transform of $G(s)$ is given as

$$g(t) = \frac{d}{dt}\delta(t) + 2\delta(t) + 2e^{-t} - e^{-2t} \qquad t \geq 0-$$

Comment. Consider a function $F(s)$ that involves a quadratic factor $s^2 + as + b$ in the denominator. If this quadratic expression has a pair of complex-conjugate roots, then it is better not to factor the quadratic, in order to avoid complex numbers. For example, if $F(s)$ is given as

$$F(s) = \frac{p(s)}{s(s^2 + as + b)}$$

where $a \geq 0$ and $b > 0$, and if $s^2 + as + b = 0$ has a pair of complex-conjugate roots, then expand $F(s)$ into the following partial-fraction expansion form:

$$F(s) = \frac{c}{s} + \frac{ds + e}{s^2 + as + b}$$

(See **Example 2–3** and **Problems A–2–15, A–2–16**, and **A–2–19**.)

Example 2–3

Find the inverse Laplace transform of

$$F(s) = \frac{2s + 12}{s^2 + 2s + 5}$$

Notice that the denominator polynomial can be factored as

$$s^2 + 2s + 5 = (s + 1 + j2)(s + 1 - j2)$$

The two roots of the denominator are complex conjugates. Hence, we expand $F(s)$ into the sum of a damped sine and a damped cosine function.

Noting that $s^2 + 2s + 5 = (s + 1)^2 + 2^2$ and referring to the Laplace transforms of $e^{-at} \sin \omega t$ and $e^{-at} \cos \omega t$, rewritten as

$$\mathcal{L}[e^{-at} \sin \omega t] = \frac{\omega}{(s + a)^2 + \omega^2}$$

and

$$\mathcal{L}[e^{-at} \cos \omega t] = \frac{s + a}{(s + a)^2 + \omega^2}$$

we can write the given $F(s)$ as a sum of a damped sine and a damped cosine function:

$$F(s) = \frac{2s + 12}{s^2 + 2s + 5} = \frac{10 + 2(s + 1)}{(s + 1)^2 + 2^2}$$

$$= 5 \frac{2}{(s + 1)^2 + 2^2} + 2 \frac{s + 1}{(s + 1)^2 + 2^2}$$

It follows that

$$f(t) = \mathcal{L}^{-1}[F(s)]$$

$$= 5\mathcal{L}^{-1}\left[\frac{2}{(s + 1)^2 + 2^2}\right] + 2\mathcal{L}^{-1}\left[\frac{s + 1}{(s + 1)^2 + 2^2}\right]$$

$$= 5e^{-t} \sin 2t + 2e^{-t} \cos 2t \qquad t \geq 0$$

Partial-fraction expansion when $F(s)$ involves multiple poles. Instead of discussing the general case, we shall use an example to show how to obtain the partial-fraction expansion of $F(s)$. (See also **Problems A–2–17** and **A–2–19**.)
Consider

$$F(s) = \frac{s^2 + 2s + 3}{(s + 1)^3}$$

The partial-fraction expansion of this $F(s)$ involves three terms:

$$F(s) = \frac{B(s)}{A(s)} = \frac{b_3}{(s+1)^3} + \frac{b_2}{(s+1)^2} + \frac{b_1}{s+1}$$

where b_3, b_2, and b_1 are determined as follows: Multiplying both sides of this last equation by $(s+1)^3$, we have

$$(s+1)^3 \frac{B(s)}{A(s)} = b_3 + b_2(s+1) + b_1(s+1)^2 \qquad (2\text{-}8)$$

Then, letting $s = -1$, we find that Equation (2-8) gives

$$\left[(s+1)^3 \frac{B(s)}{A(s)} \right]_{s=-1} = b_3$$

Also, differentiating both sides of Equation (2-8) with respect to s yields

$$\frac{d}{ds} \left[(s+1)^3 \frac{B(s)}{A(s)} \right] = b_2 + 2b_1(s+1) \qquad (2\text{-}9)$$

If we let $s = -1$ in Equation (2-9), then

$$\frac{d}{ds} \left[(s+1)^3 \frac{B(s)}{A(s)} \right]_{s=-1} = b_2$$

Differentiating both sides of Equation (2-9) with respect to s, we obtain

$$\frac{d^2}{ds^2} \left[(s+1)^3 \frac{B(s)}{A(s)} \right] = 2b_1$$

From the preceding analysis, it can be seen that the values of b_3, b_2, and b_1 are found systematically as follows:

$$b_3 = \left[(s+1)^3 \frac{B(s)}{A(s)} \right]_{s=-1}$$
$$= (s^2 + 2s + 3)_{s=-1}$$
$$= 2$$

$$b_2 = \left\{ \frac{d}{ds} \left[(s+1)^3 \frac{B(s)}{A(s)} \right] \right\}_{s=-1}$$
$$= \left[\frac{d}{ds} (s^2 + 2s + 3) \right]_{s=-1}$$
$$= (2s + 2)_{s=-1}$$
$$= 0$$

$$b_1 = \frac{1}{2!} \left\{ \frac{d^2}{ds^2} \left[(s+1)^3 \frac{B(s)}{A(s)} \right] \right\}_{s=-1}$$
$$= \frac{1}{2!} \left[\frac{d^2}{ds^2} (s^2 + 2s + 3) \right]_{s=-1}$$
$$= \frac{1}{2}(2) = 1$$

We thus obtain

$$f(t) = \mathcal{L}^{-1}[F(s)]$$

$$= \mathcal{L}^{-1}\left[\frac{2}{(s + 1)^3}\right] + \mathcal{L}^{-1}\left[\frac{0}{(s + 1)^2}\right] + \mathcal{L}^{-1}\left[\frac{1}{s + 1}\right]$$

$$= t^2 e^{-t} + 0 + e^{-t}$$

$$= (t^2 + 1)e^{-t} \qquad t \geq 0$$

2–5 SOLVING LINEAR, TIME-INVARIANT DIFFERENTIAL EQUATIONS

In this section, we are concerned with the use of the Laplace transform method in solving linear, time-invariant differential equations.

The Laplace transform method yields the complete solution (complementary solution and particular solution) of linear, time-invariant differential equations. Classical methods for finding the complete solution of a differential equation require the evaluation of the integration constants from the initial conditions. In the case of the Laplace transform method, however, this requirement is unnecessary because the initial conditions are automatically included in the Laplace transform of the differential equation.

If all initial conditions are zero, then the Laplace transform of the differential equation is obtained simply by replacing d/dt with s, d^2/dt^2 with s^2, and so on.

In solving linear, time-invariant differential equations by the Laplace transform method, two steps are followed:

1. By taking the Laplace transform of each term in the given differential equation, convert the differential equation into an algebraic equation in s and obtain the expression for the Laplace transform of the dependent variable by rearranging the algebraic equation.
2. The time solution of the differential equation is obtained by finding the inverse Laplace transform of the dependent variable.

In the discussion that follows, two examples are used to demonstrate the solution of linear, time-invariant differential equations by the Laplace transform method.

Example 2–4

Find the solution $x(t)$ of the differential equation

$$\ddot{x} + 3\dot{x} + 2x = 0, \qquad x(0) = a, \qquad \dot{x}(0) = b$$

where a and b are constants.

Writing the Laplace transform of $x(t)$ as $X(s)$, or

$$\mathcal{L}[x(t)] = X(s)$$

we obtain

$$\mathcal{L}[\dot{x}] = sX(s) - x(0)$$

$$\mathcal{L}[\ddot{x}] = s^2 X(s) - sx(0) - \dot{x}(0)$$

The Laplace transform of the given differential equation becomes

$$[s^2X(s) - sx(0) - \dot{x}(0)] + 3[sX(s) - x(0)] + 2X(s) = 0$$

Substituting the given initial conditions into the preceding equation yields

$$[s^2X(s) - as - b] + 3[sX(s) - a] + 2X(s) = 0$$

or

$$(s^2 + 3s + 2)X(s) = as + b + 3a$$

Solving this last equation for $X(s)$, we have

$$X(s) = \frac{as + b + 3a}{s^2 + 3s + 2} = \frac{as + b + 3a}{(s + 1)(s + 2)} = \frac{2a + b}{s + 1} - \frac{a + b}{s + 2}$$

The inverse Laplace transform of $X(s)$ produces

$$x(t) = \mathcal{L}^{-1}[X(s)] = \mathcal{L}^{-1}\left[\frac{2a + b}{s + 1}\right] - \mathcal{L}^{-1}\left[\frac{a + b}{s + 2}\right]$$
$$= (2a + b)e^{-t} - (a + b)e^{-2t} \qquad t \geq 0$$

which is the solution of the given differential equation. Notice that the initial conditions a and b appear in the solution. Thus, $x(t)$ has no undetermined constants.

Example 2–5

Find the solution $x(t)$ of the differential equation

$$\ddot{x} + 2\dot{x} + 5x = 3, \qquad x(0) = 0, \qquad \dot{x}(0) = 0$$

Noting that $\mathcal{L}[3] = 3/s$, $x(0) = 0$, and $\dot{x}(0) = 0$, we see that the Laplace transform of the differential equation becomes

$$s^2X(s) + 2sX(s) + 5X(s) = \frac{3}{s}$$

Solving this equation for $X(s)$, we obtain

$$X(s) = \frac{3}{s(s^2 + 2s + 5)}$$
$$= \frac{3}{5}\frac{1}{s} - \frac{3}{5}\frac{s + 2}{s^2 + 2s + 5}$$
$$= \frac{3}{5}\frac{1}{s} - \frac{3}{10}\frac{2}{(s + 1)^2 + 2^2} - \frac{3}{5}\frac{s + 1}{(s + 1)^2 + 2^2}$$

Hence, the inverse Laplace transform becomes

$$x(t) = \mathcal{L}^{-1}[X(s)]$$
$$= \frac{3}{5}\mathcal{L}^{-1}\left[\frac{1}{s}\right] - \frac{3}{10}\mathcal{L}^{-1}\left[\frac{2}{(s + 1)^2 + 2^2}\right] - \frac{3}{5}\mathcal{L}^{-1}\left[\frac{s + 1}{(s + 1)^2 + 2^2}\right]$$
$$= \frac{3}{5} - \frac{3}{10}e^{-t}\sin 2t - \frac{3}{5}e^{-t}\cos 2t \qquad t \geq 0$$

which is the solution of the given differential equation.

EXAMPLE PROBLEMS AND SOLUTIONS

Problem A–2–1

Obtain the real and imaginary parts of

$$\frac{2 + j1}{3 + j4}$$

Also, obtain the magnitude and angle of this complex quantity.

Solution

$$\frac{2 + j1}{3 + j4} = \frac{(2 + j1)(3 - j4)}{(3 + j4)(3 - j4)} = \frac{6 + j3 - j8 + 4}{9 + 16} = \frac{10 - j5}{25}$$

$$= \frac{2}{5} - j\frac{1}{5}$$

Hence,

$$\text{real part} = \frac{2}{5}, \qquad \text{imaginary part} = -j\frac{1}{5}$$

The magnitude and angle of this complex quantity are obtained as follows:

$$\text{magnitude} = \sqrt{\left(\frac{2}{5}\right)^2 + \left(\frac{-1}{5}\right)^2} = \sqrt{\frac{5}{25}} = \frac{1}{\sqrt{5}} = 0.447$$

$$\text{angle} = \tan^{-1}\frac{-1/5}{2/5} = \tan^{-1}\frac{-1}{2} = -26.565°$$

Problem A–2–2

Find the Laplace transform of

$$f(t) = 0 \qquad\qquad t < 0$$
$$= te^{-3t} \qquad\quad t \geq 0$$

Solution   Since

$$\mathscr{L}[t] = G(s) = \frac{1}{s^2}$$

referring to Equation (2–2), we obtain

$$F(s) = \mathscr{L}[te^{-3t}] = G(s + 3) = \frac{1}{(s + 3)^2}$$

Problem A–2–3

What is the Laplace transform of

$$f(t) = 0 \qquad\qquad\quad t < 0$$
$$= \sin(\omega t + \theta) \qquad t \geq 0$$

where θ is a constant?

Solution   Noting that

$$\sin(\omega t + \theta) = \sin \omega t \cos \theta + \cos \omega t \sin \theta$$

we have

$$\mathcal{L}[\sin(\omega t + \theta)] = \cos\theta\,\mathcal{L}[\sin\omega t] + \sin\theta\,\mathcal{L}[\cos\omega t]$$

$$= \cos\theta\frac{\omega}{s^2 + \omega^2} + \sin\theta\frac{s}{s^2 + \omega^2}$$

$$= \frac{\omega\cos\theta + s\sin\theta}{s^2 + \omega^2}$$

Problem A–2–4

Find the Laplace transform $F(s)$ of the function $f(t)$ shown in Figure 2–9. Also, find the limiting value of $F(s)$ as a approaches zero.

Solution   The function $f(t)$ can be written

$$f(t) = \frac{1}{a^2}1(t) - \frac{2}{a^2}1(t - a) + \frac{1}{a^2}1(t - 2a)$$

Then

$$F(s) = \mathcal{L}[f(t)]$$

$$= \frac{1}{a^2}\mathcal{L}[1(t)] - \frac{2}{a^2}\mathcal{L}[1(t - a)] + \frac{1}{a^2}\mathcal{L}[1(t - 2a)]$$

$$= \frac{1}{a^2}\frac{1}{s} - \frac{2}{a^2}\frac{1}{s}e^{-as} + \frac{1}{a^2}\frac{1}{s}e^{-2as}$$

$$= \frac{1}{a^2 s}(1 - 2e^{-as} + e^{-2as})$$

As a approaches zero, we have

$$\lim_{a\to 0}F(s) = \lim_{a\to 0}\frac{1 - 2e^{-as} + e^{-2as}}{a^2 s} = \lim_{a\to 0}\frac{\dfrac{d}{da}(1 - 2e^{-as} + e^{-2as})}{\dfrac{d}{da}(a^2 s)}$$

$$= \lim_{a\to 0}\frac{2se^{-as} - 2se^{-2as}}{2as} = \lim_{a\to 0}\frac{e^{-as} - e^{-2as}}{a}$$

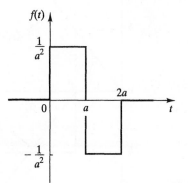

Figure 2–9   Function $f(t)$.

$$= \lim_{a \to 0} \frac{\dfrac{d}{da}(e^{-as} - e^{-2as})}{\dfrac{d}{da}(a)} = \lim_{a \to 0} \frac{-se^{-as} + 2se^{-2as}}{1}$$

$$= -s + 2s = s$$

Problem A–2–5

Obtain the Laplace transform of the function $f(t)$ shown in Figure 2–10.

Solution The given function $f(t)$ can be defined as follows:

$$
\begin{aligned}
f(t) &= 0 & t \le 0 \\
&= \frac{b}{a}t & 0 < t \le a \\
&= 0 & a < t
\end{aligned}
$$

Notice that $f(t)$ can be considered a sum of the three functions $f_1(t)$, $f_2(t)$, and $f_3(t)$ shown in Figure 2–11. Hence, $f(t)$ can be written as

$$f(t) = f_1(t) + f_2(t) + f_3(t)$$
$$= \frac{b}{a}t \cdot 1(t) - \frac{b}{a}(t - a) \cdot 1(t - a) - b \cdot 1(t - a)$$

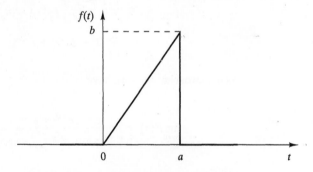

Figure 2–10 Function $f(t)$.

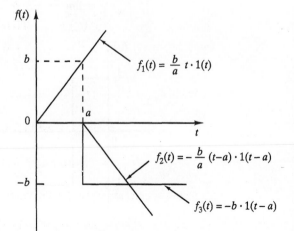

Figure 2–11 Functions $f_1(t)$, $f_2(t)$, and $f_3(t)$.

Then the Laplace transform of $f(t)$ becomes

$$F(s) = \frac{b}{a}\frac{1}{s^2} - \frac{b}{a}\frac{1}{s^2}e^{-as} - b\frac{1}{s}e^{-as}$$

$$= \frac{b}{as^2}(1 - e^{-as}) - \frac{b}{s}e^{-as}$$

The same $F(s)$ can, of course, be obtained by performing the following Laplace integration:

$$\mathcal{L}[f(t)] = \int_0^a \frac{b}{a}te^{-st}\,dt + \int_a^\infty 0\,e^{-st}\,dt$$

$$= \frac{b}{a}t\frac{e^{-st}}{-s}\Big|_0^a - \int_0^a \frac{b}{a}\frac{e^{-st}}{-s}\,dt$$

$$= b\frac{e^{-as}}{-s} + \frac{b}{as}\frac{e^{-st}}{-s}\Big|_0^a$$

$$= b\frac{e^{-as}}{-s} - \frac{b}{as^2}(e^{-as} - 1)$$

$$= \frac{b}{as^2}(1 - e^{-as}) - \frac{b}{s}e^{-as}$$

Problem A–2–6

Prove that if the Laplace transform of $f(t)$ is $F(s)$, then, except at poles of $F(s)$,

$$\mathcal{L}[tf(t)] = -\frac{d}{ds}F(s)$$

$$\mathcal{L}[t^2f(t)] = \frac{d^2}{ds^2}F(s)$$

and in general,

$$\mathcal{L}[t^nf(t)] = (-1)^n\frac{d^n}{ds^n}F(s) \qquad n = 1, 2, 3, \ldots$$

Solution

$$\mathcal{L}[tf(t)] = \int_0^\infty tf(t)e^{-st}\,dt = -\int_0^\infty f(t)\frac{d}{ds}(e^{-st})\,dt$$

$$= -\frac{d}{ds}\int_0^\infty f(t)e^{-st}\,dt = -\frac{d}{ds}F(s)$$

Similarly, by defining $tf(t) = g(t)$, the result is

$$\mathcal{L}[t^2f(t)] = \mathcal{L}[tg(t)] = -\frac{d}{ds}G(s) = -\frac{d}{ds}\left[-\frac{d}{ds}F(s)\right]$$

$$= (-1)^2\frac{d^2}{ds^2}F(s) = \frac{d^2}{ds^2}F(s)$$

Repeating the same process, we obtain

$$\mathcal{L}[t^nf(t)] = (-1)^n\frac{d^n}{ds^n}F(s) \qquad n = 1, 2, 3, \ldots$$

Problem A–2–7

Find the Laplace transform of

$$f(t) = 0 \qquad\qquad t < 0$$
$$= t^2 \sin \omega t \qquad t \geq 0$$

Solution   Since

$$\mathcal{L}[\sin \omega t] = \frac{\omega}{s^2 + \omega^2}$$

referring to **Problem A–2–6**, we have

$$\mathcal{L}[f(t)] = \mathcal{L}[t^2 \sin \omega t] = \frac{d^2}{ds^2}\left[\frac{\omega}{s^2 + \omega^2}\right] = \frac{-2\omega^3 + 6\omega s^2}{(s^2 + \omega^2)^3}$$

Problem A–2–8

Prove that if the Laplace transform of $f(t)$ is $F(s)$, then

$$\mathcal{L}\left[f\left(\frac{t}{a}\right)\right] = aF(as) \qquad a > 0$$

Solution   If we define $t/a = \tau$ and $as = s_1$, then

$$\mathcal{L}\left[f\left(\frac{t}{a}\right)\right] = \int_0^\infty f\left(\frac{t}{a}\right)e^{-st}\,dt = \int_0^\infty f(\tau)e^{-as\tau}a\,d\tau$$

$$= a\int_0^\infty f(\tau)e^{-s_1\tau}\,d\tau = aF(s_1) = aF(as)$$

Problem A–2–9

Prove that if $f(t)$ is of exponential order and if $\int_0^\infty f(t)\,dt$ exists [which means that $\int_0^\infty f(t)\,dt$ assumes a definite value], then

$$\int_0^\infty f(t)\,dt = \lim_{s\to 0} F(s)$$

where $F(s) = \mathcal{L}[f(t)]$.

Solution   Note that

$$\int_0^\infty f(t)\,dt = \lim_{t\to\infty}\int_0^t f(t)\,dt$$

Referring to Equation (2–5), we have

$$\mathcal{L}\left[\int_0^t f(t)\,dt\right] = \frac{F(s)}{s}$$

Since $\int_0^\infty f(t)\,dt$ exists, by applying the final-value theorem to this case, we obtain

$$\lim_{t\to\infty}\int_0^t f(t)\,dt = \lim_{s\to 0} s\frac{F(s)}{s}$$

or

$$\int_0^\infty f(t)\,dt = \lim_{s\to 0} F(s)$$

Problem A–2–10

The convolution of two time functions is defined by

$$\int_0^t f_1(\tau)f_2(t - \tau)\, d\tau$$

A commonly used notation for the convolution is $f_1(t)*f_2(t)$, which is defined as

$$f_1(t)*f_2(t) = \int_0^t f_1(\tau)f_2(t - \tau)\, d\tau = \int_0^t f_1(t - \tau)f_2(\tau)\, d\tau$$

Show that if $f_1(t)$ and $f_2(t)$ are both Laplace transformable, then

$$\mathscr{L}\left[\int_0^t f_1(\tau)f_2(t - \tau)\, d\tau\right] = F_1(s)F_2(s)$$

where $F_1(s) = \mathscr{L}[f_1(t)]$ and $F_2(s) = \mathscr{L}[f_2(t)]$.

Solution   Noting that $1(t - \tau) = 0$ for $t < \tau$, we have

$$\mathscr{L}\left[\int_0^t f_1(\tau)f_2(t - \tau)\, d\tau\right] = \mathscr{L}\left[\int_0^\infty f_1(\tau)f_2(t - \tau)1(t - \tau)\, d\tau\right]$$

$$= \int_0^\infty e^{-st}\left[\int_0^\infty f_1(\tau)f_2(t - \tau)1(t - \tau)\, d\tau\right] dt$$

$$= \int_0^\infty f_1(\tau)\, d\tau \int_0^\infty f_2(t - \tau)1(t - \tau)e^{-st}\, dt$$

Changing the order of integration is valid here, since $f_1(t)$ and $f_2(t)$ are both Laplace transformable, giving convergent integrals. If we substitute $\lambda = t - \tau$ into this last equation, the result is

$$\mathscr{L}\left[\int_0^t f_1(\tau)f_2(t - \tau)\, d\tau\right] = \int_0^\infty f_1(\tau)e^{-s\tau}\, d\tau \int_0^\infty f_2(\lambda)e^{-s\lambda}\, d\lambda$$

$$= F_1(s)F_2(s)$$

or

$$\mathscr{L}[f_1(t)*f_2(t)] = F_1(s)F_2(s)$$

Thus, the Laplace transform of the convolution of two time functions is the product of their Laplace transforms.

Problem A–2–11

Determine the Laplace transform of $f_1(t)*f_2(t)$, where

$$\begin{aligned}
f_1(t) = f_2(t) = 0 && \text{for } t < 0 \\
f_1(t) = t && \text{for } t \geq 0 \\
f_2(t) = 1 - e^{-t} && \text{for } t \geq 0
\end{aligned}$$

Solution   Note that

$$\mathscr{L}[t] = F_1(s) = \frac{1}{s^2}$$

$$\mathscr{L}[1 - e^{-t}] = F_2(s) = \frac{1}{s} - \frac{1}{s + 1}$$

The Laplace transform of the convolution integral is given by

$$\mathscr{L}[f_1(t)*f_2(t)] = F_1(s)F_2(s) = \frac{1}{s^2}\left(\frac{1}{s} - \frac{1}{s+1}\right)$$

$$= \frac{1}{s^3} - \frac{1}{s^2(s+1)} = \frac{1}{s^3} - \frac{1}{s^2} + \frac{1}{s} - \frac{1}{s+1}$$

To verify that the expression after the rightmost equal sign is indeed the Laplace transform of the convolution integral, let us first integrate the convolution integral and then take the Laplace transform of the result. We have

$$f_1(t)*f_2(t) = \int_0^t \tau[1 - e^{-(t-\tau)}]\, d\tau$$

$$= \int_0^t (t - \tau)(1 - e^{-\tau})\, d\tau$$

$$= \int_0^t (t - \tau - te^{-\tau} + \tau e^{-\tau})\, d\tau$$

Noting that

$$\int_0^t (t - \tau)\, d\tau = \frac{t^2}{2}$$

$$\int_0^t te^{-\tau}\, d\tau = -te^{-t} + t$$

$$\int_0^t \tau e^{-\tau}\, d\tau = -\tau e^{-\tau}\Big|_0^t + \int_0^t e^{-\tau}\, d\tau = -te^{-t} - e^{-t} + 1$$

we have

$$f_1(t)*f_2(t) = \frac{t^2}{2} - t + 1 - e^{-t}$$

Thus,

$$\mathscr{L}\left[\frac{t^2}{2} - t + 1 - e^{-t}\right] = \frac{1}{s^3} - \frac{1}{s^2} + \frac{1}{s} - \frac{1}{s+1}$$

Problem A–2–12

Prove that if $f(t)$ is a periodic function with period T, then

$$\mathscr{L}[f(t)] = \frac{\displaystyle\int_0^T f(t)e^{-st}\, dt}{1 - e^{-Ts}}$$

Solution

$$\mathscr{L}[f(t)] = \int_0^\infty f(t)e^{-st}\, dt = \sum_{n=0}^\infty \int_{nT}^{(n+1)T} f(t)e^{-st}\, dt$$

By changing the independent variable from t to $\tau = t - nT$, we obtain

$$\mathscr{L}[f(t)] = \sum_{n=0}^\infty e^{-nTs} \int_0^T f(\tau + nT)e^{-s\tau}\, d\tau$$

Since $f(t)$ is a periodic function with period T, $f(\tau + nT) = f(\tau)$. Hence,

$$\mathcal{L}[f(t)] = \sum_{n=0}^{\infty} e^{-nTs} \int_0^T f(\tau)e^{-st}\,d\tau$$

Noting that

$$\sum_{n=0}^{\infty} e^{-nTs} = 1 + e^{-Ts} + e^{-2Ts} + \cdots$$

$$= 1 + e^{-Ts}(1 + e^{-Ts} + e^{-2Ts} + \cdots)$$

$$= 1 + e^{-Ts}\left(\sum_{n=0}^{\infty} e^{-nTs}\right)$$

we obtain

$$\sum_{n=0}^{\infty} e^{-nTs} = \frac{1}{1 - e^{-Ts}}$$

It follows that

$$\mathcal{L}[f(t)] = \frac{\displaystyle\int_0^T f(t)e^{-st}\,dt}{1 - e^{-Ts}}$$

Problem A–2–13

What is the Laplace transform of the periodic function shown in Figure 2–12?

Solution   Note that

$$\int_0^T f(t)e^{-st}\,dt = \int_0^{T/2} e^{-st}\,dt + \int_{T/2}^T (-1)e^{-st}\,dt$$

$$= \frac{e^{-st}}{-s}\Big|_0^{T/2} - \frac{e^{-st}}{-s}\Big|_{T/2}^T$$

$$= \frac{e^{-(1/2)Ts} - 1}{-s} + \frac{e^{-Ts} - e^{-(1/2)Ts}}{s}$$

$$= \frac{1}{s}[e^{-Ts} - 2e^{-(1/2)Ts} + 1]$$

$$= \frac{1}{s}[1 - e^{-(1/2)Ts}]^2$$

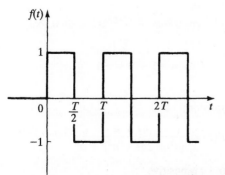

Figure 2–12   Periodic function (square wave).

Consequently,

$$F(s) = \frac{\displaystyle\int_0^T f(t)e^{-st}\, dt}{1 - e^{-Ts}} = \frac{(1/s)[1 - e^{-(1/2)Ts}]^2}{1 - e^{-Ts}}$$

$$= \frac{1 - e^{-(1/2)Ts}}{s[1 + e^{-(1/2)Ts}]} = \frac{1}{s}\tanh\frac{Ts}{4}$$

Problem A–2–14

Find the initial value of $df(t)/dt$, where the Laplace transform of $f(t)$ is given by

$$F(s) = \mathcal{L}[f(t)] = \frac{2s + 1}{s^2 + s + 1}$$

Solution   Using the initial-value theorem, we obtain

$$\lim_{t\to 0+} f(t) = \lim_{s\to\infty} sF(s) = \lim_{s\to\infty}\frac{s(2s + 1)}{s^2 + s + 1} = 2$$

Since the $\mathcal{L}_+$ transform of $df(t)/dt = g(t)$ is given by

$$\mathcal{L}_+[g(t)] = sF(s) - f(0+)$$

$$= \frac{s(2s + 1)}{s^2 + s + 1} - 2 = \frac{-s - 2}{s^2 + s + 1}$$

the initial value of $df(t)/dt$ is obtained as

$$\lim_{t\to 0+}\frac{df(t)}{dt} = g(0+) = \lim_{s\to\infty} s[sF(s) - f(0+)]$$

$$= \lim_{s\to\infty}\frac{-s^2 - 2s}{s^2 + s + 1} = -1$$

To verify this result, notice that

$$F(s) = \frac{2(s + 0.5)}{(s + 0.5)^2 + (0.866)^2} = \mathcal{L}[2e^{-0.5t}\cos 0.866t]$$

Hence,

$$f(t) = 2e^{-0.5t}\cos 0.866t$$

and

$$\dot{f}(t) = -e^{-0.5t}\cos 0.866t + 2e^{-0.5t}0.866\sin 0.866t$$

Thus,

$$\dot{f}(0) = -1 + 0 = -1$$

Problem A–2–15

Obtain the inverse Laplace transform of

$$F(s) = \frac{cs + d}{(s^2 + 2as + a^2) + b^2}$$

where $a, b, c,$ and d are real and a is positive.

Solution   Since $F(s)$ can be written as

$$F(s) = \frac{c(s + a) + d - ca}{(s + a)^2 + b^2}$$

$$= \frac{c(s + a)}{(s + a)^2 + b^2} + \frac{d - ca}{b} \frac{b}{(s + a)^2 + b^2}$$

we obtain

$$f(t) = ce^{-at} \cos bt + \frac{d - ca}{b} e^{-at} \sin bt$$

Problem A–2–16

Find the inverse Laplace transform of

$$F(s) = \frac{1}{s(s^2 + 2s + 2)}$$

Solution   Since

$$s^2 + 2s + 2 = (s + 1 + j1)(s + 1 - j1)$$

it follows that $F(s)$ involves a pair of complex-conjugate poles, so we expand $F(s)$ into the form

$$F(s) = \frac{1}{s(s^2 + 2s + 2)} = \frac{a_1}{s} + \frac{a_2 s + a_3}{s^2 + 2s + 2}$$

where a_1, a_2, and a_3 are determined from

$$1 = a_1(s^2 + 2s + 2) + (a_2 s + a_3)s$$

By comparing corresponding coefficients of the s^2, s, and s^0 terms on both sides of this last equation respectively, we obtain

$$a_1 + a_2 = 0, \qquad 2a_1 + a_3 = 0, \qquad 2a_1 = 1$$

from which it follows that

$$a_1 = \frac{1}{2}, \qquad a_2 = -\frac{1}{2}, \qquad a_3 = -1$$

Therefore,

$$F(s) = \frac{1}{2} \frac{1}{s} - \frac{1}{2} \frac{s + 2}{s^2 + 2s + 2}$$

$$= \frac{1}{2} \frac{1}{s} - \frac{1}{2} \frac{1}{(s + 1)^2 + 1^2} - \frac{1}{2} \frac{s + 1}{(s + 1)^2 + 1^2}$$

The inverse Laplace transform of $F(s)$ is

$$f(t) = \frac{1}{2} - \frac{1}{2} e^{-t} \sin t - \frac{1}{2} e^{-t} \cos t \qquad t \geq 0$$

Problem A–2–17

Derive the inverse Laplace transform of

$$F(s) = \frac{5(s + 2)}{s^2(s + 1)(s + 3)}$$

Solution

$$F(s) = \frac{5(s + 2)}{s^2(s + 1)(s + 3)} = \frac{b_2}{s^2} + \frac{b_1}{s} + \frac{a_1}{s + 1} + \frac{a_2}{s + 3}$$

where

$$a_1 = \frac{5(s + 2)}{s^2(s + 3)}\bigg|_{s=-1} = \frac{5}{2}$$

$$a_2 = \frac{5(s + 2)}{s^2(s + 1)}\bigg|_{s=-3} = \frac{5}{18}$$

$$b_2 = \frac{5(s + 2)}{(s + 1)(s + 3)}\bigg|_{s=0} = \frac{10}{3}$$

$$b_1 = \frac{d}{ds}\left[\frac{5(s + 2)}{(s + 1)(s + 3)}\right]_{s=0}$$

$$= \frac{5(s + 1)(s + 3) - 5(s + 2)(2s + 4)}{(s + 1)^2(s + 3)^2}\bigg|_{s=0} = -\frac{25}{9}$$

Thus,

$$F(s) = \frac{10}{3}\frac{1}{s^2} - \frac{25}{9}\frac{1}{s} + \frac{5}{2}\frac{1}{s + 1} + \frac{5}{18}\frac{1}{s + 3}$$

The inverse Laplace transform of $F(s)$ is

$$f(t) = \frac{10}{3}t - \frac{25}{9} + \frac{5}{2}e^{-t} + \frac{5}{18}e^{-3t} \qquad t \geq 0$$

Problem A–2–18

Find the inverse Laplace transform of

$$F(s) = \frac{s^4 + 2s^3 + 3s^2 + 4s + 5}{s(s + 1)}$$

Solution   Since the numerator polynomial is of higher degree than the denominator polynomial, by dividing the numerator by the denominator until the remainder is a fraction, we obtain

$$F(s) = s^2 + s + 2 + \frac{2s + 5}{s(s + 1)} = s^2 + s + 2 + \frac{a_1}{s} + \frac{a_2}{s + 1}$$

where

$$a_1 = \frac{2s + 5}{s + 1}\bigg|_{s=0} = 5$$

$$a_2 = \frac{2s + 5}{s}\bigg|_{s=-1} = -3$$

It follows that

$$F(s) = s^2 + s + 2 + \frac{5}{s} - \frac{3}{s+1}$$

The inverse Laplace transform of $F(s)$ is

$$f(t) = \mathcal{L}^{-1}[F(s)] = \frac{d^2}{dt^2}\delta(t) + \frac{d}{dt}\delta(t) + 2\delta(t) + 5 - 3e^{-t} \qquad t \geq 0-$$

Problem A–2–19

Obtain the inverse Laplace transform of

$$F(s) = \frac{2s^2 + 4s + 6}{s^2(s^2 + 2s + 10)} \tag{2–10}$$

Solution   Since the quadratic term in the denominator involves a pair of complex-conjugate roots, we expand $F(s)$ into the following partial-fraction form:

$$F(s) = \frac{a_1}{s^2} + \frac{a_2}{s} + \frac{bs + c}{s^2 + 2s + 10}$$

The coefficient a_1 can be obtained as

$$a_1 = \frac{2s^2 + 4s + 6}{s^2 + 2s + 10}\bigg|_{s=0} = 0.6$$

Hence, we obtain

$$F(s) = \frac{0.6}{s^2} + \frac{a_2}{s} + \frac{bs + c}{s^2 + 2s + 10}$$

$$= \frac{(a_2 + b)s^3 + (0.6 + 2a_2 + c)s^2 + (1.2 + 10a_2)s + 6}{s^2(s^2 + 2s + 10)} \tag{2–11}$$

By equating corresponding coefficients in the numerators of Equations (2–10) and (2–11), respectively, we obtain

$$a_2 + b = 0$$
$$0.6 + 2a_2 + c = 2$$
$$1.2 + 10a_2 = 4$$

from which we get

$$a_2 = 0.28, \qquad b = -0.28, \qquad c = 0.84$$

Hence,

$$F(s) = \frac{0.6}{s^2} + \frac{0.28}{s} + \frac{-0.28s + 0.84}{s^2 + 2s + 10}$$

$$= \frac{0.6}{s^2} + \frac{0.28}{s} + \frac{-0.28(s + 1) + (1.12/3) \times 3}{(s + 1)^2 + 3^2}$$

The inverse Laplace transform of $F(s)$ gives

$$f(t) = 0.6t + 0.28 - 0.28e^{-t}\cos 3t + \frac{1.12}{3}e^{-t}\sin 3t$$

Problem A–2–20

Derive the inverse Laplace transform of

$$F(s) = \frac{1}{s(s^2 + \omega^2)}$$

Solution

$$F(s) = \frac{1}{s(s^2 + \omega^2)} = \frac{1}{\omega^2}\left(\frac{1}{s} - \frac{s}{s^2 + \omega^2}\right)$$

$$= \frac{1}{\omega^2}\frac{1}{s} - \frac{1}{\omega^2}\frac{s}{s^2 + \omega^2}$$

Thus, the inverse Laplace transform of $F(s)$ is obtained as

$$f(t) = \mathcal{L}^{-1}[F(s)] = \frac{1}{\omega^2}(1 - \cos \omega t) \qquad t \geq 0$$

Problem A–2–21

Obtain the solution of the differential equation

$$\dot{x} + ax = A \sin \omega t, \qquad x(0) = b$$

Solution Laplace transforming both sides of this differential equation, we have

$$[sX(s) - x(0)] + aX(s) = A\frac{\omega}{s^2 + \omega^2}$$

or

$$(s + a)X(s) = \frac{A\omega}{s^2 + \omega^2} + b$$

Solving this last equation for $X(s)$, we obtain

$$X(s) = \frac{A\omega}{(s + a)(s^2 + \omega^2)} + \frac{b}{s + a}$$

$$= \frac{A\omega}{a^2 + \omega^2}\left(\frac{1}{s + a} - \frac{s - a}{s^2 + \omega^2}\right) + \frac{b}{s + a}$$

$$= \left(b + \frac{A\omega}{a^2 + \omega^2}\right)\frac{1}{s + a} + \frac{Aa}{a^2 + \omega^2}\frac{\omega}{s^2 + \omega^2} - \frac{A\omega}{a^2 + \omega^2}\frac{s}{s^2 + \omega^2}$$

The inverse Laplace transform of $X(s)$ then gives

$$x(t) = \mathcal{L}^{-1}[X(s)]$$

$$= \left(b + \frac{A\omega}{a^2 + \omega^2}\right)e^{-at} + \frac{Aa}{a^2 + \omega^2}\sin \omega t - \frac{A\omega}{a^2 + \omega^2}\cos \omega t \qquad t \geq 0$$

PROBLEMS

Problem B–2–1

Derive the Laplace transform of the function

$$f(t) = 0 \qquad t < 0$$
$$= te^{-2t} \qquad t \geq 0$$

Problem B–2–2

Find the Laplace transforms of the following functions:

(a) $\qquad f_1(t) = 0 \qquad\qquad\qquad t < 0$
$\qquad\qquad = 3\sin(5t + 45°) \qquad t \geq 0$

(b) $\qquad f_2(t) = 0 \qquad\qquad\qquad t < 0$
$\qquad\qquad = 0.03(1 - \cos 2t) \qquad t \geq 0$

Problem B–2–3

Obtain the Laplace transform of the function defined by

$$f(t) = 0 \qquad t < 0$$
$$= t^2 e^{-at} \qquad t \geq 0$$

Problem B–2–4

Obtain the Laplace transform of the function

$$f(t) = 0 \qquad\qquad t < 0$$
$$= \cos 2\omega t \cdot \cos 3\omega t \qquad t \geq 0$$

Problem B–2–5

What is the Laplace transform of the function $f(t)$ shown in Figure 2–13?

Problem B–2–6

Obtain the Laplace transform of the pulse function $f(t)$ shown in Figure 2–14.

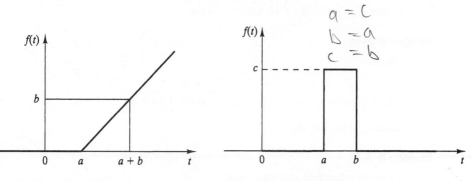

Figure 2–13 Function $f(t)$.

Figure 2–14 Pulse function.

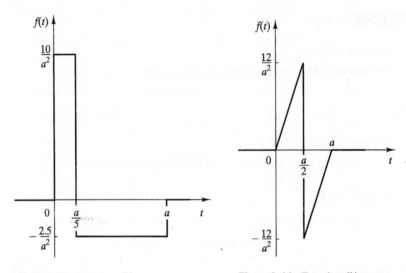

Figure 2–15   Function $f(t)$.   **Figure 2–16**   Function $f(t)$.

Problem B–2–7

What is the Laplace transform of the function $f(t)$ shown in Figure 2–15? Also, what is the limiting value of $\mathscr{L}[f(t)]$ as a approaches zero?

Problem B–2–8

Find the Laplace transform of the function $f(t)$ shown in Figure 2–16. Also, find the limiting value of $\mathscr{L}[f(t)]$ as a approaches zero.

Problem B–2–9

Given

$$F(s) = \frac{5(s + 2)}{s(s + 1)}$$

obtain $f(\infty)$. Use the final-value theorem.

Problem B–2–10

Given

$$F(s) = \frac{2(s + 2)}{s(s + 1)(s + 3)}$$

obtain $f(0+)$. Use the initial-value theorem.

Problem B–2–11

Consider a function $x(t)$. Show that

$$\dot{x}(0+) = \lim_{s \to \infty} [s^2 X(s) - sx(0+)]$$

Problem B–2–12

Derive the Laplace transform of the third derivative of $f(t)$.

Problem B–2–13

What are the inverse Laplace transforms of the following functions?

(a)
$$F_1(s) = \frac{s + 5}{(s + 1)(s + 3)}$$

(b)
$$F_2(s) = \frac{3(s + 4)}{s(s + 1)(s + 2)}$$

Problem B–2–14

Find the inverse Laplace transforms of the following functions:

(a)
$$F_1(s) = \frac{6s + 3}{s^2}$$

(b)
$$F_2(s) = \frac{5s + 2}{(s + 1)(s + 2)^2}$$

Problem B–2–15

Find the inverse Laplace transform of

$$F(s) = \frac{2s^2 + 4s + 5}{s(s + 1)}$$

Problem B–2–16

Obtain the inverse Laplace transform of

$$F(s) = \frac{s^2 + 2s + 4}{s^2}$$

Problem B–2–17

Obtain the inverse Laplace transform of

$$F(s) = \frac{s}{s^2 + 2s + 10}$$

Problem B–2–18

Obtain the inverse Laplace transform of

$$F(s) = \frac{s^2 + 2s + 5}{s^2(s + 1)}$$

Problem B–2–19

Obtain the inverse Laplace transform of

$$F(s) = \frac{2s + 10}{(s + 1)^2(s + 4)}$$

Problem B–2–20

Derive the inverse Laplace transform of

$$F(s) = \frac{1}{s^2(s^2 + \omega^2)}$$

Problem B–2–21

Obtain the inverse Laplace transform of

$$F(s) = \frac{c}{s^2}(1 - e^{-as}) - \frac{b}{s}e^{-as}$$

where $a > 0$.

Problem B–2–22

Find the solution $x(t)$ of the differential equation

$$\ddot{x} + 4x = 0, \qquad x(0) = 5, \qquad \dot{x}(0) = 0$$

Problem B–2–23

Obtain the solution $x(t)$ of the differential equation

$$\ddot{x} + \omega_n^2 x = t, \qquad x(0) = 0, \qquad \dot{x}(0) = 0$$

Problem B–2–24

Determine the solution $x(t)$ of the differential equation

$$2\ddot{x} + 2\dot{x} + x = 1, \qquad x(0) = 0, \qquad \dot{x}(0) = 2$$

Problem B–2–25

Obtain the solution $x(t)$ of the differential equation

$$\ddot{x} + x = \sin 3t, \qquad x(0) = 0, \qquad \dot{x}(0) = 0$$

Mechanical Systems

3–1 INTRODUCTION

This chapter is an introductory account of mechanical systems. Details of mathematical modeling and response analyses of various mechanical systems are given in Chapters 4, 5, 7, 8, and 9.

We begin with a review of systems of units; a clear understanding of which is necessary for the quantitative study of system dynamics.

Systems of units. Most engineering calculations in the United States are based on the *International System* (abbreviated SI)[1] *of units* and the *British engineering system* (BES) of measurement. The International System is a modified metric system, and, as such, it differs from conventional metric absolute or metric gravitational systems of units. Table 3–1 lists some units of measure from each of the International System, conventional metric systems, and British systems of units. (The table presents only those units necessary to describe the behavior of mechanical systems. Units used in describing the behaviors of electrical systems are given in Chapter 6. For additional details on systems of units, refer to Appendix A.)

The chief difference between "absolute" systems of units and "gravitational" systems of units lies in the choice of mass or force as a primary dimension. In the

---

[1]This "backward" abbreviation is for the French *Système International.*

TABLE 3–1　Systems of Units

Systems of units \ Quantity	Absolute systems				Gravitational systems	
	Metric			British	Metric engineering	British engineering
	SI	mks	cgs			
Length	m	m	cm	ft	m	ft
Mass	kg	kg	g	lb	$\dfrac{\text{kg}_f\text{-s}^2}{\text{m}}$	$\text{slug} = \dfrac{\text{lb}_f\text{-s}^2}{\text{ft}}$
Time	s	s	s	s	s	s
Force	$\dfrac{\text{N}}{= \dfrac{\text{kg-m}}{\text{s}^2}}$	$\dfrac{\text{N}}{= \dfrac{\text{kg-m}}{\text{s}^2}}$	$\dfrac{\text{dyn}}{= \dfrac{\text{g-cm}}{\text{s}^2}}$	$\dfrac{\text{poundal}}{= \dfrac{\text{lb-ft}}{\text{s}^2}}$	kg_f	lb_f
Energy	J = N-m	J = N-m	$\dfrac{\text{erg}}{= \text{dyn-cm}}$	ft-poundal	$\text{kg}_f\text{-m}$	ft-lb$_f$ or Btu
Power	$\text{W} = \dfrac{\text{N-m}}{\text{s}}$	$\text{W} = \dfrac{\text{N-m}}{\text{s}}$	$\dfrac{\text{dyn-cm}}{\text{s}}$	$\dfrac{\text{ft-poundal}}{\text{s}}$	$\dfrac{\text{kg}_f\text{-m}}{\text{s}}$	$\dfrac{\text{ft-lb}_f}{\text{s}}$ or hp

absolute systems (SI and the metric and British absolute systems), mass is chosen as a primary dimension and force is a derived quantity. Conversely, in gravitational systems (metric engineering and British engineering systems) of units, force is a primary dimension and mass is a derived quantity. In gravitational systems, the mass of a body is defined as the ratio of the magnitude of the force to that of acceleration. (Thus, the dimension of mass is force/acceleration.)

Mass.　The *mass* of a body is the quantity of matter in it, which is assumed to be constant. Physically, mass is the property of a body that gives it inertia, that is, resistance to starting and stopping. A body is attracted by the earth, and the magnitude of the force that the earth exerts on the body is called its *weight*.

In practical situations, we know the weight w of a body, but not the mass m. We calculate mass m from

$$m = \frac{w}{g}$$

where g is the gravitational acceleration constant. The value of g varies slightly from point to point on the earth's surface. As a result, the weight of a body varies slightly at different points on the earth's surface, but its mass remains constant. For engineering purposes,

$$g = 9.807 \text{ m/s}^2 = 980.7 \text{ cm/s}^2 = 32.174 \text{ ft/s}^2 = 386.1 \text{ in./s}^2$$

Far out in space, a body becomes weightless. Yet its mass remains constant, so the body possesses inertia.

The units of mass are kg, g, lb, kg_f-s^2/m, and slug, as shown in Table 3–1. If mass is expressed in units of kilograms (or pounds), we call it kilogram mass (or pound mass) to distinguish it from the unit of force, which is termed kilogram force (or pound force). In this book, kg is used to denote a kilogram mass and kg_f a kilogram force. Similarly, lb denotes a pound mass and lb_f a pound force.

A slug is a unit of mass such that, when acted on by a 1-pound force, a 1-slug mass accelerates at 1 ft/s^2 (slug = lb_f-s^2/ft). In other words, if a mass of 1 slug is acted on by a 32.174-pound force, it accelerates at 32.174 ft/s^2 ($= g$). Hence, the mass of a body weighing 32.174 lb_f at the earth's surface is 1 slug, or

$$m = \frac{w}{g} = \frac{32.174 \text{ lb}_f}{32.174 \text{ ft/s}^2} = 1 \text{ slug}$$

Force.   *Force* can be defined as the cause which tends to produce a change in motion of a body on which it acts. To move a body, force must be applied to it. Two types of forces are capable of acting on a body: *contact* forces and *field* forces. Contact forces are those which come into direct contact with a body, whereas field forces, such as gravitational force and magnetic force, act on a body, but do not come into contact with it.

The units of force are the newton (N), dyne (dyn), poundal, kg_f, and lb_f. In SI units and the mks system (a metric absolute system) of units, the force unit is the newton. One newton is the force that will give a 1-kg mass an acceleration of 1 m/s^2, or

$$1 \text{ N} = 1 \text{ kg-m/s}^2$$

This implies that 9.807 N will give a 1-kg mass an acceleration of 9.807 m/s^2. Since the gravitational acceleration constant is $g = 9.807$ m/s^2, a mass of 1 kg will produce a force of 9.807 N on its support.

The force unit in the cgs system (a metric absolute system) is the dyne, which will give a 1-g mass an acceleration of 1 cm/s^2, or

$$1 \text{ dyn} = 1 \text{ g-cm/s}^2$$

The force unit in the metric engineering (gravitational) system is kg_f, which is a primary dimension in the system. Similarly, in the British engineering system, the force unit is lb_f, a primary dimension in this system of units.

Comments.   The SI units of force, mass, and length are the newton (N), kilogram mass (kg), and meter (m). The mks units of force, mass, and length are the same as the SI units. The cgs units for force, mass, and length are the dyne (dyn), gram (g), and centimeter (cm), and those for the BES units are the pound force (lb_f), slug, and foot (ft). Each system of units is consistent in that the unit of force accelerates the unit of mass 1 unit of length per second per second.

A special effort has been made in this book to familiarize the reader with the various systems of measurement. In examples and problems, for instance, calculations are often made in SI units, conventional metric units, and BES units, in order to illustrate how to convert from one system to another. Table 3–2 shows some convenient conversion factors among different systems of units. (Other detailed conversion tables are given in Appendix B.)

TABLE 3–2   Conversion Table

Length	1	1 m = 100 cm	
	2	1 ft = 12 in.	1 in. = 2.54 cm
	3	1 m = 3.281 ft	1 ft = 0.3048 m
Mass	4	1 kg = 2.2046 lb	1 lb = 0.4536 kg
	5	1 kg = 0.10197 kg_f-s^2/m	1 kg_f-s^2/m = 9.807 kg
	6	1 slug = 14.594 kg	1 kg = 0.06852 slug
	7	1 slug = 32.174 lb	1 lb = 0.03108 slug
	8	1 slug = 1.488 kg_f-s^2/m	1 kg_f-s^2/m = 0.6720 slug
Moment of inertia	9	1 slug-ft^2 = 1.356 kg-m^2	1 kg-m^2 = 0.7376 slug-ft^2
	10	1 slug-ft^2 = 0.1383 kg_f-s^2-m	1 kg_f-s^2-m = 7.233 slug-ft^2
	11	1 slug-ft^2 = 32.174 lb-ft^2	1 lb-ft^2 = 0.03108 slug-ft^2
Force	12	1 N = 10^5 dyn	
	13	1 N = 0.10197 kg_f	1 kg_f = 9.807 N
	14	1 N = 7.233 poundals	1 poundal = 0.1383 N
	15	1 N = 0.2248 lb_f	1 lb_f = 4.4482 N
	16	1 kg_f = 2.2046 lb_f	1 lb_f = 0.4536 kg_f
	17	1 lb_f = 32.174 poundals	1 poundal = 0.03108 lb_f
Energy	18	1 N-m = 1 J = 1 W-s	1 J = 0.10197 kg_f-m
	19	1 dyn-cm = 1 erg = 10^{-7} J	1 kg_f-m = 9.807 N-m
	20	1 N-m = 0.7376 ft-lb_f	1 ft-lb_f = 1.3557 N-m
	21	1 J = 2.389 × 10^{-4} kcal	1 kcal = 4186 J
	22	1 Btu = 778 ft-lb_f	1 ft-lb_f = 1.285 × 10^{-3} Btu
Power	23	1 W = 1 J/s	
	24	1 hp = 550 ft-lb_f/s	1 ft-lb_f/s = 1.818 × 10^{-3} hp
	25	1 hp = 745.7 W	1 W = 1.341 × 10^{-3} hp

Outline of the chapter. Section 3–1 has presented a review of systems of units necessary in the discussions of dynamics of mechanical systems. Section 3–2 treats mechanical elements. Section 3–3 discusses mathematical modeling of mechanical systems and analyzes simple mechanical systems. Section 3–4 reviews the concept of work, energy, and power and then presents energy methods for deriving mathematical models of conservative systems (systems that do not dissipate energy).

3-2 MECHANICAL ELEMENTS

Any mechanical system consists of mechanical elements. There are three types of basic elements in mechanical systems: inertia elements, spring elements, and damper elements.

Inertia elements.   By *inertia elements*, we mean masses and moments of inertia. *Inertia* may be defined as the change in force (torque) required to make a unit change in acceleration (angular acceleration). That is,

$$\text{inertia (mass)} = \frac{\text{change in force}}{\text{change in acceleration}} \quad \frac{\text{N}}{\text{m/s}^2} \text{ or kg}$$

$$\text{inertia (moment of inertia)} = \frac{\text{change in torque}}{\text{change in angular acceleration}} \quad \frac{\text{N-m}}{\text{rad/s}^2} \text{ or kg-m}^2$$

Spring Elements.   A linear *spring* is a mechanical element that can be deformed by an external force or torque such that the deformation is directly proportional to the force or torque applied to the element.

Consider the spring shown in Figure 3–1(a). Here, we consider translational motion only. Suppose that the natural length of the spring is X, the spring is fixed at one end, and the other end is free. Then, when a force f is applied at the free end, the spring is stretched. The elongation of the spring is x. The force that arises in the spring is proportional to x and is given by

$$F = kx \tag{3-1}$$

where k is a proportionality constant called the *spring constant*. The dimension of the spring constant k is force/displacement. At point P, this spring force F acts opposite to the direction of the force f applied at point P.

Figure 3–1(b) shows the case where both ends (denoted by points P and Q) of the spring are deflected due to the forces f applied at each end. (The forces at each

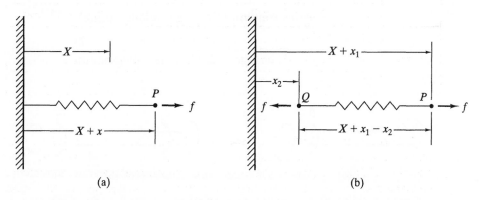

| (a) | (b) |

Figure 3-1   (a) One end of the spring is deflected; (b) both ends of the spring are deflected. (X is the natural length of the spring.)

end of the spring are on the same line and are equal in magnitude but opposite in direction.) The natural length of the spring is X. The net elongation of the spring is $x_1 - x_2$. The force acting in the spring is then

$$F = k(x_1 - x_2) \tag{3-2}$$

At point P, the spring force F acts to the left. At point Q, F acts to the right. (Note that the displacements $X + x_1$ and x_2 of the ends of the spring are measured relative to the same frame of reference.)

Next, consider the torsional spring shown in Figure 3–2(a), where one end is fixed and a torque τ is applied to the other end. The angular displacement of the free end is θ. Then the torque T that arises in the torsional spring is

$$T = k\theta \tag{3-3}$$

At the free end, this torque acts in the torsional spring in the direction opposite that of the applied torque τ.

For the torsional spring shown in Figure 3–2(b), torques equal in magnitude, but opposite in direction, are applied to the ends of the spring. In this case, the torque T acting in the torsional spring is

$$T = k(\theta_1 - \theta_2) \tag{3-4}$$

At each end, the spring torque acts in the direction opposite that of the applied torque at that end. The dimension of the torsional spring constant k is torque/angular displacement, where angular displacement is measured in radians.

When a linear spring is stretched, a point is reached in which the force per unit displacement begins to change and the spring becomes a nonlinear spring. If the spring is stretched farther, a point is reached at which the material will either break or yield. For practical springs, therefore, the assumption of linearity may be good only for relatively small net displacements. Figure 3–3 shows the force–displacement characteristic curves for linear and nonlinear springs.

For linear springs, the spring constant k may be defined as follows:

spring constant k (for translational spring)

$$= \frac{\text{change in force}}{\text{change in displacement of spring}} \frac{\text{N}}{\text{m}}$$

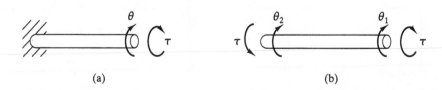

(a)   (b)

Figure 3–2   (a) A torque τ is applied at one end of torsional spring, and the other end is fixed; (b) a torque τ is applied at one end, and a torque τ, in the opposite direction, is applied at the other end.

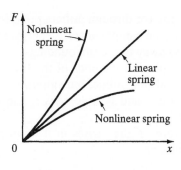

Figure 3–3   Force–displacement characteristic curves for linear and nonlinear springs.

spring constant k (for torsional spring)

$$= \frac{\text{change in torque}}{\text{change in angular displacement of spring}} \quad \frac{\text{N-m}}{\text{rad}}$$

Spring constants indicate stiffness; a large value of k corresponds to a hard spring, a small value of k to a soft spring. The reciprocal of the spring constant k is called *compliance* or *mechanical capacitance* C. Thus, $C = 1/k$. Compliance or mechanical capacitance indicates the softness of a spring.

Practical spring versus ideal spring.   All practical springs have inertia and damping. In our analysis in this book, however, we assume that the effect of the mass of a spring is negligibly small; that is, the inertia force due to acceleration of the spring is negligibly small compared with the spring force. Also, we assume that the damping effect of the spring is negligibly small.

An ideal linear spring, in comparison to a practical spring, will have neither mass nor damping and will obey the linear force–displacement law as given by Equations (3–1) and (3–2) or the linear torque–angular displacement law as given by Equations (3–3) and (3–4).

Damper elements.   A *damper* is a mechanical element that dissipates energy in the form of heat instead of storing it. Figure 3–4(a) shows a schematic diagram of a translational damper, or dashpot. It consists of a piston and an oil-filled cylinder. Any relative motion between the piston rod and the cylinder is resisted by oil

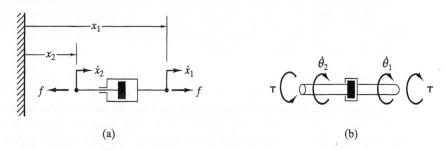

(a)   (b)

Figure 3–4   (a) Translational damper; (b) torsional (or rotational) damper.

because oil must flow around the piston (or through orifices provided in the piston) from one side to the other. Essentially, the damper absorbs energy, and the absorbed energy is dissipated as heat that flows away to the surroundings.

In Figure 3–4(a), the forces f applied at the ends of the translational damper are on the same line and are of equal magnitude, but opposite in direction. The velocities of the ends of the damper are $\dot{x}_1$ and $\dot{x}_2$. Velocities $\dot{x}_1$ and $\dot{x}_2$ are taken relative to the same frame of reference.

In the damper, the damping force F that arises in it is proportional to the velocity difference $\dot{x}_1 - \dot{x}_2$ of the ends, or

$$F = b(\dot{x}_1 - \dot{x}_2) = b\dot{x} \qquad (3\text{--}5)$$

where $\dot{x} = \dot{x}_1 - \dot{x}_2$ and the proportionality constant b relating the damping force F to the velocity difference $\dot{x}$ is called the *viscous friction coefficient* or *viscous friction constant*. The dimension of b is force/velocity. Note that the initial positions of both ends of the damper do not appear in the equation.

For the torsional damper shown in Figure 3–4(b), the torques τ applied to the ends of the damper are of equal magnitude, but opposite in direction. The angular velocities of the ends are $\dot{\theta}_1$ and $\dot{\theta}_2$ and they are taken relative to the same frame of reference. The damping torque T that arises in the damper is proportional to the angular velocity difference $\dot{\theta}_1 - \dot{\theta}_2$ of the ends, or

$$T = b(\dot{\theta}_1 - \dot{\theta}_2) = b\dot{\theta} \qquad (3\text{--}6)$$

where, analogous to the translational case, $\dot{\theta} = \dot{\theta}_1 - \dot{\theta}_2$ and the proportionality constant b relating the damping torque T to the angular velocity difference $\dot{\theta}$ is called the viscous friction coefficient or viscous friction constant. The dimension of b is torque/angular velocity. Note that the initial angular positions of both ends of the damper do not appear in the equation.

A damper is an element that provides resistance in mechanical motion, and, as such, its effect on the dynamic behavior of a mechanical system is similar to that of an electrical resistor on the dynamic behavior of an electrical system. Consequently, a damper is often referred to as a *mechanical resistance element* and the viscous friction coefficient as the *mechanical resistance*.

Practical damper versus ideal damper.   All practical dampers produce inertia and spring effects. In this book, however, we assume that these effects are negligible.

An ideal damper is massless and springless, dissipates all energy, and obeys the linear force–velocity law or linear torque–angular velocity law as given by Equation (3–5) or Equation (3–6), respectively.

Nonlinear friction.   Friction that obeys a linear law is called *linear friction*, whereas friction that does not is described as *nonlinear*. Examples of nonlinear friction include static friction, sliding friction, and square-law friction. Square-law friction occurs when a solid body moves in a fluid medium. Figure 3–5 shows a characteristic curve for square-law friction. In this book, we shall not discuss nonlinear friction any further.

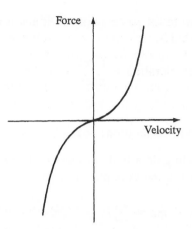

Figure 3–5   Characteristic curve for square-law friction.

3–3 MATHEMATICAL MODELING OF SIMPLE MECHANICAL SYSTEMS

A mathematical model of any mechanical system can be developed by applying Newton's laws to the system. In this section, we shall deal with the problem of deriving mathematical models of simple mechanical systems. More on deriving mathematical models of various mechanical systems and response analyses is presented in Chapters 4, 5, 7, 8, and 9.

Rigid body.   When any real body is accelerated, internal elastic deflections are always present. If these internal deflections are negligibly small relative to the gross motion of the entire body, the body is called a *rigid body*. Thus, a rigid body does not deform.

Newton's laws.   There are three well-known laws called *Newton's laws*. Newton's first law, which concerns the conservation of momentum, states that the total momentum of a mechanical system is constant in the absence of external forces. Momentum is the product of mass m and velocity v, or mv, for translational or linear motion. For rotational motion, momentum is the product of moment of inertia J and angular velocity ω, or $J\omega$, and is called angular momentum.

Newton's second law gives the force–acceleration relationship of a rigid translating body or the torque–angular acceleration relationship of a rigid rotating body. The third law concerns action and reaction and, in effect, states that every action is always opposed by an equal reaction.

Newton's second law (for translational motion).   For translational motion, Newton's second law says that if a force is acting on a rigid body through the center of mass in a given direction, the acceleration of the rigid body in the same direction is directly proportional to the force acting on it and is inversely proportional to the mass of the body. That is,

$$\text{acceleration} = \frac{\text{force}}{\text{mass}}$$

or

$$(\text{mass})(\text{acceleration}) = \text{force}$$

Suppose that forces are acting on a body of mass m. If ΣF is the sum of all forces acting on mass m through the center of mass in a given direction, then

$$ma = \sum F \qquad\qquad (3\text{--}7)$$

where a is the resulting absolute acceleration in that direction. The line of action of the force acting on a body must pass through the center of mass of the body. Otherwise, rotational motion will also be involved. Rotational motion is not defined by Equation (3–7).

Newton's second law (for rotational motion). For a rigid body in pure rotation about a fixed axis, Newton's second law states that

$$(\text{moment of inertia})(\text{angular acceleration}) = \text{torque}$$

or

$$J\alpha = \sum T \qquad\qquad (3\text{--}8)$$

where ΣT is the sum of all torques acting about a given axis, J is the moment of inertia of a body about that axis, and α is the angular acceleration of the body.

Torque or moment of force. *Torque*, or moment of force, is defined as any cause that tends to produce a change in the rotational motion of a body on which it acts. Torque is the product of a force and the perpendicular distance from a point of rotation to the line of action of the force. The units of torque are force times length, such as N-m, dyn-cm, kg_f-m, and lb_f-ft.

Moments of inertia. The *moment of inertia J* of a rigid body about an axis is defined by

$$J = \int r^2 dm$$

where dm is an element of mass, r is distance from the axis to dm, and integration is performed over the body. In considering moments of inertia, we assume that the rotating body is perfectly rigid. Physically, the moment of inertia of a body is a measure of the resistance of the body to angular acceleration.

Table 3–3 gives the moments of inertia of rigid bodies with common shapes.

TABLE 3–3 Moments of Inertia

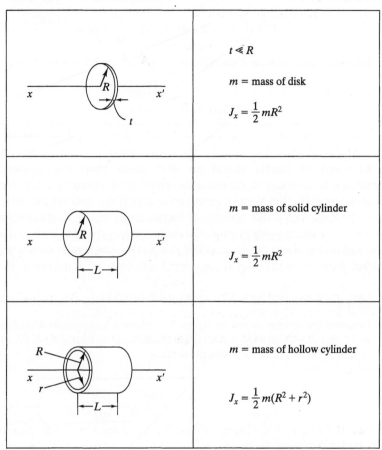

	$t \ll R$ m = mass of disk $J_x = \dfrac{1}{2} mR^2$
	m = mass of solid cylinder $J_x = \dfrac{1}{2} mR^2$
	m = mass of hollow cylinder $J_x = \dfrac{1}{2} m(R^2 + r^2)$

Example 3–1

Figure 3–6 shows a homogeneous cylinder of radius R and length L. The moment of inertia J of this cylinder about axis AA' can be obtained as follows: Consider a ring-shaped mass element of infinitesimal width dr at radius r. The mass of this ring-shaped element is $2\pi r L\rho \, dr$, where ρ is the density of the cylinder. Thus,

$$dm = 2\pi r L\rho \, dr$$

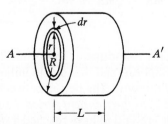

Figure 3–6 Homogeneous cylinder.

Consequently,

$$J = \int_0^R r^2 2\pi r L\rho \, dr = 2\pi L\rho \int_0^R r^3 \, dr = \frac{\pi L\rho R^4}{2}$$

Since the entire mass m of the cylinder body is $m = \pi R^2 L\rho$, we obtain

$$J = \frac{1}{2}mR^2$$

Moment of inertia about an axis other than the geometrical axis.
Sometimes it is necessary to calculate the moment of inertia of a homogeneous rigid
body about an axis other than its geometrical axis. If the axes are parallel, the calculation can be done easily. The moment of inertia about an axis that is a distance x from
the geometrical axis passing through the center of gravity of the body is the sum of the
moment of inertia about the geometrical axis and the moment of inertia about the new
axis when the mass of the body is considered concentrated at the center of gravity.

Example 3–2

Consider the system shown in Figure 3–7, where a homogeneous cylinder of mass m
and radius R rolls on a flat surface. Find the moment of inertia, J_x, of the cylinder about
its line of contact (axis xx') with the surface.

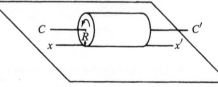

Figure 3–7 Homogeneous cylinder
rolling on a flat surface.

The moment of inertia of the cylinder about axis CC' is

$$J_C = \frac{1}{2}mR^2$$

The moment of inertia of the cylinder about axis xx' when mass m is considered concentrated at the center of gravity is mR^2. Thus, the moment of inertia J_x of the cylinder
about axis xx' is

$$J_x = J_C + mR^2 = \frac{1}{2}mR^2 + mR^2 = \frac{3}{2}mR^2$$

Forced response and natural response.   The behavior determined by a
forcing function is called a *forced response*, and that due to initial conditions (initial
energy storages) is called a *natural response*. The period between the initiation of a
response and the ending is referred to as the *transient period*. After the response has
become negligibly small, conditions are said to have reached a *steady state*.

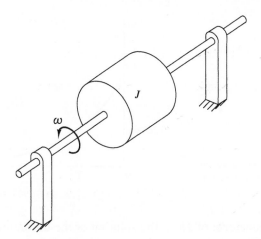

Figure 3–8 Rotor mounted in bearings.

Rotational system.   A schematic diagram of a rotor mounted in bearings is shown in Figure 3–8. The moment of inertia of the rotor about the axis of rotation is J. Let us assume that at $t = 0$ the rotor is rotating at the angular velocity $\omega(0) = \omega_0$. We also assume that the friction in the bearings is viscous friction and that no external torque is applied to the rotor. Then the only torque acting on the rotor is the friction torque $b\omega$ in the bearings.

Applying Newton's second law, Equation (3–8), we obtain the equation of motion,

$$J\dot{\omega} = -b\omega, \qquad \omega(0) = \omega_0$$

or

$$J\dot{\omega} + b\omega = 0 \tag{3–9}$$

Equation (3–9) is a mathematical model of the system. (Here, the output of the system is considered to be the angular velocity ω, not the angular displacement.)

To find $\omega(t)$, we take the Laplace transform of Equation (3–9); that is,

$$J[s\Omega(s) - \omega(0)] + b\Omega(s) = 0$$

where $\Omega(s) = \mathscr{L}[\omega(t)]$. Simplifying, we obtain

$$(Js + b)\Omega(s) = J\omega(0) = J\omega_0$$

Hence,

$$\Omega(s) = \frac{\omega_0}{s + \dfrac{b}{J}} \tag{3–10}$$

The denominator, $s + (b/J)$, is called the *characteristic polynomial*, and

$$s + \frac{b}{J} = 0$$

is called the *characteristic equation.*

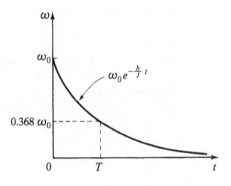

Figure 3–9 Curve of angular velocity ω versus time t for the rotor system shown in Figure 3-8.

The inverse Laplace transform of $\Omega(s)$, the solution of the differential equation given by Equation (3–9), is

$$\omega(t) = \omega_0 e^{-(b/J)t}$$

The angular velocity decreases exponentially, as shown in Figure 3–9.

Since the exponential factor $e^{-(b/J)t}$ approaches zero as t increases without limit, mathematically the response lasts forever. In dealing with such an exponentially decaying response, it is convenient to depict the response in terms of a *time constant*: that value of time which makes the exponent equal to -1. For this system, the time constant T is equal to J/b, or $T = J/b$. When $t = T$, the value of the exponential factor is

$$e^{-T/T} = e^{-1} = 0.368$$

In other words, when the time t in seconds is equal to the time constant, the exponential factor is reduced to approximately 36.8% of its initial value, as shown in Figure 3–9.

Spring–mass system. Figure 3–10 depicts a system consisting of a mass and a spring. Here, the mass is suspended by the spring. For the vertical motion, two

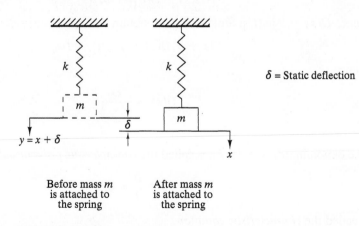

Figure 3–10 Spring–mass system.

forces are acting on the mass: the spring force ky and the gravitational force mg. (In the diagram, the positive direction of displacement y is defined downward.) If the mass is pulled downward by an external force and then released, the spring force acts upward and tends to pull the mass upward. The gravitational force pulls the mass downward. So, by applying Newton's second law to this system, we obtain the equation of motion

$$m\ddot{y} = \sum \text{forces} = -ky + mg$$

or

$$m\ddot{y} + ky = mg \tag{3–11}$$

The gravitational force is opposed statically by the equilibrium spring deflection δ. If we measure the displacement from this equilibrium position, then the term mg can be dropped from the equation of motion. By substituting $y = x + \delta$ into Equation (3–11) and noting that $\delta = $ constant, we have

$$m\ddot{x} + k(x + \delta) = mg \tag{3–12}$$

Since the spring force $k\delta$ and the gravitational force mg balance, or $k\delta = mg$, Equation (3–12) simplifies to

$$m\ddot{x} + kx = 0 \tag{3–13}$$

which is a mathematical model of the system.

In this book, unless otherwise stated, when writing equations of motion for systems involving the gravitational force, we measure the displacement of the mass from the equilibrium position in order to eliminate the term mg and simplify the mathematical model.

Free vibration.   For the spring–mass system of Figure 3–10, suppose that the mass is pulled downward and then released with arbitrary initial conditions $x(0)$ and $\dot{x}(0)$. In this case, the mass will oscillate and the motion will be periodic. (We assume that the magnitude of the displacement is such that the spring remains a linear spring.) The periodic motion that is observed as the system is displaced from its static equilibrium position is called *free vibration*. It is a natural response due to the initial condition.

To find the mathematical form of the periodic motion, let us solve Equation (3–13). By taking the Laplace transforms of both sides of that equation, we obtain

$$m[s^2 X(s) - sx(0) - \dot{x}(0)] + kX(s) = 0$$

or

$$(ms^2 + k)X(s) = m\dot{x}(0) + msx(0)$$

Hence,

$$X(s) = \frac{\dot{x}(0)}{s^2 + \dfrac{k}{m}} + \frac{sx(0)}{s^2 + \dfrac{k}{m}}$$

This last equation may be rewritten so that the inverse Laplace transform of each term can be easily identified:

$$X(s) = \sqrt{\frac{m}{k}} \dot{x}(0) \frac{\sqrt{k/m}}{s^2 + (\sqrt{k/m})^2} + x(0) \frac{s}{s^2 + (\sqrt{k/m})^2}$$

Noting that

$$\mathscr{L}[\sin \sqrt{k/m}\, t] = \frac{\sqrt{k/m}}{s^2 + (\sqrt{k/m})^2}$$

$$\mathscr{L}[\cos \sqrt{k/m}\, t] = \frac{s}{s^2 + (\sqrt{k/m})^2}$$

we obtain the inverse Laplace transform of $X(s)$ as

$$x(t) = \sqrt{\frac{m}{k}} \dot{x}(0) \sin \sqrt{\frac{k}{m}} t + x(0) \cos \sqrt{\frac{k}{m}} t \qquad (3\text{–}14)$$

Periodic motion such as that described by Equation (3–14) is called *simple harmonic motion*.

If the initial conditions were given as $x(0) = x_0$ and $\dot{x}(0) = 0$, then, by substituting these initial conditions into Equation (3–14), the displacement of the mass would be given by

$$x(t) = x_0 \cos \sqrt{\frac{k}{m}} t$$

The period and frequency of simple harmonic motion can now be defined as follows: The *period T* is the time required for a periodic motion to repeat itself. In the present case,

$$\text{period } T = \frac{2\pi}{\sqrt{\dfrac{k}{m}}} \text{ seconds}$$

The *frequency f* of periodic motion is the number of cycles per second (cps), and the standard unit of frequency is the hertz (Hz); that is, 1 Hz is 1 cps. In the present case of harmonic motion,

$$\text{frequency } f = \frac{1}{T} = \frac{\sqrt{\dfrac{k}{m}}}{2\pi} \text{ Hz}$$

The natural frequency, or undamped natural frequency, is the frequency in the free vibration of a system having no damping. If the natural frequency is measured in Hz or cps, it is denoted by f_n. If it is measured in radians per second (rad/s), it is denoted by ω_n. In the present system,

$$\omega_n = 2\pi f_n = \sqrt{\frac{k}{m}} \text{ rad/s}$$

It is important to remember that, when Equation (3–13) is written in the form

$$\ddot{x} + \frac{k}{m}x = 0$$

where the coefficient of the $\ddot{x}$ term is unity, the square root of the coefficient of the x term is the natural frequency ω_n. This means that a mathematical model for the system shown in Figure 3–10 can be put in the form

$$\ddot{x} + \omega_n^2 x = 0$$

where $\omega_n = \sqrt{k/m}$.

Experimental determination of moment of inertia.   It is possible to calculate moments of inertia for homogeneous bodies having geometrically simple shapes. However, for rigid bodies with complicated shapes or those consisting of materials of various densities, such calculation may be difficult or even impossible; moreover, calculated values may not be accurate. In these instances, experimental determination of moments of inertia is preferable. The process is as follows: We mount a rigid body in frictionless bearings so that it can rotate freely about the axis of rotation around which the moment of inertia is to be determined. Next, we attach a torsional spring with known spring constant k to the rigid body. (See Figure 3–11.) The spring is then twisted slightly and released, and the period of the resulting simple harmonic motion is measured. Since the equation of motion for this system is

$$J\ddot{\theta} + k\theta = 0$$

or

$$\ddot{\theta} + \frac{k}{J}\theta = 0$$

the natural frequency is

$$\omega_n = \sqrt{\frac{k}{J}}$$

and the period of vibration is

$$T = \frac{2\pi}{\omega_n} = \frac{2\pi}{\sqrt{\dfrac{k}{J}}}$$

The moment of inertia is then determined as

$$J = \frac{kT^2}{4\pi^2}$$

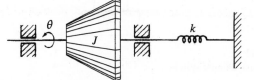

Figure 3–11 Setup for the experimental determination of moment of inertia.

Similarly, in the spring–mass system of Figure 3–10, if the spring constant k is known and the period T of the free vibration is measured, then the mass m can be calculated from

$$m = \frac{kT^2}{4\pi^2}$$

Spring–mass–damper system. Most physical systems involve some type of damping—viscous damping, magnetic damping, and so on. Such damping not only slows the motion of (a part of) the system, but also causes the motion to stop eventually. In the discussion that follows, we shall consider a simple mechanical system involving viscous damping. Note that a typical viscous damping element is a damper or dashpot.

Figure 3–12 is a schematic diagram of a spring–mass–damper system. Suppose that the mass is pulled downward and then released. If the damping is light, vibratory motion will occur. (The system is then said to be underdamped.) If the damping is heavy, vibratory motion will not occur. (The system is then said to be overdamped.) A critically damped system is a system in which the degree of damping is such that the resultant motion is on the borderline between the underdamped and overdamped cases. Regardless of whether a system is underdamped, overdamped, or critically damped, the free vibration or free motion will diminish with time because of the presence of damper. This free vibration is called *transient motion*.

In the system shown in Figure 3–12, for the vertical motion, three forces are acting on the mass: the spring force, the damping force, and the gravitational force. As noted earlier, if we measure the displacement of the mass from a static equilibrium position (so that the gravitational force is balanced by the equilibrium spring deflection), the gravitational force will not enter into the equation of motion. So, by measuring the displacement x from the static equilibrium position, we obtain the equation of motion,

$$m\ddot{x} = \sum \text{forces} = -kx - b\dot{x}$$

or

$$m\ddot{x} + b\dot{x} + kx = 0 \tag{3-15}$$

Equation (3–15), which describes the motion of the system, is a mathematical model of the system.

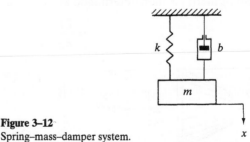

Figure 3–12
Spring–mass–damper system.

Only the underdamped case is considered in our present analysis. (A more complete analysis of this system for the underdamped, overdamped, and critically damped cases is given in Chapter 8.)

Let us solve Equation (3–15) for a particular case. Suppose that $m = 0.1$ slug, $b = 0.4$ lb$_f$-s/ft, and $k = 4$ lb$_f$/ft. Then Equation (3–15) becomes

$$0.1\ddot{x} + 0.4\dot{x} + 4x = 0$$

or

$$\ddot{x} + 4\dot{x} + 40x = 0 \qquad (3\text{–}16)$$

Let us obtain the motion $x(t)$ when the mass is pulled downward at $t = 0$, so that $x(0) = x_0$, and is released with zero velocity, or $\dot{x}(0) = 0$. (We assume that the magnitude of the downward displacement is such that the system remains a linear system.) Taking the Laplace transform of Equation (3–16), we obtain

$$[s^2 X(s) - sx(0) - \dot{x}(0)] + 4[sX(s) - x(0)] + 40X(s) = 0$$

Simplifying this last equation and noting that $x(0) = x_0$ and $\dot{x}(0) = 0$, we get

$$(s^2 + 4s + 40)X(s) = sx_0 + 4x_0$$

or

$$X(s) = \frac{(s + 4)x_0}{s^2 + 4s + 40} \qquad (3\text{–}17)$$

The characteristic equation for the system

$$s^2 + 4s + 40 = 0$$

has a pair of complex-conjugate roots. This implies that the inverse Laplace transform of $X(s)$ is a damped sinusoidal function. Hence, we may rewrite $X(s)$ in Equation (3–17) as a sum of the Laplace transforms of a damped sine function and a damped cosine function:

$$X(s) = \frac{2x_0}{s^2 + 4s + 40} + \frac{(s + 2)x_0}{s^2 + 4s + 40}$$

$$= \frac{1}{3}x_0\frac{6}{(s + 2)^2 + 6^2} + x_0\frac{s + 2}{(s + 2)^2 + 6^2}$$

Noting that

$$\frac{6}{(s + 2)^2 + 6^2} = \mathcal{L}[e^{-2t} \sin 6t], \qquad \frac{s + 2}{(s + 2)^2 + 6^2} = \mathcal{L}[e^{-2t} \cos 6t]$$

we can obtain the inverse Laplace transform of $X(s)$ as

$$x(t) = \frac{1}{3}x_0 e^{-2t} \sin 6t + x_0 e^{-2t} \cos 6t$$

$$= e^{-2t}\left(\frac{1}{3} \sin 6t + \cos 6t\right)x_0 \qquad (3\text{–}18)$$

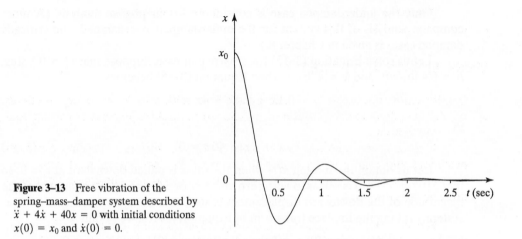

Figure 3–13 Free vibration of the
spring–mass–damper system described by
$\ddot{x} + 4\dot{x} + 40x = 0$ with initial conditions
$x(0) = x_0$ and $\dot{x}(0) = 0$.

Equation (3–18) depicts the free vibration of the spring–mass–damper system with
the given numerical values. The free vibration here is a damped sinusoidal vibration,
as shown in Figure 3–13.

Comments.   The numerical values in the preceding problem were stated in
BES units. Let us convert these values into units of other systems.

  1. *SI units (refer to Tables 3–1 and 3–2):*

$$m = 0.1 \text{ slug} = 1.459 \text{ kg}$$
$$b = 0.4 \text{ lb}_f\text{-s/ft} = 0.4 \times 4.448/0.3048 \text{ N-s/m} = 5.837 \text{ N-s/m}$$
$$k = 4 \text{ lb}_f/\text{ft} = 4 \times 4.448/0.3048 \text{ N/m} = 58.37 \text{ N/m}$$

Hence, Equation (3–15) becomes

$$1.459\ddot{x} + 5.837\dot{x} + 58.37x = 0$$

or

$$\ddot{x} + 4\dot{x} + 40x = 0$$

which is the same as Equation (3–16).

  2. *Metric engineering (gravitational) units (refer to Tables 3–1 and 3–2):*

$$m = 0.1 \text{ slug} = 0.1488 \text{ kg}_f\text{-s}^2/\text{m}$$
$$b = 0.4 \text{ lb}_f\text{-s/ft} = 0.4 \times 0.4536/0.3048 \text{ kg}_f\text{-s/m} = 0.5953 \text{ kg}_f\text{-s/m}$$
$$k = 4 \text{ lb}_f/\text{ft} = 4 \times 0.4536/0.3048 \text{ kg}_f/\text{m} = 5.953 \text{ kg}_f/\text{m}$$

Therefore, Equation (3–15) becomes

$$0.1488\ddot{x} + 0.5953\dot{x} + 5.953x = 0$$

or

$$\ddot{x} + 4\dot{x} + 40x = 0$$

which, again, is the same as Equation (3–16).

Note that as long as we use consistent units, the differential equation (mathematical model) of the system remains the same.

3–4 WORK, ENERGY, AND POWER

In this section, we discuss work, energy, and power. We also discuss energy methods for deriving equations of motion or undamped natural frequencies of certain conservative systems.

If force is considered a measure of effort, then *work* is a measure of accomplishment and *energy* is the ability to do work. The concept of work makes no allowance for a time factor. When a time factor is considered, the concept of power must be introduced. *Power* is work per unit time.

Work.   The *work* done in a mechanical system is the product of a force and the distance (or a torque and the angular displacement) through which the force is exerted, with both force and distance measured in the same direction. For instance, if a body is pushed with a horizontal force of F newtons along a horizontal floor for a distance of x meters, the work done in pushing the body is

$$W = Fx \text{ N-m}$$

Units of work.   Different systems have different units of work.

SI units and mks (metric absolute) system of units.   Force is measured in newtons and distance in meters. Thus, the unit of work is the N-m. Note that

$$1 \text{ N-m} = 1 \text{ joule} = 1 \text{ J}$$

British engineering system of units.   In this system, force is measured in pounds and distance in feet. Hence, the unit of work is the ft-lb$_f$, and

$$1 \text{ ft-lb}_f = 1.3557 \text{ J} = 1.285 \times 10^{-3} \text{ Btu}$$

$$1 \text{ Btu} = 778 \text{ ft-lb}_f$$

cgs (metric absolute) system of units.   Here, the unit of work is the dyn-cm, or erg. Note that

$$10^7 \text{ erg} = 10^7 \text{ dyn-cm} = 1 \text{ J}$$

Metric engineering (gravitational) system of units.   The unit of work in the metric engineering system is the kg$_f$-m. Note that

$$1 \text{ kg}_f\text{-m} = 9.807 \times 10^7 \text{ dyn-cm} = 9.807 \text{ J}$$

$$1 \text{ J} = 0.10197 \text{ kg}_f\text{-m}$$

Energy.   In a general way, *energy* can be defined as the capacity or ability to do work. Energy is found in many different forms and can be converted from one form into another. For instance, an electric motor converts electrical energy into mechanical energy, a battery converts chemical energy into electrical energy, and so forth.

A system is said to possess energy when it can do work. When a system does mechanical work, the system's energy decreases by the amount equal to the energy required for the work done. Units of energy are the same as units for work, that is, newton-meter, joule, kcal, Btu, and so on.

According to the law of conservation of energy, energy can be neither created nor destroyed. This means that the increase in the total energy within a system is equal to the net energy input to the system. So if there is no energy input, there is no change in total energy of the system.

The energy that a body possesses because of its position is called *potential energy*, whereas the energy that a body has as a result of its velocity is called *kinetic energy*.

Potential energy. In a mechanical system, only mass and spring elements can store potential energy. The change in the potential energy stored in a system equals the work required to change the system's configuration. Potential energy is always measured with reference to some chosen level and is relative to that level.

Potential energy is the work done by an external force. For a body of mass m in the gravitational field of the earth, the potential energy U measured from some reference level is mg times the altitude h measured from the same reference level, or

$$U = \int_0^h mg\, dx = mgh$$

Notice that the body, if dropped, has the capacity to do work, since the weight mg of the body causes it to travel a distance h when released. (The weight is a force.) Once the body is released, the potential energy decreases. The lost potential energy is converted into kinetic energy.

For a translational spring, the potential energy U is equal to the net work done on the spring by the forces acting on its ends as it is compressed or stretched. Since the spring force F is equal to kx, where x is the net displacement of the ends of the spring, the total energy stored is

$$U = \int_0^x F\, dx = \int_0^x kx\, dx = \frac{1}{2}kx^2$$

If the initial and final values of x are x_1 and x_2, respectively, then

$$\text{change in potential energy } \Delta U = \int_{x_1}^{x_2} F\, dx = \int_{x_1}^{x_2} kx\, dx = \frac{1}{2}kx_2^2 - \frac{1}{2}kx_1^2$$

Note that the potential energy stored in a spring does not depend on whether it is compressed or stretched.

Similarly, for a torsional spring,

$$\text{change in potential energy } \Delta U = \int_{\theta_1}^{\theta_2} T\, d\theta = \int_{\theta_1}^{\theta_2} k\theta\, d\theta = \frac{1}{2}k\theta_2^2 - \frac{1}{2}k\theta_1^2$$

Kinetic energy. Only inertial elements can store kinetic energy in mechanical systems. A mass m in pure translation with velocity v has kinetic energy $T = \frac{1}{2}mv^2$, whereas a moment of inertia J in pure rotation with angular velocity $\dot{\theta}$

has kinetic energy $T = \frac{1}{2}J\dot{\theta}^2$. The change in kinetic energy of the mass is equal to the work done on it by an applied force as the mass accelerates or decelerates. Thus, the change in kinetic energy T of a mass m moving in a straight line is

$$\text{change in kinetic energy} = \Delta T = \Delta W = \int_{x_1}^{x_2} F\, dx = \int_{t_1}^{t_2} F\frac{dx}{dt}\, dt$$

$$= \int_{t_1}^{t_2} Fv\, dt = \int_{t_1}^{t_2} m\dot{v}v\, dt = \int_{v_1}^{v_2} mv\, dv$$

$$= \frac{1}{2}mv_2^2 - \frac{1}{2}mv_1^2$$

where $x(t_1) = x_1$, $x(t_2) = x_2$, $v(t_1) = v_1$, and $v(t_2) = v_2$. Notice that the kinetic energy stored in the mass does not depend on the sign of the velocity v.

The change in kinetic energy of a moment of inertia in pure rotation at angular velocity $\dot{\theta}$ is

$$\text{change in kinetic energy } \Delta T = \frac{1}{2}J\dot{\theta}_2^2 - \frac{1}{2}J\dot{\theta}_1^2$$

where J is the moment of inertia about the axis of rotation, $\dot{\theta}_1 = \dot{\theta}(t_1)$, and $\dot{\theta}_2 = \dot{\theta}(t_2)$.

Dissipated energy. Consider the damper shown in Figure 3–14, in which one end is fixed and the other end is moved from x_1 to x_2. The dissipated energy ΔW of the damper is equal to the net work done on it:

$$\Delta W = \int_{x_1}^{x_2} F\, dx = \int_{x_1}^{x_2} b\dot{x}\, dx = b\int_{t_1}^{t_2} \dot{x}\frac{dx}{dt}\, dt = b\int_{t_1}^{t_2} \dot{x}^2\, dt$$

The energy of the damper element is always dissipated, regardless of the sign of $\dot{x}$.

Power. *Power* is the time rate of doing work. That is,

$$\text{power} = P = \frac{dW}{dt}$$

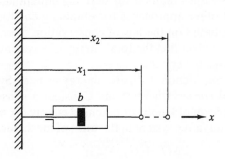

Figure 3–14 Damper.

where dW denotes work done during time interval dt. The average power during a duration of $t_2 - t_1$ seconds can be determined by measuring the work done in $t_2 - t_1$ seconds, or

$$\text{average power} = \frac{\text{work done in } (t_2 - t_1) \text{ seconds}}{(t_2 - t_1) \text{ seconds}}$$

In SI units or the mks (metric absolute) system of units, the work done is measured in newton-meters and the time in seconds. The unit of power is the newton-meter per second, or watt:

$$1 \text{ N-m/s} = 1 \text{ W}$$

In the British engineering system of units, the work done is measured in ft-lb$_f$ and the time in seconds. The unit of power is the ft-lb$_f$/s. The power 550 ft-lb$_f$/s is called 1 horsepower (hp). Thus,

$$1 \text{ hp} = 550 \text{ ft-lb}_f/\text{s} = 33\,000 \text{ ft-lb}_f/\text{min} = 745.7 \text{ W}$$

In the metric engineering system of units, the work done is measured in kg$_f$-m and the time in seconds. The unit of power is the kg$_f$-m/s, where

$$1 \text{ kg}_f\text{-m/s} = 9.807 \text{ W}$$

$$1 \text{ W} = 1 \text{ J/s} = 0.10197 \text{ kg}_f\text{-m/s}$$

Example 3–3

Find the power required to raise a body of mass 500 kg at a rate of 20 m/min.
Let us define displacement per second as x. Then

$$\text{work done in 1 second} = mgx = 500 \times 9.807 \times \frac{20}{60} \frac{\text{kg-m}^2}{\text{s}^2} = 1635 \text{ N-m}$$

and

$$\text{power} = \frac{\text{work done in 1 second}}{1 \text{ second}} = \frac{1635 \text{ N-m}}{1 \text{ s}} = 1635 \text{ W}$$

Thus, the power required is 1635 W.

An energy method for deriving equations of motion.   Earlier in this chapter, we presented Newton's method for deriving equations of motion of mechanical systems. Several other approaches for obtaining equations of motion are available, one of which is based on the law of conservation of energy. Here we derive such equations from the fact that the total energy of a system remains the same if no energy enters or leaves the system.

In mechanical systems, friction dissipates energy as heat. Systems that do not involve friction are called *conservative* systems. Consider a conservative system in which the energy is in the form of kinetic or potential energy (or both). Since energy enters and leaves the conservative system in the form of mechanical work, we obtain

$$\Delta(T + U) = \Delta W$$

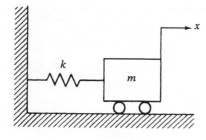

Figure 3–15   Mechanical system.

where $\Delta(T + U)$ is the change in the total energy and ΔW is the net work done on the system by an external force. If no external energy enters the system, then

$$\Delta(T + U) = 0$$

which results in

$$T + U = \text{constant}$$

If we assume no friction, then the mechanical system shown in Figure 3–15 can be considered conservative. The kinetic energy T and potential energy U are given by

$$T = \frac{1}{2}m\dot{x}^2, \qquad U = \frac{1}{2}kx^2$$

Consequently, in the absence of any external energy input,

$$T + U = \frac{1}{2}m\dot{x}^2 + \frac{1}{2}kx^2 = \text{constant}$$

The equation of motion for the system can be obtained by differentiating the total energy with respect to t:

$$\frac{d}{dt}(T + U) = m\dot{x}\ddot{x} + kx\dot{x} = (m\ddot{x} + kx)\dot{x} = 0$$

Since $\dot{x}$ is not always zero, we have

$$m\ddot{x} + kx = 0$$

which is the equation of motion for the system.

Let us look next at the mechanical system of Figure 3–16. Here, no damping is involved; therefore, the system is conservative. In this case, since the mass is suspended by a spring, the potential energy includes that due to the position of the mass element. At the equilibrium position, the potential energy of the system is

$$U_0 = mgx_0 + \frac{1}{2}k\delta^2$$

where x_0 is the equilibrium position of the mass element above an arbitrary datum line and δ is the static deflection of the spring when the system is in the equilibrium position, or $k\delta = mg$. (For the definition of δ, see Figure 3–10.)

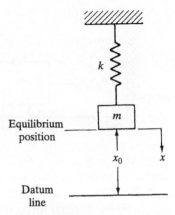

Figure 3–16 Mechanical system.

The instantaneous potential energy U is the instantaneous potential energy of the weight of the mass element, plus the instantaneous elastic energy stored in the spring. Thus,

$$U = mg(x_0 - x) + \frac{1}{2}k(\delta + x)^2$$

$$= mgx_0 - mgx + \frac{1}{2}k\delta^2 + k\delta x + \frac{1}{2}kx^2$$

$$= mgx_0 + \frac{1}{2}k\delta^2 - (mg - k\delta)x + \frac{1}{2}kx^2$$

Since $mg = k\delta$, it follows that

$$U = U_0 + \frac{1}{2}kx^2$$

Note that the increase in the total potential energy of the system is due to the increase in the elastic energy of the spring that results from its deformation from the equilibrium position. Note also that, since x_0 is the displacement measured from an arbitrary datum line, it is possible to choose the datum line such that $U_0 = 0$. Finally, note that an increase (decrease) in the potential energy is offset by a decrease (increase) in the kinetic energy.

The kinetic energy of the system is $T = \frac{1}{2}m\dot{x}^2$. Since the total energy is constant, we obtain

$$T + U = \frac{1}{2}m\dot{x}^2 + U_0 + \frac{1}{2}kx^2 = \text{constant}$$

By differentiating the total energy with respect to t and noting that U_0 is a constant, we have

$$\frac{d}{dt}(T + U) = m\dot{x}\ddot{x} + kx\dot{x} = 0$$

or

$$(m\ddot{x} + kx)\dot{x} = 0$$

Since $\dot{x}$ is not always zero, it follows that

$$m\ddot{x} + kx = 0$$

This is the equation of motion for the system.

Example 3–4

Figure 3–17 shows a homogeneous cylinder of radius R and mass m that is free to rotate about its axis of rotation and that is connected to the wall through a spring. Assuming that the cylinder rolls on a rough surface without sliding, obtain the kinetic energy and potential energy of the system. Then derive the equations of motion from the fact that the total energy is constant. Assume that x and θ are measured from respective equilibrium positions.

The kinetic energy of the cylinder is the sum of the translational kinetic energy of the center of mass and the rotational kinetic energy about the axis of rotation:

$$\text{kinetic energy} = T = \frac{1}{2}m\dot{x}^2 + \frac{1}{2}J\dot{\theta}^2$$

The potential energy of the system is due to the deflection of the spring:

$$\text{potential energy} = U = \frac{1}{2}kx^2$$

Since the total energy $T + U$ is constant in this conservation system (which means that the loss in potential energy equals the gain in kinetic energy), it follows that

$$T + U = \frac{1}{2}m\dot{x}^2 + \frac{1}{2}J\dot{\theta}^2 + \frac{1}{2}kx^2 = \text{constant} \qquad (3\text{–}19)$$

The cylinder rolls without sliding, which means that $x = R\theta$. Rewriting Equation (3–19) and noting that the moment of inertia J is equal to $\frac{1}{2}mR^2$, we have

$$\frac{3}{4}m\dot{x}^2 + \frac{1}{2}kx^2 = \text{constant}$$

Differentiating both sides of this last equation with respect to t yields

$$\frac{3}{2}m\dot{x}\ddot{x} + kx\dot{x} = 0$$

or

$$\left(m\ddot{x} + \frac{2}{3}kx\right)\dot{x} = 0$$

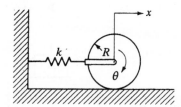

Figure 3–17   Homogeneous cylinder connected to a wall through a spring.

Note that $\dot{x}$ is not always zero, so $m\ddot{x} + \frac{2}{3}kx$ must be identically zero. Therefore,

$$m\ddot{x} + \frac{2}{3}kx = 0$$

or

$$\ddot{x} + \frac{2k}{3m}x = 0$$

This equation describes the horizontal motion of the cylinder. For the rotational motion, we substitute $x = R\theta$ to get

$$\ddot{\theta} + \frac{2k}{3m}\theta = 0$$

In either of the equations of motion, the natural frequency of vibration is the same, $\omega_n = \sqrt{2k/(3m)}$ rad/s.

An energy method for determining natural frequencies. The natural frequency of a conservative system can be obtained from a consideration of the kinetic energy and the potential energy of the system.

Let us assume that we choose the datum line so that the potential energy at the equilibrium state is zero. Then, in such a conservative system, the maximum kinetic energy equals the maximum potential energy, or

$$T_{max} = U_{max}$$

Using this relationship, we are able to determine the natural frequency of a conservative system, as presented in **Example 3–5**.

Example 3–5

Consider the system shown in Figure 3–18. The displacement x is measured from the equilibrium position. The kinetic energy of this system is

$$T = \frac{1}{2}m\dot{x}^2$$

If we choose the datum line so that the potential energy U_0 at the equilibrium state is zero, then the potential energy of the system is given by

$$U = \frac{1}{2}kx^2$$

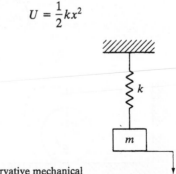

Figure 3–18   Conservative mechanical system.

Let us assume that the system is vibrating about the equilibrium position. Then the displacement is given by

$$x = A \sin \omega t$$

where A is the amplitude of vibration. Consequently,

$$T = \frac{1}{2}m\dot{x}^2 = \frac{1}{2}mA^2\omega^2(\cos \omega t)^2$$

$$U = \frac{1}{2}kx^2 = \frac{1}{2}kA^2(\sin \omega t)^2$$

Hence, the maximum values of T and U are given by

$$T_{\max} = \frac{1}{2}mA^2\omega^2, \qquad U_{\max} = \frac{1}{2}kA^2$$

Since $T_{\max} = U_{\max}$, we have

$$\frac{1}{2}mA^2\omega^2 = \frac{1}{2}kA^2$$

from which we get

$$\omega = \sqrt{\frac{k}{m}}$$

EXAMPLE PROBLEMS AND SOLUTIONS

Problem A–3–1

Calculate the moment of inertia about axis xx' of the hollow cylinder shown in Figure 3–19.

Solution The moment of inertia about axis xx' of the solid cylinder of radius R is

$$J_R = \frac{1}{2}m_1R^2$$

where

$$m_1 = \pi R^2 L \rho \qquad (\rho = \text{density})$$

The moment of inertia about axis xx' of the solid cylinder of radius r is

$$J_r = \frac{1}{2}m_2r^2$$

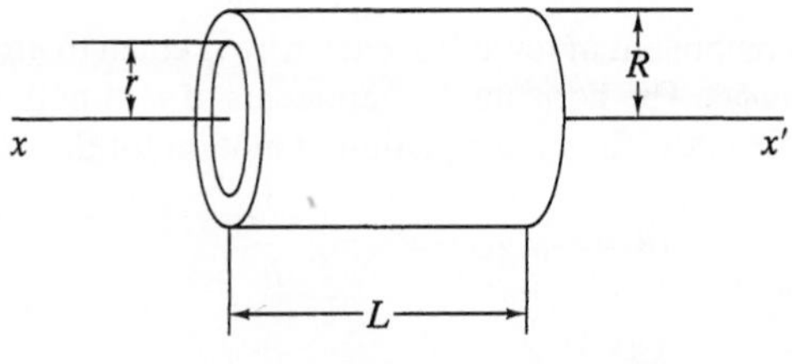

Figure 3–19 Hollow cylinder.

where

$$m_2 = \pi r^2 L \rho$$

Then the moment of inertia about axis xx' of the hollow cylinder shown in the figure is

$$J = J_R - J_r = \frac{1}{2} m_1 R^2 - \frac{1}{2} m_2 r^2$$

$$= \frac{1}{2} [(\pi R^2 L \rho) R^2 - (\pi r^2 L \rho) r^2]$$

$$= \frac{1}{2} \pi L \rho (R^4 - r^4)$$

$$= \frac{1}{2} \pi L \rho (R^2 + r^2)(R^2 - r^2)$$

The mass of the hollow cylinder is

$$m = \pi (R^2 - r^2) L \rho$$

Hence,

$$J = \frac{1}{2} (R^2 + r^2) \pi (R^2 - r^2) L \rho$$

$$= \frac{1}{2} m (R^2 + r^2)$$

(See the third item of Table 3–3.)

Problem A–3–2

A rotating body whose mass is m is suspended by two vertical wires, each of length h, a distance $2a$ apart. The center of gravity is on the vertical line that passes through the midpoint between the points of attachment of the wires. (See Figure 3–20.)

Assume that the body is turned through a small angle about the vertical axis through the center of gravity and is then released. Define the period of oscillation as T. Show that moment of inertia J of the body about the vertical axis that passes through the center of gravity is

$$J = \left(\frac{T}{2\pi}\right)^2 \frac{a^2 mg}{h}$$

Solution   Let us assume that, when the body rotates a small angle θ from the equilibrium position, the force in each wire is F. Then, from Figure 3–20, the angle ϕ that each wire makes with the vertical is small. Angles θ and ϕ are related by

$$a\theta = h\phi$$

Thus,

$$\phi = \frac{a\theta}{h}$$

Notice that the vertical component of force F in each wire is equal to $mg/2$. The horizontal component of F is $mg\phi/2$. The horizontal components of F of both wires produce a torque $mg\phi a$ to rotate the body. Thus, the equation of motion for the oscillation is

$$J\ddot{\theta} = -mg\phi a = -mg\frac{a^2\theta}{h}$$

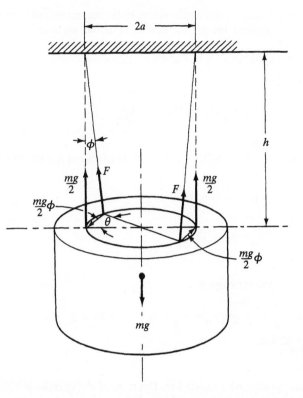

Figure 3–20 Experimental setup for measuring the moment of inertia of a rotating body.

or

$$\ddot{\theta} + \frac{a^2 mg}{Jh}\theta = 0$$

from which the period of the oscillation is found to be

$$T = \frac{2\pi}{\sqrt{\dfrac{a^2 mg}{Jh}}}$$

Solving this last equation for J gives

$$J = \left(\frac{T}{2\pi}\right)^2 \frac{a^2 mg}{h}$$

Problem A–3–3

A brake is applied to a car traveling at a constant speed of 90 km/h. If the deceleration α caused by the braking action is 5 m/s^2, find the time and distance before the car stops.

Solution   Note that

$$90 \text{ km/h} = 25 \text{ m/s}$$

The equation of motion for the car is

$$m\ddot{x} = -m\alpha$$

where m is the mass of the car and x is the displacement of the car, measured from the point where the brake is first applied. Integrating this last equation, we have

$$\dot{x}(t) = -\alpha t + v(0)$$

and

$$x(t) = -\frac{1}{2}\alpha t^2 + v(0)t + x(0)$$

where $x(0) = 0$ and $v(0) = 25$ m/s.

Assume that the car stops at $t = t_1$. Then $\dot{x}(t_1) = 0$. The value of t_1 is determined from

$$\dot{x}(t_1) = -\alpha t_1 + v(0) = 0$$

or

$$t_1 = \frac{v(0)}{\alpha} = \frac{25}{5} = 5 \text{ s}$$

The distance traveled before the car stops is

$$x(t_1) = -\frac{1}{2}\alpha t_1^2 + v(0)t_1 = -\frac{1}{2} \times 5 \times 5^2 + 25 \times 5$$
$$= 62.5 \text{ m}$$

Problem A–3–4

Consider a homogeneous cylinder with radius 1 m. The mass of the cylinder is 100 kg. What will be the angular acceleration of the cylinder if it is acted on by an external torque of 10 N-m about its axis? Assume no friction in the system.

Solution   The moment of inertia is

$$J = \frac{1}{2}mR^2 = \frac{1}{2} \times 100 \times 1^2 = 50 \text{ kg-m}^2$$

The equation of motion for this system is

$$J\ddot{\theta} = T$$

where $\ddot{\theta}$ is the angular acceleration. Therefore,

$$\ddot{\theta} = \frac{T}{J} = \frac{10 \text{ N-m}}{50 \text{ kg-m}^2} = 0.2 \text{ rad/s}^2$$

(Note that, in examining the units of this last equation, we see that the unit of $\ddot{\theta}$ is not s^{-2}, but rad/s^2. This usage occurs because writing rad/s^2 indicates that the angle θ is measured in radians. The radian is the ratio of a length of arc to the radius of a circle. That is, in radian measure, the angle is a pure number. In the algebraic handling of units, the radian is added as necessary.)

Problem A–3–5

Suppose that a disk is rotated at a constant speed of 100 rad/s and we wish to stop it in 2 min. Assuming that the moment of inertia J of the disk is 6 kg-m^2, determine the torque T necessary to stop the rotation. Assume no friction in the system.

Solution The necessary torque T must act so as to reduce the speed of the disk. Thus, the equation of motion is

$$J\dot{\omega} = -T, \qquad \omega(0) = 100$$

Noting that the torque T is a constant and taking the Laplace transform of this last equation, we obtain

$$J[s\Omega(s) - \omega(0)] = -\frac{T}{s}$$

Substituting $J = 6$ and $\omega(0) = 100$ into this equation and solving for $\Omega(s)$, we get

$$\Omega(s) = \frac{100}{s} - \frac{T}{6s^2}$$

The inverse Laplace transform of $\Omega(s)$ gives

$$\omega(t) = 100 - \frac{T}{6}t$$

At $t = 2$ min $= 120$ s, we want to stop, so $\omega(120)$ must equal zero. Therefore,

$$\omega(120) = 0 = 100 - \frac{T}{6} \times 120$$

Solving for T, we get

$$T = \frac{600}{120} = 5 \text{ N-m}$$

Problem A–3–6

Obtain the equivalent spring constant for the system shown in Figure 3–21.

Solution For the springs in parallel, the equivalent spring constant k_{eq} is obtained from

$$k_1 x + k_2 x = F = k_{eq} x$$

or

$$k_{eq} = k_1 + k_2$$

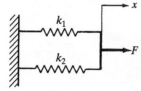

Figure 3–21 System consisting of two springs in parallel.

Problem A–3–7

Find the equivalent spring constant for the system shown in Figure 3–22(a), and show that it can also be obtained graphically as in Figure 3–22(b).

Solution For the springs in series, the force in each spring is the same. Thus,

$$k_1 y = F, \qquad k_2(x - y) = F$$

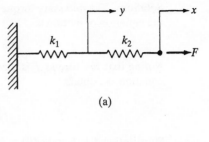

(a)

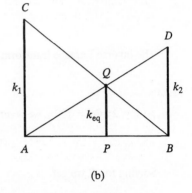

Figure 3–22   (a) System consisting of
two springs in series; (b) diagram showing the equivalent spring constant.

(b)

Eliminating y from these two equations yields

$$k_2\left(x - \frac{F}{k_1}\right) = F$$

or

$$k_2 x = F + \frac{k_2}{k_1}F = \frac{k_1 + k_2}{k_1}F$$

The equivalent spring constant for this case is then found to be

$$k_{eq} = \frac{F}{x} = \frac{k_1 k_2}{k_1 + k_2} = \frac{1}{\dfrac{1}{k_1} + \dfrac{1}{k_2}}$$

For the graphical solution, notice that

$$\frac{\overline{AC}}{\overline{PQ}} = \frac{\overline{AB}}{\overline{PB}}, \qquad \frac{\overline{BD}}{\overline{PQ}} = \frac{\overline{AB}}{\overline{AP}}$$

from which it follows that

$$\overline{PB} = \frac{\overline{AB} \cdot \overline{PQ}}{\overline{AC}}, \qquad \overline{AP} = \frac{\overline{AB} \cdot \overline{PQ}}{\overline{BD}}$$

Since $\overline{AP} + \overline{PB} = \overline{AB}$, we have

$$\frac{\overline{AB} \cdot \overline{PQ}}{\overline{BD}} + \frac{\overline{AB} \cdot \overline{PQ}}{\overline{AC}} = \overline{AB}$$

or

$$\frac{\overline{PQ}}{\overline{BD}} + \frac{\overline{PQ}}{\overline{AC}} = 1$$

Solving for $\overline{PQ}$, we obtain

$$\overline{PQ} = \frac{1}{\dfrac{1}{\overline{AC}} + \dfrac{1}{\overline{BD}}}$$

So if lengths $\overline{AC}$ and $\overline{BD}$ represent the spring constants k_1 and k_2, respectively, then length $\overline{PQ}$ represents the equivalent spring constant k_{eq}. That is,

$$\overline{PQ} = \frac{1}{\dfrac{1}{k_1} + \dfrac{1}{k_2}} = k_{eq}$$

Problem A–3–8

In Figure 3–23, the simple pendulum shown consists of a sphere of mass m suspended by a string of negligible mass. Neglecting the elongation of the string, find a mathematical model of the pendulum. In addition, find the natural frequency of the system when θ is small. Assume no friction.

Solution   The gravitational force mg has the tangential component $mg \sin \theta$ and the normal component $mg \cos \theta$. The torque due to the tangential component is $-mgl \sin \theta$, so the equation of motion is

$$J\ddot{\theta} = -mgl \sin \theta$$

where $J = ml^2$. Therefore,

$$ml^2\ddot{\theta} + mgl \sin \theta = 0$$

For small θ, $\sin \theta \doteq \theta$, and the equation of motion simplifies to

$$\ddot{\theta} + \frac{g}{l}\theta = 0$$

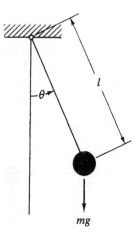

mg **Figure 3–23** Simple pendulum.

This is a mathematical model of the system. The natural frequency is then obtained as

$$\omega_n = \sqrt{\frac{g}{l}}$$

Problem A-3-9

Consider the spring-loaded pendulum system shown in Figure 3–24. Assume that the spring force acting on the pendulum is zero when the pendulum is vertical ($\theta = 0$). Assume also that the friction involved is negligible and the angle of oscillation, θ, is small. Obtain a mathematical model of the system.

Solution   Two torques are acting on this system, one due to the gravitational force and the other due to the spring force. Applying Newton's second law, we find that the equation of motion for the system becomes

$$J\ddot{\theta} = -mgl \sin \theta - 2(ka \sin \theta)(a \cos \theta)$$

where $J = ml^2$. Rewriting this last equation, we obtain

$$ml^2\ddot{\theta} + mgl \sin \theta + 2\, ka^2 \sin \theta \cos \theta = 0$$

For small θ, we have $\sin \theta = \theta$ and $\cos \theta = 1$. So the equation of motion can be simplified to

$$ml^2\ddot{\theta} + (mgl + 2\, ka^2)\theta = 0$$

or

$$\ddot{\theta} + \left(\frac{g}{l} + 2\frac{ka^2}{ml^2}\right)\theta = 0$$

This is a mathematical model of the system. The natural frequency of the system is

$$\omega_n = \sqrt{\frac{g}{l} + 2\frac{ka^2}{ml^2}}$$

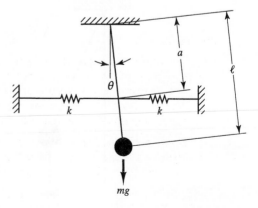

Figure 3–24   Spring-loaded pendulum system.

Problem A–3–10

Consider the rolling motion of the ship shown in Figure 3–25. The force due to buoyancy is $-w$ and that due to gravity is w. These two forces produce a couple that causes rolling motion of the ship. The point where the vertical line through the center of buoyancy, C, intersects the symmetrical line through the center of gravity, which is in the ship's centerline plane, is called the *metacenter* (point M). Define

R = distance of the metacenter to the center of gravity of the ship = $\overline{MG}$

J = moment of inertia of the ship about its longitudinal centroidal axis

Derive the equation of rolling motion of the ship when the rolling angle θ is small.

Solution   From Figure 3–25, we obtain

$$J\ddot{\theta} = -wR \sin \theta$$

or

$$J\ddot{\theta} + wR \sin \theta = 0$$

For small θ, we have $\sin \theta \doteq \theta$. Hence, the equation of rolling motion of the ship is

$$J\ddot{\theta} + wR\theta = 0$$

The natural frequency of the rolling motion is $\sqrt{wR/J}$. Note that the distance $R(=\overline{MG})$ is considered positive when the couple of weight and buoyancy tends to rotate the ship toward the upright position. That is, R is positive if point M is above point G, and R is negative if point M is below point G.

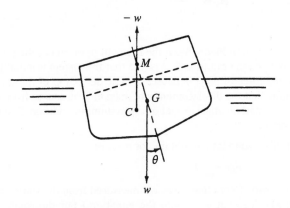

Figure 3–25   Rolling motion of a ship.

Problem A–3–11

In Figure 3–26, a homogeneous disk of radius R and mass m that can rotate about the center of mass of the disk is hung from the ceiling and is spring preloaded. (Two springs are connected by a wire that passes over a pulley as shown.) Each spring is prestretched by an amount x. Assuming that the disk is initially rotated by a small angle θ and then released, obtain both a mathematical model of the system and the natural frequency of the system.

Solution   If the disk is rotated by an angle θ as shown in Figure 3–26, then the right spring is stretched by $x + R\theta$ and the left spring is stretched by $x - R\theta$. So, applying Newton's second law to the rotational motion of the disk gives

$$J\ddot{\theta} = -k(x + R\theta)R + k(x - R\theta)R$$

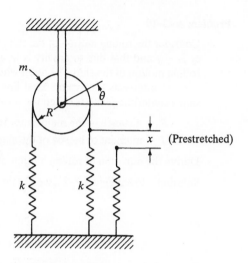

Figure 3–26   Spring–pulley system.

where the moment of inertia J is $\frac{1}{2}mR^2$. Simplifying the equation of motion, we have

$$\ddot{\theta} + \frac{4k}{m}\theta = 0$$

This is a mathematical model of the system. The natural frequency of the system is

$$\omega_n = \sqrt{\frac{4k}{m}}$$

Problem A–3–12

For the spring–mass–pulley system of Figure 3–27, the moment of inertia of the pulley about the axis of rotation is J and the radius is R. Assume that the system is initially at equilibrium. The gravitational force of mass m causes a static deflection of the spring such that $k\delta = mg$. Assuming that the displacement x of mass m is measured from the equilibrium position, obtain a mathematical model of the system. In addition, find the natural frequency of the system.

Solution   Applying Newton's second law, we obtain, for mass m,

$$m\ddot{x} = -T \tag{3–20}$$

where T is the tension in the wire. (Note that since x is measured from the static equilibrium position the term mg does not enter into the equation.) For the rotational motion of the pulley,

$$J\ddot{\theta} = TR - kxR \tag{3–21}$$

If we eliminate the tension T from Equations (3–20) and (3–21), the result is

$$J\ddot{\theta} = -m\ddot{x}R - kxR \tag{3–22}$$

Noting that $x = R\theta$, we can simplify Equation (3–22) to

$$(J + mR^2)\ddot{\theta} + kR^2\theta = 0$$

or

$$\ddot{\theta} + \frac{kR^2}{J + mR^2}\theta = 0$$

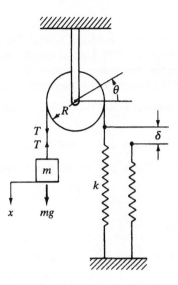

Figure 3–27
Spring–mass–pulley system.

This is a mathematical model of the system. The natural frequency is

$$\omega_n = \sqrt{\frac{kR^2}{J + mR^2}}$$

Problem A–3–13

In the mechanical system of Figure 3–28, one end of the lever is connected to a spring and a damper, and a force F is applied to the other end of the lever. Derive a mathematical model of the system. Assume that the displacement x is small and the lever is rigid and massless.

Solution From Newton's second law, for small displacement x, the rotational motion about pivot P is given by

$$Fl_1 - (b\dot{x} + kx)l_2 = 0$$

or

$$b\dot{x} + kx = \frac{l_1}{l_2}F$$

which is a mathematical model of the system.

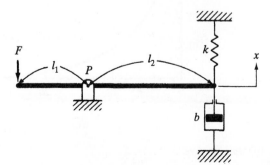

Figure 3–28 Lever system.

Problem A–3–14

Consider the mechanical system shown in Figure 3–29(a). The massless bar AA' is displaced 0.05 m by a constant force of 100 N. Suppose that the system is at rest before the force is abruptly released. The time-response curve when the force is abruptly released at $t = 0$ is shown in Figure 3–29(b). Determine the numerical values of b and k.

Solution   Since the system is at rest before the force is abruptly released, the equation of motion is

$$kx = F \qquad t \leq 0$$

Note that the effect of the force F is to give the initial condition

$$x(0) = \frac{F}{k}$$

Since $x(0) = 0.05$ m, we have

$$k = \frac{F}{x(0)} = \frac{100}{0.05} = 2000 \text{ N/m}$$

At $t = 0$, F is abruptly released, so, for $t > 0$, the equation of motion becomes

$$b\dot{x} + kx = 0 \qquad t > 0$$

Taking the Laplace transform of this last equation, we have

$$b[sX(s) - x(0)] + kX(s) = 0$$

Substituting $x(0) = 0.05$ and solving the resulting equation for $X(s)$, we get

$$X(s) = \frac{0.05}{s + \dfrac{k}{b}}$$

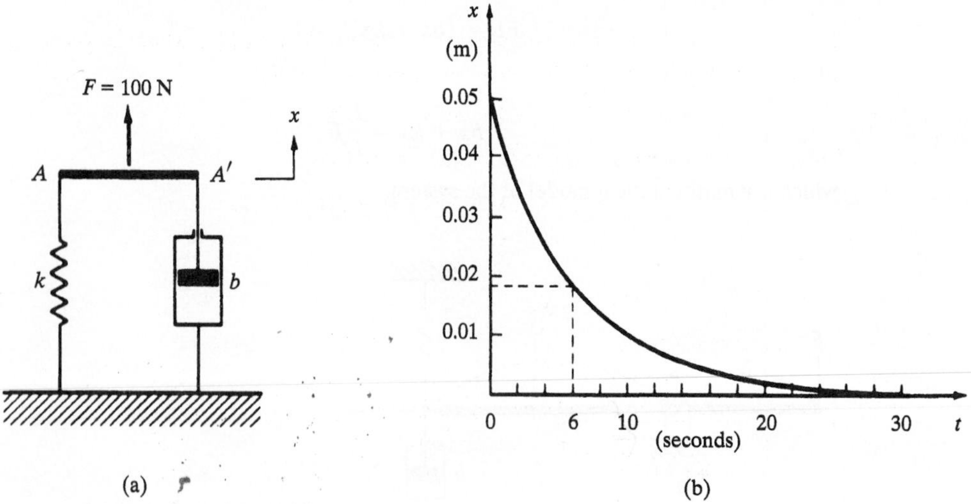

(a)

(b)

Figure 3–29   (a) Mechanical system; (b) response curve.

The inverse Laplace transform of $X(s)$, using the value of $k = 2000$ just obtained, is

$$x(t) = 0.05e^{-(2000/b)t}$$

Since the solution is an exponential function, at $t =$ time constant $= b/2000$ the response becomes

$$x\left(\frac{b}{2000}\right) = 0.05 \times 0.368 = 0.0184 \text{ m}$$

From Figure 3–29(b), $x = 0.0184$ m occurs at $t = 6$ s. Hence,

$$\frac{b}{2000} = 6$$

from which it follows that

$$b = 12,000 \text{ N-s/m}$$

Problem A–3–15

In the rotating system shown in Figure 3–30, assume that the torque T applied to the rotor is of short duration, but large amplitude, so that it can be considered an impulse input. Assume also that initially the angular velocity is zero, or $\omega(0-) = 0$. Given the numerical values

$$J = 10 \text{ kg-m}^2$$

and

$$b = 2 \text{ N-s/m}$$

find the response $\omega(t)$ of the system. Assume that the amplitude of torque T is 300 N-m/s and that the duration of T is 0.1 s; that is, the magnitude of the impulse input is $300 \times 0.1 = 30$ N-m. Show that the effect of an impulse input on a first-order system that is at rest is to generate a nonzero initial condition at $t = 0+$.

Solution   The equation of motion for the system is

$$J\dot{\omega} + b\omega = T, \qquad \omega(0-) = 0$$

Let us define the impulsive torque of magnitude 1 N-m as $\delta(t)$. Then, by substituting the given numerical values into this last equation, we obtain

$$10\dot{\omega} + 2\omega = 30\delta(t)$$

Taking the $\mathscr{L}_-$ transform of this last equation, we have

$$10[s\Omega(s) - \omega(0-)] + 2\Omega(s) = 30$$

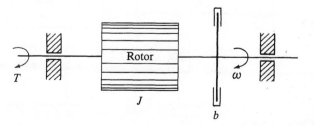

T

Rotor

J

ω

b

Figure 3–30   Mechanical rotating system.

or

$$\Omega(s) = \frac{30}{10s + 2} = \frac{3}{s + 0.2}$$

The inverse Laplace transform of $\Omega(s)$ is

$$\omega(t) = 3e^{-0.2t} \qquad\qquad\qquad\qquad (3\text{--}23)$$

Note that $\omega(0+) = 3$ rad/s. The angular velocity of the rotor is thus changed instantaneously from $\omega(0-) = 0$ to $\omega(0+) = 3$ rad/s.

If the system is subjected only to the initial condition $\omega(0) = 3$ rad/s and there is no external torque $(T = 0)$, then the equation of motion becomes

$$10\dot{\omega} + 2\omega = 0, \qquad \omega(0) = 3$$

Taking the Laplace transform of this last equation, we obtain

$$10[s\Omega(s) - \omega(0)] + 2\Omega(s) = 0$$

or

$$\Omega(s) = \frac{10\omega(0)}{10s + 2} = \frac{30}{10s + 2} = \frac{3}{s + 0.2}$$

The inverse Laplace transform of $\Omega(s)$ gives

$$\omega(t) = 3e^{-0.2t}$$

which is identical to Equation (3–23).

From the preceding analysis, we see that the response of a first-order system that is initially at rest to an impulse input is identical to the motion from the initial condition at $t = 0+$. That is, the effect of the impulse input on a first-order system that is initially at rest is to generate a nonzero initial condition at $t = 0+$.

Problem A–3–16

A mass $M = 8$ kg is supported by a spring with spring constant $k = 400$ N/m and a damper with $b = 40$ N-s/m, as shown in Figure 3–31. When a mass $m = 2$ kg is gently placed on the top of mass M, the system exhibits vibrations. Assuming that the displacement x of the masses is measured from the equilibrium position before mass m is placed on mass M, determine the response $x(t)$ of the system. Determine also the static deflection δ—the deflection of the spring when the transient response died out. Assume that $x(0) = 0$ and $\dot{x}(0) = 0$.

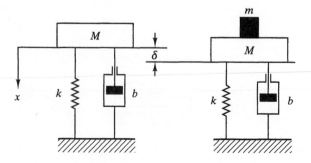

Figure 3–31 Mechanical system.

Notice that the numerical values of $M, m, b,$ and k are given in the SI system of units. If the units are changed to BES units, how does the mathematical model change? How will the solution be changed?

Solution   We shall first solve this problem using SI units. The input to the system is a constant force mg that acts as a step input to the system. The system is at rest before $t = 0$, and at $t = 0+$ the masses start to move up and down. A mathematical model, or equation of motion, is

$$(M + m)\ddot{x} + b\dot{x} + kx = mg$$

where $M + m = 10$ kg, $b = 40$ N-s/m, $k = 400$ N/m, and $g = 9.807$ m/s^2.

Substituting the numerical values into the equation of motion, we find that

$$10\ddot{x} + 40\dot{x} + 400x = 2 \times 9.807$$

or

$$\ddot{x} + 4\dot{x} + 40x = 1.9614 \tag{3-24}$$

Equation (3–24) is a mathematical model for the system when the units used are SI units. To obtain the response $x(t)$, we take the Laplace transform of Equation (3–24) and substitute the initial conditions $x(0) = 0$ and $\dot{x}(0) = 0$ into the Laplace-transformed equation as follows:

$$s^2X(s) + 4sX(s) + 40X(s) = \frac{1.9614}{s}$$

Solving for $X(s)$ yields

$$\begin{aligned}
X(s) &= \frac{1.9614}{(s^2 + 4s + 40)s} \\
&= \frac{1.9614}{40}\left(\frac{1}{s} - \frac{s + 4}{s^2 + 4s + 40}\right) \\
&= 0.04904\left[\frac{1}{s} - \frac{2}{6}\frac{6}{(s + 2)^2 + 6^2} - \frac{s + 2}{(s + 2)^2 + 6^2}\right]
\end{aligned}$$

The inverse Laplace transform of this last equation gives

$$x(t) = 0.04904\left(1 - \frac{1}{3}e^{-2t}\sin 6t - e^{-2t}\cos 6t\right)\,\text{m}$$

This solution gives the up-and-down motion of the total mass $(M + m)$. The static deflection δ is 0.04904 m.

Next, we shall solve the same problem using BES units. If we change the numerical values of $M, m, b,$ and k given in the SI system of units to BES units, we obtain

$$M = 8\text{ kg} = 0.54816\text{ slug}$$
$$m = 2\text{ kg} = 0.13704\text{ slug}$$
$$b = 40\text{ N-s/m} = 2.74063\text{ lb}_f\text{-s/ft}$$
$$k = 400\text{ N/m} = 27.4063\text{ lb}_f\text{/ft}$$
$$mg = 0.13704\text{ slug} \times 32.174\text{ ft/s}^2 = 4.4091\text{ slug-ft/s}^2$$
$$= 4.4091\text{ lb}_f$$

Then the equation of motion for the system becomes

$$0.6852\ddot{x} + 2.74063\dot{x} + 27.4063x = 4.4091$$

which can be simplified to

$$\ddot{x} + 4\dot{x} + 40x = 6.4348 \tag{3-25}$$

Equation (3–25) is a mathematical model for the system. Comparing Equations (3–24) and (3–25), we notice that the left-hand sides of the equations are the same, which means that the characteristic equation remains the same. The solution of Equation (3–25) is

$$x(t) = 0.1609 \left(1 - \frac{1}{3}e^{-2t} \sin 6t - e^{-2t} \cos 6t \right) \text{ft}$$

The static deflection δ is 0.1609 ft. (Note that 0.1609 ft = 0.04904 m.) Notice that, whenever consistent systems of units are used, the results carry the same information.

Problem A–3–17

Consider the spring-loaded inverted pendulum shown in Figure 3–32. Assume that the spring force acting on the pendulum is zero when the pendulum is vertical ($\theta = 0$). Assume also that the friction involved is negligible. Obtain a mathematical model of the system when the angle θ is small, that is, when $\sin \theta \doteq \theta$ and $\cos \theta \doteq 1$. Also, obtain the natural frequency ω_n of the system.

Solution   Suppose that the inverted pendulum is given an initial angular displacement $\theta(0)$ and released with zero initial angular velocity. Then, from Figure 3–32, for small θ such that $\sin \theta \doteq \theta$ and $\cos \theta \doteq 1$, the left-hand side spring is stretched by $h\theta$ and the right-hand side spring is compressed by $h\theta$. Hence, the torque acting on the pendulum in the counterclockwise direction is $2kh^2\theta$. The torque due to the gravitational force is $mgl\theta$, which acts in the clockwise direction. The moment of inertia of the pendulum is ml^2. Thus, the equation of motion of the system for small θ is

$$ml^2\ddot{\theta} = mgl\theta - 2kh^2\theta$$

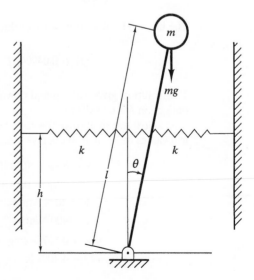

Figure 3–32   Spring-loaded inverted pendulum.

or

$$\ddot{\theta} + \left(\frac{2kh^2}{ml^2} - \frac{g}{l} \right) \theta = 0$$

This is a mathematical model of the system for small θ. If $2kh^2 > mgl$, the torques acting in the system cause it to vibrate. The undamped natural frequency of the system is

$$\omega_n = \sqrt{\frac{2kh^2}{ml^2} - \frac{g}{l}}$$

If, however, $2kh^2 < mgl$, then, starting with a small disturbance, the angle θ increases and the pendulum will fall down or hit the vertical wall and stop. The vibration will not occur.

Problem A–3–18

Consider the spring–mass–pulley system of Figure 3–33(a). If the mass m is pulled downward a short distance and released, it will vibrate. Obtain the natural frequency of the system by applying the law of conservation of energy.

Solution Define x, y, and θ as the displacement of mass m, the displacement of the pulley, and the angle of rotation of the pulley, measured respectively from their corresponding equilibrium positions. Note that $x = 2y$, $R\theta = x - y = y$, and $J = \frac{1}{2}MR^2$.

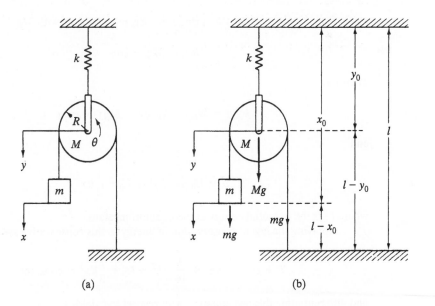

(a)   (b)

Figure 3–33 (a) Spring–mass–pulley system; (b) diagram for figuring out potential energy of the system.

The kinetic energy of the system is

$$T = \frac{1}{2}m\dot{x}^2 + \frac{1}{2}M\dot{y}^2 + \frac{1}{2}J\dot{\theta}^2$$

$$= \frac{1}{2}m\dot{x}^2 + \frac{1}{8}M\dot{x}^2 + \frac{1}{4}MR^2\left(\frac{\dot{y}}{R}\right)^2$$

$$= \frac{1}{2}m\dot{x}^2 + \frac{3}{16}M\dot{x}^2$$

The potential energy U of the system can be obtained from Figure 3–33(b). At the equilibrium state, the potential energy is

$$U_0 = \frac{1}{2}ky_\delta^2 + Mg(l - y_0) + mg(l - x_0)$$

where y_δ is the static deflection of the spring due to the hanging masses M and m. When masses m and M are displaced by x and y, respectively, the instantaneous potential energy can be obtained as

$$U = \frac{1}{2}k(y_\delta + y)^2 + Mg(l - y_0 - y) + mg(l - x_0 - x)$$

$$= \frac{1}{2}ky_\delta^2 + ky_\delta y + \frac{1}{2}ky^2 + Mg(l - y_0) - Mgy + mg(l - x_0) - mgx$$

$$= U_0 + \frac{1}{2}ky^2 + ky_\delta y - Mgy - mgx$$

Again from Figure 3–33(b), the spring force ky_δ must balance with $Mg + 2mg$, or

$$ky_\delta = Mg + 2mg$$

Therefore,

$$ky_\delta y = Mgy + 2mgy = Mgy + mgx$$

and

$$U = U_0 + \frac{1}{2}ky^2 = U_0 + \frac{1}{8}kx^2$$

where U_0 is the potential energy at the equilibrium state.

Applying the law of conservation of energy to this conservative system gives

$$T + U = \frac{1}{2}m\dot{x}^2 + \frac{3}{16}M\dot{x}^2 + U_0 + \frac{1}{8}kx^2 = \text{constant}$$

and differentiating this last equation with respect to t yields

$$m\dot{x}\ddot{x} + \frac{3}{8}M\dot{x}\ddot{x} + \frac{1}{4}kx\dot{x} = 0$$

or

$$\left[\left(m + \frac{3}{8}M\right)\ddot{x} + \frac{1}{4}kx\right]\dot{x} = 0$$

Since $\dot{x}$ is not always zero, we must have

$$\left(m + \frac{3}{8}M\right)\ddot{x} + \frac{1}{4}kx = 0$$

or

$$\ddot{x} + \frac{2k}{8m + 3M}x = 0$$

The natural frequency of the system, therefore, is

$$\omega_n = \sqrt{\frac{2k}{8m + 3M}}$$

Problem A–3–19

If, for the spring–mass system of Figure 3–34, the mass m_s of the spring is small, but not negligibly small, compared with the suspended mass m, show that the inertia of the spring can be allowed for by adding one-third of its mass m_s to the suspended mass m and then treating the spring as a massless spring.

Solution Consider the free vibration of the system. The displacement x of the mass is measured from the static equilibrium position. In free vibration, the displacement can be written as

$$x = A \cos \omega t$$

Since the mass of the spring is comparatively small, we can assume that the spring is stretched uniformly. Then the displacement of a point in the spring at a distance ξ from the top is given by $(\xi/l)A \cos \omega t$.

In the mean position, where $x = 0$ and the velocity of mass m is maximum, the velocity of the suspended mass is $A\omega$ and that of the spring at the distance ξ from the

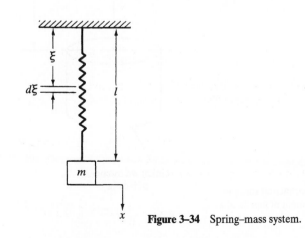

Figure 3–34 Spring–mass system.

top is $(\xi/l)A\omega$. The maximum kinetic energy is

$$
\begin{aligned}
T_{\max} &= \frac{1}{2}m(A\omega)^2 + \int_0^l \frac{1}{2}\left(\frac{m_s}{l}\right)\left(\frac{\xi}{l}A\omega\right)^2 d\xi \\
&= \frac{1}{2}mA^2\omega^2 + \frac{1}{2}\left(\frac{m_s}{l}\right)\left(\frac{A^2\omega^2}{l^2}\right)\frac{1}{3}l^3 \\
&= \frac{1}{2}\left(m + \frac{m_s}{3}\right)A^2\omega^2
\end{aligned}
$$

Note that the mass of the spring does not affect the change in the potential energy of the system and that, if the spring were massless, the maximum kinetic energy would have been $\frac{1}{2}mA^2\omega^2$. Therefore, we conclude that the inertia of the spring can be allowed for simply by adding one-third of mass m_s to the suspended mass m and then treating the spring as a massless spring, provided that m_s is small compared with m.

PROBLEMS

Problem B–3–1

A homogeneous disk has a diameter of 1 m and mass of 100 kg. Obtain the moment of inertia of the disk about the axis perpendicular to the disk and passing through its center.

Problem B–3–2

Figure 3–35 shows an experimental setup for measuring the moment of inertia of a rotating body. Suppose that the moment of inertia of a rotating body about axis AA' is known. Describe a method to determine the moment of inertia of any rotating body, using this experimental setup.

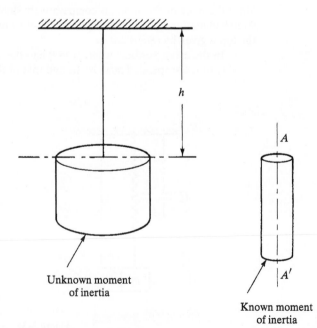

h

A

Unknown moment of inertia

A'

Figure 3–35   Experimental setup for measuring the moment of inertia of a rotating body.

Known moment of inertia

Problem B–3–3

A ball is dropped from a point 100 m above the ground with zero initial velocity. How long will it take until the ball hits the ground? What is the velocity when the ball hits the ground?

Problem B–3–4

A flywheel of $J = 50$ kg-m^2 initially standing still is subjected to a constant torque. If the angular velocity reaches 20 Hz in 5 s, find the torque given to the flywheel.

Problem B–3–5

A brake is applied to a flywheel rotating at an angular velocity of 100 rad/s. If the angular velocity reduces to 20 rad/s in 15 s, find (a) the deceleration produced by the brake and (b) the total angle the flywheel rotates in the 15-s period.

Problem B–3–6

Consider the series-connected springs shown in Figure 3–36(a). Referring to Figure 3–36(b), show that the equivalent spring constant k_{eq} can be graphically obtained as the length $\overline{OC}$ if lengths $\overline{OA}$ and $\overline{OB}$ represent k_1 and k_2, respectively.

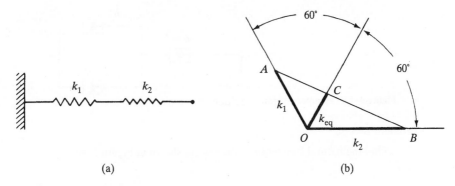

(a)   (b)

Figure 3–36   (a) System consisting of two springs in series; (b) diagram showing the equivalent spring constant.

Problem B–3–7

Obtain the equivalent spring constant k_{eq} for the system shown in Figure 3–37.

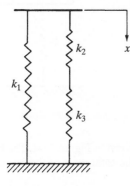

Figure 3–37   System consisting of three springs.

Problem B–3–8

Obtain the equivalent viscous-friction coefficient b_{eq} for each of the systems shown in Figure 3–38 (a) and (b).

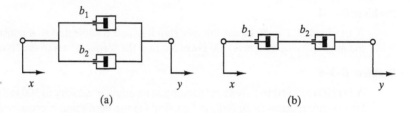

(a)   (b)

Figure 3–38   (a) Two dampers connected in parallel; (b) two dampers connected in series.

Problem B–3–9

Obtain the equivalent viscous-friction coefficient b_{eq} of the system shown in Figure 3–39.

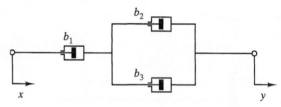

Figure 3–39   Damper system. x

Problem B–3–10

Find the natural frequency of the system shown in Figure 3–40.

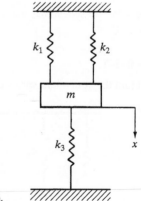

Figure 3–40   Mechanical system.

Problem B–3–11

Consider the U-shaped manometer shown in Figure 3–41. The liquid partially fills the U-shaped glass tube. Assuming that the total mass of the liquid in the tube is m, the

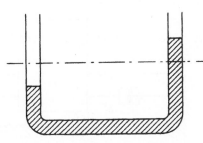

Figure 3–41 U-shaped manometer system.

total length of liquid in the tube is L, and the viscosity of the liquid is negligible, what is the equation of motion of the liquid? Find the frequency of oscillation.

Problem B–3–12

In the mechanical system shown in Figure 3–42, assume that the rod is massless, perfectly rigid, and pivoted at point P. The displacement x is measured from the equilibrium position. Assuming that x is small, that the weight mg at the end of the rod is 5 N, and that the spring constant k is 400 N/m, find the natural frequency of the system.

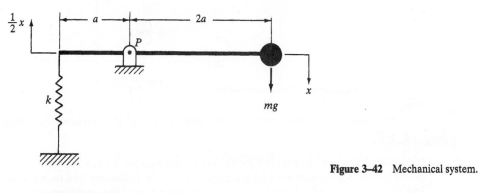

Figure 3–42 Mechanical system.

Problem B–3–13

Obtain a mathematical model of the system shown in Figure 3–43. The input to the system is the angle θ_i and the output is the angle θ_o.

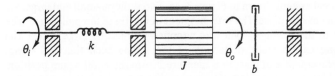

Figure 3–43 Mechanical system.

Problem B–3–14

Obtain a mathematical model for the system shown in Figure 3–44.

Problem B–3–15

Consider the system shown in Figure 3–45, where $m = 2$ kg, $b = 4$ N-s/m, and $k = 20$ N/m. Assume that $x(0) = 0.1$ m and $\dot{x}(0) = 0$. [The displacement $x(t)$ is measured from the equilibrium position.]

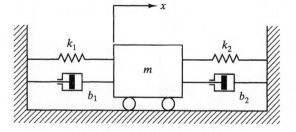

Figure 3–44 Mechanical system.

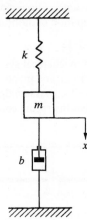

Figure 3–45 Mechanical system.

Derive a mathematical model of the system. Then find $x(t)$ as a function of time t.

Problem B–3–16

By applying Newton's second law to the spring–mass–pulley system of Figure 3–33(a), obtain the motion of mass m when it is pulled down a short distance and then released. The displacement x of a hanging mass m is measured from the equilibrium position. (The mass, the radius, and the moment of inertia of the pulley are M, R, and $J = \frac{1}{2}MR^2$, respectively.)

Problem B–3–17

Consider the mechanical system shown in Figure 3–46. Two pulleys, small and large, are bolted together and act as one piece. The total moment of inertia of the pulleys is J. The mass m is connected to the spring k_1 by a wire wrapped around the large pulley. The gravitational force mg causes static deflection of the spring such that $k_1\delta = mg$. Assume that the displacement x of mass m is measured from the equilibrium position. Two springs (denoted by k_2) are connected by a wire that passes over the small pulley as shown in the figure. Each of the two springs is prestretched by an amount y.

Obtain a mathematical model of the system. Also, obtain the natural frequency of the system.

Problem B–3–18

A disk of radius 0.5 m is subjected to a tangential force of 50 N at its periphery and is rotating at an angular velocity of 100 rad/s. Calculate the torque and power of the disk shaft.

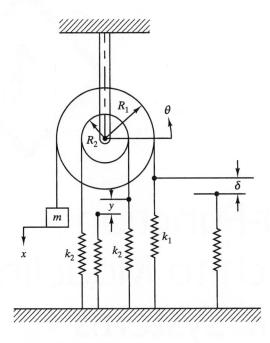

Figure 3–46 Mechanical system.

Problem B–3–19

Referring to the spring-loaded inverted pendulum system shown in Figure 3–32, obtain the natural frequency ω_n of the system, using the energy method that equates the maximum kinetic energy T_{max} and the maximum potential energy U_{max}. (Choose the potential energy at the equilibrium state to be zero.)

Problem B–3–20

Assuming that mass m of the rod of the pendulum shown in Figure 3–47 is small, but not negligible, compared with mass M, find the natural frequency of the pendulum when the angle θ is small. (Include the effect of m in the expression of the natural frequency.)

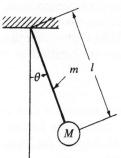

Figure 3–47 Pendulum system.

Transfer-Function Approach to Modeling Dynamic Systems

4–1 INTRODUCTION

In this chapter, we present the transfer-function approach to modeling and analyzing dynamic systems. We first define the transfer function and then introduce block diagrams. Since MATLAB plays an important role in obtaining computational solutions of transient response problems, we present a detailed introduction to writing MATLAB programs to obtain response curves for time-domain inputs such as the step, impulse, ramp, and others.

In the field of system dynamics, transfer functions are frequently used to characterize the input–output relationships of components or systems that can be described by linear, time-invariant differential equations. We begin this section by defining the transfer function and deriving the transfer function of a mechanical system. Then we discuss the impulse response function, or the weighting function, of the system.

Transfer Function. The *transfer function* of a linear, time-invariant differential-equation system is defined as the ratio of the Laplace transform of the output (response function) to the Laplace transform of the input (driving function) under the assumption that all initial conditions are zero.

106

Consider the linear time-invariant system defined by the differential equation

$$a_0 \overset{(n)}{y} + a_1 \overset{(n-1)}{y} + \cdots + a_{n-1}\dot{y} + a_n y$$

$$= b_0 \overset{(m)}{x} + b_1 \overset{(m-1)}{x} + \cdots + b_{m-1}\dot{x} + b_m x \qquad (n \geq m)$$

where y is the output of the system and x is the input. The transfer function of this system is the ratio of the Laplace-transformed output to the Laplace-transformed input when all initial conditions are zero, or

$$\text{Transfer function} = G(s) = \frac{\mathscr{L}[\text{output}]}{\mathscr{L}[\text{input}]}\Bigg|_{\text{zero initial conditions}}$$

$$= \frac{Y(s)}{X(s)} = \frac{b_0 s^m + b_1 s^{m-1} + \cdots + b_{m-1}s + b_m}{a_0 s^n + a_1 s^{n-1} + \cdots + a_{n-1}s + a_n} \qquad (4\text{--}1)$$

By using the concept of a transfer function, it is possible to represent system dynamics by algebraic equations in s. If the highest power of s in the denominator of the transfer function is equal to n, the system is called an *nth-order system*.

Comments on the Transfer Function.   The applicability of the concept of the transfer function is limited to linear, time-invariant differential-equation systems. Still, the transfer-function approach is used extensively in the analysis and design of such systems. The following list gives some important comments concerning the transfer function of a system described by a linear, time-invariant differential equation:

1. The transfer function of a system is a mathematical model of that system, in that it is an operational method of expressing the differential equation that relates the output variable to the input variable.
2. The transfer function is a property of a system itself, unrelated to the magnitude and nature of the input or driving function.
3. The transfer function includes the units necessary to relate the input to the output; however, it does not provide any information concerning the physical structure of the system. (The transfer functions of many physically different systems can be identical.)
4. If the transfer function of a system is known, the output or response can be studied for various forms of inputs with a view toward understanding the nature of the system.
5. If the transfer function of a system is unknown, it may be established experimentally by introducing known inputs and studying the output of the system. Once established, a transfer function gives a full description of the dynamic characteristics of the system, as distinct from its physical description.

Example 4–1

Consider the mechanical system shown in Figure 4–1. The displacement x of the mass m is measured from the equilibrium position. In this system, the external force $f(t)$ is the input and x is the output.

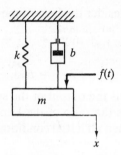

Figure 4–1 Mechanical system. x

The equation of motion for the system is

$$m\ddot{x} + b\dot{x} + kx = f(t)$$

Taking the Laplace transform of both sides of this equation and assuming that all initial conditions are zero yields

$$(ms^2 + bs + k)X(s) = F(s)$$

where $X(s) = \mathcal{L}[x(t)]$ and $F(s) = \mathcal{L}[f(t)]$. From Equation (4–1), the transfer function for the system is

$$\frac{X(s)}{F(s)} = \frac{1}{ms^2 + bs + k}$$

Impulse-Response Function.   The transfer function of a linear, time-invariant system is

$$G(s) = \frac{Y(s)}{X(s)}$$

where $X(s)$ is the Laplace transform of the input and $Y(s)$ is the Laplace transform of the output and where we assume that all initial conditions involved are zero. It follows that the output $Y(s)$ can be written as the product of $G(s)$ and $X(s)$, or

$$Y(s) = G(s)X(s) \tag{4–2}$$

· Now, consider the output (response) of the system to a unit-impulse input when the initial conditions are zero. Since the Laplace transform of the unit-impulse function is unity, or $X(s) = 1$, the Laplace transform of the output of the system is

$$Y(s) = G(s) \tag{4–3}$$

The inverse Laplace transform of the output given by Equation (4–3) yields the impulse response of the system. The inverse Laplace transform of $G(s)$, or

$$\mathcal{L}^{-1}[G(s)] = g(t)$$

is called the *impulse-response function*, or the *weighting function*, of the system.

The impulse-response function $g(t)$ is thus the response of a linear system to a unit-impulse input when the initial conditions are zero. The Laplace transform of $g(t)$ gives the transfer function. Therefore, the transfer function and impulse-response function of a linear, time-invariant system contain the same information about the system dynamics. It is hence possible to obtain complete information about the dynamic characteristics of a system by exciting it with an impulse input and measuring the response. (In practice, a large pulse input with a very short duration compared with the significant time constants of the system may be considered an impulse.)

Outline of the Chapter.   Section 4–1 has presented the concept of the transfer function and impulse-response function. Section 4–2 discusses the block diagram. Section 4–3 sets forth the MATLAB approach to the partial-fraction expansion of a ratio of two polynomials, $B(s)/A(s)$. Section 4–4 details the MATLAB approach to the transient response analysis of transfer-function systems.

4–2 BLOCK DIAGRAMS

Block diagrams of dynamic systems.   A block diagram of a dynamic system is a pictorial representation of the functions performed by each component of the system and of the flow of signals within the system. Such a diagram depicts the interrelationships that exist among the various components. Differing from a purely abstract mathematical representation, a block diagram has the advantage of indicating the signal flows of the actual system more realistically.

In a block diagram, all system variables are linked to each other through functional blocks. The *functional block*, or simply *block*, is a symbol for the mathematical operation on the input signal to the block that produces the output. The transfer functions of the components are usually entered in the corresponding blocks, which are connected by arrows to indicate the direction of the flow of signals. Note that a signal can pass only in the direction of the arrows. Thus, a block diagram of a dynamic system explicitly shows a unilateral property.

Figure 4–2 shows an element of a block diagram. The arrowhead pointing toward the block indicates the input to the block, and the arrowhead leading away from the block represents the output of the block. As mentioned, such arrows represent signals.

Note that the dimension of the output signal from a block is the dimension of the input signal multiplied by the dimension of the transfer function in the block.

The advantages of the block diagram representation of a system lie in the fact that it is easy to form the overall block diagram for the entire system merely by connecting the blocks of the components according to the signal flow and that it is possible to evaluate the contribution of each component to the overall performance of the system.

In general, the functional operation of a system can be visualized more readily by examining a block diagram of the system than by examining the physical system itself. A block diagram contains information concerning dynamic behavior, but it

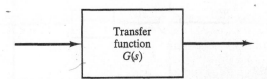

Figure 4–2   Element of a block diagram.

does not include any information about the physical construction of the system. Consequently, many dissimilar and unrelated systems can be represented by the same block diagram.

Note that in a block diagram the main source of energy is not explicitly shown and that the block diagram of a given system is not unique. A number of different block diagrams can be drawn for a system, depending on the point of view of the analysis. (See **Example 4–2.**)

Summing point. Figure 4–3 shows a circle with a cross, the symbol that stands for a summing operation. The plus or minus sign at each arrowhead indicates whether the associated signal is to be added or subtracted. It is important that the quantities being added or subtracted have the same dimensions and the same units.

Branch point.   A *branch point* is a point from which the signal from a block goes concurrently to other blocks or summing points.

Block diagram of a closed-loop system.   Figure 4–4 is a block diagram of a closed-loop system. The output $C(s)$ is fed back to the summing point, where it is compared with the input $R(s)$. The closed-loop nature of the system is indicated clearly by the figure. The output $C(s)$ of the block is obtained by multiplying the transfer function $G(s)$ by the input to the block, $E(s)$.

Any linear system can be represented by a block diagram consisting of blocks, summing points, and branch points. When the output is fed back to the summing point for comparison with the input, it is necessary to convert the form of the output signal to that of the input signal. This conversion is accomplished by the feedback element whose transfer function is $H(s)$, as shown in Figure 4–5. Another important role of the feedback element is to modify the output before it is compared with the input. In the figure, the feedback signal that is fed back to the summing point for comparison with the input is $B(s) = H(s)C(s)$.

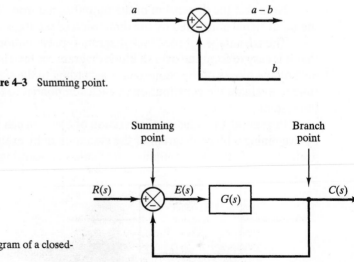

Figure 4–3 Summing point.

Figure 4–4 Block diagram of a closed-loop system.

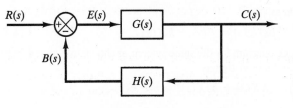

Figure 4–5   Block diagram of a closed-loop system with feedback element.

Simplifying complex block diagrams and obtaining overall transfer functions from such block diagrams are discussed in Chapter 5.

Example 4–2

Consider again the mechanical system shown in Figure 4–1. The transfer function of this system (see **Example 4–1**) is

$$\frac{X(s)}{F(s)} = \frac{1}{ms^2 + bs + k} \tag{4-4}$$

A block diagram representation of the system is shown in Figure 4–6(a).

Notice that Equation (4–4) can be written as

$$(ms^2 + bs + k)X(s) = F(s) \tag{4-5}$$

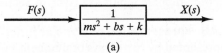

(a)

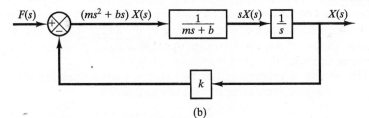

(b)

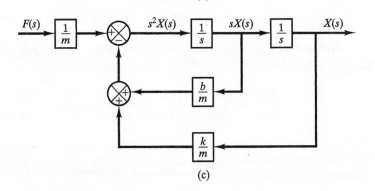

(c)

Figure 4–6   Block diagrams of the system shown in Figure 4–1. (a) Block diagram based on Equation (4–4); (b) block diagram based on Equation (4–6); (c) block diagram based on Equation (4–7).

Rewriting the latter equation as

$$(ms^2 + bs)X(s) = F(s) - kX(s) \tag{4-6}$$

we can obtain a different block diagram for the same system, as shown in Figure 4–6(b). Equation (4–5) can also be rewritten as

$$F(s) - [kX(s) + bsX(s)] = ms^2X(s)$$

or

$$\frac{1}{m}F(s) - \frac{k}{m}X(s) - \frac{b}{m}sX(s) = s^2X(s) \tag{4-7}$$

A block diagram for the system based on Equation (4–7) is shown in Figure 4–6(c).

Figures 4–6(a), (b), and (c) are thus block diagrams for the same system—that shown in Figure 4–1. (Many different block diagrams are possible for any given system.)

4–3 PARTIAL-FRACTION EXPANSION WITH MATLAB

We begin this section, with an examination of the partial-fraction expansion of the transfer function $B(s)/A(s)$ with MATLAB. Then we discuss how to obtain the system response analytically. Computational solutions (response curves) for the system responses to time-domain inputs are given in Section 4–4.

MATLAB representation of transfer functions. The transfer function of a system is represented by two arrays of numbers. For example, consider a system defined by

$$\frac{Y(s)}{U(s)} = \frac{25}{s^2 + 4s + 25}$$

This system is represented as two arrays, each containing the coefficients of the polynomials in decreasing powers of s as follows:

$$\text{num} = [25]$$
$$\text{den} = [1 \quad 4 \quad 25]$$

Partial-fraction expansion with MATLAB. MATLAB allows us to obtain the partial-fraction expansion of the ratio of two polynomials,

$$\frac{B(s)}{A(s)} = \frac{\text{num}}{\text{den}} = \frac{b(1)s^h + b(2)s^{h-1} + \cdots + b(h)}{a(1)s^n + a(2)s^{n-1} + \cdots + a(n)}$$

where $a(1) \neq 0$, some of $a(i)$ and $b(j)$ may be zero, and num and den are row vectors that specify the coefficients of the numerator and denominator of $B(s)/A(s)$. That is,

$$\text{num} = [b(1) \quad b(2) \quad \cdots \quad b(h)]$$
$$\text{den} = [a(1) \quad a(2) \quad \cdots \quad a(n)]$$

The command

$$[r,p,k] = residue(num,den)$$

finds the residues, poles, and direct terms of a partial-fraction expansion of the ratio of the two polynomials $B(s)$ and $A(s)$. The partial-fraction expansion of $B(s)/A(s)$ is given by

$$\frac{B(s)}{A(s)} = k(s) + \frac{r(1)}{s - p(1)} + \frac{r(2)}{s - p(2)} + \cdots + \frac{r(n)}{s - p(n)}$$

As an example, consider the function

$$\frac{B(s)}{A(s)} = \frac{s^4 + 8s^3 + 16s^2 + 9s + 6}{s^3 + 6s^2 + 11s + 6}$$

For this function,

$$num = [1 \quad 8 \quad 16 \quad 9 \quad 6]$$
$$den = [1 \quad 6 \quad 11 \quad 6]$$

Entering the command

$$[r,p,k] = residue(num,den)$$

as shown in MATLAB Program 4–1, we obtain the residues (r), poles (p), and direct terms (k).

MATLAB Program 4–1

```
>> num = [1     8     16     9     6];
>> den = [1     6     11     6];
>> [r,p,k] = residue(num,den)

r =

      -6.0000
      -4.0000
       3.0000

p =

      -3.0000
      -2.0000
      -1.0000

k =

       1       2
```

MATLAB Program 4–1 is the MATLAB representation of the partial-fraction expansion of $B(s)/A(s)$:

$$\frac{B(s)}{A(s)} = \frac{s^4 + 8s^3 + 16s^2 + 9s + 6}{s^3 + 6s^2 + 11s + 6}$$

$$= s + 2 + \frac{-6}{s + 3} + \frac{-4}{s + 2} + \frac{3}{s + 1}$$

Note that MATLAB first divides the numerator by the denominator and produces a polynomial in s (denoted as row vector k) plus a remainder (a ratio of polynomials in s, where the numerator is of lower degree than the denominator). Then MATLAB expands this remainder into partial fractions and returns the residues as column vector r and the pole locations as column vector p.

The command

$$[num,den] = residue(r,p,k)$$

where r, p, and k are outputs in MATLAB Program 4–1, converts the partial-fraction expansion back to the polynomial ratio $B(s)/A(s)$, as shown in MATLAB Program 4–2.

MATLAB Program 4–2

```
>> r = [−6   −4   3];
>> p = [−3   −2   −1];
>> k = [1   2];
>> [num, den] = residue(r,p,k)

num =

        1      8      16      9      6

den =

        1      6      11      6
```

Example 4–3

Consider the spring–mass–dashpot system mounted on a massless cart as shown in Figure 4–7. A dashpot is a device that provides viscous friction, or damping. It consists of a piston and oil-filled cylinder. Any relative motion between the piston rod and the cylinder is resisted by the oil because the oil must flow around the piston (or through orifices provided in the piston) from one side of the piston to the other. The dashpot essentially absorbs energy, which is dissipated as heat. The dashpot, also called a *damper*, does not store any kinetic or potential energy.

Let us obtain a mathematical model of this system by assuming that both the cart and the spring–mass–dashpot system on it are standing still for $t < 0$. In this system, $u(t)$ is the displacement of the cart and the input to the system. The displacement $y(t)$ of the mass relative to the ground is the output. Also, m denotes the mass, b denotes the viscous friction coefficient, and k denotes the spring constant. We assume that the friction force of the dashpot is proportional to $\dot{y} - \dot{u}$ and that the spring is linear; that is, the spring force is proportional to $y - u$.

After a mathematical model of the system is obtained, we determine the output $y(t)$ analytically when $m = 10$ kg, $b = 20$ N-s/m, and $k = 100$ N/m. The input is assumed to be a unit-step input.

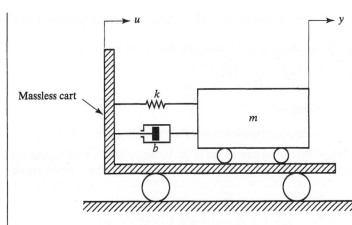

Figure 4–7 Spring–mass–dashpot system mounted on a cart.

For translational systems, Newton's second law states that

$$ma = \sum F$$

where m is a mass, a is the acceleration of the mass, and $\sum F$ is the sum of the forces acting on the mass in the direction of the acceleration. Applying Newton's second law to the present system and noting that the cart is massless, we obtain

$$m\frac{d^2y}{dt^2} = -b\left(\frac{dy}{dt} - \frac{du}{dt}\right) - k(y - u)$$

or

$$m\frac{d^2y}{dt^2} + b\frac{dy}{dt} + ky = b\frac{du}{dt} + ku$$

The latter equation represents a mathematical model of the system under consideration. Taking the Laplace transform of the equation, assuming zero initial conditions, gives

$$(ms^2 + bs + k)Y(s) = (bs + k)U(s)$$

Taking the ratio of $Y(s)$ to $U(s)$, we find the transfer function of the system to be

$$\text{Transfer function} = \frac{Y(s)}{U(s)} = \frac{bs + k}{ms^2 + bs + k} \tag{4–8}$$

Next, we shall obtain an analytical solution of the response to the unit-step input. Substituting the given numerical values into Equation (4–8) gives

$$\frac{Y(s)}{U(s)} = \frac{20s + 100}{10s^2 + 20s + 100} = \frac{2s + 10}{s^2 + 2s + 10}$$

Since the input u is a unit-step function,

$$U(s) = \frac{1}{s}$$

Then the output $Y(s)$ becomes

$$Y(s) = \frac{2s + 10}{s^2 + 2s + 10}\frac{1}{s} = \frac{2s + 10}{s^3 + 2s^2 + 10s}$$

To obtain the inverse Laplace transform of $Y(s)$, we need to expand $Y(s)$ into partial fractions.

Applying MATLAB and noting that num and den of the system are

$$\text{num} = [2 \quad 10]$$
$$\text{den} = [1 \quad 2 \quad 10 \quad 0]$$

we may use the residue command

$$[r,p,k] = \text{residue(num,den)}$$

to find the residues (r), poles (p), and direct term (k) as shown in MATLAB Program 4–3. MATLAB Program 4–3 is the MATLAB representation of the partial-fraction expansion of $Y(s)$:

$$Y(s) = \frac{-0.5 - j0.1667}{s + 1 - j3} + \frac{-0.5 + j0.1667}{s + 1 + j3} + \frac{1}{s}$$

MATLAB Program 4–3

```
>> num = [2   10];
>> den = [1    2    10    0];
>> [r,p,k] = residue(num,den)

r =

     -0.5000 - 0.1667i
     -0.5000 + 0.1667i
      1.0000

p =

     -1.0000 + 3.0000i
     -1.0000 - 3.0000i
            0

k =

     []
```

Since $Y(s)$ involves complex-conjugate poles, it is convenient to combine two complex-conjugate terms into one as follows:

$$\frac{-0.5 - j0.1667}{s + 1 - j3} + \frac{-0.5 + j0.1667}{s + 1 + j3} = \frac{-s}{(s + 1)^2 + 3^2}$$

Then $Y(s)$ can be expanded as

$$Y(s) = \frac{1}{s} - \frac{s}{(s + 1)^2 + 3^2}$$
$$= \frac{1}{s} - \frac{s + 1 - 1}{(s + 1)^2 + 3^2}$$
$$= \frac{1}{s} - \frac{s + 1}{(s + 1)^2 + 3^2} + \frac{1}{3}\frac{3}{(s + 1)^2 + 3^2}$$

The inverse Laplace transform of $Y(s)$ is obtained as

$$y(t) = 1 - e^{-t} \cos 3t + \frac{1}{3} e^{-t} \sin 3t$$

where $y(t)$ is measured in meters and t in seconds. This equation is an analytical solution to the problem.

Note that a plot of $y(t)$ versus t can be obtained easily with MATLAB from the information on num, den, and $u(t)$ without using the partial-fraction expansion. (See **Example 4-5**.)

Example 4-4

Consider the mechanical system shown in Figure 4-8. The system is at rest initially. The displacements x and y are measured from their respective equilibrium positions. Assuming that $p(t)$ is a step force input and the displacement $x(t)$ is the output, obtain the transfer function of the system. Then, assuming that $m = 0.1$ kg, $b_2 = 0.4$ N-s/m, $k_1 = 6$ N/m, $k_2 = 4$ N/m, and $p(t)$ is a step force of magnitude 10 N, obtain an analytical solution $x(t)$.

The equations of motion for the system are

$$m\ddot{x} + k_1 x + k_2(x - y) = p$$
$$k_2(x - y) = b_2 \dot{y}$$

Laplace transforming these two equations, assuming zero initial conditions, we obtain

$$(ms^2 + k_1 + k_2)X(s) = k_2 Y(s) + P(s) \qquad (4\text{-}9)$$
$$k_2 X(s) = (k_2 + b_2 s)Y(s) \qquad (4\text{-}10)$$

Solving Equation (4-10) for $Y(s)$ and substituting the result into Equation (4-9), we get

$$(ms^2 + k_1 + k_2)X(s) = \frac{k_2^2}{k_2 + b_2 s} X(s) + P(s)$$

or

$$[(ms^2 + k_1 + k_2)(k_2 + b_2 s) - k_2^2]X(s) = (k_2 + b_2 s)P(s)$$

from which we obtain the transfer function

$$\frac{X(s)}{P(s)} = \frac{b_2 s + k_2}{mb_2 s^3 + mk_2 s^2 + (k_1 + k_2)b_2 s + k_1 k_2} \qquad (4\text{-}11)$$

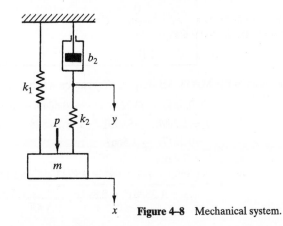

x **Figure 4-8**   Mechanical system.

Substituting the given numerical values for m, k_1, k_2, and b_2 into Equation (4–11), we have

$$\frac{X(s)}{P(s)} = \frac{0.4s + 4}{0.04s^3 + 0.4s^2 + 4s + 24}$$

$$= \frac{10s + 100}{s^3 + 10s^2 + 100s + 600} \tag{4–12}$$

Since $P(s)$ is a step force of magnitude 10 N,

$$P(s) = \frac{10}{s}$$

Then, from Equation (4–12), $X(s)$ can be written as

$$X(s) = \frac{10s + 100}{s^3 + 10s^2 + 100s + 600} \frac{10}{s}$$

To find an analytical solution, we need to expand $X(s)$ into partial fractions. For this purpose, we may use MATLAB Program 4–4, which produces the residues, poles, and direct term.

MATLAB Program 4–4

```
>> num = [100   1000];
>> den = [1   10   100   600   0];
>> [r,p,k] = residue(num, den)

r =

   -0.6845 + 0.2233i
   -0.6845 - 0.2233i
   -0.2977
    1.6667

p =

   -1.2898 + 8.8991i
   -1.2898 - 8.8991i
   -7.4204
        0

k =

   []
```

On the basis of the MATLAB output, $X(s)$ can be written as

$$X(s) = \frac{-0.6845 + j0.2233}{s + 1.2898 - j8.8991} + \frac{-0.6845 - j0.2233}{s + 1.2898 + j8.8991}$$

$$+ \frac{-0.2977}{s + 7.4204} + \frac{1.6667}{s}$$

$$= \frac{-1.3690(s + 1.2898) - 3.9743}{(s + 1.2898)^2 + 8.8991^2} - \frac{0.2977}{s + 7.4204} + \frac{1.6667}{s}$$

The inverse Laplace transform of $X(s)$ gives

$$x(t) = -1.3690e^{-1.2898t} \cos(8.8991t)$$
$$-0.4466e^{-1.2898t} \sin(8.8991t) - 0.2977e^{-7.4204t} + 1.6667$$

where $x(t)$ is measured in meters and time t in seconds. This is the analytical solution to the problem. [For the response curve $x(t)$ versus t, see **Example 4–6**.]

From the preceding examples, we have seen that once the transfer function $X(s)/U(s) = G(s)$ of a system is obtained, the response of the system to any input can be determined by taking the inverse Laplace transform of $X(s)$, or

$$\mathcal{L}^{-1}[X(s)] = \mathcal{L}^{-1}[G(s)U(s)]$$

Finding the inverse Laplace transform of $G(s)U(s)$ may be time consuming if the transfer function $G(s)$ of the system is complicated, even though the input $U(s)$ may be a simple function of time. Unless, for some reason, an analytical solution is needed, we should use a computer to get a numerical solution. Throughout this book, we use MATLAB to obtain numerical solutions to many problems. Obtaining numerical solutions and presenting them in the form of response curves is the subject discussed in the next section.

4–4 TRANSIENT-RESPONSE ANALYSIS WITH MATLAB

This section presents the MATLAB approach to obtaining system responses when the inputs are time-domain inputs such as the step, impulse, and ramp functions. The system response to the frequency-domain input (e.g., a sinusoidal input) is presented in Chapters 9 and 11.

MATLAB representation of transfer-function systems.   Figure 4–9 shows a block with a transfer function. Such a block represents a system or an element of a system. To simplify our presentation, we shall call the block with a transfer function a system. MATLAB uses sys to represent such a system. The statement

$$\text{sys} = \text{tf(num, den)} \tag{4–13}$$

represents the system. For example, consider the system

$$\frac{Y(s)}{X(s)} = \frac{2s + 25}{s^2 + 4s + 25}$$

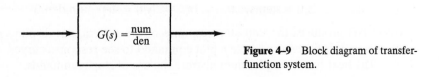

Figure 4–9   Block diagram of transfer-function system.

This system can be represented as two arrays, each containing the coefficients of the polynomials in decreasing powers of s as follows:

$$num = [2 \quad 25]$$
$$den = [1 \quad 4 \quad 25]$$

Entering MATLAB Program 4–5 into a computer produces the transfer function of the system.

MATLAB Program 4–5

```
>> num = [2   25];
>> den = [1   4   25];
>> sys = tf(num,den)

Transfer function:
  2 s + 25
--------------------
s^2 + 4 s + 25
```

In this book, we shall use Equation (4–13) to represent the transfer function system.

    Step response.   If num and den (the numerator and denominator of a transfer function) are known, we may define the system by

$$sys = tf(num,den)$$

Then, a command such as

$$step(sys) \qquad or \qquad step(num,den)$$

will generate a plot of a unit-step response and will display a response curve on the screen. The computation interval Δt and the time span of the response are determined by MATLAB.

    If we wish MATLAB to compute the response every Δt seconds and plot the response curve for $0 \leq t \leq T$ (where T is an integer multiple of Δt), we enter the statement

$$t = 0 : \Delta t : T;$$

in the program and use the command

$$step(sys,t) \qquad or \qquad step(num,den,t)$$

where t is the user-specified time.

    If step commands have left-hand arguments, such as

$$y = step(sys,t) \qquad or \qquad y = step(num,den,t)$$

and

$$[y,t] = step(sys,t) \qquad or \qquad [y,t] = step(num,den,t)$$

MATLAB produces the unit-step response of the system, but displays no plot on the screen. It is necessary to use a plot command to see response curves.

    The next two examples demonstrate the use of step commands.

Example 4–5

Consider again the spring–mass–dashpot system mounted on a cart as shown in Figure 4–7. (See **Example 4–3**.) The transfer function of the system is

$$\frac{Y(s)}{U(s)} = \frac{bs + k}{ms^2 + bs + k}$$

Assuming that $m = 10$ kg, $b = 20$ N-s/m, $k = 100$ N/m, and the input $u(t)$ is a unit-step input (a step input of 1 m), obtain the response curve $y(t)$.

Substituting the given numerical values into the transfer function, we have

$$\frac{Y(s)}{U(s)} = \frac{20s + 100}{10s^2 + 20s + 100} = \frac{2s + 10}{s^2 + 2s + 10}$$

MATLAB Program 4–6 will produce the unit-step response $y(t)$. The resulting unit-step response curve is shown in Figure 4–10.

MATLAB Program 4–6
``` >> num = [2   10]; >> den = [1   2   10]; >> sys = tf(num,den); >> step(sys) >> grid ```

In this plot, the duration of the response is automatically determined by MATLAB. The title and axis labels are also automatically determined by MATLAB.

If we wish to compute and plot the curve every 0.01 sec over the interval $0 \le t \le 8$, we need to enter the following statement in the MATLAB program:

$$t = 0:0.01:8;$$

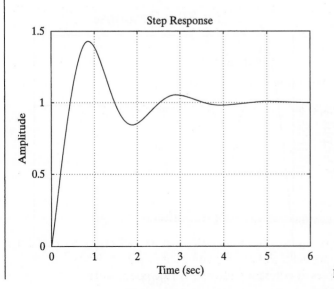

Figure 4–10   Unit-step response curve.

Also, if we wish to change the title and axis labels, we enter the desired title and desired labels as shown in MATLAB Program 4–7.

MATLAB Program 4–7

```
>> t = 0:0.01:8;
>> num = [2   10];
>> den = [1   2   10];
>> sys = tf(num,den);
>> step(sys,t)
>> grid
>> title('Unit-Step Response','Fontsize',20')
>> xlabel('t','Fontsize',20')
>> ylabel ('Output y','Fontsize',20')
```

Note that if we did not enter the desired title and desired axis labels in the program, the title, *x*-axis label, and *y*-axis label on the plot would have been "Step Response", "Time (sec)", and "Amplitude", respectively. (This statement applies to MATLAB version 6 and not to versions 3, 4, and 5.) When we enter the desired title and axis labels as shown in MATLAB Program 4–7, MATLAB erases the predetermined title and axis labels, except "(sec)" in the *x*-axis label, and replaces them with the ones we have specified. If the font sizes are too small, they can be made larger. For example, entering

'Fontsize', 20

in the title, xlabel, and ylabel variables as shown in MATLAB Program 4–7 results in that size text appearing in those places. Figure 4–11 is a plot of the response curve obtained with MATLAB Program 4–7.

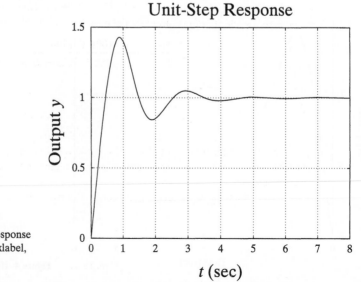

Figure 4–11 Unit-step response curve. Font sizes for title, xlabel, and ylabel are enlarged.

Example 4-6

Consider again the mechanical system shown in Figure 4-8. (See **Example 4-4.**) The transfer function $X(s)/P(s)$ was found to be

$$\frac{X(s)}{P(s)} = \frac{b_2 s + k_2}{m b_2 s^3 + m k_2 s^2 + (k_1 + k_2) b_2 s + k_1 k_2} \tag{4-14}$$

The transfer function $Y(s)/X(s)$ is obtained from Equation (4-10):

$$\frac{Y(s)}{X(s)} = \frac{k_2}{b_2 s + k_2}$$

Hence,

$$\frac{Y(s)}{P(s)} = \frac{Y(s)}{X(s)} \frac{X(s)}{P(s)} = \frac{k_2}{m b_2 s^3 + m k_2 s^2 + (k_1 + k_2) b_2 s + k_1 k_2} \tag{4-15}$$

Assuming that $m = 0.1$ kg, $b_2 = 0.4$ N-s/m, $k_1 = 6$ N/m, $k_2 = 4$ N/m, and $p(t)$ is a step force of magnitude 10 N, obtain the responses $x(t)$ and $y(t)$.

Substituting the numerical values for m, b_2, k_1, and k_2 into the transfer functions given by Equations (4-14) and (4-15), we obtain

$$\frac{X(s)}{P(s)} = \frac{0.4s + 4}{0.04s^3 + 0.4s^2 + 4s + 24}$$
$$= \frac{10s + 100}{s^3 + 10s^2 + 100s + 600} \tag{4-16}$$

and

$$\frac{Y(s)}{P(s)} = \frac{4}{0.04s^3 + 0.4s^2 + 4s + 24}$$
$$= \frac{100}{s^3 + 10s^2 + 100s + 600} \tag{4-17}$$

Since $p(t)$ is a step force of magnitude 10 N, we may define $p(t) = 10u(t)$, where $u(t)$ is a unit-step input of magnitude 1 N. Then Equations (4-16) and (4-17) can be written as

$$\frac{X(s)}{U(s)} = \frac{100s + 1000}{s^3 + 10s^2 + 100s + 600} \tag{4-18}$$

and

$$\frac{Y(s)}{U(s)} = \frac{1000}{s^3 + 10s^2 + 100s + 600} \tag{4-19}$$

Since $u(t)$ is a unit-step input, $x(t)$ and $y(t)$ can be obtained from Equations (4-18) and (4-19) with the use of a step command. (Step commands assume that the input is the unit-step input.)

In this example, we shall demonstrate the use of the commands

$$y = \text{step(sys,t)}$$

and

$$\text{plot(t,y)}$$

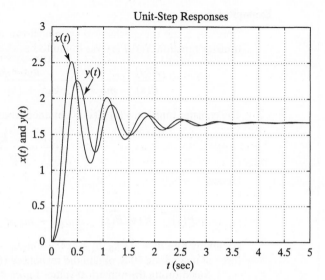

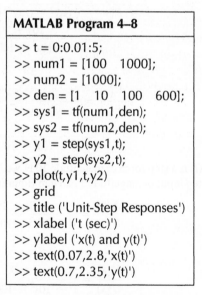

Figure 4–12 Step-response curves
$x(t)$ and $y(t)$.

MATLAB Program 4–8 produces the responses $x(t)$ and $y(t)$ of the system on one diagram.

MATLAB Program 4–8
| ```
>> t = 0:0.01:5;
>> num1 = [100   1000];
>> num2 = [1000];
>> den = [1   10   100   600];
>> sys1 = tf(num1,den);
>> sys2 = tf(num2,den);
>> y1 = step(sys1,t);
>> y2 = step(sys2,t);
>> plot(t,y1,t,y2)
>> grid
>> title ('Unit-Step Responses')
>> xlabel ('t (sec)')
>> ylabel ('x(t) and y(t)')
>> text(0.07,2.8,'x(t)')
>> text(0.7,2.35,'y(t)')
``` |

The response curves $x(t)$ and $y(t)$ are shown in Figure 4–12.

Writing text on the graph. When we plot two or more curves on one diagram, we may need to write text on the graph to distinguish the curves. For example, to write the text 'x(t)' horizontally, beginning at the point $(0.07, 2.8)$ on the graph, we use the command

$$text(0.07,2.8,'x(t)')$$

Impulse response.   The unit-impulse response of a dynamic system defined in the form of the transfer function may be obtained by use of one of the following
```

MATLAB commands:

$$
\begin{array}{lll}
\text{impulse(sys)} & \text{or} & \text{impulse(num,den)} \\
\text{impulse(sys,t)} & \text{or} & \text{impulse(num,den,t)} \\
\text{y = impulse(sys)} & \text{or} & \text{y = impulse(num,den)} \\
\text{[y,t] = impulse (sys,t)} & \text{or} & \text{[y,t] = impulse(num,den,t)}
\end{array}
$$

The command impulse(sys) will generate a plot of the unit-impulse response and will display the impulse-response curve on the screen. If the command has a left-hand argument, such as y = impulse(sys), no plot is shown on the screen. It is then necessary to use a plot command to see the response curve on the screen.

Before discussing computational solutions of problems involving impulse inputs, we present some necessary background material.

Impulse input.   The impulse response of a mechanical system can be observed when the system is subjected to a very large force for a very short time, for instance, when the mass of a spring–mass–dashpot system is hit by a hammer or a bullet. Mathematically, such an impulse input can be expressed by an impulse function.

The impulse function is a mathematical function without any actual physical counterpart. However, as shown in Figure 4–13(a), if the actual input lasts for a short time (Δt s) but has a large amplitude (h), so that the area ($h\Delta t$) in a time plot is not negligible, it can be approximated by an impulse function. The impulse input is usually denoted by a vertical arrow, as shown in Figure 4–13(b), to indicate that it has a very short duration and a very large height.

In handling impulse functions, only the magnitude (or area) of the function is important; its actual shape is immaterial. In other words, an impulse of amplitude $2h$ and duration $\Delta t/2$ can be considered the same as an impulse of amplitude h and duration Δt, as long as Δt approaches zero and $h\Delta t$ is finite.

We next briefly discuss a review of the law of conservation of momentum, which is useful in determining the impulse responses of mechanical systems.

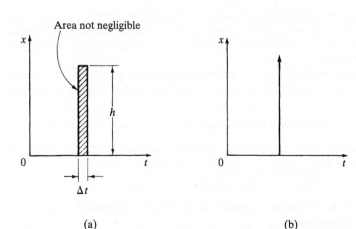

(a)   (b)   **Figure 4–13**   Impulse inputs.

Law of conservation of momentum.   The momentum of a mass m moving at a velocity v is mv. According to Newton's second law,

$$F = ma = m\frac{dv}{dt} = \frac{d}{dt}(mv)$$

Hence,

$$F \, dt = d(mv) \tag{4-20}$$

Integrating both sides of Equation (4–20), we have

$$\int_{t_1}^{t_2} F \, dt = \int_{v_1}^{v_2} d(mv) = mv_2 - mv_1 \tag{4-21}$$

where $v_1 = v(t_1)$ and $v_2 = v(t_2)$. Equation (4–21) states that a change in momentum equals the time integral of force between $t = t_1$ and $t = t_2$.

Momentum is a vector quantity, with magnitude, direction, and sense. The direction of the change in momentum is the direction of the force.

In the absence of any external force, Equation (4–20) becomes

$$d(mv) = 0$$

or

$$mv = \text{constant}$$

Thus, the total momentum of a system remains unchanged by any action that may take place within the system, provided that no external force is acting on the system. This principle is called the *law of conservation of momentum.*

The angular momentum of a rotating system is $J\omega$, where J is the moment of inertia of a body and ω is the angular velocity of the body. In the absence of an external torque, the angular momentum of a body remains unchanged. This principle is the *law of conservation of angular momentum.*

Example 4–7

A bullet is fired horizontally into a wood block resting on a horizontal, frictionless surface. If the mass m_1 of the bullet is 0.02 kg and the velocity is 600 m/s, what is the velocity of the wood block after the bullet is embedded in it? Assume that the wood block has a mass m_2 of 50 kg.

If we consider the bullet and wood block as constituting a system, no external force is acting on the system. Consequently, its total momentum remains unchanged. Thus, we have

$$\text{momentum before impact} = m_1 v_1 + m_2 v_2$$

where v_1, the velocity of the bullet before the impact, is equal to 600 m/s and v_2, the velocity of the wood block before the impact, is equal to zero. Also,

$$\text{momentum after impact} = (m_1 + m_2)v$$

where v is the velocity of the wood block after the bullet is embedded. (Velocities v_1 and v are in the same direction.)

The law of conservation of momentum states that

$$m_1 v_1 + m_2 v_2 = (m_1 + m_2)v$$

Substituting the given numerical values into this last equation, we obtain

$$0.02 \times 600 + 50 \times 0 = (0.02 + 50)v$$

or

$$v = 0.24 \text{ m/s}$$

Hence, the wood block after the bullet is embedded will move at the velocity of 0.24 m/s in the same direction as the original velocity v_1 of the bullet.

Example 4–8

Consider the mechanical system shown in Figure 4–14. A bullet of mass m is shot into a block of mass M (where $M \gg m$). Assume that when the bullet hits the block, it becomes embedded there. Determine the response (displacement x) of the block after it is hit by the bullet. The displacement x of the block is measured from the equilibrium position before the bullet hits it. Suppose that the bullet is shot at $t = 0-$ and that the initial velocity of the bullet is $v(0-)$. Assuming the following numerical values for M, m, b, k, and $v(0-)$, draw a curve $x(t)$ versus t:

$$M = 50 \text{ kg}, \qquad m = 0.01 \text{ kg}, \qquad b = 100 \text{ N-s/m},$$
$$k = 2500 \text{ N/m}, \qquad v(0-) = 800 \text{ m/s}$$

The input to the system in this case can be considered an impulse, the magnitude of which is equal to the rate of change of momentum of the bullet. At the instant the bullet hits the block, the velocity of the bullet becomes the same as that of the block, since the bullet is assumed to be embedded in it. As a result, there is a sudden change in the velocity of the bullet. [See Figure 4–15(a).] Since the change in the velocity of the bullet occurs instantaneously, $\dot{v}$ has the form of an impulse. (Note that $\dot{v}$ is negative.)

For $t > 0$, the block and the bullet move as a combined mass $M + m$. The equation of motion for the system is

$$(M + m)\ddot{x} + b\dot{x} + kx = F(t) \tag{4–22}$$

where $F(t)$, an impulse force, is equal to $-m\dot{v}$. [Note that $-m\dot{v}$ is positive; the impulse force $F(t)$ is in the positive direction of x.] From Figure 4–15(b), the impulse force can be written as

$$F(t) = -m\dot{v} = A \, \Delta t \, \delta(t)$$

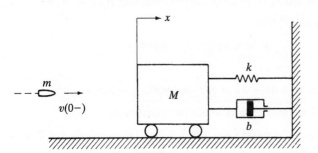

Figure 4–14 Mechanical system subjected to an impulse input.

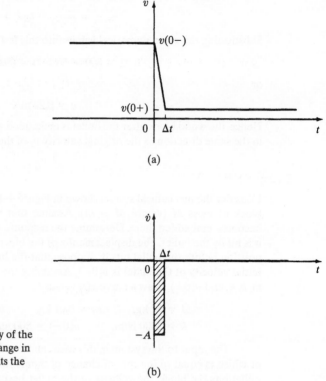

Figure 4–15 (a) Change in velocity of the bullet when it hits the block; (b) change in acceleration of the bullet when it hits the block.

where $A\,\Delta t$ is the magnitude of the impulse input. Thus,

$$\int_{0-}^{0+} A\,\Delta t\,\delta(t)\,dt = -m \int_{0-}^{0+} \dot{v}\,dt$$

or

$$A\,\Delta t = mv(0-) - mv(0+) \tag{4–23}$$

The momentum of the bullet is changed from $mv(0-)$ to $mv(0+)$. Since

$$v(0+) = \dot{x}(0+) = \text{initial velocity of combined mass } M + m$$

we can write Equation (4–23) as

$$A\,\Delta t = mv(0-) - m\dot{x}(0+)$$

Then Equation (4–22) becomes

$$(M + m)\ddot{x} + b\dot{x} + kx = F(t) = [mv(0-) - m\dot{x}(0+)]\delta(t)$$

Taking the $\mathcal{L}_-$ transform of both sides of this last equation, we see that

$$(M + m)[s^2X(s) - sx(0-) - \dot{x}(0-)] + b[sX(s) - x(0-)] + kX(s)$$
$$= mv(0-) - m\dot{x}(0+)$$

Also, noting that $x(0-) = 0$ and $\dot{x}(0-) = 0$, we have

$$X(s) = \frac{mv(0-) - m\dot{x}(0+)}{(M + m)s^2 + bs + k} \tag{4-24}$$

To determine the value of $\dot{x}(0+)$, we apply the initial-value theorem:

$$\dot{x}(0+) = \lim_{t \to 0+} \dot{x}(t) = \lim_{s \to \infty} s[sX(s)]$$

$$= \lim_{s \to \infty} \frac{s^2[mv(0-) - m\dot{x}(0+)]}{(M + m)s^2 + bs + k}$$

$$= \frac{mv(0-) - m\dot{x}(0+)}{M + m}$$

from which we get

$$mv(0-) - m\dot{x}(0+) = (M + m)\dot{x}(0+)$$

or

$$\dot{x}(0+) = \frac{m}{M + 2m}v(0-)$$

So Equation (4-24) becomes

$$X(s) = \frac{(M + m)\dot{x}(0+)}{(M + m)s^2 + bs + k}$$

$$= \frac{1}{(M + m)s^2 + bs + k} \frac{(M + m)m\,v(0-)}{M + 2m} \tag{4-25}$$

The inverse Laplace transform of Equation (4-25) gives the impulse response $x(t)$. Substituting the given numerical values into Equation (4-25), we obtain

$$X(s) = \frac{1}{50.01s^2 + 100s + 2500} \frac{50.01 \times 0.01 \times 800}{50.02}$$

$$= \frac{7.9984}{50.01s^2 + 100s + 2500}$$

$$= 0.02285 \frac{6.9993}{(s + 0.9998)^2 + (6.9993)^2}$$

Taking the inverse Laplace transform of this last equation yields

$$x(t) = 0.02285e^{-0.9998t} \sin 6.9993t$$

Thus, the response $x(t)$ is a damped sinusoidal motion.

Example 4-9

Referring to **Example 4-8**, obtain the impulse response of the system shown in Figure 4-14 with MATLAB. Use the same numerical values for M, m, b, k, and $v(0-)$ as in **Example 4-8**.

The response $X(s)$ was obtained in **Example 4-8**, as given by Equation (4-25). This is the response to the impulse input $[mv(0-) - m\dot{x}(0+)]\delta(t)$. Note that the magnitude of the impulse input is

$$mv(0-) - m\dot{x}(0+) = (M + m)\dot{x}(0+) = \frac{m(M + m)}{M + 2m}v(0-) \qquad .$$

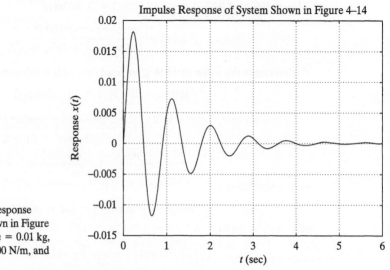

Figure 4–16   Impulse-response curve of the system shown in Figure 4–14 with $M = 50$ kg, $m = 0.01$ kg, $b = 100$ N-s/m, $k = 2500$ N/m, and $v(0-) = 800$ m/s.

Hence, the impulse input can be written as

$$F(t) = \frac{m(M + m)}{M + 2m} v(0-)\delta(t) = 7.9984\,\delta(t)$$

The system equation is

$$(M + m)\ddot{x} + b\dot{x} + kx = F(t) = 7.9984\,\delta(t)$$

so that

$$\frac{X(s)}{F(s)} = \frac{1}{(M + m)s^2 + bs + k} \tag{4–26}$$

To find the response of the system to $F(t)$ (which is an impulse input whose magnitude is not unity), we modify Equation (4–26) to the following form:

$$\frac{X(s)}{\mathcal{L}[\delta(t)]} = \frac{1}{(M + m)s^2 + bs + k} \frac{m(M + m)\,v(0-)}{M + 2m}$$

$$= \frac{7.9984}{50.01s^2 + 100s + 2500} \tag{4–27}$$

If we define

$$\text{num} = [7.9984];$$
$$\text{den} = [50.01 \quad 100 \quad 2500];$$
$$\text{sys} = \text{tf(num,den)}$$

then the command

$$\text{impulse(sys)}$$

will produce the unit-impulse response of the system defined by Equation (4–27), which is the same as the response of the system of Equation (4–26) to the impulse input $F(t) = 7.9984\,\delta(t)$. MATLAB Program 4–9 produces the response of the system subjected to the impulse input $F(t)$. The impulse response obtained is shown in Figure 4–16.

MATLAB Program 4-9

```
>> num = [7.9984];
>> den = [50.01   100   2500];
>> sys = tf(num,den);
>> impulse (sys)
>> grid
>> title ('Impulse Response of System Shown in Figure 4-14')
>> xlabel('t')
>> ylabel('Response x(t)')
```

Obtaining response to arbitrary input. The command lsim produces the response of linear, time-invariant systems to arbitrary inputs. If the initial conditions of the system are zero, then

$$\text{lsim(sys,u,t)} \qquad \text{or} \qquad \text{lsim(num,den,u,t)}$$

produces the response of the system to the input u. Here, u is the input and t represents the times at which responses to u are to be computed. (The response time span and the time increment are stated in t; an example of how t is specified is $t = 0:0.01:10$). If the initial conditions are nonzero, use the state-space approach presented in Section 5-2.

If the initial conditions of the system are zero, then any of the commands

$$y = \text{lsim(sys,u,t)} \qquad \text{or} \qquad y = \text{lsim(num,den,u,t)}$$

and

$$[y,t] = \text{lsim(sys,u,t)} \qquad \text{or} \qquad [y,t] = \text{lsim(num,den,u,t)}$$

returns the output response y. No plot is drawn. To plot the response curve, it is necessary to use the command plot(t,y).

Note that the command

$$\text{lsim(sys1,sys2, ...,u,t)}$$

plots the responses of systems sys1, sys2, ... on a single diagram. Note also that, by using lsim commands, we are able to obtain the response of the system to ramp inputs, acceleration inputs, and any other time functions that we can generate with MATLAB.

Ramp response. The next example plots the unit-ramp response curve with the use of the lsim command

$$\text{lsim(sys,u,t)}$$

where $u = t$.

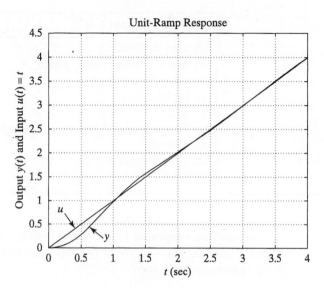

Figure 4–17 Plots of unit-ramp response curve $y(t)$ and input ramp function $u(t)$.

Example 4–10

Consider once again the system shown in Figure 4–7. (See **Example 4–3.**) Assume that $m = 10\ \text{kg}$, $b = 20\ \text{N-s/m}$, $k = 100\ \text{N/m}$, and $u(t)$ is a unit-ramp input—that is, the displacement u increases linearly, or $u = \alpha t$, where $\alpha = 1$. We shall obtain the unit-ramp response using the command

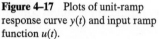

The transfer function of the system, derived in **Example 4–3**, is

$$\frac{Y(s)}{U(s)} = \frac{2s + 10}{s^2 + 2s + 10}$$

MATLAB Program 4–10 produces the unit-ramp response. The resulting response curve $y(t)$ versus t and the input ramp function $u(t)$ versus t are shown in Figure 4–17.

MATLAB Program 4–10
| ```
>> num = [2   10];
>> den = [1   2   10];
>> sys = tf(num, den);
>> t = 0:0.01:4;
>> u = t;
>> lsim(sys,u,t)
>> grid
>> title('Unit-Ramp Response')
>> xlabel('t')
>> ylabel('Output y(t) and Input u(t) = t')
>> text(0.8,0.25,'y')
>> text(0.15,0.8,'u')
``` |
```

[Note that the command lsim(sys,u,t) produces plots of both $y(t)$ versus t and $u(t)$ versus t.]

In some cases it is desired to plot multiple curves on one graph. This can be done by using a plot command with multiple arguments, for example,

$$plot(t,y1, t,y2, \ldots, t,yn)$$

MATLAB Program 4–11 uses the command

$$plot(t,y,t,u)$$

to plot a curve $y(t)$ versus t and a line $u(t)$ versus t. The resulting plots are shown in Figure 4–18.

MATLAB Program 4–11

```
>> num = [2    10];
>> den = [1    2    10];
>> sys = tf(num,den);
>> t = 0:0.01:4;
>> u = t;
>> y = lsim(sys,u,t);
>> plot(t,y,t,u)
>> grid
>> title('Unit-Ramp Response')
>> xlabel('t (sec)')
>> ylabel('Output y(t) and Input u(t) = t')
>> text(0.85,0.25,'y')
>> text(0.15,0.8,'u')
```

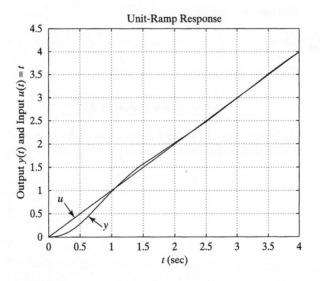

Figure 4–18 Plots of unit-ramp response curve $y(t)$ and input ramp function $u(t)$. (Plots are obtained with the use of the command plot(t,y,t,u).)

Response to initial condition (transfer-function approach). The next example obtains the response of a transfer-function system subjected to an initial condition.

Example 4–11

Consider the mechanical system shown in Figure 4–19, where $m = 1$ kg, $b = 3$ N-s/m, and $k = 2$ N/m. Assume that the displacement x of mass m is measured from the equilibrium position and that at $t = 0$ the mass m is pulled downward such that $x(0) = 0.1$ m and $\dot{x}(0) = 0.05$ m/s. Obtain the motion of the mass subjected to the initial condition. (Assume no external forcing function.)

The system equation is

$$m\ddot{x} + b\dot{x} + kx = 0$$

with the initial conditions $x(0) = 0.1$ m and $\dot{x}(0) = 0.05$ m/s. The Laplace transform of the system equation gives

$$m[s^2 X(s) - sx(0) - \dot{x}(0)] + b[sX(s) - x(0)] + kX(s) = 0$$

or

$$(ms^2 + bs + k)X(s) = mx(0)s + m\dot{x}(0) + bx(0)$$

Solving this last equation for $X(s)$ and substituting the given numerical values into $x(0)$ and $\dot{x}(0)$, we obtain

$$X(s) = \frac{mx(0)s + m\dot{x}(0) + bx(0)}{ms^2 + bs + k}$$

$$= \frac{0.1s + 0.35}{s^2 + 3s + 2}$$

This equation can be written as

$$X(s) = \frac{0.1s^2 + 0.35s}{s^2 + 3s + 2} \frac{1}{s}$$

Hence, the motion of the mass m is the unit-step response of the following system:

$$G(s) = \frac{0.1s^2 + 0.35s}{s^2 + 3s + 2}$$

MATLAB Program 4–12 produces a plot of the motion of the mass when the system is subjected to the initial condition. The plot is shown in Figure 4–20.

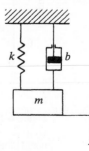

Figure 4–19   Mechanical system.

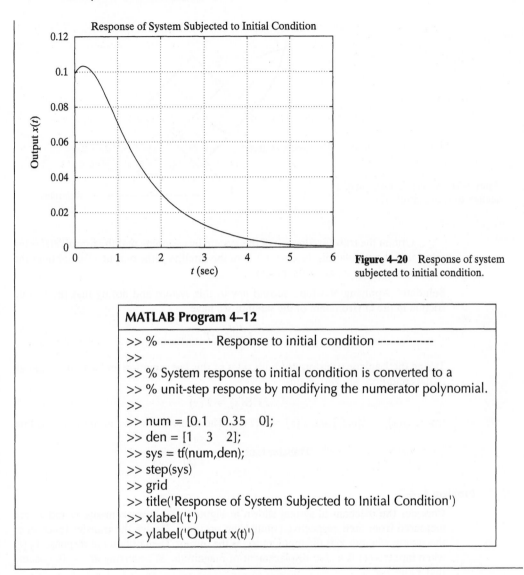

Figure 4–20 Response of system subjected to initial condition.

MATLAB Program 4–12

```
>> % ------------ Response to initial condition -------------
>>
>> % System response to initial condition is converted to a
>> % unit-step response by modifying the numerator polynomial.
>>
>> num = [0.1   0.35   0];
>> den = [1   3   2];
>> sys = tf(num,den);
>> step(sys)
>> grid
>> title('Response of System Subjected to Initial Condition')
>> xlabel('t')
>> ylabel('Output x(t)')
```

EXAMPLE PROBLEMS AND SOLUTIONS

Problem A–4–1

Consider the satellite attitude control system depicted in Figure 4–21. The diagram shows the control of only the yaw angle θ. (In the actual system, there are controls about three axes.) Small jets apply reaction forces to rotate the satellite body into the desired attitude. The two skew symmetrically placed jets denoted by A and B operate in pairs. Assume that each jet thrust is $F/2$ and a torque $T = Fl$ is applied to the system. The jets are turned on for a certain length of time, so the torque can be written as $T(t)$. The moment of inertia about the axis of rotation at the center of mass is J.

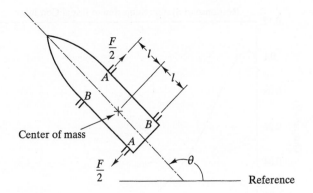

Figure 4–21 Schematic diagram of a
satellite attitude control system.

Obtain the transfer function of this system by assuming that the torque $T(t)$ is the input, and the angular displacement $\theta(t)$ of the satellite is the output. (We consider the motion only in the plane of the page.)

Solution Applying Newton's second law to this system and noting that there is no friction in the environment of the satellite, we have

$$J\frac{d^2\theta}{dt^2} = T$$

Taking the Laplace transform of both sides of this last equation and assuming that all initial conditions are zero yields

$$Js^2\Theta(s) = T(s)$$

where $\Theta(s) = \mathcal{L}[\theta(t)]$ and $T(s) = \mathcal{L}[T(t)]$. The transfer function of the system is thus

$$\text{Transfer function} = \frac{\Theta(s)}{T(s)} = \frac{1}{Js^2}$$

Problem A–4–2

Consider the mechanical system shown in Figure 4–22. Displacements x_i and x_o are measured from their respective equilibrium positions. Derive the transfer function of the system wherein x_i is the input and x_o is the output. Then obtain the response $x_o(t)$ when input $x_i(t)$ is a step displacement of magnitude X_i occurring at $t = 0$. Assume that $x_o(0-) = 0$.

Solution The equation of motion for the system is

$$b_1(\dot{x}_i - \dot{x}_o) + k_1(x_i - x_o) = b_2\dot{x}_o$$

Taking the $\mathcal{L}_-$ transform of this equation and noting that $x_i(0-) = 0$ and $x_o(0-) = 0$, we have

$$(b_1s + k_1)X_i(s) = (b_1s + k_1 + b_2s)X_o(s)$$

The transfer function $X_o(s)/X_i(s)$ is

$$\frac{X_o(s)}{X_i(s)} = \frac{b_1s + k_1}{(b_1 + b_2)s + k_1} \tag{4-28}$$

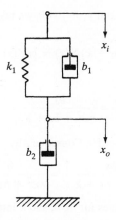

Figure 4–22　Mechanical system.

The response $x_o(t)$ when the input $x_i(t)$ is a step displacement of magnitude X_i occurring at $t = 0$ can be obtained from Equation (4–28). First we have

$$X_o(s) = \frac{b_1 s + k_1}{(b_1 + b_2)s + k_1} \frac{X_i}{s} = \left\{ \frac{1}{s} - \frac{b_2}{b_1 + b_2} \frac{1}{s + [k_1/(b_1 + b_2)]} \right\} X_i$$

Then the inverse Laplace transform of $X_o(s)$ gives

$$x_o(t) = \left[1 - \frac{b_2}{b_1 + b_2} e^{-k_1 t/(b_1 + b_2)} \right] X_i$$

Notice that $x_o(0+) = [b_1/(b_1 + b_2)]X_i$.

Problem A–4–3

The mechanical system shown in Figure 4–23 is initially at rest. At $t = 0$, a unit-step displacement input is applied to point A. Assume that the system remains linear throughout the response period. The displacement x is measured from the equilibrium position. If $m = 1$ kg, $b = 10$ N-s/m, and $k = 50$ N/m, find the response $x(t)$ as well as the values of $x(0+)$, $\dot{x}(0+)$, and $x(\infty)$.

Solution　The equation of motion for the system is

$$m\ddot{x} + b(\dot{x} - \dot{y}) + kx = 0$$

or

$$m\ddot{x} + b\dot{x} + kx = b\dot{y}$$

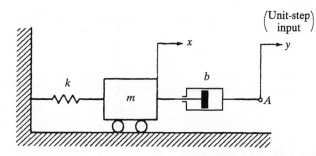

Figure 4–23　Mechanical system.

Noting that $x(0-) = 0$, $\dot{x}(0-) = 0$, and $y(0-) = 0$, we take the $\mathcal{L}_-$ transform of this last equation and obtain

$$(ms^2 + bs + k)X(s) = bsY(s)$$

Thus,

$$\frac{X(s)}{Y(s)} = \frac{bs}{ms^2 + bs + k}$$

Since the input y is a unit step, $Y(s) = 1/s$. Consequently,

$$X(s) = \frac{bs}{ms^2 + bs + k} \frac{1}{s} = \frac{b}{ms^2 + bs + k}$$

Substituting the given numerical values for m, b, and k into this last equation, we get

$$X(s) = \frac{10}{s^2 + 10s + 50} = \frac{10}{(s + 5)^2 + 5^2}$$

The inverse Laplace transform of $X(s)$ is

$$x(t) = 2e^{-5t} \sin 5t$$

The values of $x(0+)$, $\dot{x}(0+)$, and $x(\infty)$ are found from the preceding equation and are

$$x(0+) = 0, \qquad \dot{x}(0+) = 10, \qquad x(\infty) = 0$$

Thus, the mass m returns to the original position as time elapses.

Problem A–4–4

Find the transfer function $X_o(s)/X_i(s)$ of the mechanical system shown in Figure 4–24. Obtain the response $x_o(t)$ when the input $x_i(t)$ is a step displacement of magnitude X_i occurring at $t = 0$. Assume that the system is initially at rest [$x_o(0-) = 0$ and $y(0-) = 0$]. Assume also that x_i and x_o are measured from their respective equilibrium positions. The numerical values of b_1, b_2, k_1, and k_2 are as follows:

$$b_1 = 5 \text{ N-s/m}, \qquad b_2 = 20 \text{ N-s/m}, \qquad k_1 = 5 \text{ N/m}, \qquad k_2 = 10 \text{ N/m}$$

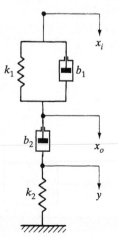

Figure 4–24   Mechanical system.

Solution   The equations of motion for the mechanical system are

$$b_1(\dot{x}_i - \dot{x}_o) + k_1(x_i - x_o) = b_2(\dot{x}_o - \dot{y})$$
$$b_2(\dot{x}_o - \dot{y}) = k_2 y$$

Taking the $\mathcal{L}_-$ transform of these two equations, with the initial conditions $x_i(0-) = 0$, $x_o(0-) = 0$ and $y(0-) = 0$, we get

$$b_1[sX_i(s) - sX_o(s)] + k_1[X_i(s) - X_o(s)] = b_2[sX_o(s) - sY(s)]$$
$$b_2[sX_o(s) - sY(s)] = k_2 Y(s)$$

If we eliminate $Y(s)$ from the last two equations, the transfer function $X_o(s)/X_i(s)$ becomes

$$\frac{X_o(s)}{X_i(s)} = \frac{\left(\dfrac{b_1}{k_1}s + 1\right)\left(\dfrac{b_2}{k_2}s + 1\right)}{\left(\dfrac{b_1}{k_1}s + 1\right)\left(\dfrac{b_2}{k_2}s + 1\right) + \dfrac{b_2}{k_1}s}$$

Substitution of the given numerical values into the transfer function yields

$$\frac{X_o(s)}{X_i(s)} = \frac{(s + 1)(2s + 1)}{(s + 1)(2s + 1) + 4s} = \frac{s^2 + 1.5s + 0.5}{s^2 + 3.5s + 0.5}$$

For an input $x_i(t) = X_i \cdot 1(t)$, the response $x_o(t)$ can be obtained as follows: Since

$$X_o(s) = \frac{s^2 + 1.5s + 0.5}{s^2 + 3.5s + 0.5} \frac{X_i}{s}$$

$$= \left(\frac{0.6247}{s + 3.3508} - \frac{0.6247}{s + 0.1492} + \frac{1}{s}\right)X_i$$

we find that

$$x_o(t) = (0.6247e^{-3.3508t} - 0.6247e^{-0.1492t} + 1)X_i$$

Notice that $x_o(0+) = X_i$.

Problem A–4–5

Obtain the transfer function $X(s)/U(s)$ of the system shown in Figure 4–25, where u is the force input. The displacement x is measured from the equilibrium position.

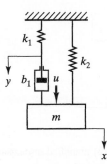

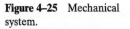

Figure 4–25   Mechanical system.

Solution The equations of motion for the system are

$$m\ddot{x} = -k_2 x - b_1(\dot{x} - \dot{y}) + u$$
$$b_1(\dot{x} - \dot{y}) = k_1 y$$

Laplace transforming these two equations and assuming initial conditions equal to zero, we obtain

$$ms^2 X(s) = -k_2 X(s) - b_1 s X(s) + b_1 s Y(s) + U(s)$$
$$b_1 s X(s) - b_1 s Y(s) = k_1 Y(s)$$

Eliminating $Y(s)$ from the last two equations yields

$$(ms^2 + b_1 s + k_2)X(s) = b_1 s \frac{b_1 s}{b_1 s + k_1} X(s) + U(s)$$

Simplifying, we obtain

$$[(ms^2 + b_1 s + k_2)(b_1 s + k_1) - b_1^2 s^2]X(s) = (b_1 s + k_1)U(s)$$

from which we get the transfer function $X(s)/U(s)$ as

$$\frac{X(s)}{U(s)} = \frac{b_1 s + k_1}{mb_1 s^3 + mk_1 s^2 + b_1(k_1 + k_2)s + k_1 k_2}$$

Problem A–4–6

Figure 4–26(a) shows a schematic diagram of an automobile suspension system. As the car moves along the road, the vertical displacements at the tires excite the automobile suspension system, whose motion consists of a translational motion of the center of mass and a rotational motion about the center of mass. Mathematical modeling of the complete system is quite complicated.

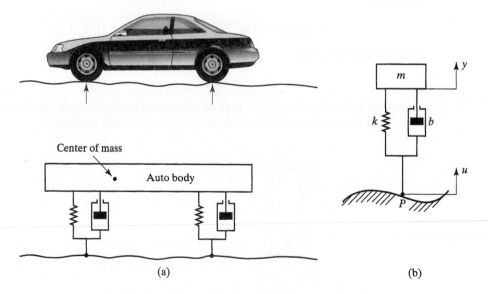

(a)   (b)

Figure 4–26 (a) Automobile suspension system; (b) simplified suspension system.

A highly simplified version of the suspension system is shown in Figure 4–26(b). Assuming that the motion u at point P is the input to the system and the vertical motion y of the body is the output, obtain the transfer function $Y(s)/U(s)$. (Consider the motion of the body only in the vertical direction.) The displacement y is measured from the equilibrium position in the absence of the input u.

Solution The equation of motion for the system shown in Figure 4–26(b) is

$$m\ddot{y} + b(\dot{y} - \dot{u}) + k(y - u) = 0$$

or

$$m\ddot{y} + b\dot{y} + ky = b\dot{u} + ku$$

Taking the Laplace transform of this last equation, assuming zero initial conditions, we obtain

$$(ms^2 + bs + k)Y(s) = (bs + k)U(s)$$

Hence, the transfer function $Y(s)/U(s)$ is

$$\frac{Y(s)}{U(s)} = \frac{bs + k}{ms^2 + bs + k}$$

Problem A–4–7

Obtain the transfer function $Y(s)/U(s)$ of the system shown in Figure 4–27. The vertical motion u at point P is the input. (Similar to the system of **Problem A–4–6**, this system is also a simplified version of an automobile or motorcycle suspension system. In Figure 4–27, m_1 and k_1 represent the wheel mass and tire stiffness, respectively.) Assume that the displacements x and y are measured from their respective equilibrium positions in the absence of the input u.

Solution Applying Newton's second law to the system, we get

$$m_1\ddot{x} = k_2(y - x) + b(\dot{y} - \dot{x}) + k_1(u - x)$$
$$m_2\ddot{y} = -k_2(y - x) - b(\dot{y} - \dot{x})$$

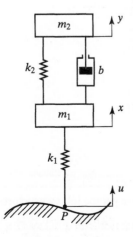

Figure 4–27 Suspension system.

Hence, we have

$$m_1\ddot{x} + b\dot{x} + (k_1 + k_2)x = b\dot{y} + k_2 y + k_1 u$$
$$m_2\ddot{y} + b\dot{y} + k_2 y = b\dot{x} + k_2 x$$

Taking the Laplace transforms of these two equations, assuming zero initial conditions, we obtain

$$[m_1 s^2 + bs + (k_1 + k_2)]X(s) = (bs + k_2)Y(s) + k_1 U(s)$$
$$[m_2 s^2 + bs + k_2]Y(s) = (bs + k_2)X(s)$$

Eliminating $X(s)$ from the last two equations, we have

$$(m_1 s^2 + bs + k_1 + k_2)\frac{m_2 s^2 + bs + k_2}{bs + k_2}Y(s) = (bs + k_2)Y(s) + k_1 U(s)$$

which yields

$$\frac{Y(s)}{U(s)} = \frac{k_1(bs + k_2)}{m_1 m_2 s^4 + (m_1 + m_2)bs^3 + [k_1 m_2 + (m_1 + m_2)k_2]s^2 + k_1 bs + k_1 k_2}$$

Problem A–4–8

Expand the function

$$\frac{B(s)}{A(s)} = \frac{3s^3 + 5s^2 + 10s + 40}{s^4 + 16s^3 + 69s^2 + 94s + 40}$$

into partial fractions with MATLAB.

Solution   A MATLAB program for obtaining the partial-fraction expansion is given in MATLAB Program 4–13.

MATLAB Program 4–13

```
>> num = [3   5   10   40];
>> den = [1   16   69   94   40];
>> [r, p, k] = residue(num,den)

r =

       5.2675
      -2.0741
      -0.1934
       1.1852

p =

     -10.0000
      -4.0000
      -1.0000
      -1.0000
k =
     []
```

From the results of the program, we get the following expression:

$$\frac{B(s)}{A(s)} = \frac{5.2675}{s + 10} + \frac{-2.0741}{s + 4} + \frac{-0.1934}{s + 1} + \frac{1.1852}{(s + 1)^2}$$

Note that the row vector k is zero, because the degree of the numerator is lower than that of the denominator.

Problem A–4–9

Expand the function

$$\frac{B(s)}{A(s)} = \frac{2s^2 + 5s + 7}{s^3 + 3s^2 + 7s + 5}$$

into partial fractions with MATLAB.

Solution A MATLAB program for obtaining the partial-fraction expansion is shown in MATLAB Program 4–14.

MATLAB Program 4–14

```
>> num = [2   5   7];
>> den = [1   3   7   5];
>> [r, p, k] = residue(num,den)

r =

    0.5000 – 0.2500i
    0.5000 + 0.2500i
    1.0000

p =

   –1.0000 + 2.0000i
   –1.0000 – 2.0000i
   –1.0000

k =

    []
```

From the MATLAB output, we get the following expression:

$$\frac{B(s)}{A(s)} = \frac{0.5 - j0.25}{s + 1 - j2} + \frac{0.5 + j0.25}{s + 1 + j2} + \frac{1}{s + 1}$$

$$= \frac{(0.5 - j0.25)(s + 1 + j2) + (0.5 + j0.25)(s + 1 - j2)}{(s + 1 - j2)(s + 1 + j2)} + \frac{1}{s + 1}$$

$$= \frac{s + 2}{s^2 + 2s + 5} + \frac{1}{s + 1}$$

Note that, because the row vector k is zero, there is no constant term in this partial-fraction expansion.

Problem A–4–10

Consider the mechanical system shown in Figure 4–28. The system is initially at rest, and the displacement x is measured from the equilibrium position. Assume that $m = 1$ kg, $b = 12$ N-s/m, and $k = 100$ N/m.

Obtain the response of the system when 10 N of force (a step input) is applied to the mass m. Also, plot a response curve with the use of MATLAB.

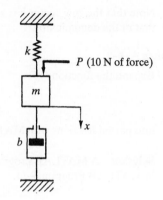

Figure 4–28   Mechanical system.

Solution   The equation of motion for the system is

$$m\ddot{x} + b\dot{x} + kx = P$$

Substituting the numerical values into this last equation, we get

$$\ddot{x} + 12\dot{x} + 100x = 10$$

Taking the Laplace transform of this last equation and substituting the initial conditions $[x(0) = 0$ and $\dot{x}(0) = 0]$ yields

$$(s^2 + 12s + 100)X(s) = \frac{10}{s}$$

Solving for $X(s)$, we obtain

$$
\begin{aligned}
X(s) &= \frac{10}{s(s^2 + 12s + 100)} \\
&= \frac{0.1}{s} - \frac{0.1s + 1.2}{s^2 + 12s + 100} \\
&= \frac{0.1}{s} - \frac{0.1(s + 6)}{(s + 6)^2 + 8^2} - \left(\frac{0.6}{8}\right)\frac{8}{(s + 6)^2 + 8^2}
\end{aligned}
$$

The inverse Laplace transform of this last equation gives

$$x(t) = 0.1 - 0.1e^{-6t}\cos 8t - 0.075e^{-6t}\sin 8t$$

The response exhibits damped vibration.

A MATLAB program to plot the response curve is given in MATLAB Program 4–15. The resulting response curve is shown in Figure 4–29.

MATLAB Program 4–15
``` >> t = 0:0.01:2; >> num = [10]; >> den = [1   12   100]; >> sys = tf(num,den); >> step(sys,t) >> grid >> title ('Step Response') >> xlabel('t') >> ylabel('Output x(t)') ```

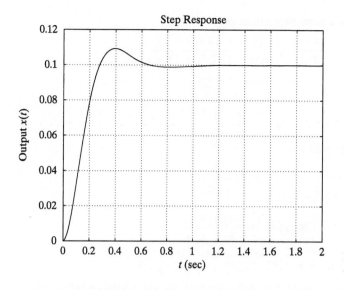

Figure 4–29 Step response of mechanical system.

Problem A–4–11

Consider the mechanical system shown in Figure 4–30, where $b_1 = 0.5$ N-s/m, $b_2 = 1$ N-s/m, $k_1 = 1$ N/m, and $k_2 = 2$ N/m. Assume that the system is initially at rest. The displacements x_i and x_o are measured from their respective equilibrium positions. Obtain the response $x_o(t)$ when $x_i(t)$ is a step input of magnitude 0.1 m.

Solution From **Problem A–4–4**, the transfer function $X_o(s)/X_i(s)$ is

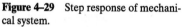

$$\frac{X_o(s)}{X_i(s)} = \frac{\left(\dfrac{b_1}{k_1}s + 1\right)\left(\dfrac{b_2}{k_2}s + 1\right)}{\left(\dfrac{b_1}{k_1}s + 1\right)\left(\dfrac{b_2}{k_2}s + 1\right) + \dfrac{b_2}{k_1}s}$$

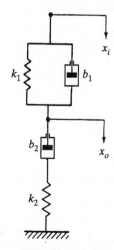

Figure 4–30 Mechanical system.

Substitution of the given numerical values yields

$$\frac{X_o(s)}{X_i(s)} = \frac{(0.5s + 1)(0.5s + 1)}{(0.5s + 1)(0.5s + 1) + s}$$

$$= \frac{0.25s^2 + s + 1}{0.25s^2 + 2s + 1}$$

$$= \frac{s^2 + 4s + 4}{s^2 + 8s + 4}$$

Since $x_i(t) = (0.1)1(t)$, we have

$$X_i(s) = \frac{0.1}{s}$$

Hence,

$$X_o(s) = \frac{s^2 + 4s + 4}{s^2 + 8s + 4} \frac{0.1}{s}$$

$$= \frac{0.1s^2 + 0.4s + 0.4}{s^2 + 8s + 4} \frac{1}{s}$$

MATLAB Program 4–16 is used to obtain the step response, which is shown in Figure 4–31.

MATLAB Program 4–16

```
>> t = 0:0.02:12;
>> num = [0.1   0.4   0.4];
>> den = [1   8   4];
>> sys = tf(num,den);
>> [x_o,t] = step(sys,t);
>> plot(t,x_o)
>> grid
>> title('Step Response of (0.1s^2 + 0.4s + 0.4) / (s^2 + 8s + 4)')
>> xlabel('t (sec)')
>> ylabel('x_o(t)')
```

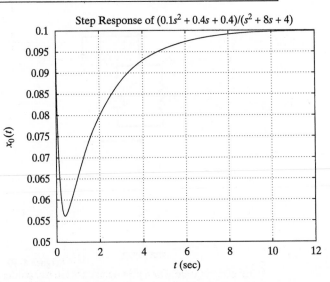

Figure 4–31 Step response of system shown in Figure 4–30.

Figure 4–31 shows the response curve from $t = 0+$ to $t = 12$. Note that $x_o(0+) = 0.1$. If we wish to plot the curve from $x_o(0-) = 0$ to $x_o(12) = 0.1$, we may add the axis command

$$v = [-2 \quad 12 \quad -0.02 \quad 0.12]; \text{axis(v)}$$

to the program, as shown in MATLAB Program 4–17. Then the xy domain of the plot becomes $-2 \le x \le 12$, $-0.02 \le y \le 0.12$. The plot of the response curve produced by MATLAB Program 4–17 is shown in Figure 4–32.

MATLAB Program 4–17

```
>> t = 0:0.02:12;
>> num = [0.1   0.4   0.4];
>> den = [1   8   4];
>> sys = tf(num,den);
>> [x_o, t] = step(sys,t);
>> plot(t,x_o)
>> v = [-2   12   -0.02   0.12]; axis(v)
>> grid
>> title('Step Response of (0.1s^2 + 0.4s + 0.4) / (s^2 + 8s + 4)')
>> xlabel('t (sec)')
>> ylabel('x_o(t)')
>> text (1.5, 0.007, 'These two lines are manually drawn.')
```

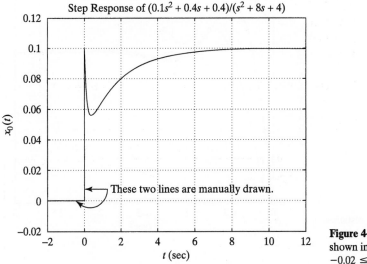

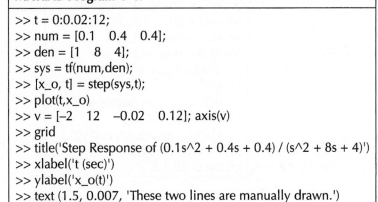

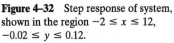

Figure 4–32 Step response of system, shown in the region $-2 \le x \le 12$, $-0.02 \le y \le 0.12$.

Problem A–4–12

Plot the unit-step response curves of the two systems defined by the transfer functions

$$\frac{X(s)}{U(s)} = \frac{25}{s^2 + 5s + 25}$$

and

$$\frac{Y(s)}{U(s)} = \frac{5s + 25}{s^2 + 5s + 25}$$

in one diagram. Then plot each curve in a separate diagram, using the subplot command.

Solution   MATLAB Program 4–18 produces the unit-step response curves of the two systems. The curves are shown in Figure 4–33.

Multiple curves on one diagram can be split into multiple windows with the use of the subplot command. MATLAB Program 4–19 uses the subplot command to plot two curves in two subwindows, one curve per subwindow. Figure 4–34 shows the resulting plot.

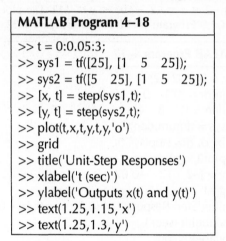

MATLAB Program 4–18

```
>> t = 0:0.05:3;
>> sys1 = tf([25], [1   5   25]);
>> sys2 = tf([5   25], [1   5   25]);
>> [x, t] = step(sys1,t);
>> [y, t] = step(sys2,t);
>> plot(t,x,t,y,t,y,'o')
>> grid
>> title('Unit-Step Responses')
>> xlabel('t (sec)')
>> ylabel('Outputs x(t) and y(t)')
>> text(1.25,1.15,'x')
>> text(1.25,1.3,'y')
```

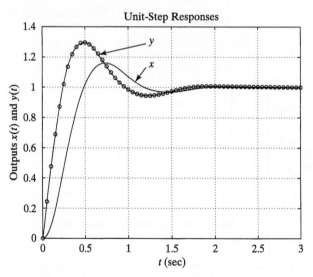

Figure 4–33   Two unit-step response curves shown in one diagram.

Problem A–4–13

Consider the system shown in Figure 4–35. The system is initially at rest. Suppose that the cart is set into motion by an impulsive force whose strength is unity. Can it be stopped by another such impulsive force?

Solution   When the mass m is set into motion by a unit-impulse force, the system equation becomes

$$m\ddot{x} + kx = \delta(t)$$

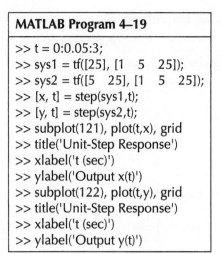

MATLAB Program 4–19

```
>> t = 0:0.05:3;
>> sys1 = tf([25], [1   5   25]);
>> sys2 = tf([5   25], [1   5   25]);
>> [x, t] = step(sys1,t);
>> [y, t] = step(sys2,t);
>> subplot(121), plot(t,x), grid
>> title('Unit-Step Response')
>> xlabel('t (sec)')
>> ylabel('Output x(t)')
>> subplot(122), plot(t,y), grid
>> title('Unit-Step Response')
>> xlabel('t (sec)')
>> ylabel('Output y(t)')
```

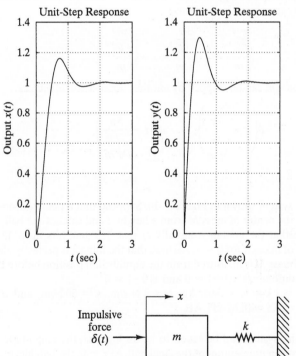

Figure 4–34 Plots of two unit-step response curves in two subwindows, one in each subwindow.

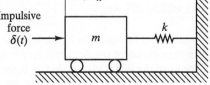

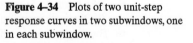

Figure 4–35 Mechanical system.

We define another impulse force to stop the motion as $A\delta(t - T)$, where A is the undetermined magnitude of the impulse force and $t = T$ is the undetermined instant that this impulse is to be given to the system to stop the motion. Then, the equation for the system when the two impulse forces are given is

$$m\ddot{x} + kx = \delta(t) + A\delta(t - T), \qquad x(0-) = 0, \qquad \dot{x}(0-) = 0$$

The $\mathcal{L}_-$ transform of this last equation with $x(0-) = 0$ and $\dot{x}(0-) = 0$ gives

$$(ms^2 + k)X(s) = 1 + Ae^{-sT}$$

Solving for $X(s)$, we obtain

$$X(s) = \frac{1}{ms^2 + k} + \frac{Ae^{-sT}}{ms^2 + k}$$

$$= \frac{1}{\sqrt{km}} \frac{\sqrt{\dfrac{k}{m}}}{s^2 + \dfrac{k}{m}} + \frac{A}{\sqrt{km}} \frac{\sqrt{\dfrac{k}{m}}\,e^{-sT}}{s^2 + \dfrac{k}{m}}$$

The inverse Laplace transform of $X(s)$ is

$$x(t) = \frac{1}{\sqrt{km}} \sin\sqrt{\frac{k}{m}}\,t + \frac{A}{\sqrt{km}}\left[\sin\sqrt{\frac{k}{m}}(t - T)\right]1(t - T)$$

If the motion of the mass m is to be stopped at $t = T$, then $x(t)$ must be identically zero for $t \geq T$, a condition we can achieve if we choose

$$A = 1, \qquad T = \frac{\pi}{\sqrt{\dfrac{k}{m}}}, \qquad \frac{3\pi}{\sqrt{\dfrac{k}{m}}}, \qquad \frac{5\pi}{\sqrt{\dfrac{k}{m}}}, \qquad \cdots$$

Thus, the motion of the mass m can be stopped by another impulse force, such as

$$\delta\left(t - \frac{\pi}{\sqrt{\dfrac{k}{m}}}\right), \qquad \delta\left(t - \frac{3\pi}{\sqrt{\dfrac{k}{m}}}\right), \qquad \delta\left(t - \frac{5\pi}{\sqrt{\dfrac{k}{m}}}\right), \qquad \cdots$$

Problem A–4–14

Consider the mechanical system shown in Figure 4–36. Suppose that a person drops a steel ball of mass m onto the center of mass M from a height d and catches the ball on the first bounce. Assume that the system is initially at rest. The ball hits mass M at $t = 0$. Obtain the motion of mass M for $0 < t$. Assume that the impact is perfectly elastic. The displacement x of mass M is measured from the equilibrium position before the ball hits it. The initial conditions are $x(0-) = 0$ and $\dot{x}(0-) = 0$.

Assuming that $M = 1$ kg, $m = 0.015$ kg, $b = 2$ N-s/m, $k = 50$ N/m, and $d = 1.45$ m, plot the response curve with MATLAB.

Solution   The input to the system can be taken to be an impulse, the magnitude of which is equal to the change in momentum of the steel ball. At $t = 0$, the ball hits mass M. Assume that the initial velocity of the ball is $v(0-)$. At $t = 0$, the ball bounces back with velocity $v(0+)$. Since the impact is assumed to be perfectly elastic, $v(0+) = -v(0-)$. Figure 4–37(a) shows a sudden change in the velocity of the ball. Define the downward velocity to be positive. Since the change in velocity of the ball occurs instantaneously, $\dot{v}$ has the form of an impulse, as shown in Figure 4–37(b). Note that $\dot{v}(0+)$ is negative.

The equation of motion for the system is

$$M\ddot{x} + b\dot{x} + kx = F(t) \tag{4-29}$$

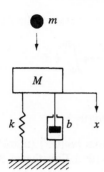

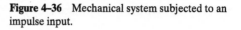

Figure 4-36 Mechanical system subjected to an impulse input.

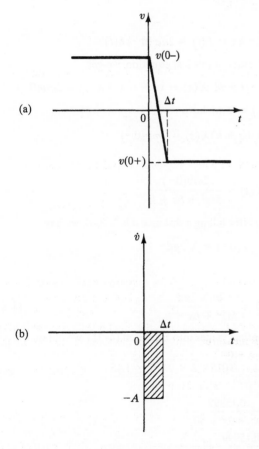

Figure 4-37 (a) Sudden change in the velocity v of steel ball; (b) plot of $\dot{v}$ versus t.

where $F(t)$, an impulse force, is equal to $-m\dot{v}$. [Note that $-m\dot{v}$ is positive; the impulse force $F(t)$ is in the positive direction of x.] From Figure 4-37(b), the impulse force can be written as

$$F(t) = A \, \Delta t \, \delta(t)$$

where $A \, \Delta t$ is the magnitude of the impulse input. Thus,

$$F(t) = A \, \Delta t \, \delta(t) = -m\dot{v}$$

from which we can get

$$\int_{0-}^{0+} A \, \Delta t \, \delta(t) dt = -m \int_{0-}^{0+} \dot{v} \, dt$$

or

$$A\Delta t = mv(0-) - mv(0+) \tag{4-30}$$

The momentum of the steel ball is changed from $mv(0-)$ (downward at $t = 0-$) to $mv(0+)$ (upward at $t = 0+$). Since $v(0+) = -v(0-)$, Equation (4-30) can be written as

$$A \, \Delta t = 2mv(0-)$$

Then Equation (4-29) becomes

$$M\ddot{x} + b\dot{x} + kx = F(t) = 2mv(0-) \, \delta(t)$$

Taking the $\mathcal{L}_{-}$ transforms of both sides of this last equation, we get

$$M[s^2X(s) - sx(0-) - \dot{x}(0-)] + b[sX(s) - x(0-)] + kX(s) = 2mv(0-)$$

Noting that $x(0-) = 0$ and $\dot{x}(0-) = 0$, we have

$$(Ms^2 + bs + k)X(s) = 2mv(0-)$$

Solving for $X(s)$, we obtain

$$X(s) = \frac{2mv(0-)}{Ms^2 + bs + k}$$

Since the velocity of the steel ball after falling a distance d is $\sqrt{2gd}$, we have

$$v(0-) = \sqrt{2gd}$$

It follows that

$$X(s) = \frac{2m\sqrt{2gd}}{Ms^2 + bs + k} \tag{4-31}$$

Substituting the given numerical values into Equation (4-31), we obtain

$$X(s) = \frac{2 \times 0.015\sqrt{2 \times 9.807 \times 1.45}}{s^2 + 2s + 50}$$

$$= \frac{0.15999}{s^2 + 2s + 50}$$

$$= \frac{0.15999}{7} \frac{7}{(s + 1)^2 + 7^2}$$

The inverse Laplace transform of $X(s)$ gives

$$x(t) = 0.02286e^{-t} \sin 7t \text{ m}$$

Thus, the response of the mass M is a damped sinusoidal motion.

A MATLAB program to produce the response curve is given in MATLAB Program 4-20. The resulting curve is shown in Figure 4-38.

MATLAB Program 4–20

```
>> num = [0.15999];
>> den = [1   2   50];
>> sys = tf(num,den);
>> impulse(sys)
>> grid
>> title('Impulse Response of Mechanical System')
>> xlabel('t')
>> ylabel('Output x(t)')
```

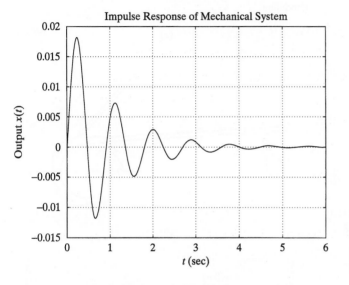

Figure 4–38 Response of mass M subjected to impulse input.

Problem A–4–15

Figure 4–39 shows a mechanism used for a safety seat belt system. Under normal operating conditions, the reel rotates freely and it is possible to let out more slack in the belt, allowing the passenger to move forward even with the belt fastened. However, if the car decelerates rapidly in a collision or sudden stop, the pendulum is subjected to an impulsive torque that causes it to swing forward and also causes the bar to engage the ratchet, locking the reel and safety belt. Thus, the passenger is restrained in place.

Referring to Figure 4–40, assume that the car is moving at a speed of 10 m/s before a sudden stop. The stopping time Δt is 0.3 s. The pendulum length is 0.05 m. Find the time needed for the pendulum to swing forward by 20°.

Solution From Figure 4–41, the moment of inertia of the pendulum about the pivot is $J = ml^2$. The angle of rotation of the pendulum is θ rad. Define the force that acts on the pendulum at the instant the car stops suddenly as $F(t)$. Then the torque that acts on the pendulum due to the force $F(t)$ is $F(t)l \cos \theta$. The equation for the pendulum system is

$$ml^2\ddot{\theta} = F(t)l \cos \theta - mgl \sin \theta \tag{4-32}$$

We linearize this nonlinear equation by assuming that the angle θ is small. (Although $\theta = 20°$ is not quite small, the resulting linearized equation will give an

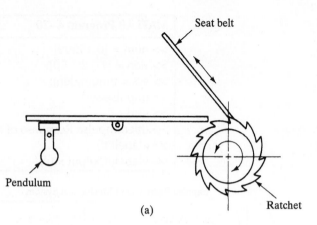

(a)

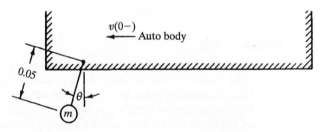

Figure 4–39 Mechanism used for a safety seat belt system. (a) Normal operating condition; (b) emergency condition.

(b)

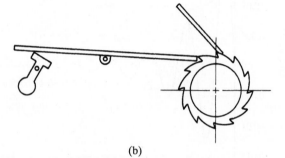

Figure 4–40 Pendulum attached to auto body.

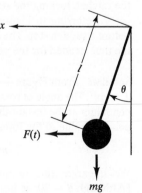

Figure 4–41 Pendulum system.

approximate solution.) Approximating $\cos\theta \doteq 1$ and $\sin\theta \doteq \theta$, we can write Equation (4–32) as

$$ml^2\ddot{\theta} = F(t)l - mgl\,\theta$$

or

$$ml\ddot{\theta} + mg\theta = F(t) \tag{4–33}$$

Since the velocity of the car at $t = 0-$ is 10 m/s and the car stops in 0.3 s, the average deceleration is 33.3 m/s^2.

Under the assumption that a constant acceleration of magnitude 33.3 m/s^2 acts on the pendulum mass for 0.3 sec, $F(t)$ may be given by

$$F(t) = m\ddot{x} = 33.3m[1(t) - 1(t - 0.3)]$$

Then, Equation (4–33) may be written as

$$ml\ddot{\theta} + mg\theta = 33.3m[1(t) - 1(t - 0.3)]$$

or

$$\ddot{\theta} + \frac{g}{l}\theta = \frac{33.3}{l}[1(t) - 1(t - 0.3)]$$

Since $l = 0.05$ m, this last equation becomes

$$\ddot{\theta} + 196.14\theta = 666[1(t) - 1(t - 0.3)]$$

Taking $\mathcal{L}_-$ transforms of both sides of the preceding equation, we obtain

$$(s^2 + 196.14)\Theta(s) = 666\left(\frac{1}{s} - \frac{1}{s}e^{-0.3s}\right) \tag{4–34}$$

where we used the initial conditions that $\theta(0-) = 0$ and $\dot{\theta}(0-) = 0$. Solving Equation (4–34) for $\Theta(s)$ yields

$$\Theta(s) = \frac{666}{s(s^2 + 196.14)}(1 - e^{-0.3s})$$

$$= \left(\frac{1}{s} - \frac{s}{s^2 + 196.14}\right)\frac{666}{196.14}(1 - e^{-0.3s})$$

The inverse Laplace transform of $\Theta(s)$ gives

$$\theta(t) = 3.3955(1 - \cos 14t)$$
$$- 3.3955\{1(t - 0.3) - [\cos 14(t - 0.3)]1(t - 0.3)\} \tag{4–35}$$

Note that $1(t - 0.3) = 0$ for $0 \le t < 0.3$.

Now assume that at $t = t_1$, $\theta = 20° = 0.3491$ rad. Then, tentatively assuming that t_1 occurs before $t = 0.3$, we seek to solve the following equation for t_1:

$$0.3491 = 3.3955(1 - \cos 14t_1)$$

Simplifying yields

$$\cos 14t_1 = 0.8972$$

and the solution is

$$t_1 = 0.0326 \text{ s}$$

Since $t_1 = 0.0326 < 0.3$, our assumption was correct. The terms involving $1(t - 0.3)$ in Equation (4–35) do not affect the value of t_1. It thus takes approximately 33 milliseconds for the pendulum to swing 20°.

Problem A–4–16

Consider the mechanical system shown in Figure 4–42(a). The cart has the mass of m kg. Assume that the wheels have negligible masses and there is no friction involved in the system. The force $u(t)$ applied to the cart is increased linearly from 0 to 5 N for the period $0 \leq t \leq 10$, as shown in Figure 4–42(b). At $t = 10+$, the force $u(t)$ is disengaged, or

$$u(t) = 0 \qquad \text{for } 10 < t$$

Assuming that $m = 100$ kg, obtain the displacement $x(t)$ of the cart for $0 \leq t \leq 30$ with MATLAB. The cart is at rest for $t < 0$, and the displacement x is measured from the rest position.

Solution   The equation of motion for the system is

$$m\ddot{x} = u$$

Hence, the transfer function of the system is

$$\frac{X(s)}{U(s)} = \frac{1}{ms^2} = \frac{1}{100s^2}$$

The input $u(t)$ is a ramp function for $0 \leq t \leq 10$ and is zero for $10 < t$, as shown in Figure 4–42(b). (At $t = 10$, $u = 5$ N.) Thus, in MATLAB, we define

$$\text{u1} = 0.5*[0:0.02:10] \qquad \text{for } 0 \leq t \leq 10$$

$$\text{u2} = 0*[10.02:0.02:30] \qquad \text{for } 10 < t \leq 30$$

where u is either u1 or u2, depending on which interval u is in. Then the input force $u(t)$ for $0 \leq t \leq 30$ can be given by the MATLAB array

$$\text{u} = [\text{u1} \quad \text{u2}]$$

MATLAB Program 4–21

```
>> t = 0:0.02:30;
>> u1 = 0.5*[0:0.02:10];
>> u2 = 0*[10.02:0.02:30];
>> u = [u1    u2];
>> num = [1]; den = [100  0  0];
>> sys = tf(num,den);
>> x = lsim(sys,u,t);
>> subplot(211),plot(t,x)
>> grid
>> ylabel('Output x(t)')
>> subplot (212), plot(t,u)
>> v = [0   30  -1   6]; axis (v)
>> grid
>> xlabel('t (sec)')
>> ylabel('Input Force u(t) newton')
```

MATLAB Program 4–21 produces the response curve. The curves of $x(t)$ versus t and the input force $u(t)$ versus t are shown in Figure 4–43.

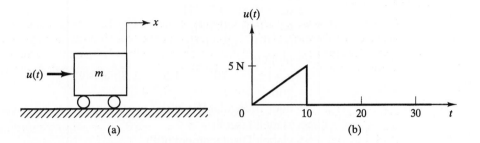

(a)   (b)

Figure 4–42   (a) Mechanical system; (b) force $u(t)$ applied to the cart.

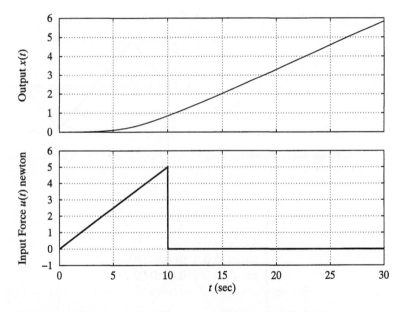

Figure 4–43   Response curve $x(t)$ versus t and input curve $u(t)$ versus t.

Problem A–4–17

Using MATLAB, generate a triangular wave as shown in Figure 4–44.

Solution   There are many ways to generate the given triangular wave. In MATLAB Program 4–22, we present one simple way to do so. The resulting wave is shown in Figure 4–45.

MATLAB Program 4–22

```
>> t = 0:1:8;
>> u1 = [0:0.5:1];
>> u2 = [0.5:-0.5:-1];
>> u3 = [-0.5:0.5:0];
>> u = [u1    u2    u3];
>> plot(t,u)
>> v = [-2    10    -1.5    1.5]; axis(v)
>> grid
>> title('Triangular Wave')
>> xlabel('t (sec)')
>> ylabel('Displacement u(t)')
```

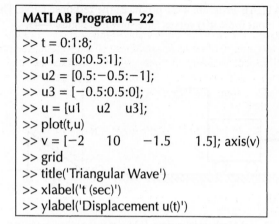

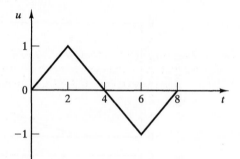

Figure 4–44   Triangular wave.

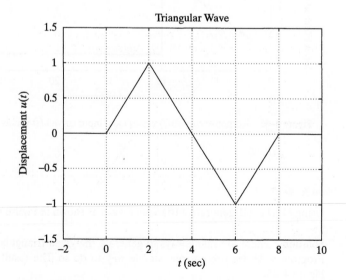

Figure 4–45   Triangular wave generated with MATLAB.

Problem A–4–18

Consider the spring–mass–dashpot system shown in Figure 4–46(a). Assume that the displacement u of point P is the input to the system. Assume also that the input $u(t)$ is a small bump, as shown in Figure 4–46(b). Obtain the response $y(t)$ of the mass m. The displacement y is measured from the equilibrium position in the absence of the input $u(t)$. To obtain the response curve, assume that $m = 100$ kg, $b = 400$ N-s/m, and $k = 800$ N/m.

Solution From **Problem A–4–6**, the transfer function of the system is

$$\frac{Y(s)}{U(s)} = \frac{bs + k}{ms^2 + bs + k}$$

Substituting the given numerical values for m, b, and k into this transfer function, we obtain

$$\frac{Y(s)}{U(s)} = \frac{400s + 800}{100s^2 + 400s + 800}$$

$$= \frac{4s + 8}{s^2 + 4s + 8}$$

The input $u(t)$ is a triangular wave for $0 \le t \le 4$ and is zero for $4 < t \le 8$. (For the generation of a triangular wave, see **Problem A–4–17**.)

As in **Problem A–4–17**, the input $u(t)$ can be generated by first defining, in MAT-LAB,

$$u1 = [0:0.02:1];$$
$$u2 = [0.98:-0.02:-1];$$
$$u3 = [-0.98:0.02:0];$$
$$u4 = 0*[4.02:0.02:8];$$

and then defining

$$u = [u1 \quad u2 \quad u3 \quad u4]$$

MATLAB Program 4–23 produces the response $y(t)$ of the system. The response curve $y(t)$ versus t and the input curve $u(t)$ versus t are shown in Figure 4–47.

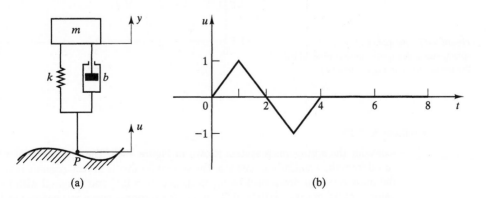

(a) (b)

Figure 4–46 (a) Spring–mass–dashpot system; (b) input $u(t)$ versus t.

MATLAB Program 4–23

```
>> t = 0:0.02:8;
>> num = [4   8];
>> den = [1   4   8];
>> sys = tf(num,den);
>> u1 = [0:0.02:1];
>> u2 = [0.98:–0.02:–1];
>> u3 = [–0.98:0.02:0];
>> u4 = 0*[4.02:0.02:8];
>> u = [u1   u2   u3   u4];
>> y = lsim(sys,u,t);
>> plot(t,y,t,u)
>> grid
>> title('Response of Spring-Mass-Dashpot System and Input u(t)')
>> xlabel('t (sec)')
>> ylabel('Output y(t) and Input u(t)')
>> text(2.2,0.72,'y')
>> text(1.05,0.1,'u')
```

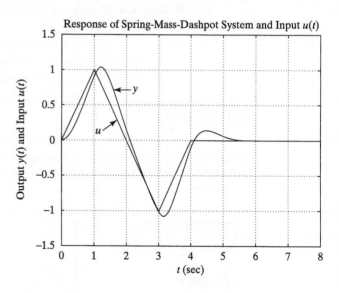

Figure 4–47 Response of spring–mass–dashpot system subjected to the input shown in Figure 4–46(b).

Problem A–4–19

Consider the spring–mass system shown in Figure 4–48. The displacement x is measured from the equilibrium position. The system is initially at rest. Assume that at $t = 0$ the mass is pulled downward by 0.1 m [i.e., $x(0) = 0.1$] and released with the initial velocity of 0.5 m/s [i.e., $\dot{x}(0) = 0.5$]. Obtain the response curve $x(t)$ versus t with MATLAB. Assume that $m = 1$ kg and $k = 9$ N/m.

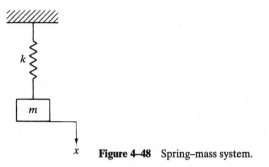

x **Figure 4-48** Spring–mass system.

Solution The equation of motion for the system is

$$m\ddot{x} = -kx$$

Taking the Laplace transform of the preceding equation, we obtain

$$m[s^2 X(s) - sx(0) - \dot{x}(0)] = -kX(s)$$

Substituting the given numerical values for m, k, $x(0)$, and $\dot{x}(0)$ into this last equation, we have

$$s^2 X(s) - 0.1s - 0.5 + 9X(s) = 0$$

Solving for $X(s)$, we get

$$X(s) = \frac{0.1s + 0.5}{s^2 + 9} = \frac{0.1s^2 + 0.5s}{s^2 + 9}\frac{1}{s}$$

Hence, the response $x(t)$ can be obtained as the unit-step response of

$$G(s) = \frac{0.1s^2 + 0.5s}{s^2 + 9}$$

MATLAB Program 4–24 produces the response curve $x(t)$ versus t. The curve is shown in Figure 4–49.

MATLAB Program 4–24

```
>> t = 0:0.001:4;
>> num = [0.1   0.5   0];
>> den = [1   0   9];
>> sys = tf(num,den);
>> x = step(sys,t);
>> plot(t,x)
>> v = [-1   4   -0.4   0.4]; axis(v)
>> grid
>> title ('Response of System Subjected to Initial Condition')
>> xlabel('t (sec)')
>> ylabel('x(t) meter')
```

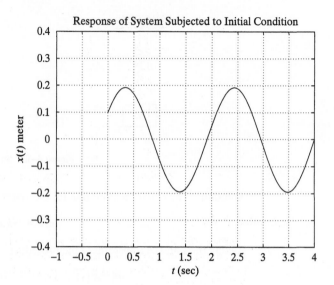

Figure 4–49 Response of spring–mass system subjected to the initial condition $x(0) = 0.1$ m and $\dot{x}(0) = 0.5$ m/s.

PROBLEMS

Problem B–4–1

Find the transfer function $X_o(s)/X_i(s)$ of the mechanical system shown in Figure 4–50. The displacements x_i and x_o are measured from their respective equilibrium positions. Obtain the displacement $x_o(t)$ when the input $x_i(t)$ is a step displacement of magnitude X_i occurring at $t = 0$. Assume that $x_o(0-) = 0$.

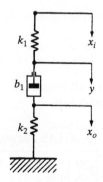

Figure 4–50 Mechanical system.

Problem B–4–2

Derive the transfer function $X_o(s)/X_i(s)$ of the mechanical system shown in Figure 4–51. The displacements x_i and x_o are measured from their respective equilibrium positions. Obtain the response $x_o(t)$ when the input $x_i(t)$ is the pulse

$$x_i(t) = X_i \qquad 0 < t < t_1$$
$$= 0 \qquad \text{elsewhere}$$

Assume that $x_o(0-) = 0$.

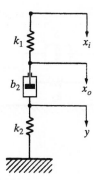

Figure 4–51 Mechanical system.

Problem B–4–3

Consider the mechanical system shown in Figure 4–52. Assume that $u(t)$ is the force applied to the cart and is the input to the system. The displacement x is measured from the equilibrium position and is the output of the system. Obtain the transfer function $X(s)/U(s)$ of the system.

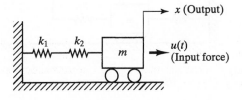

Figure 4–52 Mechanical system.

Problem B–4–4

In the mechanical system shown in Figure 4–53, the force u is the input to the system and the displacement x, measured from the equilibrium position, is the output of the system, which is initially at rest. Obtain the transfer function $X(s)/U(s)$.

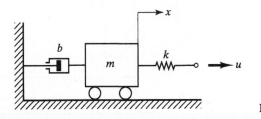

Figure 4–53 Mechanical system.

Problem B–4–5

The system shown in Figure 4–54 is initially at rest, and the displacement x is measured from the equilibrium position. At $t = 0$, a force u is applied to the system. If u is the input to the system and x is the output, obtain the transfer function $X(s)/U(s)$.

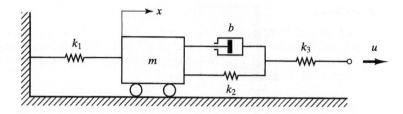

Figure 4–54   Mechanical system.

Problem B–4–6

Consider the mechanical system shown in Figure 4–55. The system is at rest for $t < 0$. The input force u is given at $t = 0$. The displacement x is the output of the system and is measured from the equilibrium position. Obtain the transfer function $X(s)/U(s)$.

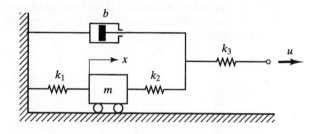

Figure 4–55   Mechanical system.

Problem B–4–7

In the system of Figure 4–56, $x(t)$ is the input displacement and $\theta(t)$ is the output angular displacement. Assume that the masses involved are negligibly small and that all motions are restricted to be small; therefore, the system can be considered linear. The initial conditions for x and θ are zeros, or $x(0-) = 0$ and $\theta(0-) = 0$. Show that this system is a differentiating element. Then obtain the response $\theta(t)$ when $x(t)$ is a unit-step input.

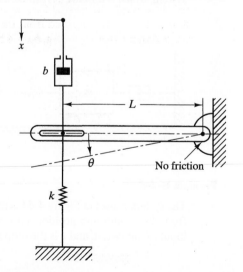

Figure 4–56   Mechanical system.

Problem B–4–8

Consider the mechanical system shown in Figure 4–57. The system is initially at rest. Assume that u is the displacement of point P and x is the displacement of mass m. The displacement x is measured from the equilibrium position when $u = 0$. Draw four different block diagrams for the system.

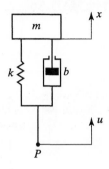

Figure 4–57 Mechanical system.

Problem B–4–9

Using MATLAB, obtain the partial-fraction expansion of

$$\frac{B(s)}{A(s)} = \frac{1}{s^4 + s^3 + 81s^2 + 81s}$$

Problem B–4–10

Using MATLAB, obtain the partial-fraction expansion of

$$\frac{B(s)}{A(s)} = \frac{5(s + 2)}{s^5 + 5s^4 + 19s^3 + 12s^2}$$

Problem B–4–11

Consider the mechanical system shown in Figure 4–58. Plot the response curve $x(t)$ versus t with MATLAB when the mass m is pulled slightly downward, generating the initial conditions $x(0) = 0.05$ m and $\dot{x}(0) = 1$ m/s, and released at $t = 0$. The displacement x is measured from the equilibrium position before m is pulled downward. Assume that $m = 1$ kg, $b_1 = 4$ N-s/m, $k_1 = 6$ N/m, and $k_2 = 10$ N/m.

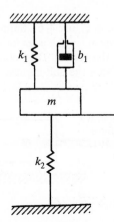

Figure 4–58 Mechanical system.

Problem B–4–12

Consider the mechanical system shown in Figure 4–59. The system is initially at rest. The displacements x_1 and x_2 are measured from their respective equilibrium positions before the input u is applied. Assume that $b_1 = 1$ N-s/m, $b_2 = 10$ N-s/m, $k_1 = 4$ N/m, and $k_2 = 20$ N/m. Obtain the displacement $x_2(t)$ when u is a step force input of 2 N. Plot the response curve $x_2(t)$ versus t with MATLAB.

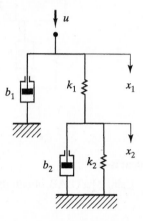

Figure 4–59   Mechanical system.

Problem B–4–13

Figure 4–60 shows a mechanical system that consists of a mass and a damper. The system is initially at rest. Find the response $x(t)$ when the system is set into motion by an impulsive force whose strength is unity. Determine the initial velocity of mass m. Plot the response curve $x(t)$ versus t when $m = 100$ kg and $b = 200$ N-s/m.

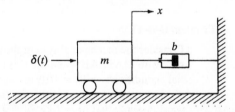

Figure 4–60   Mechanical system.

Problem B–4–14

Consider the mechanical system shown in Figure 4–61. Suppose that the system is initially at rest [$x(0-) = 0$, $\dot{x}(0-) = 0$] and at $t = 0$ it is set into motion by a unit-impulse force. Obtain the transfer function of the system. Then obtain an analytical solution $x(t)$. What is the initial velocity $\dot{x}(0+)$ after the unit-impulse force is given to the cart?

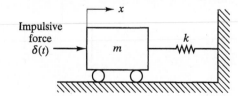

Figure 4–61   Mechanical system.

Problem B–4–15

The mechanical system shown in Figure 4–62 is initially at rest. The displacement x of mass m is measured from the rest position. At $t = 0$, mass m is set into motion by an impulsive force whose strength is unity. Using MATLAB, plot the response curve $x(t)$ versus t when $m = 10$ kg, $b = 20$ N-s/m, and $k = 50$ N/m.

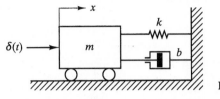

Figure 4–62 Mechanical system.

Problem B–4–16

A mass m of 1 kg is vibrating initially in the mechanical system shown in Figure 4–63. At $t = 0$, we hit the mass with an impulsive force $p(t)$ whose strength is 10 N. Assuming that the spring constant k is 100 N/m, that $x(0-) = 0.1$ m, and that $\dot{x}(0-) = 1$ m/s, find the displacement $x(t)$ as a function of time t. The displacement $x(t)$ is measured from the equilibrium position in the absence of the excitation force.

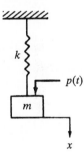

Figure 4–63 Mechanical system.

Problem B–4–17

Consider the system shown in Figure 4–64. The system is at rest for $t < 0$. Assume that the displacement x is the output of the system and is measured from the equilibrium position. At $t = 0$, the cart is given initial conditions $x(0) = x_o$ and $\dot{x}(0) = v_o$. Obtain the output motion $x(t)$. Assume that $m = 10$ kg, $b_1 = 50$ N-s/m, $b_2 = 70$ N-s/m, $k_1 = 400$ N/m, and $k_2 = 600$ N/m.

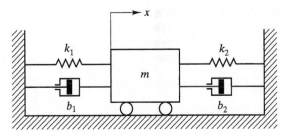

Figure 4–64 Mechanical system.

Problem B–4–18

Referring to **Problem B–4–17**, assume that $m = 100$ kg, $b_1 = 120$ N-s/m, $b_2 = 80$ N-s/m, $k_1 = 200$ N/m, and $k_2 = 300$ N/m. The initial conditions are $x(0) = 0$ m and $\dot{x}(0) = 0.5$ m/s. Obtain the response curve $x(t)$ versus t with MATLAB.

State-Space Approach to Modeling Dynamic Systems

5–1 INTRODUCTION

The modern trend in dynamic systems is toward greater complexity, due mainly to the twin requirements of complex tasks and high accuracy. Complex systems may have multiple inputs and multiple outputs. Such systems may be linear or nonlinear and may be time invariant or time varying. A very powerful approach to treating such systems is the state-space approach, based on the concept of state. This concept, by itself, is not new; it has been in existence for a long time in the field of classical dynamics and in other fields. What is new is the combination of the concept of state and the capability of high-speed solution of differential equations with the use of the digital computer.

This chapter presents an introductory account of modeling dynamic systems in state space and analyzing simple dynamic systems with MATLAB. (More on the state-space analysis of dynamic systems is given in Chapter 8.) If the dynamic system is formulated in the state space, it is very easy to simulate it on the computer and find the computer solution of the system's differential equations, because the state-space formulation is developed precisely with such computer solution in mind. Although we treat only linear, time-invariant systems in this chapter, the state-space approach can be applied to both linear and nonlinear systems and to both time-invariant and time-varying systems.

In what follows, we shall first give definitions of state, state variables, state vector, and state space. Then we shall present the outline of the chapter.

State. The *state* of a dynamic system is the smallest set of variables (called *state variables*) such that knowledge of these variables at $t = t_0$, together with knowledge of the input for $t \geq t_0$, completely determines the behavior of the system for any time $t \geq t_0$.

Thus, the state of a dynamic system at time t is uniquely determined by the state at time t_0 and the input $t \geq t_0$ and is independent of the state and input before t_0. In dealing with linear time-invariant systems, we usually choose the reference time t_0 to be zero.

State variables. The *state variables* of a dynamic system are the variables making up the smallest set of variables that determines the state of the dynamic system. If at least n variables $x_1, x_2, \ldots, x_n$ are needed to completely describe the behavior of a dynamic system (so that, once the input is given for $t \geq t_0$ and the initial state at $t = t_0$ is specified, the future state of the system is completely determined), then those n variables are a set of state variables. It is important to note that variables that do not represent physical quantities can be chosen as state variables.

State vector. If n state variables are needed to completely describe the behavior of a given system, then those state variables can be considered the n components of a vector **x** called a *state vector*. A state vector is thus a vector that uniquely determines the system state $\mathbf{x}(t)$ for any time $t \geq t_0$, once the state at $t = t_0$ is given and the input $\mathbf{u}(t)$ for $t \geq t_0$ is specified.

State space. The n-dimensional space whose coordinate axes consist of the x_1-axis, x_2-axis, $\ldots$, x_n-axis is called a *state space*. Any state can be represented by a point in the state space.

State-space equations. In state-space analysis, we are concerned with three types of variables that are involved in the modeling of dynamic systems: input variables, output variables, and state variables. As we shall see later, the state-space representation for a given system is not unique, except that the number of state variables is the same for any of the different state-space representations of the same system.

If a system is linear and time invariant and if it is described by n state variables, r input variables, and m output variables, then the state equation will have the form

$$\dot{x}_1 = a_{11}x_1 + a_{12}x_2 + \cdots + a_{1n}x_n + b_{11}u_1 + b_{12}u_2 + \cdots + b_{1r}u_r$$
$$\dot{x}_2 = a_{21}x_1 + a_{22}x_2 + \cdots + a_{2n}x_n + b_{21}u_1 + b_{22}u_2 + \cdots + b_{2r}u_r$$
$$\vdots$$
$$\dot{x}_n = a_{n1}x_1 + a_{n2}x_2 + \cdots + a_{nn}x_n + b_{n1}u_1 + b_{n2}u_2 + \cdots + b_{nr}u_r$$

and the output equation will have the form

$$y_1 = c_{11}x_1 + c_{12}x_2 + \cdots + c_{1n}x_n + d_{11}u_1 + d_{12}u_2 + \cdots + d_{1r}u_r$$
$$y_2 = c_{21}x_1 + c_{22}x_2 + \cdots + c_{2n}x_n + d_{21}u_1 + d_{22}u_2 + \cdots + d_{2r}u_r$$
$$\vdots$$
$$y_m = c_{m1}x_1 + c_{m2}x_2 + \cdots + c_{mn}x_n + d_{m1}u_1 + d_{m2}u_2 + \cdots + d_{mr}u_r$$

where the coefficients a_{ij}, b_{ij}, c_{ij}, and d_{ij} are constants, some of which may be zero. If we use vector–matrix expressions, these equations can be written as

$$\dot{\mathbf{x}} = \mathbf{A}\mathbf{x} + \mathbf{B}\mathbf{u} \tag{5–1}$$

$$\mathbf{y} = \mathbf{C}\mathbf{x} + \mathbf{D}\mathbf{u} \tag{5–2}$$

where

$$\mathbf{x} = \begin{bmatrix} x_1 \\ x_2 \\ \vdots \\ x_n \end{bmatrix}, \mathbf{A} = \begin{bmatrix} a_{11} & a_{12} & \cdots & a_{1n} \\ a_{21} & a_{22} & \cdots & a_{2n} \\ \vdots & \vdots & & \vdots \\ a_{n1} & a_{n2} & \cdots & a_{nn} \end{bmatrix}, \mathbf{B} = \begin{bmatrix} b_{11} & b_{12} & \cdots & b_{1r} \\ b_{21} & b_{22} & \cdots & b_{2r} \\ \vdots & \vdots & & \vdots \\ b_{n1} & b_{n2} & \cdots & b_{nr} \end{bmatrix}, \mathbf{u} = \begin{bmatrix} u_1 \\ u_2 \\ \vdots \\ u_r \end{bmatrix}$$

$$\mathbf{y} = \begin{bmatrix} y_1 \\ y_2 \\ \vdots \\ y_m \end{bmatrix}, \mathbf{C} = \begin{bmatrix} c_{11} & c_{12} & \cdots & c_{1n} \\ c_{21} & c_{22} & \cdots & c_{2n} \\ \vdots & \vdots & & \vdots \\ c_{m1} & c_{m2} & \cdots & c_{mn} \end{bmatrix}, \mathbf{D} = \begin{bmatrix} d_{11} & d_{12} & \cdots & d_{1r} \\ d_{21} & d_{22} & \cdots & d_{2r} \\ \vdots & \vdots & & \vdots \\ d_{m1} & d_{m2} & \cdots & d_{mr} \end{bmatrix}$$

Matrices **A**, **B**, **C**, and **D** are called the *state matrix, input matrix, output matrix,* and *direct transmission matrix,* respectively. Vectors **x**, **u**, and **y** are the *state vector, input vector,* and *output vector,* respectively. (In control systems analysis and design, the input matrix **B** and input vector **u** are called the control matrix and control vector, respectively.) The elements of the state vector are the state variables. The elements of the input vector **u** are the input variables. (If the system involves only one input variable, then u is a scalar.) The elements of the output vector **y** are the output variables. (The system may involve one or more output variables.) Equation (5–1) is called the *state equation,* and Equation (5–2) is called the *output equation.* [In this book, whenever we discuss state-space equations, they are described by Equations (5–1) and (5–2).]

A block diagram representation of Equations (5–1) and (5–2) is shown in Figure 5–1. (In the figure, double-line arrows are used to indicate that the signals are vector quantities.)

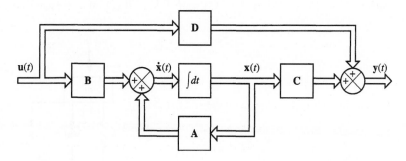

Figure 5–1 Block diagram of the linear, continuous-time system represented in state space.

Example 5–1

Consider the mechanical system shown in Figure 5–2. The displacement y of the mass is the output of the system, and the external force u is the input to the system. The displacement y is measured from the equilibrium position in the absence of the external force. Obtain a state-space representation of the system.

From the diagram, the system equation is

$$m\ddot{y} + b\dot{y} + ky = u \tag{5–3}$$

This system is of second order. (This means that the system involves two integrators.) Thus, we need two state variables to describe the system dynamics. Since $y(0)$, $\dot{y}(0)$, and $u(t) \geq 0$ completely determine the system behavior for $t \geq 0$, we choose $y(t)$ and $\dot{y}(t)$ as state variables, or define

$$x_1 = y$$
$$x_2 = \dot{y}$$

Then we obtain

$$\dot{x}_1 = x_2$$
$$\dot{x}_2 = \ddot{y} = \frac{1}{m}(-ky - b\dot{y}) + \frac{1}{m}u$$

or

$$\dot{x}_1 = x_2 \tag{5–4}$$
$$\dot{x}_2 = -\frac{k}{m}x_1 - \frac{b}{m}x_2 + \frac{1}{m}u \tag{5–5}$$

The output equation is

$$y = x_1 \tag{5–6}$$

In vector–matrix form, Equations (5–4) and (5–5) can be written as

$$\begin{bmatrix} \dot{x}_1 \\ \dot{x}_2 \end{bmatrix} = \begin{bmatrix} 0 & 1 \\ -\dfrac{k}{m} & -\dfrac{b}{m} \end{bmatrix} \begin{bmatrix} x_1 \\ x_2 \end{bmatrix} + \begin{bmatrix} 0 \\ \dfrac{1}{m} \end{bmatrix} u \tag{5–7}$$

The output equation, Equation (5–6), can be written as

$$y = \begin{bmatrix} 1 & 0 \end{bmatrix} \begin{bmatrix} x_1 \\ x_2 \end{bmatrix} \tag{5–8}$$

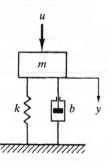

Figure 5–2 Mechanical system.

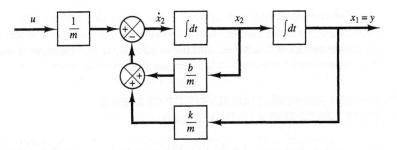

Figure 5–3   Block diagram of the mechanical system shown in Figure 5–2.

Equation (5–7) is a state equation, and Equation (5–8) is an output equation for the system. Equations (5–7) and (5–8) are in the standard form

$$\dot{\mathbf{x}} = \mathbf{A}\mathbf{x} + \mathbf{B}u$$
$$y = \mathbf{C}\mathbf{x} + Du$$

where

$$\mathbf{A} = \begin{bmatrix} 0 & 1 \\ -\dfrac{k}{m} & -\dfrac{b}{m} \end{bmatrix}, \qquad \mathbf{B} = \begin{bmatrix} 0 \\ \dfrac{1}{m} \end{bmatrix}, \qquad \mathbf{C} = [1 \quad 0], \qquad D = 0$$

Note that Equation (5–3) can be modified to

$$\frac{u}{m} - \frac{k}{m}y - \frac{b}{m}\dot{y} = \ddot{y}$$

or

$$\frac{1}{m}u - \frac{k}{m}x_1 - \frac{b}{m}x_2 = \dot{x}_2$$

On the basis of this last equation, we can draw the block diagram shown in Figure 5–3. Notice that the outputs of the integrators are state variables.

In a state-space representation, a system is represented by a state equation and an output equation. In this representation, the internal structure of the system is described by a first-order vector–matrix differential equation. This fact indicates that the state-space representation is fundamentally different from the transfer-function representation, in which the dynamics of the system are described by the input and the output, but the internal structure is put in a black box.

Outline of the chapter.   Section 5–1 has defined some terms that are necessary for the modeling of dynamic systems in state space and has derived a state-space model of a simple dynamic system. Section 5–2 gives a transient-response analysis of systems in state-space form with MATLAB. Section 5–3 discusses the state-space modeling of systems wherein derivative terms of the input function do not appear in the system differential equations. Numerical response analysis is done with MATLAB. Section 5–4 presents two methods for obtaining state-space models of systems in which derivative terms of the input function appear explicitly in the system differential equations. Section 5–5 treats the transformation of system models from transfer-function representation to state-space representation and vice versa. The section also examines the transformation of one state-space representation to another.

5–2 TRANSIENT-RESPONSE ANALYSIS OF SYSTEMS IN STATE-SPACE FORM WITH MATLAB

This section presents the MATLAB approach to obtaining transient-response curves of systems that are written in state-space form.

Step response. We first define the system with

$$sys = ss(A,B,C,D)$$

For a unit-step input, the MATLAB command

$$step(sys) \quad or \quad step(A,B,C,D)$$

will generate plots of unit-step responses. The time vector is automatically determined when t is not explicitly included in the step commands.

Note that when step commands have left-hand arguments, such as

$$y = step(sys,t), \qquad [y,t,x] = step(sys,t),$$
$$[y,x,t] = step(A,B,C,D,iu), \qquad [y,x,t] = step(A,B,C,D,iu,t)$$

no plot is shown on the screen. Hence, it is necessary to use a plot command to see the response curves. The matrices y and x contain the output and state response of the system, respectively, evaluated at the computation time points t. (Matrix y has as many columns as outputs and one row for each element in t. Matrix x has as many columns as states and one row for each element in t.)

Note also that the scalar iu is an index into the inputs of the system and specifies which input is to be used for the response; t is the user-specified time. If the system involves multiple inputs and multiple outputs, the step commands produces a series of step response plots, one for each input and output combination of

$$\dot{x} = Ax + Bu$$
$$y = Cx + Du$$

(For details, see **Example 5–2.**)

Transfer matrix. Next, consider a multiple-input–multiple-output system. Assume that there are r inputs $u_1, u_2, \ldots, u_r$, and m outputs $y_1, y_2, \ldots, y_m$. Define

$$\mathbf{u} = \begin{bmatrix} u_1 \\ u_2 \\ \vdots \\ u_r \end{bmatrix}, \qquad \mathbf{y} = \begin{bmatrix} y_1 \\ y_2 \\ \vdots \\ y_m \end{bmatrix}$$

The transfer matrix $\mathbf{G}(s)$ relates the output $\mathbf{Y}(s)$ to the input $\mathbf{U}(s)$, or

$$\mathbf{Y}(s) = \mathbf{G}(s)\mathbf{U}(s)$$

where

$$\mathbf{G}(s) = \mathbf{C}(s\mathbf{I} - \mathbf{A})^{-1}\mathbf{B} + \mathbf{D} \qquad (5\text{-}9)$$

[The derivation of Equation (5–9) is given in **Example 5–2**, to follow.] Since the input vector $\mathbf{u}$ is r dimensional and the output vector $\mathbf{y}$ is m dimensional, the transfer matrix $\mathbf{G}$(s) is an $m \times r$ matrix.

Example 5–2

Consider the following system:

$$\begin{bmatrix} \dot{x}_1 \\ \dot{x}_2 \end{bmatrix} = \begin{bmatrix} -1 & -1 \\ 6.5 & 0 \end{bmatrix} \begin{bmatrix} x_1 \\ x_2 \end{bmatrix} + \begin{bmatrix} 1 & 1 \\ 1 & 0 \end{bmatrix} \begin{bmatrix} u_1 \\ u_2 \end{bmatrix}$$

$$\begin{bmatrix} y_1 \\ y_2 \end{bmatrix} = \begin{bmatrix} 1 & 0 \\ 0 & 1 \end{bmatrix} \begin{bmatrix} x_1 \\ x_2 \end{bmatrix} + \begin{bmatrix} 0 & 0 \\ 0 & 0 \end{bmatrix} \begin{bmatrix} u_1 \\ u_2 \end{bmatrix}$$

Obtain the unit-step response curves.

Although it is not necessary to obtain the transfer-function expression for the system in order to obtain the unit-step response curves with MATLAB, we shall derive such an expression for reference purposes. For the system defined by

$$\dot{\mathbf{x}} = \mathbf{A}\mathbf{x} + \mathbf{B}\mathbf{u}$$
$$\mathbf{y} = \mathbf{C}\mathbf{x} + \mathbf{D}\mathbf{u}$$

the transfer matrix $\mathbf{G}(s)$ is a matrix that relates $\mathbf{Y}(s)$ and $\mathbf{U}(s)$ through the formula

$$\mathbf{Y}(s) = \mathbf{G}(s)\mathbf{U}(s) \qquad (5\text{-}10)$$

Taking Laplace transforms of the state-space equations, we obtain

$$s\mathbf{X}(s) - \mathbf{x}(0) = \mathbf{A}\mathbf{X}(s) + \mathbf{B}\mathbf{U}(s) \qquad (5\text{-}11)$$
$$\mathbf{Y}(s) = \mathbf{C}\mathbf{X}(s) + \mathbf{D}\mathbf{U}(s) \qquad (5\text{-}12)$$

In deriving the transfer matrix, we assume that $\mathbf{x}(0) = \mathbf{0}$. Then, from Equation (5–11), we get

$$\mathbf{X}(s) = (s\mathbf{I} - \mathbf{A})^{-1}\mathbf{B}\mathbf{U}(s)$$

Substituting this equation into Equation (5–12) yields

$$\mathbf{Y}(s) = [\mathbf{C}(s\mathbf{I} - \mathbf{A})^{-1}\mathbf{B} + \mathbf{D}]\mathbf{U}(s)$$

Upon comparing this last equation with Equation (5–10), we see that

$$\mathbf{G}(s) = \mathbf{C}(s\mathbf{I} - \mathbf{A})^{-1}\mathbf{B} + \mathbf{D}$$

The transfer matrix $\mathbf{G}(s)$ for the given system becomes

$$\mathbf{G}(s) = \mathbf{C}(s\mathbf{I} - \mathbf{A})^{-1}\mathbf{B}$$

$$= \begin{bmatrix} 1 & 0 \\ 0 & 1 \end{bmatrix} \begin{bmatrix} s+1 & 1 \\ -6.5 & s \end{bmatrix}^{-1} \begin{bmatrix} 1 & 1 \\ 1 & 0 \end{bmatrix}$$

$$= \frac{1}{s^2 + s + 6.5} \begin{bmatrix} s & -1 \\ 6.5 & s+1 \end{bmatrix} \begin{bmatrix} 1 & 1 \\ 1 & 0 \end{bmatrix}$$

$$= \frac{1}{s^2 + s + 6.5} \begin{bmatrix} s-1 & s \\ s+7.5 & 6.5 \end{bmatrix}$$

Hence,

$$\begin{bmatrix} Y_1(s) \\ Y_2(s) \end{bmatrix} = \begin{bmatrix} \dfrac{s-1}{s^2+s+6.5} & \dfrac{s}{s^2+s+6.5} \\[3mm] \dfrac{s+7.5}{s^2+s+6.5} & \dfrac{6.5}{s^2+s+6.5} \end{bmatrix} \begin{bmatrix} U_1(s) \\ U_2(s) \end{bmatrix}$$

Since this system involves two inputs and two outputs, four transfer functions can be defined, depending on which signals are considered as input and output. Note that, when considering the signal u_1 as the input, we assume that signal u_2 is zero, and vice versa. The four transfer functions are

$$\frac{Y_1(s)}{U_1(s)} = \frac{s-1}{s^2+s+6.5}, \qquad \frac{Y_1(s)}{U_2(s)} = \frac{s}{s^2+s+6.5}$$

$$\frac{Y_2(s)}{U_1(s)} = \frac{s+7.5}{s^2+s+6.5}, \qquad \frac{Y_2(s)}{U_2(s)} = \frac{6.5}{s^2+s+6.5}$$

The four individual step-response curves can be plotted with the use of the command

step(A,B,C,D)

or

sys = ss(A,B,C,D);   step(sys)

MATLAB Program 5–1 produces four individual unit-step response curves, shown in Figure 5–4.

MATLAB Program 5–1

```
>> A = [−1   −1;6.5   0];
>> B = [1   1;1   0];
>> C = [1   0;0   1];
>> D = [0   0;0   0];
>> sys = ss(A,B,C,D);
>> step(sys)
>> grid
>> title('Unit-Step Responses')
>> xlabel('t')
>> ylabel('Outputs')
```

To plot two step-response curves for the input u_1 in one diagram and two step-response curves for the input u_2 in another diagram, we may use the commands

step(A,B,C,D,1)

and

step(A,B,C,D,2)

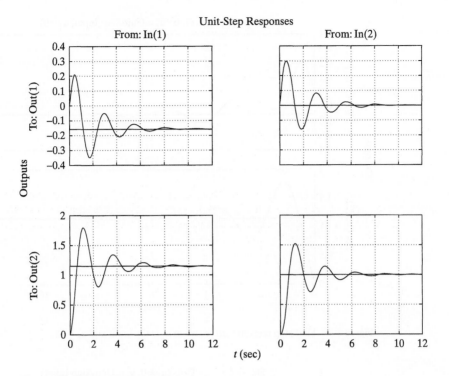

Figure 5–4   Unit-step response curves.

MATLAB Program 5–2

```
>> % ----- In this program, we first plot step-response curves
>> % when the input is u1. Then we plot response curves when
>> % the input is u2. -----
>>
>> A = [−1   −1;6.5   0];
>> B = [1   1;1   0];
>> C = [1   0;0   1];
>> D = [0   0;0   0];
>>
>> step(A,B,C,D,1)
>> grid
>> title('Step-Response Plots (u_1 = Unit-Step Input, u_2 = 0)')
>> xlabel('t'); ylabel('Outputs')
>>
>> step(A,B,C,D,2)
>> grid
>> title('Step-Response Plots (u_1 = 0, u_2 = Unit-Step Input)')
>> xlabel('t'); ylabel('Outputs')
```

---

respectively. MATLAB Program 5–2 does just that. Figures 5–5 and 5–6 show the two diagrams produced, each consisting of two unit-step response curves.

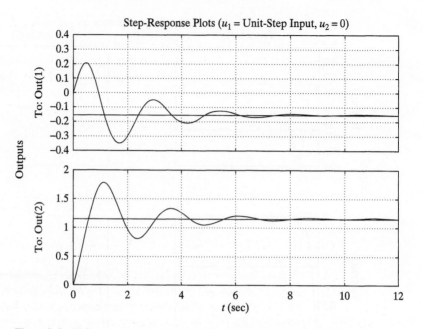

Figure 5–5 Unit-step response curves when u_1 is the input and $u_2 = 0$.

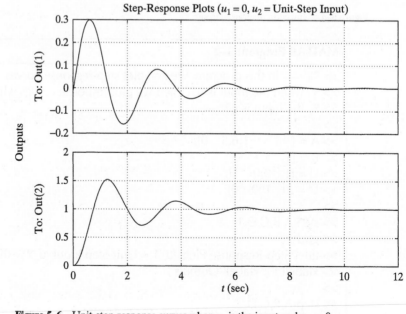

Figure 5–6 Unit-step response curves when u_2 is the input and $u_1 = 0$.

Impulse response.   The unit-impulse response of a dynamic system defined in a state space may be obtained with the use of one of the following MATLAB commands:

sys = ss(A,B,C,D);   impulse(sys),   y = impulse(sys, t),

[y,t,x] = impulse(sys),   [y,t,x] = impulse(sys,t),

impulse(A,B,C,D),   [y,x,t] = impulse(A,B,C,D),

[y,x,t] = impulse(A,B,C,D,iu),   [y,x,t] = impulse(A,B,C,D,iu,t)

The command impulse(sys) or impulse(A,B,C,D) produces a series of unit-impulse response plots, one for each input–output combination of the system

$$\dot{x} = Ax + Bu$$
$$y = Cx + Du$$

with the time vector automatically determined. If the right-hand side of a command includes the scalar iu (an index into the inputs of the system), then that scalar specifies which input to use for the impulse response.

Note that if a command includes t, it is the user-supplied time vector, which specifies the times at which the impulse response is to be computed.

If MATLAB is invoked with the left-hand argument [y,t,x], as in the case of [y,t,x] = impulse(sys,t), the command returns the output and state responses of the system and the time vector t. No plot is drawn on the screen. The matrices y and x contain the output and state responses of the system, evaluated at the time points t. (Matrix y has as many columns as outputs and one row for each element in t. Matrix x has as many columns as state variables and one row for each element in t.)

Response to arbitrary input.   The command lsim produces the response of linear time-invariant systems to arbitrary inputs. If the initial conditions of the system in state-space form are zero, then

lsim(sys,u,t)

produces the response of the system to an arbitrary input u with user-specified time t.

If the initial conditions are nonzero in a state-space model, the command

lsim(sys,u,t,x₀)

where x_0 is the initial state, produces the response of the system, subject to the input u and the initial condition x_0.

The command

[y,t] = lsim(sys,u,t,x₀)

returns the output response y. No plot is drawn. To plot the response curves, it is necessary to use the command plot(t,y).

Response to initial condition.   To find the response to the initial condition x_0 given to a system in a state-space form, the following command may be used:

[y,t] = lsim(sys,u,t,x₀)

Here, u is a vector consisting of zeros having length size(t). Alternatively, if we choose **B = 0** and **D = 0**, then u can be *any* input having length size(t).

Another way to obtain the response to the initial condition given to a system in a state-space form is to use the command

$$\text{initial}(A,B,C,D,x_0,t)$$

Example 5–3 is illustrative.

Example 5–3

Consider the system shown in Figure 5–7. The system is at rest for $t < 0$. At $t = 0$, the mass is pulled downward by 0.1 m and is released with an initial velocity of 0.05 m/s. That is, $x(0) = 0.1$ m and $\dot{x}(0) = 0.05$ m/s. The displacement x is measured from the equilibrium position. There is no external input to this system.

Assuming that $m = 1$ kg, $b = 3$ N-s/m, and $k = 2$ N/m, obtain the response curves $x(t)$ versus t and $\dot{x}(t)$ versus t with MATLAB. Use the command initial.

The system equation is

$$m\ddot{x} + b\dot{x} + kx = 0$$

Substituting the given numerical values for $m, b,$ and k yields

$$\ddot{x} + 3\dot{x} + 2x = 0$$

If we define the state variables as

$$x_1 = x$$
$$x_2 = \dot{x}$$

and the output variables as

$$y_1 = x_1$$
$$y_2 = x_2$$

then the state equation becomes

$$\begin{bmatrix} \dot{x}_1 \\ \dot{x}_2 \end{bmatrix} = \begin{bmatrix} 0 & 1 \\ -2 & -3 \end{bmatrix} \begin{bmatrix} x_1 \\ x_2 \end{bmatrix} + \begin{bmatrix} 0 \\ 0 \end{bmatrix} u$$

The output equation is

$$\begin{bmatrix} y_1 \\ y_2 \end{bmatrix} = \begin{bmatrix} 1 & 0 \\ 0 & 1 \end{bmatrix} \begin{bmatrix} x_1 \\ x_2 \end{bmatrix} + \begin{bmatrix} 0 \\ 0 \end{bmatrix} u$$

Thus,

$$A = \begin{bmatrix} 0 & 1 \\ -2 & -3 \end{bmatrix}, \quad B = \begin{bmatrix} 0 \\ 0 \end{bmatrix}, \quad C = \begin{bmatrix} 1 & 0 \\ 0 & 1 \end{bmatrix}, \quad D = \begin{bmatrix} 0 \\ 0 \end{bmatrix}, \quad x_0 = \begin{bmatrix} 0.1 \\ 0.05 \end{bmatrix}$$

Using the command

$$\text{initial}(A,B,C,D,x_0,t)$$

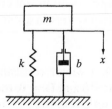

Figure 5–7   Mechanical system.

we can obtain the responses $x(t) = y_1(t)$ versus t and $\dot{x}(t) = y_2(t)$ versus t. MATLAB Program 5–3 produces the response curves, which are shown in Figure 5–8.

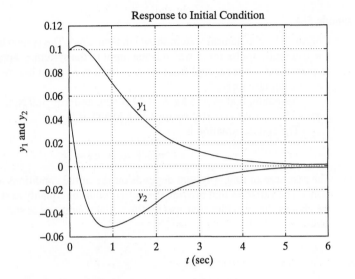

Figure 5–8   Response curves to initial condition.

MATLAB Program 5–3

```
>> t = 0:0.01:6;
>> A = [0   1;−2   −3];
>> B = [0;0];
>> C = [1   0;0   1];
>> D = [0;0];
>> [y, x, t] = initial(A,B,C,D,[0.1; 0.05],t);
>> y1 = [1      0] *y';
>> y2 = [0      1] *y';
>> plot(t,y1,t,y2)
>> grid
>> title('Response to Initial Condition')
>> xlabel('t (sec)')
>> ylabel('y_1 and y_2')
>> text(1.6, 0.05,'y_1')
>> text(1.6, −0.026,'y_2')
```

5–3 STATE-SPACE MODELING OF SYSTEMS WITH NO INPUT DERIVATIVES

In this section, we present two examples of the modeling of dynamic systems in state-space form. The systems used are limited to the case where derivatives of the input functions do not appear explicitly in the equations of motion. In each example, we first derive state-space models and then find the response curves with MATLAB, given the numerical values of all of the variables and the details of the input functions.

Example 5–4

Consider the mechanical system shown in Figure 5–9. The system is at rest for $t < 0$. At $t = 0$, a unit-impulse force, which is the input to the system, is applied to the mass. The displacement x is measured from the equilibrium position before the mass m is hit by the unit-impulse force.

Assuming that $m = 5$ kg, $b = 20$ N-s/m, and $k = 100$ N/m, obtain the response curves $x(t)$ versus t and $\dot{x}(t)$ versus t with MATLAB.

The system equation is

$$m\ddot{x} + b\dot{x} + kx = u$$

The response of such a system depends on the initial conditions and the forcing function u. The variables that provide the initial conditions qualify as state variables. Hence, we choose the variables that specify the initial conditions as state variables x_1 and x_2. Thus,

$$x_1 = x$$
$$x_2 = \dot{x}$$

The state equation then becomes

$$\dot{x}_1 = x_2$$
$$\dot{x}_2 = \frac{1}{m}(u - kx - b\dot{x}) = -\frac{k}{m}x_1 - \frac{b}{m}x_2 + \frac{1}{m}u$$

For the output variables, we choose

$$y_1 = x$$
$$y_2 = \dot{x}$$

Rewriting the state equation and output equation, we obtain

$$\begin{bmatrix} \dot{x}_1 \\ \dot{x}_2 \end{bmatrix} = \begin{bmatrix} 0 & 1 \\ -\dfrac{k}{m} & -\dfrac{b}{m} \end{bmatrix} \begin{bmatrix} x_1 \\ x_2 \end{bmatrix} + \begin{bmatrix} 0 \\ \dfrac{1}{m} \end{bmatrix} u$$

and

$$\begin{bmatrix} y_1 \\ y_2 \end{bmatrix} = \begin{bmatrix} 1 & 0 \\ 0 & 1 \end{bmatrix} \begin{bmatrix} x_1 \\ x_2 \end{bmatrix} + \begin{bmatrix} 0 \\ 0 \end{bmatrix} u$$

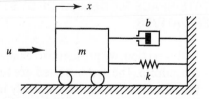

Figure 5–9 Mechanical system.

Substituting the given numerical values for m, b, and k into the state space equations yields

$$A = \begin{bmatrix} 0 & 1 \\ -20 & -4 \end{bmatrix}, \qquad B = \begin{bmatrix} 0 \\ 0.2 \end{bmatrix}, \qquad C = \begin{bmatrix} 1 & 0 \\ 0 & 1 \end{bmatrix}, \qquad D = \begin{bmatrix} 0 \\ 0 \end{bmatrix}$$

MATLAB Program 5–4 produces the impulse-response curves $x(t)$ versus t and $\dot{x}(t)$ versus t, shown in Figure 5–10.

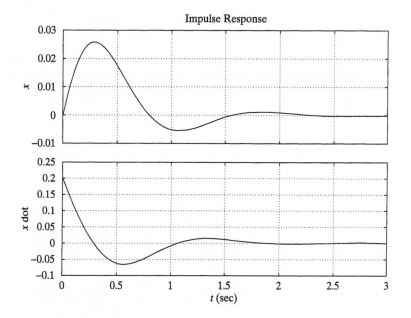

Figure 5–10   Impulse-response curves.

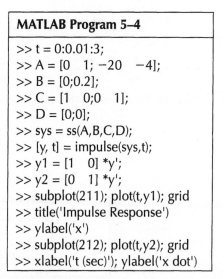

MATLAB Program 5–4

```
>> t = 0:0.01:3;
>> A = [0   1; -20  -4];
>> B = [0;0.2];
>> C = [1   0;0   1];
>> D = [0;0];
>> sys = ss(A,B,C,D);
>> [y, t] = impulse(sys,t);
>> y1 = [1   0] *y';
>> y2 = [0   1] *y';
>> subplot(211); plot(t,y1); grid
>> title('Impulse Response')
>> ylabel('x')
>> subplot(212); plot(t,y2); grid
>> xlabel('t (sec)'); ylabel('x dot')
```

Example 5–5

Consider the mechanical system shown in Figure 5–11. The system is at rest for $t < 0$. At $t = 0$, a step force f of α newtons is applied to mass m_2. [The force $f = \alpha u$, where u is a step force of 1 newton.] The displacements z_1 and z_2 are measured from the respective equilibrium positions of the carts before f is applied. Derive a state-space representation of the system. Assuming that $m_1 = 10$ kg, $m_2 = 20$ kg, $b = 20$ N-s/m, $k_1 = 30$ N/m, $k_2 = 60$ N/m, and $\alpha = 10$, obtain the response curves $z_1(t)$ versus t, $z_2(t)$ versus t, and $z_1(t) - z_2(t)$ versus t with MATLAB. Also, obtain $z_1(\infty)$ and $z_2(\infty)$.

The equations of motion for this system are

$$m_1\ddot{z}_1 = -k_1 z_1 - k_2(z_1 - z_2) - b(\dot{z}_1 - \dot{z}_2) \tag{5-13}$$
$$m_2\ddot{z}_2 = -k_2(z_2 - z_1) - b(\dot{z}_2 - \dot{z}_1) + \alpha u \tag{5-14}$$

In the absence of a forcing function, the initial conditions of any system determine the response of the system. The initial conditions for this system are $z_1(0)$, $\dot{z}_1(0)$, $z_2(0)$, and $\dot{z}_2(0)$. Hence, we choose z_1, $\dot{z}_1$, z_2, and $\dot{z}_2$ as state variables for the system and thus define

$$x_1 = z_1$$
$$x_2 = \dot{z}_1$$
$$x_3 = z_2$$
$$x_4 = \dot{z}_2$$

Then Equation (5–13) can be rewritten as

$$\dot{x}_2 = -\frac{k_1 + k_2}{m_1} x_1 - \frac{b}{m_1} x_2 + \frac{k_2}{m_1} x_3 + \frac{b}{m_1} x_4$$

and Equation (5–14) can be rewritten as

$$\dot{x}_4 = \frac{k_2}{m_2} x_1 + \frac{b}{m_2} x_2 - \frac{k_2}{m_2} x_3 - \frac{b}{m_2} x_4 + \frac{1}{m_2}\alpha u$$

The state equation now becomes

$$\dot{x}_1 = x_2$$
$$\dot{x}_2 = -\frac{k_1 + k_2}{m_1} x_1 - \frac{b}{m_1} x_2 + \frac{k_2}{m_1} x_3 + \frac{b}{m_1} x_4$$
$$\dot{x}_3 = x_4$$
$$\dot{x}_4 = \frac{k_2}{m_2} x_1 + \frac{b}{m_2} x_2 - \frac{k_2}{m_2} x_3 - \frac{b}{m_2} x_4 + \frac{1}{m_2}\alpha u$$

Note that z_1 and z_2 are the outputs of the system; hence, the output equations are

$$y_1 = z_1$$
$$y_2 = z_2$$

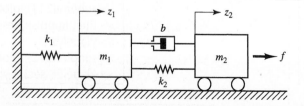

Figure 5–11   Mechanical system.

In terms of vector–matrix equations, we have

$$
\begin{bmatrix} \dot{x}_1 \\ \dot{x}_2 \\ \dot{x}_3 \\ \dot{x}_4 \end{bmatrix} = \begin{bmatrix} 0 & 1 & 0 & 0 \\ -\dfrac{k_1+k_2}{m_1} & -\dfrac{b}{m_1} & \dfrac{k_2}{m_1} & \dfrac{b}{m_1} \\ 0 & 0 & 0 & 1 \\ \dfrac{k_2}{m_2} & \dfrac{b}{m_2} & -\dfrac{k_2}{m_2} & -\dfrac{b}{m_2} \end{bmatrix} \begin{bmatrix} x_1 \\ x_2 \\ x_3 \\ x_4 \end{bmatrix} + \begin{bmatrix} 0 \\ 0 \\ 0 \\ \dfrac{\alpha}{m_2} \end{bmatrix} u \tag{5-15}
$$

$$
\begin{bmatrix} y_1 \\ y_2 \end{bmatrix} = \begin{bmatrix} 1 & 0 & 0 & 0 \\ 0 & 0 & 1 & 0 \end{bmatrix} \begin{bmatrix} x_1 \\ x_2 \\ x_3 \\ x_4 \end{bmatrix} + \begin{bmatrix} 0 \\ 0 \end{bmatrix} u \tag{5-16}
$$

Equations (5–15) and (5–16) represent the system in state-space form.

Next, we substitute the given numerical values for m_1, m_2, b, k_1, and k_2 into Equation (5–15). The result is

$$
\begin{bmatrix} \dot{x}_1 \\ \dot{x}_2 \\ \dot{x}_3 \\ \dot{x}_4 \end{bmatrix} = \begin{bmatrix} 0 & 1 & 0 & 0 \\ -9 & -2 & 6 & 2 \\ 0 & 0 & 0 & 1 \\ 3 & 1 & -3 & -1 \end{bmatrix} \begin{bmatrix} x_1 \\ x_2 \\ x_3 \\ x_4 \end{bmatrix} + \begin{bmatrix} 0 \\ 0 \\ 0 \\ 0.5 \end{bmatrix} u \tag{5-17}
$$

From Equations (5–17) and (5–16), we have

$$
A = \begin{bmatrix} 0 & 1 & 0 & 0 \\ -9 & -2 & 6 & 2 \\ 0 & 0 & 0 & 1 \\ 3 & 1 & -3 & -1 \end{bmatrix}, \quad B = \begin{bmatrix} 0 \\ 0 \\ 0 \\ 0.5 \end{bmatrix}, \quad C = \begin{bmatrix} 1 & 0 & 0 & 0 \\ 0 & 0 & 1 & 0 \end{bmatrix}, \quad D = \begin{bmatrix} 0 \\ 0 \end{bmatrix}
$$

MATLAB Program 5–5 produces the response curves $z_1(t)$ versus t, $z_2(t)$ versus t, and $z_1(t) - z_2(t)$ versus t. The curves are shown in Figure 5–12.

MATLAB Program 5–5

```
>> t = 0:0.1:200;
>> A = [0  1  0  0;-9  -2  6  2;0  0  0  1;3  1  -3  -1];
>> B = [0;0;0;0.5];
>> C = [1  0  0  0;0  0  1  0];
>> D = [0;0];
>> sys = ss(A,B,C,D);
>> [y,t] = step(sys,t);
>> y1 = [1 0]*y';
>> y2 = [0 1]*y';
>> z1 = y1; subplot(311); plot(t,z1); grid
>> title('Responses z_1 Versus t, z_2 Versus t, and z_1 - z_2 Versus t')
>> ylabel('Output z_1')
>> z2 = y2; subplot(312); plot(t,z2); grid
>> ylabel('Output z_2')
>> subplot(313); plot(t,z1-z2); grid
>> xlabel('t (sec)'); ylabel('z_1 - z_2')
```

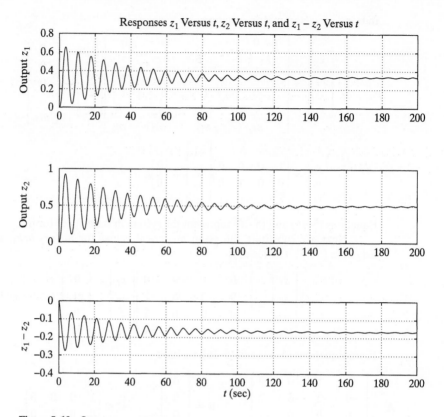

Figure 5–12   Step-response curves.

To obtain $z_1(\infty)$ and $z_2(\infty)$, we set all derivatives of z_1 and z_2 in Equations (5–13) and (5–14) equal to zero, because all derivative terms must approach zero at steady state in the system. Then, from Equation (5–14), we get

$$k_2[z_2(\infty) - z_1(\infty)] = \alpha u$$

from which it follows that

$$z_2(\infty) - z_1(\infty) = \frac{\alpha u}{k_2} = \frac{10}{60} = \frac{1}{6}$$

From Equation (5–13), we have

$$k_1 z_1(\infty) = k_2[z_2(\infty) - z_1(\infty)]$$

Hence,

$$z_1(\infty) = \frac{k_2}{k_1}[z_2(\infty) - z_1(\infty)] = \frac{60}{30}\frac{1}{6} = \frac{1}{3}$$

and

$$z_2(\infty) = \frac{1}{6} + z_1(\infty) = \frac{1}{2}$$

Thus,

$$z_1(\infty) = \frac{1}{3}\,\text{m}, \qquad z_2(\infty) = \frac{1}{2}\,\text{m}$$

The final values of $z_1(t)$ and $z_2(t)$ obtained with MATLAB (see the response curves in Figure 5–12) agree, of course, with the result obtained here.

5–4 STATE-SPACE MODELING OF SYSTEMS WITH INPUT DERIVATIVES

In this section, we take up the case where the equations of motion of a system involve one or more derivatives of the input function. In such a case, the variables that specify the initial conditions do not qualify as state variables. The main problem in defining the state variables is that they must be chosen such that they will eliminate the derivatives of the input function u in the state equation.

For example, consider the mechanical system shown in Figure 5–13. The displacements y and u are measured from their respective equilibrium positions. The equation of motion for this system is

$$m\ddot{y} = -ky - b(\dot{y} - \dot{u})$$

or

$$\ddot{y} = -\frac{k}{m}y - \frac{b}{m}\dot{y} + \frac{b}{m}\dot{u}$$

If we choose the state variables

$$x_1 = y$$
$$x_2 = \dot{y}$$

then we get

$$\dot{x}_1 = x_2$$
$$\dot{x}_2 = -\frac{k}{m}x_1 - \frac{b}{m}x_2 + \frac{b}{m}\dot{u} \tag{5–18}$$

The right-hand side of Equation (5–18) involves the derivative term $\dot{u}$. Note that, in formulating state-space representations of dynamic systems, we constrain the input function to be any function of time of order up to the impulse function, but not any higher order impulse functions, such as $d\delta(t)/dt$, $d^2\delta(t)/dt^2$, etc.

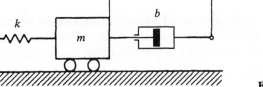

Figure 5–13   Mechanical system.

To explain why the right-hand side of the state equation should not involve the derivative of the input function u, suppose that u is the unit-impulse function $\delta(t)$. Then the integral of Equation (5–18) becomes

$$x_2 = -\frac{k}{m}\int y\,dt - \frac{b}{m}y + \frac{k}{m}\delta(t)$$

Notice that x_2 includes the term $(k/m)\,\delta(t)$. This means that $x_2(0) = \infty$, which is not acceptable as a state variable. We should choose the state variables such that the state equation will not include the derivative of u.

Suppose that we try to eliminate the term involving $\dot{u}$ from Equation (5–18). One possible way to accomplish this is to define

$$x_1 = y$$

$$x_2 = \dot{y} - \frac{b}{m}u$$

Then

$$\dot{x}_2 = \ddot{y} - \frac{b}{m}\dot{u}$$

$$= -\frac{k}{m}y - \frac{b}{m}\dot{y} + \frac{b}{m}\dot{u} - \frac{b}{m}\dot{u}$$

$$= -\frac{k}{m}x_1 - \frac{b}{m}\left(x_2 + \frac{b}{m}u\right)$$

$$= -\frac{k}{m}x_1 - \frac{b}{m}x_2 - \left(\frac{b}{m}\right)^2 u$$

Thus, we have eliminated the term that involves $\dot{u}$. The acceptable state equation can now be given by

$$\begin{bmatrix} \dot{x}_1 \\ \dot{x}_2 \end{bmatrix} = \begin{bmatrix} 0 & 1 \\ -\dfrac{k}{m} & -\dfrac{b}{m} \end{bmatrix}\begin{bmatrix} x_1 \\ x_2 \end{bmatrix} + \begin{bmatrix} \dfrac{b}{m} \\ -\left(\dfrac{b}{m}\right)^2 \end{bmatrix} u$$

If equations of motion involve $u, \dot{u}, \ddot{u}$, etc., the choice of state variables becomes more complicated. Fortunately, there are systematic methods for choosing state variables for a general case of equations of motion that involve derivatives of the input function u. In what follows we shall present two systematic methods for eliminating derivatives of the input function from the state equations. Note that MATLAB can also be used to obtain state-space representations of systems involving derivatives of the input function u. (See Section 5–5.)

State-space representation of dynamic systems in which derivatives of the input function appear in the system differential equations. We consider the case where the input function u is a scalar. (That is, only one input function u is involved in the system.)

The differential equation of a system that involves derivatives of the input function has the general form

$$\overset{(n)}{y} + a_1 \overset{(n-1)}{y} + \cdots + a_{n-1} \dot{y} + a_n y = b_0 \overset{(n)}{u} + b_1 \overset{(n-1)}{u} + \cdots + b_{n-1} \dot{u} + b_n u \qquad (5\text{-}19)$$

To apply the methods presented in this section, it is necessary that the system be written as a differential equation in the form of Equation (5-19) or its equivalent transfer function

$$\frac{Y(s)}{U(s)} = \frac{b_0 s^n + b_1 s^{n-1} + \cdots + b_{n-1} s + b_n}{s^n + a_1 s^{n-1} + \cdots + a_{n-1} s + a_n}$$

We examine two methods when $n = 2$; for an arbitrary $n = 1, 2, 3, \ldots,$ see **Problems A-5-12** and **A-5-13**.

Method 1.   Consider the second-order system

$$\ddot{y} + a_1 \dot{y} + a_2 y = b_0 \ddot{u} + b_1 \dot{u} + b_2 u \qquad (5\text{-}20)$$

As a set of state variables, suppose that we choose

$$x_1 = y - \beta_0 u \qquad (5\text{-}21)$$
$$x_2 = \dot{x}_1 - \beta_1 u \qquad (5\text{-}22)$$

where

$$\beta_0 = b_0 \qquad (5\text{-}23)$$
$$\beta_1 = b_1 - a_1 \beta_0 \qquad (5\text{-}24)$$

Then, from Equation (5-21), we have

$$y = x_1 + \beta_0 u \qquad (5\text{-}25)$$

Substituting this last equation into Equation (5-20), we obtain

$$\ddot{x}_1 + \beta_0 \ddot{u} + a_1 (\dot{x}_1 + \beta_0 \dot{u}) + a_2 (x_1 + \beta_0 u) = b_0 \ddot{u} + b_1 \dot{u} + b_2 u$$

Noting that $\beta_0 = b_0$ and $\beta_1 = b_1 - a_1 \beta_0$, we can simplify the preceding equation to

$$\ddot{x}_1 + a_1 \dot{x}_1 + a_2 x_1 = \beta_1 \dot{u} + (b_2 - a_2 \beta_0) u \qquad (5\text{-}26)$$

From Equation (5-22), we have

$$\dot{x}_1 = x_2 + \beta_1 u \qquad (5\text{-}27)$$

Substituting Equation (5-27) into Equation (5-26), we obtain

$$\dot{x}_2 + \beta_1 \dot{u} + a_1 (x_2 + \beta_1 u) + a_2 x_1 = \beta_1 \dot{u} + (b_2 - a_2 \beta_0) u$$

which can be simplified to

$$\dot{x}_2 = -a_2 x_1 - a_1 x_2 + \beta_2 u \qquad (5\text{-}28)$$

where

$$\beta_2 = b_2 - a_1 \beta_1 - a_2 \beta_0 \qquad (5\text{-}29)$$

From Equations (5-27) and (5-28), we obtain the state equation:

$$\begin{bmatrix} \dot{x}_1 \\ \dot{x}_2 \end{bmatrix} = \begin{bmatrix} 0 & 1 \\ -a_2 & -a_1 \end{bmatrix} \begin{bmatrix} x_1 \\ x_2 \end{bmatrix} + \begin{bmatrix} \beta_1 \\ \beta_2 \end{bmatrix} u \qquad (5\text{-}30)$$

From Equation (5–25), we get the output equation:

$$y = \begin{bmatrix} 1 & 0 \end{bmatrix} \begin{bmatrix} x_1 \\ x_2 \end{bmatrix} + \beta_0 u \tag{5–31}$$

Equations (5–30) and (5–31) represent the system in a state space.

Note that if $\beta_0 = b_0 = 0$, then the state variable x_1 is the output signal y, which can be measured, and, in this case, the state variable x_2 is the output velocity $\dot{y}$ minus $b_1 u$.

Note that, for the case of the nth-order differential-equation system

$$\overset{(n)}{y} + a_1 \overset{(n-1)}{y} + \cdots + a_{n-1}\dot{y} + a_n y = b_0 \overset{(n)}{u} + b_1 \overset{(n-1)}{u} + \cdots + b_{n-1}\dot{u} + b_n u$$

the state equation and output equation can be given by

$$\begin{bmatrix} \dot{x}_1 \\ \dot{x}_2 \\ \vdots \\ \dot{x}_{n-1} \\ \dot{x}_n \end{bmatrix} = \begin{bmatrix} 0 & 1 & 0 & \cdots & 0 \\ 0 & 0 & 1 & \cdots & 0 \\ \vdots & \vdots & \vdots & & \vdots \\ 0 & 0 & 0 & \cdots & 1 \\ -a_n & -a_{n-1} & -a_{n-2} & \cdots & -a_1 \end{bmatrix} \begin{bmatrix} x_1 \\ x_2 \\ \vdots \\ x_{n-1} \\ x_n \end{bmatrix} + \begin{bmatrix} \beta_1 \\ \beta_2 \\ \vdots \\ \beta_{n-1} \\ \beta_n \end{bmatrix} u$$

and

$$y = \begin{bmatrix} 1 & 0 & \cdots & 0 \end{bmatrix} \begin{bmatrix} x_1 \\ x_2 \\ \vdots \\ x_n \end{bmatrix} + \beta_0 u$$

where $\beta_0, \beta_1, \beta_2, \ldots, \beta_n$ are determined from

$$\beta_0 = b_0$$
$$\beta_1 = b_1 - a_1\beta_0$$
$$\beta_2 = b_2 - a_1\beta_1 - a_2\beta_0$$
$$\beta_3 = b_3 - a_1\beta_2 - a_2\beta_1 - a_3\beta_0$$
$$\vdots$$
$$\beta_n = b_n - a_1\beta_{n-1} - \cdots - a_{n-1}\beta_1 - a_n\beta_0$$

Method 2. Consider the second-order system

$$\ddot{y} + a_1\dot{y} + a_2 y = b_0\ddot{u} + b_1\dot{u} + b_2 u$$

or its equivalent transfer function

$$\frac{Y(s)}{U(s)} = \frac{b_0 s^2 + b_1 s + b_2}{s^2 + a_1 s + a_2} \tag{5–32}$$

Equation (5–32) can be split into two equations as follows:

$$\frac{Z(s)}{U(s)} = \frac{1}{s^2 + a_1 s + a_2}, \qquad \frac{Y(s)}{Z(s)} = b_0 s^2 + b_1 s + b_2$$

We then have

$$\ddot{z} + a_1\dot{z} + a_2 z = u \tag{5-33}$$
$$b_0\ddot{z} + b_1\dot{z} + b_2 z = y \tag{5-34}$$

If we define

$$x_1 = z$$
$$x_2 = \dot{z} \tag{5-35}$$

then Equation (5-33) can be written as

$$\dot{x}_2 = -a_2 x_1 - a_1 x_2 + u \tag{5-36}$$

and Equation (5-34) can be written as

$$b_0\dot{x}_2 + b_1 x_2 + b_2 x_1 = y$$

Substituting Equation (5-36) into this last equation, we obtain

$$b_0(-a_2 x_1 - a_1 x_2 + u) + b_1 x_2 + b_2 x_1 = y$$

which can be rewritten as

$$y = (b_2 - a_2 b_0)x_1 + (b_1 - a_1 b_0)x_2 + b_0 u \tag{5-37}$$

From Equations (5-35) and (5-36), we get

$$\dot{x}_1 = x_2$$
$$\dot{x}_2 = -a_2 x_1 - a_1 x_2 + u$$

These two equations can be combined into the vector–matrix differential equation

$$\begin{bmatrix} \dot{x}_1 \\ \dot{x}_2 \end{bmatrix} = \begin{bmatrix} 0 & 1 \\ -a_2 & -a_1 \end{bmatrix} \begin{bmatrix} x_1 \\ x_2 \end{bmatrix} + \begin{bmatrix} 0 \\ 1 \end{bmatrix} u \tag{5-38}$$

Equation (5-37) can be rewritten as

$$y = [b_2 - a_2 b_0 \;\vdots\; b_1 - a_1 b_0] \begin{bmatrix} x_1 \\ x_2 \end{bmatrix} + b_0 u \tag{5-39}$$

Equations (5-38) and (5-39) are the state equation and output equation, respectively. Note that the state variables x_1 and x_2 in this case may not correspond to any physical signals that can be measured.

If the system equation is given by

$$\overset{(n)}{y} + a_1 \overset{(n-1)}{y} + \cdots + a_{n-1}\dot{y} + a_n y = b_0 \overset{(n)}{u} + b_1 \overset{(n-1)}{u} + \cdots + b_{n-1}\dot{u} + b_n u$$

or its equivalent transfer function

$$\frac{Y(s)}{U(s)} = \frac{b_0 s^n + b_1 s^{n-1} + \cdots + b_{n-1}s + b_n}{s^n + a_1 s^{n-1} + \cdots + a_{n-1}s + a_n}$$

then the state equation and the output equation obtained with the use of Method 2 are given by

$$
\begin{bmatrix} \dot{x}_1 \\ \dot{x}_2 \\ \cdot \\ \cdot \\ \cdot \\ \dot{x}_{n-1} \\ \dot{x}_n \end{bmatrix} = \begin{bmatrix} 0 & 1 & 0 & \cdots & 0 \\ 0 & 0 & 1 & \cdots & 0 \\ \cdot & \cdot & \cdot & & \cdot \\ \cdot & \cdot & \cdot & & \cdot \\ \cdot & \cdot & \cdot & & \cdot \\ 0 & 0 & 0 & \cdots & 1 \\ -a_n & -a_{n-1} & -a_{n-2} & \cdots & -a_1 \end{bmatrix} \begin{bmatrix} x_1 \\ x_2 \\ \cdot \\ \cdot \\ \cdot \\ x_{n-1} \\ x_n \end{bmatrix} + \begin{bmatrix} 0 \\ 0 \\ \cdot \\ \cdot \\ \cdot \\ 0 \\ 1 \end{bmatrix} u \quad (5\text{-}40)
$$

and

$$
y = [b_n - a_n b_0 \;\vdots\; b_{n-1} - a_{n-1} b_0 \;\vdots\; \cdots \;\vdots\; b_1 - a_1 b_0] \begin{bmatrix} x_1 \\ x_2 \\ \cdot \\ \cdot \\ \cdot \\ x_n \end{bmatrix} + b_0 u \quad (5\text{-}41)
$$

   Examples 5–6 and **5–7** illustrate the use of the preceding two analytical methods for obtaining state-space representations of a differential-equation system involving derivatives of the input signal.

Example 5–6

   Consider the spring–mass–dashpot system mounted on a cart as shown in Figure 5–14. Assume that the cart is standing still for $t < 0$. In this system, $u(t)$ is the displacement of the cart and is the input to the system. At $t = 0$, the cart is moved at a constant speed, or $\dot{u}$ = constant. The displacement y of the mass is the output. (y is measured

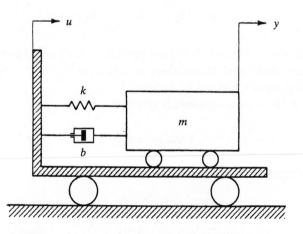

Figure 5–14   Spring–mass–dashpot system mounted on a cart.

from the rest position and is relative to the ground.) In this system, m denotes the mass of the small cart on the large cart (assume that the large cart is massless), b denotes the viscous-friction coefficient, and k is the spring constant. We assume that the entire system is a linear system.

Obtain a state-space representations of the system based on methods 1 and 2 just presented. Assuming that $m = 10$ kg, $b = 20$ N-s/m, $k = 100$ N/m, and the input is a ramp function such that $\dot{u} = 1$ m/s, obtain the response curve $y(t)$ versus t with MATLAB.

First, we shall obtain the system equation. Applying Newton's second law, we obtain

$$m\frac{d^2y}{dt^2} = -b\left(\frac{dy}{dt} - \frac{du}{dt}\right) - k(y - u)$$

or

$$m\frac{d^2y}{dt^2} + b\frac{dy}{dt} + ky = b\frac{du}{dt} + ku \tag{5–42}$$

Equation (5–42) is the differential equation (mathematical model) of the system. The transfer function is

$$\frac{Y(s)}{U(s)} = \frac{bs + k}{ms^2 + bs + k}$$

Method 1. We shall obtain a state-space model of this system based on Method 1. We first compare the differential equation of the system,

$$\ddot{y} + \frac{b}{m}\dot{y} + \frac{k}{m}y = \frac{b}{m}\dot{u} + \frac{k}{m}u$$

with the standard form

$$\ddot{y} + a_1\dot{y} + a_2y = b_0\ddot{u} + b_1\dot{u} + b_2u$$

and identify

$$a_1 = \frac{b}{m}, \qquad a_2 = \frac{k}{m}, \qquad b_0 = 0, \qquad b_1 = \frac{b}{m}, \qquad b_2 = \frac{k}{m}$$

From Equations (5–23), (5–24), and (5–29), we have

$$\beta_0 = b_0 = 0$$

$$\beta_1 = b_1 - a_1\beta_0 = \frac{b}{m}$$

$$\beta_2 = b_2 - a_1\beta_1 - a_2\beta_0 = \frac{k}{m} - \left(\frac{b}{m}\right)^2$$

From Equations (5–21) and (5–22), we define

$$x_1 = y - \beta_0u = y$$

$$x_2 = \dot{x}_1 - \beta_1u = \dot{x}_1 - \frac{b}{m}u \tag{5–43}$$

From Equations (5–43) and (5–28), we obtain

$$\dot{x}_1 = x_2 + \beta_1u = x_2 + \frac{b}{m}u \tag{5–44}$$

$$\dot{x}_2 = -a_2x_1 - a_1x_2 + \beta_2u = -\frac{k}{m}x_1 - \frac{b}{m}x_2 + \left[\frac{k}{m} - \left(\frac{b}{m}\right)^2\right]u \tag{5–45}$$

and the output equation is

$$y = x_1 \tag{5-46}$$

Combining Equations (5–44) and (5–45) yields the state equation, and from Equation (5–46), we get the output equation:

$$\begin{bmatrix} \dot{x}_1 \\ \dot{x}_2 \end{bmatrix} = \begin{bmatrix} 0 & 1 \\ -\dfrac{k}{m} & -\dfrac{b}{m} \end{bmatrix} \begin{bmatrix} x_1 \\ x_2 \end{bmatrix} + \begin{bmatrix} \dfrac{b}{m} \\ \dfrac{k}{m} - \left(\dfrac{b}{m}\right)^2 \end{bmatrix} u$$

$$y = \begin{bmatrix} 1 & 0 \end{bmatrix} \begin{bmatrix} x_1 \\ x_2 \end{bmatrix}$$

These two equations give a state-space representation of the system.

Next, we shall obtain the response curve $y(t)$ versus t for the unit-ramp input $\dot{u} = 1$ m/s. Substituting the given numerical values for m, b, and k into the state equation, we obtain

$$\begin{bmatrix} \dot{x}_1 \\ \dot{x}_2 \end{bmatrix} = \begin{bmatrix} 0 & 1 \\ -10 & -2 \end{bmatrix} \begin{bmatrix} x_1 \\ x_2 \end{bmatrix} + \begin{bmatrix} 2 \\ 6 \end{bmatrix} u$$

and the output equation is

$$y = \begin{bmatrix} 1 & 0 \end{bmatrix} \begin{bmatrix} x_1 \\ x_2 \end{bmatrix}$$

MATLAB Program 5–6 produces the response $y(t)$ of the system to the ramp input $\dot{u} = 1$ m/s. The response curve $y(t)$ versus t and the unit-ramp input are shown in Figure 5–15.

MATLAB Program 5–6

```
>> % ----- The response y(t) is obtained by use of the
>> % state-space equation obtained by Method 1. -----
>>
>> t = 0:0.01:4;
>> A = [0   1;−10   −2];
>> B = [2;6];
>> C = [1   0];
>> D = 0;
>> sys = ss(A,B,C,D);
>> u = t;
>> lsim(sys,u,t)
>> grid
>> title('Unit-Ramp Response (Method 1)')
>> xlabel('t')
>> ylabel('Output y and Unit-Ramp Input u')
>> text(0.85, 0.25,'y')
>> text(0.15,0.8,'u')
```

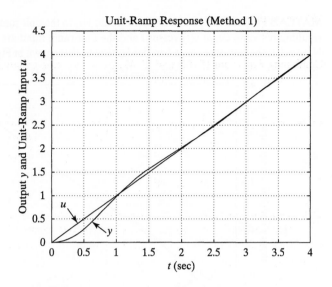

Figure 5–15   Unit-ramp response obtained with the use of Method 1.

Method 2. Since

$$b_0 = 0$$

$$b_2 - a_2 b_0 = \frac{k}{m} - \frac{k}{m} \times 0 = \frac{k}{m}$$

$$b_1 - a_1 b_0 = \frac{b}{m} - \frac{b}{m} \times 0 = \frac{b}{m}$$

from Equations (5–38) and (5–39), we obtain

$$\begin{bmatrix} \dot{x}_1 \\ \dot{x}_2 \end{bmatrix} = \begin{bmatrix} 0 & 1 \\ -\dfrac{k}{m} & -\dfrac{b}{m} \end{bmatrix} \begin{bmatrix} x_1 \\ x_2 \end{bmatrix} + \begin{bmatrix} 0 \\ 1 \end{bmatrix} u$$

$$y = \begin{bmatrix} \dfrac{k}{m} & \dfrac{b}{m} \end{bmatrix} \begin{bmatrix} x_1 \\ x_2 \end{bmatrix}$$

The last two equations give another state-space representation of the same system.

Substituting the given numerical values for m, b, and k into the state equation, we get

$$\begin{bmatrix} \dot{x}_1 \\ \dot{x}_2 \end{bmatrix} = \begin{bmatrix} 0 & 1 \\ -10 & -2 \end{bmatrix} \begin{bmatrix} x_1 \\ x_2 \end{bmatrix} + \begin{bmatrix} 0 \\ 1 \end{bmatrix} u$$

and the output equation is

$$y = \begin{bmatrix} 10 & 2 \end{bmatrix} \begin{bmatrix} x_1 \\ x_2 \end{bmatrix}$$

MATLAB Program 5–7 produces the response $y(t)$ to the unit-ramp input $\dot{u} = 1$ m/s. The resulting response curve $y(t)$ versus t and the unit-ramp input are shown in Figure 5–16. Notice that the response curve here is identical to that shown in Figure 5–15.

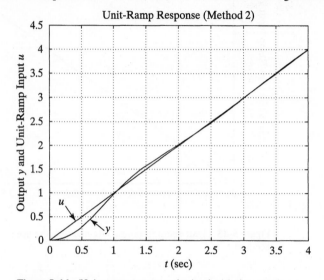

Figure 5–16   Unit-ramp response obtained with the use of Method 2.

MATLAB Program 5–7

```
>> % ----- The response y(t) is obtained by use of the
>> % state-space equation obtained by Method 2. -----
>>
>> t = 0:0.01:4;
>> A = [0   1;−10   −2];
>> B = [0;1];
>> C = [10   2];
>> D = 0;
>> sys = ss(A,B,C,D);
>> u = t;
>> lsim(sys,u,t)
>> grid
>> title('Unit-Ramp Response (Method 2)')
>> xlabel('t')
>> ylabel('Output y and Unit-Ramp Input u')
>> text(0.85,0.25,'y')
>> text(0.15,0.8,'u')
```

Example 5–7

Consider the front suspension system of a motorcycle. A simplified version is shown in Figure 5–17(a). Point P is the contact point with the ground. The vertical displacement u of point P is the input to the system. The displacements x and y are measured from their respective equilibrium positions before the input u is given to the system. Assume that

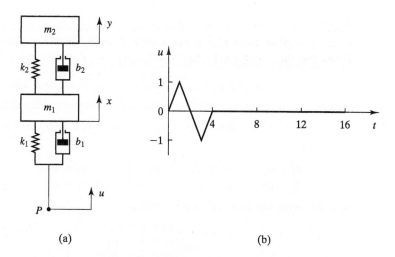

(a)   (b)

Figure 5–17   (a) Mechanical system; (b) triangular bump input u.

m_1, b_1, and k_1 represent the front tire and shock absorber assembly and m_2, b_2, and k_2 represent half of the body of the vehicle. Assume also that the system is at rest for $t < 0$. At $t = 0$, P is given a triangular bump input as shown in Figure 5–17(b). Point P moves only in the vertical direction. Assume that $m_1 = 10$ kg, $m_2 = 100$ kg, $b_1 = 50$ N-s/m, $b_2 = 100$ N-s/m, $k_1 = 50$ N/m, and $k_2 = 200$ N/m. (These numerical values are chosen to simplify the computations involved.) Obtain a state-space representation of the system. Plot the response curve $y(t)$ versus t with MATLAB.

   Method 1. Applying Newton's second law to the system, we obtain

$$m_1\ddot{x} = -k_1(x - u) - b_1(\dot{x} - \dot{u})$$
$$m_2\ddot{y} = -k_2(y - x) - b_2(\dot{y} - \dot{x})$$

which can be rewritten as

$$m_1\ddot{x} + b_1\dot{x} + k_1x = b_1\dot{u} + k_1u$$
$$m_2\ddot{y} + b_2\dot{y} + k_2y = b_2\dot{x} + k_2x$$

If we substitute the given numerical values for m_1, m_2, b_1, b_2, k_1, and k_2, the equations of motion become

$$10\ddot{x} + 50\dot{x} + 50x = 50\dot{u} + 50u$$
$$100\ddot{y} + 100\dot{y} + 200y = 100\dot{x} + 200x$$

which can be simplified to

$$\ddot{x} + 5\dot{x} + 5x = 5\dot{u} + 5u \tag{5-47}$$
$$\ddot{y} + \dot{y} + 2y = \dot{x} + 2x \tag{5-48}$$

Laplace transforming Equations (5–47) and (5–48), assuming the zero initial conditions, we obtain

$$(s^2 + 5s + 5)X(s) = (5s + 5)U(s)$$
$$(s^2 + s + 2)Y(s) = (s + 2)X(s)$$

Eliminating $X(s)$ from these two equations, we get

$$(s^2 + 5s + 5)(s^2 + s + 2)Y(s) = 5(s + 1)(s + 2)U(s)$$

or

$$(s^4 + 6s^3 + 12s^2 + 15s + 10)Y(s) = (5s^2 + 15s + 10)U(s) \qquad (5\text{--}49)$$

Equation (5–49) corresponds to the differential equation

$$\ddddot{y} + 6\dddot{y} + 12\ddot{y} + 15\dot{y} + 10y = 5\ddot{u} + 15\dot{u} + 10u$$

Comparing this last equation with the standard fourth-order differential equation

$$\ddddot{y} + a_1\dddot{y} + a_2\ddot{y} + a_3\dot{y} + a_4y = b_0\ddddot{u} + b_1\dddot{u} + b_2\ddot{u} + b_3\dot{u} + b_4u$$

we find that

$$a_1 = 6, \qquad a_2 = 12, \qquad a_3 = 15, \qquad a_4 = 10$$
$$b_0 = 0, \qquad b_1 = 0, \qquad b_2 = 5, \qquad b_3 = 15, \qquad b_4 = 10$$

Next, we define the state variables as follows:

$$x_1 = y - \beta_0 u$$
$$x_2 = \dot{x}_1 - \beta_1 u$$
$$x_3 = \dot{x}_2 - \beta_2 u$$
$$x_4 = \dot{x}_3 - \beta_3 u$$

where

$$\beta_0 = b_0 = 0$$
$$\beta_1 = b_1 - a_1\beta_0 = 0$$
$$\beta_2 = b_2 - a_1\beta_1 - a_2\beta_0 = 5$$
$$\beta_3 = b_3 - a_1\beta_2 - a_2\beta_1 - a_3\beta_0 = 15 - 6 \times 5 = -15$$

Hence,

$$\dot{x}_1 = x_2$$
$$\dot{x}_2 = x_3 + 5u$$
$$\dot{x}_3 = x_4 - 15u$$
$$\dot{x}_4 = -a_4x_1 - a_3x_2 - a_2x_3 - a_1x_4 + \beta_4 u$$
$$\quad = -10x_1 - 15x_2 - 12x_3 - 6x_4 + \beta_4 u$$

where

$$\beta_4 = b_4 - a_1\beta_3 - a_2\beta_2 - a_3\beta_1 - a_4\beta_0$$
$$\quad = 10 + 6 \times 15 - 12 \times 5 - 15 \times 0 - 10 \times 0 = 40$$

Thus,

$$\dot{x}_4 = -10x_1 - 15x_2 - 12x_3 - 6x_4 + 40u$$

and the state equation and output equation become

$$\begin{bmatrix} \dot{x}_1 \\ \dot{x}_2 \\ \dot{x}_3 \\ \dot{x}_4 \end{bmatrix} = \begin{bmatrix} 0 & 1 & 0 & 0 \\ 0 & 0 & 1 & 0 \\ 0 & 0 & 0 & 1 \\ -10 & -15 & -12 & -6 \end{bmatrix} \begin{bmatrix} x_1 \\ x_2 \\ x_3 \\ x_4 \end{bmatrix} + \begin{bmatrix} 0 \\ 5 \\ -15 \\ 40 \end{bmatrix} u$$

$$y = \begin{bmatrix} 1 & 0 & 0 & 0 \end{bmatrix} \begin{bmatrix} x_1 \\ x_2 \\ x_3 \\ x_4 \end{bmatrix} + 0u$$

MATLAB Program 5–8 produces the response $y(t)$ to the triangular bump input shown in Figure 5–17(b). The resulting response curve $y(t)$ versus t, as well as the input $u(t)$ versus t, is shown in Figure 5–18.

MATLAB Program 5–8

```
>> t = 0:0.01:16;
>> A = [0  1  0  0;0  0  1  0;0  0  0  1;−10  −15  −12  −6];
>> B = [0;5;−15;40];
>> C = [1  0  0  0];
>> D = 0;
>> sys = ss(A,B,C,D);
>> u1 = [0:0.01:1];
>> u2 = [0.99:−0.01:−1];
>> u3 = [−0.99:0.01:0];
>> u4 = 0*[4.01:0.01:16];
>> u = [u1  u2  u3  u4];
>> y = lsim(sys,u,t);
>> plot(t,y,t,u)
>> v = [0  16  −1.5  1.5]; axis(v)
>> grid
>> title('Response to Triangular Bump (Method 1)')
>> xlabel('t (sec)')
>> ylabel('Triangular Bump and Response')
```

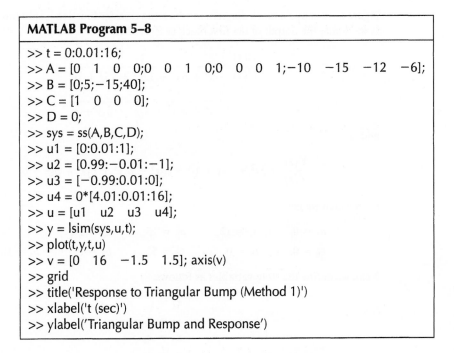

Figure 5–18   Response curve $y(t)$ and triangular bump input $u(t)$.

Method 2. From Equation (5–49), the transfer function of the system is given by

$$\frac{Y(s)}{U(s)} = \frac{5s^2 + 15s + 10}{s^4 + 6s^3 + 12s^2 + 15s + 10}$$

Figure 5–19 shows a block diagram in which the transfer function is split into two parts. If we define the output of the first block as $Z(s)$, then

$$\frac{Z(s)}{U(s)} = \frac{1}{s^4 + 6s^3 + 12s^2 + 15s + 10}$$

$$= \frac{1}{s^4 + a_1 s^3 + a_2 s^2 + a_3 s + a_4}$$

and

$$\frac{Y(s)}{Z(s)} = 5s^2 + 15s + 10 = b_0 s^4 + b_1 s^3 + b_2 s^2 + b_3 s + b_4$$

from which we get

$$a_1 = 6, \quad a_2 = 12, \quad a_3 = 15, \quad a_4 = 10,$$
$$b_0 = 0, \quad b_1 = 0, \quad b_2 = 5, \quad b_3 = 15, \quad b_4 = 10$$

Next, we define the state variables as follows:

$$x_1 = z$$
$$x_2 = \dot{x}_1$$
$$x_3 = \dot{x}_2$$
$$x_4 = \dot{x}_3$$

From Equation (5–40), noting that $a_1 = 6$, $a_2 = 12$, $a_3 = 15$, and $a_4 = 10$, we obtain

$$\begin{bmatrix} \dot{x}_1 \\ \dot{x}_2 \\ \dot{x}_3 \\ \dot{x}_4 \end{bmatrix} = \begin{bmatrix} 0 & 1 & 0 & 0 \\ 0 & 0 & 1 & 0 \\ 0 & 0 & 0 & 1 \\ -10 & -15 & -12 & -6 \end{bmatrix} \begin{bmatrix} x_1 \\ x_2 \\ x_3 \\ x_4 \end{bmatrix} + \begin{bmatrix} 0 \\ 0 \\ 0 \\ 1 \end{bmatrix} u$$

Similarly, from the output equation given by Equation (5–41), we have

$$y = [b_4 - a_4 b_0 \ \vdots \ b_3 - a_3 b_0 \ \vdots \ b_2 - a_2 b_0 \ \vdots \ b_1 - a_1 b_0] \begin{bmatrix} x_1 \\ x_2 \\ x_3 \\ x_4 \end{bmatrix} + b_0 u$$

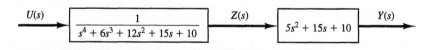

Figure 5–19   Block diagram of $Y(s)/U(s)$.

or

$$y = [10 \quad 15 \quad 5 \quad 0] \begin{bmatrix} x_1 \\ x_2 \\ x_3 \\ x_4 \end{bmatrix} + 0u$$

MATLAB Program 5–9 produces the response $y(t)$ to the triangular bump input. The response curve is shown in Figure 5–20. (This response curve is identical to that shown in Figure 5–19.)

MATLAB Program 5–9

```
>> t = 0:0.01:16;
>> A = [0  1  0  0;0  0  1  0;0  0  0  1;−10  −15  −12  −6];
>> B = [0;0;0;1];
>> C = [10    15    5    0];
>> D = 0;
>> sys = ss(A,B,C,D);
>> u1 = [0:0.01:1];
>> u2 = [0.99:−0.01:−1];
>> u3 = [−0.99:0.01:0];
>> u4 = 0*[4.01:0.01:16];
>> u = [u1  u2  u3  u4];
>> y = lsim(sys,u,t);
>> plot(t,y,t,u)
>> v = [0    16    −1.5    1.5]; axis(v)
>> grid
>> title('Response to Triangular Bump (Method 2)')
>> xlabel('t (sec)')
>> ylabel('Triangular Bump and Response')
```

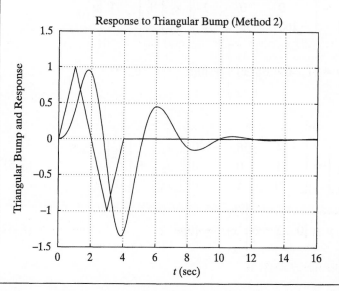

Response to Triangular Bump (Method 2)

Figure 5–20   Response $y(t)$ to the triangular bump input $u(t)$.

5–5 TRANSFORMATION OF MATHEMATICAL MODELS WITH MATLAB

MATLAB is quite useful in transforming a system model from transfer function to state space and vice versa. We shall begin our discussion with the transformation from transfer function to state space.

Let us write the transfer function $Y(s)/U(s)$ as

$$\frac{Y(s)}{U(s)} = \frac{\text{numerator polynomial in } s}{\text{denominator polynomial in } s} = \frac{\text{num}}{\text{den}}$$

Once we have this transfer-function expression, the MATLAB command

$$[A, B, C, D] = \text{tf2ss(num,den)}$$

will give a state-space representation. Note that the command can be used when the system equation involves one or more derivatives of the input function. (In such a case, the transfer function of the system involves a numerator polynomial in s.)

It is important to note that the state-space representation of any system is not unique. There are many (indeed, infinitely many) state-space representations of the same system. The MATLAB command gives one possible such representation.

Transformation from transfer function to state space. Consider the transfer function system

$$\frac{Y(s)}{U(s)} = \frac{s}{s^3 + 14s^2 + 56s + 160} \tag{5–50}$$

Of the infinitely many possible state-space representations of this system, one is

$$\begin{bmatrix} \dot{x}_1 \\ \dot{x}_2 \\ \dot{x}_3 \end{bmatrix} = \begin{bmatrix} 0 & 1 & 0 \\ 0 & 0 & 1 \\ -160 & -56 & -14 \end{bmatrix} \begin{bmatrix} x_1 \\ x_2 \\ x_3 \end{bmatrix} + \begin{bmatrix} 0 \\ 1 \\ -14 \end{bmatrix} u$$

$$y = \begin{bmatrix} 1 & 0 & 0 \end{bmatrix} \begin{bmatrix} x_1 \\ x_2 \\ x_3 \end{bmatrix} + [0]u$$

Another is

$$\begin{bmatrix} \dot{x}_1 \\ \dot{x}_2 \\ \dot{x}_3 \end{bmatrix} = \begin{bmatrix} -14 & -56 & -160 \\ 1 & 0 & 0 \\ 0 & 1 & 0 \end{bmatrix} \begin{bmatrix} x_1 \\ x_2 \\ x_3 \end{bmatrix} + \begin{bmatrix} 1 \\ 0 \\ 0 \end{bmatrix} u \tag{5–51}$$

$$y = \begin{bmatrix} 0 & 1 & 0 \end{bmatrix} \begin{bmatrix} x_1 \\ x_2 \\ x_3 \end{bmatrix} + [0]u \tag{5–52}$$

MATLAB transforms the transfer function given by Equation (5–50) into the state-space representation given by Equations (5–51) and (5–52). For the system considered here, MATLAB Program 5–10 will produce matrices $\mathbf{A}, \mathbf{B}, \mathbf{C}$, and D.

MATLAB Program 5–10

```
>> % ----- Transforming transfer-function model to
>> %        state-space model -----
>>
>> num = [0   0   1   0];
>> den = [1   14   56   160];
>>
>> % ----- Enter the following transformation command -----
>>
>> [A, B, C, D] = tf2ss(num,den)

A =

    -14      -56      160
      1        0        0
      0        1        0

B =

      1
      0
      0

C =

      0        1        0

D =

      0
```

Transformation from state space to transfer function.   To obtain the transfer function from state-space equations, use the command

$$[num,den] = ss2tf(A,B,C,D,iu)$$

Note that iu must be specified for systems with more than one input. For example, if the system has three inputs ($u1, u2, u3$), then iu must be either 1, 2, or 3, where 1 implies $u1$, 2 implies $u2$, and 3 implies $u3$.

If the system has only one input, then either

$$[num,den] = ss2tf(A,B,C,D)$$

or

$$[num,den] = ss2tf(A,B,C,D,1)$$

may be used. (For the case where the system has multiple inputs and multiple outputs, see **Example 5–9**.)

Example 5–8

Obtain the transfer function of the system defined by the following state-space equations:

$$\begin{bmatrix} \dot{x}_1 \\ \dot{x}_2 \\ \dot{x}_3 \\ \dot{x}_4 \end{bmatrix} = \begin{bmatrix} 0 & 1 & 0 & 0 \\ 0 & 0 & 1 & 0 \\ 0 & 0 & 0 & 1 \\ -10 & -15 & -12 & -6 \end{bmatrix} \begin{bmatrix} x_1 \\ x_2 \\ x_3 \\ x_4 \end{bmatrix} + \begin{bmatrix} 0 \\ 5 \\ -15 \\ 40 \end{bmatrix} u$$

$$Y = \begin{bmatrix} 1 & 0 & 0 & 0 \end{bmatrix} \begin{bmatrix} x_1 \\ x_2 \\ x_3 \\ x_4 \end{bmatrix} + 0u$$

MATLAB Program 5–11 produces the transfer function of the system, namely,

$$\frac{Y(s)}{U(s)} = \frac{5s^2 + 15s + 10}{s^4 + 6s^3 + 12s^2 + 15s + 10}$$

MATLAB Program 5–11

```
>> % ----- Transforming state-space model to
>> %      transfer function model -----
>>
>> A = [0  1  0  0;0  0  1  0;0  0  0  1;-10  -15  -12  -6];
>> B = [0;5;-15;40];
>> C = [1  0  0  0];
>> D = 0;
>>
>> % ----- Enter the following transformation command ——
>>
>> [num,den] = ss2tf(A,B,C,D)
num =
         0         0    5.0000    15.0000    10.0000

den =
    1.0000    6.0000    12.0000    15.0000    10.0000
```

Example 5–9

Consider a system with multiple inputs and multiple outputs. When the system has more than one output, the command

$$[NUM,den] = ss2tf(A,B,C,D,iu)$$

produces transfer functions for all outputs to each input. (The numerator coefficients are returned to matrix NUM with as many rows as there are outputs.)

Let the system be defined by

$$\begin{bmatrix} \dot{x}_1 \\ \dot{x}_2 \end{bmatrix} = \begin{bmatrix} 0 & 1 \\ -25 & -4 \end{bmatrix} \begin{bmatrix} x_1 \\ x_2 \end{bmatrix} + \begin{bmatrix} 1 & 1 \\ 0 & 1 \end{bmatrix} \begin{bmatrix} u_1 \\ u_2 \end{bmatrix}$$

$$\begin{bmatrix} y_1 \\ y_2 \end{bmatrix} = \begin{bmatrix} 1 & 0 \\ 0 & 1 \end{bmatrix} \begin{bmatrix} x_1 \\ x_2 \end{bmatrix} + \begin{bmatrix} 0 & 0 \\ 0 & 0 \end{bmatrix} \begin{bmatrix} u_1 \\ u_2 \end{bmatrix}$$

This system involves two inputs and two outputs. Four transfer functions are involved: $Y_1(s)/U_1(s)$, $Y_2(s)/U_1(s)$, $Y_1(s)/U_2(s)$, and $Y_2(s)/U_2(s)$. (When considering input u_1, we assume that input u_2 is zero, and vice versa.)

MATLAB Program 5–12 produces representations of the following four transfer functions:

$$\frac{Y_1(s)}{U_1(s)} = \frac{s + 4}{s^2 + 4s + 25}, \qquad \frac{Y_1(s)}{U_2(s)} = \frac{s + 5}{s^2 + 4s + 25}$$

$$\frac{Y_2(s)}{U_1(s)} = \frac{-25}{s^2 + 4s + 25}, \qquad \frac{Y_2(s)}{U_2(s)} = \frac{s - 25}{s^2 + 4s + 25}$$

MATLAB Program 5–12

```
>> A = [0     1;-25     -4];
>> B = [1     1;0     1];
>> C = [1     0;0     1];
>> D = [0     0;0     0];
>> [NUM,den] = ss2tf(A,B,C,D,1)

NUM =

          0      1.0000      4.0000
          0           0    -25.0000

den =

     1.0000      4.0000     25.0000
>> [NUM,den] = ss2tf(A,B,C,D,2)

NUM =

          0      1.0000      5.0000
          0      1.0000    -25.0000

den =

     1.0000      4.0000     25.0000
```

Nonuniqueness of a set of state variables.   A set of state variables is not unique for a given system. Suppose that $x_1, x_2, \ldots, x_n$ are a set of state variables. Then we may take as another set of state variables any set of functions

$$\hat{x}_1 = X_1(x_1, x_2, \ldots, x_n)$$
$$\hat{x}_2 = X_2(x_1, x_2, \ldots, x_n)$$
$$\vdots$$
$$\hat{x}_n = X_n(x_1, x_2, \ldots, x_n)$$

provided that, for every set of values $\hat{x}_1, \hat{x}_2, \ldots, \hat{x}_n$, there corresponds a unique set of values $x_1, x_2, \ldots, x_n$, and vice versa. Thus, if $\mathbf{x}$ is a state vector, then

$$\hat{\mathbf{x}} = \mathbf{Px}$$

is also a state vector, provided that the matrix $\mathbf{P}$ is nonsingular. (Note that a square matrix $\mathbf{P}$ is nonsingular if the determinant $|\mathbf{P}|$ is nonzero.) Different state vectors convey the same information about the system behavior.

Transformation of a state-space model into another state-space model. A state-space model

$$\dot{\mathbf{x}} = \mathbf{Ax} + \mathbf{Bu} \tag{5–53}$$
$$\mathbf{y} = \mathbf{Cx} + \mathbf{Du} \tag{5–54}$$

can be transformed into another state-space model by transforming the state vector $\mathbf{x}$ into state vector $\hat{\mathbf{x}}$ by means of the transformation

$$\mathbf{x} = \mathbf{P}\hat{\mathbf{x}}$$

where $\mathbf{P}$ is nonsingular. Then Equations (5–53) and (5–54) can be written as

$$\mathbf{P}\dot{\hat{\mathbf{x}}} = \mathbf{AP}\hat{\mathbf{x}} + \mathbf{Bu}$$
$$\mathbf{y} = \mathbf{CP}\hat{\mathbf{x}} + \mathbf{Du}$$

or

$$\dot{\hat{\mathbf{x}}} = \mathbf{P}^{-1}\mathbf{AP}\hat{\mathbf{x}} + \mathbf{P}^{-1}\mathbf{Bu} \tag{5–55}$$
$$\mathbf{y} = \mathbf{CP}\hat{\mathbf{x}} + \mathbf{Du} \tag{5–56}$$

Equations (5–55) and (5–56) represent another state-space model of the same system. Since infinitely many $n \times n$ nonsingular matrices can be used as a transformation matrix $\mathbf{P}$, there are infinitely many state-space models for a given system.

Eigenvalues of an $n \times n$ matrix A.   The *eigenvalues* of an $n \times n$ matrix $\mathbf{A}$ are the roots of the characteristic equation

$$|\lambda\mathbf{I} - \mathbf{A}| = 0 \tag{5–57}$$

The eigenvalues are also called the *characteristic roots*.

Consider, for example, the matrix

$$\mathbf{A} = \begin{bmatrix} 0 & 1 & 0 \\ 0 & 0 & 1 \\ -6 & -11 & -6 \end{bmatrix}$$

The characteristic equation is

$$|\lambda\mathbf{I} - \mathbf{A}| = \begin{vmatrix} \lambda & -1 & 0 \\ 0 & \lambda & -1 \\ 6 & 11 & \lambda + 6 \end{vmatrix}$$
$$= \lambda^3 + 6\lambda^2 + 11\lambda + 6$$
$$= (\lambda + 1)(\lambda + 2)(\lambda + 3) = 0$$

The eigenvalues of $\mathbf{A}$ are the roots of the characteristic equation, or -1, -2, and -3.

It is sometimes desirable to transform the state matrix into a diagonal matrix. This may be done by choosing an appropriate transformation matrix **P**. In what follows, we shall discuss the diagonalization of a state matrix.

Diagonalization of state matrix A.   Consider an $n \times n$ state matrix

$$\mathbf{A} = \begin{bmatrix} 0 & 1 & 0 & \cdots & 0 \\ 0 & 0 & 1 & \cdots & 0 \\ \vdots & \vdots & \vdots & & \vdots \\ 0 & 0 & 0 & \cdots & 1 \\ -a_n & -a_{n-1} & -a_{n-2} & \cdots & -a_1 \end{bmatrix} \tag{5-58}$$

We first consider the case where matrix **A** has distinct eigenvalues only. If the state vector **x** is transformed into another state vector **z** with the use of a transformation matrix **P**, or

$$\mathbf{x} = \mathbf{Pz}$$

where

$$\mathbf{P} = \begin{bmatrix} 1 & 1 & \cdots & 1 \\ \lambda_1 & \lambda_2 & \cdots & \lambda_n \\ \lambda_1^2 & \lambda_2^2 & \cdots & \lambda_n^2 \\ \vdots & \vdots & & \vdots \\ \lambda_1^{n-1} & \lambda_2^{n-1} & \cdots & \lambda_n^{n-1} \end{bmatrix} \tag{5-59}$$

in which $\lambda_1, \lambda_2, \ldots,$ and λ_n are n distinct eigenvalues of **A**, then $\mathbf{P}^{-1}\mathbf{AP}$ becomes a diagonal matrix, or

$$\mathbf{P}^{-1}\mathbf{AP} = \begin{bmatrix} \lambda_1 & & & & 0 \\ & \lambda_2 & & & \\ & & \cdot & & \\ & & & \cdot & \\ 0 & & & & \lambda_n \end{bmatrix} \tag{5-60}$$

Note that each column of the transformation matrix **P** in Equation (5–59) is an eigenvector of the matrix **A** given by Equation (5–58). (See **Problem A–5–18** for details.)

Next, consider the case where matrix **A** involves multiple eigenvalues. In this case, diagonalization is not possible, but matrix **A** can be transformed into a Jordan canonical form. For example, consider the 3×3 matrix

$$\mathbf{A} = \begin{bmatrix} 0 & 1 & 0 \\ 0 & 0 & 1 \\ -a_3 & -a_2 & -a_1 \end{bmatrix}$$

Assume that $\mathbf{A}$ has eigenvalues λ_1, λ_1, and λ_3, where $\lambda_1 \neq \lambda_3$. In this case, the transformation $\mathbf{x} = \mathbf{S}\mathbf{z}$, where

$$\mathbf{S} = \begin{bmatrix} 1 & 0 & 1 \\ \lambda_1 & 1 & \lambda_3 \\ \lambda_1^2 & 2\lambda_1 & \lambda_3^2 \end{bmatrix} \tag{5-61}$$

will yield

$$\mathbf{S}^{-1}\mathbf{A}\mathbf{S} = \begin{bmatrix} \lambda_1 & 1 & 0 \\ 0 & \lambda_1 & 0 \\ 0 & 0 & \lambda_3 \end{bmatrix} \tag{5-62}$$

This matrix is in Jordan canonical form.

Example 5–10

Consider a system with the state-space representation

$$\begin{bmatrix} \dot{x}_1 \\ \dot{x}_2 \\ \dot{x}_3 \end{bmatrix} = \begin{bmatrix} 0 & 1 & 0 \\ 0 & 0 & 1 \\ -6 & -11 & -6 \end{bmatrix} \begin{bmatrix} x_1 \\ x_2 \\ x_3 \end{bmatrix} + \begin{bmatrix} 0 \\ 0 \\ 6 \end{bmatrix} u$$

$$y = \begin{bmatrix} 1 & 0 & 0 \end{bmatrix} \begin{bmatrix} x_1 \\ x_2 \\ x_3 \end{bmatrix}$$

or

$$\dot{\mathbf{x}} = \mathbf{A}\mathbf{x} + \mathbf{B}u \tag{5-63}$$
$$y = \mathbf{C}\mathbf{x} + Du$$

where

$$\mathbf{A} = \begin{bmatrix} 0 & 1 & 0 \\ 0 & 0 & 1 \\ -6 & -11 & -6 \end{bmatrix}, \qquad \mathbf{B} = \begin{bmatrix} 0 \\ 0 \\ 6 \end{bmatrix}, \qquad \mathbf{C} = \begin{bmatrix} 1 & 0 & 0 \end{bmatrix}, \qquad D = 0$$

The eigenvalues of the state matrix $\mathbf{A}$ are -1, -2, and -3, or

$$\lambda_1 = -1, \qquad \lambda_2 = -2, \qquad \lambda_3 = -3$$

We shall show that Equation (5–63) is not the only possible state equation for the system. Suppose we define a set of new state variables z_1, z_2, and z_3 by the transformation

$$\begin{bmatrix} x_1 \\ x_2 \\ x_3 \end{bmatrix} = \begin{bmatrix} 1 & 1 & 1 \\ \lambda_1 & \lambda_2 & \lambda_3 \\ \lambda_1^2 & \lambda_2^2 & \lambda_3^2 \end{bmatrix} \begin{bmatrix} z_1 \\ z_2 \\ z_3 \end{bmatrix}$$

$$= \begin{bmatrix} 1 & 1 & 1 \\ -1 & -2 & -3 \\ 1 & 4 & 9 \end{bmatrix} \begin{bmatrix} z_1 \\ z_2 \\ z_3 \end{bmatrix}$$

or

$$\mathbf{x} = \mathbf{P}\mathbf{z} \tag{5-64}$$

where

$$\mathbf{P} = \begin{bmatrix} 1 & 1 & 1 \\ -1 & -2 & -3 \\ 1 & 4 & 9 \end{bmatrix} \tag{5-65}$$

Then, substituting Equation (5–64) into Equation (5–63), we obtain

$$\mathbf{P}\dot{\mathbf{z}} = \mathbf{APz} + \mathbf{B}u$$

Premultiplying both sides of this last equation by $\mathbf{P}^{-1}$, we get

$$\dot{\mathbf{z}} = \mathbf{P}^{-1}\mathbf{APz} + \mathbf{P}^{-1}\mathbf{B}u \tag{5-66}$$

or

$$\begin{bmatrix} \dot{z}_1 \\ \dot{z}_2 \\ \dot{z}_3 \end{bmatrix} = \begin{bmatrix} 3 & 2.5 & 0.5 \\ -3 & -4 & -1 \\ 1 & 1.5 & 0.5 \end{bmatrix} \begin{bmatrix} 0 & 1 & 0 \\ 0 & 0 & 1 \\ -6 & -11 & -6 \end{bmatrix} \begin{bmatrix} 1 & 1 & 1 \\ -1 & -2 & -3 \\ 1 & 4 & 9 \end{bmatrix} \begin{bmatrix} z_1 \\ z_2 \\ z_3 \end{bmatrix}$$
$$+ \begin{bmatrix} 3 & 2.5 & 0.5 \\ -3 & -4 & -1 \\ 1 & 1.5 & 0.5 \end{bmatrix} \begin{bmatrix} 0 \\ 0 \\ 6 \end{bmatrix} u$$

Simplifying gives

$$\begin{bmatrix} \dot{z}_1 \\ \dot{z}_2 \\ \dot{z}_3 \end{bmatrix} = \begin{bmatrix} -1 & 0 & 0 \\ 0 & -2 & 0 \\ 0 & 0 & -3 \end{bmatrix} \begin{bmatrix} z_1 \\ z_2 \\ z_3 \end{bmatrix} + \begin{bmatrix} 3 \\ -6 \\ 3 \end{bmatrix} u \tag{5-67}$$

Equation (5–67) is a state equation that describes the system defined by Equation (5–63).

The output equation is modified to

$$y = \begin{bmatrix} 1 & 0 & 0 \end{bmatrix} \begin{bmatrix} 1 & 1 & 1 \\ -1 & -2 & -3 \\ 1 & 4 & 9 \end{bmatrix} \begin{bmatrix} z_1 \\ z_2 \\ z_3 \end{bmatrix}$$
$$= \begin{bmatrix} 1 & 1 & 1 \end{bmatrix} \begin{bmatrix} z_1 \\ z_2 \\ z_3 \end{bmatrix} \tag{5-68}$$

Notice that the transformation matrix $\mathbf{P}$ defined by Equation (5–65) changes the coefficient matrix of $\mathbf{z}$ into the diagonal matrix. As is clearly seen from Equation (5–67), the three separate state equations are uncoupled. Notice also that the diagonal elements of the matrix $\mathbf{P}^{-1}\mathbf{AP}$ in Equation (5–66) are identical to the three eigenvalues of $\mathbf{A}$. (For a proof, see **Problem A–5–20**.)

EXAMPLE PROBLEMS AND SOLUTIONS

Problem A–5–1

Consider the pendulum system shown in Figure 5–21. Assuming angle θ to be the output of the system, obtain a state-space representation of the system.

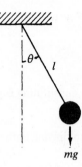

Figure 5–21   Pendulum system.

Solution   The equation for the pendulum system is

$$ml^2\ddot{\theta} = -mgl \sin \theta$$

or

$$\ddot{\theta} + \frac{g}{l} \sin \theta = 0$$

This is a second-order system; accordingly, we need two state variables, x_1 and x_2, to completely describe the system dynamics. If we define

$$x_1 = \theta$$
$$x_2 = \dot{\theta}$$

then we get

$$\dot{x}_1 = x_2$$
$$\dot{x}_2 = -\frac{g}{l} \sin x_1$$

(There is no input u to this system.) The output y is angle θ. Thus,

$$y = \theta = x_1$$

A state-space representation of the system is

$$\begin{bmatrix} \dot{x}_1 \\ \dot{x}_2 \end{bmatrix} = \begin{bmatrix} 0 & 1 \\ -\dfrac{g}{l}\dfrac{\sin x_1}{x_1} & 0 \end{bmatrix} \begin{bmatrix} x_1 \\ x_2 \end{bmatrix}$$

$$y = \begin{bmatrix} 1 & 0 \end{bmatrix} \begin{bmatrix} x_1 \\ x_2 \end{bmatrix}$$

Note that the state equation just obtained is a nonlinear differential equation.
  If the angle θ is limited to be small, then the system can be linearized. For small angle θ, we have $\sin \theta = \sin x_1 \doteq x_1$ and $(\sin x_1)/x_1 \doteq 1$. A state-space representation

of the linearized model is then given by

$$\begin{bmatrix} \dot{x}_1 \\ \dot{x}_2 \end{bmatrix} = \begin{bmatrix} 0 & 1 \\ -\dfrac{g}{l} & 0 \end{bmatrix} \begin{bmatrix} x_1 \\ x_2 \end{bmatrix}$$

$$y = \begin{bmatrix} 1 & 0 \end{bmatrix} \begin{bmatrix} x_1 \\ x_2 \end{bmatrix}$$

Problem A–5–2

Obtain a state-space representation of the mechanical system shown in Figure 5–22. The external force $u(t)$ applied to mass m_2 is the input to the system. The displacements y and z are measured from their respective equilibrium positions and are the outputs of the system.

Solution   Applying Newton's second law to this system, we obtain

$$m_2\ddot{y} + b_1(\dot{y} - \dot{z}) + k_1(y - z) + k_2y = u \tag{5–69}$$
$$m_1\ddot{z} + b_1(\dot{z} - \dot{y}) + k_1(z - y) = 0 \tag{5–70}$$

If we define the state variables

$$x_1 = y$$
$$x_2 = \dot{y}$$
$$x_3 = z$$
$$x_4 = \dot{z}$$

then, from Equation (5–69), we get

$$m_2\dot{x}_2 = -(k_1 + k_2)x_1 - b_1x_2 + k_1x_3 + b_1x_4 + u$$

Also, from Equation (5–70), we obtain

$$m_1\dot{x}_4 = k_1x_1 + b_1x_2 - k_1x_3 - b_1x_4$$

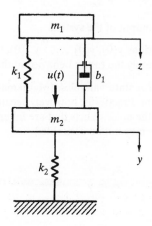

Figure 5–22   Mechanical system.

Hence, the state equation is

$$
\begin{bmatrix} \dot{x}_1 \\ \dot{x}_2 \\ \dot{x}_3 \\ \dot{x}_4 \end{bmatrix} = \begin{bmatrix} 0 & 1 & 0 & 0 \\ -\dfrac{k_1 + k_2}{m_2} & -\dfrac{b_1}{m_2} & \dfrac{k_1}{m_2} & \dfrac{b_1}{m_2} \\ 0 & 0 & 0 & 1 \\ \dfrac{k_1}{m_1} & \dfrac{b_1}{m_1} & -\dfrac{k_1}{m_1} & -\dfrac{b_1}{m_1} \end{bmatrix} \begin{bmatrix} x_1 \\ x_2 \\ x_3 \\ x_4 \end{bmatrix} + \begin{bmatrix} 0 \\ \dfrac{1}{m_2} \\ 0 \\ 0 \end{bmatrix} u \tag{5-71}
$$

The outputs of the system are y and z. Consequently, if we define the output variables as

$$
\begin{aligned}
y_1 &= y \\
y_2 &= z
\end{aligned}
$$

then we have

$$
\begin{aligned}
y_1 &= x_1 \\
y_2 &= x_3
\end{aligned}
$$

The output equation can now be put in the form

$$
\begin{bmatrix} y_1 \\ y_2 \end{bmatrix} = \begin{bmatrix} 1 & 0 & 0 & 0 \\ 0 & 0 & 1 & 0 \end{bmatrix} \begin{bmatrix} x_1 \\ x_2 \\ x_3 \\ x_4 \end{bmatrix} \tag{5-72}
$$

Equations (5–71) and (5–72) give a state-space representation of the mechanical system shown in Figure 5–22.

Problem A–5–3

Obtain a state-space representation of the system defined by

$$
\overset{(n)}{y} + a_1 \overset{(n-1)}{y} + \cdots + a_{n-1}\dot{y} + a_n y = u \tag{5-73}
$$

where u is the input and y is the output of the system.

Solution   Since the initial conditions $y(0), \dot{y}(0), \ldots, \overset{(n-1)}{y}(0)$, together with the input $u(t)$ for $t \geq 0$, determines completely the future behavior of the system, we may take $y(t), \dot{y}(t), \ldots, \overset{(n-1)}{y}(t)$ as a set of n state variables. (Mathematically, such a choice of state variables is quite convenient. Practically, however, because higher order derivative terms are inaccurate due to the noise effects that are inherent in any practical system, this choice of state variables may not be desirable.)

Let us define

$$
\begin{aligned}
x_1 &= y \\
x_2 &= \dot{y} \\
&\vdots \\
x_n &= \overset{(n-1)}{y}
\end{aligned}
$$

Then Equation (5–73) can be written as

$$\dot{x}_1 = x_2$$
$$\dot{x}_2 = x_3$$
$$\vdots$$
$$\dot{x}_{n-1} = x_n$$
$$\dot{x}_n = -a_n x_1 - \cdots - a_1 x_n + u$$

or

$$\dot{\mathbf{x}} = \mathbf{A}\mathbf{x} + \mathbf{B}u \qquad (5\text{–}74)$$

where

$$\mathbf{x} = \begin{bmatrix} x_1 \\ x_2 \\ \vdots \\ x_n \end{bmatrix}, \qquad \mathbf{A} = \begin{bmatrix} 0 & 1 & 0 & \cdots & 0 \\ 0 & 0 & 1 & \cdots & 0 \\ \vdots & \vdots & \vdots & & \vdots \\ 0 & 0 & 0 & \cdots & 1 \\ -a_n & -a_{n-1} & -a_{n-2} & \cdots & -a_1 \end{bmatrix}, \qquad \mathbf{B} = \begin{bmatrix} 0 \\ 0 \\ \vdots \\ 0 \\ 1 \end{bmatrix}$$

The output can be given by

$$y = \begin{bmatrix} 1 & 0 & \cdots & 0 \end{bmatrix} \begin{bmatrix} x_1 \\ x_2 \\ \vdots \\ x_n \end{bmatrix}$$

or

$$y = \mathbf{C}\mathbf{x} \qquad (5\text{–}75)$$

where

$$\mathbf{C} = \begin{bmatrix} 1 & 0 & \cdots & 0 \end{bmatrix}$$

Equation (5–74) is the state equation and Equation (5–75) is the output equation.
Note that the state-space representation of the transfer function of the system,

$$\frac{Y(s)}{U(s)} = \frac{1}{s^n + a_1 s^{n-1} + \cdots + a_{n-1}s + a_n}$$

is also given by Equations (5–74) and (5–75).

Problem A–5–4

Consider a system described by the state equation

$$\dot{\mathbf{x}} = \mathbf{A}\mathbf{x} + \mathbf{B}u$$

and output equation

$$y = \mathbf{C}\mathbf{x} + Du$$

where

$$\mathbf{A} = \begin{bmatrix} 0 & 1 \\ -0.125 & -1.375 \end{bmatrix}, \qquad \mathbf{B} = \begin{bmatrix} -0.25 \\ 0.34375 \end{bmatrix}, \qquad \mathbf{C} = \begin{bmatrix} 1 & 0 \end{bmatrix}, \qquad D = 1$$

Obtain the transfer function of this system.

Solution From Equation (5–9), the transfer function $G(s)$ can be given in terms of matrices $\mathbf{A}, \mathbf{B}, \mathbf{C}$, and D as

$$G(s) = \mathbf{C}(s\mathbf{I} - \mathbf{A})^{-1}\mathbf{B} + D$$

Since

$$s\mathbf{I} - \mathbf{A} = \begin{bmatrix} s & -1 \\ 0.125 & s + 1.375 \end{bmatrix}$$

we have

$$(s\mathbf{I} - \mathbf{A})^{-1} = \frac{1}{s^2 + 1.375s + 0.125} \begin{bmatrix} s + 1.375 & 1 \\ -0.125 & s \end{bmatrix}$$

Therefore, the transfer function of the system is

$$
\begin{aligned}
G(s) &= \begin{bmatrix} 1 & 0 \end{bmatrix} \frac{1}{s^2 + 1.375s + 0.125} \begin{bmatrix} s + 1.375 & 1 \\ -0.125 & s \end{bmatrix} \begin{bmatrix} -0.25 \\ 0.34375 \end{bmatrix} + 1 \\
&= \frac{-0.25(s + 1.375) + 0.34375}{s^2 + 1.375s + 0.125} + 1 \\
&= \frac{s^2 + 1.125s + 0.125}{s^2 + 1.375s + 0.125} \\
&= \frac{8s^2 + 9s + 1}{8s^2 + 11s + 1}
\end{aligned}
$$

Problem A–5–5

Consider the following state equation and output equation:

$$\begin{bmatrix} \dot{x}_1 \\ \dot{x}_2 \end{bmatrix} = \begin{bmatrix} -1 & -1 \\ 6.5 & 0 \end{bmatrix} \begin{bmatrix} x_1 \\ x_2 \end{bmatrix} + \begin{bmatrix} 1 & 1 \\ 1 & 0 \end{bmatrix} \begin{bmatrix} u_1 \\ u_2 \end{bmatrix}$$

$$\begin{bmatrix} y_1 \\ y_2 \end{bmatrix} = \begin{bmatrix} 1 & 0 \\ 0 & 1 \end{bmatrix} \begin{bmatrix} x_1 \\ x_2 \end{bmatrix} + \begin{bmatrix} 0 & 0 \\ 0 & 0 \end{bmatrix} \begin{bmatrix} u_1 \\ u_2 \end{bmatrix}$$

The system involves two inputs and two outputs, so there are four input–output combinations. Obtain the impulse-response curves of the four combinations. (When u_1 is a unit-impulse input, we assume that $u_2 = 0$, and vice versa.)

Next, find the outputs y_1 and y_2 when both inputs, u_1 and u_2, are given at the same time (i.e., $u_1 = u_2 =$ unit-impulse function occurring at the same time $t = 0$).

Solution The command

$$\text{sys} = \text{ss(A,B,C,D1),} \qquad \text{impulse(sys,t)}$$

produces the impulse-response curves for the four input–output combinations. (See MATLAB Program 5–13; when u_1 is a unit-impulse function, we assume that $u_2 = 0$, and vice versa.) The resulting curves are shown in Figure 5–23.

When both unit-impulse inputs $u_1(t)$ and $u_2(t)$ are given at the same time $t = 0$, the responses are

$$y_1(t) = y_{11}(t) + y_{21}(t)$$
$$y_2(t) = y_{12}(t) + y_{22}(t)$$

MATLAB Program 5–13

```
>> t = 0:0.01:10;
>> A = [−1    −1;6.5    0];
>> B = [1    1;1    0];
>> C = [1    0;0    1];
>> D = [0    0;0    0];
>> sys = ss(A,B,C,D);
>> impulse(sys,t)
>> grid
>> title('Impulse-Response Curves')
>> xlabel('t'); ylabel('Outputs')
```

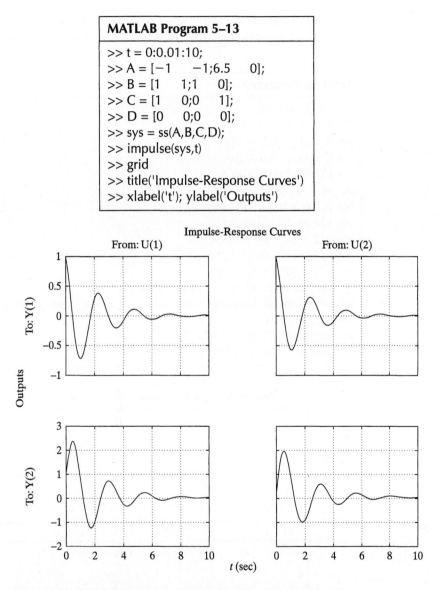

Figure 5–23 Unit-impulse response curves. (The left column corresponds to u_1 = unit-impulse input and u_2 = 0. The right column corresponds to u_1 = 0 and u_2 = unit-impulse input.)

where

$$
\begin{aligned}
y_{11} &= y_1 && \text{when } u_1 = \delta(t),\ u_2 = 0 \\
y_{12} &= y_2 && \text{when } u_1 = \delta(t),\ u_2 = 0 \\
y_{21} &= y_1 && \text{when } u_1 = 0,\ u_2 = \delta(t) \\
y_{22} &= y_2 && \text{when } u_1 = 0,\ u_2 = \delta(t)
\end{aligned}
$$

MATLAB Program 5–14 produces the responses $y_1(t) = y_{11}(t) + y_{21}(t)$ and $y_2(t) = y_{12}(t) + y_{22}(t)$. The resulting response curves are shown in Figure 5–24.

MATLAB Program 5–14

```
>> t = 0:0.01:10;
>> A = [-1    -1;6.5    0];
>> B = [1    1;1    0];
>> C = [1    0;0    1];
>> D = [0    0;0    0];
>> sys = ss(A,B,C,D);
>> [y,t,x] = impulse(sys,t);
>> y11 = [1    0]*y(:,:,1)';
>> y12 = [0    1]*y(:,:,1)';
>> y21 = [1    0]*y(:,:,2)';
>> y22 = [0    1]*y(:,:,2)';
>> subplot(211); plot(t,y11+y21); grid
>> title('Impulse Response when Both u_1 and u_2 are given at t = 0')
>> ylabel('y_1')
>> subplot(212); plot(t,y12+y22); grid
>> xlabel('t (sec)'); ylabel('y_2')
```

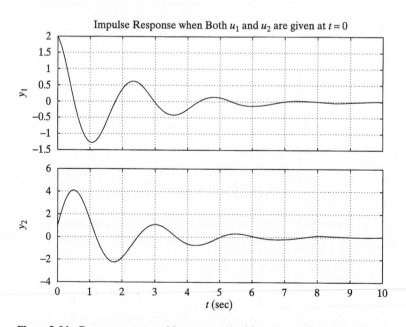

Figure 5–24 Response curves $y_1(t)$ versus t and $y_2(t)$ versus t when $u_1(t)$ and $u_2(t)$ are given at the same time. [Both $u_1(t)$ and $u_2(t)$ are unit-impulse inputs occurring at $t = 0$.]

Problem A–5–6

Obtain the unit-step response and unit-impulse response of the following system with MATLAB:

$$\begin{bmatrix} \dot{x}_1 \\ \dot{x}_2 \\ \dot{x}_3 \\ \dot{x}_4 \end{bmatrix} = \begin{bmatrix} 0 & 1 & 0 & 0 \\ 0 & 0 & 1 & 0 \\ 0 & 0 & 0 & 1 \\ -0.01 & -0.1 & -0.5 & -1.5 \end{bmatrix} \begin{bmatrix} x_1 \\ x_2 \\ x_3 \\ x_4 \end{bmatrix} + \begin{bmatrix} 0 \\ 0.04 \\ -0.012 \\ 0.008 \end{bmatrix} u$$

$$y = \begin{bmatrix} 1 & 0 & 0 & 0 \end{bmatrix} \begin{bmatrix} x_1 \\ x_2 \\ x_3 \\ x_4 \end{bmatrix}$$

The initial conditions are zeros.

Solution To obtain the unit-step response of this system, the following command may be used:

$$[y, x, t] = step(A, B, C, D)$$

Since the unit-impulse response is the derivative of the unit-step response, the derivative of the output ($y = x1$) will give the unit-impulse response. From the state equation, we see that the derivative of y is

$$x2 = [0 \quad 1 \quad 0 \quad 0]*x'$$

Hence, $x2$ versus t will give the unit-impulse response.

MATLAB Program 5–15 produces both the unit-step and unit-impulse responses. The resulting unit-step response curve and unit-impulse curve are shown in Figure 5–25.

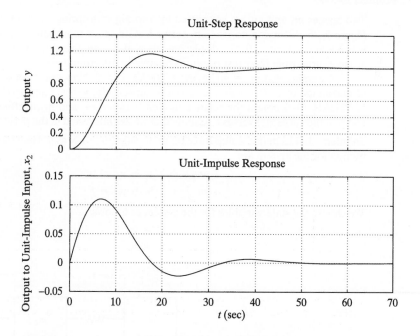

Figure 5–25 Unit-step response curve and unit-impulse response curve.

MATLAB Program 5–15

```
>> A = [0  1  0  0;0  0  1  0;0  0  0  1;−0.01  −0.1  −0.5  −1.5];
>> B = [0;0.04;−0.012;0.008];
>> C = [1  0  0  0];
>> D = 0;
>>
>> % To get the step response, enter, for example, the following
>> % command:
>>
>> [y,x,t] = step(A,B,C,D);
>> subplot(211); plot(t,y); grid
>> title('Unit-Step Response')
>> ylabel('Output y')
>>
>> % The unit-impulse response of the system is the same as the
>> % derivative of the unit-step response. (Note that x_1dot
>> % = x_2 in this system.) Hence, the unit-impulse response
>> % of this system is given by ydot = x_2. To plot the unit-
>> % impulse response curve, enter the following command:
>>
>> x2 = [0  1  0  0]*x'; subplot(212); plot(t,x2); grid
>> title('Unit-Impulse Response')
>> xlabel('t (sec)'); ylabel('Output to Unit-Impulse Input, x_2')
```

Problem A–5–7

Two masses m_1 and m_2 are connected by a spring with spring constant k, as shown in Figure 5–26. Assuming no friction, derive a state-space representation of the system, which is at rest for $t < 0$. The displacements y_1 and y_2 are the outputs of the system and are measured from their rest positions relative to the ground.

Assuming that $m_1 = 40$ kg, $m_2 = 100$ kg, $k = 40$ N/m, and f is a step force input of magnitude of 10 N, obtain the response curves $y_1(t)$ versus t and $y_2(t)$ versus t with MATLAB. Also, obtain the relative motion between m_1 and m_2. Define $y_2 - y_1 = x$ and plot the curve $x(t)$ versus t. Assume that we are interested in the period $0 \le t \le 20$.

Solution   Let us define a step force input of magnitude 1 N as u. Then the equations of motion for the system are

$$m_1\ddot{y}_1 + k(y_1 - y_2) = 0$$
$$m_2\ddot{y}_2 + k(y_2 - y_1) = f$$

We choose the state variables for the system as follows:

$$x_1 = y_1$$
$$x_2 = \dot{y}_1$$

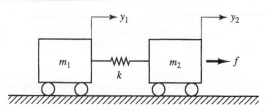

Figure 5–26   Mechanical system.

$$x_3 = y_2$$
$$x_4 = \dot{y}_2$$

Then we obtain

$$\dot{x}_1 = x_2$$

$$\dot{x}_2 = -\frac{k}{m_1}x_1 + \frac{k}{m_1}x_3$$

$$\dot{x}_3 = x_4$$

$$\dot{x}_4 = \frac{k}{m_2}x_1 - \frac{k}{m_2}x_3 + \frac{1}{m_2}f$$

Noting that $f = 10u$ and substituting the given numerical values for m_1, m_2, and k, we obtain the state equation

$$
\begin{bmatrix} \dot{x}_1 \\ \dot{x}_2 \\ \dot{x}_3 \\ \dot{x}_4 \end{bmatrix} =
\begin{bmatrix} 0 & 1 & 0 & 0 \\ -1 & 0 & 1 & 0 \\ 0 & 0 & 0 & 1 \\ 0.4 & 0 & -0.4 & 0 \end{bmatrix}
\begin{bmatrix} x_1 \\ x_2 \\ x_3 \\ x_4 \end{bmatrix} +
\begin{bmatrix} 0 \\ 0 \\ 0 \\ 0.1 \end{bmatrix} u
$$

The output equation is

$$
\begin{bmatrix} y_1 \\ y_2 \end{bmatrix} =
\begin{bmatrix} 1 & 0 & 0 & 0 \\ 0 & 0 & 1 & 0 \end{bmatrix}
\begin{bmatrix} x_1 \\ x_2 \\ x_3 \\ x_4 \end{bmatrix} + 0u
$$

MATLAB Program 5–16 produces the outputs y_1 and y_2 and the relative motion $x(= y_2 - y_1 = x_3 - x_1)$. The resulting response curves $y_1(t)$ versus t, $y_2(t)$ versus t, and $x(t)$ versus t are shown in Figure 5–27. Notice that the vibration between m_1 and m_2 continues forever.

MATLAB Program 5–16

```
>> t = 0:0.02:20;
>> A = [0  1  0  0;−1  0  1  0;0  0  0  1;0.4  0  −0.4  0];
>> B = [0;0;0;0.1];
>> C = [1  0  0  0;0  0  1  0];
>> D = 0;
>> sys = ss(A,B,C,D);
>> [y,t,x] = step(sys,t);
>> y1 = [1    0]*y';
>> y2 = [0    1]*y';
>> subplot(311); plot(t,y1), grid
>> title('Step Response')
>> ylabel('Output y_1')
>> subplot(312); plot(t,y2), grid
>> ylabel('Output y_2')
>> subplot(313); plot(t,y2 − y1), grid
>> xlabel('t (sec)'); ylabel('x = y_2 − y_1')
```

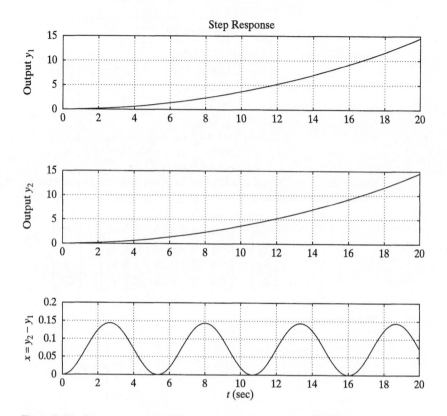

Figure 5–27   Response curves $y_1(t)$ versus t, $y_2(t)$ versus t, and $x(t)$ versus t.

Problem A–5–8

Obtain the unit-ramp response of the following system:

$$\begin{bmatrix} \dot{x}_1 \\ \dot{x}_2 \end{bmatrix} = \begin{bmatrix} 0 & 1 \\ -1 & -0.4 \end{bmatrix} \begin{bmatrix} x_1 \\ x_2 \end{bmatrix} + \begin{bmatrix} 0 \\ 1 \end{bmatrix} u$$

$$y = \begin{bmatrix} 1 & 0 \end{bmatrix} \begin{bmatrix} x_1 \\ x_2 \end{bmatrix} + [0]u$$

The system is initially at rest.

Solution   Noting that the unit-ramp input is defined by

$$u = t \qquad (0 \le t)$$

we may use the command

$$\text{lsim(sys, u, t)}$$

as shown in MATLAB Program 5–17. The unit-ramp response curve and the unit-ramp input are shown in Figure 5–28.

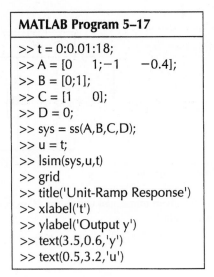

MATLAB Program 5–17

```
>> t = 0:0.01:18;
>> A = [0    1;−1    −0.4];
>> B = [0;1];
>> C = [1    0];
>> D = 0;
>> sys = ss(A,B,C,D);
>> u = t;
>> lsim(sys,u,t)
>> grid
>> title('Unit-Ramp Response')
>> xlabel('t')
>> ylabel('Output y')
>> text(3.5,0.6,'y')
>> text(0.5,3.2,'u')
```

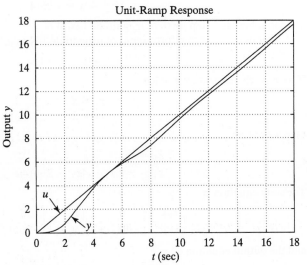

Figure 5–28 Plot of unit-ramp response curve, together with unit-ramp input.

Problem A–5–9

A mass M (where $M = 8$ kg) is supported by a spring (where $k = 400$ N/m) and a damper (where $b = 40$ N-s/m), as shown in Figure 5–29. At $t = 0$, a mass $m = 2$ kg is gently placed on the top of mass M, causing the system to exhibit vibrations. Assuming that the displacement x of the combined mass is measured from the equilibrium position before m is placed on M, obtain a state-space representation of the system. Then plot the response curve $x(t)$ versus t. (For an analytical solution, see **Problem A–3–16**.)

Solution   The equation of motion for the system is

$$(M + m)\ddot{x} + b\dot{x} + kx = mg \qquad (0 < t)$$

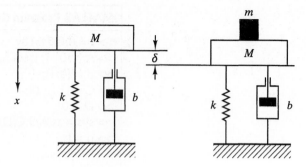

Figure 5–29   Mechanical system.

Substituting the given numerical values for M, m, b, k, and g = 9.807 m/s^2 into this last equation, we obtain

$$10\ddot{x} + 40\dot{x} + 400x = 2 \times 9.807$$

or

$$\ddot{x} + 4\dot{x} + 40x = 1.9614$$

The input here is a step force of magnitude 1.9614 N.

Let us define a step force input of magnitude 1 N as u. Then we have

$$\ddot{x} + 4\dot{x} + 40x = 1.9614u$$

If we now choose state variables

$$x_1 = x$$
$$x_2 = \dot{x}$$

then we obtain

$$\dot{x}_1 = x_2$$
$$\dot{x}_2 = -40x_1 - 4x_2 + 1.9614u$$

The state equation is

$$\begin{bmatrix} \dot{x}_1 \\ \dot{x}_2 \end{bmatrix} = \begin{bmatrix} 0 & 1 \\ -40 & -4 \end{bmatrix} \begin{bmatrix} x_1 \\ x_2 \end{bmatrix} + \begin{bmatrix} 0 \\ 1.9614 \end{bmatrix} u$$

and the output equation is

$$y = \begin{bmatrix} 1 & 0 \end{bmatrix} \begin{bmatrix} x_1 \\ x_2 \end{bmatrix} + 0u$$

MATLAB Program 5–18 produces the response curve $y(t)$ [$= x(t)$] versus t, shown in Figure 5–30. Notice that the static deflection $x(\infty) = y(\infty) \doteq y(600)$ is 0.049035 m.

Problem A–5–10

Consider the system shown in Figure 5–31. The system is at rest for $t < 0$. The displacements z_1 and z_2 are measured from their respective equilibrium positions relative to the ground. Choosing z_1, $\dot{z}_1$, z_2, and $\dot{z}_2$ as state variables, derive a state-space representation of the system. Assuming that $m_1 = 10$ kg, $m_2 = 20$ kg, $b = 20$ N-s/m, $k = 60$ N/m, and f is a step force input of magnitude 10 N, plot the response curves $z_1(t)$ versus t, $z_2(t)$ versus t, $z_2(t) - z_1(t)$ versus t, and $\dot{z}_2(t) - \dot{z}_1(t)$ versus t. Also, obtain the steady-state values of $\ddot{z}_1$, $\ddot{z}_2$, and $z_2 - z_1$.

MATLAB Program 5–18

```
>> t = 0:0.01:6;
>> A = [0    1;−40    −4];
>> B = [0;1.9614];
>> C = [1    0];
>> D = 0;
>> sys = ss(A,B,C,D);
>> [y,t] = step(sys,t);
>> plot(t,y)
>> grid
>> title('Step Response')
>> xlabel('t (sec)'); ylabel('Output y')
>>
>> format long;
>> y(600)

ans =

    0.04903515818520
```

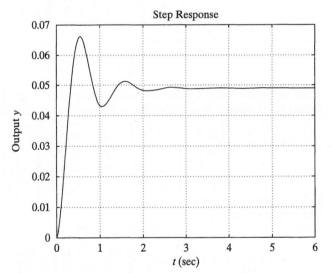

Figure 5–30   Step-response curve.

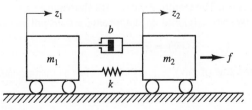

Figure 5–31   Mechanical system.

Solution   The equations of motion for the system are

$$m_1\ddot{z}_1 = k(z_2 - z_1) + b(\dot{z}_2 - \dot{z}_1) \tag{5-76}$$
$$m_2\ddot{z}_2 = -k(z_2 - z_1) - b(\dot{z}_2 - \dot{z}_1) + f \tag{5-77}$$

Since we chose state variables as

$$x_1 = z_1$$
$$x_2 = \dot{z}_1$$
$$x_3 = z_2$$
$$x_4 = \dot{z}_2$$

Equations (5–76) and (5–77) can be written as

$$m_1\dot{x}_2 = k(x_3 - x_1) + b(x_4 - x_2)$$
$$m_2\dot{x}_4 = -k(x_3 - x_1) - b(x_4 - x_2) + f$$

We thus have

$$\dot{x}_1 = x_2$$
$$\dot{x}_2 = -\frac{k}{m_1}x_1 - \frac{b}{m_1}x_2 + \frac{k}{m_1}x_3 + \frac{b}{m_1}x_4$$
$$\dot{x}_3 = x_4$$
$$\dot{x}_4 = \frac{k}{m_2}x_1 + \frac{b}{m_2}x_2 - \frac{k}{m_2}x_3 - \frac{b}{m_2}x_4 + \frac{1}{m_2}f$$

Let us define z_1 and z_2 as the system outputs. Then

$$y_1 = z_1 = x_1$$
$$y_2 = z_2 = x_3$$

After substitution of the given numerical values and $f = 10u$ (where u is a step force input of magnitude 1 N occurring at $t = 0$), the state equation becomes

$$\begin{bmatrix} \dot{x}_1 \\ \dot{x}_2 \\ \dot{x}_3 \\ \dot{x}_4 \end{bmatrix} = \begin{bmatrix} 0 & 1 & 0 & 0 \\ -6 & -2 & 6 & 2 \\ 0 & 0 & 0 & 1 \\ 3 & 1 & -3 & -1 \end{bmatrix} \begin{bmatrix} x_1 \\ x_2 \\ x_3 \\ x_4 \end{bmatrix} + \begin{bmatrix} 0 \\ 0 \\ 0 \\ 0.5 \end{bmatrix} u$$

The output equation is

$$\begin{bmatrix} y_1 \\ y_2 \end{bmatrix} = \begin{bmatrix} 1 & 0 & 0 & 0 \\ 0 & 0 & 1 & 0 \end{bmatrix} \begin{bmatrix} x_1 \\ x_2 \\ x_3 \\ x_4 \end{bmatrix} + \begin{bmatrix} 0 \\ 0 \end{bmatrix} u$$

MATLAB Program 5–19 produces the response curves z_1 versus t, z_2 versus t, $z_2 - z_1$ versus t, and $\dot{z}_2 - \dot{z}_1$ versus t. The resulting curves are shown in Figure 5–32. Note that at steady state $\ddot{z}_1(t)$ and $\ddot{z}_2(t)$ approach a constant value, or

$$\ddot{z}_1(\infty) = \ddot{z}_2(\infty) = \alpha$$

Also, at steady state the value of $z_2(t) - z_1(t)$ approaches a constant value, or

$$z_2(\infty) - z_1(\infty) = \beta$$

MATLAB Program 5–19

```
>> t = 0:0.01:15;
>> A = [0   1   0   0;−6  −2   6   2;0   0   0   1;3   1  −3  −1];
>> B = [0;0;0;0.5];
>> C = [1   0   0   0;0   0   1   0];
>> D = [0;0];
>> sys = ss(A,B,C,D);
>> [y,t,x] = step(sys, t);
>> x1 = [1   0   0   0]*x';
>> x2 = [0   1   0   0]*x';
>> x3 = [0   0   1   0]*x';
>> x4 = [0   0   0   1]*x';
>> subplot(221); plot(t,x1); grid
>> xlabel('t (sec)'); ylabel('Output z_1')
>> subplot(222); plot(t,x3); grid
>> xlabel('t (sec)'); ylabel('Output z_2')
>> subplot(223); plot(t,x3 − x1); grid
>> xlabel('t (sec)'); ylabel('Output z_2 − z_1')
>> subplot(224); plot(t,x4 − x2); grid
>> xlabel('t (sec)'); ylabel('z_2dot − z_1dot')
```

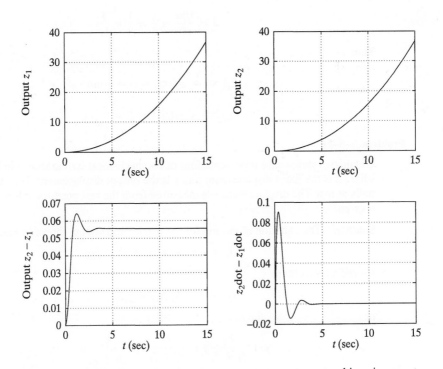

Figure 5–32 Response curves z_1 versus t, z_2 versus t, $z_2 - z_1$ versus t, and $\dot{z}_2 - \dot{z}_1$ versus t.

The steady-state value of $\dot{z}_2(t) - \dot{z}_1(t)$ is zero, or

$$\dot{z}_2(\infty) - \dot{z}_1(\infty) = 0$$

For $t = \infty$, Equation (5–76) becomes

$$m_1\ddot{z}_1(\infty) = k[z_2(\infty) - z_1(\infty)] + b[\dot{z}_2(\infty) - \dot{z}_1(\infty)]$$

or

$$10\alpha = k\beta + b \times 0$$

Also, Equation (5–77) becomes

$$m_2\ddot{z}_2(\infty) = -k[z_2(\infty) - z_1(\infty)] - b[\dot{z}_2(\infty) - \dot{z}_1(\infty)] + f$$

or

$$20\alpha = -k\beta - b \times 0 + f$$

Hence,

$$10\alpha = 60\beta$$
$$20\alpha = -60\beta + f$$

from which we get

$$\alpha = \frac{f}{30} = \frac{10}{30} = \frac{1}{3}$$

and

$$\beta = \frac{10\alpha}{60} = \frac{1}{6} \times \frac{1}{3} = \frac{1}{18}$$

Thus,

$$\ddot{z}_1(\infty) = \ddot{z}_2(\infty) = \alpha = \frac{1}{3}\,\text{m/s}^2$$

$$z_2(\infty) - z_1(\infty) = \beta = \frac{1}{18}\,\text{m}$$

Problem A–5–11

Obtain two state-space representations of the mechanical system shown in Figure 5–33 where u is the input displacement and y is the output displacement. The system is initially at rest. The displacement y is measured from the rest position before the input u is given.

Solution The equation of motion for the mechanical system shown in Figure 5–33 is

$$f_1(\dot{u} - \dot{y}) + k_1(u - y) = f_2\dot{y}$$

Rewriting, we obtain

$$(f_1 + f_2)\dot{y} + k_1y = f_1\dot{u} + k_1u$$

or

$$\dot{y} + \frac{k_1}{f_1 + f_2}y = \frac{f_1}{f_1 + f_2}\dot{u} + \frac{k_1}{f_1 + f_2}u \qquad (5\text{–}78)$$

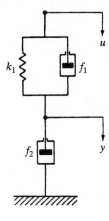

Figure 5-33 Mechanical system.

Comparing this last equation with

$$\dot{y} + a_1 y = b_0 \dot{u} + b_1 u \qquad (5\text{--}79)$$

we get

$$a_1 = \frac{k_1}{f_1 + f_2}, \qquad b_0 = \frac{f_1}{f_1 + f_2}, \qquad b_1 = \frac{k_1}{f_1 + f_2}$$

We shall obtain two state-space representations of the system, based on Methods 1 and 2 presented in Section 5–4.

Method 1. First calculate β_0 and β_1:

$$\beta_0 = b_0 = \frac{f_1}{f_1 + f_2}$$

$$\beta_1 = b_1 - a_1 \beta_0 = \frac{k_1 f_2}{(f_1 + f_2)^2}$$

Define the state variable x by

$$x = y - \beta_0 u = y - \frac{f_1}{f_1 + f_2} u$$

Then the state equation can be obtained from Equation (5–78) as follows:

$$\dot{x} = -\frac{k_1}{f_1 + f_2} x + \frac{k_1 f_2}{(f_1 + f_2)^2} u \qquad (5\text{--}80)$$

The output equation is

$$y = x + \frac{f_1}{f_1 + f_2} u \qquad (5\text{--}81)$$

Equations (5–80) and (5–81) give a state-space representation of the system.

Method 2. From Equation (5–79), we have

$$\frac{Y(s)}{U(s)} = \frac{b_0 s + b_1}{s + a_1}$$

If we define

$$\frac{Z(s)}{U(s)} = \frac{1}{s + a_1}, \qquad \frac{Y(s)}{Z(s)} = b_0 s + b_1$$

then we get

$$\dot{z} + a_1 z = u \qquad \qquad (5\text{--}82)$$
$$b_0 \dot{z} + b_1 z = y \qquad \qquad (5\text{--}83)$$

Next, we define the state variable x by

$$x = z$$

Then Equation (5–82) can be written as

$$\dot{x} = -a_1 x + u$$

or

$$\dot{x} = -\frac{k_1}{f_1 + f_2} x + u \qquad \qquad (5\text{--}84)$$

and Equation (5–83) becomes

$$b_0 \dot{x} + b_1 x = y$$

or

$$y = \frac{k_1}{f_1 + f_2} x + \frac{f_1}{f_1 + f_2} \dot{x} \qquad \qquad (5\text{--}85)$$

Substituting Equation (5–84) into Equation (5–85), we get

$$y = \frac{k_1 f_2}{(f_1 + f_2)^2} x + \frac{f_1}{f_1 + f_2} u \qquad \qquad (5\text{--}86)$$

Equations (5–84) and (5–86) give a state-space representation of the system.

Problem A–5–12

Show that, for the differential-equation system

$$\dddot{y} + a_1 \ddot{y} + a_2 \dot{y} + a_3 y = b_0 \dddot{u} + b_1 \ddot{u} + b_2 \dot{u} + b_3 u \qquad \qquad (5\text{--}87)$$

state and output equations can be given, respectively, by

$$\begin{bmatrix} \dot{x}_1 \\ \dot{x}_2 \\ \dot{x}_3 \end{bmatrix} = \begin{bmatrix} 0 & 1 & 0 \\ 0 & 0 & 1 \\ -a_3 & -a_2 & -a_1 \end{bmatrix} \begin{bmatrix} x_1 \\ x_2 \\ x_3 \end{bmatrix} + \begin{bmatrix} \beta_1 \\ \beta_2 \\ \beta_3 \end{bmatrix} u \qquad \qquad (5\text{--}88)$$

and

$$y = \begin{bmatrix} 1 & 0 & 0 \end{bmatrix} \begin{bmatrix} x_1 \\ x_2 \\ x_3 \end{bmatrix} + \beta_0 u \qquad \qquad (5\text{--}89)$$

where the state variables are defined by

$$x_1 = y - \beta_0 u$$
$$x_2 = \dot{y} - \beta_0 \dot{u} - \beta_1 u = \dot{x}_1 - \beta_1 u$$
$$x_3 = \ddot{y} - \beta_0 \ddot{u} - \beta_1 \dot{u} - \beta_2 u = \dot{x}_2 - \beta_2 u$$

The constants, β_0, β_1, β_2, and β_3 are defined by

$$\beta_0 = b_0$$
$$\beta_1 = b_1 - a_1\beta_0$$
$$\beta_2 = b_2 - a_1\beta_1 - a_2\beta_0$$
$$\beta_3 = b_3 - a_1\beta_2 - a_2\beta_1 - a_3\beta_0$$

Solution   From the definition of the state variables x_2 and x_3, we have

$$\dot{x}_1 = x_2 + \beta_1 u \qquad\qquad (5\text{–}90)$$
$$\dot{x}_2 = x_3 + \beta_2 u \qquad\qquad (5\text{–}91)$$

To derive the equation for $\dot{x}_3$, we note that

$$\dddot{y} = -a_1\ddot{y} - a_2\dot{y} - a_3 y + b_0\dddot{u} + b_1\ddot{u} + b_2\dot{u} + b_3 u$$

Since

$$x_3 = \ddot{y} - \beta_0\ddot{u} - \beta_1\dot{u} - \beta_2 u$$

we have

$$
\begin{aligned}
\dot{x}_3 &= \dddot{y} - \beta_0\dddot{u} - \beta_1\ddot{u} - \beta_2\dot{u} \\
&= (-a_1\ddot{y} - a_2\dot{y} - a_3 y) + b_0\dddot{u} + b_1\ddot{u} + b_2\dot{u} + b_3 u - \beta_0\dddot{u} - \beta_1\ddot{u} - \beta_2\dot{u} \\
&= -a_1(\ddot{y} - \beta_0\ddot{u} - \beta_1\dot{u} - \beta_2 u) - a_1\beta_0\ddot{u} - a_1\beta_1\dot{u} - a_1\beta_2 u \\
&\quad - a_2(\dot{y} - \beta_0\dot{u} - \beta_1 u) - a_2\beta_0\dot{u} - a_2\beta_1 u - a_3(y - \beta_0 u) - a_3\beta_0 u \\
&\quad + b_0\dddot{u} + b_1\ddot{u} + b_2\dot{u} + b_3 u - \beta_0\dddot{u} - \beta_1\ddot{u} - \beta_2\dot{u} \\
&= -a_1 x_3 - a_2 x_2 - a_3 x_1 + (b_0 - \beta_0)\dddot{u} + (b_1 - \beta_1 - a_1\beta_0)\ddot{u} \\
&\quad + (b_2 - \beta_2 - a_1\beta_1 - a_2\beta_0)\dot{u} + (b_3 - a_1\beta_2 - a_2\beta_1 - a_3\beta_0)u \\
&= -a_1 x_3 - a_2 x_2 - a_3 x_1 + (b_3 - a_1\beta_2 - a_2\beta_1 - a_3\beta_0)u \\
&= -a_1 x_3 - a_2 x_2 - a_3 x_1 + \beta_3 u
\end{aligned}
$$

Hence, we get

$$\dot{x}_3 = -a_3 x_1 - a_2 x_2 - a_1 x_3 + \beta_3 u \qquad\qquad (5\text{–}92)$$

Combining Equations (5–90), (5–91), and (5–92) into a vector-matrix differential equation, we obtain Equation (5–88). Also, from the definition of state variable x_1, we get the output equation given by Equation (5–89).

Note that the derivation presented here can be easily extended to the general case of an nth-order system.

Problem A–5–13

Show that, for the system

$$\dddot{y} + a_1\ddot{y} + a_2\dot{y} + a_3 y = b_0\dddot{u} + b_1\ddot{u} + b_2\dot{u} + b_3 u$$

or

$$\frac{Y(s)}{U(s)} = \frac{b_0 s^3 + b_1 s^2 + b_2 s + b_3}{s^3 + a_1 s^2 + a_2 s + a_3}$$

state and output equations may be given, respectively, by

$$
\begin{bmatrix} \dot{x}_1 \\ \dot{x}_2 \\ \dot{x}_3 \end{bmatrix} =
\begin{bmatrix} 0 & 1 & 0 \\ 0 & 0 & 1 \\ -a_3 & -a_2 & -a_1 \end{bmatrix}
\begin{bmatrix} x_1 \\ x_2 \\ x_3 \end{bmatrix} +
\begin{bmatrix} 0 \\ 0 \\ 1 \end{bmatrix} u
$$

and

$$y = [b_3 - a_3 b_0 \vdots b_2 - a_2 b_0 \vdots b_1 - a_1 b_0] \begin{bmatrix} x_1 \\ x_2 \\ x_3 \end{bmatrix} + b_0 u$$

Solution   Let us define

$$\frac{Z(s)}{U(s)} = \frac{1}{s^3 + a_1 s^2 + a_2 s + a_3}, \qquad \frac{Y(s)}{Z(s)} = b_0 s^3 + b_1 s^2 + b_2 s + b_3$$

Then we obtain

$$\dddot{z} + a_1 \ddot{z} + a_2 \dot{z} + a_3 z = u$$
$$b_0 \dddot{z} + b_1 \ddot{z} + b_2 \dot{z} + b_3 z = y$$

Now we define

$$x_1 = z$$
$$x_2 = \dot{x}_1 \tag{5-93}$$
$$x_3 = \dot{x}_2 \tag{5-94}$$

Then, noting that $\dot{x}_3 = \ddot{x}_2 = \dddot{x}_1 = \dddot{z}$, we obtain

$$\dot{x}_3 = -a_3 z - a_2 \dot{z} - a_1 \ddot{z} + u$$

or

$$\dot{x}_3 = -a_3 x_1 - a_2 x_2 - a_1 x_3 + u \tag{5-95}$$

Also,

$$\begin{aligned} y &= b_0 \dddot{z} + b_1 \ddot{z} + b_2 \dot{z} + b_3 z \\ &= b_0 \dot{x}_3 + b_1 x_3 + b_2 x_2 + b_3 x_1 \\ &= b_0 [(-a_3 x_1 - a_2 x_2 - a_1 x_3) + u] + b_1 x_3 + b_2 x_2 + b_3 x_1 \\ &= (b_3 - a_3 b_0) x_1 + (b_2 - a_2 b_0) x_2 + (b_1 - a_1 b_0) x_3 + b_0 u \end{aligned} \tag{5-96}$$

From Equations (5–93), (5–94), and (5–95), we obtain

$$\begin{bmatrix} \dot{x}_1 \\ \dot{x}_2 \\ \dot{x}_3 \end{bmatrix} = \begin{bmatrix} 0 & 1 & 0 \\ 0 & 0 & 1 \\ -a_3 & -a_2 & -a_1 \end{bmatrix} \begin{bmatrix} x_1 \\ x_2 \\ x_3 \end{bmatrix} + \begin{bmatrix} 0 \\ 0 \\ 1 \end{bmatrix} u$$

which is the state equation. From Equation (5–96), we get

$$y = [b_3 - a_3 b_0 \vdots b_2 - a_2 b_0 \vdots b_1 - a_1 b_0] \begin{bmatrix} x_1 \\ x_2 \\ x_3 \end{bmatrix} + b_0 u$$

which is the output equation.

Note that the derivation presented here can be easily extended to the general case of an nth-order system.

Problem A–5–14

Consider the mechanical system shown in Figure 5–34. The system is initially at rest. The displacements u, y, and z are measured from their respective rest positions.

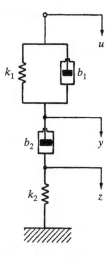

Figure 5–34 Mechanical system.

Assuming that u is the input and y is the output, obtain the transfer function $Y(s)/U(s)$ of the system. Then obtain a state-space representation of the system.

Solution The equations of motion for the system are

$$b_1(\dot{u} - \dot{y}) + k_1(u - y) = b_2(\dot{y} - \dot{z})$$
$$b_2(\dot{y} - \dot{z}) = k_2 z$$

Laplace transforming these two equations, assuming zero initial conditions, we obtain

$$b_1[sU(s) - sY(s)] + k_1[U(s) - Y(s)] = b_2[sY(s) - sZ(s)]$$
$$b_2[sY(s) - sZ(s)] = k_2 Z(s)$$

Eliminating $Z(s)$ from the last two equations yields

$$(b_1 s + k_1)U(s) = \left(b_1 s + k_1 + b_2 s - \frac{b_2^2 s^2}{b_2 s + k_2}\right)Y(s)$$

Multiplying both sides of this last equation by $(b_2 s + k_2)$, we get

$$(b_1 s + k_1)(b_2 s + k_2)U(s) = [(b_1 s + k_1)(b_2 s + k_2) + b_2 k_2 s]Y(s)$$

The transfer function of the system then becomes

$$\frac{Y(s)}{U(s)} = \frac{(b_1 s + k_1)(b_2 s + k_2)}{(b_1 s + k_1)(b_2 s + k_2) + b_2 k_2 s}$$

$$= \frac{s^2 + \left(\dfrac{k_1}{b_1} + \dfrac{k_2}{b_2}\right)s + \dfrac{k_1 k_2}{b_1 b_2}}{s^2 + \left(\dfrac{k_1}{b_1} + \dfrac{k_2}{b_2} + \dfrac{k_2}{b_1}\right)s + \dfrac{k_1 k_2}{b_1 b_2}} \qquad (5\text{--}97)$$

Next, we shall obtain a state-space representation of the system. The differential equation corresponding to Equation (5–97) is

$$\ddot{y} + \left(\frac{k_1}{b_1} + \frac{k_2}{b_2} + \frac{k_2}{b_1}\right)\dot{y} + \frac{k_1 k_2}{b_1 b_2}y = \ddot{u} + \left(\frac{k_1}{b_1} + \frac{k_2}{b_2}\right)\dot{u} + \frac{k_1 k_2}{b_1 b_2}u$$

Comparing this equation with the standard second-order differential equation given by Equation (5–20), namely,

$$\ddot{y} + a_1 \dot{y} + a_2 y = b_0 \ddot{u} + b_1 \dot{u} + b_2 u$$

we find that

$$a_1 = \frac{k_1}{b_1} + \frac{k_2}{b_2} + \frac{k_2}{b_1}, \qquad a_2 = \frac{k_1 k_2}{b_1 b_2}$$

$$b_0 = 1, \qquad b_1 = \frac{k_1}{b_1} + \frac{k_2}{b_2}, \qquad b_2 = \frac{k_1 k_2}{b_1 b_2}$$

From Equations (5–23), (5–24), and (5–29), we have

$$\beta_0 = b_0 = 1$$

$$\beta_1 = b_1 - a_1 \beta_0 = -\frac{k_2}{b_1}$$

$$\beta_2 = b_2 - a_1 \beta_1 - a_2 \beta_0 = \frac{k_1 k_2}{b_1^2} + \frac{k_2^2}{b_1 b_2} + \frac{k_2^2}{b_1^2}$$

From Equations (5–21) and (5–22), we define the state variables x_1 and x_2 as

$$x_1 = y - \beta_0 u = y - u$$

$$x_2 = \dot{x}_1 - \beta_1 u = \dot{x}_1 + \frac{k_2}{b_1} u$$

The state equation is given by Equation (5–30) as

$$\begin{bmatrix} \dot{x}_1 \\ \dot{x}_2 \end{bmatrix} = \begin{bmatrix} 0 & 1 \\ -a_2 & -a_1 \end{bmatrix} \begin{bmatrix} x_1 \\ x_2 \end{bmatrix} + \begin{bmatrix} \beta_1 \\ \beta_2 \end{bmatrix} u$$

or

$$\begin{bmatrix} \dot{x}_1 \\ \dot{x}_2 \end{bmatrix} = \begin{bmatrix} 0 & 1 \\ -\dfrac{k_1 k_2}{b_1 b_2} & -\left(\dfrac{k_1}{b_1} + \dfrac{k_2}{b_2} + \dfrac{k_2}{b_1} \right) \end{bmatrix} \begin{bmatrix} x_1 \\ x_2 \end{bmatrix} + \begin{bmatrix} -\dfrac{k_2}{b_1} \\ \dfrac{k_1 k_2}{b_1^2} + \dfrac{k_2^2}{b_1 b_2} + \dfrac{k_2^2}{b_1^2} \end{bmatrix} u \qquad (5\text{–}98)$$

The output equation is given by Equation (5–31) as

$$y = \begin{bmatrix} 1 & 0 \end{bmatrix} \begin{bmatrix} x_1 \\ x_2 \end{bmatrix} + \beta_0 u$$

or

$$y = \begin{bmatrix} 1 & 0 \end{bmatrix} \begin{bmatrix} x_1 \\ x_2 \end{bmatrix} + u \qquad (5\text{–}99)$$

Equations (5–98) and (5–99) constitute a state-space representation of the system.

Problem A–5–15

Consider the mechanical system shown in Figure 5–35, in which $m = 0.1$ kg, $b = 0.4$ N-s/m, $k_1 = 6$ N/m, and $k_2 = 4$ N/m. The displacements y and z are measured from their respective equilibrium positions. Assume that force u is the input to the system. Considering that displacement y is the output, obtain the transfer function $Y(s)/U(s)$. Also, obtain a state-space representation of the system.

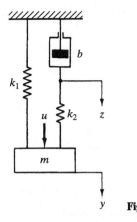

y   **Figure 5–35**   Mechanical system.

Solution   The equations of motion for the system are

$$m\ddot{y} + k_1 y + k_2(y - z) = u \tag{5–100}$$
$$k_2(y - z) = b\dot{z} \tag{5–101}$$

Taking the Laplace transforms of Equations (5–100) and (5–101), assuming zero initial conditions, we obtain

$$[ms^2 + (k_1 + k_2)]Y(s) = k_2 Z(s) + U(s)$$
$$k_2 Y(s) = (k_2 + bs)Z(s)$$

Eliminating $Z(s)$ from these two equations yields

$$\frac{Y(s)}{U(s)} = \frac{k_2 + bs}{mbs^3 + mk_2 s^2 + (k_1 + k_2)bs + k_1 k_2}$$

$$= \frac{\dfrac{1}{m}s + \dfrac{k_2}{mb}}{s^3 + \dfrac{k_2}{b}s^2 + \dfrac{k_1 + k_2}{m}s + \dfrac{k_1 k_2}{mb}}$$

Substituting numerical values for m, b, k_1, and k_2 into this last equation results in

$$\frac{Y(s)}{U(s)} = \frac{10s + 100}{s^3 + 10s^2 + 100s + 600} \tag{5–102}$$

This is the transfer function of the system.

Next, we shall obtain a state-space representation of the system using Method 1 presented in Section 5–4. From Equation (5–102), we obtain

$$\dddot{y} + 10\ddot{y} + 100\dot{y} + 600y = 10\dot{u} + 100u$$

Comparing this equation with the standard third-order differential equation, namely,

$$\dddot{y} + a_1\ddot{y} + a_2\dot{y} + a_3 y = b_0\dddot{u} + b_1\ddot{u} + b_2\dot{u} + b_3 u$$

we find that

$$a_1 = 10, \qquad a_2 = 100, \qquad a_3 = 600$$
$$b_0 = 0, \qquad b_1 = 0, \qquad b_2 = 10, \qquad b_3 = 100$$

Referring to **Problem A–5–12**, define

$$x_1 = y - \beta_0 u$$
$$x_2 = \dot{x}_1 - \beta_1 u$$
$$x_3 = \dot{x}_2 - \beta_2 u$$

where

$$\beta_0 = b_0 = 0$$
$$\beta_1 = b_1 - a_1\beta_0 = 0$$
$$\beta_2 = b_2 - a_1\beta_1 - a_2\beta_0 = 10$$

Also, note that

$$\beta_3 = b_3 - a_1\beta_2 - a_2\beta_1 - a_3\beta_0 = 100 - 10 \times 10 = 0$$

Then the state equation for the system becomes

$$\begin{bmatrix} \dot{x}_1 \\ \dot{x}_2 \\ \dot{x}_3 \end{bmatrix} = \begin{bmatrix} 0 & 1 & 0 \\ 0 & 0 & 1 \\ -600 & -100 & -10 \end{bmatrix} \begin{bmatrix} x_1 \\ x_2 \\ x_3 \end{bmatrix} + \begin{bmatrix} 0 \\ 10 \\ 0 \end{bmatrix} u \tag{5–103}$$

and the output equation becomes

$$y = \begin{bmatrix} 1 & 0 & 0 \end{bmatrix} \begin{bmatrix} x_1 \\ x_2 \\ x_3 \end{bmatrix} \tag{5–104}$$

Equations (5–103) and (5–104) give a state-space representation of the system.

Problem A–5–16

Consider the system defined by

$$\dddot{y} + 6\ddot{y} + 11\dot{y} + 6y = 6u \tag{5–105}$$

Obtain a state-space representation of the system by the partial-fraction expansion technique.

Solution   First, rewrite Equation (5–105) in the form of a transfer function:

$$\frac{Y(s)}{U(s)} = \frac{6}{s^3 + 6s^2 + 11s + 6} = \frac{6}{(s + 1)(s + 2)(s + 3)}$$

Next, expanding this transfer function into partial fractions, we get

$$\frac{Y(s)}{U(s)} = \frac{3}{s + 1} + \frac{-6}{s + 2} + \frac{3}{s + 3}$$

from which we obtain

$$Y(s) = \frac{3}{s + 1}U(s) + \frac{-6}{s + 2}U(s) + \frac{3}{s + 3}U(s) \tag{5–106}$$

Let us define

$$X_1(s) = \frac{3}{s + 1}U(s)$$
$$X_2(s) = \frac{-6}{s + 2}U(s)$$
$$X_3(s) = \frac{3}{s + 3}U(s)$$

Then, rewriting these three equations, we have

$$sX_1(s) = -X_1(s) + 3U(s)$$
$$sX_2(s) = -2X_2(s) - 6U(s)$$
$$sX_3(s) = -3X_3(s) + 3U(s)$$

The inverse Laplace transforms of the last three equations give

$$\dot{x}_1 = -x_1 + 3u \qquad (5\text{--}107)$$
$$\dot{x}_2 = -2x_2 - 6u \qquad (5\text{--}108)$$
$$\dot{x}_3 = -3x_3 + 3u \qquad (5\text{--}109)$$

Since Equation (5–106) can be written as

$$Y(s) = X_1(s) + X_2(s) + X_3(s)$$

we obtain

$$y = x_1 + x_2 + x_3 \qquad (5\text{--}110)$$

Combining Equations (5–107), (5–108), and (5–109) into a vector–matrix differential equation yields the following state equation:

$$\begin{bmatrix} \dot{x}_1 \\ \dot{x}_2 \\ \dot{x}_3 \end{bmatrix} = \begin{bmatrix} -1 & 0 & 0 \\ 0 & -2 & 0 \\ 0 & 0 & -3 \end{bmatrix} \begin{bmatrix} x_1 \\ x_2 \\ x_3 \end{bmatrix} + \begin{bmatrix} 3 \\ -6 \\ 3 \end{bmatrix} u \qquad (5\text{--}111)$$

From Equation (5–110), we get the following output equation:

$$y = \begin{bmatrix} 1 & 1 & 1 \end{bmatrix} \begin{bmatrix} x_1 \\ x_2 \\ x_3 \end{bmatrix} \qquad (5\text{--}112)$$

Equations (5–111) and (5–112) constitute a state-space representation of the system given by Equation (5–105). (Note that this representation is the same as that obtained in **Example 5–10**.)

Problem A–5–17

Show that the 2×2 matrix

$$\mathbf{A} = \begin{bmatrix} 1 & 1 \\ 0 & 2 \end{bmatrix}$$

has two distinct eigenvalues and that the eigenvectors are linearly independent of each other.

Solution The eigenvalues, obtained from

$$|\lambda \mathbf{I} - \mathbf{A}| = \begin{vmatrix} \lambda - 1 & -1 \\ 0 & \lambda - 2 \end{vmatrix} = (\lambda - 1)(\lambda - 2) = 0$$

are

$$\lambda_1 = 1 \qquad \text{and} \qquad \lambda_2 = 2$$

Thus, matrix $\mathbf{A}$ has two distinct eigenvalues.

There are two eigenvectors $\mathbf{x}_1$ and $\mathbf{x}_2$ associated with λ_1 and λ_2, respectively. If we define

$$\mathbf{x}_1 = \begin{bmatrix} x_{11} \\ x_{21} \end{bmatrix}, \qquad \mathbf{x}_2 = \begin{bmatrix} x_{12} \\ x_{22} \end{bmatrix}$$

then the eigenvector $\mathbf{x}_1$ can be found from

$$\mathbf{A}\mathbf{x}_1 = \lambda_1\mathbf{x}_1$$

or

$$(\lambda_1\mathbf{I} - \mathbf{A})\mathbf{x}_1 = \mathbf{0}$$

Noting that $\lambda_1 = 1$, we have

$$\begin{bmatrix} 1 - 1 & -1 \\ 0 & 1 - 2 \end{bmatrix}\begin{bmatrix} x_{11} \\ x_{21} \end{bmatrix} = \begin{bmatrix} 0 \\ 0 \end{bmatrix}$$

which gives

$$x_{11} = \text{arbitrary constant} \qquad \text{and} \qquad x_{21} = 0$$

Hence, eigenvector $\mathbf{x}_1$ may be written as

$$\mathbf{x}_1 = \begin{bmatrix} x_{11} \\ x_{21} \end{bmatrix} = \begin{bmatrix} c_1 \\ 0 \end{bmatrix}$$

where $c_1 \neq 0$ is an arbitrary constant.

Similarly, for the eigenvector $\mathbf{x}_2$, we have

$$\mathbf{A}\mathbf{x}_2 = \lambda_2\mathbf{x}_2$$

or

$$(\lambda_2\mathbf{I} - \mathbf{A})\mathbf{x}_2 = \mathbf{0}$$

Noting that $\lambda_2 = 2$, we obtain

$$\begin{bmatrix} 2 - 1 & -1 \\ 0 & 2 - 2 \end{bmatrix}\begin{bmatrix} x_{12} \\ x_{22} \end{bmatrix} = \begin{bmatrix} 0 \\ 0 \end{bmatrix}$$

from which we get

$$x_{12} - x_{22} = 0$$

Thus, the eigenvector associated with $\lambda_2 = 2$ may be selected as

$$\mathbf{x}_2 = \begin{bmatrix} x_{12} \\ x_{22} \end{bmatrix} = \begin{bmatrix} c_2 \\ c_2 \end{bmatrix}$$

where $c_2 \neq 0$ is an arbitrary constant.

The two eigenvectors are therefore given by

$$\mathbf{x}_1 = \begin{bmatrix} c_1 \\ 0 \end{bmatrix} \qquad \text{and} \qquad \mathbf{x}_2 = \begin{bmatrix} c_2 \\ c_2 \end{bmatrix}$$

That eigenvectors $\mathbf{x}_1$ and $\mathbf{x}_2$ are linearly independent can be seen from the fact that the determinant of the matrix $[\mathbf{x}_1\,\mathbf{x}_2]$ is nonzero:

$$\begin{vmatrix} c_1 & c_2 \\ 0 & c_2 \end{vmatrix} \neq 0$$

Problem A–5–18

Obtain the eigenvectors of the matrix

$$\mathbf{A} = \begin{bmatrix} 0 & 1 & 0 \\ 0 & 0 & 1 \\ -a_3 & -a_2 & -a_1 \end{bmatrix}$$

Assume that the eigenvalues are λ_1, λ_2, and λ_3; that is,

$$|\lambda\mathbf{I} - \mathbf{A}| = \begin{vmatrix} \lambda & -1 & 0 \\ 0 & \lambda & -1 \\ a_3 & a_2 & \lambda + a_1 \end{vmatrix}$$

$$= \lambda^3 + a_1\lambda^2 + a_2\lambda + a_3$$

$$= (\lambda - \lambda_1)(\lambda - \lambda_2)(\lambda - \lambda_3)$$

Assume also that λ_1, λ_2, and λ_3 are distinct.

Solution The eigenvector $\mathbf{x}_i$ associated with an eigenvalue λ_i is a vector that satisfies the equation

$$\mathbf{A}\mathbf{x}_i = \lambda_i\mathbf{x}_i \tag{5-113}$$

which can be written as

$$\begin{bmatrix} 0 & 1 & 0 \\ 0 & 0 & 1 \\ -a_3 & -a_2 & -a_1 \end{bmatrix}\begin{bmatrix} x_{i1} \\ x_{i2} \\ x_{i3} \end{bmatrix} = \lambda_i\begin{bmatrix} x_{i1} \\ x_{i2} \\ x_{i3} \end{bmatrix}$$

Simplifying this last equation, we obtain

$$x_{i2} = \lambda_i x_{i1}$$

$$x_{i3} = \lambda_i x_{i2}$$

$$-a_3 x_{i1} - a_2 x_{i2} - a_1 x_{i3} = \lambda_i x_{i3}$$

Thus,

$$\begin{bmatrix} x_{i1} \\ x_{i2} \\ x_{i3} \end{bmatrix} = \begin{bmatrix} x_{i1} \\ \lambda_i x_{i1} \\ \lambda_i^2 x_{i1} \end{bmatrix} = \begin{bmatrix} 1 \\ \lambda_i \\ \lambda_i^2 \end{bmatrix} x_{i1}$$

Hence, the eigenvectors are

$$\begin{bmatrix} x_{11} \\ \lambda_1 x_{11} \\ \lambda_1^2 x_{11} \end{bmatrix}, \qquad \begin{bmatrix} x_{21} \\ \lambda_2 x_{21} \\ \lambda_2^2 x_{21} \end{bmatrix}, \qquad \begin{bmatrix} x_{31} \\ \lambda_3 x_{31} \\ \lambda_3^2 x_{31} \end{bmatrix} \tag{5-114}$$

Note that if $\mathbf{x}_i$ is an eigenvector, then $a\mathbf{x}_i$ (where $a = $ scalar $\neq 0$) is also an eigenvector, because Equation (5-113) can be written as

$$a(\mathbf{A}\mathbf{x}_i) = a(\lambda_i\mathbf{x}_i)$$

or

$$\mathbf{A}(a\mathbf{x}_i) = \lambda_i(a\mathbf{x}_i)$$

Thus, by dividing the eigenvectors given by (5-114) by x_{11}, x_{21}, and x_{31}, respectively, we obtain

$$\begin{bmatrix} 1 \\ \lambda_1 \\ \lambda_1^2 \end{bmatrix}, \qquad \begin{bmatrix} 1 \\ \lambda_2 \\ \lambda_2^2 \end{bmatrix}, \qquad \begin{bmatrix} 1 \\ \lambda_3 \\ \lambda_3^2 \end{bmatrix}$$

These are also a set of eigenvectors.

Problem A-5-19

Consider a matrix

$$\mathbf{A} = \begin{bmatrix} 0 & 1 & 0 \\ 0 & 0 & 1 \\ -a_3 & -a_2 & -a_1 \end{bmatrix}$$

Assume that λ_1, λ_2, and λ_3 are distinct eigenvalues of matrix $\mathbf{A}$.
Show that if a transformation matrix $\mathbf{P}$ is defined by

$$\mathbf{P} = \begin{bmatrix} 1 & 1 & 1 \\ \lambda_1 & \lambda_2 & \lambda_3 \\ \lambda_1^2 & \lambda_2^2 & \lambda_3^2 \end{bmatrix}$$

then

$$\mathbf{P}^{-1}\mathbf{AP} = \begin{bmatrix} \lambda_1 & 0 & 0 \\ 0 & \lambda_2 & 0 \\ 0 & 0 & \lambda_3 \end{bmatrix}$$

Solution   First note that

$$
\mathbf{AP} = \begin{bmatrix} 0 & 1 & 0 \\ 0 & 0 & 1 \\ -a_3 & -a_2 & -a_1 \end{bmatrix}\begin{bmatrix} 1 & 1 & 1 \\ \lambda_1 & \lambda_2 & \lambda_3 \\ \lambda_1^2 & \lambda_2^2 & \lambda_3^2 \end{bmatrix}
$$

$$
= \begin{bmatrix} \lambda_1 & \lambda_2 & \lambda_3 \\ \lambda_1^2 & \lambda_2^2 & \lambda_3^2 \\ -a_3 - a_2\lambda_1 - a_1\lambda_1^2 & -a_3 - a_2\lambda_2 - a_1\lambda_2^2 & -a_3 - a_2\lambda_3 - a_1\lambda_3^2 \end{bmatrix} \qquad (5\text{–}115)
$$

Since λ_1, λ_2, and λ_3 are eigenvalues, they satisfy the characteristic equation, or

$$\lambda_i^3 + a_1\lambda_i^2 + a_2\lambda_i + a_3 = 0$$

Thus,

$$\lambda_i^3 = -a_3 - a_2\lambda_i - a_1\lambda_i^2$$

Hence,

$$-a_3 - a_2\lambda_1 - a_1\lambda_1^2 = \lambda_1^3$$
$$-a_3 - a_2\lambda_2 - a_1\lambda_2^2 = \lambda_2^3$$
$$-a_3 - a_2\lambda_3 - a_1\lambda_3^2 = \lambda_3^3$$

Consequently, Equation (5–115) can be written as

$$\mathbf{AP} = \begin{bmatrix} \lambda_1 & \lambda_2 & \lambda_3 \\ \lambda_1^2 & \lambda_2^2 & \lambda_3^2 \\ \lambda_1^3 & \lambda_2^3 & \lambda_3^3 \end{bmatrix} \qquad (5\text{–}116)$$

Next, define

$$\mathbf{D} = \begin{bmatrix} \lambda_1 & 0 & 0 \\ 0 & \lambda_2 & 0 \\ 0 & 0 & \lambda_3 \end{bmatrix}$$

Then

$$PD = \begin{bmatrix} 1 & 1 & 1 \\ \lambda_1 & \lambda_2 & \lambda_3 \\ \lambda_1^2 & \lambda_2^2 & \lambda_3^2 \end{bmatrix} \begin{bmatrix} \lambda_1 & 0 & 0 \\ 0 & \lambda_2 & 0 \\ 0 & 0 & \lambda_3 \end{bmatrix} = \begin{bmatrix} \lambda_1 & \lambda_2 & \lambda_3 \\ \lambda_1^2 & \lambda_2^2 & \lambda_3^2 \\ \lambda_1^3 & \lambda_2^3 & \lambda_3^3 \end{bmatrix} \qquad (5\text{--}117)$$

Comparing Equations (5–116) and (5–117), we have

$$AP = PD$$

Thus, we have shown that

$$P^{-1}AP = D = \begin{bmatrix} \lambda_1 & 0 & 0 \\ 0 & \lambda_2 & 0 \\ 0 & 0 & \lambda_3 \end{bmatrix}$$

Problem A–5–20

Prove that the eigenvalues of a square matrix A are invariant under a linear transformation.

Solution To prove the invariance of the eigenvalues under a linear transformation, we must show that the characteristic polynomials $|\lambda I - A|$ and $|\lambda I - P^{-1}AP|$ are identical.

Since the determinant of a product is the product of the determinants, we obtain

$$\begin{aligned} |\lambda I - P^{-1}AP| &= |\lambda P^{-1}P - P^{-1}AP| \\ &= |P^{-1}(\lambda I - A)P| \\ &= |P^{-1}||\lambda I - A||P| \\ &= |P^{-1}||P||\lambda I - A| \end{aligned}$$

Noting that the product of the determinants $|P^{-1}|$ and $|P|$ is the determinant of the product $|P^{-1}P|$, we obtain

$$\begin{aligned} |\lambda I - P^{-1}AP| &= |P^{-1}P||\lambda I - A| \\ &= |\lambda I - A| \end{aligned}$$

Thus, we have proven that the eigenvalues of A are invariant under a linear transformation.

PROBLEMS

Problem B–5–1

Obtain state-space representations of the mechanical systems shown in Figures 5–36(a) and (b).

Problem B–5–2

For the spring–mass–pulley system of Figure 5–37, the moment of inertia of the pulley about the axis of rotation is J and the radius is R. Assume that the system is initially in equilibrium. The gravitational force of mass m causes a static deflection of the spring such that $k\delta = mg$. Assuming that the displacement y of mass m is measured from the equilibrium position, obtain a state-space representation of the system. The external force u applied to mass m is the input and the displacement y is the output of the system.

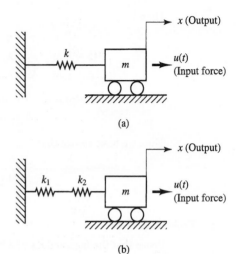

(a)

(b)

Figure 5–36   (a) and (b) Mechanical systems.

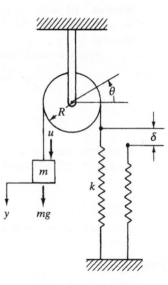

Figure 5–37   Spring–mass–pulley system.

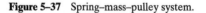

Problem B–5–3

Obtain a state-space representation of the mechanical system shown in Figure 5–38. The force $u(t)$ applied to mass m_1 is the input to the system. The displacements y and z are the outputs of the system. Assume that y and z are measured from their respective equilibrium positions.

Problem B–5–4

Obtain a state-space representation of the mechanical system shown in Figure 5–39, where u_1 and u_2 are the inputs and y_1 and y_2 are the outputs. The displacements y_1 and y_2 are measured from their respective equilibrium positions.

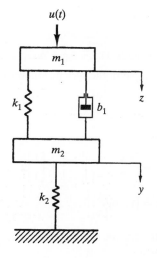

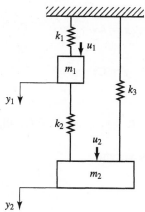

Figure 5–38 Mechanical system.

Figure 5–39 Mechanical system.

Problem B–5–5

Given the state equation

$$\begin{bmatrix} \dot{x}_1 \\ \dot{x}_2 \\ \dot{x}_3 \end{bmatrix} = \begin{bmatrix} 0 & 1 & 0 \\ 0 & 0 & 1 \\ 1 & -3 & 3 \end{bmatrix} \begin{bmatrix} x_1 \\ x_2 \\ x_3 \end{bmatrix} + \begin{bmatrix} 1 \\ 1 \\ 1 \end{bmatrix} u$$

and output equation

$$y = \begin{bmatrix} 1 & 0 & 0 \end{bmatrix} \begin{bmatrix} x_1 \\ x_2 \\ x_3 \end{bmatrix}$$

obtain the corresponding scalar differential equation in terms of y and u.

Problem B-5-6

Consider the system defined by

$$\dddot{y} + 6\ddot{y} + 11\dot{y} + 6y = 6u$$

where y is the output and u is the input of the system. Obtain a state-space representation for the system.

Problem B-5-7

Consider the system described by

$$\begin{bmatrix} \dot{x}_1 \\ \dot{x}_2 \end{bmatrix} = \begin{bmatrix} -4 & -1 \\ 3 & -1 \end{bmatrix} \begin{bmatrix} x_1 \\ x_2 \end{bmatrix} + \begin{bmatrix} 1 \\ 1 \end{bmatrix} u$$

$$y = \begin{bmatrix} 1 & 0 \end{bmatrix} \begin{bmatrix} x_1 \\ x_2 \end{bmatrix}$$

Obtain the transfer function of the system.

Problem B-5-8

Consider a system described by the state equation

$$\dot{x} = Ax + Bu$$

and the output equation

$$y = Cx + Du$$

where

$$A = \begin{bmatrix} 0 & 1 \\ -1 & -2 \end{bmatrix}, \qquad B = \begin{bmatrix} 1 \\ 1 \end{bmatrix}, \qquad C = \begin{bmatrix} 1 & 0 \end{bmatrix}, \qquad D = 1$$

Obtain the transfer function of the system.

Problem B-5-9

Consider the system

$$\dot{x} = Ax + Bu$$
$$y = Cx + Du$$

where

$$A = \begin{bmatrix} 0 & 1 & 0 \\ 0 & 0 & 1 \\ -600 & -100 & -10 \end{bmatrix}, \qquad B = \begin{bmatrix} 0 \\ 10 \\ 0 \end{bmatrix}, \qquad C = \begin{bmatrix} 1 & 0 & 0 \end{bmatrix}, \qquad D = 0$$

Obtain the transfer function of the system.

Problem B-5-10

Consider the following system:

$$\begin{bmatrix} \dot{x}_1 \\ \dot{x}_2 \end{bmatrix} = \begin{bmatrix} -1 & -1 \\ 5 & 0 \end{bmatrix} \begin{bmatrix} x_1 \\ x_2 \end{bmatrix} + \begin{bmatrix} 1 & 1 \\ 1 & 0 \end{bmatrix} \begin{bmatrix} u_1 \\ u_2 \end{bmatrix}$$

$$\begin{bmatrix} y_1 \\ y_2 \end{bmatrix} = \begin{bmatrix} 1 & 0 \\ 0 & 1 \end{bmatrix} \begin{bmatrix} x_1 \\ x_2 \end{bmatrix} + \begin{bmatrix} 1 & 0 \\ 0 & 0 \end{bmatrix} \begin{bmatrix} u_1 \\ u_2 \end{bmatrix}$$

Obtain the unit-step response curves with MATLAB.

Problem B–5–11

Obtain the unit-step response curve and unit-impulse response curve of the following system with MATLAB:

$$
\begin{bmatrix} \dot{x}_1 \\ \dot{x}_2 \\ \dot{x}_3 \end{bmatrix} = \begin{bmatrix} -5 & -25 & -5 \\ 1 & 0 & 0 \\ 0 & 1 & 0 \end{bmatrix} \begin{bmatrix} x_1 \\ x_2 \\ x_3 \end{bmatrix} + \begin{bmatrix} 1 \\ 0 \\ 0 \end{bmatrix} u
$$

$$
y = \begin{bmatrix} 0 & 25 & 5 \end{bmatrix} \begin{bmatrix} x_1 \\ x_2 \\ x_3 \end{bmatrix} + [0]u
$$

Problem B–5–12

Consider the system defined by

$$
\dot{x} = Ax + Bu, \qquad x(0) = x_0
$$
$$
y = Cx + Du
$$

where

$$
A = \begin{bmatrix} 0 & 1 \\ -10 & -5 \end{bmatrix}, \qquad B = \begin{bmatrix} 0 \\ 0 \end{bmatrix}, \qquad C = [1 \quad 0], \qquad D = 0
$$

Obtain the response to the initial condition

$$
x_0 = \begin{bmatrix} 2 \\ 1 \end{bmatrix}
$$

Use MATLAB command initial(A,B,C,D,[initial condition],t).

Problem B–5–13

Consider the system

$$
\dddot{y} + 8\ddot{y} + 17\dot{y} + 10y = 0
$$

subjected to the initial condition

$$
y(0) = 2, \qquad \dot{y}(0) = 1, \qquad \ddot{y}(0) = 0.5
$$

(No external forcing function is present.) Obtain the response curve $y(t)$ to the given initial condition with MATLAB. Use command lsim.

Problem B–5–14

Consider the mechanical system shown in Figure 5–40(a). The system is at rest for $t < 0$. The displacement y is measured from the equilibrium position for $t < 0$. At $t = 0$, an input force

$$
\begin{aligned}
u(t) &= 1\,\text{N} && \text{for } 0 \le t \le 5 \\
&= 0 && \text{for } 5 < t
\end{aligned}
$$

is given to the system. [See Figure 5–40(b).] Derive a state-space representation of the system. Plot the response curve $y(t)$ versus t (where $0 < t < 10$) with MATLAB. Assume that $m = 5$ kg, $b = 8$ N-s/m, and $k = 20$ N/m.

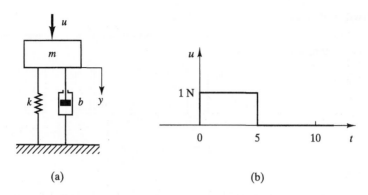

Figure 5–40   (a) Mechanical system; (b) input force u.

Problem B–5–15

Consider the mechanical system shown in Figure 5–41(a). Assume that at $t = 0$ mass m is placed on the massless bar AA'. [See Figure 5–41(b).] Neglecting the mass of the spring–damper device, what is the subsequent motion $y(t)$ of the bar AA'? The displacement $y(t)$ is measured from the equilibrium position before the mass is placed on the bar AA'. Assume that $m = 1$ kg, the viscous-friction coefficient $b = 4$ N-s/m, and the spring constant $k = 40$ N/m. Derive a state-space representation of the system, and plot the response curve $y(t)$ versus t with MATLAB.

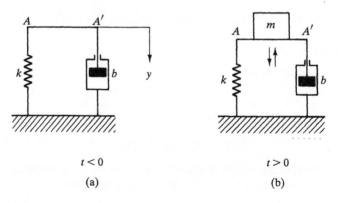

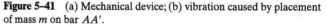

Figure 5–41   (a) Mechanical device; (b) vibration caused by placement of mass m on bar AA'.

Problem B–5–16

Consider the system shown in Figure 5–42. The system is at rest for $t < 0$. Assume that the input and output are the displacements u and y, respectively, measured from the rest positions. Assume that $m = 10$ kg, $b = 20$ N-s/m, and $k = 40$ N/m. The input u is a step displacement input of 0.2 m. Assume also that the system remains linear throughout the transient period. Obtain a state-space representation of the system. Plot the response curve $y(t)$ versus t with MATLAB.

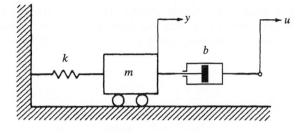

Figure 5–42 Mechanical system.

Problem B–5–17

Referring to **Problem A–5–10**, consider the system shown in Figure 5–31. The system is at rest for $t < 0$. The displacements z_1 and z_2 are measured from their respective equilibrium positions relative to the ground. Define $z_2 - z_1 = z$. Derive a state-space equation when $z, \dot{z}, z_1,$ and $\dot{z}_1$ are chosen as state variables. Assuming that $m_1 = 10$ kg, $m_2 = 20$ kg, $b = 20$ N-s/m, $k = 60$ N/m, and f is a step force input of magnitude 10 N, plot the response curves $z_1(t)$ versus t, $z_2(t)$ versus t, and $z(t)$ versus t.

Problem B–5–18

Consider the system shown in Figure 5–43(a). The system is at rest for $t < 0$. The displacements z_1 and z_2 are measured from their respective equilibrium positions before the input force

$$f = t \text{ N} \qquad (0 < t \le 10)$$
$$= 0 \qquad (10 < t)$$

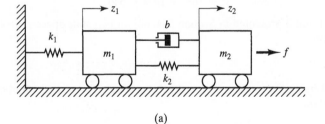

(a)

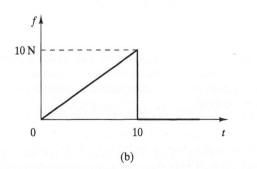

(b)

Figure 5–43 (a) Mechanical system; (b) input force f.

[see Figure 5–43(b)] is given to the system. Assume that $m_1 = 10$ kg, $m_2 = 20$ kg, $b = 20$ N-s/m. $k_1 = 30$ N/m, and $k_2 = 60$ N/m, and derive a state-space representation of the system. Then plot the response curves $z_1(t)$ versus t, $z_2(t)$ versus t, and $z_2(t) - z_1(t)$ versus t.

Problem B–5–19

Consider the system equation given by

$$\overset{(n)}{y} + a_1 \overset{(n-1)}{y} + \cdots + a_{n-1}\dot{y} + a_n y = b_0 \overset{(n)}{u} + b_1 \overset{(n-1)}{u} + \cdots + b_{n-1}\dot{u} + b_n u$$

By choosing appropriate state variables, derive the state equation

$$
\begin{bmatrix} \dot{x}_1 \\ \dot{x}_2 \\ \vdots \\ \dot{x}_n \end{bmatrix}
=
\begin{bmatrix}
0 & 0 & \cdots & 0 & -a_n \\
1 & 0 & \cdots & 0 & -a_{n-1} \\
\vdots & \vdots & & \vdots & \vdots \\
0 & 0 & \cdots & 1 & -a_1
\end{bmatrix}
\begin{bmatrix} x_1 \\ x_2 \\ \vdots \\ x_n \end{bmatrix}
+
\begin{bmatrix}
b_n - a_n b_0 \\
b_{n-1} - a_{n-1} b_0 \\
\vdots \\
b_1 - a_1 b_0
\end{bmatrix} u
\qquad (5\text{–}118)
$$

and output equation

$$
y = \begin{bmatrix} 0 & 0 & \cdots & 0 & 1 \end{bmatrix}
\begin{bmatrix} x_1 \\ x_2 \\ \vdots \\ x_{n-1} \\ x_n \end{bmatrix}
+ b_0 u
\qquad (5\text{–}119)
$$

Problem B–5–20

Consider the system defined by the following transfer function:

$$\frac{Y(s)}{U(s)} = \frac{160(s + 4)}{s^3 + 18s^2 + 192s + 640}$$

Using Methods 1 and 2 presented in Section 5–4, obtain two state-space representations of the system.

Problem B–5–21

Using the partial-fraction expansion approach, obtain a state-space representation for the following system:

$$\frac{Y(s)}{U(s)} = \frac{5}{(s + 1)^2(s + 2)}$$

Problem B–5–22

Consider the mechanical system shown in Figure 5–44. The system is at rest for $t < 0$. The force u is the input to the system and the displacement y, measured from the equilibrium position before u is given at $t = 0$, is the output of the system. Obtain a state-space representation of the system.

Problem B–5–23

Consider the system shown in Figure 5–45. The system is at rest for $t < 0$. The displacements y_1 and y_2 are measured from their respective equilibrium positions before the input force u is given at $t = 0$. Obtain a state-space representation of the system.

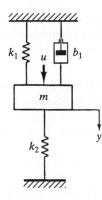

Figure 5–44 Mechanical system.

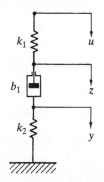

Figure 5–45 Mechanical system.

Assuming that $m_1 = 10$ kg, $m_2 = 5$ kg, $b = 10$ N-s/m, $k_1 = 40$ N/m, and $k_2 = 20$ N/m and that input force u is a constant force of 5 N, obtain the response of the system. Plot the response curves $y_1(t)$ versus t and $y_2(t)$ versus t with MATLAB.

Problem B–5–24

Consider the mechanical system shown in Figure 5–46. The system is initially at rest. The displacement u is the input to the system, and the displacements y and z, measured

Figure 5–46 Mechanical system.

from their respective rest positions before the input displacement u is given to the system, are the outputs of the system. Obtain a state-space representation of the system.

Problem B–5–25

Consider the mechanical system shown in Figure 5–47. The system is at rest for $t < 0$. The force u is the input to the system and the displacements z_1 and z_2, measured from their respective equilibrium positions before u is applied at $t = 0$, are the outputs of the system. Obtain a state-space representation of the system.

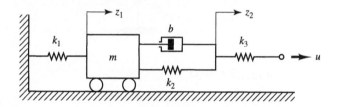

Figure 5–47   Mechanical system.

Problem B–5–26

Consider the system shown in Figure 5–48. The system is at rest for $t < 0$. The force u is the input to the system and the displacements z_1 and z_2, measured from their respective equilibrium positions before u is applied at $t = 0$, are the outputs of the system. Obtain a state-space representation of the system,

      Assume that $m_1 = 100$ kg, $m_2 = 200$ kg, $b = 25$ N-s/m, $k_1 = 50$ N/m, and $k_2 = 100$ N/m. The input force u is a step force of magnitude 10 N. Plot the response curves $z_1(t)$ versus t, $z_2(t)$ versus t, and $z_2(t) - z_1(t)$ versus t.

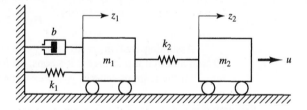

Figure 5–48   Mechanical system.

Problem B–5–27

Consider the system

$$\frac{Y(s)}{U(s)} = \frac{25s + 5}{s^3 + 5s^2 + 25s + 5}$$

Obtain a state-space representation of the system with MATLAB.

Problem B–5–28

Consider the system

$$\frac{Y(s)}{U(s)} = \frac{s^3 + 2s^2 + 15s + 10}{s^3 + 4s^2 + 8s + 10}$$

Obtain a state-space representation of the system with MATLAB.

Problem B–5–29

Consider the system defined by

$$\begin{bmatrix} \dot{x}_1 \\ \dot{x}_2 \end{bmatrix} = \begin{bmatrix} 0 & 1 \\ -25 & -4 \end{bmatrix} \begin{bmatrix} x_1 \\ x_2 \end{bmatrix} + \begin{bmatrix} 1 & 1 \\ 0 & 1 \end{bmatrix} \begin{bmatrix} u_1 \\ u_2 \end{bmatrix}$$

$$\begin{bmatrix} y_1 \\ y_2 \end{bmatrix} = \begin{bmatrix} 1 & 0 \\ 0 & 1 \end{bmatrix} \begin{bmatrix} x_1 \\ x_2 \end{bmatrix} + \begin{bmatrix} 1 & 0 \\ 0 & 1 \end{bmatrix} \begin{bmatrix} u_1 \\ u_2 \end{bmatrix}$$

This system involves two inputs and two outputs. Four transfer functions are involved: $Y_1(s)/U_1(s)$, $Y_2(s)/U_1(s)$, $Y_1(s)/U_2(s)$, and $Y_2(s)/U_2(s)$. (When considering input u_1, we assume that input u_2 is zero, and vice versa.)

Obtain the transfer matrix (consisting of the preceding four transfer functions) of the system.

Problem B–5–30

Obtain the transfer matrix of the system defined by

$$\begin{bmatrix} \dot{x}_1 \\ \dot{x}_2 \\ \dot{x}_3 \end{bmatrix} = \begin{bmatrix} 0 & 1 & 0 \\ 0 & 0 & 1 \\ -2 & -4 & -6 \end{bmatrix} \begin{bmatrix} x_1 \\ x_2 \\ x_3 \end{bmatrix} + \begin{bmatrix} 0 & 0 \\ 0 & 1 \\ 1 & 0 \end{bmatrix} \begin{bmatrix} u_1 \\ u_2 \end{bmatrix}$$

$$\begin{bmatrix} y_1 \\ y_2 \end{bmatrix} = \begin{bmatrix} 1 & 0 & 0 \\ 0 & 1 & 0 \end{bmatrix} \begin{bmatrix} x_1 \\ x_2 \\ x_3 \end{bmatrix}$$

Problem B–5–31

Consider a 3×3 matrix having a triple eigenvalue of λ_1. Then any one of the following Jordan canonical forms is possible:

$$\begin{bmatrix} \lambda_1 & 1 & 0 \\ 0 & \lambda_1 & 1 \\ 0 & 0 & \lambda_1 \end{bmatrix}, \quad \begin{bmatrix} \lambda_1 & 1 & 0 \\ 0 & \lambda_1 & 0 \\ 0 & 0 & \lambda_1 \end{bmatrix}, \quad \begin{bmatrix} \lambda_1 & 0 & 0 \\ 0 & \lambda_1 & 0 \\ 0 & 0 & \lambda_1 \end{bmatrix}$$

Each of the three matrices has the same characteristic equation $(\lambda - \lambda_1)^3 = 0$. The first corresponds to the case where there exists only one linearly independent eigenvector. This fact can be seen by denoting the first matrix by $\mathbf{A}$ and solving the following equation for $\mathbf{x}$:

$$\mathbf{A}\mathbf{x} = \lambda_1 \mathbf{x}$$

That is,

$$\begin{bmatrix} \lambda_1 & 1 & 0 \\ 0 & \lambda_1 & 1 \\ 0 & 0 & \lambda_1 \end{bmatrix} \begin{bmatrix} x_1 \\ x_2 \\ x_3 \end{bmatrix} = \lambda_1 \begin{bmatrix} x_1 \\ x_2 \\ x_3 \end{bmatrix}$$

which can be rewritten as

$$\lambda_1 x_1 + x_2 = \lambda_1 x_1$$
$$\lambda_1 x_2 + x_3 = \lambda_1 x_2$$
$$\lambda_1 x_3 = \lambda_1 x_3$$

which, in turn, gives

$$x_1 = \text{arbitrary constant}, \qquad x_2 = 0, \qquad x_3 = 0$$

Hence,

$$\mathbf{x} = \begin{bmatrix} a \\ 0 \\ 0 \end{bmatrix}$$

where a is a nonzero constant. Thus, there is only one linearly independent eigenvector.

Show that the second and third of the three matrices have, respectively, two and three linearly independent eigenvectors.

Electrical Systems and Electromechanical Systems

6–1 INTRODUCTION

This chapter is concerned with mathematical modeling and the response analysis of electrical systems and electromechanical systems. Electrical systems and mechanical systems (as well as other systems, such as fluid systems) are very often described by analogous mathematical models. Therefore, we present brief discussions on analogous systems in the chapter.

In this section, we first review three types of basic elements of electrical systems: resistance, capacitance, and inductance elements. (These elements are passive elements, because, although they can store or dissipate energy that is already present in the circuit, they cannot introduce additional energy into the circuit.) Then we briefly discuss voltage and current sources. (These are active elements, because they can introduce energy into the circuit.) Finally, we provide an outline of the chapter.

Resistance elements.   The *resistance R* of a linear resistor is given by

$$R = \frac{e_R}{i}$$

where e_R is the voltage across the resistor and i is the current through the resistor. The unit of resistance is the ohm (Ω), where

$$\text{ohm} = \frac{\text{volt}}{\text{ampere}}$$

Resistors do not store electric energy in any form, but instead dissipate it as heat. Note that real resistors may not be linear and may also exhibit some capacitance and inductance effects.

Capacitance elements. Two conductors separated by a nonconducting medium form a capacitor, so two metallic plates separated by a very thin dielectric material form a capacitor. The *capacitance* C is a measure of the quantity of charge that can be stored for a given voltage across the plates. The capacitance C of a capacitor can thus be given by

$$C = \frac{q}{e_C}$$

where q is the quantity of charge stored and e_C is the voltage across the capacitor. The unit of capacitance is the farad (F), where

$$\text{farad} = \frac{\text{ampere-second}}{\text{volt}} = \frac{\text{coulomb}}{\text{volt}}$$

Note that, since $i = dq/dt$ and $e_C = q/C$, we have

$$i = C \frac{de_C}{dt}$$

or

$$de_C = \frac{1}{C} i \, dt$$

Therefore,

$$e_C(t) = \frac{1}{C} \int_0^t i \, dt + e_C(0)$$

Although a pure capacitor stores energy and can release all of it, real capacitors exhibit various losses. These energy losses are indicated by a *power factor*, which is the ratio of the energy lost per cycle of ac voltage to the energy stored per cycle. Thus, a small-valued power factor is desirable.

Inductance elements. If a circuit lies in a time-varying magnetic field, an electromotive force is induced in the circuit. The inductive effects can be classified as self-inductance and mutual inductance.

Self-inductance is that property of a single coil that appears when the magnetic field set up by the current in the coil links to the coil itself. The magnitude of the induced voltage is proportional to the rate of change of flux linking the circuit. If the circuit does not contain ferromagnetic elements (such as an iron core), the rate of change of flux is proportional to di/dt. Self-inductance, or simply inductance, L, is the proportionality constant between the induced voltage e_L volts and the rate of change of current (or change in current per second) di/dt amperes per second; that is,

$$L = \frac{e_L}{di/dt}$$

The unit of inductance is the henry (H). An electrical circuit has an inductance of 1 henry when a rate of change of 1 ampere per second will induce an emf of 1 volt:

$$\text{henry} = \frac{\text{volt}}{\text{ampere/second}} = \frac{\text{weber}}{\text{ampere}}$$

The voltage e_L across the inductor L is given by

$$e_L = L\frac{di_L}{dt}$$

where i_L is the current through the inductor. The current $i_L(t)$ can thus be given by

$$i_L(t) = \frac{1}{L}\int_0^t e_L dt + i_L(0)$$

Because most inductors are coils of wire, they have considerable resistance. The energy loss due to the presence of resistance is indicated by the *quality factor Q*, which denotes the ratio of stored to dissipated energy. A high value of Q generally means that the inductor contains small resistance.

Mutual inductance refers to the influence between inductors that results from the interaction of their fields. If two inductors are involved in an electrical circuit, each may come under the influence of the magnetic field of the other inductor. Then the voltage drop in the first inductor is related to the current flowing through the first inductor, as well as to the current flowing through the second inductor, whose magnetic field influences the first. The second inductor is also influenced by the first in exactly the same manner. When a change in current of 1 ampere per second in either of the two inductors induces an electromotive force of 1 volt in the other inductor, their mutual inductance M is 1 henry. (Note that it is customary to use the symbol M to denote mutual inductance, to distinguish it from self-inductance L.)

Voltage and current sources.   A *voltage source* is a device that causes a specified voltage to exist between two points in a circuit. The voltage may be time varying or time invariant (for a sufficiently long time). Figure 6–1(a) is a schematic diagram of a voltage source. Figure 6–1(b) shows a voltage source that has a constant value for an indefinite time. Often the voltage is denoted by E. A battery is an example of this type of voltage source.

A *current source* causes a specified current to flow through a wire containing this source. Figure 6–1(c) is a schematic diagram of a current source.

Outline of the chapter.   Section 6–1 has presented introductory material. Section 6–2 reviews the fundamentals of electrical circuits that are presented in college physics courses. Section 6–3 deals with mathematical modeling and the response analysis of electrical systems. The complex-impedance approach is included. Section 6–4 discusses analogous systems. Section 6–5 offers brief discussions of electromechanical systems. Finally, Section 6–6 treats operational-amplifier systems.

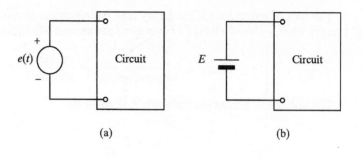

(a) (b)

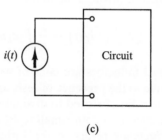

(c)

Figure 6–1 (a) Voltage source; (b) constant voltage source; (c) current source.

6–2 FUNDAMENTALS OF ELECTRICAL CIRCUITS

In this section, we review Ohm's law, series and parallel circuits, and Kirchhoff's current and voltage laws.

Ohm's law. *Ohm's law* states that the current in a circuit is proportional to the total electromotive force (emf) acting in the circuit and inversely proportional to the total resistance of the circuit. Ohm's law can be expressed as

$$i = \frac{e}{R}$$

where i is the current (amperes), e is the emf (volts), and R is the resistance (ohms).

Series circuits. The combined resistance of series-connected resistors is the sum of the separate resistances. Figure 6–2 shows a simple series circuit. The voltage between points A and B is

$$e = e_1 + e_2 + e_3$$

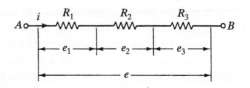

Figure 6–2 Series circuit.

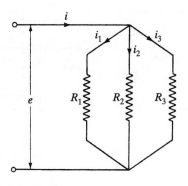

Figure 6–3   Parallel circuit.

where

$$e_1 = iR_1, \qquad e_2 = iR_2, \qquad e_3 = iR_3$$

Thus,

$$\frac{e}{i} = R_1 + R_2 + R_3$$

The combined resistance is then given by

$$R = R_1 + R_2 + R_3$$

Parallel circuits.   For the parallel circuit shown in Figure 6–3,

$$i_1 = \frac{e}{R_1}, \qquad i_2 = \frac{e}{R_2}, \qquad i_3 = \frac{e}{R_3}$$

Since $i = i_1 + i_2 + i_3$, it follows that

$$i = \frac{e}{R_1} + \frac{e}{R_2} + \frac{e}{R_3} = \frac{e}{R}$$

where R is the combined resistance. Hence,

$$\frac{1}{R} = \frac{1}{R_1} + \frac{1}{R_2} + \frac{1}{R_3}$$

or

$$R = \frac{1}{\dfrac{1}{R_1} + \dfrac{1}{R_2} + \dfrac{1}{R_3}} = \frac{R_1 R_2 R_3}{R_1 R_2 + R_2 R_3 + R_3 R_1}$$

Resistance of combined series and parallel resistors.   Consider the circuit shown in Figure 6–4(a). The combined resistance between points B and C is

$$R_{BC} = \frac{R_2 R_3}{R_2 + R_3}$$

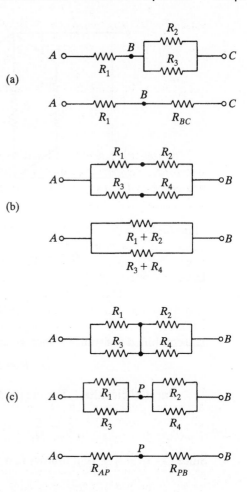

Figure 6–4 Combined series and parallel resistors.

The combined resistance R between points A and C is

$$R = R_1 + R_{BC} = R_1 + \frac{R_2 R_3}{R_2 + R_3}$$

The circuit shown in Figure 6–4(b) can be considered a parallel circuit consisting of resistances $(R_1 + R_2)$ and $(R_3 + R_4)$. So the combined resistance R between points A and B is

$$\frac{1}{R} = \frac{1}{R_1 + R_2} + \frac{1}{R_3 + R_4}$$

or

$$R = \frac{(R_1 + R_2)(R_3 + R_4)}{R_1 + R_2 + R_3 + R_4}$$

Next, consider the circuit shown in Figure 6–4(c). Here, R_1 and R_3 are parallel and R_2 and R_4 are parallel, and the two parallel pairs of resistances are connected in series. Redrawing this circuit as shown in Figure 6–4(c), therefore, we obtain

$$R_{AP} = \frac{R_1 R_3}{R_1 + R_3}, \qquad R_{PB} = \frac{R_2 R_4}{R_2 + R_4}$$

As a result, the combined resistance R becomes

$$R = R_{AP} + R_{PB} = \frac{R_1 R_3}{R_1 + R_3} + \frac{R_2 R_4}{R_2 + R_4}$$

Kirchhoff's laws.   In solving circuit problems that involve many electromotive forces, resistances, capacitances, inductances, and so on, it is often necessary to use *Kirchhoff's laws*, of which there are two: the *current law* (node law) and the *voltage law* (loop law).

Kirchhoff's current law (node law).   A *node* in an electrical circuit is a point where three or more wires are joined together. Kirchhoff's current law (node law) states that the algebraic sum of all currents entering and leaving a node is zero. (This law can also be stated as follows: The sum of all the currents entering a node is equal to the sum of all the currents leaving the same node.) In applying the law to circuit problems, the following rules should be observed: Currents going toward a node should be preceded by a plus sign; currents going away from a node should be preceded by a minus sign. As applied to Figure 6–5, Kirchhoff's current law states that

$$i_1 + i_2 + i_3 - i_4 - i_5 = 0$$

Kirchhoff's voltage law (loop law).   Kirchhoff's voltage law states that at any given instant of time the algebraic sum of the voltages around any loop in an electrical circuit is zero. This law can also be stated as follows: The sum of the voltage drops is equal to the sum of the voltage rises around a loop. In applying the law to circuit problems, the following rules should be observed: A rise in voltage [which occurs in going through a source of electromotive force from the negative to the positive terminal, as shown in Figure 6–6(a), or in going through a resistance in opposition to the current flow, as shown in Figure 6–6(b)] should be preceded by a plus sign. A drop in voltage [which occurs in going through a source of electromotive force from the positive to the negative terminal, as shown in Figure 6–6(c), or in going through a resistance in the direction of the current flow, as shown in Figure 6–6(d)] should be preceded by a minus sign.

Figure 6–7 shows a circuit that consists of a battery and an external resistance. Here, E is the electromotive force, r is the internal resistance of the battery, R is the external resistance, and i is the current. If we follow the loop in the clockwise direction

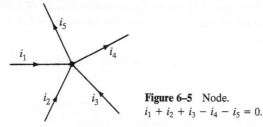

Figure 6–5   Node.
$i_1 + i_2 + i_3 - i_4 - i_5 = 0.$

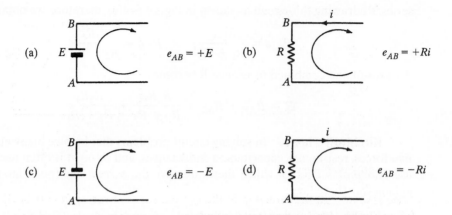

(a) $\quad E$ $e_{AB} = +E$ (b) $\quad R$ $e_{AB} = +Ri$

(c) $\quad E$ $e_{AB} = -E$ (d) $\quad R$ $e_{AB} = -Ri$

Figure 6–6   Diagrams showing voltage rises and voltage drops in circuits. (*Note*: Each circular arrow shows the direction one follows in analyzing the respective circuit.)

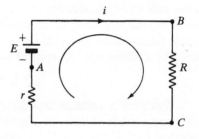

Figure 6–7   Electrical circuit.

$(A \rightarrow B \rightarrow C \rightarrow A)$ as shown, then we have

$$e_{\overrightarrow{AB}} + e_{\overrightarrow{BC}} + e_{\overrightarrow{CA}} = 0$$

or

$$E - iR - ir = 0$$

from which it follows that

$$i = \frac{E}{R + r}$$

A circuit consisting of two batteries and an external resistance appears in Figure 6–8(a), where E_1 and r_1 (E_2 and r_2) are the electromotive force and internal resistance of battery 1 (battery 2), respectively, and R is the external resistance. By assuming the direction of the current i as shown and following the loop clockwise as shown, we obtain

$$E_1 - iR - E_2 - ir_2 - ir_1 = 0$$

or

$$i = \frac{E_1 - E_2}{r_1 + r_2 + R} \tag{6-1}$$

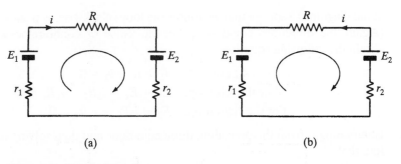

(a)   (b)

Figure 6–8   Electrical circuits.

If we assume that the direction of the current i is reversed [Figure 6–8(b)], then, by following the loop clockwise, we obtain

$$E_1 + iR - E_2 + ir_2 + ir_1 = 0$$

or

$$i = \frac{E_2 - E_1}{r_1 + r_2 + R} \qquad (6\text{–}2)$$

Note that, in solving circuit problems, if we assume that the current flows to the right and if the value of i is calculated and found to be positive, then the current i actually flows to the right. If the value of i is found to be negative, the current i actually flows to the left. For the circuits shown in Figure 6–8, suppose that $E_1 > E_2$. Then Equation (6–1) gives $i > 0$, which means that the current i flows in the direction assumed. Equation (6–2), however, yields $i < 0$, which means that the current i flows opposite to the assumed direction.

Note that the direction used to follow the loop is arbitrary, just as the direction of current flow can be assumed to be arbitrary. That is, the direction used in following the loop can be clockwise or counterclockwise; the final result is the same in either case.

Circuits with two or more loops.   For circuits with two or more loops, both Kirchhoff's current law and voltage law may be applied. The first step in writing the circuit equations is to define the directions of the currents in each wire. The second is to determine the directions that we follow in each loop.

Consider the circuit shown in Figure 6–9, which has two loops. Let us find the current in each wire. Here, we can assume the directions of currents as shown in the diagram. (Note that these directions are arbitrary and could differ from those shown

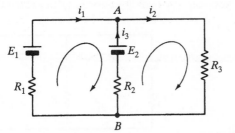

Figure 6–9   Electrical circuit.

in the diagram.) Suppose that we follow the loops clockwise, as is shown in the figure. (Again, the directions could be either clockwise or counterclockwise.) Then we obtain the following equations:

$$\text{At point } A: \quad i_1 + i_3 - i_2 = 0$$
$$\text{For the left loop:} \quad E_1 - E_2 + i_3R_2 - i_1R_1 = 0$$
$$\text{For the right loop:} \quad E_2 - i_2R_3 - i_3R_2 = 0$$

Eliminating i_2 from the preceding three equations and then solving for i_1 and i_3, we find that

$$i_1 = \frac{E_1(R_2 + R_3) - E_2R_3}{R_1R_2 + R_2R_3 + R_3R_1}$$
$$i_3 = \frac{E_2(R_1 + R_3) - E_1R_3}{R_1R_2 + R_2R_3 + R_3R_1}$$

Hence,

$$i_2 = i_1 + i_3 = \frac{E_1R_2 + E_2R_1}{R_1R_2 + R_2R_3 + R_3R_1}$$

Writing equations for loops by using cyclic currents. In this approach, we assume that a cyclic current exists in each loop. For instance, in Figure 6–10, we assume that clockwise cyclic currents i_1 and i_2 exist in the left and right loops, respectively, of the circuit.

Applying Kirchhoff's voltage law to the circuit results in the following equations:

$$\text{For left loop:} \quad E_1 - E_2 - R_2(i_1 - i_2) - R_1i_1 = 0$$
$$\text{For right loop:} \quad E_2 - R_3i_2 - R_2(i_2 - i_1) = 0$$

Note that the net current through resistance R_2 is the difference between i_1 and i_2. Solving for i_1 and i_2 gives

$$i_1 = \frac{E_1(R_2 + R_3) - E_2R_3}{R_1R_2 + R_2R_3 + R_3R_1}$$
$$i_2 = \frac{E_1R_2 + E_2R_1}{R_1R_2 + R_2R_3 + R_3R_1}$$

(By comparing the circuits shown in Figures 6–9 and 6–10, verify that i_3 in Figure 6–9 is equal to $i_2 - i_1$ in Figure 6–10.)

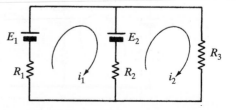

Figure 6–10   Electrical circuit.

6–3 MATHEMATICAL MODELING OF ELECTRICAL SYSTEMS

The first step in analyzing circuit problems is to obtain mathematical models for the circuits. (Although the terms *circuit* and *network* are sometimes used interchangeably, *network* implies a more complicated interconnection than *circuit*.) A mathematical model may consist of algebraic equations, differential equations, integrodifferential equations, and similar ones. Such a model may be obtained by applying one or both of Kirchhoff's laws to a given circuit. The variables of interest in the circuit analysis are voltages and currents at various points along the circuit.

In this section, we first present the mathematical modeling of electrical circuits and obtain solutions of simple circuit problems. Then we review the concept of complex impedances, followed by derivations of mathematical models of electrical circuits.

Example 6–1

Consider the circuit shown in Figure 6–11. Assume that the switch S is open for $t < 0$ and closed at $t = 0$. Obtain a mathematical model for the circuit and obtain an equation for the current $i(t)$.

By arbitrarily choosing the direction of the current around the loop as shown in the figure, we obtain

$$E - L\frac{di}{dt} - Ri = 0$$

or

$$L\frac{di}{dt} + Ri = E \tag{6–3}$$

This is a mathematical model for the given circuit. Note that at the instant switch S is closed the current $i(0)$ is zero, because the current in the inductor cannot change from zero to a finite value instantaneously. Thus, $i(0) = 0$.

Let us solve Equation (6–3) for the current $i(t)$. Taking the Laplace transforms of both sides, we obtain

$$L[sI(s) - i(0)] + RI(s) = \frac{E}{s}$$

Noting that $i(0) = 0$, we have

$$(Ls + R)I(s) = \frac{E}{s}$$

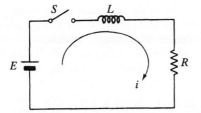

Figure 6–11   Electrical circuit.

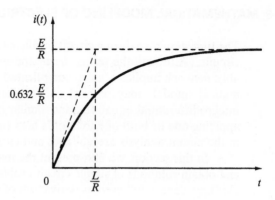

Figure 6–12 Plot of $i(t)$ versus t for the circuit shown in Figure 6–11 when switch S is closed at $t = 0$.

or

$$I(s) = \frac{E}{s(Ls + R)} = \frac{E}{R}\left[\frac{1}{s} - \frac{1}{s + (R/L)}\right]$$

The inverse Laplace transform of this last equation gives

$$i(t) = \frac{E}{R}[1 - e^{-(R/L)t}] \tag{6-4}$$

A typical plot of $i(t)$ versus t appears in Figure 6–12.

Example 6–2

Consider again the circuit shown in Figure 6–11. Assume that switch S is open for $t < 0$, it is closed at $t = 0$, and is open again at $t = t_1 > 0$. Obtain a mathematical model for the system, and find the current $i(t)$ for $t \geq 0$.

The equation for the circuit is

$$L\frac{di}{dt} + Ri = E \qquad i(0) = 0 \qquad t_1 > t \geq 0 \tag{6-5}$$

From Equation (6–4), the solution of Equation (6–5) is

$$i(t) = \frac{E}{R}[1 - e^{-(R/L)t}] \qquad t_1 > t \geq 0 \tag{6-6}$$

At $t = t_1$, the switch is opened. The equation for the circuit for $t \geq t_1$ is

$$L\frac{di}{dt} + Ri = 0 \qquad t \geq t_1 \tag{6-7}$$

where the initial condition at $t = t_1$ is given by

$$i(t_1) = \frac{E}{R}[1 - e^{-(R/L)t_1}] \tag{6-8}$$

(Note that the instantaneous value of the current at the switching instant $t = t_1$ serves as the initial condition for the transient response for $t \geq t_1$.) Equations (6–5), (6–7), and (6–8) constitute a mathematical model for the system.

Now we shall obtain the solution of Equation (6–7) with the initial condition given by Equation (6–8). The Laplace transform of Equation (6–7), with $t = t_1$ the initial time, gives

$$L[sI(s) - i(t_1)] + RI(s) = 0$$

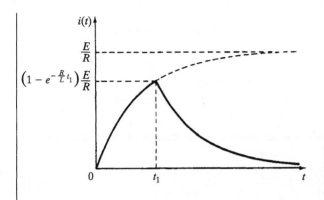

Figure 6–13 Plot of $i(t)$ versus t for the circuit shown in Figure 6–12 when switch S is closed at $t = 0$ and opened at $t = t_1$.

or

$$(Ls + R)I(s) = Li(t_1)$$

Hence,

$$I(s) = \frac{Li(t_1)}{Ls + R} = \frac{E}{R}[1 - e^{-(R/L)t_1}]\frac{1}{s + (R/L)} \tag{6–9}$$

The inverse Laplace transform of Equation (6–9) gives

$$i(t) = \frac{E}{R}[1 - e^{-(R/L)t_1}]e^{-(R/L)(t-t_1)} \qquad t \ge t_1 \tag{6–10}$$

Consequently, from Equations (6–6) and (6–10), the current $i(t)$ for $t \ge 0$ can be written

$$i(t) = \frac{E}{R}[1 - e^{-(R/L)t}] \qquad\qquad t_1 > t \ge 0$$

$$= \frac{E}{R}[1 - e^{-(R/L)t_1}]e^{-(R/L)(t-t_1)} \qquad t \ge t_1$$

A typical plot of $i(t)$ versus t for this case is given in Figure 6–13.

Example 6–3

Consider the electrical circuit shown in Figure 6–14. The circuit consists of a resistance R(in ohms) and a capacitance C (in farads). Obtain the transfer function $E_o(s)/E_i(s)$. Also, obtain a state-space representation of the system.

Applying Kirchhoff's voltage law to the system, we obtain the following equations:

$$Ri + \frac{1}{C}\int i\, dt = e_i \tag{6–11}$$

$$\frac{1}{C}\int i\, dt = e_o \tag{6–12}$$

The transfer-function model of the circuit can be obtained as follows: Taking the Laplace transforms of Equations (6–11) and (6–12), assuming zero initial conditions, we get

$$RI(s) + \frac{1}{C}\frac{1}{s}I(s) = E_i(s)$$

$$\frac{1}{C}\frac{1}{s}I(s) = E_o(s)$$

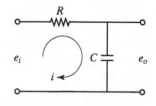

Figure 6–14 RC circuit.

Assuming that the input is e_i and the output is e_o, the transfer function of the system is

$$\frac{E_o(s)}{E_i(s)} = \frac{\dfrac{1}{C}\dfrac{1}{s}I(s)}{\left(R + \dfrac{1}{C}\dfrac{1}{s}\right)I(s)} = \frac{1}{RCs + 1} \tag{6–13}$$

This system is a first-order system.

A state-space model of the system may be obtained as follows: First, note that, from Equation (6–13), the differential equation for the circuit is

$$RC\dot{e}_o + e_o = e_i$$

If we define the state variable

$$x = e_o$$

and the input and output variables

$$u = e_i, \qquad y = e_o = x$$

then we obtain

$$\dot{x} = -\frac{1}{RC}x + \frac{1}{RC}u$$

$$y = x$$

These two equations give a state-space representation of the system.

Example 6–4

Consider the electrical circuit shown in Figure 6–15. The circuit consists of an inductance L (in henrys), a resistance R (in ohms), and a capacitance C (in farads). Obtain the transfer function $E_o(s)/E_i(s)$. Also, obtain a state-space representation of the system.

Applying Kirchhoff's voltage law to the system, we obtain the following equations:

$$L\frac{di}{dt} + Ri + \frac{1}{C}\int i\, dt = e_i \tag{6–14}$$

$$\frac{1}{C}\int i\, dt = e_o \tag{6–15}$$

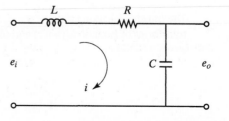

Figure 6–15 Electrical circuit.

The transfer-function model of the circuit can be obtained as follows: Taking the Laplace transforms of Equations (6–14) and (6–15), assuming zero initial conditions, we get

$$LsI(s) + RI(s) + \frac{1}{C}\frac{1}{s}I(s) = E_i(s)$$

$$\frac{1}{C}\frac{1}{s}I(s) = E_o(s)$$

Then the transfer function $E_o(s)/E_i(s)$ becomes

$$\frac{E_o(s)}{E_i(s)} = \frac{1}{LCs^2 + RCs + 1} \tag{6–16}$$

A state-space model of the system may be obtained as follows: First, note that, from Equation (6–16), the differential equation for the system is

$$\ddot{e}_o + \frac{R}{L}\dot{e}_o + \frac{1}{LC}e_o = \frac{1}{LC}e_i$$

Then, by defining state variables

$$x_1 = e_o$$
$$x_2 = \dot{e}_o$$

and the input and output variables

$$u = e_i$$
$$y = e_o = x_1$$

we obtain

$$\begin{bmatrix} \dot{x}_1 \\ \dot{x}_2 \end{bmatrix} = \begin{bmatrix} 0 & 1 \\ -\dfrac{1}{LC} & -\dfrac{R}{L} \end{bmatrix} \begin{bmatrix} x_1 \\ x_2 \end{bmatrix} + \begin{bmatrix} 0 \\ \dfrac{1}{LC} \end{bmatrix} u$$

and

$$y = \begin{bmatrix} 1 & 0 \end{bmatrix} \begin{bmatrix} x_1 \\ x_2 \end{bmatrix}$$

These two equations give a mathematical model of the system in state space.

Transfer Functions of Nonloading Cascaded Elements.   The transfer function of a system consisting of two nonloading cascaded elements can be obtained by eliminating the intermediate input and output. For example, consider the system shown in Figure 6–16(a). The transfer functions of the elements are

$$G_1(s) = \frac{X_2(s)}{X_1(s)} \qquad \text{and} \qquad G_2(s) = \frac{X_3(s)}{X_2(s)}$$

If the input impedance of the second element is infinite, the output of the first element is not affected by connecting it to the second element. Then the transfer function of the whole system becomes

$$G(s) = \frac{X_3(s)}{X_1(s)} = \frac{X_2(s)X_3(s)}{X_1(s)X_2(s)} = G_1(s)G_2(s)$$

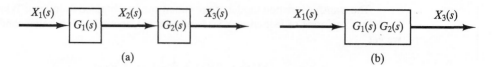

(a)   (b)

Figure 6–16   (a) System consisting of two nonloading cascaded elements; (b) an equivalent system.

The transfer function of the whole system is thus the product of the transfer functions of the individual elements. This is shown in Figure 6–16(b).

As an example, consider the system shown in Figure 6–17. The insertion of an isolating amplifier between the circuits to obtain nonloading characteristics is frequently used in combining circuits. Since amplifiers have very high input impedances, an isolation amplifier inserted between the two circuits justifies the nonloading assumption.

The two simple RC circuits, isolated by an amplifier as shown in Figure 6–17, have negligible loading effects, and the transfer function of the entire circuit equals the product of the individual transfer functions. Thus, in this case,

$$\frac{E_o(s)}{E_i(s)} = \left(\frac{1}{R_1 C_1 s + 1}\right)(K)\left(\frac{1}{R_2 C_2 s + 1}\right)$$

$$= \frac{K}{(R_1 C_1 s + 1)(R_2 C_2 s + 1)}$$

Transfer functions of cascaded elements.   Many feedback systems have components that load each other. Consider the system shown in Figure 6–18. Assume that e_i is the input and e_o is the output. The capacitances C_1 and C_2 are not charged initially. Let us show that the second stage of the circuit (the $R_2 C_2$ portion) produces a loading effect on the first stage (the $R_1 C_1$ portion). The equations for the system are

$$\frac{1}{C_1}\int (i_1 - i_2)\, dt + R_1 i_1 = e_i \tag{6–17}$$

$$\frac{1}{C_1}\int (i_2 - i_1)\, dt + R_2 i_2 + \frac{1}{C_2}\int i_2 dt = 0 \tag{6–18}$$

$$\frac{1}{C_2}\int i_2 dt = e_o \tag{6–19}$$

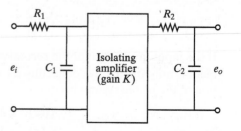

Figure 6–17   Electrical system.

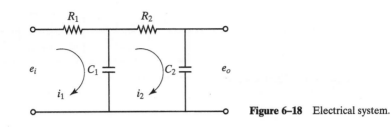

Figure 6–18   Electrical system.

Taking the Laplace transforms of Equations (6–17) through (6–19), respectively, assuming zero initial conditions, we obtain

$$\frac{1}{C_1 s}[I_1(s) - I_2(s)] + R_1 I_1(s) = E_i(s) \qquad (6\text{–}20)$$

$$\frac{1}{C_1 s}[I_2(s) - I_1(s)] + R_2 I_2(s) + \frac{1}{C_2 s}I_2(s) = 0 \qquad (6\text{–}21)$$

$$\frac{1}{C_2 s}I_2(s) = E_o(s) \qquad (6\text{–}22)$$

Eliminating $I_1(s)$ from Equations (6–20) and (6–21) and writing $E_i(s)$ in terms of $I_2(s)$, we find the transfer function between $E_o(s)$ and $E_i(s)$ to be

$$\frac{E_o(s)}{E_i(s)} = \frac{1}{(R_1 C_1 s + 1)(R_2 C_2 s + 1) + R_1 C_2 s}$$

$$= \frac{1}{R_1 C_1 R_2 C_2 s^2 + (R_1 C_1 + R_2 C_2 + R_1 C_2)s + 1} \qquad (6\text{–}23)$$

The term $R_1 C_2 s$ in the denominator of the transfer function represents the interaction of two simple RC circuits. Since $(R_1 C_1 + R_2 C_2 + R_1 C_2)^2 > 4R_1 C_1 R_2 C_2$, the two roots of the denominator of Equation (6–23) are real.

The analysis just presented shows that, if two RC circuits are connected in cascade so that the output from the first circuit is the input to the second, the overall transfer function is *not* the product of $1/(R_1 C_1 s + 1)$ and $1/(R_2 C_2 s + 1)$. The reason for this is that, when we derive the transfer function for an isolated circuit, we implicitly assume that the output is unloaded. In other words, the load impedance is assumed to be infinite, which means that no power is being withdrawn at the output. When the second circuit is connected to the output of the first, however, a certain amount of power is withdrawn, and thus the assumption of no loading is violated. Therefore, if the transfer function of this system is obtained under the assumption of no loading, then it is not valid. The degree of the loading effect determines the amount of modification of the transfer function.

Complex impedances.   In deriving transfer functions for electrical circuits, we frequently find it convenient to write the Laplace-transformed equations directly, without writing the differential equations. Consider the system shown in Figure 6–19. In this system, Z_1 and Z_2 represent complex impedances. The complex impedance

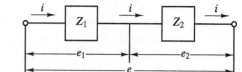

Figure 6–19 Electrical circuit.

$Z(s)$ of a two-terminal circuit is the ratio of $E(s)$, the Laplace transform of the voltage across the terminals, to $I(s)$, the Laplace transform of the current through the element, under the assumption that the initial conditions are zero, so that $Z(s) = E(s)/I(s)$. If the two-terminal element is a resistance R, a capacitance C, or an inductance L, then the complex impedance is given by R, $1/Cs$, or Ls, respectively. If complex impedances are connected in series, the total impedance is the sum of the individual complex impedances.

The general relationship

$$E(s) = Z(s)I(s)$$

corresponds to Ohm's law for purely resistive circuits. (Note that, like resistances, impedances can be combined in series and in parallel.)

Remember that the impedance approach is valid only if the initial conditions involved are all zero. Since the transfer function requires zero initial conditions, the impedance approach can be applied to obtain the transfer function of the electrical circuit. This approach greatly simplifies the derivation of transfer functions of electrical circuits.

Deriving transfer functions of electrical circuits with the use of complex impedances. The transfer function of an electrical circuit can be obtained as a ratio of complex impedances. For the circuit shown in Figure 6–20, assume that the voltages e_i and e_o are the input and output of the circuit, respectively. Then the transfer function of this circuit can be obtained as

$$\frac{E_o(s)}{E_i(s)} = \frac{Z_2(s)I(s)}{Z_1(s)I(s) + Z_2(s)I(s)} = \frac{Z_2(s)}{Z_1(s) + Z_2(s)}$$

where $I(s)$ is the Laplace transform of the current $i(t)$ in the circuit.

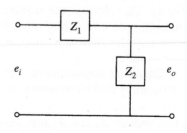

Figure 6–20 Electrical circuit.

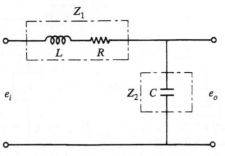

Figure 6–21   Electrical circuit.

For the circuit shown in Figure 6–21,

$$Z_1 = Ls + R, \qquad Z_2 = \frac{1}{Cs}$$

Hence, the transfer function $E_o(s)/E_i(s)$ is

$$\frac{E_o(s)}{E_i(s)} = \frac{\dfrac{1}{Cs}}{Ls + R + \dfrac{1}{Cs}} = \frac{1}{LCs^2 + RCs + 1}$$

Example 6–5

Consider the system shown in Figure 6–22. Obtain the transfer function $E_o(s)/E_i(s)$ by the complex-impedance approach. (Capacitances C_1 and C_2 are not charged initially.)

The circuit shown in Figure 6–22 can be redrawn as that shown in Figure 6–23(a), which can be further modified to Figure 6–23(b).

In the system shown in Figure 6–23(b), the current I is divided into two currents I_1 and I_2. Noting that

$$Z_2 I_1 = (Z_3 + Z_4)I_2, \qquad I_1 + I_2 = I$$

we obtain

$$I_1 = \frac{Z_3 + Z_4}{Z_2 + Z_3 + Z_4}I, \qquad I_2 = \frac{Z_2}{Z_2 + Z_3 + Z_4}I$$

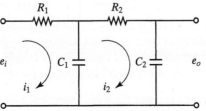

Figure 6–22   Electrical circuit.

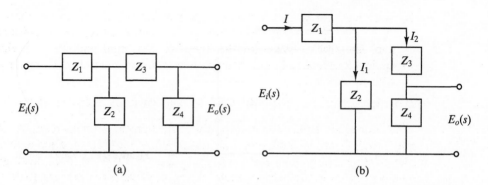

Figure 6-23 (a) The circuit of Figure 6–22 shown in terms of impedances; (b) equivalent circuit diagram.

Observing that

$$E_i(s) = Z_1 I + Z_2 I_1 = \left[Z_1 + \frac{Z_2(Z_3 + Z_4)}{Z_2 + Z_3 + Z_4} \right] I$$

$$E_o(s) = Z_4 I_2 = \frac{Z_2 Z_4}{Z_2 + Z_3 + Z_4} I$$

we get

$$\frac{E_o(s)}{E_i(s)} = \frac{Z_2 Z_4}{Z_1(Z_2 + Z_3 + Z_4) + Z_2(Z_3 + Z_4)}$$

Substituting $Z_1 = R_1$, $Z_2 = 1/(C_1 s)$, $Z_3 = R_2$, and $Z_4 = 1/(C_2 s)$ into this last equation yields

$$\frac{E_o(s)}{E_i(s)} = \frac{\dfrac{1}{C_1 s}\dfrac{1}{C_2 s}}{R_1\left(\dfrac{1}{C_1 s} + R_2 + \dfrac{1}{C_2 s}\right) + \dfrac{1}{C_1 s}\left(R_2 + \dfrac{1}{C_2 s}\right)}$$

$$= \frac{1}{R_1 C_1 R_2 C_2 s^2 + (R_1 C_1 + R_2 C_2 + R_1 C_2)s + 1}$$

which is the transfer function of the system. [Notice that it is the same as that given by Equation (6–23).]

6-4 ANALOGOUS SYSTEMS

Systems that can be represented by the same mathematical model, but that are physically different, are called *analogous* systems. Thus, analogous systems are described by the same differential or integrodifferential equations or transfer functions.

The concept of analogous systems is useful in practice, for the following reasons:

1. The solution of the equation describing one physical system can be directly applied to analogous systems in any other field.

2. Since one type of system may be easier to handle experimentally than another, instead of building and studying a mechanical system (or a hydraulic system, pneumatic system, or the like), we can build and study its electrical analog, for electrical or electronic systems are, in general, much easier to deal with experimentally.

This section presents analogies between mechanical and electrical systems.

Mechanical–electrical analogies. Mechanical systems can be studied through their electrical analogs, which may be more easily constructed than models of the corresponding mechanical systems. There are two electrical analogies for mechanical systems: the force–voltage analogy and the force–current analogy.

Force–voltage analogy. Consider the mechanical system of Figure 6–24(a) and the electrical system of Figure 6–24(b). In the mechanical system p is the external force, and in the electrical system e is the voltage source. The equation for the mechanical system is

$$m\frac{d^2x}{dt^2} + b\frac{dx}{dt} + kx = p \tag{6-24}$$

where x is the displacement of mass m, measured from the equilibrium position. The equation for the electrical system is

$$L\frac{di}{dt} + Ri + \frac{1}{C}\int i\,dt = e$$

In terms of the electric charge q, this last equation becomes

$$L\frac{d^2q}{dt^2} + R\frac{dq}{dt} + \frac{1}{C}q = e \tag{6-25}$$

Comparing Equations (6–24) and (6–25), we see that the differential equations for the two systems are of identical form. Thus, these two systems are analogous systems.

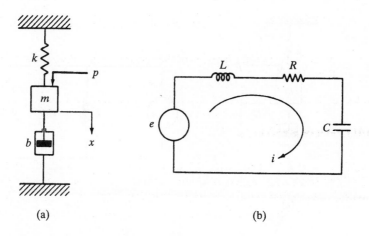

(a) (b)

Figure 6–24 Analogous mechanical and electrical systems.

TABLE 6–1   Force–Voltage Analogy

Mechanical Systems	Electrical Systems
Force p (torque T) Mass m (moment of inertia J) Viscous-friction coefficient b Spring constant k Displacement x (angular displacement θ) Velocity $\dot{x}$ (angular velocity $\dot{\theta}$)	Voltage e Inductance L Resistance R Reciprocal of capacitance, $1/C$ Charge q Current i

The terms that occupy corresponding positions in the differential equations are called *analogous quantities*, a list of which appears in Table 6–1. The analogy here is called the *force–voltage analogy* (or *mass–inductance analogy*).

Force–current analogy.   Another analogy between mechanical and electrical systems is based on the force–current analogy. Consider the mechanical system shown in Figure 6–25(a), where p is the external force. The system equation is

$$m\frac{d^2x}{dt^2} + b\frac{dx}{dt} + kx = p \tag{6–26}$$

where x is the displacement of mass m, measured from the equilibrium position.

Consider next the electrical system shown in Figure 6–25(b), where i_s is the current source. Applying Kirchhoff's current law gives

$$i_L + i_R + i_C = i_s \tag{6–27}$$

where

$$i_L = \frac{1}{L}\int e\,dt, \qquad i_R = \frac{e}{R}, \qquad i_C = C\frac{de}{dt}$$

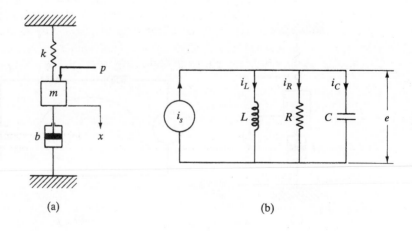

(a)   (b)

Figure 6–25   Analogous mechanical and electrical systems.

Thus, Equation (6–27) can be written as

$$\frac{1}{L}\int e \, dt + \frac{e}{R} + C\frac{de}{dt} = i_s \tag{6–28}$$

Since the magnetic flux linkage ψ is related to the voltage e by the equation

$$\frac{d\psi}{dt} = e$$

Equation (6–28) can be written in terms of ψ as

$$C\frac{d^2\psi}{dt^2} + \frac{1}{R}\frac{d\psi}{dt} + \frac{1}{L}\psi = i_s \tag{6–29}$$

Comparing Equations (6–26) and (6–29), we find that the two systems are analogous. The analogous quantities are listed in Table 6–2. The analogy here is called the *force–current analogy* (or *mass–capacitance analogy*).

Comments.   Analogies between two systems break down if the regions of operation are extended too far. In other words, since the mathematical models on which the analogies are based are only approximations to the dynamic characteristics of physical systems, the analogy may break down if the operating region of one system is very wide. Nevertheless, even if the operating region of a given mechanical system is wide, it can be divided into two or more subregions, and analogous electrical systems can be built for each subregion.

Analogy, of course, is not limited to mechanical–electrical analogy, but includes any physical or nonphysical system. Systems having an identical transfer function (or identical mathematical model) are analogous systems. (The transfer function is one of the simplest and most concise forms of mathematical models available today.)

Analogous systems exhibit the same output in response to the same input. For any given physical system, the mathematical response can be given a physical interpretation.

The concept of analogy is useful in applying well-known results in one field to another. It proves particularly useful when a given physical system (mechanical, hydraulic, pneumatic, and so on) is complicated, so that analyzing an analogous electrical circuit first is advantageous. Such an analogous electrical circuit can be built physically or can be simulated on the digital computer.

TABLE 6–2   Force–Current Analogy

Mechanical Systems	Electrical Systems
Force p (torque T)	Current i
Mass m (moment of inertia J)	Capacitance C
Viscous-friction coefficient b	Reciprocal of resistance, $1/R$
Spring constant k	Reciprocal of inductance, $1/L$
Displacement x (angular displacement θ)	Magnetic flux linkage ψ
Velocity $\dot{x}$ (angular velocity $\dot{\theta}$)	Voltage e

Example 6–6

Obtain the transfer functions of the systems shown in Figures 6–26(a) and (b), and show that these systems are analogous.

For the mechanical system shown in Figure 6–26(a), the equation of motion is

$$b(\dot{x}_i - \dot{x}_o) = kx_o$$

or

$$b\dot{x}_i = kx_o + b\dot{x}_o$$

Taking the Laplace transform of this last equation, assuming zero initial conditions, we obtain

$$bsX_i(s) = (k + bs)X_o(s)$$

Hence, the transfer function between $X_o(s)$ and $X_i(s)$ is

$$\frac{X_o(s)}{X_i(s)} = \frac{bs}{bs + k} = \frac{\dfrac{b}{k}s}{\dfrac{b}{k}s + 1}$$

For the electrical system shown in Figure 6–26(b), we have

$$\frac{E_o(s)}{E_i(s)} = \frac{RCs}{RCs + 1}$$

Comparing the transfer functions obtained, we see that the two systems are analogous. (Note that both b/k and RC have the dimension of time and are time constants of the respective systems.)

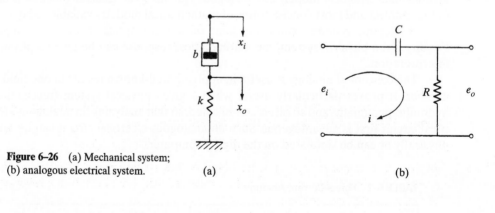

Figure 6–26   (a) Mechanical system;
(b) analogous electrical system.   (a)   (b)

6–5 MATHEMATICAL MODELING OF ELECTROMECHANICAL SYSTEMS

In this section, we obtain mathematical models of dc servomotors. To control the motion or speed of dc servomotors, we control the field current or armature current or we use a servodriver as a motor–driver combination. There are many different types of servodrivers. Most are designed to control the speed of dc servomotors, which improves the efficiency of operating servomotors. Here, however, we shall discuss only armature control of a dc servomotor and obtain its mathematical model in the form of a transfer function.

Armature control of dc servomotors.   Consider the armature-controlled dc servomotor shown in Figure 6–27, where the field current is held constant. In this system,

R_a = armature resistance, Ω

L_a = armature inductance, H

i_a = armature current, A

i_f = field current, A

e_a = applied armature voltage, V

e_b = back emf, V

θ = angular displacement of the motor shaft, rad

T = torque developed by the motor, N-m

J = moment of inertia of the motor and load referred to the motor shaft, kg-m^2

b = viscous-friction coefficient of the motor and load referred to the motor shaft, N-m/rad/s

The torque T developed by the motor is proportional to the product of the armature current i_a and the air gap flux ψ, which in turn is proportional to the field current, or

$$\psi = K_f i_f$$

where K_f is a constant. The torque T can therefore be written as

$$T = K_f i_f K_1 i_a$$

where K_1 is a constant.

For a constant field current, the flux becomes constant and the torque becomes directly proportional to the armature current, so

$$T = K i_a$$

where K is a motor-torque constant. Notice that if the sign of the current i_a is reversed, the sign of the torque T will be reversed, which will result in a reversal of the direction of rotor rotation.

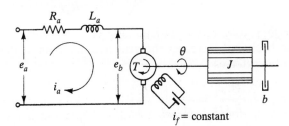

i_f = constant

Figure 6–27   Armature-controlled dc servomotor.

When the armature is rotating, a voltage proportional to the product of the flux and angular velocity is induced in the armature. For a constant flux, the induced voltage e_b is directly proportional to the angular velocity $d\theta/dt$, or

$$e_b = K_b \frac{d\theta}{dt} \tag{6-30}$$

where e_b is the back emf and K_b is a back-emf constant.

The speed of an armature-controlled dc servomotor is controlled by the armature voltage e_a. The differential equation for the armature circuit is

$$L_a \frac{di_a}{dt} + R_a i_a + e_b = e_a \tag{6-31}$$

The armature current produces the torque that is applied to the inertia and friction; hence,

$$J \frac{d^2\theta}{dt^2} + b \frac{d\theta}{dt} = T = K i_a \tag{6-32}$$

Assuming that all initial conditions are zero and taking the Laplace transforms of Equations (6–30), (6–31), and (6–32), we obtain the following equations:

$$K_b s \Theta(s) = E_b(s) \tag{6-33}$$

$$(L_a s + R_a) I_a(s) + E_b(s) = E_a(s) \tag{6-34}$$

$$(J s^2 + b s) \Theta(s) = T(s) = K I_a(s) \tag{6-35}$$

Considering $E_a(s)$ as the input and $\Theta(s)$ as the output and eliminating $I_a(s)$ and $E_b(s)$ from Equations (6–33), (6–34), and (6–35), we obtain the transfer function for the dc servomotor:

$$\frac{\Theta(s)}{E_a(s)} = \frac{K}{s[L_a J s^2 + (L_a b + R_a J)s + R_a b + K K_b]} \tag{6-36}$$

The inductance L_a in the armature circuit is usually small and may be neglected. If L_a is neglected, then the transfer function given by Equation (6–36) reduces to

$$\frac{\Theta(s)}{E_a(s)} = \frac{K}{s(R_a J s + R_a b + K K_b)} = \frac{\dfrac{K}{R_a J}}{s\left(s + \dfrac{R_a b + K K_b}{R_a J}\right)} \tag{6-37}$$

Notice that the term $(R_a b + K K_b)/(R_a J)$ in Equation (6–37) corresponds to the damping term. Thus, the back emf increases the effective damping of the system. Equation (6–37) may be rewritten as

$$\frac{\Theta(s)}{E_a(s)} = \frac{K_m}{s(T_m s + 1)} \tag{6-38}$$

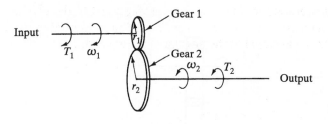

Figure 6-28   Gear train system.

where

$$K_m = K/(R_a b + K K_b) = \text{motor gain constant}$$
$$T_m = R_a J/(R_a b + K K_b) = \text{motor time constant}$$

Equation (6-38) is the transfer function of the dc servomotor when the armature voltage $e_a(t)$ is the input and the angular displacement $\theta(t)$ is the output. Since the transfer function involves the term $1/s$, this system possesses an integrating property. (Notice that the time constant T_m of the motor becomes smaller as the resistance R_a is reduced and the moment of inertia J is made smaller.)

Gear train.   Gear trains are frequently used in mechanical systems to reduce speed, to magnify torque, or to obtain the most efficient power transfer by matching the driving member to the given load. Figure 6-28 illustrates a simple gear train system in which the gear train transmits motion and torque from the input member to the output member. If the radii of gear 1 and gear 2 are r_1 and r_2, respectively, and the numbers of teeth on gear 1 and gear 2 are n_1 and n_2, respectively, then

$$\frac{r_1}{r_2} = \frac{n_1}{n_2}$$

Because the surface speeds at the point of contact of the two gears must be identical, we have

$$r_1 \omega_1 = r_2 \omega_2$$

where ω_1 and ω_2 are the angular velocities of gear 1 and gear 2, respectively. Therefore,

$$\frac{\omega_2}{\omega_1} = \frac{r_1}{r_2} = \frac{n_1}{n_2}$$

If we neglect friction loss, the gear train transmits the power unchanged. In other words, if the torque applied to the input shaft is T_1 and the torque transmitted to the output shaft is T_2, then

$$T_1 \omega_1 = T_2 \omega_2$$

Example 6-7

Consider the system shown in Figure 6-29. Here, a load is driven by a motor through the gear train. Assuming that the stiffness of the shafts of the gear train is infinite, that there is neither backlash nor elastic deformation, and that the number of teeth on each gear is

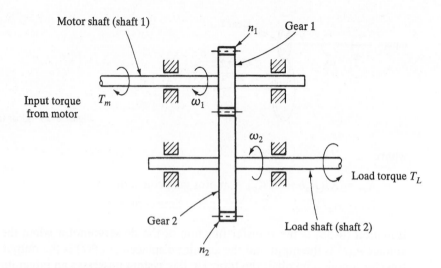

Motor shaft (shaft 1)

n_1　　Gear 1

Input torque T_m
from motor

ω_1

ω_2

Load torque T_L

Gear 2

Load shaft (shaft 2)

n_2

Figure 6–29　Gear train system.

proportional to the radius of the gear, find the equivalent inertia and equivalent friction referred to the motor shaft (shaft 1) and those referred to the load shaft (shaft 2). The numbers of teeth on gear 1 and gear 2 are n_1 and n_2, respectively, and the angular velocities of shaft 1 and shaft 2 are ω_1 and ω_2, respectively. The inertia and viscous friction coefficient of each gear train component are denoted by J_1, b_1 and J_2, b_2, respectively.

By applying Newton's second law to this system, the following two equations can be derived: For the motor shaft (shaft 1),

$$J_1\dot\omega_1 + b_1\omega_1 + T_1 = T_m \tag{6–39}$$

where T_m is the torque developed by the motor and T_1 is the load torque on gear 1 due to the rest of the gear train. For the load shaft (shaft 2),

$$J_2\dot\omega_2 + b_2\omega_2 + T_L = T_2 \tag{6–40}$$

where T_2 is the torque transmitted to gear 2 and T_L is the load torque. Since the gear train transmits the power unchanged, we have

$$T_1\omega_1 = T_2\omega_2$$

or

$$T_1 = T_2\frac{\omega_2}{\omega_1} = T_2\frac{n_1}{n_2}$$

If $n_1/n_2 < 1$, the gear ratio reduces the speed in addition to magnifying the torque. Eliminating T_1 and T_2 from Equations (6–39) and (6–40) yields

$$J_1\dot\omega_1 + b_1\omega_1 + \frac{n_1}{n_2}(J_2\dot\omega_2 + b_2\omega_2 + T_L) = T_m \tag{6–41}$$

Since $\omega_2 = (n_1/n_2)\omega_1$, eliminating ω_2 from Equation (6–41) gives

$$\left[J_1 + \left(\frac{n_1}{n_2}\right)^2 J_2\right]\dot\omega_1 + \left[b_1 + \left(\frac{n_1}{n_2}\right)^2 b_2\right]\omega_1 + \left(\frac{n_1}{n_2}\right)T_L = T_m \tag{6–42}$$

Thus, the equivalent inertia and equivalent viscous friction coefficient of the gear train referred to shaft 1 are given by

$$J_{1\,eq} = J_1 + \left(\frac{n_1}{n_2}\right)^2 J_2, \qquad b_{1\,eq} = b_1 + \left(\frac{n_1}{n_2}\right)^2 b_2$$

The effect of J_2 on the equivalent inertia $J_{1\,eq}$ is determined by the gear ratio n_1/n_2. For speed-reducing gear trains, the ratio n_1/n_2 is much smaller than unity. If $n_1/n_2 \ll 1$, then the effect of J_2 on the equivalent inertia $J_{1\,eq}$ is negligible. Similar comments apply to the equivalent friction of the gear train.

In terms of the equivalent inertia $J_{1\,eq}$ and equivalent viscous friction coefficient $b_{1\,eq}$, Equation (6–42) can be simplified to give

$$J_{1\,eq}\dot{\omega}_1 + b_{1\,eq}\omega_1 + nT_L = T_m$$

where $n = n_1/n_2$.

The equivalent inertia and equivalent viscous friction coefficient of the gear train referred to shaft 2 are

$$J_{2\,eq} = J_2 + \left(\frac{n_2}{n_1}\right)^2 J_1, \qquad b_{2\,eq} = b_2 + \left(\frac{n_2}{n_1}\right)^2 b_1$$

So the relationship between $J_{1\,eq}$ and $J_{2\,eq}$ is

$$J_{1\,eq} = \left(\frac{n_1}{n_2}\right)^2 J_{2\,eq}$$

and that between $b_{1\,eq}$ and $b_{2\,eq}$ is

$$b_{1\,eq} = \left(\frac{n_1}{n_2}\right)^2 b_{2\,eq}$$

and Equation (6–42) can be modified to give

$$J_{2\,eq}\dot{\omega}_2 + b_{2\,eq}\omega_2 + T_L = \frac{1}{n}T_m$$

Example 6–8

Consider the dc servomotor system shown in Figure 6–30. The armature inductance is negligible and is not shown in the circuit. Obtain the transfer function between the output θ_2 and the input e_a. In the diagram,

$$R_a = \text{armature resistance, } \Omega$$
$$i_a = \text{armature current, A}$$

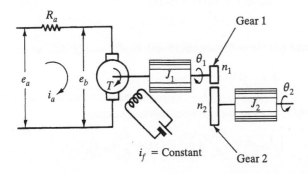

i_f = Constant

Figure 6–30   DC servomotor system.

i_f = field current, A

e_a = applied armature voltage, V

e_b = back emf, V

θ_1 = angular displacement of the motor shaft, rad

θ_2 = angular displacement of the load element, rad

T = torque developed by the motor, N-m

J_1 = moment of inertia of the rotor of the motor, kg-m^2

J_2 = moment of inertia of the load, kg-m^2

n_1 = number of teeth on gear 1

n_2 = number of teeth on gear 2

The torque T developed by the dc servomotor is

$$T = Ki_a$$

where K is the motor torque constant. The induced voltage e_b is proportional to the angular velocity $\dot{\theta}_1$, or

$$e_b = K_b \frac{d\theta_1}{dt} \tag{6-43}$$

where K_b is the back-emf constant.

The equation for the armature circuit is

$$R_a i_a + e_b = e_a \tag{6-44}$$

The equivalent moment of inertia of the motor rotor plus the load inertia referred to the motor shaft is

$$J_{1\,eq} = J_1 + \left(\frac{n_1}{n_2}\right)^2 J_2$$

The armature current produces the torque that is applied to the equivalent moment of inertia $J_{1\,eq}$. Thus,

$$J_{1\,eq} \frac{d^2\theta_1}{dt^2} = T = Ki_a \tag{6-45}$$

Assuming that all initial conditions are zero and taking the Laplace transforms of Equations (6-43), (6-44), and (6-45), we obtain

$$E_b(s) = K_b s \Theta_1(s) \tag{6-46}$$
$$R_a I_a(s) + E_b(s) = E_a(s) \tag{6-47}$$
$$J_{1\,eq} s^2 \Theta_1(s) = K I_a(s) \tag{6-48}$$

Eliminating $E_b(s)$ and $I_a(s)$ from Equations (6-46), (6-47), and (6-48), we obtain

$$\left(J_{1\,eq} s^2 + \frac{KK_b}{R_a} s\right) \Theta_1(s) = \frac{K}{R_a} E_a(s)$$

Noting that $\Theta_1(s)/\Theta_2(s) = n_2/n_1$, we can write this last equation as

$$\left(J_{1\,eq} s^2 + \frac{KK_b}{R_a} s\right) \frac{n_2}{n_1} \Theta_2(s) = \frac{K}{R_a} E_a(s)$$

Hence, the transfer function $\Theta_2(s)/E_a(s)$ is given by

$$\frac{\Theta_2(s)}{E_a(s)} = \frac{\dfrac{n_1}{n_2}K}{\left\{ R_a\left[J_1 + \left(\dfrac{n_1}{n_2}\right)^2 J_2 \right]s + KK_b \right\}s}$$

6-6 MATHEMATICAL MODELING OF OPERATIONAL-AMPLIFIER SYSTEMS

In this section, we briefly discuss operational amplifiers. We present several examples of operational-amplifier systems and obtain their mathematical models.

   Operational amplifiers, often called *op-amps*, are important building blocks in modern electronic systems. They are used in filters in control systems and to amplify signals in sensor circuits.

   Consider the operational amplifier shown in Figure 6-31. There are two terminals on the input side, one with a minus sign and the other with a plus sign, called the *inverting* and *noninverting* terminals, respectively. We choose the ground as 0 volts and measure the input voltages e_1 and e_2 relative to the ground. (The input e_1 to the minus terminal of the amplifier is inverted; the input e_2 to the plus terminal is not inverted.) The total input to the amplifier is $e_2 - e_1$. The ideal operational amplifier has the characteristic

$$e_o = K(e_2 - e_1) = -K(e_1 - e_2)$$

where the inputs e_1 and e_2 may be dc or ac signals and K is the differential gain or voltage gain. The magnitude of K is approximately 10^5 to 10^6 for dc signals and ac signals with frequencies less than approximately 10 Hz. (The differential gain K decreases with the frequency of the signal and becomes about unity for frequencies of 1 MHz to about 50 MHz.) Note that the operational amplifier amplifies the difference in voltages e_1 and e_2. Such an amplifier is commonly called a *differential amplifier*. Since the gain of the operational amplifier is very high, the device is inherently unstable. To stabilize it, it is necessary to have negative feedback from the output to the input (feedback from the output to the inverted input).

   In the ideal operational amplifier, no current flows into the input terminals and the output voltage is not affected by the load connected to the output terminal. In other words, the input impedance is infinity and the output impedance is zero. In

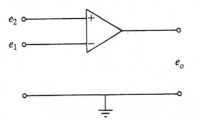

Figure 6-31 Operational amplifier.

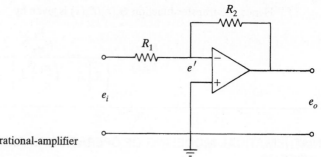

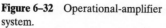

Figure 6–32 Operational-amplifier system.

an actual operational amplifier, a very small (almost negligible) current flows into an input terminal and the output cannot be loaded too much. In our analysis here, however, we make the assumption that the operational amplifiers are ideal.

Inverting amplifier. Consider the operational-amplifier system shown in Figure 6–32. Assume that the magnitudes of the resistances R_1 and R_2 are of comparable order.

Let us obtain the voltage ratio e_o/e_i. In the derivation, we assume the voltage gain to be $K \gg 1$. Let us define the voltage at the minus terminal as e'. Ignoring the current flowing into the amplifier, we have

$$\frac{e_i - e'}{R_1} + \frac{e_o - e'}{R_2} = 0$$

from which we get

$$\frac{e_i}{R_1} + \frac{e_o}{R_2} = \left(\frac{1}{R_1} + \frac{1}{R_2}\right) e'$$

Thus,

$$e' = \frac{\dfrac{e_i}{R_1} + \dfrac{e_o}{R_2}}{\dfrac{1}{R_1} + \dfrac{1}{R_2}} \tag{6–49}$$

Also,

$$e_o = -Ke' \tag{6–50}$$

Eliminating e' from Equations (6–49) and (6–50), we obtain

$$-\frac{e_o}{K} = \frac{\dfrac{e_i}{R_1} + \dfrac{e_o}{R_2}}{\dfrac{1}{R_1} + \dfrac{1}{R_2}}$$

or

$$e_o\left(-\frac{1}{KR_1} - \frac{1}{KR_2} - \frac{1}{R_2}\right) = \frac{e_i}{R_1}$$

Hence,

$$\frac{e_o}{e_i} = \frac{-\dfrac{R_2}{R_1}}{1 + \dfrac{1 + \dfrac{R_2}{R_1}}{K}}$$

Since $K \gg 1 + (R_2/R_1)$, we have

$$\frac{e_o}{e_i} = -\frac{R_2}{R_1} \qquad\qquad (6\text{--}51)$$

Equation (6–51) gives the relationship between the output voltage e_o and the input voltage e_i. From Equations (6–49) and (6–51) we have

$$e' = \frac{\dfrac{e_i}{R_1} + \dfrac{e_o}{R_2}}{\dfrac{1}{R_1} + \dfrac{1}{R_2}} = 0$$

In an operational-amplifier circuit, when the output signal is fed back to the minus terminal, the voltage at the minus terminal becomes equal to the voltage at the plus terminal. This is called an *imaginary short*. If we use the concept of an imaginary short, the ratio e_o/e_i can be obtained much more quickly than the way we just found it, as the following analysis shows:

Consider again the amplifier system shown in Figure 6–32, and define

$$i_1 = \frac{e_i - e'}{R_1}, \qquad i_2 = \frac{e' - e_o}{R_2}$$

Since only a negligible current flows into the amplifier, the current i_1 must be equal to the current i_2. Thus,

$$\frac{e_i - e'}{R_1} = \frac{e' - e_o}{R_2}$$

Because the output signal is fed back to the minus terminal, the voltage at the minus terminal and the voltage at the plus terminal become equal, or $e' = 0$. Hence, we have

$$\frac{e_i}{R_1} = \frac{-e_o}{R_2}$$

or

$$e_o = -\frac{R_2}{R_1} e_i$$

This is a mathematical model relating voltages e_o and e_i. We obtained the same result as we got in the previous analysis [see Equation (6–51)], but much more quickly.

Note that the sign of the output voltage e_o is the negative of that of the input voltage e_i. Hence, this operational amplifier is called an *inverted amplifier*. If $R_1 = R_2$, then the circuit is a sign inverter.

Obtaining mathematical models of physical operational-amplifier systems by means of equations for idealized operational-amplifier systems. In the remaining part of this section, we derive mathematical models of operational-amplifier systems, using the following three conditions that apply to idealized operational-amplifier systems:

1. From Figure 6–31, the output voltage e_o is the differential input voltage $(e_2 - e_1)$ multiplied by the differential gain K. That is,

$$e_o = K(e_2 - e_1)$$

where K is infinite. In designing active filters, we construct the circuit such that the negative feedback appears in the operational amplifier like the system shown in Figure 6–32. As a result, the differential input voltage becomes zero, and we have

Voltage at negative terminal = Voltage at positive terminal

2. The input impedance is infinite.
3. The output impedance is zero.

The use of these three conditions simplifies the derivation of transfer functions of operational-amplifier systems. The derived transfer functions are, of course, not exact, but are approximations that are sufficiently accurate.

In what follows, we shall derive the characteristics of circuits consisting of operational amplifiers, resistors, and capacitors.

Example 6–9

Consider the operational-amplifier circuit shown in Figure 6–33. Obtain the relationship between e_o and e_i.

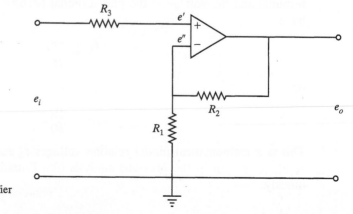

Figure 6–33 Operational-amplifier circuit.

If the operational amplifier is an ideal one, then the output voltage e_o is limited and the differential input voltage becomes zero, or voltage $e'(= e_i)$ and voltage e'', which is equal to $[R_1/(R_1 + R_2)]e_o$ are equal. Thus,

$$e_i = \frac{R_1}{R_1 + R_2} e_o$$

from which it follows that

$$e_o = \left(1 + \frac{R_2}{R_1}\right) e_i$$

This operational-amplifier circuit is a noninverting circuit. If we choose $R_1 = \infty$, then $e_o = e_i$, and the circuit is called a *voltage follower*.

Example 6–10

Consider the operational-amplifier circuit shown in Figure 6–34. Obtain the relationship between the output e_o and the inputs e_1, e_2, and e_3.

We define

$$i_1 = \frac{e_1 - e'}{R_1}, \qquad i_2 = \frac{e_2 - e'}{R_2}, \qquad i_3 = \frac{e_3 - e'}{R_3}, \qquad i_4 = \frac{e' - e_o}{R_4}$$

Noting that the current flowing into the amplifier is negligible, we have

$$\frac{e_1 - e'}{R_1} + \frac{e_2 - e'}{R_2} + \frac{e_3 - e'}{R_3} + \frac{e_o - e'}{R_4} = 0 \qquad (6\text{–}52)$$

Since the amplifier involves negative feedback, the voltage at the minus terminal and that at the plus terminal become equal. Thus, $e' = 0$, and Equation (6–52) becomes

$$\frac{e_1}{R_1} + \frac{e_2}{R_2} + \frac{e_3}{R_3} + \frac{e_o}{R_4} = 0$$

or

$$e_o = -\frac{R_4}{R_1} e_1 - \frac{R_4}{R_2} e_2 - \frac{R_4}{R_3} e_3$$

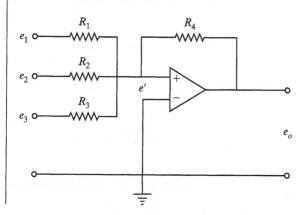

Figure 6–34   Operational-amplifier circuit.

If we choose $R_1 = R_2 = R_3 = R_4$, then

$$e_o = -(e_1 + e_2 + e_3)$$

The circuit is an *inverting adder*.

Example 6–11

Consider the operational-amplifier system shown in Figure 6–35. Letting $e_i(t)$ be the input and $e_o(t)$ be the output of the system, obtain the transfer function for the system. Then obtain the response of the system to a step input of a small magnitude.

Let us define

$$i_1 = \frac{e_i - e'}{R_1}, \qquad i_2 = C\frac{d(e' - e_o)}{dt}, \qquad i_3 = \frac{e' - e_o}{R_2}$$

Noting that the current flowing into the amplifier is negligible, we have

$$i_1 = i_2 + i_3$$

Hence,

$$\frac{e_i - e'}{R_1} = C\frac{d(e' - e_o)}{dt} + \frac{e' - e_o}{R_2} \tag{6-53}$$

Since the operational amplifier involves negative feedback, the voltage at the minus terminal and that at the plus terminal become equal. Hence, $e' = 0$. Substituting $e' = 0$ into Equation (6–53), we obtain

$$\frac{e_i}{R_1} = -C\frac{de_o}{dt} - \frac{e_o}{R_2}$$

Taking the Laplace transform of this last equation, assuming a zero initial condition, we have

$$\frac{E_i(s)}{R_1} = -\frac{R_2Cs + 1}{R_2}E_o(s)$$

which can be written as

$$\frac{E_o(s)}{E_i(s)} = -\frac{R_2}{R_1}\frac{1}{R_2Cs + 1} \tag{6-54}$$

Equation (6–54) is the transfer function for the system, which is a first-order lag system.

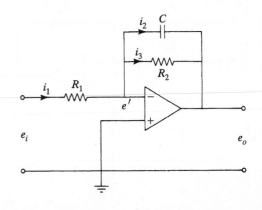

Figure 6–35 First-order lag circuit using an operational amplifier.

Next, we shall find the response of the system to a step input. Suppose that the input $e_i(t)$ is a step function of E volts; that is,

$$e_i(t) = 0 \qquad \text{for } t < 0$$
$$= E \qquad \text{for } t > 0$$

where we assume $0 < (R_2/R_1)E < 10$ V. The output $e_o(t)$ can be determined from

$$E_o(s) = -\frac{R_2}{R_1}\frac{1}{R_2Cs + 1}E_i(s)$$

$$= -\frac{R_2}{R_1}\frac{1}{R_2Cs + 1}\frac{E}{s}$$

$$= -\frac{R_2E}{R_1}\left[\frac{1}{s} - \frac{1}{s + 1/(R_2C)}\right]$$

The inverse Laplace transform of $E_o(s)$ gives

$$e_o(t) = -\frac{R_2E}{R_1}[1 - e^{-t/(R_2C)}]$$

The output voltage reaches $-(R_2/R_1)E$ volts as t increases to infinity.

Example 6–12

Consider the operational-amplifier circuit shown in Figure 6–36. Obtain the transfer function $E_o(s)/E_i(s)$ of the circuit.

The voltage at point A is

$$e_A = \frac{1}{2}(e_i + e_o)$$

The Laplace-transformed version of this last equation is

$$E_A(s) = \frac{1}{2}[E_i(s) + E_o(s)]$$

The voltage at point B is

$$E_B(s) = \frac{\dfrac{1}{Cs}}{R_2 + \dfrac{1}{Cs}}E_i(s) = \frac{1}{R_2Cs + 1}E_i(s)$$

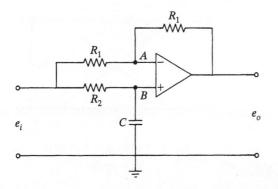

Figure 6–36 Operational-amplifier circuit.

Since the operational amplifier involves negative feedback, the voltage at the minus terminal and that at the plus terminal become equal. Thus,

$$E_A(s) = E_B(s)$$

and it follows that

$$\frac{1}{2}[E_i(s) + E_o(s)] = \frac{1}{R_2Cs + 1}E_i(s)$$

or

$$\frac{E_o(s)}{E_i(s)} = -\frac{R_2Cs - 1}{R_2Cs + 1} = -\frac{s - \dfrac{1}{R_2C}}{s + \dfrac{1}{R_2C}}$$

EXAMPLE PROBLEMS AND SOLUTIONS

Problem A–6–1

Obtain the resistance between points A and B of the circuit shown in Figure 6–37.

Solution This circuit is equivalent to the one shown in Figure 6–38(a). Since $R_1 = R_4 = 10\ \Omega$ and $R_2 = R_3 = 20\ \Omega$, the voltages at points C and D are equal, and there is no current flowing through R_5. Because resistance R_5 does not affect the value of the total resistance between points A and B, it may be removed from the circuit, as shown in Figure 6–38(b). Then

$$\frac{1}{R_{AB}} = \frac{1}{R_1 + R_4} + \frac{1}{R_2 + R_3} = \frac{1}{20} + \frac{1}{40} = \frac{3}{40}$$

and

$$R_{AB} = \frac{40}{3} = 13.3\ \Omega$$

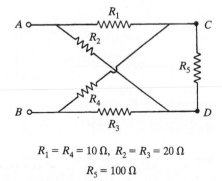

$$R_1 = R_4 = 10\ \Omega,\ R_2 = R_3 = 20\ \Omega$$
$$R_5 = 100\ \Omega$$

Figure 6–37   Electrical circuit.

(a)

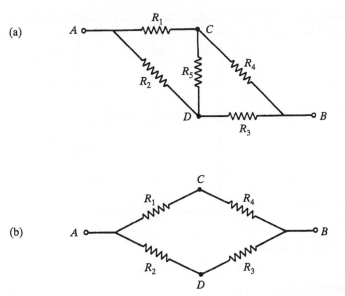

(b)

Figure 6–38 Equivalent circuits to the one shown in Figure 6–37.

Problem A–6–2

Given the circuit of Figure 6–39, calculate currents i_1, i_2, and i_3.

Solution The circuit can be redrawn as shown in Figure 6–40. The combined resistance R of the path in which current i_2 flows is

$$R = 100 + \frac{1}{\dfrac{1}{10} + \dfrac{1}{40}} + 50 = 158 \ \Omega$$

The combined resistance R_0 as seen from the battery is

$$\frac{1}{R_0} = \frac{1}{40} + \frac{1}{158}$$

or

$$R_0 = 31.92 \ \Omega$$

Consequently,

$$i_1 + i_2 = \frac{12}{R_0} = \frac{12}{31.92} = 0.376 \ \text{A}$$

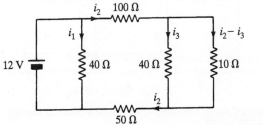

Figure 6–39 Electrical circuit.

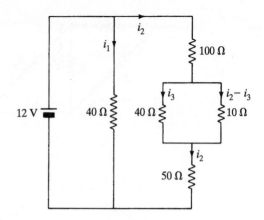

Figure 6–40 Equivalent circuit to the
one shown in Figure 6–39.

Noting that $40i_1 = 158i_2$, we obtain

$$i_1 = 0.300 \text{ A}, \qquad i_2 = 0.076 \text{ A}$$

To determine i_3, note that

$$40i_3 = 10(i_2 - i_3)$$

Then

$$i_3 = \frac{10}{50}i_2 = 0.0152 \text{ A}$$

Problem A–6–3

Obtain the combined resistance between points A and B of the circuit shown in Figure 6–41, which consists of an infinite number of resistors connected in the form of a ladder.

Solution We define the combined resistance between points A and B as R_0. Now, let us separate the first three resistors from the rest. [See Figure 6–42(a).] Since the circuit consists of an infinite number of resistors, the removal of the first three resistors does not affect the combined resistance value. Therefore, the combined resistance between points C and D is the same as R_0. Then the circuit shown in Figure 6–41 may be redrawn as shown in Figure 6–42(b), and R_0, the resistance between points A and B, can be obtained as

$$R_0 = 2R + \cfrac{1}{\cfrac{1}{R} + \cfrac{1}{R_0}} = 2R + \frac{RR_0}{R_0 + R}$$

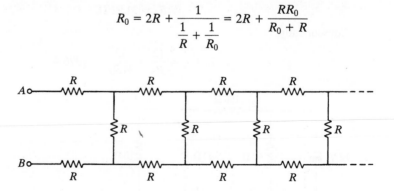

Figure 6–41 Electrical circuit consisting of an infinite number of resistors connected in the form of a ladder.

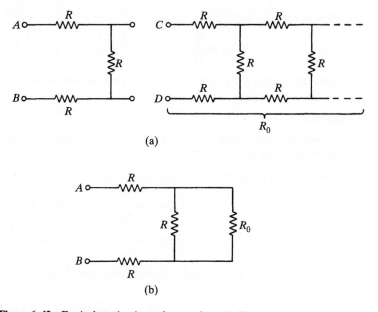

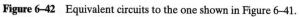

(b)

Figure 6–42 Equivalent circuits to the one shown in Figure 6–41.

Rewriting, we get

$$R_0^2 - 2RR_0 - 2R^2 = 0$$

Solving for R_0, we find that

$$R_0 = R \pm \sqrt{3}R$$

Finally, neglecting the negative value for resistance, we obtain

$$R_0 = R + \sqrt{3}R = 2.732R$$

Problem A–6–4

Find currents i_1, i_2, and i_3 for the circuit shown in Figure 6–43.

Solution Applying Kirchhoff's voltage law and current law to the circuit, we have

$$12 - 10i_1 - 5i_3 = 0$$
$$8 - 15i_2 - 5i_3 = 0$$
$$i_1 + i_2 - i_3 = 0$$

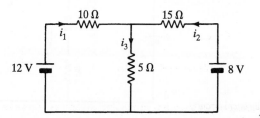

Figure 6–43 Electrical circuit.

Solving for i_1, i_2, and i_3 gives

$$i_1 = \frac{8}{11} \text{A}, \qquad i_2 = \frac{12}{55} \text{A}, \qquad i_3 = \frac{52}{55} \text{A}$$

Since all i values are found to be positive, the currents actually flow in the directions shown in the diagram.

Problem A–6–5

Given the circuit shown in Figure 6–44, obtain a mathematical model. Here, currents i_1 and i_2 are cyclic currents.

Solution   Applying Kirchhoff's voltage law gives

$$R_1 i_1 + \frac{1}{C} \int (i_1 - i_2)\, dt = E$$

$$L\frac{di_2}{dt} + R_2 i_2 + \frac{1}{C} \int (i_2 - i_1)\, dt = 0$$

These two equations constitute a mathematical model for the circuit.

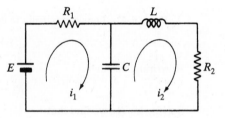

Figure 6–44   Electrical circuit.

Problem A–6–6

In the circuit of Figure 6–45, assume that, for $t < 0$, switch S is connected to voltage source E, and the current in coil L is in a steady state. At $t = 0$, S disconnects the voltage source and simultaneously short-circuits the coil. What is the current $i(t)$ for $t > 0$?

Solution   For $t > 0$, the equation for the circuit is

$$L\frac{di}{dt} + Ri = 0, \qquad i(0) = \frac{E}{R}$$

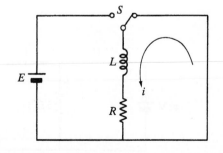

Figure 6–45   Electrical circuit.

[Note that there is a nonzero initial current $i(0-) = E/R$. Since inductance L stores energy, the current in the coil cannot be changed instantaneously. Hence, $i(0+) = i(0-) = i(0) = E/R$.]

Taking the Laplace transform of the system equation, we obtain

$$L[sI(s) - i(0)] + RI(s) = 0$$

or

$$(Ls + R)I(s) = Li(0) = \frac{LE}{R}$$

Thus,

$$I(s) = \frac{E}{R} \frac{L}{Ls + R}$$

The inverse Laplace transform of this last equation gives

$$i(t) = \frac{E}{R} e^{-(R/L)t}$$

Problem A–6–7

Consider the circuit shown in Figure 6–46, and assume that capacitor C is initially charged to q_0. At $t = 0$, switch S is disconnected from the battery and simultaneously connected to inductor L. The capacitance has a value of 50 μF. Calculate the value of the inductance L that will make the oscillation occur at a frequency of 200 Hz.

Solution The equation for the circuit for $t > 0$ is

$$L\frac{di}{dt} + \frac{1}{C}\int i\, dt = 0$$

or, by substituting $i = dq/dt$ into this last equation,

$$L\frac{d^2q}{dt^2} + \frac{1}{C}q = 0$$

where $q(0) = q_0$ and $\dot{q}(0) = 0$. The frequency of oscillation is

$$\omega_n = \sqrt{\frac{1}{LC}}$$

Since

$$200 \text{ Hz} = 200 \text{ cps} = 200 \times 6.28 \text{ rad/s} = 1256 \text{ rad/s}$$

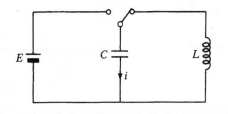

Figure 6–46 Electrical circuit.

we obtain

$$\omega_n = 1256 = \sqrt{\frac{1}{LC}} = \sqrt{\frac{1}{L \times 50 \times 10^{-6}}}$$

Thus,

$$L = \frac{1}{1256^2 \times 50 \times 10^{-6}} = 0.0127 \text{ H}$$

Problem A–6–8

In Figure 6–47(a), suppose that switch S is open for $t < 0$ and that the system is in a steady state. Switch S is closed at $t = 0$. Find the current $i(t)$ for $t \geq 0$.

Solution Notice that, for $t < 0$, the circuit resistance is $R_1 + R_2$. There is a nonzero initial current

$$i(0-) = \frac{E}{R_1 + R_2}$$

For $t \geq 0$, the circuit resistance becomes R_1. Because of the presence of inductance L, there is no instantaneous change in the current in the circuit when switch S is closed. Hence,

$$i(0+) = i(0-) = \frac{E}{R_1 + R_2} = i(0)$$

Therefore, the equation for the circuit for $t \geq 0$ is

$$L\frac{di}{dt} + R_1 i = E \qquad (6\text{–}55)$$

where

$$i(0) = \frac{E}{R_1 + R_2}$$

Taking the Laplace transforms of both sides of Equation (6–55), we obtain

$$L[sI(s) - i(0)] + R_1 I(s) = \frac{E}{s}$$

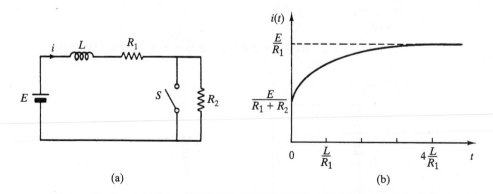

(a)   (b)

Figure 6–47   (a) Electrical circuit; (b) plot of $i(t)$ versus t of the circuit when switch S is closed at $t = 0$.

Substituting the initial condition $i(0)$ into this last equation and simplifying, we get

$$(Ls + R_1)I(s) = \frac{E}{s} + \frac{LE}{R_1 + R_2}$$

Hence,

$$I(s) = \frac{E}{s(Ls + R_1)} + \frac{E}{R_1 + R_2} \frac{L}{Ls + R_1}$$

$$= \frac{E}{R_1}\left(\frac{1}{s} - \frac{L}{Ls + R_1}\right) + \frac{E}{R_1 + R_2} \frac{L}{Ls + R_1}$$

$$= \frac{E}{R_1}\left(\frac{1}{s} - \frac{R_2}{R_1 + R_2} \frac{L}{Ls + R_1}\right)$$

Taking the inverse Laplace transform of this equation, we obtain

$$i(t) = \frac{E}{R_1}\left[1 - \frac{R_2}{R_1 + R_2}e^{-(R_1/L)t}\right]$$

A typical plot of $i(t)$ versus t is shown in Figure 6–47(b).

Problem A–6–9

In the electrical circuit shown in Figure 6–48, there is an initial charge q_0 on the capacitor just before switch S is closed at $t = 0$. Find the current $i(t)$.

Solution   The equation for the circuit when switch S is closed is

$$Ri + \frac{1}{C}\int i\, dt = E$$

Taking the Laplace transform of this last equation yields

$$RI(s) + \frac{1}{C}\frac{I(s) + \left.\int i(t)\, dt\right|_{t=0}}{s} = \frac{E}{s}$$

Since

$$\left.\int i(t)\, dt\right|_{t=0} = q(0) = q_0$$

we obtain

$$RI(s) + \frac{1}{C}\frac{I(s) + q_0}{s} = \frac{E}{s}$$

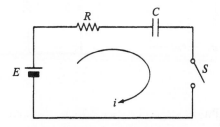

Figure 6–48   Electrical circuit.

or

$$RCsI(s) + I(s) + q_0 = CE$$

Solving for $I(s)$, we have

$$I(s) = \frac{CE - q_0}{RCs + 1} = \left(\frac{E}{R} - \frac{q_0}{RC}\right)\frac{1}{s + \dfrac{1}{RC}}$$

The inverse Laplace transform of this last equation gives

$$i(t) = \left(\frac{E}{R} - \frac{q_0}{RC}\right)e^{-t/RC}$$

Problem A–6–10

Obtain the impedances of the circuits shown in Figures 6–49(a) and (b).

Solution Consider the circuit shown in Figure 6–49(a). From

$$E(s) = E_L(s) + E_R(s) + E_C(s) = \left(Ls + R + \frac{1}{Cs}\right)I(s)$$

where $I(s)$ is the Laplace transform of the current $i(t)$ in the circuit, the complex impedance is

$$Z(s) = \frac{E(s)}{I(s)} = Ls + R + \frac{1}{Cs}$$

For the circuit shown in Figure 6–49(b),

$$I(s) = \frac{E(s)}{Ls} + \frac{E(s)}{R} + \frac{E(s)}{1/(Cs)} = E(s)\left(\frac{1}{Ls} + \frac{1}{R} + Cs\right)$$

Consequently,

$$Z(s) = \frac{E(s)}{I(s)} = \frac{1}{\dfrac{1}{Ls} + \dfrac{1}{R} + Cs}$$

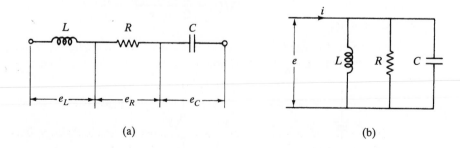

(a)   (b)

Figure 6–49 Electrical circuits.

Problem A–6–11

Find the transfer function $E_o(s)/E_i(s)$ of the electrical circuit shown in Figure 6–50. Obtain the voltage $e_o(t)$ when the input voltage $e_i(t)$ is a step change of voltage E_i occurring at $t = 0$. Assume that $e_i(0-) = 0$. Assume also that the initial charges in the capacitors are zero. [Thus, $e_o(0-) = 0.$]

Solution With the complex-impedance method, the transfer function $E_o(s)/E_i(s)$ can be obtained as

$$\frac{E_o(s)}{E_i(s)} = \frac{\dfrac{1}{(1/R_2) + C_2s}}{\dfrac{1}{C_1s} + \dfrac{1}{(1/R_2) + C_2s}} = \frac{R_2C_1s}{R_2(C_1 + C_2)s + 1}$$

Next, we determine $e_o(t)$. For the input $e_i(t) = E_i \cdot 1(t)$, we have

$$E_o(s) = \frac{R_2C_1s}{R_2(C_1 + C_2)s + 1} \frac{E_i}{s}$$

$$= \frac{R_2C_1E_i}{R_2(C_1 + C_2)s + 1}$$

Inverse Laplace transforming $E_o(s)$, we get

$$e_o(t) = \frac{C_1E_i}{C_1 + C_2}e^{-t/[R_2(C_1+C_2)]}$$

from which it follows that $e_o(0+) = C_1E_i/(C_1 + C_2)$.

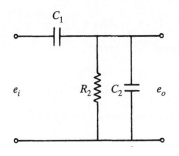

Figure 6–50 Electrical circuit.

Problem A–6–12

Derive the transfer function $E_o(s)/E_i(s)$ of the electrical circuit shown in Figure 6–51. The input voltage is a pulse signal given by

$$e_i(t) = 10 \text{ V} \qquad 0 \le t \le 5$$
$$= 0 \qquad \text{elsewhere}$$

Obtain the output $e_o(t)$. Assume that the initial charges in the capacitors C_1 and C_2 are zero. Assume also that $C_2 = 1.5\,C_1$ and $R_1C_1 = 1$ s.

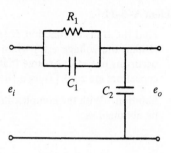

Figure 6–51 Electrical circuit.

Solution   By the use of the complex-impedance method, the transfer function $E_o(s)/E_i(s)$ can be obtained as

$$\frac{E_o(s)}{E_i(s)} = \frac{\dfrac{1}{C_2 s}}{\dfrac{R_1}{R_1 C_1 s + 1} + \dfrac{1}{C_2 s}} = \frac{R_1 C_1 s + 1}{R_1(C_1 + C_2)s + 1} = \frac{s + 1}{2.5\,s + 1}$$

For the given input $e_i(t)$, we have

$$E_i(s) = \frac{10}{s}(1 - e^{-5s})$$

Thus, the response $E_o(s)$ can be given by

$$E_o(s) = \frac{s + 1}{2.5\,s + 1}\frac{10}{s}(1 - e^{-5s})$$

$$= \left(\frac{10}{s} - \frac{15}{2.5\,s + 1}\right)(1 - e^{-5s})$$

The inverse Laplace transform of $E_o(s)$ gives

$$e_o(t) = (10 - 6\,e^{-0.4t})$$
$$- [10 - 6\,e^{-0.4(t-5)}]\,1(t - 5)$$

Figure 6–52 shows a possible response curve $e_o(t)$ versus t.

Problem A–6–13

Obtàin the transfer functions $E_o(s)/E_i(s)$ of the bridged T networks shown in Figures 6–53(a) and (b).

Solution   The bridged T networks shown can both be represented by the network of Figure 6–54(a), which uses complex impedances. This network may be modified to that shown in Figure 6–54(b), in which

$$I_1 = I_2 + I_3, \qquad I_2 Z_1 = (Z_3 + Z_4)I_3$$

Hence,

$$I_2 = \frac{Z_3 + Z_4}{Z_1 + Z_3 + Z_4}I_1, \qquad I_3 = \frac{Z_1}{Z_1 + Z_3 + Z_4}I_1$$

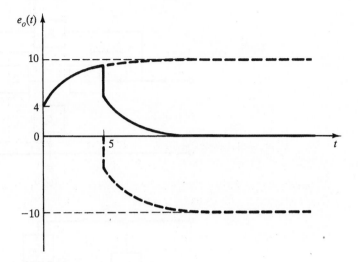

Figure 6–52   Response curve $e_o(t)$ versus t.

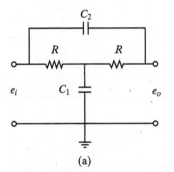

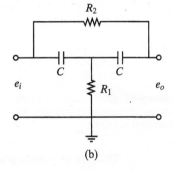

Figure 6–53   Bridged T networks.

Then the voltages $E_i(s)$ and $E_o(s)$ can be obtained as

$$E_i(s) = Z_1 I_2 + Z_2 I_1$$

$$= \left[Z_2 + \frac{Z_1(Z_3 + Z_4)}{Z_1 + Z_3 + Z_4} \right] I_1$$

$$= \frac{Z_2(Z_1 + Z_3 + Z_4) + Z_1(Z_3 + Z_4)}{Z_1 + Z_3 + Z_4} I_1$$

$$E_o(s) = Z_3 I_3 + Z_2 I_1$$

$$= \frac{Z_3 Z_1}{Z_1 + Z_3 + Z_4} I_1 + Z_2 I_1$$

$$= \frac{Z_3 Z_1 + Z_2(Z_1 + Z_3 + Z_4)}{Z_1 + Z_3 + Z_4} I_1$$

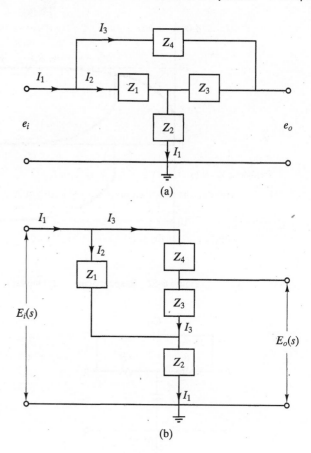

Figure 6–54 (a) Bridged T network in terms of complex impedances; (b) equivalent network.

Thus, the transfer function of the network shown in Figure 6–54(a) is

$$\frac{E_o(s)}{E_i(s)} = \frac{Z_3 Z_1 + Z_2(Z_1 + Z_3 + Z_4)}{Z_2(Z_1 + Z_3 + Z_4) + Z_1 Z_3 + Z_1 Z_4} \tag{6-56}$$

For the bridged T network shown in Figure 6–53(a), we substitute

$$Z_1 = R, \qquad Z_2 = \frac{1}{C_1 s}, \qquad Z_3 = R, \qquad Z_4 = \frac{1}{C_2 s}$$

into Equation (6–56). Then we obtain the transfer function

$$\frac{E_o(s)}{E_i(s)} = \frac{R^2 + \dfrac{1}{C_1 s}\left(R + R + \dfrac{1}{C_2 s}\right)}{\dfrac{1}{C_1 s}\left(R + R + \dfrac{1}{C_2 s}\right) + R^2 + R\dfrac{1}{C_2 s}}$$

$$= \frac{R C_1 R C_2 s^2 + 2 R C_2 s + 1}{R C_1 R C_2 s^2 + (2 R C_2 + R C_1)s + 1}$$

Similarly, for the bridged T network shown in Figure 6–53(b), we substitute

$$Z_1 = \frac{1}{Cs}, \qquad Z_2 = R_1, \qquad Z_3 = \frac{1}{Cs}, \qquad Z_4 = R_2$$

into Equation (6–56). Then we obtain the transfer function

$$\frac{E_o(s)}{E_i(s)} = \frac{\dfrac{1}{Cs}\dfrac{1}{Cs} + R_1\left(\dfrac{1}{Cs} + \dfrac{1}{Cs} + R_2\right)}{R_1\left(\dfrac{1}{Cs} + \dfrac{1}{Cs} + R_2\right) + \dfrac{1}{Cs}\dfrac{1}{Cs} + R_2\dfrac{1}{Cs}}$$

$$= \frac{R_1CR_2Cs^2 + 2R_1Cs + 1}{R_1CR_2Cs^2 + (2R_1C + R_2C)s + 1}$$

Problem A–6–14

Consider the electrical circuit shown in Figure 6–55. Assume that voltage e_i is the input and voltage e_o is the output of the circuit. Derive a state equation and an output equation.

Solution The transfer function for the system is

$$\frac{E_o(s)}{E_i(s)} = \frac{R_2 + \dfrac{1}{C_2s}}{\left(\dfrac{R_1}{R_1C_1s + 1}\right) + \left(R_2 + \dfrac{1}{C_2s}\right)}$$

$$= \frac{(R_1C_1s + 1)(R_2C_2s + 1)}{R_1C_2s + (R_1C_1s + 1)(R_2C_2s + 1)}$$

$$= \frac{R_1C_1R_2C_2s^2 + (R_1C_1 + R_2C_2)s + 1}{R_1C_1R_2C_2s^2 + (R_1C_1 + R_2C_2 + R_1C_2)s + 1} \qquad (6-57)$$

Hence, we have

$$[R_1C_1R_2C_2s^2 + (R_1C_1 + R_2C_2 + R_1C_2)s + 1]E_o(s)$$
$$= [R_1C_1R_2C_2s^2 + (R_1C_1 + R_2C_2)s + 1]E_i(s)$$

The inverse Laplace transform of this last equation gives

$$R_1C_1R_2C_2\ddot{e}_o + (R_1C_1 + R_2C_2 + R_1C_2)\dot{e}_o + e_o$$
$$= R_1C_1R_2C_2\ddot{e}_i + (R_1C_1 + R_2C_2)\dot{e}_i + e_i$$

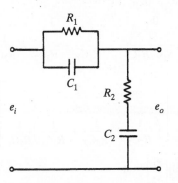

Figure 6–55 Electrical circuit.

By dividing each term of the preceding equation by $R_1 C_1 R_2 C_2$ and defining $e_o = y$ and $e_i = u$, we obtain

$$\ddot{y} + \left(\frac{1}{R_1 C_1} + \frac{1}{R_2 C_2} + \frac{1}{R_2 C_1}\right)\dot{y} + \frac{1}{R_1 C_1 R_2 C_2}y$$
$$= \ddot{u} + \left(\frac{1}{R_1 C_1} + \frac{1}{R_2 C_2}\right)\dot{u} + \frac{1}{R_1 C_1 R_2 C_2}u \tag{6-58}$$

To derive a state equation and an output equation based on Method 1 given in Section 5–4, we first compare Equation (6–58) with the following standard second-order equation:

$$\ddot{y} + a_1 \dot{y} + a_2 y = b_0 \ddot{u} + b_1 \dot{u} + b_2 u$$

We then identify a_1, a_2, b_0, b_1, and b_2 as follows:

$$a_1 = \frac{1}{R_1 C_1} + \frac{1}{R_2 C_2} + \frac{1}{R_2 C_1}$$

$$a_2 = \frac{1}{R_1 C_1 R_2 C_2}$$

$$b_0 = 1$$

$$b_1 = \frac{1}{R_1 C_1} + \frac{1}{R_2 C_2}$$

$$b_2 = \frac{1}{R_1 C_1 R_2 C_2}$$

From Equations (5–23), (5–24), and (5–29), we have

$$\beta_0 = b_0 = 1$$

$$\beta_1 = b_1 - a_1\beta_0 = b_1 - a_1 = -\frac{1}{R_2 C_1}$$

$$\beta_2 = b_2 - a_2\beta_0 - a_1\beta_1 = \left(\frac{1}{R_1 C_1} + \frac{1}{R_2 C_2} + \frac{1}{R_2 C_1}\right)\frac{1}{R_2 C_1}$$

If we define state variables x_1 and x_2 as

$$x_1 = y - \beta_0 u$$
$$x_2 = \dot{x}_1 - \beta_1 u$$

then, from Equations (5–30) and (5–31), the state-space representation for the system can be given by

$$\begin{bmatrix} \dot{x}_1 \\ \dot{x}_2 \end{bmatrix} = \begin{bmatrix} 0 & 1 \\ -\dfrac{1}{R_1 C_1 R_2 C_2} & -\dfrac{1}{R_1 C_1} - \dfrac{1}{R_2 C_2} - \dfrac{1}{R_2 C_1} \end{bmatrix}\begin{bmatrix} x_1 \\ x_2 \end{bmatrix}$$

$$+ \begin{bmatrix} -\dfrac{1}{R_2 C_1} \\ \left(\dfrac{1}{R_1 C_1} + \dfrac{1}{R_2 C_2} + \dfrac{1}{R_2 C_1}\right)\dfrac{1}{R_2 C_1} \end{bmatrix} u$$

$$y = [1 \quad 0]\begin{bmatrix} x_1 \\ x_2 \end{bmatrix} + u$$

Problem A–6–15

Show that the mechanical and electrical systems illustrated in Figure 6–56 are analogous. Assume that the displacement x in the mechanical system is measured from the equilibrium position and that mass m is released from the initial displacement $x(0) = x_0$ with zero initial velocity, or $\dot{x}(0) = 0$. Assume also that in the electrical system the capacitor has the initial charge $q(0) = q_0$ and that the switch is closed at $t = 0$. Note that $\dot{q}(0) = i(0) = 0$. Obtain $x(t)$ and $q(t)$.

Solution   The equation of motion for the mechanical system is

$$m\ddot{x} + kx = 0 \tag{6–59}$$

For the electrical system,

$$L\frac{di}{dt} + \frac{1}{C}\int i\, dt = 0$$

or, by substituting $i = dq/dt = \dot{q}$ into this last equation,

$$L\ddot{q} + \frac{1}{C}q = 0 \tag{6–60}$$

Since Equations (6–59) and (6–60) are of the same form, the two systems are analogous (i.e., they satisfy the force–voltage analogy).

The solution of Equation (6–59) with the initial condition $x(0) = x_0$, $\dot{x}(0) = 0$ is a simple harmonic motion given by

$$x(t) = x_0 \cos \sqrt{\frac{k}{m}}\, t$$

Similarly, the solution of Equation (6–60) with the initial condition $q(0) = q_0$, $\dot{q}(0) = 0$ is

$$q(t) = q_0 \cos \sqrt{\frac{1}{LC}}\, t$$

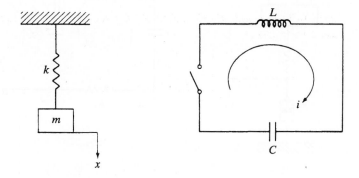

Figure 6–56   Analogous mechanical and electrical systems.

Problem A–6–16

Obtain mathematical models for the systems shown in Figures 6–57(a) and (b), and show that they are analogous systems. In the mechanical system, displacements x_1 and x_2 are measured from their respective equilibrium positions.

Solution   For the mechanical system shown in Figure 6–57(a), the equations of motion are

$$m_1\ddot{x}_1 + b_1\dot{x}_1 + k_1x_1 + k_2(x_1 - x_2) = 0$$
$$b_2\dot{x}_2 + k_2(x_2 - x_1) = 0$$

These two equations constitute a mathematical model for the mechanical system.

For the electrical system shown in Figure 6–57(b), the loop-voltage equations are

$$L_1\frac{di_1}{dt} + \frac{1}{C_2}\int (i_1 - i_2)\, dt + R_1i_1 + \frac{1}{C_1}\int i_1\, dt = 0$$

$$R_2i_2 + \frac{1}{C_2}\int (i_2 - i_1)\, dt = 0$$

Let us write $i_1 = \dot{q}_1$ and $i_2 = \dot{q}_2$. Then, in terms of q_1 and q_2, the preceding two equations can be written

$$L_1\ddot{q}_1 + R_1\dot{q}_1 + \frac{1}{C_1}q_1 + \frac{1}{C_2}(q_1 - q_2) = 0$$

$$R_2\dot{q}_2 + \frac{1}{C_2}(q_2 - q_1) = 0$$

These two equations constitute a mathematical model for the electrical system.

Comparing the two mathematical models, we see that the two systems are analogous. (i.e., they satisfy the force–voltage analogy).

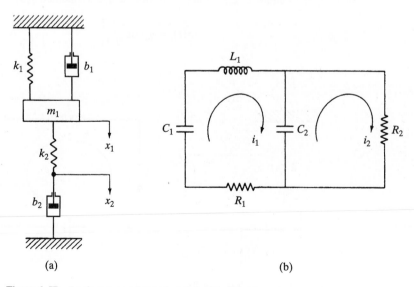

(a)   (b)

Figure 6–57   Analogous mechanical and electrical systems.

Problem A–6–17

Using the force–voltage analogy, obtain an electrical analog of the mechanical system shown in Figure 6–58. Assume that the displacements x_1 and x_2 are measured from their respective equilibrium positions.

Solution   The equations of motion for the mechanical system are

$$m_1\ddot{x}_1 + b_1\dot{x}_1 + k_1x_1 + b_2(\dot{x}_1 - \dot{x}_2) + k_2(x_1 - x_2) = 0$$
$$m_2\ddot{x}_2 + b_2(\dot{x}_2 - \dot{x}_1) + k_2(x_2 - x_1) = 0$$

With the use of the force–voltage analogy, the equations for an analogous electrical system may be written

$$L_1\ddot{q}_1 + R_1\dot{q}_1 + \frac{1}{C_1}q_1 + R_2(\dot{q}_1 - \dot{q}_2) + \frac{1}{C_2}(q_1 - q_2) = 0$$

$$L_2\ddot{q}_2 + R_2(\dot{q}_2 - \dot{q}_1) + \frac{1}{C_2}(q_2 - q_1) = 0$$

Substituting $\dot{q}_1 = i_1$ and $\dot{q}_2 = i_2$ into the last two equations gives

$$L_1\frac{di_1}{dt} + R_1i_1 + \frac{1}{C_1}\int i_1\, dt + R_2(i_1 - i_2) + \frac{1}{C_2}\int (i_1 - i_2)\, dt = 0 \qquad (6\text{–}61)$$

$$L_2\frac{di_2}{dt} + R_2(i_2 - i_1) + \frac{1}{C_2}\int (i_2 - i_1)\, dt = 0 \qquad (6\text{–}62)$$

These two equations are loop-voltage equations. From Equation (6–61), we obtain the diagram shown in Figure 6–59(a). Similarly, from Equation (6–62), we obtain the one given in Figure 6–59(b). Combining these two diagrams produces the desired analogous electrical system (Figure 6–60).

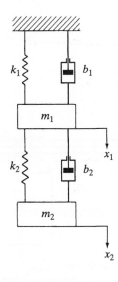

Figure 6–58   Mechanical system.

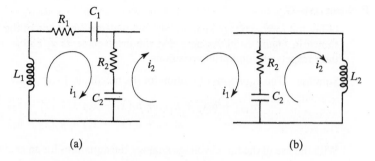

Figure 6–59   (a) Electrical circuit corresponding to Equation (6–61); (b) electrical circuit corresponding to Equation (6–62).

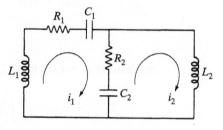

Figure 6–60   Electrical system analogous to the mechanical system shown in Figure 6–58 (force–voltage analogy).

Problem A–6–18

Figure 6–61 shows an inertia load driven by a dc servomotor by means of pulleys and a belt. Obtain the equivalent moment of intertia, J_{eq}, of the system with respect to the motor shaft axis. Assume that there is no slippage between the belt and the pulleys.

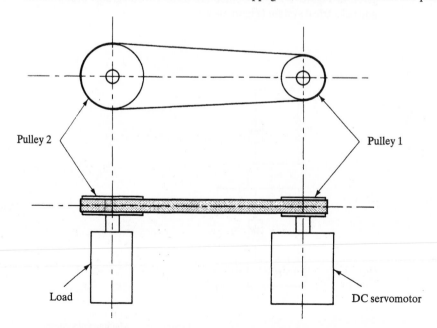

Figure 6–61   Inertia load driven by a dc servomotor by means of pulleys and belt.

Assume also that the diameters of pulleys 1 and 2 are d_1 and d_2, respectively. The moment of inertia of the rotor of the motor is J_r and that of the load element is J_L. The moments of inertia of pulleys 1 and 2 are J_1 and J_2, respectively. Neglect the moment of inertia of the belt.

Solution The given system uses a belt and two pulleys as a drive device. The system acts similarly to a gear train system. Since we assume no slippage between the belt and pulleys, the work done by the belt and pulley 1 $(T_1\theta_1)$ is equal to that done by the belt and pulley 2 $(T_2\theta_2)$, or

$$T_1\theta_1 = T_2\theta_2 \qquad (6\text{--}63)$$

where T_1 is the load torque on the motor shaft, θ_1 is the angular displacement of pulley 1, T_2 is the torque transmitted to the load shaft, and θ_2 is the angular displacement of pulley 2. Note that

$$\frac{\theta_2}{\theta_1} = \frac{d_1}{d_2} \qquad (6\text{--}64)$$

For the servomotor system,

$$(J_1 + J_r)\ddot{\theta}_1 + T_1 = T_m \qquad (6\text{--}65)$$

where T_m is the torque developed by the motor. For the load shaft,

$$(J_L + J_2)\ddot{\theta}_2 = T_2 \qquad (6\text{--}66)$$

From Equations (6–63) and (6–64), we have

$$T_2 = T_1\frac{\theta_1}{\theta_2} = T_1\frac{d_2}{d_1}$$

Then Equation (6–66) becomes

$$(J_L + J_2)\ddot{\theta}_2 = T_1\frac{d_2}{d_1} \qquad (6\text{--}67)$$

From Equations (6–65) and (6–67), we obtain

$$(J_1 + J_r)\ddot{\theta}_1 + \frac{d_1}{d_2}(J_L + J_2)\ddot{\theta}_2 = T_m$$

Since $\theta_2 = (d_1/d_2)\theta_1$, this last equation can be written as

$$(J_1 + J_r)\ddot{\theta}_1 + \left(\frac{d_1}{d_2}\right)^2(J_L + J_2)\ddot{\theta}_1 = T_m$$

or

$$\left[(J_1 + J_r) + (J_L + J_2)\left(\frac{d_1}{d_2}\right)^2\right]\ddot{\theta}_1 = T_m$$

The equivalent moment of inertia of the system with respect to the motor shaft axis is thus given by

$$J_{eq} = J_1 + J_r + (J_L + J_2)\left(\frac{d_1}{d_2}\right)^2$$

Problem A–6–19

Obtain the transfer function $E_o(s)/E_i(s)$ of the operational-amplifier circuit shown in Figure 6–62.

Solution Define the voltage at point A as e_A. Then

$$\frac{E_A(s)}{E_i(s)} = \frac{R_1}{\dfrac{1}{Cs} + R_1} = \frac{R_1 Cs}{R_1 Cs + 1}$$

Define the voltage at point B as e_B. Then

$$E_B(s) = \frac{R_3}{R_2 + R_3} E_o(s)$$

In this operational-amplifier system, negative feedback appears in the operational amplifier. As a result, the differential input voltage becomes zero, and we have $E_A(s) = E_B(s)$. Hence,

$$E_A(s) = \frac{R_1 Cs}{R_1 Cs + 1} E_i(s) = E_B(s) = \frac{R_3}{R_2 + R_3} E_o(s)$$

from which we obtain

$$\frac{E_o(s)}{E_i(s)} = \frac{R_2 + R_3}{R_3}\,\frac{R_1 Cs}{R_1 Cs + 1}$$

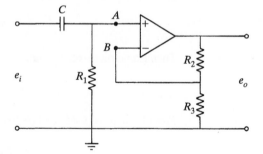

Figure 6–62 Operational-amplifier circuit.

Problem A–6–20

Consider the operational-amplifier circuit shown in Figure 6–63. Obtain the transfer function of this circuit by the complex-impedance approach.

Solution For the circuit shown, we have

$$\frac{E_i(s) - E'(s)}{Z_1} = \frac{E'(s) - E_o(s)}{Z_2}$$

Since the operational amplifier involves negative feedback, the differential input voltage becomes zero. Hence, $E'(s) = 0$. Thus,

$$\frac{E_o(s)}{E_i(s)} = -\frac{Z_2(s)}{Z_1(s)}$$

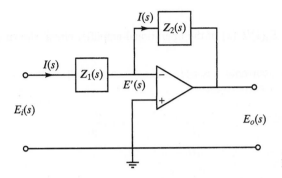

Figure 6–63 Operational-amplifier circuit.

Problem A–6–21

Obtain the transfer function $E_o(s)/E_i(s)$ of the operational-amplifier circuit shown in Figure 6–64 by the complex-impedance approach.

Solution The complex impedances for this circuit are

$$Z_1(s) = R_1 \qquad \text{and} \qquad Z_2(s) = \frac{1}{Cs + \dfrac{1}{R_2}} = \frac{R_2}{R_2 Cs + 1}$$

From **Problem A–6–20**, the transfer function of the system is

$$\frac{E_o(s)}{E_i(s)} = -\frac{Z_2(s)}{Z_1(s)} = -\frac{R_2}{R_1} \frac{1}{R_2 Cs + 1}$$

Notice that the circuit considered here is the same as that discussed in **Example 6–11**. Accordingly, the transfer function $E_o(s)/E_i(s)$ obtained here is, of course, the same as the one obtained in that example.

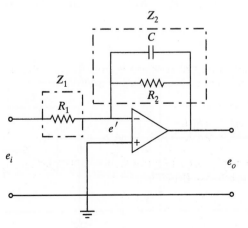

Figure 6–64 Operational-amplifier circuit.

Problem A–6–22

Obtain the transfer function $E_o(s)/E_i(s)$ of the operational-amplifier circuit shown in Figure 6–65.

Solution   We shall first obtain currents i_1, i_2, i_3, i_4, and i_5. Then we shall use the node equation at nodes A and B. The currents are

$$i_1 = \frac{e_i - e_A}{R_1}, \qquad i_2 = \frac{e_A - e_o}{R_3}, \qquad i_3 = C_1 \frac{de_A}{dt}$$

$$i_4 = \frac{e_A - 0}{R_2}, \qquad i_5 = C_2 \frac{d(0 - e_o)}{dt}$$

At node A, we have $i_1 = i_2 + i_3 + i_4$, or

$$\frac{e_i - e_A}{R_1} = \frac{e_A - e_o}{R_3} + C_1 \frac{de_A}{dt} + \frac{e_A}{R_2} \qquad (6\text{–}68)$$

At node B, we have $e_B = 0$, and no current flows into the amplifier. Thus, we get $i_4 = i_5$, or

$$\frac{e_A}{R_2} = C_2 \frac{-de_o}{dt} \qquad (6\text{–}69)$$

Rewriting Equation (6–68), we have

$$C_1 \frac{de_A}{dt} + \left(\frac{1}{R_1} + \frac{1}{R_2} + \frac{1}{R_3} \right) e_A = \frac{e_i}{R_1} + \frac{e_o}{R_3} \qquad (6\text{–}70)$$

From Equation (6–69), we get

$$e_A = -R_2 C_2 \frac{de_o}{dt} \qquad (6\text{–}71)$$

Substituting Equation (6–71) into Equation (6–70), we obtain

$$C_1 \left(-R_2 C_2 \frac{d^2 e_o}{dt^2} \right) + \left(\frac{1}{R_1} + \frac{1}{R_2} + \frac{1}{R_3} \right) (-R_2 C_2) \frac{de_o}{dt} = \frac{e_i}{R_1} + \frac{e_o}{R_3}$$

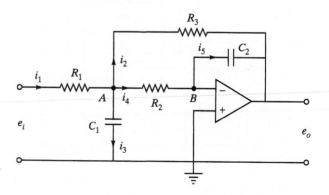

Figure 6–65 Operational-amplifier circuit.

Laplace transforming this last equation, assuming zero initial conditions, yields

$$-C_1C_2R_2s^2E_o(s) + \left(\frac{1}{R_1} + \frac{1}{R_2} + \frac{1}{R_3}\right)(-R_2C_2)sE_o(s) - \frac{1}{R_3}E_o(s) = \frac{E_i(s)}{R_1}$$

from which we get the transfer function $E_o(s)/E_i(s)$:

$$\frac{E_o(s)}{E_i(s)} = -\frac{1}{R_1C_1R_2C_2s^2 + [R_2C_2 + R_1C_2 + (R_1/R_3)R_2C_2]s + (R_1/R_3)}$$

Problem A–6–23

Obtain the transfer function $E_o(s)/E_i(s)$ of the op-amp circuit shown in Figure 6–66 in terms of the complex impedances $Z_1, Z_2, Z_3,$ and Z_4. Using the equation derived, obtain the transfer function $E_o(s)/E_i(s)$ of the op-amp circuit shown in Figure 6–36.

Solution   From Figure 6–66, we find that

$$\frac{E_i(s) - E_A(s)}{Z_3} = \frac{E_A(s) - E_o(s)}{Z_4}$$

or

$$E_i(s) - \left(1 + \frac{Z_3}{Z_4}\right)E_A(s) = -\frac{Z_3}{Z_4}E_o(s) \tag{6–72}$$

Since the system involves negative feedback, we have $E_A(s) = E_B(s)$, or

$$E_A(s) = E_B(s) = \frac{Z_1}{Z_1 + Z_2}E_i(s) \tag{6–73}$$

By substituting Equation (6–73) into Equation (6–72), we obtain

$$\left[\frac{Z_4Z_1 + Z_4Z_2 - Z_4Z_1 - Z_3Z_1}{Z_4(Z_1 + Z_2)}\right]E_i(s) = -\frac{Z_3}{Z_4}E_o(s)$$

from which we get the transfer function

$$\frac{E_o(s)}{E_i(s)} = -\frac{Z_4Z_2 - Z_3Z_1}{Z_3(Z_1 + Z_2)} \tag{6–74}$$

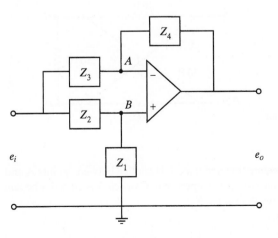

Figure 6–66   Operational-amplifier circuit.

To find the transfer function $E_o(s)/E_i(s)$ of the circuit shown in Figure 6–36, we substitute

$$Z_1 = \frac{1}{Cs}, \qquad Z_2 = R_2, \qquad Z_3 = R_1, \qquad Z_4 = R_1$$

into Equation (6–74). The result is

$$\frac{E_o(s)}{E_i(s)} = -\frac{R_1 R_2 - R_1 \dfrac{1}{Cs}}{R_1 \left(\dfrac{1}{Cs} + R_2 \right)} = -\frac{R_2 Cs - 1}{R_2 Cs + 1}$$

which is, as a matter of course, the same as that obtained in **Example 6–12**.

PROBLEMS

Problem B–6–1

Three resistors R_1, R_2, and R_3 are connected in a triangular shape (Figure 6–67). Obtain the resistance between points A and B.

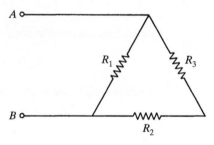

Figure 6–67 Three resistors connected in a triangular shape.

Problem B–6–2

Calculate the resistance between points A and B for the circuit shown in Figure 6–68.

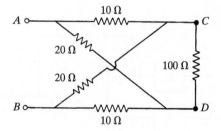

Figure 6–68 Electrical circuit.

Problem B–6–3

In the circuit of Figure 6–69, assume that a voltage E is applied between points A and B and that the current i is i_0 when switch S is open. When switch S is closed, i becomes equal to $2i_0$. Find the value of the resistance R.

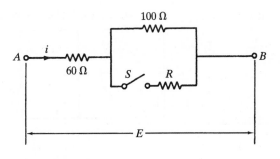

100 Ω

A $\circ$ —ww— 60 Ω   *S*   *R*   — $\circ$ B

— *E* —

Figure 6–69   Electrical circuit.

Problem B–6–4

Obtain a mathematical model of the circuit shown in Figure 6–70.

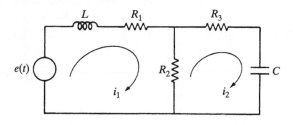

L R_1 R_3

$e(t)$ R_2 i_1 i_2 *C*

Figure 6–70   Electrical circuit.

Problem B–6–5

Consider the circuit shown in Figure 6–71. Assume that switch *S* is open for $t < 0$ and that capacitor *C* is initially charged so that the initial voltage $q(0)/C = e_0$ appears on the capacitor. Calculate cyclic currents i_1 and i_2 when switch *S* is closed at $t = 0$.

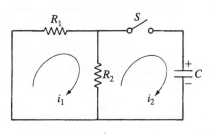

R_1 *S*

i_1 R_2 i_2 *C*

Figure 6–71   Electrical circuit.

Problem B–6–6

The circuit shown in Figure 6–72 is in a steady state with switch *S* closed. Switch *S* is then opened at $t = 0$. Obtain $i(t)$.

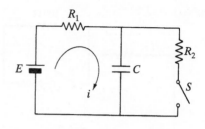

Figure 6–72   Electrical circuit.

Problem B–6–7

Obtain the transfer function $E_o(s)/E_i(s)$ of the circuit shown in Figure 6–73.

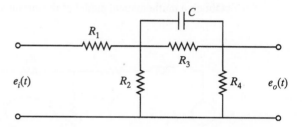

Figure 6–73   Electrical circuit.

Problem B–6–8

Obtain the transfer function $E_o(s)/E_i(s)$ of the system shown in Figure 6–74.

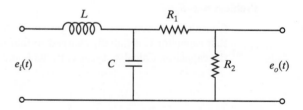

Figure 6–74   Electrical circuit.

Problem B–6–9

Obtain the transfer function $E_o(s)/E_i(s)$ of the circuit shown in Figure 6–75.

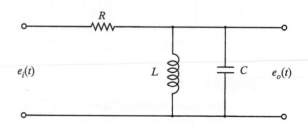

Figure 6–75   Electrical circuit.

Problem B–6–10

Obtain the transfer function $E_o(s)/E_i(s)$ of the electrical circuit shown in Figure 6–76.

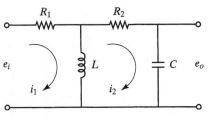

Figure 6–76 Electrical circuit.

Problem B–6–11

Determine the transfer function $E_o(s)/E_i(s)$ of the circuit shown in Figure 6–77. Use the complex-impedance method.

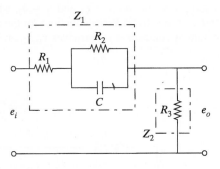

Figure 6–77 Electrical circuit.

Problem B–6–12

Obtain the transfer function $E_o(s)/E_i(s)$ of the circuit shown in Figure 6–78. Use the complex-impedance method.

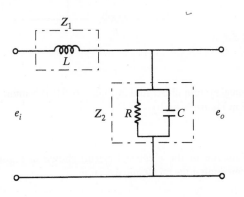

Figure 6–78 Electrical circuit.

Problem B–6–13

Obtain a state-space representation for the electrical circuit shown in Figure 6–79. Assume that voltage e_i is the input and voltage e_o is the output of the system.

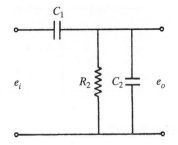

Figure 6–79   Electrical circuit.

Problem B–6–14

In the circuit shown in Figure 6–80, define $i_1 = \dot{q}_1$ and $i_2 = \dot{q}_2$, where q_1 and q_2 are charges in capacitors C_1 and C_2, respectively. Write equations for the circuit. Then obtain a state equation for the system by choosing state variables x_1, x_2, and x_3 as follows:

$$x_1 = q_1$$
$$x_2 = \dot{q}_1$$
$$x_3 = q_2$$

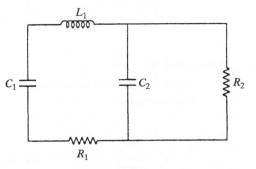

Figure 6–80   Electrical circuit.

Problem B–6–15

Show that the mechanical system illustrated in Figure 6–81(a) is analogous to the electrical system depicted in Figure 6–81(b).

Problem B–6–16

Derive the transfer function of the electrical circuit shown in Figure 6–82. Draw a schematic diagram of an analogous mechanical system.

Problem B–6–17

Obtain a mechanical system analogous to the electrical system shown in Figure 6–83.

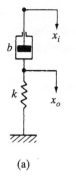

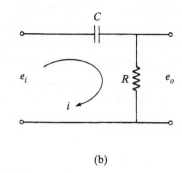

(a)

(b)

Figure 6–81 (a) Mechanical system; (b) analogous electrical system.

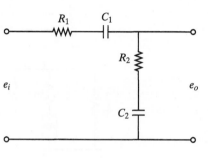

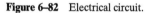

Figure 6–82 Electrical circuit.

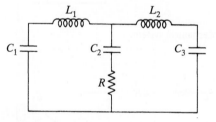

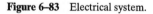

Figure 6–83 Electrical system.

Problem B–6–18

Determine an electrical system analogous to the mechanical system shown in Figure 6–84, where $p(t)$ is the force input to the system. The displacements x_1 and x_2 are measured from their respective equilibrium positions.

Problem B–6–19

Consider the dc servomotor shown in Figure 6–85. Assume that the input of the system is the applied armature voltage e_a and the output is the load shaft position θ_2. Assume also the following numerical values for the constants:

$$R_a = \text{armature winding resistance} = 0.2 \ \Omega$$
$$L_a = \text{armature winding inductance} = \text{negligible}$$
$$K_b = \text{back-emf constant} = 5.5 \times 10^{-2} \ \text{V-s/rad}$$

K = motor-torque constant = 6×10^{-5} lb$_f$-ft/A

J_r = moment of inertia of the rotor of the motor = 1×10^{-5} lb$_f$-ft-s^2

b_r = viscous-friction coefficient of the rotor of the motor = negligible

J_L = moment of inertia of the load = 4.4×10^{-3} lb$_f$-ft-s^2

b_L = viscous-friction coefficient of the load = 4×10^{-2} lb$_f$-ft/rad/s

n = gear ratio = N_1/N_2 = 0.1

Obtain the transfer function $\Theta_2(s)/E_a(s)$.

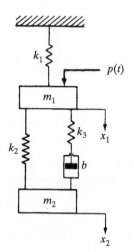

Figure 6–84 Mechanical system.

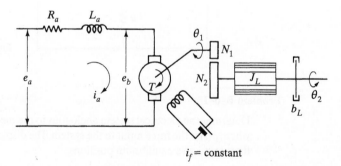

Figure 6–85 DC servomotor.

Problem B–6–20

Obtain the transfer function $E_o(s)/E_i(s)$ of the operational-amplifier circuit shown in Figure 6–86.

Problem B–6–21

Obtain the transfer function $E_o(s)/E_i(s)$ of the operational-amplifier circuit shown in Figure 6–87.

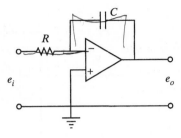

Figure 6–86 Operational-amplifier circuit.

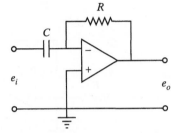

Figure 6–87 Operational-amplifier circuit.

Problem B–6–22

Obtain the transfer function $E_o(s)/E_i(s)$ of the operational-amplifier circuit shown in Figure 6–88.

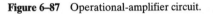
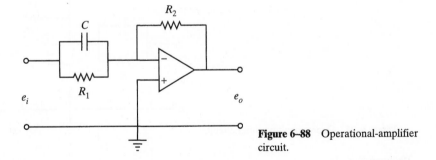

Figure 6–88 Operational-amplifier circuit.

Problem B–6–23

Obtain a state-space representation of the operational-amplifier circuit shown in Figure 6–89.

Problem B–6–24

Obtain the transfer function $E_o(s)/E_i(s)$ of the operational-amplifier circuit shown in Figure 6–90.

Problem B–6–25

Obtain the transfer function $E_o(s)/E_i(s)$ of the operational-amplifier circuit shown in Figure 6–91.

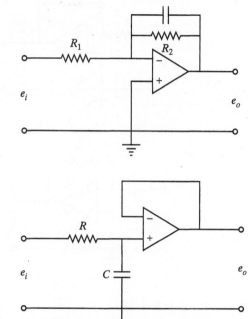

Figure 6–89 Operational-amplifier circuit.

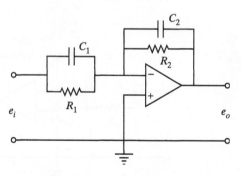

Figure 6–90 Operational-amplifier circuit.

Figure 6–91 Operational-amplifier circuit.

Problem B–6–26

Obtain the transfer function $E_o(s)/E_i(s)$ of the operational-amplifier circuit shown in Figure 6–92.

Problem B–6–27

Obtain the transfer function $E_o(s)/E_i(s)$ of the operational-amplifier circuit shown in Figure 6–93.

Problem B–6–28

Obtain the transfer function $E_o(s)/E_i(s)$ of the operational-amplifier circuit shown in Figure 6–94.

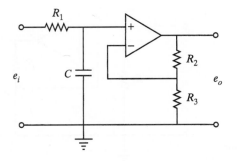

Figure 6–92 Operational-amplifier circuit.

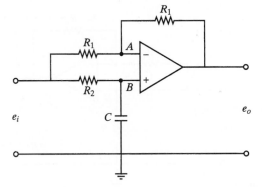

Figure 6–93 Operational-amplifier circuit.

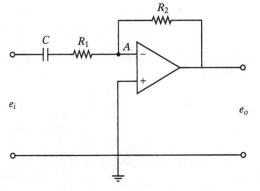

Figure 6–94 Operational-amplifier circuit.

Problem B–6–29

Obtain the transfer function $E_o(s)/E_i(s)$ of the operational-amplifier circuit shown in Figure 6–95.

Problem B–6–30

Using the impedance approach, obtain the transfer function $E_o(s)/E_i(s)$ of the operational-amplifier circuit shown in Figure 6–96.

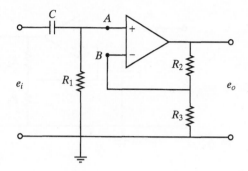

Figure 6–95 Operational-amplifier circuit.

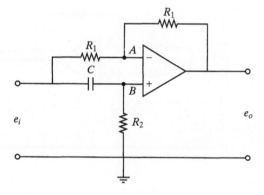

Figure 6–96 Operational-amplifier circuit.

Problem B–6–31

Obtain the output voltage e_o of the operational-amplifier circuit shown in Figure 6–97 in terms of the input voltages e_1 and e_2.

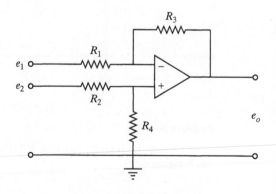

Figure 6–97 Operational-amplifier circuit.

Fluid Systems and Thermal Systems

7–1 INTRODUCTION

As the most versatile medium for transmitting signals and power, fluids—liquids or gases—have wide usage in industry. Liquids and gases can be distinguished from each other by their relative incompressibilities and from the fact that a liquid may have a free surface whereas a gas expands to fill its vessel. In the engineering field, the term *hydraulic* describes fluid systems that use liquids and *pneumatic* applies to those using air or gases.

Mathematical models of fluid systems are generally nonlinear. However, if we assume that the operation of a nonlinear system is near a normal operating point, then the system can be linearized near the operating point, and the mathematical model can be made linear.

Mathematical models of fluid systems obtained in this chapter are linearized models near normal operating points.

Thermal systems generally have distributed parameters. Mathematical models of thermal systems normally involve partial differential equations. In this chapter, however, we assume that thermal systems have lumped parameters, so that approximate mathematical models may be obtained in terms of ordinary differential equations or transfer functions. Such simplified models provide fairly good approximations to actual systems near their normal operating points.

Since fluid systems inevitably involve pressure signals, we shall briefly review units of pressure, gage pressure, and absolute pressure.

Units of pressure. *Pressure* is defined as force per unit area. The units of pressure include N/m^2, kg_f/cm^2, $lb_f/in.^2$, and so on. In the SI system, the unit of pressure is N/m^2. The name *pascal* (abbreviated Pa) has been given to this unit, so

$$1 \text{ Pa} = 1 \text{ N/m}^2$$

Kilopascals ($10^3 \text{ Pa} = \text{kPa}$) and megapascals ($10^6 \text{ Pa} = \text{MPa}$) may be used in expressing hydraulic pressure. Note that

$$1 \text{ lb}_f/\text{in.}^2 = 6895 \text{ Pa}$$

$$1 \text{ kg}_f/\text{cm}^2 = 14.22 \text{ lb}_f/\text{in.}^2 = 0.9807 \times 10^5 \text{ N/m}^2 = 0.09807 \text{ MPa}$$

Gage pressure and absolute pressure. The standard barometer reading at sea level is 760 mm of mercury at 0°C (29.92 in. of mercury at 32°F). *Gage pressure* refers to the pressure that is measured with respect to atmospheric pressure. It is the pressure indicated by a gage above atmospheric pressure. *Absolute pressure* is the sum of the gage and barometer pressures. Note that, in engineering measurement, pressure is expressed in gage pressure. In theoretical calculations, however, absolute pressure must be used. Note also that

$$760 \text{ mm Hg} = 1.0332 \text{ kg}_f/\text{cm}^2 = 1.0133 \times 10^5 \text{ N/m}^2 = 14.7 \text{ lb}_f/\text{in.}^2$$

$$0 \text{ N/m}^2 \text{ gage} = 1.0133 \times 10^5 \text{ N/m}^2 \text{ abs}$$

$$0 \text{ kg}_f/\text{cm}^2 \text{ gage} = 1.0332 \text{ kg}_f/\text{cm}^2 \text{ abs}$$

$$0 \text{ lb}_f/\text{in.}^2 \text{ gage} = 0 \text{ psig} = 14.7 \text{ lb}_f/\text{in.}^2 \text{ abs} = 14.7 \text{ psia}$$

Outline of the chapter. Section 7–1 has presented introductory material for the chapter. Section 7–2 discusses liquid-level systems and obtains their mathematical models. Section 7–3 treats pneumatic systems and derives a mathematical model for a pressure system. Section 7–4 presents a useful linearization method: Linearized models are obtained for nonlinear systems near their respective operating points. Section 7–5 deals with hydraulic systems and derives mathematical models of such systems. Finally, Section 7–6 discusses the mathematical modeling of thermal systems.

7–2 MATHEMATICAL MODELING OF LIQUID-LEVEL SYSTEMS

Industrial processes often involve systems consisting of liquid-filled tanks connected by pipes having orifices, valves, or other flow-restricting devices. Often, it is important to know the dynamic behavior of such systems. The dynamic behavior can be predicted once mathematical models of the systems are known.

In this section, we first review the Reynolds number, laminar flow, and turbulent flow. We then derive mathematical models of liquid-level systems. We shall see that, by introducing the concept of resistance and capacitance, it is possible to describe the dynamic characteristics of such systems in simple forms.

Reynolds number.   The forces that affect fluid flow are due to gravity, bouyancy, fluid inertia, viscosity, surface tension, and similar factors. In many flow situations, the forces resulting from fluid inertia and viscosity are most significant. In fact, fluid flows in many important situations are dominated by either inertia or viscosity of the fluid. The dimensionless ratio of inertia force to viscous force is called the *Reynolds number*. Thus, a large Reynolds number indicates the dominance of inertia force and a small number the dominance of viscosity. The Reynolds number R is given by

$$R = \frac{\rho v D}{\mu}$$

where ρ is the mass density of the fluid, μ is the dynamic viscosity of the fluid, v is the average velocity of flow, and D is a characteristic length. For flow in pipes, the characteristic length is the inside pipe diameter. Since the average velocity v for flow in a pipe is

$$v = \frac{Q}{A} = \frac{4Q}{\pi D^2}$$

where Q is the volumetric flow rate, A is the area of the pipe and D is the inside diameter of the pipe, the Reynolds number for flow in pipes can be given by

$$R = \frac{\rho v D}{\mu} = \frac{4 \rho Q}{\pi \mu D}$$

Laminar flow and turbulent flow.   Flow dominated by viscosity forces is called *laminar flow* and is characterized by a smooth, parallel-line motion of the fluid. When inertia forces dominate, the flow is called *turbulent flow* and is characterized by an irregular and eddylike motion of the fluid. For a Reynolds number below 2000 ($R < 2000$), the flow is always laminar. For a Reynolds number above 4000 ($R > 4000$), the flow is usually turbulent, except in special cases.

In capillary tubes, flow is laminar. If velocities are kept very low or viscosities are very high, flow in pipes of relatively large diameter may also result in laminar flow. In general, flow in a pipe is laminar if the cross section of the passage is comparatively small or the pipe length is relatively long. Otherwise, turbulent flow results. (Note that laminar flow is temperature sensitive, for it depends on viscosity.)

For laminar flow, the velocity profile in a pipe becomes parabolic, as shown in Figure 7–1(a). Figure 7–1(b) shows a velocity profile in a pipe for turbulent flow.

Industrial processes often involve the flow of liquids through connecting pipes and tanks. In hydraulic control systems, there are many cases of flow through small passages, such as flow between spool and bore and between piston and cylinder. The properties of such flow through small passages depend on the Reynolds number of flow involved in each situation.

Resistance and capacitance of liquid-level systems.   Consider the flow through a short pipe with a valve connecting two tanks, as shown in Figure 7–2. The

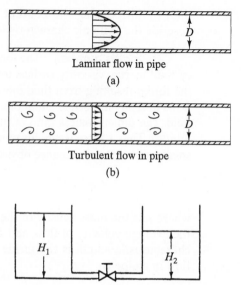

Laminar flow in pipe

(a)

Turbulent flow in pipe

(b)

Figure 7–1 (a) Velocity profile for laminar flow;
(b) velocity profile for turbulent flow.

Figure 7–2 Two tanks connected by a short
pipe with a valve.

resistance R for liquid flow in such a pipe or restriction is defined as the change in
the level difference (the difference of the liquid levels of the two tanks) necessary to
cause a unit change in flow rate; that is,

$$R = \frac{\text{change in level difference}}{\text{change in flow rate}} \quad \frac{\text{m}}{\text{m}^3/\text{s}}$$

Since the relationship between the flow rate and the level difference differs for laminar flow and turbulent flow, we shall consider both cases in what follows.

Consider the liquid-level system shown in Figure 7–3(a). In this system, the liquid spouts through the load valve in the side of the tank. If the flow through the
valve is laminar, the relationship between the steady-state flow rate and the steady-
state head at the level of the restriction is given by

$$Q = K_l H$$

where

Q = steady-state liquid flow rate, m³/s

K_l = constant, m²/s

H = steady-state head, m

For laminar flow, the resistance R_l is

$$R_l = \frac{dH}{dQ} = \frac{1}{K_l} = \frac{H}{Q}$$

The laminar-flow resistance is constant and is analogous to the electrical resistance.
(The laminar-flow resistance of the flow in a capillary tube is given by the
Hagen–Poiseuille formula; see **Problem A–7–1**.)

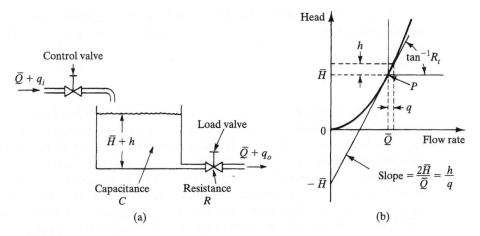

Figure 7–3   (a) Liquid-level system; (b) curve of head versus flow rate.

If the flow through the restriction is turbulent, the steady-state flow rate is given by

$$Q = K_t \sqrt{H} \qquad (7\text{–}1)$$

where

Q = steady-state liquid flow rate, m^3/s

K_t = constant, $m^{2.5}/s$

H = steady-state head, m

The resistance R_t for turbulent flow is obtained from

$$R_t = \frac{dH}{dQ}$$

From Equation (7–1), we obtain

$$dQ = \frac{K_t}{2\sqrt{H}} dH$$

Consequently, we have

$$\frac{dH}{dQ} = \frac{2\sqrt{H}}{K_t} = \frac{2\sqrt{H}\sqrt{H}}{Q} = \frac{2H}{Q}$$

Thus,

$$R_t = \frac{2H}{Q} \qquad (7\text{–}2)$$

The value of the turbulent-flow resistance R_t depends on the flow rate and the head. The value of R_t, however, may be considered constant if the changes in head and flow rate are small.

If the changes in the head and flow rate from their respective steady-state values are small, then, from Equation (7–2), the relationship between Q and H is given by

$$Q = \frac{2H}{R_t}$$

In many practical cases, the value of the constant K_t in Equation (7–1) is not known. Then the resistance may be determined by plotting the curve of head versus flow rate based on experimental data and measuring the slope of the curve at the operating condition. An example of such a plot is shown in Figure 7–3(b). In the figure, point P is the steady-state operating point. The tangent line to the curve at point P intersects the ordinate at the point (Head $= -\overline{H}$). Thus, the slope of this tangent line is $2\overline{H}/\overline{Q}$. Since the resistance R_t at the operating point P is given by $2\overline{H}/\overline{Q}$, the resistance R_t is the slope of the curve at the operating point.

Now define a small deviation of the head from the steady-state value as h and the corresponding small change in the flow rate as q. Then the slope of the curve at point P is given by

$$\text{slope of curve at point } P = \frac{h}{q} = \frac{2\overline{H}}{\overline{Q}} = R_t$$

The capacitance C of a tank is defined to be the change in quantity of stored liquid necessary to cause a unit change in the potential, or head. (The potential is the quantity that indicates the energy level of the system.) Thus,

$$C = \frac{\text{change in liquid stored}}{\text{change in head}} \quad \frac{\text{m}^3}{\text{m}} \text{ or } \text{m}^2$$

Note that the capacity (m^3) and the capacitance (m^2) are different. The capacitance of the tank is equal to its cross-sectional area. If this is constant, the capacitance is constant for any head.

Inertance.   The terms *inertance, inertia,* and *inductance* refer to the change in potential required to make a unit rate of change in flow rate, velocity, or current [change in flow rate per second, change in velocity per second (acceleration), or change in current per second], or

Inertance (inertia or inductance)

$$= \frac{\text{change in potential}}{\text{change in flow rate (velocity or current) per second}}$$

For the inertia effect of liquid flow in pipes, tubes, and similar devices, the potential may be either pressure (N/m^2) or head (m), and the change in flow rate per second may be the volumetric liquid-flow acceleration (m^3/s^2). Applying the preceding general definition of inertance, inertia, or inductance to liquid flow gives

$$\text{Inertance } I = \frac{\text{change in pressure}}{\text{change in flow rate per second}} \quad \frac{\text{N/m}^2}{\text{m}^3/\text{s}^2} \text{ or } \frac{\text{N-s}^2}{\text{m}^5}$$

or

$$\text{Inertance } I = \frac{\text{change in head}}{\text{change in flow rate per second}} \frac{\text{m}}{\text{m}^3/\text{s}^2} \text{ or } \frac{\text{s}^2}{\text{m}^2}$$

(For the computation of inertance, see **Problem A–7–2.**)

Inertia elements in mechanical systems and inductance elements in electrical systems are important in describing system dynamics. However, in deriving mathematical models of liquid-filled tanks connected by pipes with orifices, valves, and so on, only resistance and capacitance are important, and the effects of liquid-flow inertance may be negligible. Such liquid-flow inertance becomes important only in special cases. For instance, it plays a dominant role in vibration transmitted through water, such as water hammer, which results from both the inertia effects and the elastic or capacitance effects of water flow in pipes. Note that this vibration or wave propagation results from inertance–capacitance effects of hydraulic circuits—comparable to free vibration in a mechanical spring–mass system or free oscillation in an electrical LC circuit.

Mathematical modeling of liquid-level systems. In the mathematical modeling of liquid-level systems, we do not take inertance into consideration, because it is negligible. Instead, we characterize liquid-level systems in terms of resistance and capacitance. Let us now obtain a mathematical model of the liquid-level system shown in Figure 7–3(a). If the operating condition as to the head and flow rate varies little for the period considered, a mathematical model can easily be found in terms of resistance and capacitance. In the present analysis, we assume that the liquid outflow from the valve is turbulent.

Let us define

$\overline{H}$ = steady-state head (before any change has occurred), m

h = small deviation of head from its steady-state value, m

$\overline{Q}$ = steady-state flow rate (before any change has occurred), m^3/s

q_i = small deviation of inflow rate from its steady-state value, m^3/s

q_o = small deviation of outflow rate from its steady-state value, m^3/s

The change in the liquid stored in the tank during dt seconds is equal to the net inflow to the tank during the same dt seconds, so

$$C\,dh = (q_i - q_o)\,dt \tag{7–3}$$

where C is the capacitance of the tank.

Note that if the operating condition varies little (i.e., if the changes in head and flow rate are small during the period of operation considered), then the resistance R may be considered constant during the entire period of operation.

In the present system, we defined h and q_o as small deviations from steady-state head and steady-state outflow rate, respectively. Thus,

$$dH = h, \qquad dQ = q_o$$

and the resistance R may be written as

$$R = \frac{dH}{dQ} = \frac{h}{q_o}$$

Substituting $q_o = h/R$ into Equation (7–3), we obtain

$$C\frac{dh}{dt} = q_i - \frac{h}{R}$$

or

$$RC\frac{dh}{dt} + h = Rq_i \tag{7–4}$$

Note that RC has the dimension of time and is the time constant of the system. Equation (7–4) is a linearized mathematical model for the system when h is considered the system output. Such a linearized mathematical model is valid, provided that changes in the head and flow rate from their respective steady-state values are small.

   If q_o (the change in the outflow rate), rather than h (the change in head), is considered the system output, then another mathematical model may be obtained. Substituting $h = Rq_o$ into Equation (7–4) gives

$$RC\frac{dq_o}{dt} + q_o = q_i \tag{7–5}$$

which is also a linearized mathematical model for the system.

   Analogous systems.   The liquid-level system considered here is analogous to the electrical system shown in Figure 7–4(a). It is also analogous to the mechanical system shown in Figure 7–4(b). For the electrical system, a mathematical model is

$$RC\frac{de_o}{dt} + e_o = e_i \tag{7–6}$$

For the mechanical system, a mathematical model is

$$\frac{b}{k}\frac{dx_o}{dt} + x_o = x_i \tag{7–7}$$

Equations (7–5), (7–6), and (7–7) are of the same form; thus, they are analogous. Hence, the liquid-level system shown in Figure 7–3(a), the electrical system shown

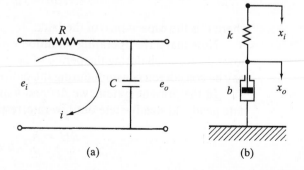

Figure 7–4 Systems analogous to the liquid-level system shown in Figure 7–3(a). (a) Electrical system; (b) mechanical system.

(a)   (b)

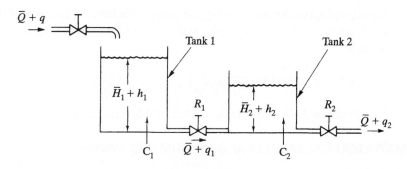

Figure 7–5 Liquid-level system with interaction.

in Figure 7–4(a), and the mechanical system shown in Figure 7–4(b) are analogous systems. [Note that there are many other electrical and mechanical systems that are analogous to the liquid-level system shown in Figure 7–3(a).]

Liquid-level system with interaction. Consider the liquid-level system shown in Figure 7–5. In this system, the two tanks interact. (Note that the transfer function for such a case is not the product of two individual first-order transfer functions.)

If the variations of the variables from their respective steady-state values are small, the resistance R_1 stays constant. Hence, at steady state,

$$\overline{Q} = \frac{\overline{H}_1 - \overline{H}_2}{R_1} \tag{7–8}$$

After small changes have occurred, we have

$$\overline{Q} + q_1 = \frac{\overline{H}_1 + h_1 - (\overline{H}_2 + h_2)}{R_1}$$

$$= \frac{\overline{H}_1 - \overline{H}_2}{R_1} + \frac{h_1 - h_2}{R_1}$$

Substituting Equation (7–8) into this last equation, we obtain

$$q_1 = \frac{h_1 - h_2}{R_1}$$

In the analysis that follows, we assume that variations of the variables from their respective steady-state values are small. Then, using the symbols as defined in Figure 7–5, we can obtain the following four equations for the system:

$$\frac{h_1 - h_2}{R_1} = q_1 \tag{7–9}$$

$$C_1 \frac{dh_1}{dt} = q - q_1 \tag{7–10}$$

$$\frac{h_2}{R_2} = q_2 \tag{7–11}$$

$$C_2 \frac{dh_2}{dt} = q_1 - q_2 \tag{7–12}$$

If q is considered the input and q_2 the output, the transfer function of the system can be obtained by eliminating q_1, h_1, and h_2 from Equations (7–9) through (7–12). The result is

$$\frac{Q_2(s)}{Q(s)} = \frac{1}{R_1C_1R_2C_2s^2 + (R_1C_1 + R_2C_2 + R_2C_1)s + 1} \tag{7–13}$$

(See **Problem A–7–5** for the derivation of this transfer function.)

7–3 MATHEMATICAL MODELING OF PNEUMATIC SYSTEMS

Pneumatic systems are fluid systems that use air as the medium for transmitting signals and power. (Although the most common fluid in these systems is air, other gases can be used as well.)

Pneumatic systems are used extensively in the automation of production machinery and in the field of automatic controllers. For instance, pneumatic circuits that convert the energy of compressed air into mechanical energy enjoy wide usage, and various types of pneumatic controllers are found in industry.

In our discussions of pneumatic systems here, we assume that the flow condition is subsonic. If the speed of air in the pneumatic system is below the velocity of sound, then, like liquid-level systems, such pneumatic systems can be described in terms of resistance and capacitance. (For numerical values of the velocity of sound, see **Problem A–7–13.**)

Before we derive a mathematical model of a pneumatic system, we examine some physical properties of air and other gases. Then we define the resistance and capacitance of pneumatic systems. Finally, we derive a mathematical model of a pneumatic system in terms of resistance and capacitance.

Physical properties of air and other gases.   Some physical properties of air and other gases at standard pressure and temperature are shown in Table 7–1. Standard pressure p and temperature t are defined as

$$p = 1.0133 \times 10^5 \text{ N/m}^2 \text{ abs} = 1.0332 \text{ kg}_f/\text{cm}^2 \text{ abs}$$
$$= 14.7 \text{ lb}_f/\text{in.}^2 \text{ abs} = 14.7 \text{ psia}$$
$$t = 0°C = 273 \text{ K} = 32°F = 492°R$$

TABLE 7–1   Properties of Gases

Gas	Molecular weight	Gas constant R_{gas}		Specific heat, kcal/kg K or Btu/lb °R		Specific heat ratio, c_p/c_v
		N-m/kg K	ft-lb$_f$/lb °R	c_p	c_v	
Air	29.0	287	53.3	0.240	0.171	1.40
Hydrogen (H$_2$)	2.02	4121	766	3.40	2.42	1.41
Nitrogen (N$_2$)	28.0	297	55.2	0.248	0.177	1.40
Oxygen (O$_2$)	32.0	260	48.3	0.218	0.156	1.40
Water vapor(H$_2$O)	18.0	462	85.8	0.444	0.334	1.33

The density ρ, specific volume v, and specific weight γ of air at standard pressure and temperature are

$$\rho = 1.293 \text{ kg/m}^3$$

$$v = 0.7733 \text{ m}^3\text{/kg}$$

$$\gamma = 12.68 \text{ N/m}^3$$

Resistance and capacitance of pneumatic systems. Many industrial processes and pneumatic controllers involve the flow of air (or some other gas) through connected pipelines and pressure vessels.

Consider the pneumatic system shown in Figure 7–6(a). Assume that at steady state the pressure in the system is $\overline{P}$. If the pressure upstream changes to $\overline{P} + p_i$, where p_i is a small quantity compared with $\overline{P}$, then the pressure downstream (the pressure in the vessel) changes to $\overline{P} + p_o$, where p_o is also a small quantity compared with $\overline{P}$. Under the condition that the flow is subsonic, $|\overline{P}| \gg |p_o|$, and $|\overline{P}| \gg |p_i|$, the airflow rate through the restriction becomes proportional to $\sqrt{p_i - p_o}$. Such a pneumatic system may be characterized in terms of a resistance and a capacitance.

Airflow resistance in pipes, orifices, valves, and any other flow-restricting devices can be defined as the change in differential pressure (existing between upstream and downstream of a flow-restricting device) (N/m^2) required to make a unit change in the mass flow rate (kg/s), or

$$\text{resistance } R = \frac{\text{change in differential pressure}}{\text{change in mass flow rate}} \frac{\text{N/m}^2}{\text{kg/s}} \text{ or } \frac{\text{N-s}}{\text{kg-m}^2}$$

Therefore, resistance R can be expressed as

$$R = \frac{d(\Delta p)}{dq}$$

where $d(\Delta p)$ is a change in the differential pressure and dq is a change in the mass flow rate. A theoretical determination of the value of the airflow resistance R is very time consuming. Experimentally, however, it can be easily determined from a plot of

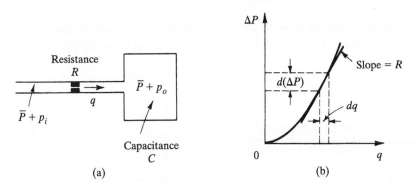

Figure 7–6 (a) Pneumatic system; (b) curve of pressure difference versus flow rate.

the pressure difference Δp versus flow rate q by calculating the slope of the curve at a given operating condition, as shown in Figure 7–6(b). Notice that the airflow resistance R is not constant, but varies with the change in the operating condition.

For a pneumatic pressure vessel, *capacitance* can be defined as the change in the mass of air (kg) [or other gas (kg)] in the vessel required to make a unit change in pressure (N/m²), or

$$\text{capacitance } C = \frac{\text{change in mass of air (or gas)}}{\text{change in pressure}} \quad \frac{\text{kg}}{\text{N/m}^2} \text{ or } \frac{\text{kg-m}^2}{\text{N}}$$

which may be expressed as

$$C = \frac{dm}{dp} = V \frac{d\rho}{dp} \frac{\text{kg}}{\text{N/m}^2} \tag{7–14}$$

where

m = mass of air (or other gas) in vessel, kg

p = absolute pressure of air (or other gas), N/m²

V = volume of vessel, m³

ρ = mass density of air (or other gas), kg/m³

Such a capacitance C may be calculated with the use of the perfect-gas law. For air, we have

$$pv = \frac{p}{\rho} = \frac{\overline{R}}{M}T = R_{\text{air}}T \tag{7–15}$$

where

p = absolute pressure of air, N/m²

v = specific volume of air, m³/kg

M = molecular weight of air per mole, kg/kg-mole

$\overline{R}$ = universal gas constant, N-m/kg-mole K

R_{air} = gas constant of air, N-m/kg K

T = absolute temperature of air, K

If the change of state of air is between isothermal and adiabatic, then the expansion process can be expressed as polytropic and can be given by

$$\frac{p}{\rho^n} = \text{constant} \tag{7–16}$$

where

n = polytropic exponent

Since $d\rho/dp$ can be obtained from Equation (7–16) as

$$\frac{d\rho}{dp} = \frac{\rho}{np}$$

by substituting Equation (7–15) into this last equation, we have

$$\frac{d\rho}{dp} = \frac{1}{nR_{air}T} \tag{7-17}$$

Then, from Equations (7–14) and (7–17), the capacitance C of a vessel is

$$C = \frac{V}{nR_{air}T} \quad \frac{\text{kg}}{\text{N/m}^2} \tag{7-18}$$

Note that if a gas other than air is used in a pressure vessel, the capacitance C is given by

$$C = \frac{V}{nR_{gas}T} \quad \frac{\text{kg}}{\text{N/m}^2} \tag{7-19}$$

where R_{gas} is the gas constant for the particular gas involved.

From the preceding analysis, it is clear that the capacitance of a pressure vessel is not constant, but depends on the expansion process involved, the nature of the gas (air, N_2, H_2, and so on) and the temperature of the gas in the vessel. The value of the polytropic exponent n is approximately constant ($n = 1.0$ to 1.2) for gases in uninsulated metal vessels.

Example 7–1

Find the capacitance C of a 2-m³ pressure vessel that contains air at 50°C. Assume that the expansion and compression of air occur slowly and that there is sufficient time for heat to transfer to and from the vessel so that the expansion process may be considered isothermal, or $n = 1$.

The capacitance C is found by substituting $V = 2\ \text{m}^3$, $R_{air} = 287\ \text{N-m/kg K}$, $T = 273 + 50 = 323\ \text{K}$, and $n = 1$ into Equation (7–18) as follows:

$$C = \frac{V}{nR_{air}T} = \frac{2}{1 \times 287 \times 323} = 2.16 \times 10^{-5}\ \text{kg-m}^2/\text{N}$$

Example 7–2

In **Example 7–1**, if hydrogen (H_2), rather than air, is used to fill the same pressure vessel, what is the capacitance? Assume that the temperature of the gas is 50°C and that the expansion process is isothermal, or $n = 1$.

The gas constant for hydrogen is

$$R_{H_2} = 4121\ \text{N-m/kg K}$$

Substituting $V = 2\ \text{m}^3$, $R_{H_2} = 4121\ \text{N-m/kg K}$, $T = 273 + 50 = 323\ \text{K}$, and $n = 1$ into Equation (7–19), we have

$$C = \frac{V}{nR_{H_2}T} = \frac{2}{1 \times 4121 \times 323} = 1.50 \times 10^{-6}\ \text{kg-m}^2/\text{N}$$

Mathematical modeling of a pneumatic system.   The pneumatic pressure system shown in Figure 7–7(a) consists of a pressure vessel and connecting pipe with

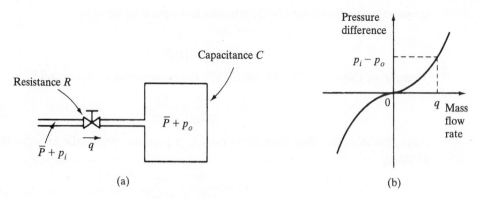

Figure 7-7   (a) Pneumatic pressure system; (b) curve of pressure difference versus mass flow rate.

a valve. If we assume only small deviations in the variables from their respective steady-state values, then this system may be considered linear. We define

$$\overline{P} = \text{steady-state pressure of the system, N/m}^2$$
$$p_i = \text{small change in inflow pressure, N/m}^2$$
$$p_o = \text{small change in air pressure in vessel, N/m}^2$$
$$V = \text{volume of vessel, m}^3$$
$$m = \text{mass of air in vessel, kg}$$
$$q = \text{mass flow rate, kg/s}$$

Let us obtain a mathematical model of this pneumatic pressure system. Assume that the system operates in such a way that the average flow through the valve is zero (i.e., the normal operating condition corresponds to $p_i - p_o = 0, q = 0$). Assume also that the flow is subsonic for the entire range of operation of the system.

As noted earlier, the resistance R is not constant. Hence, for the present system, we shall use an average resistance in the region of its operation. From Figure 7-7(b), the average resistance of the valve may be written as

$$R = \frac{p_i - p_o}{q}$$

From Equation (7-14), the capacitance of the pressure vessel can be written

$$C = \frac{dm}{dp_o}$$

or

$$C\,dp_o = dm$$

This last equation states that the capacitance C times the pressure change dp_o (during dt seconds) is equal to dm, the change in the mass of air in the vessel (during dt seconds). Now, the change in mass, dm, is equal to the mass flow during dt seconds, or $q\,dt$; hence,

$$C\,dp_o = q\,dt$$

Substituting $q = (p_i - p_o)/R$ into this last equation, we have

$$C \, dp_o = \frac{p_i - p_o}{R} dt$$

Rewriting yields

$$RC \frac{dp_o}{dt} + p_o = p_i \tag{7-20}$$

where RC has the dimension of time and is the time constant of the system. Equation (7-20) is a mathematical model for the system shown in Figure 7-7(a).

Note that the pneumatic pressure system considered here is analogous to the electrical system shown in Figure 7-4(a) and the mechanical system shown in Figure 7-4(b). It is also analogous to the liquid-level system shown in Figure 7-3(a).

7-4 LINEARIZATION OF NONLINEAR SYSTEMS

In this section, we present a linearization technique that is applicable to many nonlinear systems. The process of linearizing nonlinear systems is important, for by linearizing nonlinear equations, it is possible to apply numerous linear analysis methods that will produce information on the behavior of those systems. The linearization procedure presented here is based on the expansion of the nonlinear function into a Taylor series about the operating point and the retention of only the linear term. Because we neglect higher order terms of the Taylor series expansion, these neglected terms must be small enough; that is, the variables must deviate only slightly from the operating condition.

Linearization of $z = f(x)$ about a point $(\bar{x}, \bar{z})$.   Consider a nonlinear system whose input is x and output is z. The relationship between z and x may be written

$$z = f(x) \tag{7-21}$$

If the normal operating condition corresponds to a point $(\bar{x}, \bar{z})$, then Equation (7-21) can be expanded into a Taylor series about this point as follows:

$$z = f(x) = f(\bar{x}) + \frac{df}{dx}(x - \bar{x}) + \frac{1}{2!}\frac{d^2 f}{dx^2}(x - \bar{x})^2 + \cdots \tag{7-22}$$

Here, the derivatives df/dx, $d^2 f/dx^2$, ... are evaluated at the operating point, $x = \bar{x}, z = \bar{z}$. If the variation $x - \bar{x}$ is small, we can neglect the higher order terms in $x - \bar{x}$. Noting that $\bar{z} = f(\bar{x})$, we can write Equation (7-22) as

$$z - \bar{z} = a(x - \bar{x}) \tag{7-23}$$

where

$$a = \frac{df}{dx}\bigg|_{x=\bar{x}}$$

Equation (7-23) indicates that $z - \bar{z}$ is proportional to $x - \bar{x}$. The equation is a linear mathematical model for the nonlinear system given by Equation (7-21) near the operating point $x = \bar{x}, z = \bar{z}$.

Linearization of $z = f(x, y)$ about a point $(\bar{x}, \bar{y}, \bar{z})$.   Next, consider a non-linear system whose output z is a function of two inputs x and y, or

$$z = f(x, y) \tag{7-24}$$

To obtain a linear mathematical model for this nonlinear system about an operating point $(\bar{x}, \bar{y}, \bar{z})$, we expand Equation (7-24) into a Taylor series about that point. Then Equation (7-24) becomes

$$z = f(\bar{x}, \bar{y}) + \left[\frac{\partial f}{\partial x}(x - \bar{x}) + \frac{\partial f}{\partial y}(y - \bar{y}) \right]$$

$$+ \frac{1}{2!}\left[\frac{\partial^2 f}{\partial x^2}(x - \bar{x})^2 + 2\frac{\partial^2 f}{\partial x\,\partial y}(x - \bar{x})(y - \bar{y}) + \frac{\partial^2 f}{\partial y^2}(y - \bar{y})^2 \right] + \cdots$$

where the partial derivatives are evaluated at the operating point, $x = \bar{x}$, $y = \bar{y}$, $z = \bar{z}$. Near this point, the higher order terms may be neglected. Noting that $\bar{z} = f(\bar{x}, \bar{y})$, we find that a linear mathematical model of this nonlinear system near the operating point $x = \bar{x}$, $y = \bar{y}$, $z = \bar{z}$ is

$$z - \bar{z} = a(x - \bar{x}) + b(y - \bar{y})$$

where

$$a = \left. \frac{\partial f}{\partial x} \right|_{x=\bar{x},\, y=\bar{y}}$$

$$b = \left. \frac{\partial f}{\partial y} \right|_{x=\bar{x},\, y=\bar{y}}$$

It is important to remember that in the present linearization procedure, the deviations of the variables from the operating condition must be sufficiently small. Otherwise, the procedure does not apply.

Example 7-3

Linearize the nonlinear equation

$$z = xy$$

in the region $5 \le x \le 7, 10 \le y \le 12$. Find the error if the linearized equation is used to calculate the value of z when $x = 5$ and $y = 10$.

Since the region considered is given by $5 \le x \le 7, 10 \le y \le 12$, choose $\bar{x} = 6, \bar{y} = 11$. Then $\bar{z} = \bar{x}\bar{y} = 66$. Let us obtain a linearized equation for the nonlinear equation near a point $\bar{x} = 6, \bar{y} = 11, \bar{z} = 66$.

Expanding the nonlinear equation into a Taylor series about the point $x = \bar{x}, y = \bar{y}, z = \bar{z}$ and neglecting the higher order terms, we have

$$z - \bar{z} = a(x - \bar{x}) + b(y - \bar{y})$$

where

$$a = \frac{\partial(xy)}{\partial x}\Bigg|_{x=\bar{x},\, y=\bar{y}} = \bar{y} = 11$$

$$b = \frac{\partial(xy)}{\partial y}\Bigg|_{x=\bar{x},\, y=\bar{y}} = \bar{x} = 6$$

Hence, the linearized equation is

$$z - 66 = 11(x - 6) + 6(y - 11)$$

or

$$z = 11x + 6y - 66$$

When $x = 5$ and $y = 10$, the value of z given by the linearized equation is

$$z = 11x + 6y - 66 = 55 + 60 - 66 = 49$$

The exact value of z is $z = xy = 50$. The error is thus $50 - 49 = 1$. In terms of percentage, the error is 2%.

Example 7–4

Consider the liquid-level system shown in Figure 7–8. At steady state, the inflow rate is $Q_i = \bar{Q}$, the outflow rate is $Q_o = \bar{Q}$, and the head is $H = \bar{H}$. Assume that the flow is turbulent. Then

$$Q_o = K\sqrt{H}$$

For this system, we have

$$C\frac{dH}{dt} = Q_i - Q_o = Q_i - K\sqrt{H}$$

where C is the capacitance of the tank. Let us define

$$\frac{dH}{dt} = \frac{1}{C}Q_i - \frac{K\sqrt{H}}{C} = f(H, Q_i) \qquad (7\text{--}25)$$

Assume that the system operates near the steady-state condition $(\bar{H}, \bar{Q})$. That is, $H = \bar{H} + h$ and $Q_i = \bar{Q} + q_i$, where h and q_i are small quantities (either positive or negative). At steady-state operation, $dH/dt = 0$. Hence, $f(\bar{H}, \bar{Q}) = 0$.

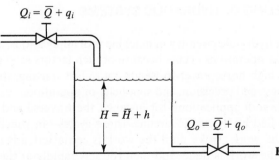

Figure 7–8 Liquid-level system.

Let us linearize Equation (7–25) near the operating point $(\overline{H}, \overline{Q})$. Using the linearization technique just presented, we obtain the linearized equation

$$\frac{dH}{dt} - f(\overline{H}, \overline{Q}) = \frac{\partial f}{\partial H}(H - \overline{H}) + \frac{\partial f}{\partial Q_i}(Q_i - \overline{Q}) \tag{7–26}$$

where

$$f(\overline{H}, \overline{Q}) = 0$$

$$\left.\frac{\partial f}{\partial H}\right|_{H=\overline{H}, Q_i=\overline{Q}} = -\frac{K}{2C\sqrt{\overline{H}}} = -\frac{\overline{Q}}{\sqrt{\overline{H}}}\frac{1}{2C\sqrt{\overline{H}}} = -\frac{\overline{Q}}{2C\overline{H}} = -\frac{1}{RC}$$

in which we used the resistance R defined by

$$R = \frac{2\overline{H}}{\overline{Q}}$$

Also,

$$\left.\frac{\partial f}{\partial Q_i}\right|_{H=\overline{H}, Q_i=\overline{Q}} = \frac{1}{C}$$

Then Equation (7–26) can be written as

$$\frac{dH}{dt} = -\frac{1}{RC}(H - \overline{H}) + \frac{1}{C}(Q_i - \overline{Q}) \tag{7–27}$$

Since $H - \overline{H} = h$ and $Q_i - \overline{Q} = q_i$, Equation (7–27) can be written as

$$\frac{dh}{dt} = -\frac{1}{RC}h + \frac{1}{C}q_i$$

or

$$RC\frac{dh}{dt} + h = Rq_i$$

which is the linearized equation for the liquid-level system and is the same as Equation (7–4). (See Section 7–2.)

7–5 MATHEMATICAL MODELING OF HYDRAULIC SYSTEMS

The widespread use of hydraulic circuitry in machine tool applications, aircraft control systems, and similar operations occurs because of such factors as dependability; accuracy; flexibility; a high horsepower-to-weight ratio; fast starting, stopping, and reversal with smoothness and precision; and simplicity of operation.

In many machine tool applications, for instance, the traverse and feed cycles required are best handled by hydraulic circuits. These cycles—in which the piston advances rapidly on the work stroke until the work is contacted, advances slowly under pressure while the work is done, and then retracts rapidly at the end of the slow tool feed stroke—are easily handled by the use of two pumps (one large-capacity, low-pressure pump and one small-capacity, high-pressure pump) and flow control devices. The large-capacity, low-pressure pump is used only during the rapid advance and return of the cylinder. The small-capacity, high-pressure pump supplies hydraulic fluid for the compression stroke. An unloading valve maintains high pressure while the low-pressure pump is unloaded to the reservoir. (The unloading valve unloads whatever is delivered by the large-capacity, low-pressure pump during the small-capacity, high-pressure phase of a cycle.) Such an unloading valve is designed for the rapid discharge of hydraulic fluid at near atmospheric pressure after permitting the buildup of pressure to a preset value.

Generally, the operating pressure in hydraulic systems is somewhere between 10^6 N/m^2 (1 MPa) and 35×10^6 N/m^2 (35 MPa) (approximately between 10 kg$_f$/cm^2 and 350 kg$_f$/cm^2, or approximately between 145 lb$_f$/in.^{2} and 5000 lb$_f$/in.^{2}). In some special applications, the operating pressure may go up to 70×10^6 N/m^2 (70 MPa, which is approximately 700 kg$_f$/cm^2 or 10,000 lb$_f$/in.^{2}). For the same power requirement, the weight and size of the hydraulic unit can be made smaller by increasing the supply pressure.

In this section, we first present some properties of hydraulic fluids and then introduce general concepts of hydraulic systems. We then model a hydraulic servo. Since this is a nonlinear device, we linearize the nonlinear equation describing the dynamics of the hydraulic servo by using the linearization technique presented in Section 7–4. Afterward, we obtain the transfer function of the hydraulic servo. Finally, we derive a mathematical model of a hydraulic damper.

Properties of hydraulic fluids.   The properties of hydraulic fluids have an important effect on the performance of hydraulic systems. Besides serving as a power-transmitting medium, a hydraulic fluid must minimize the wear of moving parts by providing satisfactory lubrication. In practice, petroleum-based oils with proper additives are the most commonly used hydraulic fluids, because they give good lubrication for the moving parts of a system and are almost imcompressible. The use of a clean, high-quality oil is required for satisfactory operation of the hydraulic system.

Viscosity, the most important property of a hydraulic fluid, is a measure of the internal friction or the resistance of the fluid to flow. Low viscosity means an increase in leakage losses, and high viscosity implies sluggish operation. In hydraulic systems, allowable viscosities are limited by the operating characteristics of the pump, motor, and valves, as well as by ambient and operating temperatures. The viscosity of a liquid decreases with temperature.

The resistance of a fluid to the relative motion of its parts is called *dynamic*, or *absolute, viscosity*. It is the ratio of the shearing stress to the rate of shear deformation of the fluid. The SI units of dynamic viscosity are N-s/m^2 and kg/m-s. The cgs unit of dynamic viscosity is the poise (P) (dyn-s/cm^2 or g/cm-s). The SI unit is 10 times larger than the poise. The centipoise (cP) is one-hundredth of a poise. The BES units of dynamic viscosity are lb$_f$-s/ft^2 and slug/ft-s. Note that

$$1 \text{ slug/ft-s} = 1 \text{ lb}_f\text{-s/ft}^2 = 47.9 \text{ kg/m-s} = 47.9 \text{ N-s/m}^2$$

$$1 \text{ P} = 100 \text{ cP} = 0.1 \text{ N-s/m}^2$$

The *kinematic viscosity* v is the dynamic viscosity μ divided by the mass density ρ, or

$$v = \frac{\mu}{\rho}$$

For petroleum-based oils, the mass density is approximately

$$\rho = 820 \text{ kg/m}^3 = 51.2 \text{ lb/ft}^3 = 1.59 \text{ slug/ft}^3$$

The SI unit of kinematic viscosity is m^2/s; the cgs unit of kinematic viscosity is the stoke(St) (cm^2/s), and one-hundredth of a stoke is called a centistoke (cSt). The BES unit of kinematic viscosity is ft^2/s. In changing from the stoke to the poise, multiply by the mass density in g/cm^3. Note that

$1 \text{ m}^2/\text{s}$ (SI unit of kinematic viscosity)

$\qquad = 10.764 \text{ ft}^2/\text{s}$ (BES unit of kinematic viscosity)

$\qquad 1 \text{ St} = 100 \text{ cSt} = 0.0001 \text{ m}^2/\text{s}$

For hydraulic oils at normal operating conditions, the kinematic viscosity is about 5 to 100 centistokes (5×10^{-6} to $100 \times 10^{-6} \text{ m}^2/\text{s}$).

Petroleum oils tend to become thin as the temperature increases and thick as the temperature decreases. If the system operates over a wide temperature range, fluid having a viscosity that is relatively less sensitive to temperature changes must be used.

Some additional remarks on hydraulic fluids are as follows:

1. The operating life of a hydraulic fluid depends on its oxidation resistance. Oxidation of hydraulic fluid is caused by air, heat, and contamination. Note that any hydraulic fluid combines with air to a certain extent, especially at high operating temperatures. Note also that the operating temperature of the hydraulic system should be kept between 30 and 60°C. For operating temperatures above 70°C, oxidation is accelerated. Premium-grade fluids usually contain inhibitors to slow down oxidation.

2. For hydraulic systems located near high-temperature sources, fire-resistant fluids should be used. These fluids are available in several general types, such as water–glycol, synthetic oil, and water–oil emulsions.

Hydraulic circuits.   Hydraulic circuits are capable of producing many different combinations of motion and force. All, however, are fundamentally the same, regardless of the application. Such circuits involve four basic components: a reservoir to hold the hydraulic fluid, a pump or pumps to force the fluid through the circuit, valves to control fluid pressure and flow, and an actuator or actuators to convert hydraulic energy into mechanical energy to do the work. Figure 7–9 shows a simple circuit that involves a reservoir, a pump, valves, a hydraulic cylinder, and so on.

High-pressure hydraulic systems enable very large forces to be derived. Moreover, these systems permit a rapid and accurate positioning of loads.

Hydraulic servomotor.   Figure 7–10 shows a hydraulic servomotor consisting of a spool valve and a power cylinder and piston. The valve admits hydraulic fluid under high pressure into a power cylinder that contains a large piston, so a large hydraulic force is established to move a load. Assume that the spool valve is

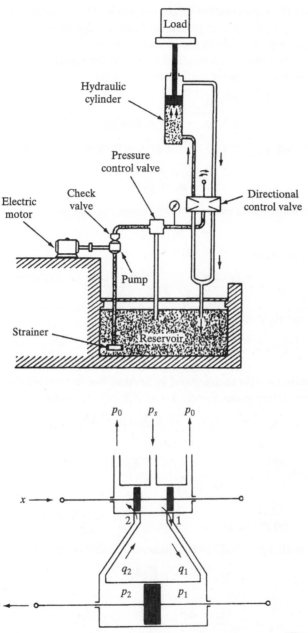

Figure 7–9   Hydraulic circuit.

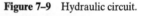

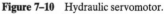

Figure 7–10   Hydraulic servomotor.

symmetrical and has zero overlapping, that the valve orifice areas are proportional to the valve displacement x, and that the orifice coefficient and the pressure drop across the orifice are constant and independent of the valve position. Assume also the following: The supply pressure is p_s, the return pressure p_0 in the return line is small and can be neglected, the hydraulic fluid is incompressible, the inertia force of the power piston and the load reactive forces are negligible compared with the hydraulic force developed by the power piston, and the leakage flow around the spool valve from the supply pressure side to the return pressure side is negligible.

Let us derive a linearized mathematical model of the spool valve near the origin. The flow rates through the valve orifices are given by

$$q_1 = C\sqrt{p_s - p_1}\, x$$
$$q_2 = C\sqrt{p_2 - p_0}\, x = C\sqrt{p_2}\, x$$

where we assumed that $p_0 = 0$ and C is a proportionality constant. Noting that $q_1 = q_2$, we have

$$p_s - p_1 = p_2$$

Let us define the pressure difference across the power piston as

$$\Delta p = p_1 - p_2$$

Then p_1 and p_2 can be written

$$p_1 = \frac{p_s + \Delta p}{2}, \qquad p_2 = \frac{p_s - \Delta p}{2}$$

The flow rate q_1 to the right side of the power piston is

$$q_1 = C\sqrt{p_s - p_1}\, x = C\sqrt{\frac{p_s - \Delta p}{2}}\, x = f(x, \Delta p)$$

Using the linearization technique discussed in Section 7–4, we obtain the linearized equation near the operating point $x = \bar{x}$, $\Delta p = \Delta \bar{p}$, $q_1 = \bar{q}_1$ to be

$$q_1 - \bar{q}_1 = a(x - \bar{x}) + b(\Delta p - \Delta \bar{p}) \tag{7–28}$$

where

$$a = \left. \frac{\partial f}{\partial x} \right|_{x=\bar{x},\,\Delta p=\Delta\bar{p}} = C\sqrt{\frac{p_s - \Delta \bar{p}}{2}}$$

$$b = \left. \frac{\partial f}{\partial \Delta p} \right|_{x=\bar{x},\,\Delta p=\Delta\bar{p}} = -\frac{C}{2\sqrt{2}\sqrt{p_s - \Delta \bar{p}}} \bar{x} \le 0$$

Near the origin ($\bar{x} = 0$, $\Delta \bar{p} = 0$, $\bar{q}_1 = 0$), Equation (7–28) becomes

$$q_1 = K_1 x - K_2 \Delta p$$

where

$$K_1 = \left. C\sqrt{\frac{p_s - \Delta \bar{p}}{2}} \right|_{\bar{x}=0,\,\Delta\bar{p}=0} = C\sqrt{\frac{p_s}{2}}$$

$$K_2 = \left. \frac{C}{2\sqrt{2}\sqrt{p_s - \Delta \bar{p}}} \bar{x} \right|_{\bar{x}=0,\,\Delta\bar{p}=0} = 0$$

Hence,

$$q_1 = K_1 x \tag{7–29}$$

This is a linearized model of the spool valve near the origin.

Mathematical model of hydraulic servomotor.   In obtaining a mathematical model of the hydraulic servomotor shown in Figure 7–10, we assume that the hydraulic fluid is incompressible and that the inertia force of the power piston and load is negligible compared with the hydraulic force at the power piston. We also assume that the pilot valve is a zero-lapped valve. As given by Equation (7–29), the oil flow rate is proportional to the pilot valve displacement.

The operation of this hydraulic servomotor is as follows: If input x moves the pilot valve to the right, port 1 is uncovered, and high-pressure oil enters the right-hand side of the power piston. Since port 2 is connected to the drain port, the oil on the left-hand side of the power piston is returned to the drain. The oil flowing into the power cylinder is at high pressure; the oil flowing out from the power cylinder into the drain is at low pressure. The resulting difference in pressure on both sides of the power piston will cause it to move to the left.

Note that the rate of flow of oil, q_1 (kg/s), times dt (s) is equal to the power piston displacement dy (m) times the piston area A (m^2) times the density of the oil, ρ (kg/m^3). That is,

$$A\rho\,dy = q_1\,dt \tag{7–30}$$

As given by Equation (7–29), the oil flow rate q_1 is proportional to the pilot valve displacement x, or

$$q_1 = K_1 x \tag{7–31}$$

where K_1 is a proportionality constant. From Equations (7–30) and (7–31), we obtain

$$A\rho\frac{dy}{dt} = K_1 x$$

The Laplace transform of this last equation, assuming a zero initial condition, gives

$$A\rho s\,Y(s) = K_1 X(s)$$

or

$$\frac{Y(s)}{X(s)} = \frac{K_1}{A\rho s} = \frac{K}{s} \tag{7–32}$$

where $K = K_1/(A\rho)$. Thus, the hydraulic servomotor shown in Figure 7–10 acts as an integral controller.

Dashpots.   The dashpot (also called a damper) shown in Figure 7–11(a) acts as a differentiating element. Suppose that we introduce a step displacement into the piston position x. Then the displacement y becomes momentarily equal to x. Because of the spring force, however, the oil will flow through the resistance R, and the cylinder will come back to the original position. The curves of x versus t and y versus t are shown in Figure 7–11(b).

Let us derive the transfer function between the displacement y and the displacement x. We define the pressures existing on the right-hand side and left-hand side of the piston as P_1 (lb$_f$/in.^{2}) and P_2 (lb$_f$/in.^{2}), respectively. Suppose that the inertia force involved is negligible. Then the force acting on the piston must balance

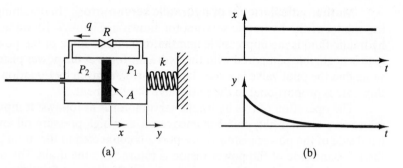

Figure 7–11 (a) Dashpot; (b) step change in x and the corresponding change in y plotted against t.

the spring force. Thus,

$$A(P_1 - P_2) = ky$$

where

$A = $ piston area, in.^{2}

$k = $ spring constant, lb$_f$/in.

The flow rate q through the restriction, in lb/s, is given by

$$q = \frac{P_1 - P_2}{R}$$

where R is the resistance to flow at the restriction, lb$_f$-s/in.^{2}-lb.

Since the flow through the restriction during dt seconds must equal the change in the mass of oil to the left of the piston during the same dt seconds, we obtain

$$q\,dt = A\rho(dx - dy) \tag{7–33}$$

where $\rho = $ density, lb/in.^{3}. (We assume that the fluid is incompressible, or $\rho = $ constant.) Equation (7–33) can be rewritten as

$$\frac{dx}{dt} - \frac{dy}{dt} = \frac{q}{A\rho} = \frac{P_1 - P_2}{RA\rho} = \frac{ky}{RA^2\rho}$$

or

$$\frac{dx}{dt} = \frac{dy}{dt} + \frac{ky}{RA^2\rho}$$

Taking the Laplace transforms of both sides of this last equation, assuming zero initial conditions, we obtain

$$sX(s) = sY(s) + \frac{k}{RA^2\rho}Y(s)$$

The transfer function of the system thus becomes

$$\frac{Y(s)}{X(s)} = \frac{s}{s + \dfrac{k}{RA^2\rho}}$$

Let us define $RA^2\rho/k = T$. Then

$$\frac{Y(s)}{X(s)} = \frac{Ts}{Ts + 1} \tag{7-34}$$

In earlier chapters, we frequently treated the spring–dashpot system as shown in Figure 7–12, which is equivalent to the system of Figure 7–11(a). A mathematical model of the system shown in Figure 7–12 is

$$b(\dot{x} - \dot{y}) = ky$$

or

$$\frac{Y(s)}{X(s)} = \frac{bs}{bs + k} = \frac{\dfrac{b}{k}s}{\dfrac{b}{k}s + 1} = \frac{Ts}{Ts + 1} \tag{7-35}$$

where b/k is the *time constant T*.

Notice that, since $T = RA^2\rho/k$ in Equation (7–34) and $T = b/k$ in Equation (7–35), we find the viscous-friction coefficient b to be equal to $RA^2\rho$ or

$$b = RA^2\rho$$

Note that the resistance R depends on the viscosity of oil.

Comments.   Since hydraulic systems are used frequently in industry, in what follows we shall list the advantages and disadvantages of using hydraulic systems over comparable electrical systems.

Advantages and disadvantages of hydraulic systems.   Some of the advantages to using hydraulic systems rather than electrical systems are as follows:

1. Hydraulic fluid acts as a lubricant, in addition to carrying away heat generated in the system to a convenient heat exchanger.
2. Comparatively small hydraulic actuators can develop large forces or torques.
3. Hydraulic actuators have a higher speed of response, with fast starts, stops, and reversals of speed.

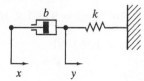

Figure 7–12   Spring-dashpot system.

4. Hydraulic actuators can be operated under continuous, intermittent, reversing, and stalled conditions without damage.

5. The availability of both linear and rotary actuators lends flexibility to design.

6. Because of low leakages in hydraulic actuators, drops in speed when loads are applied are small.

Several disadvantages, however, tend to limit the use of hydraulic systems:

1. Hydraulic power is not readily available, compared with electric power.

2. The cost of a hydraulic system may be higher than that of a comparable electrical system performing a similar function.

3. Fire and explosion hazards exist, unless fire-resistant fluids are used.

4. Because it is difficult to maintain a hydraulic system that is free from leaks, the system tends to be messy.

5. Contaminated oil may cause failure in the proper functioning of a hydraulic system.

6. As a result of the nonlinear and other complex characteristics involved, the design of sophisticated hydraulic systems is quite involved.

7. Hydraulic circuits have generally poor damping characteristics. If a hydraulic circuit is not designed properly, some unstable phenomena may appear or disappear, depending on the operating condition of the circuit.

7–6 MATHEMATICAL MODELING OF THERMAL SYSTEMS

Thermal systems involve the transfer of heat from one substance to another. Thermal systems may be analyzed in terms of resistance and capacitance, although the thermal capacitance and thermal resistance may not be represented accurately as lumped parameters, since they are usually distributed throughout the substance. (For precise analysis, distributed-parameter models must be used.) Here, however, to simplify the analysis, we shall assume that a thermal system can be represented by a lumped-parameter model, that substances characterized by resistance to heat flow have negligible heat capacitance, and that substances characterized by heat capacitance have negligible resistance to heat flow.

Before we derive mathematical models of thermal systems, let us review units of heat.

Units of heat. *Heat* is energy transferred from one body to another because of a temperature difference. The SI unit of heat is the joule (J). Other units of heat commonly used in engineering calculations are the kilocalorie (kcal) and Btu (British thermal unit). The following conversions are applicable:

$$1 \text{ J} = 1 \text{ N-m} = 2.389 \times 10^{-4} \text{ kcal} = 9.480 \times 10^{-4} \text{ Btu}$$

$$1 \text{ kcal} = 4186 \text{ J} = \frac{1}{0.860} \text{ Wh} = 1.163 \text{ Wh}$$

$$1 \text{ Btu} = 1055 \text{ J} = 778 \text{ ft-lb}_f$$

From an engineering point of view, the kilocalorie can be considered to be that amount of energy needed to raise the temperature of 1 kilogram of water from 14.5 to 15.5°C. The Btu can be considered as the energy required to raise 1 pound of water 1 degree Fahrenheit at some arbitrarily chosen temperature. (These units give roughly the same values as those previously defined.)

Heat transfer by conduction, convection, and radiation.   Heat can flow from one substance to another in three different ways: conduction, convection and radiation. In this section, we shall be concerned with systems that involve just conduction and convection; radiation heat transfer is appreciable only if the temperature of the emitter is very high compared with that of the receiver. Most thermal processes in process control systems do not involve radiation heat transfer and may be described in terms of thermal resistance and thermal capacitance.

For conduction or convection heat transfer,

$$q = K\Delta\theta$$

where

$\quad q$ = heat flow rate, kcal/s

$\Delta\theta$ = temperature difference, °C

$\quad K$ = coefficient, kcal/s °C

The coefficient K is given by

$$K = \frac{kA}{\Delta X} \quad \text{for conduction}$$

$$= HA \quad \text{for convection}$$

where

$\quad k$ = thermal conductivity, kcal/m s°C

$\quad A$ = area normal to heat flow, m^2

ΔX = thickness of conductor, m

$\quad H$ = convection coefficient, kcal/m^2 s°C

Thermal resistance and thermal capacitance.   The thermal resistance R for heat transfer between two substances may be defined as follows:

$$R = \frac{\text{change in temperature difference}}{\text{change in heat flow rate}} \quad \frac{°C}{\text{kcal/s}}$$

Thus, the thermal resistance for conduction or convection heat transfer is given by

$$R = \frac{d(\Delta\theta)}{dq} = \frac{1}{K}$$

Since the thermal conductivity and convection coefficients are almost constant, the thermal resistance for either conduction or convection is constant. The thermal capacitance C is defined by

$$C = \frac{\text{change in heat stored}}{\text{change in temperature}} \quad \frac{\text{kcal}}{°C}$$

Accordingly, the thermal capacitance is the product of the specific heat and the mass of the material. Therefore, thermal capacitance can also be written as

$$C = mc$$

where

     m = mass of substance considered, kg
     c = specific heat of substance, kcal/kg °C

Mathematical modeling of a thermal system: thermometer system.
Consider the thin, glass-walled mercury thermometer system shown in Figure 7–13. Assume that the thermometer is at a uniform temperature $\overline{\Theta}$°C (ambient temperature) and that at $t = 0$ it is immersed in a bath of temperature $\overline{\Theta} + \theta_b$°C, where θ_b is the bath temperature (which may be constant or changing), measured from the ambient temperature $\overline{\Theta}$. Let us denote the instantaneous thermometer temperature by $\overline{\Theta} + \theta$°C, so that θ is the change in the temperature of the thermometer, satisfying the condition that $\theta(0) = 0$. The dynamics of this thermometer system can be characterized in terms of a thermal resistance R (°C/kcal/s) that resists the heat flow and a thermal capacitance C (kcal/°C) that stores heat.

A mathematical model for this thermal system can be derived by considering heat balance as follows: The heat entering the thermometer during dt seconds is $q\,dt$, where q (kcal/s) is the heat flow rate to the thermometer. This heat is stored in the thermal capacitance C of the thermometer, thereby raising its temperature by $d\theta$. Thus, the heat balance equation is

$$C\,d\theta = q\,dt \tag{7–36}$$

Since the thermal resistance may be written

$$R = \frac{d(\Delta\theta)}{dq} = \frac{\Delta\theta}{q}$$

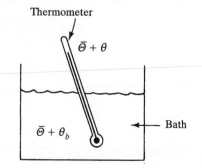

Thermometer

$\overline{\Theta} + \theta$

$\overline{\Theta} + \theta_b$

Bath

Figure 7–13 Thin, glass-walled mercury thermometer system.

the heat flow rate q may be given, in terms of R, as

$$q = \frac{(\overline{\Theta} + \theta_b) - (\overline{\Theta} + \theta)}{R} = \frac{\theta_b - \theta}{R}$$

where $\overline{\Theta} + \theta_b$ is the bath temperature and $\overline{\Theta} + \theta$ is the thermometer temperature. Consequently, we can rewrite Equation (7–36) as

$$C\frac{d\theta}{dt} = \frac{\theta_b - \theta}{R}$$

or

$$RC\frac{d\theta}{dt} + \theta = \theta_b$$

where RC is the time constant. This is a mathematical model of the thermometer system, which is analogous to the electrical system shown in Figure 7–4(a), the mechanical system of Figure 7–4(b), the liquid-level system depicted in Figure 7–3(a), and the pneumatic pressure system shown in Figure 7–7(a).

Example 7–5

Consider the air-heating system shown in Figure 7–14. Assuming small deviations from steady-state operation, let us derive a mathematical model for the system. We shall also assume that the heat loss to the surroundings and the heat capacitance of the metal parts of the heater are negligible.

To derive a mathematical model for the system, let us define

$\overline{\Theta}_i$ = steady-state temperature of inlet air, °C
$\overline{\Theta}_o$ = steady-state temperature of outlet air, °C
G = mass flow rate of air through the heating chamber, kg/s
M = mass of air contained in the heating chamber, kg
c = specific heat of air, kcal/kg °C
R = thermal resistance, °C s/kcal
C = thermal capacitance of air contained in the heating chamber = Mc, kcal/°C
$\overline{H}$ = steady-state heat input, kcal/s

Let us assume that the heat input is suddenly changed from $\overline{H}$ to $\overline{H} + h$, and at the same time, the inlet air temperature is suddenly changed from $\overline{\Theta}_i$ to $\overline{\Theta}_i + \theta_i$. Then the outlet air temperature will be changed from $\overline{\Theta}_o$ to $\overline{\Theta}_o + \theta_o$.

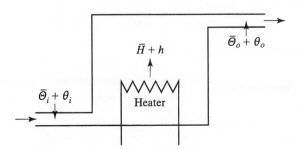

Figure 7–14   Air-heating system.

The equation describing the system behavior is

$$C \, d\theta_o = [h + Gc(\theta_i - \theta_o)] \, dt$$

or

$$C\frac{d\theta_o}{dt} = h + Gc(\theta_i - \theta_o)$$

Noting that

$$Gc = \frac{1}{R}$$

we obtain

$$C\frac{d\theta_o}{dt} = h + \frac{1}{R}(\theta_i - \theta_o)$$

or

$$RC\frac{d\theta_o}{dt} + \theta_o = Rh + \theta_i \tag{7-37}$$

Taking the Laplace transforms of both sides of this last equation and substituting the initial condition $\theta_o(0) = 0$ yields

$$\Theta_o(s) = \frac{R}{RCs + 1}H(s) + \frac{1}{RCs + 1}\Theta_i(s) \tag{7-38}$$

Equation (7–37) is a mathematical model of the system. Equation (7–38) is also a mathematical model of the system, but one in which the Laplace transform of the output $\Theta_o(s)$ is given as a sum of the responses to the inputs $H(s)$ and $\Theta_i(s)$.

EXAMPLE PROBLEMS AND SOLUTIONS

Problem A–7–1

Liquid flow resistance depends on the flow condition, either laminar or turbulent. Here, we consider the laminar-flow resistance.

For laminar flow, the flow rate Q m³/s and differential head $(H_1 - H_2)$ m are proportional, or

$$Q = K(H_1 - H_2)$$

where K is a proportionality constant. Since

$$\text{resistance } R = \frac{\text{change in differential head}}{\text{change in flow rate}} \quad \frac{\text{m}}{\text{m}^3/\text{s}}$$

$$= \frac{d(H_1 - H_2)}{dQ} \text{ s/m}^2$$

the laminar-flow resistance can be given by

$$R = \frac{d(H_1 - H_2)}{dQ} = \frac{1}{K} \, \text{s/m}^2$$

Note that the laminar flow resistance is constant.

In considering laminar flow through a cylindrical pipe, the relationship between the differential head h $(= H_1 - H_2)$ m and the flow rate Q m³/s is given by the Hagen–Poiseuille formula

$$h = \frac{128\nu L}{g\pi D^4} Q$$

where

ν = kinematic viscosity, m²/s
L = length of pipe, m
D = diameter of pipe, m

So the laminar-flow resistance R for liquid flow through cylindrical pipes is given by

$$R = \frac{dh}{dQ} = \frac{128\nu L}{g\pi D^4} \, \text{s/m}^2 \qquad (7\text{--}39)$$

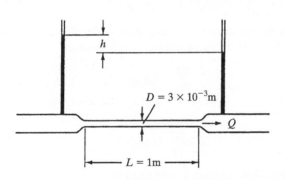

Figure 7–15 Flow of water through a capillary tube.

Now consider the flow of water through a capillary tube as shown in Figure 7–15. Assuming that the temperature of the water is 20°C and that the flow is laminar, find the resistance R of the capillary tube. The kinetic viscosity ν of water at 20°C is 1.004×10^{-6} m²/s.

Solution Substituting numerical values into Equation (7–39), we obtain

$$R = \frac{128 \times 1.004 \times 10^{-6} \times 1}{9.807 \times 3.14 \times (3 \times 10^{-3})^4} = 5.15 \times 10^4 \, \text{s/m}^2$$

Problem A–7–2

Consider a liquid flow in a pipe. The liquid-flow inertance is the potential difference (either pressure difference or head difference) between two sections in the pipe required to cause a unit rate of change in flow rate (a unit volumetric flow acceleration).

Suppose that the cross-sectional area of a pipe is constant and equal to A m² and that the pressure difference between two sections in the pipe is Δp N/m². Then the force $A\,\Delta p$ will accelerate the liquid between the two sections, or

$$M\frac{dv}{dt} = A\,\Delta p$$

where M kg is the mass of liquid in the pipe between the two sections and v m/s is the velocity of liquid flow. Note that the mass M is equal to $\rho\,AL$, where ρ kg/m^3 is the density and L m is the distance between the two sections considered. Therefore, the last equation can be written

$$\rho AL\frac{dv}{dt} = A\,\Delta p$$

Noting that Av m^3/s is the volumetric flow rate and defining $Q = Av$ m^3/s, we can rewrite the preceding equation as

$$\frac{\rho L}{A}\frac{dQ}{dt} = \Delta p \tag{7–40}$$

If pressure (N/m^2) is chosen as a measure of potential, then the liquid-flow inertance I is obtained as

$$I = \frac{\Delta p}{dQ/dt} = \frac{\rho L}{A}\ \frac{\text{N-s}^2}{\text{m}^5}$$

If head (m) is chosen as a measure of potential, then, noting that $\Delta p = \Delta h \rho g$, where Δh is the differential head, we see that Equation (7–40) becomes

$$\frac{\rho L}{A}\frac{dQ}{dt} = \Delta h \rho g$$

or

$$\frac{L}{Ag}\frac{dQ}{dt} = \Delta h$$

Consequently, the liquid-flow inertance I is obtained as

$$I = \frac{\Delta h}{dQ/dt} = \frac{L}{Ag}\ \frac{\text{s}^2}{\text{m}^2}$$

Now consider water flow through a pipe whose cross-sectional area is 1×10^{-3} m^2 and in which two sections are 15 m apart. Compute the inertance I. Assuming that the differential head between two sections is 1 m, compute the volumetric water flow acceleration dQ/dt.

Solution   The liquid-flow inertance is

$$I = \frac{\rho L}{A} = \frac{1000 \times 15}{1 \times 10^{-3}}\ \frac{\text{kg}}{\text{m}^3}\frac{\text{m}}{\text{m}^2} = 1.5 \times 10^7\ \text{N-s}^2/\text{m}^5$$

or

$$I = \frac{L}{Ag} = \frac{15}{1 \times 10^{-3} \times 9.807}\ \frac{\text{m}}{\text{m}^2}\frac{\text{s}^2}{\text{m}} = 1529.5\ \text{s}^2/\text{m}^2$$

For a differential head of 1 m between two sections that are 15 m apart, the volumetric water flow acceleration is

$$\frac{dQ}{dt} = \frac{\Delta h}{I} = \frac{\Delta h}{L/Ag} = \frac{1}{1529.5} = 0.000654 \text{ m}^3/\text{s}^2$$

Problem A–7–3

Consider the liquid-level system shown in Figure 7–16. Assume that the outflow rate Q m³/s through the outflow valve is related to the head H m by

$$Q = K\sqrt{H} = 0.01\sqrt{H}$$

Assume also that, when the inflow rate Q_i is 0.015 m³/s, the head stays constant. At $t = 0$ the inflow valve is closed, so there is no inflow for $t \geq 0$. Find the time necessary to empty the tank to half the original head. The capacitance of the tank is 2 m².

Solution When the head is stationary, the inflow rate equals the outflow rate. Thus, the head H_0 at $t = 0$ is obtained from

$$0.015 = 0.01\sqrt{H_0}$$

or

$$H_0 = 2.25 \text{ m}$$

The equation for the system for $t > 0$ is

$$-C dH = Q \, dt$$

or

$$\frac{dH}{dt} = -\frac{Q}{C} = \frac{-0.01\sqrt{H}}{2}$$

Consequently,

$$\frac{dH}{\sqrt{H}} = -0.005 \, dt$$

Assume that $H = 1.125$ m at $t = t_1$. Integrating both sides of this last equation, we have

$$\int_{2.25}^{1.125} \frac{dH}{\sqrt{H}} = \int_{0}^{t_1} (-0.005) \, dt = -0.005 t_1$$

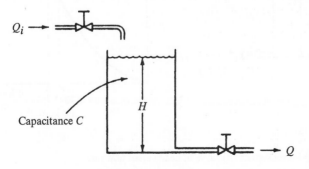

Q_i

H

Capacitance C

Q

Figure 7–16 Liquid-level system.

It follows that

$$2\sqrt{H}\Big|_{2.25}^{1.125} = 2\sqrt{1.125} - 2\sqrt{2.25} = -0.005t_1$$

or

$$t_1 = 175.7$$

Thus, the time necessary to empty the tank to half the original head is 175.7 s.

Problem A–7–4

Consider the liquid-level system of Figure 7–17(a). The curve of head versus flow rate is shown in Figure 7–17(b). Assume that at steady state the liquid flow rate is 4×10^{-4} m³/s and the steady-state head is 1 m. At $t = 0$, the inflow valve is opened further and the inflow rate is changed to 4.5×10^{-4} m³/s. Determine the average resistance R of the outflow valve. Also, determine the change in head as a function of time. The capacitance C of the tank is 0.02 m².

Solution   The flow rate through the outflow valve can be assumed to be

$$Q = K\sqrt{H}$$

Next, from the curve given in Figure 7–17(b), we see that

$$4 \times 10^{-4} = K\sqrt{1}$$

or

$$K = 4 \times 10^{-4}$$

So if the steady-state flow rate is changed to 4.5×10^{-4} m³/s, then the new steady-state head can be obtained from

$$4.5 \times 10^{-4} = 4 \times 10^{-4}\sqrt{H}$$

or

$$H = 1.266 \text{ m}$$

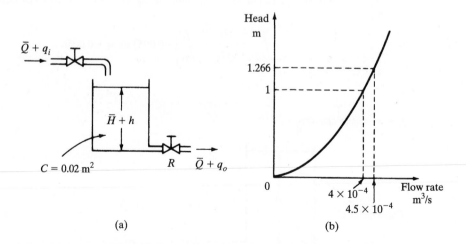

(a)   (b)

Figure 7–17   (a) Liquid-level system; (b) curve of head versus flow rate.

This means that the change in head is $1.266 - 1 = 0.266$ m. The average resistance R of the outflow valve is then

$$R = \frac{dH}{dQ} = \frac{1.266 - 1}{(4.5 - 4) \times 10^{-4}} = 0.532 \times 10^4 \text{ s/m}^2$$

Noting that the change in the liquid stored in the tank during dt seconds is equal to the net flow into the tank during the same dt seconds, we have

$$C \, dh = (q_i - q_o) \, dt$$

where q_i and q_o are the changes in the inflow rate and outflow rate of the tank, respectively, and h is the change in the head. Thus,

$$C \frac{dh}{dt} = q_i - q_o$$

Since

$$R = \frac{h}{q_o}$$

it follows that

$$C \frac{dh}{dt} = q_i - \frac{h}{R}$$

or

$$RC \frac{dh}{dt} + h = Rq_i$$

Substituting $R = 0.532 \times 10^4$ s/m^2, $C = 0.02$ m^2, and $q_i = 0.5 \times 10^{-4}$ m^3/s into this last equation yields

$$0.532 \times 10^4 \times 0.02 \frac{dh}{dt} + h = 0.532 \times 10^4 \times 0.5 \times 10^{-4}$$

or

$$106.4 \frac{dh}{dt} + h = 0.266$$

Taking Laplace transforms of both sides of this last equation, with the initial condition $h(0) = 0$, we obtain

$$(106.4s + 1)H(s) = \frac{0.266}{s}$$

or

$$H(s) = \frac{0.266}{s(106.4s + 1)} = 0.266 \left[\frac{1}{s} - \frac{1}{s + (1/106.4)} \right]$$

The inverse Laplace transform of $H(s)$ gives

$$h(t) = 0.266(1 - e^{-t/106.4}) \text{ m}$$

This equation gives the change in head as a function of time.

Problem A–7–5

For the liquid-level system shown in Figure 7–18, the steady-state flow rate through the tanks is $\overline{Q}$ and the steady-state heads of tank 1 and tank 2 are $\overline{H}_1$ and $\overline{H}_2$, respectively. At $t = 0$, the inflow rate is changed from $\overline{Q}$ to $\overline{Q} + q$, where q is small. The corresponding changes in the heads (h_1 and h_2) and changes in flow rates (q_1 and q_2) are assumed to be small as well. The capacitances of tank 1 and tank 2 are C_1 and C_2, respectively. The resistance of the valve between the tanks is R_1 and that of the outflow valve is R_2. Assuming that q is the input and q_2 the output, derive the transfer function for the system.

Solution   For tank 1, we have

$$q_1 = \frac{h_1 - h_2}{R_1}$$

$$C_1 \frac{dh_1}{dt} = q - q_1$$

Hence,

$$C_1 \frac{dh_1}{dt} + \frac{h_1}{R_1} = q + \frac{h_2}{R_1} \tag{7–41}$$

For tank 2, we get

$$q_2 = \frac{h_2}{R_2}$$

$$C_2 \frac{dh_2}{dt} = q_1 - q_2$$

Therefore,

$$C_2 \frac{dh_2}{dt} + \frac{h_2}{R_1} + \frac{h_2}{R_2} = \frac{h_1}{R_1} \tag{7–42}$$

Taking Laplace transforms of both sides of Equations (7–41) and (7–42), under the initial conditions $h_1(0) = 0$ and $h_2(0) = 0$, we obtain

$$\left(C_1 s + \frac{1}{R_1} \right) H_1(s) = Q(s) + \frac{1}{R_1} H_2(s) \tag{7–43}$$

$$\left(C_2 s + \frac{1}{R_1} + \frac{1}{R_2} \right) H_2(s) = \frac{1}{R_1} H_1(s) \tag{7–44}$$

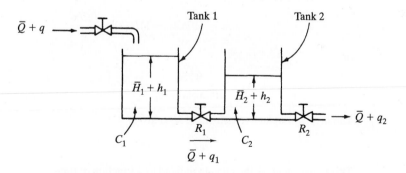

Figure 7–18   Liquid-level system.

From Equation (7–43), we have

$$(R_1C_1s + 1)H_1(s) = R_1Q(s) + H_2(s)$$

or

$$H_1(s) = \frac{R_1Q(s) + H_2(s)}{R_1C_1s + 1}$$

Substituting this last equation into Equation (7–44) yields

$$\left(C_2s + \frac{1}{R_1} + \frac{1}{R_2}\right)H_2(s) = \frac{1}{R_1}\frac{R_1Q(s) + H_2(s)}{R_1C_1s + 1}$$

Since $H_2(s) = R_2Q_2(s)$, we get

$$\left(C_2s + \frac{1}{R_1} + \frac{1}{R_2}\right)R_2Q_2(s) = \frac{Q(s)}{R_1C_1s + 1} + \frac{R_2}{R_1}\frac{Q_2(s)}{R_1C_1s + 1}$$

which can be simplified to

$$[(C_2R_2s + 1)(R_1C_1s + 1) + R_2C_1s]Q_2(s) = Q(s)$$

Thus, the transfer function $Q_2(s)/Q(s)$ can be given by

$$\frac{Q_2(s)}{Q(s)} = \frac{1}{R_1C_1R_2C_2s^2 + (R_1C_1 + R_2C_2 + R_2C_1)s + 1}$$

which is Equation (7–13).

Problem A–7–6

Consider the liquid-level system of Figure 7–19. At steady state, the inflow rate and out-flow rate are both $\overline{Q}$, the flow rate between the tanks is zero, and the heads of tank 1 and tank 2 are both $\overline{H}$. At $t = 0$, the inflow rate is changed from $\overline{Q}$ to $\overline{Q} + q$, where q is small. The resulting changes in the heads (h_1 and h_2) and flow rates (q_1 and q_2) are assumed to be small as well. The capacitances of tanks 1 and 2 are C_1 and C_2, respectively. The resistance of the valve between the tanks is R_1 and that of the outflow valve is R_2.

Derive the transfer function for the system when q is the input and h_2 is the output.

Solution　For tank 1, we have

$$C_1\,dh_1 = q_1\,dt$$

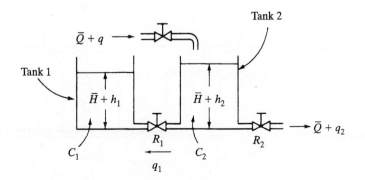

Figure 7–19　Liquid-level system.

where

$$q_1 = \frac{h_2 - h_1}{R_1}$$

Consequently,

$$R_1 C_1 \frac{dh_1}{dt} + h_1 = h_2 \tag{7-45}$$

For tank 2, we get

$$C_2 dh_2 = (q - q_1 - q_2)\, dt$$

where

$$q_1 = \frac{h_2 - h_1}{R_1}, \qquad q_2 = \frac{h_2}{R_2}$$

It follows that

$$R_2 C_2 \frac{dh_2}{dt} + \frac{R_2}{R_1} h_2 + h_2 = R_2 q + \frac{R_2}{R_1} h_1 \tag{7-46}$$

Eliminating h_1 from Equations (7–45) and (7–46), we have

$$R_1 C_1 R_2 C_2 \frac{d^2 h_2}{dt^2} + (R_1 C_1 + R_2 C_2 + R_2 C_1)\frac{dh_2}{dt} + h_2 = R_1 C_1 R_2 \frac{dq}{dt} + R_2 q \tag{7-47}$$

The transfer function $H_2(s)/Q(s)$ is then obtained from Equation (7–47) and is

$$\frac{H_2(s)}{Q(s)} = \frac{R_1 C_1 R_2 s + R_2}{R_1 C_1 R_2 C_2 s^2 + (R_1 C_1 + R_2 C_2 + R_2 C_1)s + 1}$$

Problem A–7–7

Consider the liquid-level system shown in Figure 7–20. In the system, $\overline{Q}_1$ and $\overline{Q}_2$ are steady-state inflow rates and $\overline{H}_1$ and $\overline{H}_2$ are steady-state heads. The quantities $q_{i1}, q_{i2}, h_1, h_2, q_1,$ and q_o are considered small. Obtain a state-space representation of the system when h_1 and h_2 are the outputs and q_{i1} and q_{i2} are the inputs.

Solution The equations for the system are

$$C_1\, dh_1 = (q_{i1} - q_1)\, dt \tag{7-48}$$

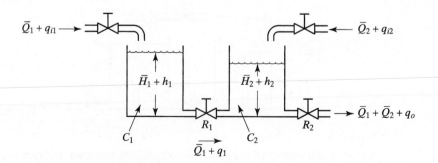

Figure 7–20 Liquid-level system.

$$\frac{h_1 - h_2}{R_1} = q_1 \tag{7-49}$$

$$C_2 \, dh_2 = (q_1 + q_{i2} - q_o) \, dt \tag{7-50}$$

$$\frac{h_2}{R_2} = q_o \tag{7-51}$$

Using Equation (7–49) to eliminate q_1 from Equation (7–48) results in

$$\frac{dh_1}{dt} = \frac{1}{C_1}\left(q_{i1} - \frac{h_1 - h_2}{R_1}\right) \tag{7-52}$$

Using Equations (7–49) and (7–51) to eliminate q_1 and q_o from Equation (7–50) gives

$$\frac{dh_2}{dt} = \frac{1}{C_2}\left(\frac{h_1 - h_2}{R_1} + q_{i2} - \frac{h_2}{R_2}\right) \tag{7-53}$$

If we define state variables

$$x_1 = h_1$$
$$x_2 = h_2$$

input variables

$$u_1 = q_{i1}$$
$$u_2 = q_{i2}$$

and output variables

$$y_1 = h_1 = x_1$$
$$y_2 = h_2 = x_2$$

then Equations (7–52) and (7–53) can be written as

$$\dot{x}_1 = -\frac{1}{R_1 C_1} x_1 + \frac{1}{R_1 C_1} x_2 + \frac{1}{C_1} u_1$$

$$\dot{x}_2 = \frac{1}{R_1 C_2} x_1 - \left(\frac{1}{R_1 C_2} + \frac{1}{R_2 C_2}\right) x_2 + \frac{1}{C_2} u_2$$

In standard vector–matrix representation, we have

$$\begin{bmatrix} \dot{x}_1 \\ \dot{x}_2 \end{bmatrix} = \begin{bmatrix} -\dfrac{1}{R_1 C_1} & \dfrac{1}{R_1 C_1} \\ \dfrac{1}{R_1 C_2} & -\left(\dfrac{1}{R_1 C_2} + \dfrac{1}{R_2 C_2}\right) \end{bmatrix} \begin{bmatrix} x_1 \\ x_2 \end{bmatrix} + \begin{bmatrix} \dfrac{1}{C_1} & 0 \\ 0 & \dfrac{1}{C_2} \end{bmatrix} \begin{bmatrix} u_1 \\ u_2 \end{bmatrix}$$

which is the state equation, and

$$\begin{bmatrix} y_1 \\ y_2 \end{bmatrix} = \begin{bmatrix} 1 & 0 \\ 0 & 1 \end{bmatrix} \begin{bmatrix} x_1 \\ x_2 \end{bmatrix}$$

which is the output equation.

Problem A–7–8

Obtain a mechanical analog of the liquid-level system shown in Figure 7–21 when q is the input and q_2 the output.

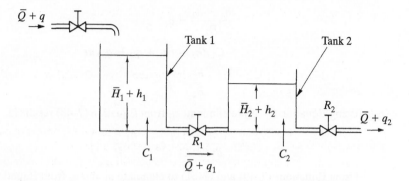

Figure 7–21   Liquid-level system.

Solution   The equations for the liquid-level system are

$$C_1 \frac{dh_1}{dt} = q - q_1 \tag{7–54}$$

$$q_1 = \frac{h_1 - h_2}{R_1} \tag{7–55}$$

$$C_2 \frac{dh_2}{dt} = q_1 - q_2 \tag{7–56}$$

$$q_2 = \frac{h_2}{R_2} \tag{7–57}$$

Analogous quantities in a mechanical–liquid-level analogy are shown in Table 7–2. (Note that other mechanical–liquid-level analogies are possible as well.) Using the analogous quantities shown in the table, Equations (7–54) through (7–57) can be modified to

$$b_1 \dot{x}_1 = F - F_1 \tag{7–58}$$
$$F_1 = k_1(x_1 - x_2) \tag{7–59}$$
$$b_2 \dot{x}_2 = F_1 - F_2 \tag{7–60}$$
$$F_2 = k_2 x_2 \tag{7–61}$$

TABLE 7–2   Mechanical–Liquid-Level Analogy

Mechanical Systems	Liquid-Level Systems
F (force)	q (flow rate)
x (displacement)	h (head)
$\dot{x}$ (velocity)	$\dot{h}$ (time change of head)
b (viscous-friction coefficient)	C (capacitance)
k (spring constant)	$\dfrac{1}{R}$ (reciprocal of resistance)

Rewriting Equations (7–58) through (7–61), we obtain

$$b_1\dot{x}_1 + k_1(x_1 - x_2) = F$$
$$b_2\dot{x}_2 + k_2x_2 = k_1(x_1 - x_2)$$

On the basis of the last two equations, we can obtain an analogous mechanical system as shown in Figure 7–22.

Problem A–7–9

In dealing with gas systems, we find it convenient to work in molar quantities, because 1 mole of any gas contains the same number of molecules. Thus, 1 mole occupies the same volume if measured under the same conditions of pressure and temperature.

   At standard pressure and temperature (1.0133×10^5 N/m² abs and 273 K, or 14.7 psia and 492°R), 1 kg mole of any gas is found to occupy 22.4 m³ (or 1 lb mole of any gas is found to occupy 359 ft³). For instance, at standard pressure and temperature, the volume occupied by 2 kg of hydrogen, 32 kg of oxygen, or 28 kg of nitrogen is the same, 22.4 m³. This volume is called the *molal* volume and is denoted by $\bar{v}$.

   For 1 mole of gas,

$$p\bar{v} = \bar{R}T \qquad (7\text{–}62)$$

The value of $\bar{R}$ is the same for all gases under all conditions. The constant $\bar{R}$ is the universal gas constant.

   Find the value of the universal gas constant in SI and BES units.

Solution   Substituting $p = 1.0133 \times 10^5$ N/m² abs, $\bar{v} = 22.4$ m³/kg-mole, and $T = 273$ K into Equation (7–62), we obtain

$$\bar{R} = \frac{p\bar{v}}{T} = \frac{1.0133 \times 10^5 \times 22.4}{273} = 8314 \text{ N-m/kg-mole K}$$

This is the universal gas constant in SI units.

   To obtain the universal gas constant in BES units, we substitute $p = 14.7$ psia $= 14.7 \times 144$ lb$_f$/ft² abs, $\bar{v} = 359$ ft³/lb-mole, and $T = 492$°R into Equation (7–62).

$$\bar{R} = \frac{p\bar{v}}{T} = \frac{14.7 \times 144 \times 359}{492} = 1545 \text{ ft-lb}_f\text{/lb-mole °R}$$
$$= 1.985 \text{ Btu/lb-mole °R}$$

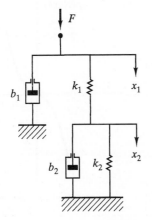

Figure 7–22   Mechanical analog of the liquid-level system shown in Figure 7–21.

Problem A–7–10

Referring to the pneumatic pressure system shown in Figure 7–23, assume that the system is at steady state for $t < 0$ and that the steady-state pressure of the system is $\overline{P} = 5 \times 10^5$ N/m² abs. At $t = 0$, the inlet pressure is suddenly changed from $\overline{P}$ to $\overline{P} + p_i$, where p_i is a step change with a magnitude equal to 2×10^4 N/m². This change causes the air to flow into the vessel until the pressure equalizes. Assume that the initial flow rate is $q(0) = 1 \times 10^{-4}$ kg/s. As air flows into the vessel, the pressure of the air in the vessel rises from $\overline{P}$ to $\overline{P} + p_o$. Determine p_o as a function of time. Assume that the expansion process is isothermal $(n = 1)$, that the temperature of the entire system is constant at $T = 293$ K, and that the vessel has a capacity of 0.1 m³.

Solution   The average resistance of the valve is

$$R = \frac{\Delta p}{q} = \frac{2 \times 10^4}{1 \times 10^{-4}} = 2 \times 10^8 \text{ N-s/kg-m}^2$$

The capacitance of the vessel is

$$C = \frac{V}{nR_{air}T} = \frac{0.1}{1 \times 287 \times 293} = 1.19 \times 10^{-6} \text{ kg-m}^2/\text{N}$$

A mathematical model for this system is obtained from

$$C \, dp_o = q \, dt$$

where

$$q = \frac{\Delta p}{R} = \frac{p_i - p_o}{R}$$

Thus,

$$RC\frac{dp_o}{dt} + p_o = p_i$$

Substituting the values of R, C, and p_i into this last equation, we have

$$2 \times 10^8 \times 1.19 \times 10^{-6}\frac{dp_o}{dt} + p_o = 2 \times 10^4$$

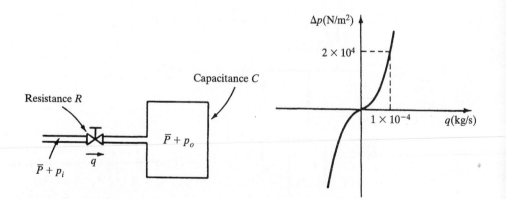

Figure 7–23   Pneumatic pressure system.

or

$$238\frac{dp_o}{dt} + p_o = 2 \times 10^4 \tag{7-63}$$

Taking Laplace transforms of both sides of Equation (7–63), with the initial condition $p_o(0) = 0$, we get

$$(238s + 1)P_o(s) = 2 \times 10^4 \frac{1}{s}$$

or

$$P_o(s) = \frac{2 \times 10^4}{s(238s + 1)}$$

$$= 2 \times 10^4 \left(\frac{1}{s} - \frac{1}{s + 0.0042} \right)$$

The inverse Laplace transform of this last equation is

$$p_o(t) = 2 \times 10^4 (1 - e^{-0.0042t})$$

which gives $p_o(t)$ as a function of time.

Problem A–7–11

Air is compressed into a tank of volume $2 \, m^3$. The compressed air pressure is $5 \times 10^5 \, N/m^2$ gage and the temperature is 20°C. Find the mass of air in the tank. Also, find the specific volume and specific weight of the compressed air.

Solution The pressure and temperature are

$$p = (5 + 1.0133) \times 10^5 \, N/m^2 \text{ abs}$$
$$T = 273 + 20 = 293 \, K$$

From Table 7–1, the gas constant of air is $R_{air} = 287 \, N\text{-m/kg K}$. Therefore, the mass of the compressed air is

$$m = \frac{pV}{R_{air}T} = \frac{6.0133 \times 10^5 \times 2}{287 \times 293} = 14.3 \, kg$$

The specific volume v is

$$v = \frac{V}{m} = \frac{2}{14.3} = 0.140 \, m^3/kg$$

The specific weight γ is

$$\gamma = \frac{mg}{V} = \frac{14.3 \times 9.807}{2} = 70.1 \, N/m^3$$

Problem A–7–12

The molecular weight of a pure substance is the weight of one molecule of the substance, compared with the weight of one oxygen atom, which is taken to be 16. That is, the molecular weight of carbon dioxide (CO_2) is $12 + (16 \times 2) = 44$. The molecular weights of (molecular) oxygen and water vapor are 32 and 18, respectively.

Determine the specific volume v of a mixture that consists of $100 \, m^3$ of oxygen, $5 \, m^3$ of carbon dioxide, and $20 \, m^3$ of water vapor when the pressure and temperature are $1.0133 \times 10^5 \, N/m^2$ abs and 294 K, respectively.

Solution   The mean molecular weight of the mixture is

$$M = \left(32 \times \frac{100}{125} \right) + \left(44 \times \frac{5}{125} \right) + \left(18 \times \frac{20}{125} \right) = 30.24$$

Thus,

$$v = \frac{\overline{R}T}{Mp} = \frac{8314 \times 294}{30.24 \times 1.0133 \times 10^5} = 0.798 \text{ m}^3/\text{kg}$$

Problem A–7–13

Sound is a longitudinal wave phenomenon representing the propagation of compressional waves in an elastic medium. The speed c of propagation of a sound wave is given by

$$c = \sqrt{\frac{dp}{d\rho}}$$

Show that the speed c of sound can also be given by

$$c = \sqrt{kRT}$$

where

> k = ratio of specific heats, c_p/c_v
>
> R = gas constant
>
> T = absolute temperature

Find the speed of sound in air when the temperature is 293 K.

Solution   Since the pressure and temperature changes due to the passage of a sound wave are negligible, the process can be considered isentropic. Then

$$\frac{p}{\rho^k} = \text{constant}$$

Therefore,

$$\frac{dp}{d\rho} = \frac{kp}{\rho}$$

Since $p = \rho RT$, we obtain

$$c = \sqrt{\frac{dp}{d\rho}} = \sqrt{\frac{kp}{\rho}} = \sqrt{kRT}$$

For a given gas, the values of k and R are constant. So the speed of sound in a gas is a function only of the absolute temperature of the gas.

Noting that, for air,

$$k = 1.40$$
$$R_{\text{air}} = 287 \text{ N-m/kg K}$$

we find the speed of sound to be

$$c = \sqrt{kR_{\text{air}}T} = \sqrt{1.40 \times 287 \times 293} = 343.1 \text{ m/s}$$
$$= 1235 \text{ km/h} = 1126 \text{ ft/s} = 768 \text{ mi/h}$$

Problem A–7–14

Find a linearized equation for

$$z = 0.4x^3 = f(x)$$

about a point $\bar{x} = 2, \bar{z} = 3.2$.

Solution   The Taylor series expansion of $f(x)$ about the point $(2, 3.2)$, neglecting the higher order terms, is

$$z - \bar{z} = a(x - \bar{x})$$

where

$$a = \left.\frac{df}{dx}\right|_{x=2} = \left.1.2x^2\right|_{x=2} = 4.8$$

So the linear approximation of the given nonlinear equation is

$$z - 3.2 = 4.8(x - 2) \tag{7-64}$$

Figure 7–24 depicts a nonlinear curve $z = 0.4x^3$ and the linear equation given by Equation (7–64). Note that the straight-line approximation of the cubic curve is valid near the point $(2, 3.2)$.

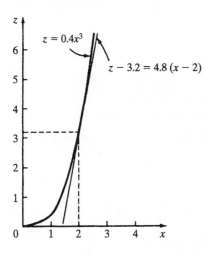

Figure 7–24   Nonlinear curve $z = 0.4x^3$ and its linear approximation at point $\bar{x} = 2$ and $\bar{z} = 3.2$.

Problem A–7–15

Linearize the nonlinear equation

$$z = xy^2$$

in the region $5 \le x \le 7, 10 \le y \le 12$. Find the error if the linearized equation is used to calculate the value of z when $x = 5, y = 10$.

Solution   Since the region considered is given by $5 \le x \le 7, 10 \le y \le 12$, choose $\bar{x} = 6, \bar{y} = 11$. Then $\bar{z} = \bar{x}\bar{y}^2 = 726$. Let us obtain a linearized equation for the nonlinear equation near a point $\bar{x} = 6, \bar{y} = 11, \bar{z} = 726$.

Expanding the nonlinear equation into a Taylor series about the point $x = \bar{x}, y = \bar{y}, z = \bar{z}$ and neglecting the higher order terms, we have

$$z - \bar{z} = a(x - \bar{x}) + b(y - \bar{y})$$

where

$$a = \frac{\partial(xy^2)}{\partial x}\Bigg|_{x=\bar{x},\, y=\bar{y}} = \bar{y}^2 = 121$$

$$b = \frac{\partial(xy^2)}{\partial y}\Bigg|_{x=\bar{x},\, y=\bar{y}} = 2\bar{x}\bar{y} = 132$$

Hence, the linearized equation is

$$z - 726 = 121(x - 6) + 132(y - 11)$$

or

$$z = 121x + 132y - 1452$$

When $x = 5$, $y = 10$, the value of z given by the linearized equation is

$$z = 121x + 132y - 1452 = 605 + 1320 - 1452 = 473$$

The exact value of z is $z = xy^2 = 500$. The error is thus $500 - 473 = 27$. In terms of percentage, the error is 5.4%.

Problem A–7–16

Linearize the nonlinear equation

$$z = \frac{x}{y}$$

in the region defined by $90 \le x \le 110, 45 \le y \le 55$.

Solution   Let us choose $\bar{x} = 100$, $\bar{y} = 50$. The given function $z = x/y$ can be expanded into a Taylor series as follows:

$$z = \frac{x}{y} = f(x, y)$$

$$= f(\bar{x}, \bar{y}) + \frac{\partial f}{\partial x}(x - \bar{x}) + \frac{\partial f}{\partial y}(y - \bar{y}) + \cdots$$

Thus, a linearized equation for the system is

$$z - \frac{\bar{x}}{\bar{y}} = a(x - \bar{x}) + b(y - \bar{y})$$

where $\bar{x} = 100, \bar{y} = 50$, and

$$a = \frac{\partial f}{\partial x}\Bigg|_{x=100,\, y=50} = \frac{1}{y}\Bigg|_{x=100,\, y=50} = \frac{1}{50}$$

$$b = \frac{\partial f}{\partial y}\Bigg|_{x=100,\, y=50} = -\frac{x}{y^2}\Bigg|_{x=100,\, y=50} = -\frac{1}{25}$$

Hence,

$$z - \frac{100}{50} = \frac{1}{50}(x - 100) - \frac{1}{25}(y - 50)$$

or

$$x - 2y - 50z + 100 = 0$$

This is a linearized equation for the nonlinear system in the given region.

Problem A-7-17

A six-pulley hoist is shown in Figure 7–25. If the piston area A is $30 \times 10^{-4} \text{ m}^2$ and the pressure difference $p_1 - p_2$ is $5 \times 10^6 \text{ N/m}^2$, find the mass m of the maximum load that can be pulled up. Neglect the friction force in the system.

Solution   The hydraulic force on the piston is

$$A(p_1 - p_2) = 30 \times 10^{-4} \times 5 \times 10^6 = 15{,}000 \text{ N}$$

Note that in this system the piston pulls six cables. Since the tension is the same on the entire length of the cable, we obtain

$$6F = 15{,}000 \text{ N}$$

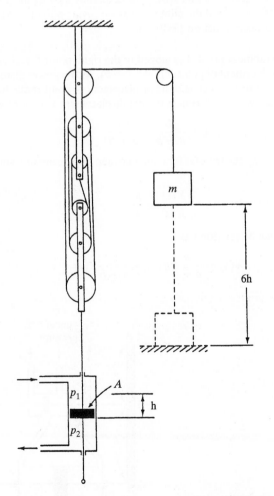

Figure 7-25   Six-pulley hoist.

where F is the tension in the cable and also is the lifting force. This force should be equal to mg; that is,

$$F = mg$$

or

$$m = \frac{15,000}{9.807 \times 6} = 254.9 \text{ kg}$$

Problem A–7–18

Consider the hydraulic system shown in Figure 7–26. The left-hand side of the pilot valve is joined to the left-hand side of the power piston by a link ABC. This link is a floating link rather than one moving about a fixed pivot. The system is a hydraulic controller.

    The system operates in the following way: If input e moves the pilot valve to the right, port I will be uncovered and high-pressure oil will flow through that port into the right-hand side of the power piston, forcing it to the left. The power piston, in moving to the left, will carry the feedback link ABC with it, thereby moving the pilot valve to the left. This action continues until the pilot valve again covers ports I and II.

    Derive the transfer function $Y(s)/E(s)$.

Solution   At the moment point A is moved to the right, point C acts as a fixed point. Therefore, the displacement of point B is $eb/(a + b)$. As the power piston moves to the left, point A acts as a fixed point, and the displacement of point B due to the motion of the power piston is $ya/(a + b)$. Hence, the net displacement x of point B is

$$x = \frac{eb}{a + b} - \frac{ya}{a + b} \tag{7–65}$$

From Equation (7–32), the transfer function between displacement y and displacement x is given by

$$\frac{Y(s)}{X(s)} = \frac{K}{s} \tag{7–66}$$

Equation (7–65) can be rewritten as

$$X(s) = \frac{b}{a + b} E(s) - \frac{a}{a + b} Y(s) \tag{7–67}$$

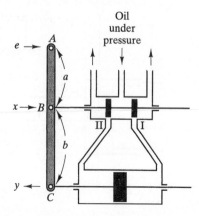

Figure 7–26   Hydraulic system.

Eliminating $X(s)$ from Equations (7–66) and (7–67), we obtain

$$\frac{s}{K}Y(s) = \frac{b}{a+b}E(s) - \frac{a}{a+b}Y(s)$$

or

$$\left(\frac{s}{K} + \frac{a}{a+b}\right)Y(s) = \frac{b}{a+b}E(s)$$

Hence,

$$\frac{Y(s)}{E(s)} = \frac{\dfrac{bK}{(a+b)s}}{1 + \dfrac{aK}{(a+b)s}} \tag{7–68}$$

Under normal operations of the system, $|Ka/[s(a+b)]| \gg 1$. Thus, Equation (7–68) can be simplified to

$$\frac{Y(s)}{E(s)} = \frac{b}{a} = K_p$$

Thus the transfer function between y and e becomes a constant. The hydraulic system shown in Figure 7–26 acts as a proportional controller, the gain of which is K_p. This gain can be adjusted by effectively changing the lever ratio b/a. (The adjusting mechanism is not shown in the diagram.)

Problem A–7–19

Consider the thermal system shown in Figure 7–27. Assume that the tank is insulated to eliminate heat loss to the surrounding air. Assume also that there is no heat storage in the insulation and that the liquid in the tank is perfectly mixed so that it is at a uniform temperature. Thus, a single temperature is used to describe both the temperature of the liquid in the tank and that of the outflowing liquid.

Let us define

$\overline{\Theta}_i$ = steady-state temperature of inflowing liquid, °C

$\overline{\Theta}_o$ = steady-state temperature of outflowing liquid, °C

G = steady-state liquid flow rate, kg/s

M = mass of liquid in tank, kg

c = specific heat of liquid, kcal/kg °C

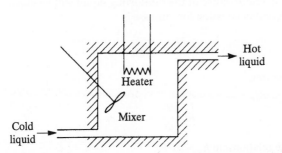

Figure 7–27 Thermal system.

R = thermal resistance, °C s/kcal

C = thermal capacitance, kcal/°C

$\overline{H}$ = steady-state heat input rate, kcal/s

Suppose that the system is subjected to changes in both the heat input rate and the temperature of the inflow liquid, while the liquid flow rate is kept constant. Define θ as the change in the temperature of the outflowing liquid when both the heat input rate and inflow liquid temperature are changed. Obtain a differential equation in θ.

Solution   The system is subjected to two inputs. In **Example 7–5**, we considered two inputs at the same time in deriving the system equation there. In the current example problem, we consider the two inputs independently. (This approach is valid for any linear system.) We shall first consider the change in the temperature of the outflowing liquid when the heat input rate is changed.

Assume that the temperature of the inflowing liquid is kept constant and that the heat input rate to the system (the heat supplied by the heater) is suddenly changed from $\overline{H}$ to $\overline{H} + h_i$, where h_i is small. The heat outflow rate will then change gradually from $\overline{H}$ to $\overline{H} + h_o$. The temperature of the outflowing liquid will also change, from $\overline{\Theta}_o$ to $\overline{\Theta}_o + \theta_1$. For this case,

$$h_o = Gc\theta_1$$
$$C = Mc$$
$$R = \frac{\theta_1}{h_o} = \frac{1}{Gc}$$

The differential equation for the system is

$$C\frac{d\theta_1}{dt} = h_i - h_o$$

which may be rewritten as

$$RC\frac{d\theta_1}{dt} + \theta_1 = Rh_i$$

Next, consider the change in the temperature of the outflowing liquid when the temperature of the inflowing liquid is changed. If the temperature of the inflowing liquid is suddenly changed from $\overline{\Theta}_i$ to $\overline{\Theta}_i + \theta_i$ while the heat input rate H and the liquid flow rate G are kept constant, then the heat outflow rate will be changed from $\overline{H}$ to $\overline{H} + h_o$, and the temperature of the outflowing liquid will be changed from $\overline{\Theta}_o$ to $\overline{\Theta}_o + \theta_2$. The differential equation for this case is

$$C\frac{d\theta_2}{dt} = Gc\theta_i - h_o$$

which may be rewritten

$$RC\frac{d\theta_2}{dt} + \theta_2 = \theta_i$$

where we used the relationship $h_o = Gc\theta_2$.

Since the present thermal system is subjected to changes in both the temperature of the inflow liquid and the heat input rate, the total change θ in the temperature of the outflowing liquid is the sum of the two individual changes, or $\theta = \theta_1 + \theta_2$. Thus, we obtain

$$RC\frac{d\theta}{dt} + \theta = \theta_i + Rh_i$$

Problem A–7–20

In the thermal system shown in Figure 7–28(a), it is assumed that the tank is insulated to eliminate heat loss to the surrounding air, that there is no heat storage in the insulation, and that the liquid in the tank is perfectly mixed so that it is at a uniform temperature. (Thus, a single temperature can be used to denote both the temperature of the liquid in the tank and that of the outflowing liquid.) It is further assumed that the flow rate of liquid into and out of the tank is constant and that the inflow temperature is constant at $\overline{\Theta}_i°$C. For $t < 0$, the system is at steady state and the heater supplies heat at the rate $\overline{H}$ J/s. At $t = 0$, the heat input rate is changed from $\overline{H}$ to $\overline{H} + h$ J/s. This change causes the outflow liquid temperature to change from $\overline{\Theta}_o$ to $\overline{\Theta}_o + \theta°$C. Suppose that the change in temperature, $\theta°$C, is the output and that the change in the heat input, h J/s, is the input to the system. Determine the transfer function $\Theta(s)/H(s)$, where $\Theta(s) = \mathcal{L}[\theta(t)]$ and $H(s) = \mathcal{L}[h(t)]$. Show that the thermal system is analogous to the electrical system of Figure 7–28(b), where voltage e_o is the output and current i is the input.

Solution Define

G = liquid flow rate, kg/s
c = specific heat of liquid, J/kg-K
M = mass of liquid in the tank, kg
R = thermal resistance, K-s/J
C = thermal capacitance, J/K
h_o = change in heat added to outflowing liquid, J/s

Then

$$C = Mc$$

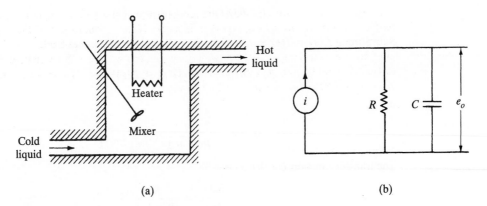

(a) (b)

Figure 7–28 (a) Thermal system; (b) analogous electrical system.

Note that

$$\overline{H} = Gc(\overline{\Theta}_o - \overline{\Theta}_i)$$

$$\overline{H} + h_o = Gc(\overline{\Theta}_o + \theta - \overline{\Theta}_i)$$

So we have

$$h_o = Gc\theta$$

Note also that

$$R = \frac{\theta}{h_o} = \frac{1}{Gc}$$

The heat balance equation is

$$C\, d\theta = (h - h_o)\, dt$$

or

$$C\frac{d\theta}{dt} = h - \frac{\theta}{R}$$

Thus,

$$RC\frac{d\theta}{dt} + \theta = Rh$$

and the transfer function is

$$\frac{\Theta(s)}{H(s)} = \frac{R}{RCs + 1}$$

For the electrical circuit shown in Figure 7–28(b), define the currents through resistance R and capacitance C as i_1 and i_2, respectively. Then the equation for the circuit becomes

$$Ri_1 = \frac{1}{C}\int i_2\, dt = e_o$$

The Laplace transform of this last equation, assuming a zero initial condition, is

$$RI_1(s) = \frac{1}{Cs}I_2(s) = E_o(s)$$

Substituting $I_2(s) = I(s) - I_1(s)$ into the preceding equation, we have

$$RI_1(s) = \frac{1}{Cs}[I(s) - I_1(s)]$$

or

$$RI_1(s) = \frac{R}{RCs + 1}I(s) = E_o(s)$$

The transfer function $E_o(s)/I(s)$ is

$$\frac{E_o(s)}{I(s)} = \frac{R}{RCs + 1}$$

Comparing the transfer function of the thermal system with that of the electrical system, we find the analogy apparent.

PROBLEMS

Problem B–7–1

For laminar flow through a cylindrical pipe, the relationship between the differential head h m and flow rate Q m³/s is given by the Hagen–Poiseuille formula

$$h = \frac{128vL}{g\pi D^4}Q$$

where

    v = kinematic viscosity, m²/s
    L = length of pipe, m
    D = diameter of pipe, m

Thus, the laminar-flow resistance R_l for the liquid flow through cylindrical pipes is given by

$$R_l = \frac{dh}{dQ} = \frac{128vL}{g\pi D^4}\ \text{s/m}^2$$

Now consider the flow of water through a capillary tube. Assuming that the temperature of the water is 20°C and that the flow is laminar, obtain the resistance R_l of the capillary tube. The kinematic viscosity v of water at a temperature of 20°C is 1.004×10^{-6} m²/s. Assume that the length L of the capillary tube is 2 m and the diameter is 4 mm.

Problem B–7–2

In the liquid-level system shown in Figure 7–29, the head is kept at 1 m for $t < 0$. The inflow valve opening is changed at $t = 0$, and the inflow rate is 0.05 m³/s for $t \geq 0$. Determine the time needed to fill the tank to a 2.5-m level. Assume that the outflow rate Q m³/s and head H m are related by

$$Q = 0.02\sqrt{H}$$

The capacitance of the tank is 2 m².

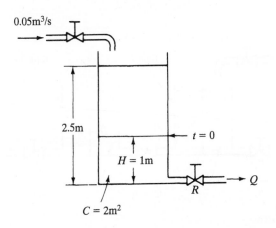

Figure 7–29 Liquid-level system.

Problem B-7-3

At steady state, the flow rate throughout the liquid-level system shown in Figure 7–30 is $\overline{Q}$, and the heads of tanks 1 and 2 are $\overline{H}_1$ and $\overline{H}_2$, respectively. At $t = 0$, the inflow rate is changed from $\overline{Q}$ to $\overline{Q} + q$, where q is small. The resulting changes in the heads (h_1 and h_2) and flow rates (q_1 and q_2) are assumed to be small as well. The capacitances of tanks 1 and 2 are C_1 and C_2, respectively. The resistance of the outflow valve of tank 1 is R_1 and that of tank 2 is R_2. Obtain the transfer function for the system when q is the input and q_2 the output.

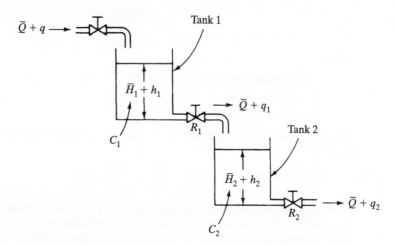

Figure 7–30   Liquid-level system.

Problem B-7-4

Consider the liquid-level system shown in Figure 7–31. At steady state, the inflow rate and outflow rate are both $\overline{Q}$ and the heads of tanks 1, 2, and 3 are $\overline{H}_1$, $\overline{H}_2$, and $\overline{H}_3$, respectively, where $\overline{H}_1 = \overline{H}_2$. At $t = 0$, the inflow rate is changed from $\overline{Q}$ to $\overline{Q} + q_i$. Assuming that h_1, h_2, and h_3 are small changes, obtain the transfer function $Q_o(s)/Q_i(s)$.

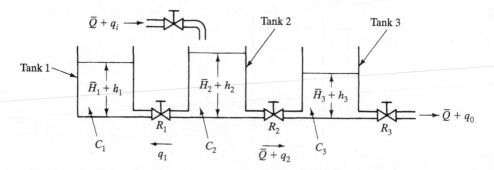

Figure 7–31   Liquid-level system.

Problem B–7–5

Consider the conical water tank system shown in Figure 7–32. The flow through the valve is turbulent and is related to the head H by

$$Q = 0.005\sqrt{H}$$

where Q is the flow rate measured in m³/s and H is in meters. Suppose that the head is 2 m at $t = 0$. What will be the head at $t = 60$ s?

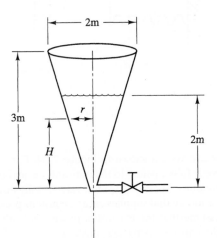

Figure 7–32   Conical water tank system.

Problem B–7–6

Obtain an electrical analog of the liquid-level system shown in Figure 7–30.

Problem B–7–7

Obtain an electrical analog of the liquid-level system shown in Figure 7–21 when q is the input and q_2 the output.

Problem B–7–8

Air is compressed into a tank of volume 10 m³. The pressure is 7×10^5 N/m² gage and the temperature is 20°C. Find the mass of air in the tank. If the temperature of the compressed air is raised to 40°C, what is the gage pressure of air in the tank in N/m², in kg$_f$/cm², and in lb$_f$/in.²?

Problem B–7–9

For the pneumatic system shown in Figure 7–33, assume that the steady-state values of the air pressure and the displacement of the bellows are $\overline{P}$ and $\overline{X}$, respectively. Assume also that the input pressure is changed from $\overline{P}$ to $\overline{P} + p_i$, where p_i is small. This change will cause the displacement of the bellows to change a small amount x. Assuming that the capacitance of the bellows is C and the resistance of the valve is R, obtain the transfer function relating x and p_i.

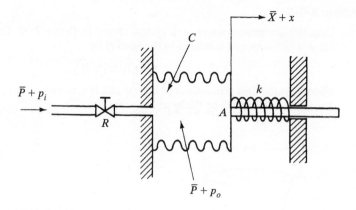

Figure 7–33   Pneumatic system.

Problem B–7–10

Consider the pneumatic pressure system shown in Figure 7–34. For $t < 0$, the inlet valve is closed, the outlet valve is fully opened to the atmosphere, and the pressure p_2 in the vessel is atmospheric pressure. At $t = 0$, the inlet valve is fully opened. The inlet pipe is connected to a pressure source that supplies air at a constant pressure p_1, where $p_1 = 0.5 \times 10^5 \text{ N/m}^2$ gage. Assume that the expansion process is isothermal ($n = 1$) and that the temperature of the entire system stays constant.

Determine the steady-state pressure p_2 in the vessel after the inlet valve is fully opened, assuming that the inlet and outlet valves are identical (i.e., both valves have identical flow characteristics).

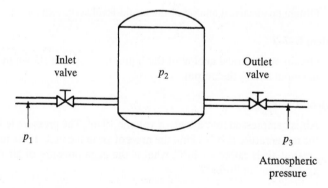

Figure 7–34   Pneumatic pressure system.

Problem B–7–11

Figure 7–35 shows a toggle joint. Show that

$$F = 2\frac{l_1}{l_2}R$$

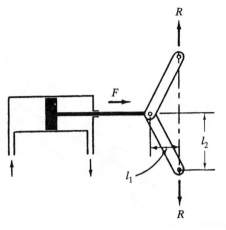

Figure 7–35 Toggle joint.

Problem B–7–12

Consider the pneumatic system shown in Figure 7–36. The load consists of a mass m and friction. The frictional force is assumed to be $\mu N = \mu mg$. If $m = 1000$ kg, $\mu = 0.3$, and $p_1 - p_2 = 5 \times 10^5$ N/m^2, find the minimum area of the piston needed if the load is to be moved. Note that the frictional force $\mu\, mg$ acts in the direction opposite to the intended direction of motion.

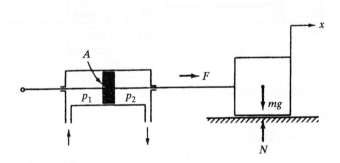

Figure 7–36 Pneumatic system.

Problem B–7–13

In the system of Figure 7–37, a mass m is to be pushed upward along the inclined plane by the pneumatic cylinder. The friction force μN is acting opposite to the direction of motion or intended motion. If the load is to be moved, show that the area A of the piston must not be smaller than

$$\frac{mg\,\sin(\theta + \alpha)}{(p_1 - p_2)\cos\theta}$$

where $\theta = \tan^{-1}\mu$ and α is the angle of inclination of the plane.

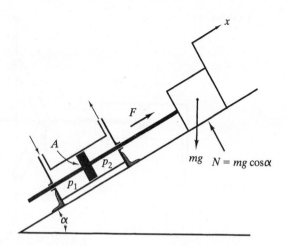

Figure 7–37 Pneumatic system.

Problem B–7–14

The system shown in Figure 7–38 consists of a power cylinder and a rack-and-pinion mechanism to drive the load. Power piston D moves rack C, which, in turn, causes pinion B to rotate on rack A. Find the displacement y of the output when the displacement of the power piston is x.

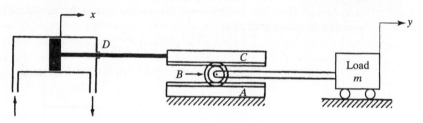

Figure 7–38 Pneumatic system.

Problem B–7–15

Obtain a linear approximation of

$$Q = 0.1\sqrt{H} = f(H)$$

about the operating point $H = 4, Q = 0.2$.

Problem B–7–16

Find a linearized equation of

$$z = 5x^2$$

about the point $x = 2, z = 20$.

Problem B–7–17

Linearize the nonlinear equation

$$z = x^2 + 2xy + 5y^2$$

in the region defined by $10 \le x \le 12, 4 \le y \le 6$.

Problem B–7–18

Pascal's law states that the pressure at any point in a static liquid is the same in every direction and exerts equal force on equal areas. Examine Figure 7–39. If a force of P_1 is applied to the left-hand-side piston, find the force P_2 acting on the right-hand-side piston. Also, find the distance x_2 m traveled by the piston on the right-hand side when the one on the left-hand side is moved by x_1 m.

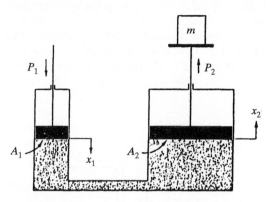

Figure 7–39 Hydraulic system.

Problem B–7–19

Figure 7–40 is a schematic diagram of an aircraft elevator control system. The input to the system is the deflection angle θ of the control lever, and the output is the elevator angle ϕ. Assume that angles θ and ϕ are relatively small. Show that, for each angle θ of the control lever, there is a corresponding (steady-state) elevator angle ϕ.

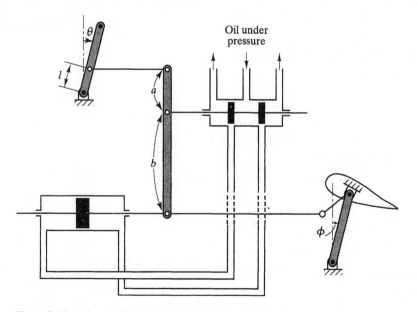

Figure 7–40 Aircraft elevator control system.

Problem B–7–20

Consider the thermal system shown in Figure 7–41. The temperature of the inflow liquid is kept constant, and the liquid inflow rate G kg/s is also kept constant. For $t < 0$, the system is at steady state, wherein the heat input from the heater is $\overline{U}$ kcal/s and the temperature of the liquid in tank 2 is $\overline{\Theta}_2°$C. At $t = 0$, the heat input is changed from $\overline{U}$ to $\overline{U} + u$, where u is small. This change will cause the temperature of the liquid in tank 2 to change from $\overline{\Theta}_2$ to $\overline{\Theta}_2 + \theta_2$. Taking the change in the heat input from the heater to be the input to the system and the change in the temperature of the liquid in tank 2 to be the output, obtain the transfer function $\Theta_2(s)/U(s)$. Assume that the thermal capacitances of tanks 1 and 2 are C_1 kcal/°C and C_2 kcal/°C, respectively, and that the specific heat of the liquid is c kcal/kg°C. The steady-state heat flow rates from tanks 1 and 2 are the same, $\overline{Q}$ kcal/s. The changes in heat flow rates from tanks 1 and 2 are q_1, and q_2, respectively.

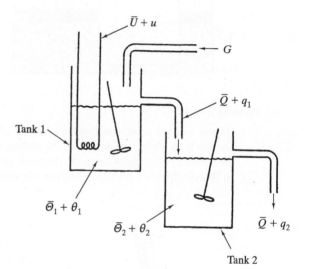

Figure 7–41 Thermal system.

Time-Domain Analysis of Dynamic Systems

8–1 INTRODUCTION

This chapter deals primarily with the transient-response analysis of dynamic systems and obtains analytical solutions giving the responses. The chapter also derives an analytical solution of the state equation when the input is a step, impulse, or ramp function. (The method can be extended to obtain an analytical solution for any time-domain input.)

Natural and forced responses. Consider a system defined by a differential equation, for instance,

$$\overset{(n)}{x} + a_1 \overset{(n-1)}{x} + \cdots + a_{n-1}\dot{x} + a_n x = p(t) \qquad (8\text{–}1)$$

where the coefficients $a_1, a_2, \ldots, a_n$ are constants, $x(t)$ is the dependent variable, t is the independent variable, and $p(t)$ is the input function.

The differential equation (8–1) has a complete solution $x(t)$ composed of two parts: the complementary solution $x_c(t)$ and the particular solution $x_p(t)$. The complementary solution $x_c(t)$ is found by equating the right-hand side of Equation (8–1) to zero and solving the associated homogeneous differential equation. The particular solution $x_p(t)$ depends on the functional form of the input function $p(t)$.

If the complementary solution $x_c(t)$ approaches zero or a constant value as time t approaches infinity and if $\lim_{t\to\infty} x_p(t)$ is a bounded function of time, the system is said to be in a *steady state*.

Customarily, engineers call the complementary solution $x_c(t)$ and the particular solution $x_p(t)$ the *natural* and *forced responses*, respectively. Although the natural behavior of a system is not itself a response to any external or input function, a study of this type of behavior will reveal characteristics that will be useful in predicting the forced response as well.

Transient response and steady-state response.   Both the natural and forced responses of a dynamic system consist of two parts: the transient response and the steady-state response. *Transient response* refers to the process generated in going from the initial state to the final state. By *steady-state response*, we mean the way in which the system output behaves as t approaches infinity. The transient response of a dynamic system often exhibits damped vibrations before reaching a steady state.

Outline of the chapter.   Section 8–1 has presented introductory material. Section 8–2 deals with the transient-response analysis of first-order systems subjected to step and ramp inputs. Section 8–3 begins with the transient-response analysis of second-order systems subjected to initial conditions only. A discussion of the transient response of such systems to step inputs then follows. Section 8–4 treats higher order systems. Finally, Section 8–5 presents an analytical solution of the state-space equation.

8–2 TRANSIENT-RESPONSE ANALYSIS OF FIRST-ORDER SYSTEMS

From time to time in Chapters 2 through 7, we analyzed the transient response of several first-order systems. Essentially, this section is a systematic review of the transient response analysis of first-order systems. In the current section, we consider a thermal system (a thin, glass-walled mercury thermometer system) as an example of a first-order system. We shall find the system's response to step and ramp inputs. Then we point out that the mathematical results obtained can be applied to any physical or nonphysical system having the same mathematical model.

Step response of first-order system.   For the thin, glass-walled mercury thermometer system shown in Figure 8–1, assume that the thermometer is at the ambient temperature $\overline{\Theta}°C$ and that at $t = 0$ it is immersed in a water bath of temperature $\overline{\Theta} + \theta_b°C$. ($\theta_b$ is the difference between the temperature of the bath and the ambient temperature.) Let us define the instantaneous thermometer temperature as $\overline{\Theta} + \theta°C$. [Note that θ is the change in the thermometer temperature satisfying the condition $\theta(0) = 0$.] We shall find the response $\theta(t)$ when the bath temperature is constant, or θ_b is constant.

We presented a mathematical model of this system in Section 7-6. The basic equation for the heat balance for this system is

$$C\, d\theta = q\, dt \tag{8–2}$$

where C (kcal/°C) is the thermal capacitance of the thermometer and q (kcal/s) is the heat input to the thermometer. The heat input q (kcal/s) can be given in terms of

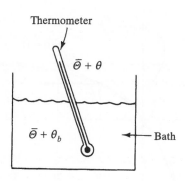

Thermometer

$\bar{\Theta} + \theta$

$\bar{\Theta} + \theta_b$

Bath

Figure 8–1   Thin, glass-walled mercury thermometer system.

the thermal resistance R (°C/kcal/s) as

$$q = \frac{\theta_b - \theta}{R} \tag{8–3}$$

Substituting Equation (8–3) into Equation (8–2), we obtain

$$C \, d\theta = \frac{\theta_b - \theta}{R} dt$$

or

$$T \frac{d\theta}{dt} + \theta = \theta_b \tag{8–4}$$

where $T = RC$ = time constant. Equation (8–4) is a mathematical model of the thermometer system.

To obtain the step response of this system, we first take the Laplace transform of Equation (8–4):

$$T[s\Theta(s) - \theta(0)] + \Theta(s) = \Theta_b(s)$$

Since $\theta(0) = 0$, this last equation simplifies to

$$\Theta(s) = \frac{1}{Ts + 1} \Theta_b(s) \tag{8–5}$$

Note that, for θ_b = constant, we have

$$\Theta_b(s) = \frac{\theta_b}{s}$$

Hence, Equation (8–5) becomes

$$\Theta(s) = \frac{1}{Ts + 1} \frac{\theta_b}{s} = \left[\frac{1}{s} - \frac{1}{s + (1/T)} \right] \theta_b$$

The inverse Laplace transform of this last equation gives

$$\theta(t) = (1 - e^{-t/T})\theta_b \tag{8–6}$$

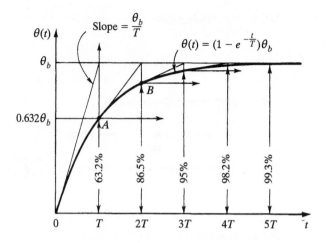

Figure 8–2 Step response curve for the first-order system.

The response curve $\theta(t)$ versus t is shown in Figure 8–2. Equation (8–6) states that, initially, the response $\theta(t)$ is zero and that it finally becomes θ_b. (There is no steady-state error.) One important characteristic of such an exponential response curve is that at $t = T$ the value of $\theta(t)$ is $0.632\theta_b$, or the response $\theta(t)$ has reached 63.2% of its total change. This fact can readily be seen by substituting $e^{-1} = 0.368$ into the equation.

Another important property of the exponential response curve is that the slope of the tangent line at $t = 0$ is θ_b/T, since

$$\left.\frac{d\theta}{dt}\right|_{t=0} = \left.\frac{\theta_b}{T}e^{-t/T}\right|_{t=0} = \frac{\theta_b}{T}$$

The response would reach the final value at $t = T$ if it maintained its initial speed. The slope of the response curve $\theta(t)$ decreases monotonically from θ_b/T at $t = 0$ to zero at $t = \infty$.

Figure 8–2 shows that in one time constant the exponential response curve has gone from zero to 63.2% of the total change. In two time constants, the response reaches 86.5% of the total change. At $t = 3T$, $4T$, and $5T$, the response $\theta(t)$ reaches 95, 98.2, and 99.3% of the total change, respectively. So, for $t \geq 4T$, the response remains within 2% of the final value. As can be seen from Equation (8–6), the steady state is reached mathematically only after an infinite time. In practice, however, a reasonable estimate of the response time is the length of time that the response curve needs to reach the 2% line of the final value, or four time constants.

Ramp response of first-order system. Consider again the thermometer system shown in Figure 8–1. Assume that, for $t < 0$, both the bath temperature and the thermometer temperature are in a steady state at the ambient temperature $\overline{\Theta}°C$ and that, for $t \geq 0$, heat is added to the bath and the bath temperature changes linearly at the rate of $r°C/s$; that is,

$$\theta_b(t) = rt$$

Let us derive the ramp response $\theta(t)$.

First, note that

$$\Theta_b(s) = \mathcal{L}[\theta_b(t)] = \mathcal{L}[rt] = \frac{r}{s^2}$$

Substituting this equation into Equation (8–5), we find that

$$\Theta(s) = \frac{1}{Ts+1}\frac{r}{s^2} = r\left[\frac{1}{s^2} - \frac{T}{s} + \frac{T}{s+(1/T)}\right]$$

The inverse Laplace transform of this last equation gives

$$\theta(t) = r(t - T + Te^{-t/T}) \qquad t \geq 0 \tag{8–7}$$

The error $e(t)$ between the actual bath temperature and the indicated thermometer temperature is

$$e(t) = rt - \theta(t) = rT(1 - e^{-t/T})$$

As t approaches infinity, $e^{-t/T}$ approaches zero. Thus, the error $e(t)$ approaches rT, or

$$e(\infty) = rT$$

The ramp input rt and response $\theta(t)$ versus t are shown in Figure 8–3. The error in following the ramp input is equal to rT for sufficiently large t. The smaller the time constant T, the smaller is the steady-state error in following the ramp input.

Comments.   Since the mathematical analysis does not depend on the physical structure of the system, the preceding results for step and ramp responses can be applied to any systems having the mathematical model

$$T\frac{dx_o}{dt} + x_o = x_i \tag{8–8}$$

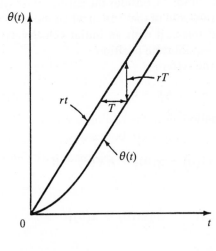

0

t

Figure 8–3   Ramp response curve for the first-order system.

where

T = time constant of the system
x_i = input or forcing function
x_o = output or response function

From Equation (8–6), for a step input $x_i(t) = r \cdot 1(t)$, any system described by Equation (8–8) will exhibit the following response:

$$x_o(t) = (1 - e^{-t/T})r$$

Similarly, for a ramp input $x_i(t) = rt \cdot 1(t)$, any system described by Equation (8–8) will exhibit the following response [see Equation (8–7)]:

$$x_o(t) = r(t - T + Te^{-t/T})$$

Many physical systems have the mathematical model given by Equation (8–8); Table 8–1 shows several such systems, which are analogous. All analogous systems exhibit the same response to the same input function.

8–3 TRANSIENT-RESPONSE ANALYSIS OF SECOND-ORDER SYSTEMS

Let us next consider the transient-response analysis of second-order systems such as a spring–mass system and a spring–mass–dashpot system. The results obtained can be applied to the response of any analogous systems.

We first discuss the free vibration of a spring–mass system and then treat the free vibration of a spring–mass–dashpot system. Since the step response of the second-order system is discussed fully in Chapter 10, we shall not present the details of such a response here. Instead, we shall treat only illustrative examples of the step responses of second-order systems with and without damping.

Free vibration without damping.   Consider the spring–mass system shown in Figure 8–4. We shall obtain the response of the system when the mass is displaced downward by a distance $x(0)$ and released with an initial velocity $\dot{x}(0)$. The displacement x is measured from the equilibrium position.

The mathematical model of the system is

$$m\ddot{x} + kx = 0$$

The solution of the preceding equation gives the response $x(t)$. To solve this differential equation, let us take Laplace transforms of both sides:

$$m[s^2X(s) - sx(0) - \dot{x}(0)] + kX(s) = 0$$

which can be rewritten as

$$(ms^2 + k)X(s) = m[sx(0) + \dot{x}(0)]$$

TABLE 8–1 Examples of Physical Systems Having a Mathematical
Model of the Form $T(dx_0/dt) + x_0 = x_i$

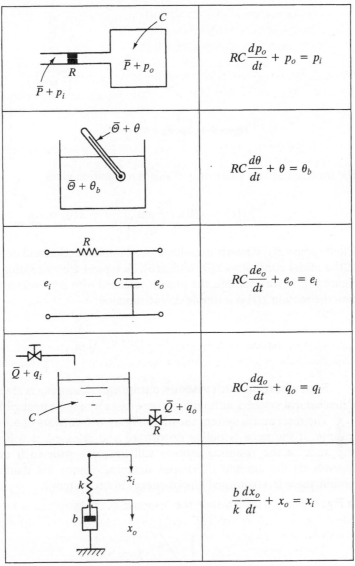

	$RC\dfrac{dp_o}{dt} + p_o = p_i$
	$RC\dfrac{d\theta}{dt} + \theta = \theta_b$
	$RC\dfrac{de_o}{dt} + e_o = e_i$
	$RC\dfrac{dq_o}{dt} + q_o = q_i$
	$\dfrac{b}{k}\dfrac{dx_o}{dt} + x_o = x_i$

Solving for $X(s)$ yields

$$X(s) = \frac{sx(0) + \dot{x}(0)}{s^2 + (k/m)}$$

$$= \frac{\dot{x}(0)}{\sqrt{k/m}} \frac{\sqrt{k/m}}{s^2 + (\sqrt{k/m})^2} + \frac{x(0)s}{s^2 + (\sqrt{k/m})^2}$$

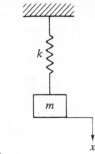

Figure 8-4   Spring–mass system.

The inverse Laplace transform of this last equation gives

$$x(t) = \dot{x}(0)\sqrt{\frac{m}{k}} \sin\sqrt{\frac{k}{m}}t + x(0)\cos\sqrt{\frac{k}{m}}t$$

The response $x(t)$ consists of a sine and a cosine function and depends on the values of the initial conditions $x(0)$ and $\dot{x}(0)$. A typical free-vibration curve is shown in Figure 8–5. If, for example, the mass is released with zero velocity, so that $\dot{x}(0) = 0$, then the motion $x(t)$ is a simple cosine function:

$$x(t) = x(0)\cos\sqrt{\frac{k}{m}}t$$

Free vibration with viscous damping.   Damping is always present in actual mechanical systems, although in some cases it may be negligibly small.

The mechanical system shown in Figure 8–6 consists of a mass, a spring, and a dashpot. If the mass is pulled downward and released, it will vibrate freely. The amplitude of the resulting motion will decrease with each cycle at a rate that depends on the amount of viscous damping. (Since the damping force opposes motion, there is a continual loss of energy in the system.)

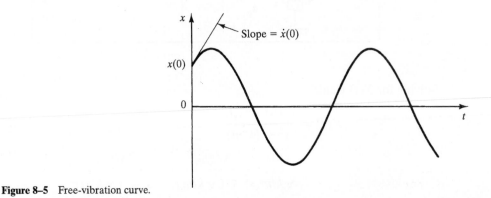

Figure 8-5   Free-vibration curve.

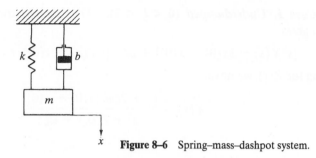

Figure 8–6 Spring–mass–dashpot system.

The mathematical model of this system is

$$m\ddot{x} + b\dot{x} + kx = 0 \qquad (8\text{–}9)$$

where the displacement x is measured from the equilibrium position.

The characteristic of the natural response of a second-order system like this one is determined by the roots of the characteristic equation

$$ms^2 + bs + k = 0$$

The two roots of this equation are

$$s = \frac{-b \pm \sqrt{b^2 - 4mk}}{2m}$$

If the damping coefficient b is small, so that $b^2 < 4mk$, the roots of the characteristic equation are complex conjugates. The natural response is then an exponentially decaying sinusoid, and the system is said to be *underdamped*.

If the damping coefficient b is increased, a point will be reached at which $b^2 = 4mk$. When the damping has reached this value ($b = 2\sqrt{mk}$), the two roots of the characteristic equation become real and equal. The system is then said to be *critically damped*.

If the damping coefficient b is increased further, so that $b^2 > 4mk$, the two roots are real and distinct. The response is the sum of two decaying exponentials, and the system is said to be *overdamped*.

In solving Equation (8–9) for the response $x(t)$, it is convenient to define

$$\omega_n = \sqrt{\frac{k}{m}} = \text{undamped natural frequency, rad/s}$$

$$\zeta = \text{damping ratio} = \frac{\text{actual damping value}}{\text{critical damping value}} = \frac{b}{2\sqrt{km}}$$

and rewrite Equation (8–9) as follows:

$$\ddot{x} + 2\zeta\omega_n\dot{x} + \omega_n^2 x = 0 \qquad (8\text{–}10)$$

In what follows, we shall use Equation (8–10) as the system equation and derive the response $x(t)$ for three cases: the underdamped case ($0 < \zeta < 1$), the overdamped case ($\zeta > 1$), and the critically damped case ($\zeta = 1$).

Case 1. Underdamped ($0 < \zeta < 1$). The Laplace transform of Equation (8–10) gives

$$[s^2 X(s) - sx(0) - \dot{x}(0)] + 2\zeta\omega_n[sX(s) - x(0)] + \omega_n^2 X(s) = 0$$

Solving for $X(s)$, we have

$$X(s) = \frac{(s + 2\zeta\omega_n)x(0) + \dot{x}(0)}{s^2 + 2\zeta\omega_n s + \omega_n^2} \tag{8–11}$$

or

$$X(s) = \frac{\zeta\omega_n x(0) + \dot{x}(0)}{\omega_n\sqrt{1 - \zeta^2}} \frac{\omega_n\sqrt{1 - \zeta^2}}{(s + \zeta\omega_n)^2 + (\omega_n\sqrt{1 - \zeta^2})^2}$$

$$+ \frac{(s + \zeta\omega_n)x(0)}{(s + \zeta\omega_n)^2 + (\omega_n\sqrt{1 - \zeta^2})^2}$$

The inverse Laplace transform of this last equation gives

$$x(t) = \frac{\zeta\omega_n x(0) + \dot{x}(0)}{\omega_n\sqrt{1 - \zeta^2}} e^{-\zeta\omega_n t} \sin \omega_n\sqrt{1 - \zeta^2}t$$

$$+ x(0)e^{-\zeta\omega_n t} \cos \omega_n\sqrt{1 - \zeta^2}t$$

Next, we define

$$\omega_d = \omega_n\sqrt{1 - \zeta^2} = \text{damped natural frequency, rad/s}$$

Then the response $x(t)$ is given by

$$x(t) = e^{-\zeta\omega_n t}\left\{\left[\frac{\zeta}{\sqrt{1 - \zeta^2}}x(0) + \frac{1}{\omega_d}\dot{x}(0)\right]\sin \omega_d t + x(0)\cos \omega_d t\right\} \tag{8–12}$$

If the initial velocity is zero, or $\dot{x}(0) = 0$, Equation (8–12) simplifies to

$$x(t) = x(0)e^{-\zeta\omega_n t}\left(\frac{\zeta}{\sqrt{1 - \zeta^2}}\sin \omega_d t + \cos \omega_d t\right) \tag{8–13}$$

or

$$x(t) = \frac{x(0)}{\sqrt{1 - \zeta^2}}e^{-\zeta\omega_n t}\sin\left(\omega_d t + \tan^{-1}\frac{\sqrt{1 - \zeta^2}}{\zeta}\right)$$

$$= \frac{x(0)}{\sqrt{1 - \zeta^2}}e^{-\zeta\omega_n t}\cos\left(\omega_d t - \tan^{-1}\frac{\zeta}{\sqrt{1 - \zeta^2}}\right) \tag{8–14}$$

Notice that, in the present case, the damping introduces the term $e^{-\zeta\omega_n t}$ as a multiplicative factor. This factor is a decreasing exponential and becomes smaller and smaller as time increases, thus causing the amplitude of the harmonic motion to decrease with time.

Case 2. Overdamped ($\zeta > 1$).   Here, the two roots of the characteristic equation are real, so Equation (8–11) can be written

$$X(s) = \frac{(s + 2\zeta\omega_n)x(0) + \dot{x}(0)}{(s + \zeta\omega_n + \omega_n\sqrt{\zeta^2 - 1})(s + \zeta\omega_n - \omega_n\sqrt{\zeta^2 - 1})}$$

$$= \frac{\hat{a}}{s + \zeta\omega_n + \omega_n\sqrt{\zeta^2 - 1}} + \frac{\hat{b}}{s + \zeta\omega_n - \omega_n\sqrt{\zeta^2 - 1}}$$

where

$$\hat{a} = \frac{(-\zeta + \sqrt{\zeta^2 - 1})x(0)}{2\sqrt{\zeta^2 - 1}} - \frac{\dot{x}(0)}{2\omega_n\sqrt{\zeta^2 - 1}}$$

$$\hat{b} = \frac{(\zeta + \sqrt{\zeta^2 - 1})x(0)}{2\sqrt{\zeta^2 - 1}} + \frac{\dot{x}(0)}{2\omega_n\sqrt{\zeta^2 - 1}}$$

The inverse Laplace transform of $X(s)$ gives the following response:

$$x(t) = \hat{a}e^{-(\zeta\omega_n + \omega_n\sqrt{\zeta^2 - 1})t} + \hat{b}e^{-(\zeta\omega_n - \omega_n\sqrt{\zeta^2 - 1})t}$$

$$= \left[\frac{(-\zeta + \sqrt{\zeta^2 - 1})x(0)}{2\sqrt{\zeta^2 - 1}} - \frac{\dot{x}(0)}{2\omega_n\sqrt{\zeta^2 - 1}}\right]e^{-(\zeta\omega_n + \omega_n\sqrt{\zeta^2 - 1})t}$$

$$+ \left[\frac{(\zeta + \sqrt{\zeta^2 - 1})x(0)}{2\sqrt{\zeta^2 - 1}} + \frac{\dot{x}(0)}{2\omega_n\sqrt{\zeta^2 - 1}}\right]e^{-(\zeta\omega_n - \omega_n\sqrt{\zeta^2 - 1})t}$$

Notice that both terms on the right-hand side of this last equation decrease exponentially. The motion of the mass in this case is a gradual creeping back to the equilibrium position.

Case 3. Critically damped ($\zeta = 1$).   In reality, all systems have a damping ratio greater or less than unity, and $\zeta = 1$ rarely occurs in practice. Nevertheless, the case $\zeta = 1$ is useful as a mathematical reference. (The response does not exhibit any vibration, but it is the fastest among such nonvibratory motions.)

In the critically damped case, the damping ratio ζ is equal to unity. So the two roots of the characteristic equation are the same and are equal to the negative of the natural frequency ω_n. Equation (8–11) can, therefore, be written

$$X(s) = \frac{(s + 2\omega_n)x(0) + \dot{x}(0)}{s^2 + 2\omega_n s + \omega_n^2}$$

$$= \frac{(s + \omega_n)x(0) + \omega_n x(0) + \dot{x}(0)}{(s + \omega_n)^2}$$

$$= \frac{x(0)}{s + \omega_n} + \frac{\omega_n x(0) + \dot{x}(0)}{(s + \omega_n)^2}$$

The inverse Laplace transform of this last equation gives

$$x(t) = x(0)e^{-\omega_n t} + [\omega_n x(0) + \dot{x}(0)]te^{-\omega_n t}$$

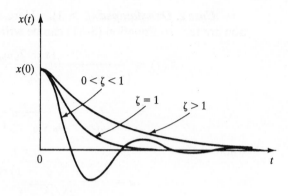

Figure 8–7 Typical response curves of the spring–mass–dashpot system.

The response $x(t)$ is similar to that found for the overdamped case. The mass, when displaced and released, will return to the equilibrium position without vibration.

Figure 8–7 shows the response $x(t)$ versus t for the three cases (underdamped, critically damped, and overdamped) with initial conditions $x(0) \neq 0$ and $\dot{x}(0) = 0$.

Experimental determination of damping ratio.   It is sometimes necessary to determine the damping ratios and damped natural frequencies of recorders and other instruments. To determine the damping ratio and damped natural frequency of a system experimentally, a record of decaying or damped oscillations, such as that shown in Figure 8–8, is needed. (Such an oscillation may be recorded by giving the system any convenient initial conditions.)

The period of oscillation, T, can be measured directly from crossing points on the zero axis, as shown in Figure 8–8.

To determine the damping ratio ζ from the rate of decay of the oscillation, we measure amplitudes; that is, at time $t = t_1$ we measure the amplitude x_1, and at time $t = t_1 + (n-1)T$ we measure the amplitude x_n. Note that it is necessary to choose n large enough so that x_n/x_1 is not near unity. Since the decay in amplitude from one cycle to the next may be represented as the ratio of the exponential multiplying factors at times t_1 and $t_1 + T$, we obtain, from Equation (8–12),

$$\frac{x_1}{x_2} = \frac{e^{-\zeta\omega_n t_1}}{e^{-\zeta\omega_n(t_1+T)}} = \frac{1}{e^{-\zeta\omega_n T}} = e^{\zeta\omega_n T}$$

Similarly,

$$\frac{x_1}{x_n} = \frac{1}{e^{-\zeta\omega_n(n-1)T}} = e^{(n-1)\zeta\omega_n T}$$

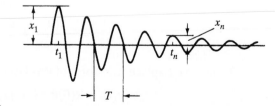

Figure 8–8   Decaying oscillation.

The logarithm of the ratio of succeeding amplitudes is called the *logarithmic decrement*. Thus,

$$\text{logarithmic decrement} = \ln\frac{x_1}{x_2} = \frac{1}{n-1}\left(\ln\frac{x_1}{x_n}\right) = \zeta\omega_n T$$

$$= \zeta\omega_n\frac{2\pi}{\omega_d} = \frac{2\pi\zeta}{\sqrt{1-\zeta^2}}$$

Once the amplitudes x_1 and x_n are measured and the logarithmic decrement is calculated, the damping ratio ζ is found from

$$\frac{1}{n-1}\left(\ln\frac{x_1}{x_n}\right) = \frac{2\pi\zeta}{\sqrt{1-\zeta^2}}$$

or

$$\zeta = \frac{\dfrac{1}{n-1}\left(\ln\dfrac{x_1}{x_n}\right)}{\sqrt{4\pi^2 + \left[\dfrac{1}{n-1}\left(\ln\dfrac{x_1}{x_n}\right)\right]^2}} \tag{8-15}$$

Note that this equation is valid only for the system described by Equation (8–10).

Example 8–1

In the system shown in Figure 8–6, assume that $m = 1\,\text{kg}$, $b = 2\,\text{N-s/m}$, and $k = 100\,\text{N/m}$. The mass is displaced 0.05 m and released without initial velocity. (The displacement x is measured from the equilibrium position.) Find the frequency observed in the vibration. In addition, find the amplitude four cycles later.

The equation of motion for the system is

$$m\ddot{x} + b\dot{x} + kx = 0$$

Substituting the numerical values for m, b, and k into this equation gives

$$\ddot{x} + 2\dot{x} + 100x = 0$$

where the initial conditions are $x(0) = 0.05$ and $\dot{x}(0) = 0$. From the system equation, the undamped natural frequency ω_n and the damping ratio ζ are respectively found to be

$$\omega_n = 10, \qquad \zeta = 0.1$$

The frequency actually observed in the vibration is the damped natural frequency ω_d:

$$\omega_d = \omega_n\sqrt{1-\zeta^2} = 10\sqrt{1-0.01} = 9.95\,\text{rad/s}$$

In the present analysis, $\dot{x}(0)$ is given as zero. So, from Equation (8–13), the solution can be written as

$$x(t) = x(0)e^{-\zeta\omega_n t}\left(\frac{\zeta}{\sqrt{1-\zeta^2}}\sin\omega_d t + \cos\omega_d t\right)$$

It follows that at $t = nT$, where $T = 2\pi/\omega_d$,

$$x(nT) = x(0)e^{-\zeta\omega_n nT}$$

Consequently, the amplitude four cycles later becomes

$$x(4T) = x(0)e^{-\zeta\omega_n 4T} = x(0)e^{-(0.1)(10)(4)(0.63)}$$
$$= 0.05e^{-2.52} = 0.05 \times 0.0804 = 0.00402 \text{ m}$$

Estimate of response time.   The mass of the mechanical system shown in Figure 8–6 is displaced $x(0)$ and released without initial velocity. The response is given by Equation (8–14), rewritten thus:

$$x(t) = \frac{x(0)}{\sqrt{1 - \zeta^2}} e^{-\zeta\omega_n t} \cos\left(\omega_d t - \tan^{-1}\frac{\zeta}{\sqrt{1 - \zeta^2}}\right)$$

A typical response curve is shown in Figure 8–9. Note that such a response curve is tangent to the envelope exponentials $\pm[x(0)/\sqrt{1 - \zeta^2}]e^{-\zeta\omega_n t}$. The time constant T of these exponential curves is $1/(\zeta\omega_n)$.

The fact that the response curve $x(t)$ is tangent to the exponential curves enables us to estimate the response time of a second-order system such as that shown in Figure 8–6 in terms of the settling time t_s defined by

$$t_s = 4T = \frac{4}{\zeta\omega_n}$$

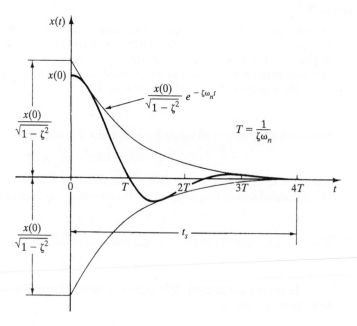

Figure 8–9   Typical response curve of the system shown in Figure 8–6 and its envelope exponentials.

The settling time t_s can be considered an approximate response time of the system, since, for $t > t_s$, the response curve remains within 2% of the final value or 2% of the total change.

Comments.   The preceding analysis, as well as the results derived, can be applied to any analogous systems having mathematical models of the form of Equation (8–10).

Step-response of second-order system.   Consider the mechanical system shown in Figure 8–10. Assume that the system is at rest for $t < 0$. At $t = 0$, force $u = a \cdot 1(t)$ [where a is a constant and $1(t)$ is a step force of magnitude 1 newton] is applied to the mass m. The displacement x is measured from the equilibrium position before the input force u is applied. Assume that the system is underdamped.

The equation of motion for the system is

$$m\ddot{x} + b\dot{x} + kx = u = a \cdot 1(t)$$

The transfer function for the system is

$$\frac{X(s)}{U(s)} = \frac{1}{ms^2 + bs + k}$$

Hence

$$\frac{X(s)}{\mathcal{L}[1(t)]} = \frac{a}{ms^2 + bs + k} = \frac{\dfrac{a}{m}}{s^2 + \dfrac{b}{m}s + \dfrac{k}{m}}$$

Let us define

$$\omega_n = \sqrt{\frac{k}{m}}, \qquad \zeta = \frac{b}{2\sqrt{km}}$$

Then

$$\frac{X(s)}{\mathcal{L}[1(t)]} = \frac{a}{m\omega_n^2}\left(\frac{\omega_n^2}{s^2 + 2\zeta\omega_n s + \omega_n^2}\right) \tag{8–16}$$

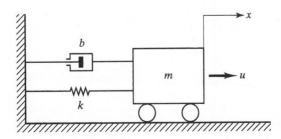

Figure 8–10   Mechanical system.

Hence,

$$X(s) = \frac{a}{m\omega_n^2}\left(\frac{\omega_n^2}{s^2 + 2\zeta\omega_n s + \omega_n^2}\frac{1}{s}\right)$$

$$= \frac{a}{m\omega_n^2}\left(\frac{1}{s} - \frac{s + 2\zeta\omega_n}{s^2 + 2\zeta\omega_n s + \omega_n^2}\right)$$

$$= \frac{a}{m\omega_n^2}\left[\frac{1}{s} - \frac{\zeta\omega_n}{(s + \zeta\omega_n)^2 + \omega_d^2} - \frac{s + \zeta\omega_n}{(s + \zeta\omega_n)^2 + \omega_d^2}\right]$$

where $\omega_d = \omega_n\sqrt{1 - \zeta^2}$. The inverse Laplace transform of this last equation gives

$$x(t) = \frac{a}{m\omega_n^2}\left(1 - \frac{\zeta}{\sqrt{1 - \zeta^2}}e^{-\zeta\omega_n t}\sin\omega_d t - e^{-\zeta\omega_n t}\cos\omega_d t\right)$$

$$= \frac{a}{m\omega_n^2}\left[1 - e^{-\zeta\omega_n t}\left(\frac{\zeta}{\sqrt{1 - \zeta^2}}\sin\omega_d t + \cos\omega_d t\right)\right]$$

$$= \frac{a}{m\omega_n^2}\left[1 - \frac{e^{-\zeta\omega_n t}}{\sqrt{1 - \zeta^2}}\sin\left(\omega_d t + \tan^{-1}\frac{\sqrt{1 - \zeta^2}}{\zeta}\right)\right] \qquad (8\text{–}17)$$

The response starts from $x(0) = 0$ and reaches $x(\infty) = a/(m\omega_n^2)$. The general shape of the response curve is shown in Figure 8–11.

Note that Equation (8–17) is an analytical solution for the step response of the system. If the numerical values of m, b, k, and a are given, an exact response curve can be plotted easily with MATLAB.

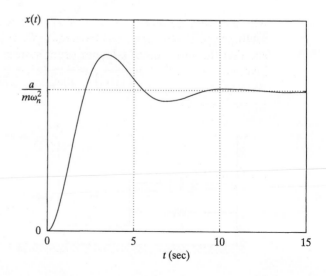

Figure 8–11 Step response of second-order system. (The response curve shown corresponds to the case where $\zeta = 0.4$ and $\omega_n = 1$ rad/s.)

8–4 TRANSIENT-RESPONSE ANALYSIS OF HIGHER ORDER SYSTEMS

Consider an nth-order $(n \geq 3)$ dynamic system defined by

$$\frac{X(s)}{U(s)} = \frac{b_0 s^m + b_1 s^{m-1} + \cdots + b_{n-1}s + b_n}{s^n + a_1 s^{n-1} + \cdots + a_{n-1}s + a_n} \tag{8–18}$$

where $m \leq n$. The transient response of this system to any given time-domain input can be obtained easily by a computer simulation. If an analytical expression for the transient response is desired, then it is necessary to expand $X(s)/U(s)$ into partial fractions. (Use MATLAB command residue for the partial-fraction expansion.)

Unit-step response of higher order systems. Let us examine the response of the system to a unit-step input. If all poles of the denominator polynomials are real and distinct, then, for a unit-step input $U(s) = 1/s$, Equation (8–18) can be written as

$$X(s) = \frac{a}{s} + \sum_{i=1}^{n} \frac{a_i}{s + p_i} \tag{8–19}$$

where a_i is the residue of the pole at $s = -p_i$. [If the system involves multiple poles, then $X(s)$ will have multiple-pole terms.]

Next, consider the case where the poles of $X(s)$ consist of real poles and pairs of complex-conjugate poles. A pair of complex-conjugate poles yields a second-order term in s. Since the factored form of the higher order characteristic equation consists of first-order and second-order terms, Equation (8–19) can be rewritten as

$$X(s) = \frac{a}{s} + \sum_{j=1}^{q} \frac{a_j}{s + p_j} + \sum_{k=1}^{r} \frac{b_k(s + \zeta_k \omega_k) + c_k \omega_k \sqrt{1 - \zeta_k^2}}{s^2 + 2\zeta_k \omega_k s + \omega_k^2} \quad (q + 2r = n) \tag{8–20}$$

where we assume that all poles of $X(s)$ are distinct. [If some of the poles of $X(s)$ are multiple poles, then $X(s)$ must have multiple-pole terms.] From this last equation, we see that the response of a higher order system is composed of a number of terms involving the simple functions found in the responses of first-order and second-order systems. From Equation (8–19), the unit-step response $x(t)$, the inverse Laplace transform of $X(s)$, is

$$x(t) = a + \sum_{i=1}^{n} a_i e^{-p_i t} \qquad \text{for } t \geq 0$$

or, from Equation (8–20),

$$x(t) = a + \sum_{j=1}^{q} a_j e^{-p_j t} + \sum_{k=1}^{r} b_k e^{-\zeta_k \omega_k t} \cos \omega_k \sqrt{1 - \zeta_k^2}\, t$$

$$+ \sum_{k=1}^{r} c_k e^{-\zeta_k \omega_k t} \sin \omega_k \sqrt{1 - \zeta_k^2}\, t, \qquad \text{for } t \geq 0$$

Thus, the response curve of a higher order system is the sum of a number of exponential curves and damped sinusoidal curves. [If the poles of $X(s)$ involve multiple poles, then $x(t)$ must have the corresponding multiple-pole terms.]

If all poles of $X(s)$ lie in the left-half s-plane, then the exponential terms [including those terms multiplied by t, t^2, etc., that occur when $X(s)$ involves multiple poles] and the damped exponential terms in $x(t)$ will approach zero as the time t increases. The steady-state output is then $x(\infty) = a$.

8–5 SOLUTION OF THE STATE EQUATION

In this section, we shall obtain the general solution of the linear time-invariant state equation. We first consider the homogeneous case and then the nonhomogeneous case. After we obtain the general solution, we shall derive the analytical expression for the step response. (For details of analytical expressions for the step response, impulse response, and ramp response, see **Problems A–8–13, A–8–14**, and **A–8–15**.)

Solution of homogeneous state equations.   Before we solve vector–matrix differential equations, let us review the solution of the scalar differential equation

$$\dot{x} = ax \tag{8–21}$$

In solving this equation, we may assume a solution of the form

$$x(t) = b_0 + b_1 t + b_2 t^2 + \cdots + b_k t^k + \cdots \tag{8–22}$$

Substituting this assumed solution into Equation (8–21), we obtain

$$b_1 + 2b_2 t + 3b_3 t^2 + \cdots + k b_k t^{k-1} + \cdots$$
$$= a(b_0 + b_1 t + b_2 t^2 + \cdots + b_k t^k + \cdots) \tag{8–23}$$

If the assumed solution is the true solution, Equation (8–23) must hold for any t. Hence, equating the coefficients of equal powers of t, we find that

$$b_1 = ab_0$$
$$b_2 = \frac{1}{2}ab_1 = \frac{1}{2}a^2 b_0$$
$$b_3 = \frac{1}{3}ab_2 = \frac{1}{3 \times 2}a^3 b_0$$
$$\vdots$$
$$b_k = \frac{1}{k!}a^k b_0$$
$$\vdots$$

The value of b_0 is determined by substituting $t = 0$ into Equation (8–22), or

$$x(0) = b_0$$

Hence, the solution can be written as

$$x(t) = \left(1 + at + \frac{1}{2!}a^2t^2 + \cdots + \frac{1}{k!}a^kt^k + \cdots \right)x(0)$$

$$= e^{at}x(0)$$

We shall now solve the vector–matrix differential equation

$$\dot{\mathbf{x}} = \mathbf{A}\mathbf{x} \tag{8–24}$$

where

$\mathbf{x} = n$-vector
$\mathbf{A} = n \times n$ constant matrix

By analogy with the scalar case, we assume that the solution is in the form of a vector power series in t, or

$$\mathbf{x}(t) = \mathbf{b}_0 + \mathbf{b}_1 t + \mathbf{b}_2 t^2 + \cdots + \mathbf{b}_k t^k + \cdots \tag{8–25}$$

Substituting this assumed solution into Equation (8–24), we obtain

$$\mathbf{b}_1 + 2\mathbf{b}_2 t + 3\mathbf{b}_3 t^2 + \cdots + k\mathbf{b}_k t^{k-1} + \cdots$$
$$= \mathbf{A}(\mathbf{b}_0 + \mathbf{b}_1 t + \mathbf{b}_2 t^2 + \cdots + \mathbf{b}_k t^k + \cdots) \tag{8–26}$$

If the assumed solution is the true solution, Equation (8–26) must hold for all t. Thus, equating the coefficients of like powers of t on both sides of Equation (8–26), we find that

$$\mathbf{b}_1 = \mathbf{A}\mathbf{b}_0$$

$$\mathbf{b}_2 = \frac{1}{2}\mathbf{A}\mathbf{b}_1 = \frac{1}{2}\mathbf{A}^2\mathbf{b}_0$$

$$\mathbf{b}_3 = \frac{1}{3}\mathbf{A}\mathbf{b}_2 = \frac{1}{3 \times 2}\mathbf{A}^3\mathbf{b}_0$$

$$\vdots$$

$$\mathbf{b}_k = \frac{1}{k!}\mathbf{A}^k\mathbf{b}_0$$

$$\vdots$$

Substituting $t = 0$ into Equation (8–25) yields

$$\mathbf{x}(0) = \mathbf{b}_0$$

Thus, the solution $\mathbf{x}(t)$ can be written as

$$\mathbf{x}(t) = \left(\mathbf{I} + \mathbf{A}t + \frac{1}{2!}\mathbf{A}^2t^2 + \cdots + \frac{1}{k!}\mathbf{A}^kt^k + \cdots \right)\mathbf{x}(0)$$

The multiterm expression in parentheses on the right-hand side of this last equation is an $n \times n$ matrix. Because of its similarity to the infinite power series for a scalar exponential, we call it the *matrix exponential* and write

$$\mathbf{I} + \mathbf{A}t + \frac{1}{2!}\mathbf{A}^2t^2 + \cdots + \frac{1}{k!}\mathbf{A}^kt^k + \cdots = e^{\mathbf{A}t}$$

In terms of the matrix exponential, the solution of Equation (8–24) can be written as

$$\mathbf{x}(t) = e^{\mathbf{A}t}\mathbf{x}(0) \tag{8–27}$$

Since the matrix exponential is very important in the state-space analysis of linear systems, we examine its properties next.

Matrix exponential.   It can be proved that the matrix exponential

$$e^{\mathbf{A}t} = \sum_{k=0}^{\infty} \frac{\mathbf{A}^k t^k}{k!}$$

of an $n \times n$ matrix $\mathbf{A}$ converges absolutely for all finite t. (Hence, computer calculations for evaluating the elements of $e^{\mathbf{A}t}$ by using the series expansion can easily be carried out.)

Because of the convergence of the infinite series $\sum_{k=0}^{\infty} \mathbf{A}^k t^k / k!$, the series can be differentiated term by term to give

$$\frac{d}{dt} e^{\mathbf{A}t} = \mathbf{A} + \mathbf{A}^2 t + \frac{\mathbf{A}^3 t^2}{2!} + \cdots + \frac{\mathbf{A}^k t^{k-1}}{(k-1)!} + \cdots$$

$$= \mathbf{A}\left[\mathbf{I} + \mathbf{A}t + \frac{\mathbf{A}^2 t^2}{2!} + \cdots + \frac{\mathbf{A}^{k-1} t^{k-1}}{(k-1)!} + \cdots \right] = \mathbf{A}e^{\mathbf{A}t}$$

$$= \left[\mathbf{I} + \mathbf{A}t + \frac{\mathbf{A}^2 t^2}{2!} + \cdots + \frac{\mathbf{A}^{k-1} t^{k-1}}{(k-1)!} + \cdots \right]\mathbf{A} = e^{\mathbf{A}t}\mathbf{A}$$

The matrix exponential has the property that

$$e^{\mathbf{A}(t+s)} = e^{\mathbf{A}t} e^{\mathbf{A}s}$$

This can be proved as follows:

$$e^{\mathbf{A}t} e^{\mathbf{A}s} = \left(\sum_{k=0}^{\infty} \frac{\mathbf{A}^k t^k}{k!} \right)\left(\sum_{h=0}^{\infty} \frac{\mathbf{A}^h s^h}{h!} \right) = \sum_{k=0}^{\infty} \sum_{h=0}^{\infty} \mathbf{A}^{k+h} \frac{t^k s^h}{k! h!}$$

Let $k + h = m$. Then

$$e^{\mathbf{A}t} e^{\mathbf{A}s} = \sum_{k=0}^{\infty} \sum_{m=k}^{\infty} \mathbf{A}^m \frac{t^k s^{m-k}}{k!(m-k)!} = \sum_{m=0}^{\infty} \frac{1}{m!} \mathbf{A}^m \sum_{k=0}^{\infty} \frac{m! t^k s^{m-k}}{k!(m-k)!}$$

$$= \sum_{m=0}^{\infty} \frac{1}{m!} \mathbf{A}^m (t+s)^m = e^{\mathbf{A}(t+s)}$$

In particular, if $s = -t$, then

$$e^{\mathbf{A}t} e^{-\mathbf{A}t} = e^{-\mathbf{A}t} e^{\mathbf{A}t} = e^{\mathbf{A}(t-t)} = \mathbf{I}$$

Thus, the inverse of $e^{\mathbf{A}t}$ is $e^{-\mathbf{A}t}$. Since the inverse of $e^{\mathbf{A}t}$ always exists, $e^{\mathbf{A}t}$ is nonsingular.
It is very important to remember that

$$e^{(\mathbf{A}+\mathbf{B})t} = e^{\mathbf{A}t} e^{\mathbf{B}t}, \qquad \text{if } \mathbf{AB} = \mathbf{BA}$$

$$e^{(\mathbf{A}+\mathbf{B})t} \neq e^{\mathbf{A}t} e^{\mathbf{B}t}, \qquad \text{if } \mathbf{AB} \neq \mathbf{BA}$$

To prove this, note that

$$e^{(\mathbf{A}+\mathbf{B})t} = \mathbf{I} + (\mathbf{A}+\mathbf{B})t + \frac{(\mathbf{A}+\mathbf{B})^2}{2!}t^2 + \frac{(\mathbf{A}+\mathbf{B})^3}{3!}t^3 + \cdots$$

$$e^{\mathbf{A}t}e^{\mathbf{B}t} = \left(\mathbf{I} + \mathbf{A}t + \frac{\mathbf{A}^2 t^2}{2!} + \frac{\mathbf{A}^3 t^3}{3!} + \cdots\right)\left(\mathbf{I} + \mathbf{B}t + \frac{\mathbf{B}^2 t^2}{2!} + \frac{\mathbf{B}^3 t^3}{3!} + \cdots\right)$$

$$= \mathbf{I} + (\mathbf{A}+\mathbf{B})t + \frac{\mathbf{A}^2 t^2}{2!} + \mathbf{A}\mathbf{B}t^2 + \frac{\mathbf{B}^2 t^2}{2!} + \frac{\mathbf{A}^3 t^3}{3!}$$

$$+ \frac{\mathbf{A}^2\mathbf{B}t^3}{2!} + \frac{\mathbf{A}\mathbf{B}^2 t^3}{2!} + \frac{\mathbf{B}^3 t^3}{3!} + \cdots$$

Hence,

$$e^{(\mathbf{A}+\mathbf{B})t} - e^{\mathbf{A}t}e^{\mathbf{B}t} = \frac{\mathbf{B}\mathbf{A} - \mathbf{A}\mathbf{B}}{2!}t^2$$

$$+ \frac{\mathbf{B}\mathbf{A}^2 + \mathbf{A}\mathbf{B}\mathbf{A} + \mathbf{B}^2\mathbf{A} + \mathbf{B}\mathbf{A}\mathbf{B} - 2\mathbf{A}^2\mathbf{B} - 2\mathbf{A}\mathbf{B}^2}{3!}t^3 + \cdots$$

The difference between $e^{(\mathbf{A}+\mathbf{B})t}$ and $e^{\mathbf{A}t}e^{\mathbf{B}t}$ vanishes if $\mathbf{A}$ and $\mathbf{B}$ commute.

Laplace transform approach to the solution of homogeneous state equations.   Let us first consider the scalar case:

$$\dot{x} = ax \qquad\qquad\qquad (8\text{–}28)$$

Taking the Laplace transform of Equation (8–28), we obtain

$$sX(s) - x(0) = aX(s) \qquad\qquad\qquad (8\text{–}29)$$

where $X(s) = \mathscr{L}[x]$. Solving Equation (8–29) for $X(s)$ gives

$$X(s) = \frac{x(0)}{s - a} = (s - a)^{-1}x(0)$$

The inverse Laplace transform of this last equation produces the solution:

$$x(t) = e^{at}x(0)$$

The foregoing approach to the solution of the homogeneous scalar differential equation can be extended to the homogeneous state equation

$$\dot{\mathbf{x}}(t) = \mathbf{A}\mathbf{x}(t) \qquad\qquad\qquad (8\text{–}30)$$

Taking the Laplace transform of both sides of Equation (8–30), we obtain

$$s\mathbf{X}(s) - \mathbf{x}(0) = \mathbf{A}\mathbf{X}(s)$$

where $\mathbf{X}(s) = \mathscr{L}[\mathbf{x}]$. Hence,

$$(s\mathbf{I} - \mathbf{A})\mathbf{X}(s) = \mathbf{x}(0)$$

Premultiplying both sides of this last equation by $(s\mathbf{I} - \mathbf{A})^{-1}$, we obtain

$$\mathbf{X}(s) = (s\mathbf{I} - \mathbf{A})^{-1}\mathbf{x}(0)$$

The inverse Laplace transform of $\mathbf{X}(s)$ gives the solution

$$\mathbf{x}(t) = \mathcal{L}^{-1}[(s\mathbf{I} - \mathbf{A})^{-1}]\mathbf{x}(0) \tag{8-31}$$

Note that

$$(s\mathbf{I} - \mathbf{A})^{-1} = \frac{\mathbf{I}}{s} + \frac{\mathbf{A}}{s^2} + \frac{\mathbf{A}^2}{s^3} + \cdots$$

Hence, the inverse Laplace transform of $(s\mathbf{I} - \mathbf{A})^{-1}$ gives

$$\mathcal{L}^{-1}[(s\mathbf{I} - \mathbf{A})^{-1}] = \mathbf{I} + \mathbf{A}t + \frac{\mathbf{A}^2 t^2}{2!} + \frac{\mathbf{A}^3 t^3}{3!} + \cdots = e^{\mathbf{A}t} \tag{8-32}$$

(The inverse Laplace transform of a matrix is the matrix consisting of the inverse Laplace transforms of all of the elements of the matrix.) From Equations (8–31) and (8–32), the solution of Equation (8–30) is

$$\mathbf{x}(t) = e^{\mathbf{A}t}\mathbf{x}(0)$$

The importance of Equation (8–32) lies in the fact that it provides a convenient means for finding the closed solution of the matrix exponential.

State-transition matrix. We can write the solution of the homogeneous state equation

$$\dot{\mathbf{x}} = \mathbf{A}\mathbf{x} \tag{8-33}$$

as

$$\mathbf{x}(t) = \mathbf{\Phi}(t)\mathbf{x}(0) \tag{8-34}$$

where $\mathbf{\Phi}(t)$ is an $n \times n$ matrix and is the unique solution of

$$\dot{\mathbf{\Phi}}(t) = \mathbf{A}\mathbf{\Phi}(t), \qquad \mathbf{\Phi}(0) = \mathbf{I}$$

To verify this, note that

$$\mathbf{x}(0) = \mathbf{\Phi}(0)\mathbf{x}(0) = \mathbf{x}(0)$$

and

$$\dot{\mathbf{x}}(t) = \dot{\mathbf{\Phi}}(t)\mathbf{x}(0) = \mathbf{A}\mathbf{\Phi}(t)\mathbf{x}(0) = \mathbf{A}\mathbf{x}(t)$$

We thus confirm that Equation (8–34) is the solution of Equation (8–33). From Equations (8–27), (8–32), and (8–34), we obtain

$$\mathbf{\Phi}(t) = e^{\mathbf{A}t} = \mathcal{L}^{-1}[(s\mathbf{I} - \mathbf{A})^{-1}]$$

Note that

$$\mathbf{\Phi}^{-1}(t) = e^{-\mathbf{A}t} = \mathbf{\Phi}(-t)$$

From Equation (8–34), we see that the solution of Equation (8–33) is simply a transformation of the initial condition. Hence, the unique matrix $\mathbf{\Phi}(t)$ is called the *state-transition matrix*. This matrix contains all the information about the free motions of the system defined by Equation (8–33).

Example 8–2

Obtain the state-transition matrix $\Phi(t)$ of the following system:

$$\begin{bmatrix} \dot{x}_1 \\ \dot{x}_2 \end{bmatrix} = \begin{bmatrix} 0 & 1 \\ -2 & -3 \end{bmatrix} \begin{bmatrix} x_1 \\ x_2 \end{bmatrix}$$

Obtain also the inverse of the state-transition matrix, $\Phi^{-1}(t)$.
For this system,

$$\mathbf{A} = \begin{bmatrix} 0 & 1 \\ -2 & -3 \end{bmatrix}$$

The state-transition matrix is given by

$$\Phi(t) = e^{\mathbf{A}t} = \mathcal{L}^{-1}[(s\mathbf{I} - \mathbf{A})^{-1}]$$

Since

$$s\mathbf{I} - \mathbf{A} = \begin{bmatrix} s & 0 \\ 0 & s \end{bmatrix} - \begin{bmatrix} 0 & 1 \\ -2 & -3 \end{bmatrix} = \begin{bmatrix} s & -1 \\ 2 & s+3 \end{bmatrix}$$

the inverse of $(s\mathbf{I} - \mathbf{A})$ is given by

$$(s\mathbf{I} - \mathbf{A})^{-1} = \frac{1}{(s+1)(s+2)} \begin{bmatrix} s+3 & 1 \\ -2 & s \end{bmatrix}$$

$$= \begin{bmatrix} \dfrac{s+3}{(s+1)(s+2)} & \dfrac{1}{(s+1)(s+2)} \\ \dfrac{-2}{(s+1)(s+2)} & \dfrac{s}{(s+1)(s+2)} \end{bmatrix}$$

$$= \begin{bmatrix} \dfrac{2}{s+1} - \dfrac{1}{s+2} & \dfrac{1}{s+1} - \dfrac{1}{s+2} \\ \dfrac{-2}{s+1} + \dfrac{2}{s+2} & \dfrac{-1}{s+1} + \dfrac{2}{s+2} \end{bmatrix}$$

Hence,

$$\Phi(t) = e^{\mathbf{A}t} = \mathcal{L}^{-1}[(s\mathbf{I} - \mathbf{A})^{-1}]$$

$$= \begin{bmatrix} 2e^{-t} - e^{-2t} & e^{-t} - e^{-2t} \\ -2e^{-t} + 2e^{-2t} & -e^{-t} + 2e^{-2t} \end{bmatrix}$$

Noting that $\Phi^{-1}(t) = \Phi(-t)$, we obtain the inverse of the state-transition matrix as follows:

$$\Phi^{-1}(t) = e^{-\mathbf{A}t} = \begin{bmatrix} 2e^{t} - e^{2t} & e^{t} - e^{2t} \\ -2e^{t} + 2e^{2t} & -e^{t} + 2e^{2t} \end{bmatrix}$$

Solution of nonhomogeneous state equations. We shall begin by considering the scalar case

$$\dot{x} = ax + bu$$

Let us rewrite this equation as

$$\dot{x} - ax = bu$$

Multiplying both sides of the latter equation by e^{-at}, we obtain

$$e^{-at}[\dot{x}(t) - ax(t)] = \frac{d}{dt}[e^{-at}x(t)] = e^{-at}bu(t)$$

Integrating this equation between 0 and t gives

$$e^{-at}x(t) = x(0) + \int_0^t e^{-a\tau}bu(\tau)\,d\tau$$

or

$$x(t) = e^{at}x(0) + e^{at}\int_0^t e^{-a\tau}bu(\tau)\,d\tau$$

The first term on the right-hand side is the response to the initial condition, and the second term is the response to the input $u(t)$.

Let us now consider the nonhomogeneous state equation defined by

$$\dot{\mathbf{x}} = \mathbf{Ax} + \mathbf{Bu} \tag{8-35}$$

where

     $\mathbf{x} = n$-vector
     $\mathbf{u} = r$-vector
     $\mathbf{A} = n \times n$ constant matrix
     $\mathbf{B} = n \times r$ constant matrix

Writing Equation (8-35) as

$$\dot{\mathbf{x}}(t) - \mathbf{Ax}(t) = \mathbf{Bu}(t)$$

and premultiplying both sides of this equation by $e^{-\mathbf{A}t}$, we obtain

$$e^{-\mathbf{A}t}[\dot{\mathbf{x}}(t) - \mathbf{Ax}(t)] = \frac{d}{dt}[e^{-\mathbf{A}t}\mathbf{x}(t)] = e^{-\mathbf{A}t}\mathbf{Bu}(t)$$

Integrating the preceding equation between 0 and t gives

$$e^{-\mathbf{A}t}\mathbf{x}(t) = \mathbf{x}(0) + \int_0^t e^{-\mathbf{A}\tau}\mathbf{Bu}(\tau)\,d\tau$$

or

$$\mathbf{x}(t) = e^{\mathbf{A}t}\mathbf{x}(0) + \int_0^t e^{\mathbf{A}(t-\tau)}\mathbf{Bu}(\tau)\,d\tau \tag{8-36}$$

Equation (8–36) can also be written as

$$\mathbf{x}(t) = \mathbf{\Phi}(t)\mathbf{x}(0) + \int_0^t \mathbf{\Phi}(t - \tau)\mathbf{Bu}(\tau)\,d\tau \tag{8–37}$$

where $\mathbf{\Phi}(t) = e^{\mathbf{A}t}$. Equation (8–36) or Equation (8–37) is the solution of Equation (8–35). The solution $\mathbf{x}(t)$ is clearly the sum of a term consisting of the transition of the initial state and a term arising from the input vector.

Laplace transform approach to the solution of nonhomogeneous state equations. The solution of the nonhomogeneous state equation

$$\dot{\mathbf{x}} = \mathbf{Ax} + \mathbf{Bu}$$

can also be obtained by the Laplace transform approach. The Laplace transform of this last equation yields

$$s\mathbf{X}(s) - \mathbf{x}(0) = \mathbf{AX}(s) + \mathbf{BU}(s)$$

or

$$(s\mathbf{I} - \mathbf{A})\mathbf{X}(s) = \mathbf{x}(0) + \mathbf{BU}(s)$$

Premultiplying both sides of the foregoing equation by $(s\mathbf{I} - \mathbf{A})^{-1}$, we obtain

$$\mathbf{X}(s) = (s\mathbf{I} - \mathbf{A})^{-1}\mathbf{x}(0) + (s\mathbf{I} - \mathbf{A})^{-1}\mathbf{BU}(s)$$

Using the relationship given by Equation (8–32) yields

$$\mathbf{X}(s) = \mathscr{L}[e^{\mathbf{A}t}]\mathbf{x}(0) + \mathscr{L}[e^{\mathbf{A}t}]\mathbf{BU}(s)$$

The inverse Laplace transform of this last equation can be obtained with the use of the convolution integral as follows:

$$\mathbf{x}(t) = e^{\mathbf{A}t}\mathbf{x}(0) + \int_0^t e^{\mathbf{A}(t-\tau)}\mathbf{Bu}(\tau)\,d\tau$$

Solution in terms of x(t_0). Thus far, we have assumed the initial time to be zero. If, however, the initial time is given by t_0 instead of 0, then Equation (8–36), the solution of Equation (8–35), must be modified to

$$\mathbf{x}(t) = e^{\mathbf{A}(t-t_0)}\mathbf{x}(t_0) + \int_{t_0}^t e^{\mathbf{A}(t-\tau)}\mathbf{Bu}(\tau)\,d\tau \tag{8–38}$$

Example 8–3

Obtain the time response of the system

$$\begin{bmatrix} \dot{x}_1 \\ \dot{x}_2 \end{bmatrix} = \begin{bmatrix} 0 & 1 \\ -2 & -3 \end{bmatrix}\begin{bmatrix} x_1 \\ x_2 \end{bmatrix} + \begin{bmatrix} 0 \\ 1 \end{bmatrix}u$$

where $u(t)$ is the unit-step function occurring at $t = 0$, or

$$u(t) = 1(t)$$

For this system,

$$\mathbf{A} = \begin{bmatrix} 0 & 1 \\ -2 & -3 \end{bmatrix}, \qquad \mathbf{B} = \begin{bmatrix} 0 \\ 1 \end{bmatrix}$$

The state-transition matrix was obtained in Example 8–2 as

$$\Phi(t) = e^{\mathbf{A}t} = \begin{bmatrix} 2e^{-t} - e^{-2t} & e^{-t} - e^{-2t} \\ -2e^{-t} + 2e^{-2t} & -e^{-t} + 2e^{-2t} \end{bmatrix}$$

The response to the unit-step input is then obtained as

$$\mathbf{x}(t) = e^{\mathbf{A}t}\mathbf{x}(0) + \int_0^t \begin{bmatrix} 2e^{-(t-\tau)} - e^{-2(t-\tau)} & e^{-(t-\tau)} - e^{-2(t-\tau)} \\ -2e^{-(t-\tau)} + 2e^{-2(t-\tau)} & -e^{-(t-\tau)} + 2e^{-2(t-\tau)} \end{bmatrix} \begin{bmatrix} 0 \\ 1 \end{bmatrix} [1]\, d\tau$$

or

$$\begin{bmatrix} x_1(t) \\ x_2(t) \end{bmatrix} = \begin{bmatrix} 2e^{-t} - e^{-2t} & e^{-t} - e^{-2t} \\ -2e^{-t} + 2e^{-2t} & -e^{-t} + 2e^{-2t} \end{bmatrix} \begin{bmatrix} x_1(0) \\ x_2(0) \end{bmatrix} + \begin{bmatrix} 0.5 - e^{-t} + 0.5e^{-2t} \\ e^{-t} - e^{-2t} \end{bmatrix}$$

In the special case where the initial state is zero, or $\mathbf{x}(0) = \mathbf{0}$, the solution $\mathbf{x}(t)$ can be simplified to

$$\begin{bmatrix} x_1(t) \\ x_2(t) \end{bmatrix} = \begin{bmatrix} 0.5 - e^{-t} + 0.5e^{-2t} \\ e^{-t} - e^{-2t} \end{bmatrix}$$

Analytical solution for step response of system in state-space form.
Consider the system described by

$$\dot{\mathbf{x}} = \mathbf{A}\mathbf{x} + \mathbf{B}u$$

where matrix $\mathbf{A}$ is nonsingular. Let us show that the response $\mathbf{x}(t)$ when the input u is a unit-step function $1(t)$ can be given by

$$\mathbf{x}(t) = e^{\mathbf{A}t}\mathbf{x}(0) + \mathbf{A}^{-1}(e^{\mathbf{A}t} - \mathbf{I})\mathbf{B}$$

From Equation (8–37), the response $\mathbf{x}(t)$ is

$$\mathbf{x}(t) = e^{\mathbf{A}t}\mathbf{x}(0) + \int_0^t e^{\mathbf{A}(t-\tau)}\mathbf{B}u(\tau)\, d\tau$$

$$= e^{\mathbf{A}t}\mathbf{x}(0) + e^{\mathbf{A}t}\left[\int_0^t e^{-\mathbf{A}\tau}1(\tau)\, d\tau\right]\mathbf{B}$$

$$= e^{\mathbf{A}t}\mathbf{x}(0) + e^{\mathbf{A}t}\left[\int_0^t \left(\mathbf{I} - \mathbf{A}\tau + \frac{1}{2!}\mathbf{A}^2\tau^2 - \cdots\right)d\tau\right]\mathbf{B}$$

$$= e^{\mathbf{A}t}\mathbf{x}(0) + e^{\mathbf{A}t}\left(\mathbf{I}t - \frac{1}{2!}\mathbf{A}t^2 + \frac{1}{3!}\mathbf{A}^2t^3 - \cdots\right)\mathbf{B}$$

Since

$$e^{\mathbf{A}t} = \mathbf{I} + \mathbf{A}t + \frac{1}{2!}\mathbf{A}^2t^2 + \frac{1}{3!}\mathbf{A}^3t^3 + \cdots$$

we have

$$e^{-\mathbf{A}t} = \mathbf{I} - \mathbf{A}t + \frac{1}{2!}\mathbf{A}^2t^2 - \frac{1}{3!}\mathbf{A}^3t^3 + \cdots$$

Thus,

$$It - \frac{1}{2!} \mathbf{A}t^2 + \frac{1}{3!} \mathbf{A}^2 t^3 - \cdots$$

$$= \mathbf{A}^{-1} \left(\mathbf{A}t - \frac{1}{2!} \mathbf{A}^2 t^2 + \frac{1}{3!} \mathbf{A}^3 t^3 - \cdots \right)$$

$$= -\mathbf{A}^{-1} \left(-\mathbf{A}t + \frac{1}{2!} \mathbf{A}^2 t^2 - \frac{1}{3!} \mathbf{A}^3 t^3 + \cdots \right)$$

$$= -\mathbf{A}^{-1} \left(\mathbf{I} - \mathbf{A}t + \frac{1}{2!} \mathbf{A}^2 t^2 - \frac{1}{3!} \mathbf{A}^3 t^3 + \cdots - \mathbf{I} \right)$$

$$= -\mathbf{A}^{-1} (e^{-\mathbf{A}t} - \mathbf{I})$$

$$= \mathbf{A}^{-1} (\mathbf{I} - e^{-\mathbf{A}t})$$

Therefore, we get

$$\mathbf{x}(t) = e^{\mathbf{A}t}\mathbf{x}(0) + e^{\mathbf{A}t}\mathbf{A}^{-1}(\mathbf{I} - e^{-\mathbf{A}t})\mathbf{B}$$

$$= e^{\mathbf{A}t}\mathbf{x}(0) + \mathbf{A}^{-1}(e^{\mathbf{A}t} - \mathbf{I})\mathbf{B} \qquad (8\text{–}39)$$

More on analytical solutions for responses of systems in state-space form is given in **Problems A–8–13, A–8–14,** and **A–8–15.**

EXAMPLE PROBLEMS AND SOLUTIONS

Problem A–8–1

Consider the electrical circuit shown in Figure 8–12. Assume that there is an initial charge q_0 on the capacitor just before switch S is closed at $t = 0$. Find the current $i(t)$.

Solution The equation for the circuit when switch S is closed is

$$Ri + \frac{1}{C} \int i \, dt = E$$

Taking the Laplace transform of this last equation yields

$$RI(s) + \frac{1}{C} \frac{I(s) + \int i(t) \, dt \Big|_{t=0}}{s} = \frac{E}{s}$$

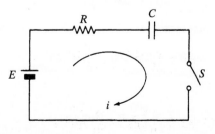

Figure 8–12 Electrical circuit.

Since

$$\int i(t)\, dt \Big|_{t=0} = q(0) = q_0$$

we obtain

$$RI(s) + \frac{1}{C}\frac{I(s) + q_0}{s} = \frac{E}{s}$$

or

$$RCsI(s) + I(s) + q_0 = CE$$

Solving for $I(s)$, we have

$$I(s) = \frac{CE - q_0}{RCs + 1} = \left(\frac{E}{R} - \frac{q_0}{RC}\right)\frac{1}{s + \dfrac{1}{RC}}$$

The inverse Laplace transform of this equation gives the current $i(t)$:

$$i(t) = \left(\frac{E}{R} - \frac{q_0}{RC}\right)e^{-t/RC}$$

Problem A–8–2

Suppose that a disk is rotated at a constant speed of 100 rad/s and we wish to stop it in 2 min. Assuming that the moment of inertia J of the disk is 6 kg-m^2, find the constant torque T necessary to stop the rotation.

Solution   The necessary torque T must act so as to reduce the speed of the disk. Thus, the equation of motion is

$$J\dot\omega = -T \qquad \omega(0) = 100$$

Taking the Laplace transform of this last equation, under the condition that the torque T is a constant, we obtain

$$J[s\Omega(s) - \omega(0)] = -\frac{T}{s}$$

Substituting $J = 6$ and $\omega(0) = 100$ into this equation and solving for $\Omega(s)$, we get

$$\Omega(s) = \frac{100}{s} - \frac{T}{6s^2}$$

The inverse Laplace transform of the latter equation gives

$$\omega(t) = 100 - \frac{T}{6}t$$

At $t = 2$ min $= 120$ s, we want to stop, or $\omega(120)$ must equal zero. Therefore,

$$\omega(120) = 0 = 100 - \frac{T}{6} \times 120$$

Solving for T, we get

$$T = \frac{600}{120} = 5 \text{ N-m}$$

Problem A–8–3

Find the transfer function $E_o(s)/E_i(s)$ of the electrical circuit shown in Figure 8–13. Obtain the voltage $e_o(t)$ when the input voltage $e_i(t)$ is a step change of voltage E_i occurring at $t = 0$ $[e_i(0-) = 0]$. Assume that the initial charges in the capacitors are zero. [Thus, $e_o(0-) = 0$.]

Solution Using the complex-impedance method, we obtain the transfer function $E_o(s)/E_i(s)$ as

$$\frac{E_o(s)}{E_i(s)} = \frac{\dfrac{1}{(1/R_2) + C_2 s}}{\dfrac{1}{C_1 s} + \dfrac{1}{(1/R_2) + C_2 s}} = \frac{R_2 C_1 s}{R_2(C_1 + C_2)s + 1}$$

Next, we determine $e_o(t)$. For the input $e_i(t) = E_i \cdot 1(t)$, we have

$$E_o(s) = \frac{R_2 C_1 s}{R_2(C_1 + C_2)s + 1} \frac{E_i}{s}$$

$$= \frac{R_2 C_1 E_i}{R_2(C_1 + C_2)s + 1}$$

Then, inverse Laplace transforming $E_o(s)$, we get

$$e_o(t) = \frac{C_1 E_i}{C_1 + C_2} e^{-t/[R_2(C_1 + C_2)]}$$

from which we see that $e_o(0+) = C_1 E_i/(C_1 + C_2)$. Since $e_0(0-) = 0$, there is a sudden change in $e_0(t)$ at $t = 0$.

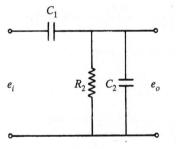

Figure 8–13 Electrical circuit.

Problem A–8–4

Derive the transfer function $E_o(s)/E_i(s)$ of the electrical circuit shown in Figure 8–14. The input voltage is a pulse signal given by

$$e_i(t) = E_i \qquad 0 \le t \le t_1$$
$$= 0 \qquad \text{elsewhere}$$

Obtain the output $e_o(t)$. Assume that the initial charges in the capacitors C_1 and C_2 are zero.

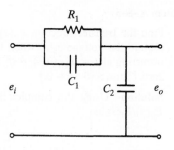

Figure 8–14 Electrical circuit.

Solution By using the complex-impedance method, the transfer function $E_o(s)/E_i(s)$ can be obtained as

$$\frac{E_o(s)}{E_i(s)} = \frac{\dfrac{1}{C_2 s}}{\dfrac{R_1}{R_1 C_1 s + 1} + \dfrac{1}{C_2 s}} = \frac{R_1 C_1 s + 1}{R_1 (C_1 + C_2) s + 1}$$

For the given input $e_i(t)$, we have

$$E_i(s) = \frac{E_i}{s}(1 - e^{-st_1})$$

Thus, the response $E_o(s)$ can be given by

$$E_o(s) = \frac{R_1 C_1 s + 1}{R_1 (C_1 + C_2) s + 1} \frac{E_i}{s}(1 - e^{-st_1})$$

$$= \left[\frac{-R_1 C_2}{R_1 (C_1 + C_2) s + 1} + \frac{1}{s} \right] E_i (1 - e^{-st_1})$$

The inverse Laplace transform of $E_o(s)$ gives

$$e_o(t) = \left\{ 1 - \frac{C_2}{C_1 + C_2} e^{-t/[R_1(C_1 + C_2)]} \right\} E_i$$

$$- \left\{ 1 - \frac{C_2}{C_1 + C_2} e^{-(t - t_1)/[R_1(C_1 + C_2)]} \right\} E_i \cdot 1(t - t_1)$$

Figure 8–15 shows a possible response curve $e_o(t)$ versus t.

Problem A–8–5

A mass m is attached to a string that is under tension T in the mechanical system of Figure 8–16(a). We assume that tension T is to remain constant for small displacement x. (The displacement x is that of mass m perpendicular to the string.) Neglecting gravity, find the natural frequency of the vertical motion of mass m. What is the displacement $x(t)$ when the mass is given initial conditions $x(0) = x_o$ and $\dot{x}(0) = 0$?

Solution From Figure 8–16(b), the vertical component of the force due to tension is

$$-T \sin \theta_1 - T \sin \theta_2$$

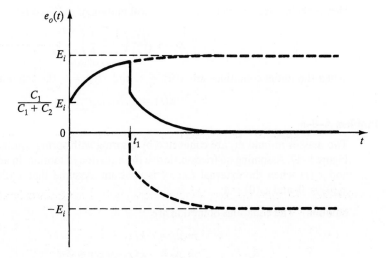

Figure 8–15   Response curve $e_o(t)$ versus t (solid curve).

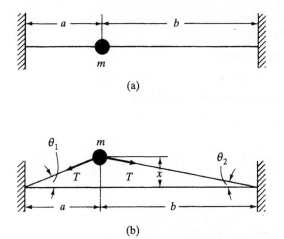

(a)

(b)

Figure 8–16   (a) Mechanical vibratory system;
(b) diagram showing tension forces.

For small x, angles θ_1 and θ_2 are small, and

$$\sin \theta_1 = \tan \theta_1 = \frac{x}{a}$$

$$\sin \theta_2 = \tan \theta_2 = \frac{x}{b}$$

The equation of motion for the system when x is small is

$$m\ddot{x} = -T \sin \theta_1 - T \sin \theta_2 = -T\frac{x}{a} - T\frac{x}{b}$$

or

$$m\ddot{x} + T\left(\frac{1}{a} + \frac{1}{b}\right)x = 0$$

Hence, the natural frequency of the vertical motion of the mass is

$$\omega_n = \sqrt{\frac{T}{m}\left(\frac{1}{a} + \frac{1}{b}\right)}$$

When the initial conditions are $x(0) = x_o$ and $\dot{x}(0) = 0$, the solution $x(t)$ is given by

$$x(t) = x_0 \cos \omega_n t$$

Problem A–8–6

Two masses m_1 and m_2 are connected by a spring with spring constant k, as shown in Figure 8–17. Assuming no friction, derive the equation of motion. In addition, find $x_1(t)$ and $x_2(t)$ when the external force f is constant. Assume that $x_1(0) = 0$, $\dot{x}_1(0) = 0$, $x_2(0) = 0$, and $\dot{x}_2(0) = 0$.

Solution   The equations of motion are

$$m_1 \ddot{x}_1 = -k(x_1 - x_2)$$
$$m_2 \ddot{x}_2 = -k(x_2 - x_1) + f$$

Rewriting yields

$$m_1 \ddot{x}_1 + k(x_1 - x_2) = 0 \tag{8–40}$$
$$m_2 \ddot{x}_2 + k(x_2 - x_1) = f \tag{8–41}$$

From Equations (8–40) and (8–41), we obtain

$$m_1 m_2 (\ddot{x}_2 - \ddot{x}_1) + (km_2 + km_1)(x_2 - x_1) = m_1 f$$

If we define $x_2 - x_1 = x$, then this last equation simplifies to

$$m_1 m_2 \ddot{x} + k(m_1 + m_2)x = m_1 f$$

It follows that

$$\ddot{x} + \frac{k(m_1 + m_2)}{m_1 m_2}x = \frac{f}{m_2} \tag{8–42}$$

Let us define

$$\omega_n^2 = \frac{k(m_1 + m_2)}{m_1 m_2}$$

Then Equation (8–42) becomes

$$\ddot{x} + \omega_n^2 x = \frac{f}{m_2}$$

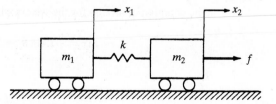

Figure 8–17   Mechanical system.

Taking the Laplace transform of this last equation, substituting the initial conditions $x(0) = 0$ and $\dot{x}(0) = 0$, and noting that f is a constant, we have

$$(s^2 + \omega_n^2)X(s) = \frac{f}{m_2 s}$$

or

$$X(s) = \frac{f}{m_2 s}\frac{1}{s^2 + \omega_n^2} = \frac{f}{m_2 \omega_n^2}\left(\frac{1}{s} - \frac{s}{s^2 + \omega_n^2}\right)$$

The inverse Laplace transform of $X(s)$ gives

$$x(t) = \frac{f}{m_2 \omega_n^2}(1 - \cos \omega_n t) \tag{8–43}$$

Now we shall determine $x_1(t)$. From Equations (8–40) and (8–43), we find that

$$m_1 \ddot{x}_1 = kx = \frac{kf}{m_2 \omega_n^2}(1 - \cos \omega_n t)$$

Since f is a constant, we can easily integrate the right-hand side of this last equation. Noting that $x_1(0) = 0$ and $\dot{x}_1(0) = 0$, we get

$$m_1 \dot{x}_1 = \frac{kf}{m_2 \omega_n^2}\left(t - \frac{1}{\omega_n}\sin \omega_n t\right)$$

and

$$m_1 x_1 = \frac{kf}{m_2 \omega_n^2}\left(\frac{t^2}{2} + \frac{1}{\omega_n^2}\cos \omega_n t\right) - \frac{kf}{m_2 \omega_n^4}$$

Thus,

$$x_1(t) = \frac{f}{m_1 + m_2}\frac{t^2}{2} - \frac{fm_1 m_2}{k(m_1 + m_2)^2}\left[1 - \cos\sqrt{\frac{k(m_1 + m_2)}{m_1 m_2}}t\right] \tag{8–44}$$

and the solution $x_2(t)$ is obtained from

$$x_2(t) = x(t) + x_1(t)$$

Substituting Equations (8–43) and (8–44) into this last equation and simplifying yields

$$x_2(t) = \frac{f}{m_1 + m_2}\frac{t^2}{2} + \frac{fm_1^2}{k(m_1 + m_2)^2}\left[1 - \cos\sqrt{\frac{k(m_1 + m_2)}{m_1 m_2}}t\right]$$

Problem A–8–7

The step response of a second-order system may be described by

$$\frac{Y(s)}{U(s)} = \frac{\omega_n^2}{s^2 + 2\zeta\omega_n s + \omega_n^2}$$

To see this, refer to Equation (8–16), rewritten thus:

$$\frac{X(s)}{\mathcal{L}[1(t)]} = \frac{a}{m\omega_n^2}\frac{\omega_n^2}{s^2 + 2\zeta\omega_n s + \omega_n^2} \tag{8–45}$$

If we define

$$\frac{m\omega_n^2}{a}X(s) = Y(s), \qquad \mathcal{L}[1(t)] = U(s)$$

then Equation (8–45) can be written as

$$\frac{Y(s)}{U(s)} = \frac{\omega_n^2}{s^2 + 2\zeta\omega_n s + \omega_n^2} \tag{8–46}$$

The maximum overshoot in the step response depends on ζ, and the time taken for the response to reach 2% of the final value depends on ζ and ω_n.

Obtain unit-step response curves of the system defined by Equation (8–46) for the following three cases:

1. Case 1: $\zeta = 0.3$, $\omega_n = 1$
2. Case 2: $\zeta = 0.5$, $\omega_n = 2$
3. Case 3: $\zeta = 0.7$, $\omega_n = 4$

Solution In writing a MATLAB program, we use a "for loop." Define $\omega_n^2 = a$ and $2\zeta\omega_n = b$. Then, a and b each have three elements as follows:

$$a = [1 \quad 4 \quad 16]$$
$$b = [0.6 \quad 2 \quad 5.6]$$

Using vectors a and b, MATLAB Program 8–1 will produce the unit-step response curves as shown in Figure 8–18.

MATLAB Program 8–1

```
>> a = [1       4       16];
>> b = [0.6     2       5.6];
>> t = 0:0.1:10;
>> y = zeros(101,3);
>>      for i = 1:3;
              num = [0   0   a(i)];
              den = [1   b(i)   a(i)];
              y(:,i) = step(num,den,t);
            end
>> plot(t,y(:,1),'o',t,y(:,2),'x',t,y(:,3),'-')
>> grid
>> title('Unit-Step Response Curves for Three Cases')
>> xlabel('t (sec)')
>> ylabel('Outputs')
>> text(4.5, 1.28, '1')
>> text(2.8, 1.1, '2')
>> text(0.35, 0.93, '3')
```

Problem A–8–8

As mentioned in **Problem A–8–7**, the step response of a second-order system may be described by

$$\frac{Y(s)}{U(s)} = \frac{\omega_n^2}{s^2 + 2\zeta\omega_n s + \omega_n^2} \tag{8–47}$$

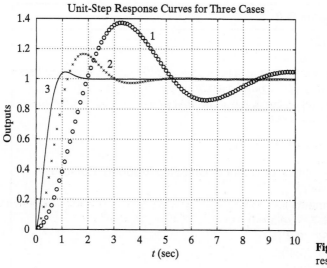

Unit-Step Response Curves for Three Cases

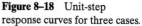

Figure 8–18 Unit-step response curves for three cases.

This equation involves two parameters: ζ and ω_n. If we normalize the system equation by defining $t = (1/\omega_n)\tau$ and writing the system equation in terms of τ, then Equation (8–47) may be modified to

$$\frac{Y(s)}{U(s)} = \frac{1}{s^2 + 2\zeta s + 1} \tag{8–48}$$

This normalized equation involves only one parameter: ζ. It is easy to see the effect of ζ on the unit-step response of the second-order system if it is defined by Equation (8–48).

Obtain the unit-step response curves of the system defined by Equation (8–48), where $\zeta = 0, 0.2, 0.4, 0.6, 0.8,$ and 1.0. Write a MATLAB program that uses a "for loop" to obtain the two-dimensional and three-dimensional plots of the system output.

Solution MATLAB Program 8–2 obtains two-dimensional and three-dimensional plots. Figure 8–19 is a two-dimensional plot of the unit-step response curves for the specified values of ζ. Figure 8–20 is a three-dimensional plot obtained with the use of the command "mesh(y')". [Note that if we use the command "mesh(y)", we get a similar three-dimensional plot, but the x-axis and y-axis are interchanged.]

Problem A–8–9

Obtain $e^{\mathbf{A}t}$, where

$$\mathbf{A} = \begin{bmatrix} 0 & 1 \\ 0 & -2 \end{bmatrix}$$

Solution Since

$$s\mathbf{I} - \mathbf{A} = \begin{bmatrix} s & 0 \\ 0 & s \end{bmatrix} - \begin{bmatrix} 0 & 1 \\ 0 & -2 \end{bmatrix} = \begin{bmatrix} s & -1 \\ 0 & s + 2 \end{bmatrix}$$

MATLAB Program 8–2

```
>> t = 0:0.2:12;
>>     for n = 1:6;
            num = [0    0    1];
            den = [1    2*(n−1)*0.2    1];
            [y(1:61,n),x,t] = step(num,den,t);
        end
>> plot(t,y)
>> grid
>> title('Two-Dimensional Plot of Unit-Step Response Curves')
>> xlabel('t (sec)')
>> ylabel('Outputs')
>> text(3.5, 1.7, '\zeta = 0')
>> text(3, 1.52, '0.2')
>> text(3,1.23, '0.4')
>> text(3,1.05, '0.6')
>> text(3,0.93, '0.8')
>> text(3,0.8, '1.0')
>>
>> % To draw a three-dimensional plot, enter the command mesh(y').
>>
>> mesh(y')
>> title('Three-Dimensional Plot of Unit-Step Response Curves')
>> xlabel('Computation Time Points')
>> ylabel('n  [\zeta = 0.2 (n−1)]')
>> zlabel('Outputs')
```

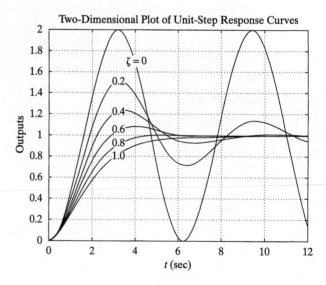

Figure 8–19 Two-dimensional plot of unit-step response curves.

Three-Dimensional Plot of Unit-Step Response Curves

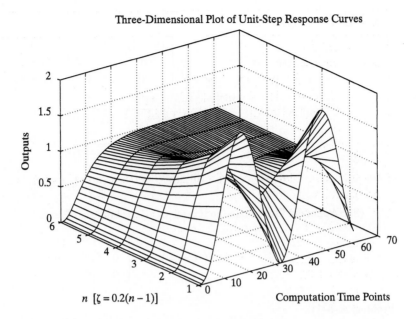

Figure 8–20   Three-dimensional plot of unit-step response curves. [In the plot,
$\zeta = 0.2(n - 1)$ and the incremental computation time is 0.2 s.]

it follows that

$$(s\mathbf{I} - \mathbf{A})^{-1} = \frac{1}{s(s + 2)}\begin{bmatrix} s + 2 & 1 \\ 0 & s \end{bmatrix} = \begin{bmatrix} \dfrac{1}{s} & \dfrac{1}{s(s + 2)} \\ 0 & \dfrac{1}{s + 2} \end{bmatrix}$$

Hence,

$$e^{\mathbf{A}t} = \mathscr{L}^{-1}[(s\mathbf{I} - \mathbf{A})^{-1}] = \begin{bmatrix} 1 & \frac{1}{2}(1 - e^{-2t}) \\ 0 & e^{-2t} \end{bmatrix}$$

Problem A–8–10

Obtain $e^{\mathbf{A}t}$, where

$$\mathbf{A} = \begin{bmatrix} 2 & 1 & 0 \\ 0 & 2 & 1 \\ 0 & 0 & 2 \end{bmatrix}$$

Solution   Since

$$s\mathbf{I} - \mathbf{A} = \begin{bmatrix} s & 0 & 0 \\ 0 & s & 0 \\ 0 & 0 & s \end{bmatrix} - \begin{bmatrix} 2 & 1 & 0 \\ 0 & 2 & 1 \\ 0 & 0 & 2 \end{bmatrix} = \begin{bmatrix} s - 2 & -1 & 0 \\ 0 & s - 2 & -1 \\ 0 & 0 & s - 2 \end{bmatrix}$$

we have

$$(s\mathbf{I} - \mathbf{A})^{-1} = \frac{1}{(s-2)^3} \begin{bmatrix} (s-2)^2 & (s-2) & 1 \\ 0 & (s-2)^2 & (s-2) \\ 0 & 0 & (s-2)^2 \end{bmatrix}$$

$$= \begin{bmatrix} \dfrac{1}{s-2} & \dfrac{1}{(s-2)^2} & \dfrac{1}{(s-2)^3} \\ 0 & \dfrac{1}{s-2} & \dfrac{1}{(s-2)^2} \\ 0 & 0 & \dfrac{1}{s-2} \end{bmatrix}$$

(See **Appendix C** to obtain the inverse of a 3 × 3 matrix.) Hence,

$$e^{\mathbf{A}t} = \mathcal{L}^{-1}[(s\mathbf{I} - \mathbf{A})^{-1}] = \begin{bmatrix} e^{2t} & te^{2t} & \frac{1}{2}t^2 e^{2t} \\ 0 & e^{2t} & te^{2t} \\ 0 & 0 & e^{2t} \end{bmatrix}$$

Problem A–8–11

Obtain the response of the system

$$\begin{bmatrix} \dot{x}_1 \\ \dot{x}_2 \end{bmatrix} = \begin{bmatrix} 0 & 1 \\ 0 & -2 \end{bmatrix} \begin{bmatrix} x_1 \\ x_2 \end{bmatrix} + \begin{bmatrix} 0 \\ 1 \end{bmatrix} u$$

when the input u is a unit-step function. Assume that $\mathbf{x}(0) = \mathbf{0}$.

Solution   From **Problem A–8–9**, we have

$$e^{\mathbf{A}t} = \begin{bmatrix} 1 & \frac{1}{2}(1 - e^{-2t}) \\ 0 & e^{-2t} \end{bmatrix}$$

From Equation (8–36),

$$\mathbf{x}(t) = e^{\mathbf{A}t}\mathbf{x}(0) + \int_0^t e^{\mathbf{A}(t-\tau)}\mathbf{B}u(\tau)\, d\tau$$

Since $\mathbf{x}(0) = \mathbf{0}$ and $u(t) = 1(t)$, we get

$$\mathbf{x}(t) = \int_0^t e^{\mathbf{A}(t-\tau)}\mathbf{B}1(\tau)\, d\tau$$

$$= e^{\mathbf{A}t}\int_0^t e^{-\mathbf{A}\tau}\,\mathbf{B}\, d\tau$$

$$= \begin{bmatrix} 1 & \frac{1}{2}(1 - e^{-2t}) \\ 0 & e^{-2t} \end{bmatrix} \int_0^t \begin{bmatrix} 1 & \frac{1}{2}(1 - e^{2\tau}) \\ 0 & e^{2\tau} \end{bmatrix} \begin{bmatrix} 0 \\ 1 \end{bmatrix} d\tau$$

$$= \begin{bmatrix} 1 & \frac{1}{2}(1 - e^{-2t}) \\ 0 & e^{-2t} \end{bmatrix} \begin{bmatrix} \frac{1}{2}t - \frac{1}{4}e^{2t} + \frac{1}{4} \\ \frac{1}{2}e^{2t} - \frac{1}{2} \end{bmatrix}$$

$$= \begin{bmatrix} \frac{1}{2}t - \frac{1}{4} + \frac{1}{4}e^{-2t} \\ \frac{1}{2} - \frac{1}{2}e^{-2t} \end{bmatrix}$$

Thus,

$$\begin{bmatrix} x_1(t) \\ x_2(t) \end{bmatrix} = \begin{bmatrix} \frac{1}{2}t - \frac{1}{4} + \frac{1}{4}e^{-2t} \\ \frac{1}{2} - \frac{1}{2}e^{-2t} \end{bmatrix}$$

Problem A–8–12

Obtain the response $y(t)$ of the system

$$\begin{bmatrix} \dot{x}_1 \\ \dot{x}_1 \end{bmatrix} = \begin{bmatrix} -1 & -0.5 \\ 1 & 0 \end{bmatrix}\begin{bmatrix} x_1 \\ x_2 \end{bmatrix} + \begin{bmatrix} 0.5 \\ 0 \end{bmatrix}u, \qquad \begin{bmatrix} x_1(0) \\ x_2(0) \end{bmatrix} = \begin{bmatrix} 0 \\ 0 \end{bmatrix}$$

$$y = \begin{bmatrix} 1 & 0 \end{bmatrix}\begin{bmatrix} x_1 \\ x_2 \end{bmatrix}$$

where $u(t)$ is the unit-step input occurring at $t = 0$, or

$$u(t) = 1(t)$$

Solution   For this system,

$$\mathbf{A} = \begin{bmatrix} -1 & -0.5 \\ 1 & 0 \end{bmatrix}, \qquad \mathbf{B} = \begin{bmatrix} 0.5 \\ 0 \end{bmatrix}$$

The state transition matrix $\Phi(t) = e^{\mathbf{A}t}$ can be obtained as follows:

$$\Phi(t) = e^{\mathbf{A}t} = \mathscr{L}^{-1}[(s\mathbf{I} - \mathbf{A})^{-1}]$$

Since

$$(s\mathbf{I} - \mathbf{A})^{-1} = \begin{bmatrix} s+1 & 0.5 \\ -1 & s \end{bmatrix}^{-1} = \frac{1}{s^2 + s + 0.5}\begin{bmatrix} s & -0.5 \\ 1 & s+1 \end{bmatrix}$$

$$= \begin{bmatrix} \dfrac{s + 0.5 - 0.5}{(s + 0.5)^2 + 0.5^2} & \dfrac{-0.5}{(s + 0.5)^2 + 0.5^2} \\[3mm] \dfrac{1}{(s + 0.5)^2 + 0.5^2} & \dfrac{s + 0.5 + 0.5}{(s + 0.5)^2 + 0.5^2} \end{bmatrix}$$

we have

$$\Phi(t) = e^{\mathbf{A}t} = \mathscr{L}^{-1}[(s\mathbf{I} - \mathbf{A})^{-1}]$$

$$= \begin{bmatrix} e^{-0.5t}(\cos 0.5t - \sin 0.5t) & -e^{-0.5t}\sin 0.5t \\ 2e^{-0.5t}\sin 0.5t & e^{-0.5t}(\cos 0.5t + \sin 0.5t) \end{bmatrix}$$

Since $\mathbf{x}(0) = \mathbf{0}$ and $u(t) = 1(t)$, referring to Equation (8–39), we get

$$\mathbf{x}(t) = e^{\mathbf{A}t}\mathbf{x}(0) + \mathbf{A}^{-1}(e^{\mathbf{A}t} - \mathbf{I})\mathbf{B}$$

$$= \mathbf{A}^{-1}(e^{\mathbf{A}t} - \mathbf{I})\mathbf{B}$$

$$= \begin{bmatrix} 0 & 1 \\ -2 & -2 \end{bmatrix}\begin{bmatrix} 0.5e^{-0.5t}(\cos 0.5t - \sin 0.5t) - 0.5 \\ e^{-0.5t}\sin 0.5t \end{bmatrix}$$

$$= \begin{bmatrix} e^{-0.5t}\sin 0.5t \\ -e^{-0.5t}(\cos 0.5t + \sin 0.5t) + 1 \end{bmatrix}$$

Hence, the output $y(t)$ is given by

$$y(t) = \begin{bmatrix} 1 & 0 \end{bmatrix}\begin{bmatrix} x_1 \\ x_2 \end{bmatrix} = x_1 = e^{-0.5t}\sin 0.5t$$

Problem A–8–13

Consider the system defined by

$$\dot{\mathbf{x}} = \mathbf{A}\mathbf{x} + \mathbf{B}\mathbf{u} \tag{8–49}$$

where

$\mathbf{x}$ = state vector (n-vector)
$\mathbf{u}$ = input vector (r-vector)
$\mathbf{A}$ = $n \times n$ constant matrix
$\mathbf{B}$ = $n \times r$ constant matrix

Obtain the response of the system to the input $\mathbf{u}$ whose r components are step functions of various magnitudes, or

$$\mathbf{u} = \begin{bmatrix} u_1 \\ u_2 \\ \vdots \\ u_r \end{bmatrix} = \begin{bmatrix} k_1 \cdot 1(t) \\ k_2 \cdot 1(t) \\ \vdots \\ k_r \cdot 1(t) \end{bmatrix} = \mathbf{k} \cdot 1(t)$$

Solution   The response to the step input $\mathbf{u} = \mathbf{k} \cdot 1(t)$ given at $t = 0$ is

$$\mathbf{x}(t) = e^{\mathbf{A}t}\mathbf{x}(0) + \int_0^t e^{\mathbf{A}(t-\tau)}\mathbf{B}\mathbf{k}\, d\tau$$

$$= e^{\mathbf{A}t}\mathbf{x}(0) + e^{\mathbf{A}t}\left[\int_0^t \left(\mathbf{I} - \mathbf{A}\tau + \frac{\mathbf{A}^2\tau^2}{2!} - \cdots\right)d\tau\right]\mathbf{B}\mathbf{k}$$

$$= e^{\mathbf{A}t}\mathbf{x}(0) + e^{\mathbf{A}t}\left(\mathbf{I}t - \frac{\mathbf{A}t^2}{2!} + \frac{\mathbf{A}^2t^3}{3!} - \cdots\right)\mathbf{B}\mathbf{k}$$

If $\mathbf{A}$ is nonsingular, then this last equation can be simplified to give

$$\mathbf{x}(t) = e^{\mathbf{A}t}\mathbf{x}(0) + e^{\mathbf{A}t}[-(\mathbf{A}^{-1})(e^{-\mathbf{A}t} - \mathbf{I})]\mathbf{B}\mathbf{k}$$
$$= e^{\mathbf{A}t}\mathbf{x}(0) + \mathbf{A}^{-1}(e^{\mathbf{A}t} - \mathbf{I})\mathbf{B}\mathbf{k}$$

This is the analytical expression of the step response of the system defined by Equation (8–49).

Problem A–8–14

Consider the system defined by

$$\dot{\mathbf{x}} = \mathbf{A}\mathbf{x} + \mathbf{B}\mathbf{u}$$

where

$\mathbf{x}$ = state vector (n-vector)
$\mathbf{u}$ = input vector (r-vector)
$\mathbf{A}$ = $n \times n$ constant matrix
$\mathbf{B}$ = $n \times r$ constant matrix

Obtain the response of the system to the input $\mathbf{u}$ whose r components are impulse functions of various magnitudes occuring at $t = 0$, or

$$\mathbf{u} = \mathbf{w}\delta(t)$$

where **w** is a vector whose components are the magnitudes of r impulse functions applied at $t = 0$.

Solution   From Equation (8–38), the solution of the given state equation is

$$\mathbf{x}(t) = e^{\mathbf{A}(t-t_0)}\mathbf{x}(t_0) + \int_{t_0}^{t} e^{\mathbf{A}(t-\tau)}\mathbf{B}\mathbf{u}(\tau)\, d\tau$$

Substituting $t_0 = 0-$ into this solution, we obtain

$$\mathbf{x}(t) = e^{\mathbf{A}t}\mathbf{x}(0-) + \int_{0-}^{t} e^{\mathbf{A}(t-\tau)}\mathbf{B}\mathbf{u}(\tau)\, d\tau$$

The solution of the state equation when the impulse input $\mathbf{w}\,\delta(t)$ is given at $t = 0$ is

$$\mathbf{x}(t) = e^{\mathbf{A}t}\mathbf{x}(0-) + \int_{0-}^{t} e^{\mathbf{A}(t-\tau)}\mathbf{B}\delta(\tau)\mathbf{w}\, d\tau$$

$$= e^{\mathbf{A}t}\mathbf{x}(0-) + e^{\mathbf{A}t}\mathbf{B}\mathbf{w}$$

This last equation gives the response to the impulse input $\mathbf{w}\,\delta(t)$.

Problem A–8–15

Consider the system defined by

$$\dot{\mathbf{x}} = \mathbf{A}\mathbf{x} + \mathbf{B}\mathbf{u}$$

where

> $\mathbf{x}$ = state vector (n-vector)
> $\mathbf{u}$ = input vector (r-vector)
> $\mathbf{A}$ = $n \times n$ constant matrix
> $\mathbf{B}$ = $n \times r$ constant matrix

Obtain the response of the system to the input $\mathbf{u}$ whose r components are ramp functions of various magnitudes, or

$$\mathbf{u} = \mathbf{v}t$$

where **v** is a vector whose components are magnitudes of ramp functions applied at $t = 0$.

Solution   The response to the ramp input $\mathbf{v}t$ given at $t = 0$ is

$$\mathbf{x}(t) = e^{\mathbf{A}t}\mathbf{x}(0) + \int_{0}^{t} e^{\mathbf{A}(t-\tau)}\mathbf{B}\tau\mathbf{v}\, d\tau$$

$$= e^{\mathbf{A}t}\mathbf{x}(0) + e^{\mathbf{A}t}\int_{0}^{t} e^{-\mathbf{A}\tau}\tau\, d\tau\,\mathbf{B}\mathbf{v}$$

$$= e^{\mathbf{A}t}\mathbf{x}(0) + e^{\mathbf{A}t}\left(\frac{\mathbf{I}}{2}t^2 - \frac{2\mathbf{A}}{3!}t^3 + \frac{3\mathbf{A}^2}{4!}t^4 - \frac{4\mathbf{A}^3}{5!}t^5 + \cdots\right)\mathbf{B}\mathbf{v}$$

If **A** is nonsingular, then this last equation can be simplified to

$$\mathbf{x}(t) = e^{\mathbf{A}t}\mathbf{x}(0) + (\mathbf{A}^{-2})(e^{\mathbf{A}t} - \mathbf{I} - \mathbf{A}t)\mathbf{B}\mathbf{v}$$

$$= e^{\mathbf{A}t}\mathbf{x}(0) + [\mathbf{A}^{-2}(e^{\mathbf{A}t} - \mathbf{I}) - \mathbf{A}^{-1}t]\mathbf{B}\mathbf{v}$$

The latter equation gives the response to the ramp input $\mathbf{v}t$.

PROBLEMS

Problem B–8–1

In the electrical system of Figure 8–21, switch S is closed at $t = 0$. Find the voltage $e_o(t)$. Assume that the capacitor is initially uncharged.

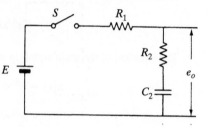

Figure 8–21   Electrical system.

Problem B–8–2

Consider the electrical system shown in Figure 8–22. The voltage source E is suddenly connected by means of switch S at $t = 0$. Assume that capacitor C is initially uncharged and that inductance L carries no initial current. What is the current $i(t)$ for $t > 0$?

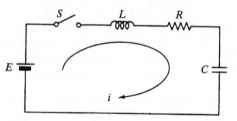

Figure 8–22   Electrical system.

Problem B–8–3

Derive the transfer function $E_o(s)/E_i(s)$ of the electrical circuit shown in Figure 8–23. Then obtain the response $e_o(t)$ when the input $e_i(t)$ is a step function of magnitude E_i, or

$$e_i(t) = E_i \cdot 1(t)$$

Assume that the initial charge in the capacitor is zero.

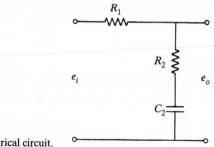

Figure 8–23   Electrical circuit.

Problem B–8–4

Find the transfer function $X_o(s)/X_i(s)$ of the mechanical system shown in Figure 8–24. Obtain the displacement $x_o(t)$ when the input $x_i(t)$ is a step displacement of magnitude X_i occurring at $t = 0$. Assume that $x_o(0-) = 0$. The displacement $x_o(t)$ is measured from the equilibrium position before the input $x_i(t)$ is given.

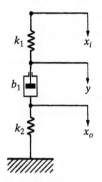

Figure 8–24   Mechanical system.

Problem B–8–5

Derive the transfer function $X_o(s)/X_i(s)$ of the mechanical system shown in Figure 8–25. Then obtain the response $x_o(t)$ when the input $x_i(t)$ is a pulse signal given by

$$x_i(t) = X_i \qquad 0 < t < t_1$$
$$= 0 \qquad \text{elsewhere}$$

Assume that $x_o(0-) = 0$. The displacement $x_o(t)$ is measured from the rest position before the input $x_i(t)$ is given.

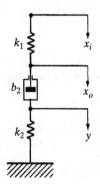

Figure 8–25   Mechanical system.

Problem B–8–6

Find the transfer function $E_o(s)/E_i(s)$ of the electrical circuit shown in Figure 8–26. Suppose that the input $e_i(t)$ is a pulse signal given by

$$e_i(t) = E_i \qquad 0 < t < t_1$$
$$= 0 \qquad \text{elsewhere}$$

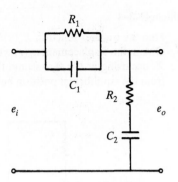

Figure 8–26 Electrical circuit.

Obtain the output $e_o(t)$. Assume that $R_2 = 1.5R_1$, $C_2 = C_1$, and $R_1C_1 = 1$ s. Assume also that the initial charges in the capacitors are zero.

Problem B–8–7

A free vibration of the mechanical system shown in Figure 8–27(a) indicates that the amplitude of vibration decreases to 25% of the value at $t = t_o$ after four consecutive cycles of motion, as Figure 8–27(b) shows. Determine the viscous-friction coefficient b of the system if $m = 1$ kg and $k = 500$ N/m.

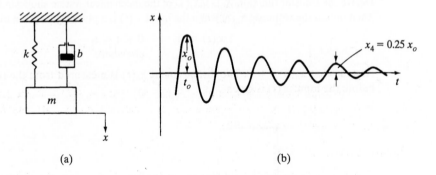

(a)   (b)

Figure 8–27 (a) Mechanical system; (b) portion of a free-vibration curve.

Problem B–8–8

A mass of 20 kg is supported by a spring and damper as shown in Figure 8–28(a). The system is at rest for $t < 0$. At $t = 0$, a mass of 2 kg is added to the 20-kg mass. The system vibrates as shown in Figure 8–28(b). Determine the spring constant k and the viscous-friction coefficient b. [Note that $(0.02/0.08) \times 100 = 25\%$ maximum overshoot corresponds to $\zeta = 0.4$.]

Problem B–8–9

Consider the mechanical system shown in Figure 8–29. It is at rest for $t < 0$. The pendulum m_2 is supported by mass m_1, which vibrates because of an elastic connection. Derive the equations of motion for the system. The displacement x is measured from the equilibrium position for $t < 0$. The angular displacement θ is measured from the vertical axis passing through the pivot of the pendulum.

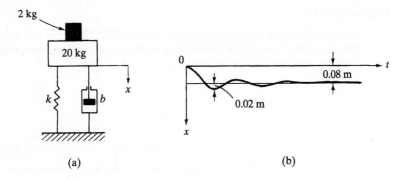

Figure 8–28 (a) Mechanical system; (b) step-response curve.

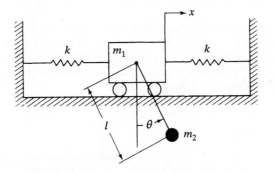

Figure 8–29 Mechanical system.

Assuming the initial conditions to be $x(0) = 0.1$ m, $\dot{x}(0) = 0$ m/s, $\theta(0) = 0$ rad, and $\dot{\theta}(0) = 0$ rad/s, obtain the motion of the pendulum. Assume also that $m_1 = 10$ kg, $m_2 = 1$ kg, $k = 250$ N/m, and $l = 1$ m.

Problem B–8–10

Mass $m = 1$ kg is vibrating initially in the mechanical system shown in Figure 8–30. At $t = 0$, we hit the mass with an impulsive force $p(t)$ whose strength is 10 N. Assuming that the spring constant k is 100 N/m and that $x(0-) = 0.1$ m and $\dot{x}(0-) = 1$ m/s, find the displacement $x(t)$ as a function of time t. The displacement $x(t)$ is measured from the equilibrium position in the absence of an excitation force.

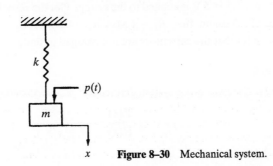

Figure 8–30 Mechanical system.

Problem B-8-11

Figure 8-31 shows a mechanical system that consists of a mass and a damper. The system, initially at rest, is set into motion by an impulsive force whose strength is unity. Find the response $x(t)$ and the initial velocity of mass m.

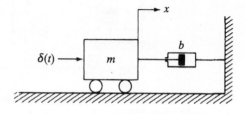

Figure 8-31 Mechanical system.

Problem B-8-12

Consider the mechanical system shown in Figure 8-32. The system is at rest for $t < 0$. Assume that $k_1 = 4$ N/m, $k_2 = 20$ N/m, $b_1 = 1$ N-s/m, and $b_2 = 10$ N-s/m. Obtain the displacement $x_2(t)$ when u is a step force input of 2 N. Plot the response curve $x_2(t)$ versus t with MATLAB. The displacements x_1 and x_2 are measured from their respective equilibrium positions before the input u is given.

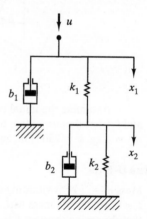

Figure 8-32 Mechanical system.

Problem B-8-13

Consider the electrical circuit shown in Figure 8-33. Obtain the response $e_o(t)$ when a step input $e_i(t) = 5$ V is applied to the system. Plot the response curve $e_o(t)$ versus t with MATLAB. Assume that $R_1 = 1$ MΩ, $R_2 = 0.5$ MΩ, $C_1 = 0.5$ μF, and $C_2 = 0.1$ μF. Assume also that the capacitors are not charged initially.

Problem B-8-14

Consider a second-order system defined by

$$\frac{Y(s)}{U(s)} = \frac{1}{s^2 + s + 1}$$

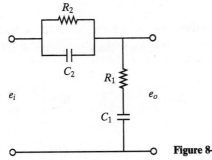

Figure 8–33 Electrical circuit.

Obtain the response $y(t)$ when the input $u(t)$ is a unit acceleration input $[u(t) = \frac{1}{2}t^2]$. Obtain the response curve with MATLAB.

Problem B–8–15

Consider a second-order system defined by

$$\frac{Y(s)}{U(s)} = \frac{1}{s^2 + 2\zeta s + 1}$$

Obtain the unit-impulse response curves of the system for $\zeta = 0, 0.2, 0.4, 0.6, 0.8,$ and 1.0. Plot the six response curves in a two-dimensional diagram and a three-dimensional diagram.

Problem B–8–16

Obtain $e^{\mathbf{A}t}$, where

$$\mathbf{A} = \begin{bmatrix} -2 & -1 \\ 2 & -5 \end{bmatrix}$$

Problem B–8–17

Obtain $e^{\mathbf{A}t}$, where

$$\mathbf{A} = \begin{bmatrix} 0 & 1 & 0 \\ 0 & 0 & 1 \\ 1 & -3 & 3 \end{bmatrix}$$

Problem B–8–18

Consider the system defined by

$$\dot{\mathbf{x}} = \mathbf{A}\mathbf{x} + \mathbf{B}u, \qquad \mathbf{x}(0) = \mathbf{0}$$

where

$$\mathbf{A} = \begin{bmatrix} 0 & 1 \\ -6 & -5 \end{bmatrix}, \qquad \mathbf{B} = \begin{bmatrix} 0 \\ 1 \end{bmatrix}$$

Obtain the response $\mathbf{x}(t)$ analytically when u is a unit-step function.

Problem B–8–19

Consider the system defined by

$$\dot{x} = Ax + Bu, \qquad x(0) = 0$$

where

$$A = \begin{bmatrix} 0 & 1 \\ -6 & -5 \end{bmatrix}, \qquad B = \begin{bmatrix} 0 & 1 \\ 1 & 1 \end{bmatrix}$$

and

$$u = \begin{bmatrix} u_1 \\ u_2 \end{bmatrix} = \begin{bmatrix} 2 \cdot 1(t) \\ 5 \cdot 1(t) \end{bmatrix}$$

Obtain the response $x(t)$ analytically.

Problem B–8–20

For the system of **Problem B–8–19**, obtain the response curves $x_1(t)$ versus t and $x_2(t)$ versus t with MATLAB.

Frequency-Domain Analysis of Dynamic Systems

9–1 INTRODUCTION

Responses of linear, time-invariant systems to sinusoidal inputs are the major subject of this chapter. First we define the sinusoidal transfer function and explain its use in the steady-state sinusoidal response. Then we treat vibrations in rotating mechanical systems, present some vibration isolation problems, and examine dynamic vibration absorbers. Finally, we deal with vibrations in multi-degrees-of-freedom systems.

Outline of the chapter. Section 9–1 gives introductory material. Section 9–2 begins with forced vibrations of mechanical systems and then derives the sinusoidal transfer function for the linear, time-invariant dynamic system. Section 9–3 treats vibrations in rotating mechanical systems. Section 9–4 examines vibration isolation problems that arise in rotating mechanical systems. In this regard, transmissibility for force excitation and that for motion excitation are discussed. Section 9–5 presents a way to reduce vibrations caused by rotating unbalance and treats a dynamic vibration absorber commonly used in industries. Section 9–6 analyzes free vibrations in multi-degrees-of-freedom systems and discusses modes of vibration.

9-2 SINUSOIDAL TRANSFER FUNCTION

When a sinusoidal input is applied to a linear, time-invariant system, the system will tend to vibrate at its own natural frequency, as well as follow the frequency of the input. In the presence of damping, that portion of motion not sustained by the sinusoidal input will gradually die out. As a result, the response at steady state is sinusoidal at the same frequency as the input. The steady-state output differs from the input only in the amplitude and phase angle. Thus, the output–input amplitude ratio and the phase angle between the output and input sinusoids are the only two parameters needed to predict the steady-state output of a linear, time-invariant system when the input is a sinusoid. In general, the amplitude ratio and the phase angle depend on the input frequency.

Frequency response. The term *frequency response* refers to the steady-state response of a system to a sinusoidal input. For all frequencies from zero to infinity, the frequency-response characteristics of a system can be completely described by the output–input amplitude ratio and the phase angle between the output and input sinusoids. In this method of systems analysis, we vary the frequency of the input signal over a wide range and study the resulting response. (We shall present detailed discussions of frequency response in Chapter 11.)

Forced vibration without damping. Figure 9–1 illustrates a spring–mass system in which the mass is subjected to a sinusoidal input force $P \sin \omega t$. Let us find the response of the system when it is initially at rest.

If we measure the displacement x from the equilibrium position, the equation of motion for the system becomes

$$m\ddot{x} + kx = P \sin \omega t$$

or

$$\ddot{x} + \frac{k}{m}x = \frac{P}{m} \sin \omega t \tag{9-1}$$

Note that the solution of this equation consists of the vibration at its own natural frequency (the complementary solution) and that at the forcing frequency (the particular solution). Thus, the solution $x(t)$ can be written as

$$x(t) = \text{complementary solution} + \text{particular solution}$$

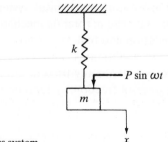

Figure 9–1 Spring–mass system.

Now we shall obtain the solution of Equation (9–1) under the condition that the system is initially at rest. Taking the Laplace transform of Equation (9–1) and using the initial conditions $x(0) = 0$ and $\dot{x}(0) = 0$, we obtain

$$\left(s^2 + \frac{k}{m}\right)X(s) = \frac{P}{m}\frac{\omega}{s^2 + \omega^2}$$

Solving for $X(s)$ yields

$$X(s) = \frac{P}{m}\frac{\omega}{s^2 + \omega^2}\frac{1}{s^2 + (k/m)}$$

$$= \frac{-P\omega\sqrt{m/k}}{k - m\omega^2}\frac{\sqrt{k/m}}{s^2 + (k/m)} + \frac{P}{k - m\omega^2}\frac{\omega}{s^2 + \omega^2}$$

The inverse Laplace transform of this last equation gives

$$x(t) = -\frac{P\omega\sqrt{m/k}}{k - m\omega^2}\sin\sqrt{\frac{k}{m}}t + \frac{P}{k - m\omega^2}\sin\omega t \qquad (9\text{--}2)$$

This is the complete solution. The first term is the complementary solution (the natural frequency vibration does not decay in this system), and the second term is the particular solution. [Note that if we need only a steady-state solution (particular solution) of a stable system, the use of the sinusoidal transfer function simplifies the solution. The sinusoidal transfer function is discussed in detail later in this section.]

Let us examine Equation (9–2). As the forcing frequency ω approaches zero, the amplitude of the vibration at the natural frequency $\sqrt{k/m}$ approaches zero and the amplitude of the vibration at the forcing frequency ω approaches P/k. This value P/k is the deflection of the mass that would result if the force P were applied steadily (at zero frequency). That is, P/k is the static deflection. As the frequency ω increases from zero, the denominator $k - m\omega^2$ of the solution becomes smaller and the amplitudes become larger. As the frequency ω is further increased and becomes equal to the natural frequency of the system (i.e., $\omega = \omega_n = \sqrt{k/m}$), resonance occurs. At resonance, the denominator $k - m\omega^2$, becomes zero, and the amplitude of vibration will increase without bound. (When the sinusoidal input is applied at the natural frequency and in phase with the motion—that is, in the same direction as the velocity—the input force is actually doing work on the system and is adding energy to it that will appear as an increase in amplitudes.) As ω continues to increase past resonance, the denominator $k - m\omega^2$ becomes negative and assumes increasingly larger values, approaching negative infinity. Therefore, the amplitudes of vibration (at the natural frequency and at the forcing frequency) approach zero from the negative side, starting at negative infinity when $\omega = \omega_n+$. In other words, if ω is below resonance, that part of the vibration at the forcing frequency (particular solution) is in phase with the forcing sinusoid. If ω is above resonance, this vibration becomes 180° out of phase.

Sinusoidal transfer function.   The *sinusoidal transfer function* is defined as the transfer function $G(s)$ in which s is replaced by $j\omega$. When only the steady-state solution (the particular solution) is wanted, the sinusoidal transfer function $G(j\omega)$

can simplify the solution. In the discussion that follows, we shall consider the behavior of stable linear, time-invariant systems under steady-state conditions—that is, after the initial transients have died out. We shall see that sinusoidal inputs produce sinusoidal outputs in the steady state, with the amplitude and phase angle at each frequency ω determined by the magnitude and angle of $G(j\omega)$, respectively.

Deriving steady-state output caused by sinusoidal input.   We shall show how the frequency-response characteristics of a stable system can be derived directly from the sinusoidal transfer function. For the linear, time-invariant system $G(s)$ shown in Figure 9–2, the input and output are denoted by $p(t)$ and $x(t)$, respectively. The input $p(t)$ is sinusoidal and is given by

$$p(t) = P \sin \omega t$$

We shall show that the output $x(t)$ at steady state is given by

$$x(t) = |G(j\omega)| \, P \sin(\omega t + \phi)$$

where $|G(j\omega)|$ and ϕ are the magnitude and angle of $G(j\omega)$, respectively.

Suppose that the transfer function $G(s)$ can be written as a ratio of two polynomials in s; that is,

$$G(s) = \frac{K(s + z_1)(s + z_2)\cdots(s + z_m)}{(s + s_1)(s + s_2)\cdots(s + s_n)}$$

The Laplace-transformed output $X(s)$ is

$$X(s) = G(s)P(s) \tag{9–3}$$

where $P(s)$ is the Laplace transform of the input $p(t)$.

Let us limit our discussion to stable systems. For such systems, the real parts of the $-s_i$ are negative. The steady-state response of a stable linear system to a sinusoidal input does not depend on the initial conditions, so they can be ignored.

If $G(s)$ has only distinct poles, then the partial-fraction expansion of Equation (9–3) yields

$$X(s) = G(s)\frac{P\omega}{s^2 + \omega^2}$$

$$= \frac{a}{s + j\omega} + \frac{\bar{a}}{s - j\omega} + \frac{b_1}{s + s_1} + \frac{b_2}{s + s_2} + \cdots + \frac{b_n}{s + s_n} \tag{9–4}$$

where a and $b_i(i = 1, 2, \ldots, n)$ are constants and $\bar{a}$ is the complex conjugate of a. The inverse Laplace transform of Equation (9–4) gives

$$x(t) = ae^{-j\omega t} + \bar{a}e^{j\omega t} + b_1 e^{-s_1 t} + b_2 e^{-s_2 t} + \cdots + b_n e^{-s_n t}$$

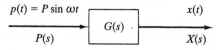

Figure 9–2   Linear, time-invariant system.

For a stable system, as t approaches infinity, the terms $e^{-s_1 t}, e^{-s_2 t}, \ldots, e^{-s_n t}$ approach zero, since $-s_1, -s_2, \ldots, -s_n$ have negative real parts. Thus, all terms on the right-hand side of this last equation, except the first two, drop out at steady state.

If $G(s)$ involves k multiple poles s_j, then $x(t)$ will involve such terms as $t^h e^{-s_j t}$ (where $h = 0, 1, \ldots, k - 1$). Since the real part of the $-s_j$ is negative for a stable system, the terms $t^h e^{-s_j t}$ approach zero as t approaches infinity.

Regardless of whether the system involves multiple poles, the steady-state response thus becomes

$$x(t) = ae^{-j\omega t} + \bar{a}e^{j\omega t} \tag{9-5}$$

where the constants a and $\bar{a}$ can be evaluated from Equation (9–4):

$$a = G(s)\frac{P\omega}{s^2 + \omega^2}(s + j\omega)\bigg|_{s=-j\omega} = -\frac{P}{2j}G(-j\omega)$$

$$\bar{a} = G(s)\frac{P\omega}{s^2 + \omega^2}(s - j\omega)\bigg|_{s=j\omega} = \frac{P}{2j}G(j\omega)$$

(Note that $\bar{a}$ is the complex conjugate of a.) Referring to Figure 9–3, we can write

$$
\begin{aligned}
G(j\omega) &= G_x + jG_y \\
&= |G(j\omega)|\cos\phi + j|G(j\omega)|\sin\phi \\
&= |G(j\omega)|(\cos\phi + j\sin\phi) \\
&= |G(j\omega)|e^{j\phi}
\end{aligned}
$$

(Note that $\underline{/G(j\omega)} = \underline{/e^{j\phi}} = \phi$.) Similarly,

$$G(-j\omega) = |G(-j\omega)|e^{-j\phi} = |G(j\omega)|e^{-j\phi}$$

It follows that

$$a = -\frac{P}{2j}|G(j\omega)|e^{-j\phi}$$

$$\bar{a} = \frac{P}{2j}|G(j\omega)|e^{j\phi}$$

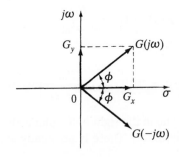

Figure 9–3   Complex function and its complex conjugate.

Then Equation (9–5) can be written as

$$x(t) = |G(j\omega)|P\frac{e^{j(\omega t+\phi)} - e^{-j(\omega t+\phi)}}{2j}$$

$$= |G(j\omega)|P\sin(\omega t + \phi)$$

$$= X\sin(\omega t + \phi) \tag{9–6}$$

where $X = |G(j\omega)|P$ and $\phi = \underline{/G(j\omega)}$. We see that a stable linear system subjected to a sinusoidal input will, at steady state, have a sinusoidal output of the same frequency as the input. But the amplitude and phase angle of the output will, in general, differ from the input's. In fact, the output's amplitude is given by the product of the amplitude of the input and $|G(j\omega)|$, whereas the phase angle differs from that of the input by the amount $\phi = \underline{/G(j\omega)}$.

On the basis of the preceding analysis, we are able to derive the following important result: For sinusoidal inputs,

$$|G(j\omega)| = \left|\frac{X(j\omega)}{P(j\omega)}\right| = \begin{array}{l}\text{amplitude ratio of the output}\\ \text{sinusoid to the input sinusoid}\end{array} \tag{9–7}$$

$$\underline{/G(j\omega)} = \underline{/\frac{X(j\omega)}{P(j\omega)}} = \tan^{-1}\left[\frac{\text{imaginary part of } G(j\omega)}{\text{real part of } G(j\omega)}\right]$$

$$= \begin{array}{l}\text{phase shift of the output sinusoid}\\ \text{with respect to the input sinusoid}\end{array} \tag{9–8}$$

Thus, the steady-state response characteristics of a linear system to a sinusoidal input can be found directly from $G(j\omega)$, the ratio of $X(j\omega)$ to $P(j\omega)$.

Note that the sinusoidal transfer function $G(j\omega)$ is a complex quantity that can be represented by the magnitude and phase angle with the frequency ω as a parameter. To characterize a linear system completely by its frequency-response curves, we must specify both the amplitude ratio and the phase angle as a function of the frequency ω.

Comments.   Equation (9–6) is valid only if $G(s) = X(s)/P(s)$ is a stable system, that is, if all poles of $G(s)$ lie in the left half s-plane. If a pole is at the origin and/or poles of $G(s)$ lie on the $j\omega$-axis (any poles on the $j\omega$-axis, except that at the origin, must occur as a pair of complex conjugates), the output $x(t)$ may be obtained by taking the inverse Laplace transform of the equation

$$X(s) = G(s)P(s) = G(s)\frac{P\omega}{s^2 + \omega^2}$$

or

$$x(t) = \mathcal{L}^{-1}[X(s)] = \mathcal{L}^{-1}\left[G(s)\frac{P\omega}{s^2 + \omega^2}\right]$$

Note that if one or more poles of $G(s)$ lie in the right half s-plane, then the system is unstable and the response grows indefinitely. There is no steady state for such an unstable system.

Example 9–1

Consider the transfer-function system

$$\frac{X(s)}{P(s)} = G(s) = \frac{1}{Ts + 1}$$

For the sinusoidal input $p(t) = P \sin \omega t$, what is the steady-state output $x(t)$?
Substituting $j\omega$ for s in $G(s)$ yields

$$G(j\omega) = \frac{1}{Tj\omega + 1}$$

The output–input amplitude ratio is

$$|G(j\omega)| = \frac{1}{\sqrt{T^2\omega^2 + 1}}$$

whereas the phase angle ϕ is

$$\phi = \underline{/G(j\omega)} = -\tan^{-1} T\omega$$

So, for the input $p(t) = P \sin \omega t$, the steady-state output $x(t)$ can be found as

$$x(t) = \frac{P}{\sqrt{T^2\omega^2 + 1}} \sin(\omega t - \tan^{-1} T\omega) \tag{9–9}$$

From this equation, we see that, for small ω, the amplitude of the output $x(t)$ is almost equal to the amplitude of the input. For large ω, the amplitude of the output is small and almost inversely proportional to ω. The phase angle is $0°$ at $\omega = 0$ and approaches $-90°$ as ω increases indefinitely.

Example 9–2

Suppose that a sinusoidal force $p(t) = P \sin \omega t$ is applied to the mechanical system shown in Figure 9–4. Assuming that the displacement x is measured from the equilibrium position, find the steady-state output.

The equation of motion for the system is

$$m\ddot{x} + b\dot{x} + kx = p(t)$$

The Laplace transform of this equation, assuming zero initial conditions, is

$$(ms^2 + bs + k)X(s) = P(s)$$

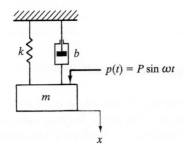

Figure 9–4 Mechanical system.

where $X(s) = \mathcal{L}[x(t)]$ and $P(s) = \mathcal{L}[p(t)]$. (Note that the initial conditions do not affect the steady-state output and so can be taken to be zero.) The transfer function between the displacement $X(s)$ and the input force $P(s)$ is, therefore, obtained as

$$\frac{X(s)}{P(s)} = G(s) = \frac{1}{ms^2 + bs + k}$$

Since the input is a sinusoidal function $p(t) = P \sin \omega t$, we can use the sinusoidal transfer function to obtain the steady-state solution. The sinusoidal transfer function is

$$\frac{X(j\omega)}{P(j\omega)} = G(j\omega) = \frac{1}{-m\omega^2 + bj\omega + k} = \frac{1}{(k - m\omega^2) + jb\omega}$$

From Equation (9–6), the steady-state output $x(t)$ can be written

$$x(t) = |G(j\omega)|P \sin(\omega t + \phi)$$

where

$$|G(j\omega)| = \frac{1}{\sqrt{(k - m\omega^2)^2 + b^2\omega^2}}$$

and

$$\phi = \angle G(j\omega) = \angle \frac{1}{(k - m\omega^2) + jb\omega} = -\tan^{-1}\frac{b\omega}{k - m\omega^2}$$

Thus,

$$x(t) = \frac{P}{\sqrt{(k - m\omega^2)^2 + b^2\omega^2}} \sin\left(\omega t - \tan^{-1}\frac{b\omega}{k - m\omega^2}\right)$$

Since $k/m = \omega_n^2$ and $b/k = 2\zeta/\omega_n$, the equation for $x(t)$ can be written

$$x(t) = \frac{x_{st}}{\sqrt{[1 - (\omega^2/\omega_n^2)]^2 + (2\zeta\omega/\omega_n)^2}} \sin\left[\omega t - \tan^{-1}\frac{2\zeta\omega/\omega_n}{1 - (\omega^2/\omega_n^2)}\right] \qquad (9\text{–}10)$$

where $x_{st} = P/k$ is the static deflection.

Writing the amplitude of $x(t)$ as X, we find that the amplitude ratio X/x_{st} is

$$\frac{X}{x_{st}} = \frac{1}{\sqrt{[1 - (\omega^2/\omega_n^2)]^2 + (2\zeta\omega/\omega_n)^2}}$$

and the phase shift ϕ is

$$\phi = -\tan^{-1}\frac{2\zeta\omega/\omega_n}{1 - (\omega^2/\omega_n^2)}$$

9–3 VIBRATIONS IN ROTATING MECHANICAL SYSTEMS

Vibration is, in general, undesirable because it may cause parts to break down, generate noise, transmit forces to foundations, and so on. To reduce the amount of force transmitted to the foundation as a result of a machine's vibration (a technique known as *force isolation*) as much as possible, machines are usually mounted on vibration isolators that consist of springs and dampers. Similarly, to reduce the amount of motion transmitted to a delicate instrument by the motion of its foundation (a technique called *motion isolation*), instruments are mounted on isolators. In this section, centripetal force, centrifugal force, and force due to a rotating unbalance are described first. Afterward, vibrations caused by the excitatory force resulting from unbalance are discussed. Vibration isolation is examined in Section 9–4.

Centripetal force and centrifugal force.   Suppose that a point mass m is moving in a circular path with a constant speed, as shown in Figure 9–5(a). The magnitudes of the velocities of the mass m at point A and point B are the same, but the directions are different. Referring to Figure 9–5(b), the direction $\overrightarrow{PQ}$ becomes perpendicular to the direction $\overrightarrow{AP}$ (the direction of the velocity vector at point A) if points A and B are close to each other. This means that the point mass must be subjected to a force that acts *toward* the center of rotation, point O. Such a force is called a *centripetal force*. For example, if a mass is attached to the end of a cord and is rotated at an angular speed ω in a horizontal plane like a conical pendulum, then the horizontal component of the tension in the cord is the centripetal force acting to keep the rotating configuration.

The force ma acting toward the center of rotation is derived as follows: Noting that triangles OAB and APQ are similar, we have

$$\frac{|\Delta v|}{|v_A|} = \frac{r\,\Delta\theta}{r}$$

where $|\Delta v|$ and $|v_A|$ represent the magnitudes of velocity Δv and velocity v_A, respectively. Observing that $|v_A| = \omega r$ and $\omega = \lim_{\Delta t \to 0} (\Delta\theta/\Delta t)$, we obtain

$$a = \lim_{\Delta t \to 0} \frac{|\Delta v|}{\Delta t} = \lim_{\Delta t \to 0} \frac{|v_A| r\,\Delta\theta}{r\,\Delta t} = \omega^2 r$$

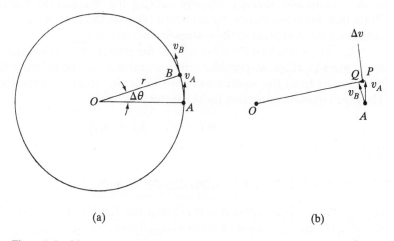

(a)   (b)

Figure 9–5   (a) Point mass moving in a circular path; (b) velocity vector diagram.

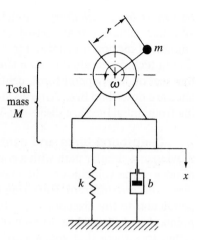

Figure 9–6 Unbalanced machine resting on shock mounts.

This acceleration acts toward the center of rotation, and the centripetal force is $ma = m\omega^2 r$. The *centrifugal force* is the opposing inertia force that acts outward. Its magnitude is also $m\omega^2 r$.

Vibration due to rotating unbalance. Force inputs that excite vibratory motion often arise from rotating unbalance, a condition that arises when the mass center of a rotating rigid body and the center of rotation do not coincide. Figure 9–6 shows an unbalanced machine resting on shock mounts. Assume that the rotor is rotating at a constant speed ω rad/s and that the unbalanced mass m is located a distance r from the center of rotation. Then the unbalanced mass will produce a centrifugal force of magnitude $m\omega^2 r$.

In the present analysis, we limit the motion to the vertical direction only, even though the rotating unbalance produces a horizontal component of force. The vertical component of this force, $m\omega^2 r \sin \omega t$, acts on the bearings and is thus transmitted to the foundation, thereby possibly causing the machine to vibrate excessively. [Note that, for convenience, we arbitrarily choose the time origin $t = 0$, so that the unbalance force applied to the system is $m\omega^2 r \sin \omega t$.]

Let us assume that the total mass of the system is M, which includes the unbalanced mass m. Here, we consider only vertical motion and measure the vertical displacement x from the equilibrium position in the absence of the forcing function. Then the equation of motion for the system becomes

$$M\ddot{x} + b\dot{x} + kx = p(t) \tag{9–11}$$

where

$$p(t) = m\omega^2 r \sin \omega t$$

is the force applied to the system. Taking the Laplace transform of both sides of Equation (9–11), assuming zero initial conditions, we have

$$(Ms^2 + bs + k)X(s) = P(s)$$

or

$$\frac{X(s)}{P(s)} = \frac{1}{Ms^2 + bs + k}$$

The sinusoidal transfer function is

$$\frac{X(j\omega)}{P(j\omega)} = G(j\omega) = \frac{1}{-M\omega^2 + bj\omega + k}$$

For the sinusoidal forcing function $p(t)$, the steady-state output is obtained from Equation (9-6) as

$$x(t) = X\sin(\omega t + \phi)$$

$$= |G(j\omega)|m\omega^2 r \sin\left(\omega t - \tan^{-1}\frac{b\omega}{k - M\omega^2}\right)$$

$$= \frac{m\omega^2 r}{\sqrt{(k - M\omega^2)^2 + b^2\omega^2}} \sin\left(\omega t - \tan^{-1}\frac{b\omega}{k - M\omega^2}\right)$$

In this last equation, if we divide the numerator and denominator of the amplitude and those of the phase angle by k and substitute $k/M = \omega_n^2$ and $b/M = 2\zeta\omega_n$ into the result, the steady-state output becomes

$$x(t) = \frac{m\omega^2 r/k}{\sqrt{[1 - (\omega^2/\omega_n^2)]^2 + (2\zeta\omega/\omega_n)^2}} \sin\left[\omega t - \tan^{-1}\frac{2\zeta\omega/\omega_n}{1 - (\omega^2/\omega_n^2)}\right]$$

Thus, the steady-state output is a sinusoidal motion whose amplitude becomes large when the damping ratio ζ is small and the forcing frequency ω is close to the natural frequency ω_n.

9-4 VIBRATION ISOLATION

Vibration isolation is a process by which vibratory effects are minimized or eliminated. The function of a vibration isolator is to reduce the magnitude of force transmitted from a machine to its foundation or to reduce the magnitude of motion transmitted from a vibratory foundation to a machine.

The concept is illustrated in Figures 9-7(a) and (b). The system consists of a rigid body representing a machine connected to a foundation by an isolator that consists of a spring and a damper. Figure 9-7(a) illustrates the case in which the source of vibration is a vibrating force originating within the machine (force excitation). The isolator reduces the force transmitted to the foundation. In Figure 9-7(b), the source of vibration is a vibrating motion of the foundation (motion excitation). The isolator reduces the vibration amplitude of the machine.

The isolator essentially consists of a resilient load-supporting means (such as a spring) and an energy-dissipating means (such as a damper). A typical vibration isolator appears in Figure 9-8. (In a simple vibration isolator, a single element like synthetic rubber can perform the functions of both the load-supporting means and the energy-dissipating means.) In the analysis given here, the machine and the foundation are assumed rigid and the isolator is assumed massless.

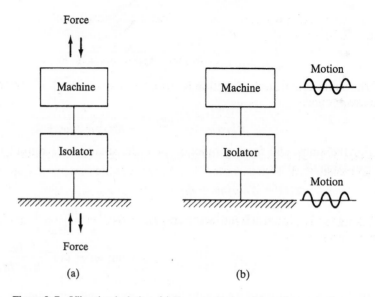

Figure 9–7   Vibration isolation. (a) Force excitation; (b) motion excitation.

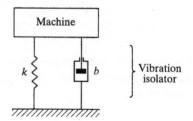

Figure 9–8   Vibration isolator.

Transmissibility.   *Transmissibility* is a measure of the reduction of a transmitted force or of motion afforded by an isolator. If the source of vibration is a vibrating force due to the unbalance of the machine (force excitation), transmissibility is the ratio of the amplitude of the force transmitted to the foundation to the amplitude of the excitatory force. If the source of vibration is a vibratory motion of the foundation (motion excitation), transmissibility is the ratio of the vibration amplitude of the machine to the vibration amplitude of the foundation.

Transmissibility for force excitation.   For the system shown in Figure 9–6, the source of vibration is a vibrating force resulting from the unbalance of the machine. The transmissibility in this case is the force amplitude ratio and is given by

$$\text{transmissibility} = \text{TR} = \frac{F_t}{F_0} = \frac{\text{amplitude of the transmitted force}}{\text{amplitude of the excitatory force}}$$

Let us find the transmissibility of this system in terms of the damping ratio ζ and the frequency ratio $\beta = \omega/\omega_n$.

The excitatory force (in the vertical direction) is caused by the unbalanced mass of the machine and is

$$p(t) = m\omega^2 r \sin \omega t = F_0 \sin \omega t$$

The equation of motion for the system is Equation (9–11), rewritten here for convenience:

$$M\ddot{x} + b\dot{x} + kx = p(t) \tag{9–12}$$

where M is the total mass of the machine including the unbalance mass m. The force $f(t)$ transmitted to the foundation is the sum of the damper and spring forces, or

$$f(t) = b\dot{x} + kx = F_t \sin(\omega t + \phi) \tag{9–13}$$

Taking the Laplace transforms of Equations (9–12) and (9–13), assuming zero initial conditions, gives

$$(Ms^2 + bs + k)X(s) = P(s)$$
$$(bs + k)X(s) = F(s)$$

where $X(s) = \mathcal{L}[x(t)]$, $P(s) = \mathcal{L}[p(t)]$, and $F(s) = \mathcal{L}[f(t)]$. Hence,

$$\frac{X(s)}{P(s)} = \frac{1}{Ms^2 + bs + k}$$

$$\frac{F(s)}{X(s)} = bs + k$$

Eliminating $X(s)$ from the last two equations yields

$$\frac{F(s)}{P(s)} = \frac{F(s)}{X(s)}\frac{X(s)}{P(s)} = \frac{bs + k}{Ms^2 + bs + k}$$

The sinusoidal transfer function is thus

$$\frac{F(j\omega)}{P(j\omega)} = \frac{bj\omega + k}{-M\omega^2 + bj\omega + k} = \frac{(b/M)j\omega + (k/M)}{-\omega^2 + (b/M)j\omega + (k/M)}$$

Substituting $k/M = \omega_n^2$ and $b/M = 2\zeta\omega_n$ into this last equation and simplifying, we have

$$\frac{F(j\omega)}{P(j\omega)} = \frac{1 + j(2\zeta\omega/\omega_n)}{1 - (\omega^2/\omega_n^2) + j(2\zeta\omega/\omega_n)}$$

from which it follows that

$$\left|\frac{F(j\omega)}{P(j\omega)}\right| = \frac{\sqrt{1 + (2\zeta\omega/\omega_n)^2}}{\sqrt{[1 - (\omega^2/\omega_n^2)]^2 + (2\zeta\omega/\omega_n)^2}} = \frac{\sqrt{1 + (2\zeta\beta)^2}}{\sqrt{(1 - \beta^2)^2 + (2\zeta\beta)^2}}$$

where $\beta = \omega/\omega_n$.

Noting that the amplitude of the excitatory force is $F_0 = |P(j\omega)|$ and that the amplitude of the transmitted force is $F_t = |F(j\omega)|$, we obtain the transmissibility:

$$\text{TR} = \frac{F_t}{F_0} = \frac{|F(j\omega)|}{|P(j\omega)|} = \frac{\sqrt{1 + (2\zeta\beta)^2}}{\sqrt{(1 - \beta^2)^2 + (2\zeta\beta)^2}} \tag{9-14}$$

From Equation (9-14), we see that the transmissibility depends on both β and ζ. When $\beta = \sqrt{2}$, however, the transmissibility is equal to unity, regardless of the value of the damping ratio ζ.

Figure 9-9 shows some curves of transmissibility versus $\beta(= \omega/\omega_n)$. We see that all of the curves pass through a critical point where $\text{TR} = 1$ and $\beta = \sqrt{2}$. For $\beta < \sqrt{2}$, as the damping ratio ζ increases, the transmissibility at resonance decreases. For $\beta > \sqrt{2}$, as ζ increases, the transmissibility increases. Therefore, for $\beta < \sqrt{2}$, or $\omega < \sqrt{2}\,\omega_n$ (the forcing frequency ω is smaller than $\sqrt{2}$ times the undamped natural frequency ω_n), increasing damping improves the vibration isolation. For $\beta > \sqrt{2}$, or $\omega > \sqrt{2}\,\omega_n$, increasing damping adversely affects the vibration isolation.

Note that, since $|P(j\omega)| = F_0 = m\omega^2 r$, the amplitude of the force transmitted to the foundation is

$$F_t = |F(j\omega)| = \frac{m\omega^2 r\sqrt{1 + (2\zeta\beta)^2}}{\sqrt{(1 - \beta^2)^2 + (2\zeta\beta)^2}} \tag{9-15}$$

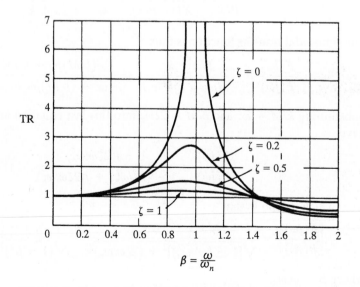

$$\beta = \frac{\omega}{\omega_n}$$

Figure 9-9   Curves of transmissibility TR versus $\beta(= \omega/\omega_n)$.

Example 9–3

In the system shown in Figure 9–6, if $M = 15$ kg, $b = 450$ N-s/m, $k = 6000$ N/m, $m = 0.005$ kg, $r = 0.2$ m, and $\omega = 16$ rad/s, what is the force transmitted to the foundation?

The equation of motion for the system is

$$15\ddot{x} + 450\dot{x} + 6000x = (0.005)(16)^2(0.2) \sin 16t$$

Consequently,

$$\omega_n = 20 \text{ rad/s}, \qquad \zeta = 0.75$$

and we find that $\beta = \omega/\omega_n = 16/20 = 0.8$. From Equation (9–15), we have

$$F_t = \frac{m\omega^2 r \sqrt{1 + (2\zeta\beta)^2}}{\sqrt{(1 - \beta^2)^2 + (2\zeta\beta)^2}}$$

$$= \frac{(0.005)(16)^2(0.2)\sqrt{1 + (2 \times 0.75 \times 0.8)^2}}{\sqrt{(1 - 0.8^2)^2 + (2 \times 0.75 \times 0.8)^2}} = 0.319 \text{ N}$$

The force transmitted to the foundation is sinusoidal with an amplitude of 0.319 N.

Automobile suspension system.   Figure 9–10(a) shows an automobile system. Figure 9–10(b) is a schematic diagram of an automobile suspension system. As the car moves along the road, the vertical displacements at the tires act as motion excitation to the automobile suspension system. The motion of this system consists of a translational motion of the center of mass and a rotational motion about the center of mass. A complete analysis of the suspension system would be very involved. A highly simplified version appears in Figure 9–11. Let us analyze this simple model when the motion input is sinusoidal. We shall derive the transmissibility for the motion excitation system. (As a related problem, see **Problem B–9–13**.)

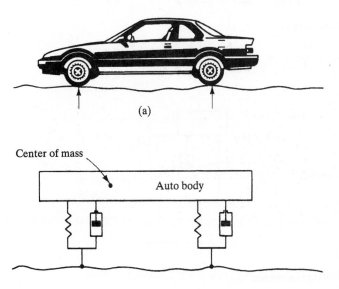

(a)

Center of mass

Auto body

(b)

Figure 9–10   (a) Automobile system; (b) schematic diagram of an automobile suspension system.

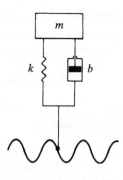

Figure 9–11 Simplified version of the automobile suspension system of Figure 9–10.

Transmissibility for motion excitation.   In the mechanical system shown
in Figure 9–12, the motion of the body is in the vertical direction only. The motion
$p(t)$ at point A is the input to the system; the vertical motion $x(t)$ of the body is the
output. The displacement $x(t)$ is measured from the equilibrium position in the absence of input $p(t)$. We assume that $p(t)$ is sinusoidal, or $p(t) = P \sin \omega t$.
    The equation of motion for the system is

$$m\ddot{x} + b(\dot{x} - \dot{p}) + k(x - p) = 0$$

or

$$m\ddot{x} + b\dot{x} + kx = b\dot{p} + kp$$

The Laplace transform of this last equation, assuming zero initial conditions, gives

$$(ms^2 + bs + k)X(s) = (bs + k)P(s)$$

Hence,

$$\frac{X(s)}{P(s)} = \frac{bs + k}{ms^2 + bs + k}$$

The sinusoidal transfer function is

$$\frac{X(j\omega)}{P(j\omega)} = \frac{bj\omega + k}{-m\omega^2 + bj\omega + k}$$

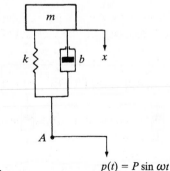

Figure 9–12 Mechanical system. $p(t) = P \sin \omega t$

The steady-state output $x(t)$ has the amplitude $|X(j\omega)|$. The input amplitude is $|P(j\omega)|$. The transmissibility TR in this case is the displacement amplitude ratio and is given by

$$TR = \frac{\text{amplitude of the output displacement}}{\text{amplitude of the input displacement}}$$

Thus,

$$TR = \frac{|X(j\omega)|}{|P(j\omega)|} = \frac{\sqrt{b^2\omega^2 + k^2}}{\sqrt{(k - m\omega^2)^2 + b^2\omega^2}}$$

Noting that $k/m = \omega_n^2$ and $b/m = 2\zeta\omega_n$, we see that the transmissibility is given, in terms of the damping ratio ζ and the undamped natural frequency ω_n, by

$$TR = \frac{\sqrt{1 + (2\zeta\beta)^2}}{\sqrt{(1 - \beta^2)^2 + (2\zeta\beta)^2}} \tag{9–16}$$

where $\beta = \omega/\omega_n$. This equation is identical to Equation (9–14).

Example 9–4

A rigid body is mounted on an isolator to reduce vibratory effects. Assume that the mass of the rigid body is 500 kg, the damping coefficient of the isolator is very small ($\zeta = 0.01$), and the effective spring constant of the isolator is 12,500 N/m. Find the percentage of motion transmitted to the body if the frequency of the motion excitation of the base of the isolator is 20 rad/s.

The undamped natural frequency ω_n of the system is

$$\omega_n = \sqrt{\frac{12,500}{500}} = 5 \text{ rad/s}$$

so

$$\beta = \frac{\omega}{\omega_n} = \frac{20}{5} = 4$$

Substituting $\zeta = 0.01$ and $\beta = 4$ into Equation (9–16), we have

$$TR = \frac{\sqrt{1 + (2\zeta\beta)^2}}{\sqrt{(1 - \beta^2)^2 + (2\zeta\beta)^2}} = \frac{\sqrt{1 + (2 \times 0.01 \times 4)^2}}{\sqrt{(1 - 4^2)^2 + (2 \times 0.01 \times 4)^2}} = 0.0669$$

The isolator thus reduces the vibratory motion of the rigid body to 6.69% of the vibratory motion of the base of the isolator.

9–5 DYNAMIC VIBRATION ABSORBERS

If a mechanical system operates near a critical frequency, the amplitude of vibration increases to a degree that cannot be tolerated, because the machine might break down or might transmit too much vibration to the surrounding machines. This section discusses a way to reduce vibrations near a specified operating frequency that is close to the natural frequency (i.e., the critical frequency) of the system by the use of a dynamic vibration absorber.

Basically, the dynamic vibration absorber adds one degree of freedom to the system. If the original system is a one-degree-of-freedom system (meaning that the system has only one critical frequency ω_c), the addition of the dynamic vibration absorber increases the number of degrees of freedom to two and thereby increases the number of critical frequencies to two. This means that it is possible to shift the critical frequencies from the operating frequency. One of the two new critical frequencies will be well below the original critical frequency ω_c, and the other will be well above ω_c. Therefore, operation at the given frequency (near ω_c) is possible.

Before we present vibration absorbers, we shall discuss systems with two or more degrees of freedom.

Mechanical systems with two or more degrees of freedom. In real-life situations, the motion of a mechanical system may be simultaneously translational and rotational in three-dimensional space, and parts of the system may have constraints on where they can move. The geometrical description of such motions can become complicated, but the fundamental physical laws still apply.

For some simple systems, only one coordinate may be necessary to specify the motion of the system. However, more than one coordinate is necessary to describe the motion of complicated systems. The term used to describe the minimum number of independent coordinates required to specify this motion is *degrees of freedom*.

Degrees of freedom. The number of degrees of freedom that a mechanical system possesses is the minimum number of independent coordinates required to specify the positions of all of the elements of the system. For instance, if only one independent coordinate is needed to completely specify the geometric location of the mass of a system in space, the system has a one degree of freedom. Thus, a rigid body rotating on an axis has one degree of freedom, whereas a rigid body in space has six degrees of freedom—three translational and three rotational.

It is important to note that, in general, neither the number of masses nor any other obvious quantity will always lead to a correct assessment of the number of degrees of freedom.

In terms of the number of equations of motion and the number of constraints, we have

$$\text{number of degrees of freedom} = (\text{number of equations of motion})$$
$$- (\text{number of equations of constraint})$$

Example 9–5

Let us find the degrees of freedom of each of the systems shown in Figure 9–13.

   (a) We begin with the system shown in Figure 9–13(a). If the mass m is constrained to move vertically, only one coordinate x is required to define the location of the mass at any time. Thus, the system shown in Figure 9–13(a) has one degree of freedom.

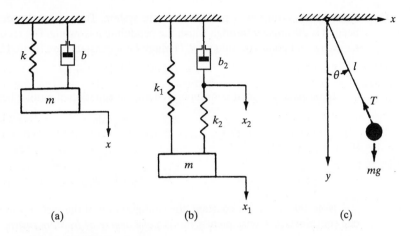

Figure 9–13   Mechanical systems.

We can verify this statement by counting the number of equations of motion and the number of equations of constraint. This system has one equation of motion, namely,

$$m\ddot{x} + b\dot{x} + kx = 0$$

and no equation of constraint. Consequently,

$$\text{degree of freedom} = 1 - 0 = 1$$

(b) Next, consider the system shown in Figure 9–13(b). The equations of motion here are

$$m\ddot{x}_1 + k_1x_1 + k_2(x_1 - x_2) = 0$$
$$k_2(x_1 - x_2) = b_2\dot{x}_2$$

so the number of equations of motion is two. There is no equation of constraint. Therefore,

$$\text{degrees of freedom} = 2 - 0 = 2$$

(c) Finally, consider the pendulum system shown in Figure 9–13(c). If we define the coordinates of the pendulum mass as (x, y), then the equations of motion are

$$m\ddot{x} = -T \sin\theta$$
$$m\ddot{y} = mg - T \cos\theta$$

where T is the tension in the wire. Thus, the number of equations of motion is two. The constraint equation for this system is

$$x^2 + y^2 = l^2$$

The number of equations of constraint is one, so

$$\text{degree of freedom} = 2 - 1 = 1$$

Note that when physical constraints are present, the most convenient coordinate system may not be a rectangular one. In the pendulum system of Figure 9–13(c), the pendulum is constrained to move in a circular path. The most convenient coordinate system here would be a polar coordinate system. Then the only coordinate that is needed is the angle θ through which the pendulum has swung. The rectangular coordinates x, y and polar coordinates θ, l (where l is a constant) are related by

$$x = l \sin \theta, \qquad y = l \cos \theta$$

In terms of the polar coordinate system, the equation of motion becomes

$$ml^2\ddot{\theta} = -mgl \sin \theta$$

or

$$\ddot{\theta} + \frac{g}{l} \sin \theta = 0$$

Note that, since l is constant, the configuration of the system can be specified by one coordinate, θ. Consequently, this is a *one-degree-of-freedom system.*

Dynamic vibration absorber.   In many situations, rotating machines (such as turbines and compressors) cause vibrations and transmit large vibratory forces to the machines' foundations. Vibratory forces may be caused by an unbalanced mass of the rotor. If the excitatory frequency ω is equal to or nearly equal to the undamped natural frequency of the rotating machine on its mounts, then resonance occurs and large forces are transmitted to the foundation.

If the machine operates at nearly constant speed, a device called a *dynamic vibration absorber* can be attached to it to eliminate the large transmitted force. This device is usually in the form of a spring–mass system tuned to have a natural frequency equal to the operating frequency ω. When a vibration absorber is added to a one-degree-of-freedom vibratory system, the entire system becomes a two-degrees-of-freedom system with two natural frequencies. To reduce or nearly eliminate the transmitted force, one of the natural frequencies is set above the operating frequency, the other below it.

Our discussion here focuses on a simple dynamic vibration absorber that will reduce the vertical force transmitted to the foundation. Note that only vertical motions are discussed.

Reducing vibrations by means of a dynamic vibration absorber.   If the mass of the rotor of a rotating machine is unbalanced, the machine transmits a large vibratory force to its foundation. Let us assume that the machine is supported by a spring and a damper as shown in Figure 9–14(a). The unbalanced rotor is represented by mass M, which includes the unbalanced mass, and is rotating at frequency ω. The excitatory force is $p(t) = P \sin \omega t$, where $P = m\omega^2 r$. (Here, m is the unbalanced mass and r is the distance of the unbalanced mass from the center of rotation.) Because of this force excitation, a sinusoidal force of amplitude

$$\frac{m\omega^2 r \sqrt{k^2 + b^2\omega^2}}{\sqrt{(k - M\omega^2)^2 + b^2\omega^2}}$$

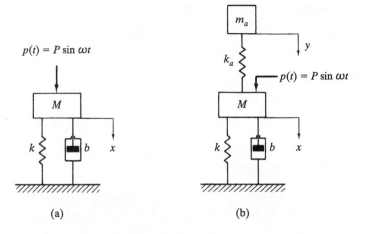

Figure 9–14   (a) Machine supported by a spring and damper; (b) machine with a dynamic vibration absorber.

is transmitted to the foundation. [To obtain this amplitude, substitute $\beta = \omega/\omega_n = \omega/\sqrt{k/M}$ and $\zeta = b/(2\sqrt{kM})$ into Equation (9–15).]

If the viscous damping coefficient b is small and the natural frequency $\sqrt{k/M}$ of the system is equal to the excitation frequency, then resonance occurs, the machine is subjected to excessive vibration, and the transmitted force becomes extremely large.

In the analysis which follows, we assume that b is very small and that the natural frequency $\sqrt{k/M}$ is very close to the excitation frequency ω. In such a case, to reduce the transmitted force, a dynamic vibration absorber consisting of a mass (m_a) and a spring (k_a) may be added to the machine as shown in Figure 9–14(b).

The equations of motion for the system of Figure 9–14(b) are

$$M\ddot{x} + b\dot{x} + kx + k_a(x - y) = p(t) = P \sin \omega t$$
$$m_a\ddot{y} + k_a(y - x) = 0$$

where x and y, the displacements of mass M and mass m_a, respectively, are measured from the respective equilibrium positions of these masses in the absence of excitation force $p(t)$. Taking the Laplace transforms of the last two equations, assuming zero initial conditions, we obtain

$$(Ms^2 + bs + k + k_a)X(s) - k_aY(s) = P(s)$$
$$(m_as^2 + k_a)Y(s) - k_aX(s) = 0$$

Eliminating $Y(s)$ from these equations yields

$$\left(Ms^2 + bs + k + k_a - \frac{k_a^2}{m_as^2 + k_a}\right)X(s) = P(s)$$

It follows that

$$\frac{X(s)}{P(s)} = \frac{m_a s^2 + k_a}{(Ms^2 + bs + k + k_a)(m_a s^2 + k_a) - k_a^2}$$

The sinusoidal transfer function is

$$\frac{X(j\omega)}{P(j\omega)} = \frac{-m_a\omega^2 + k_a}{(-M\omega^2 + bj\omega + k + k_a)(-m_a\omega^2 + k_a) - k_a^2}$$

If the viscous damping coefficient b is negligibly small, we may substitute $b = 0$ into this last equation. Then

$$\frac{X(j\omega)}{P(j\omega)} \doteq \frac{-m_a\omega^2 + k_a}{(-M\omega^2 + k + k_a)(-m_a\omega^2 + k_a) - k_a^2}$$

[Note that, in the actual system, free vibrations eventually die out due to damping (even though it may be negligibly small), and the forced vibration at steady state can be represented by the preceding equation.] The force $f(t)$ transmitted to the foundation is

$$f(t) = kx + b\dot{x} \doteq kx$$

The amplitude of this transmitted force is $k|X(j\omega)|$, where

$$
\begin{aligned}
|X(j\omega)| &= \left| \frac{k_a - m_a\omega^2}{(k + k_a - M\omega^2)(k_a - m_a\omega^2) - k_a^2} \right| |P(j\omega)| \\
&= \left| \frac{m\omega^2 r(k_a - m_a\omega^2)}{(k + k_a - M\omega^2)(k_a - m_a\omega^2) - k_a^2} \right|
\end{aligned}
\tag{9-17}
$$

[Note that $|P(j\omega)| = P = m\omega^2 r$.]

In examining Equation (9–17), observe that if m_a and k_a are chosen such that

$$k_a - m_a\omega^2 = 0$$

or $k_a/m_a = \omega^2$, then $|X(j\omega)| = 0$ and the force transmitted to the foundation is zero. So if the natural frequency $\sqrt{k_a/m_a}$ of the dynamic vibration absorber is made equal to the excitation frequency ω, it is possible to eliminate the force transmitted to the foundation. In general, such a dynamic vibration absorber is used only when the natural frequency $\sqrt{k/M}$ of the original system is very close to the excitation frequency ω. (Without this device, the system would be in near resonance.)

Physically, the effect of the dynamic vibration absorber is to produce a spring force $k_a y$ that cancels the excitation force $p(t)$. To see this point, note that if the viscous damping coefficient b is negligibly small, then

$$\frac{Y(j\omega)}{P(j\omega)} = \frac{X(j\omega)}{P(j\omega)} \frac{Y(j\omega)}{X(j\omega)}$$

$$= \frac{k_a}{(-M\omega^2 + k + k_a)(-m_a\omega^2 + k_a) - k_a^2} \tag{9-18}$$

If m_a and k_a are chosen so that $k_a = m_a\omega^2$, we find that

$$\frac{Y(j\omega)}{P(j\omega)} = \frac{k_a}{-k_a^2} = -\frac{1}{k_a}$$

Consequently,

$$y(t) = \left|-\frac{1}{k_a}\right| P \sin\left(\omega t + \left/-\frac{1}{k_a}\right.\right)$$

$$= \frac{P}{k_a} \sin(\omega t - 180°)$$

$$= -\frac{P}{k_a} \sin \omega t$$

This means that the spring k_a transmits a force $k_a y = -P \sin \omega t$ to mass M. The magnitude of this force is equal to that of the excitation force, and the phase angle lags 180° from that of the excitation force (mass m_a is vibrating in phase opposition to the excitation force), with the result that the spring force $k_a y$ and the excitation force $p(t)$ cancel each other and mass M stays stationary.

We have shown that the addition of a dynamic vibration absorber will reduce the vibration of a machine and the force transmitted to the foundation to zero when the machine is excited by an unbalanced mass at a frequency ω. It can also be shown that there will now be two frequencies at which mass M will be in resonance. These two frequencies are the natural frequencies of the two-degrees-of-freedom system and can be found from the characteristic equation for the sinusoidal transfer function $Y(j\omega)/P(j\omega)$ given by Equation (9–18):

$$(k + k_a - M\omega_i^2)(k_a - m_a\omega_i^2) - k_a^2 = 0 \qquad i = 1, 2$$

The two values of frequency, ω_1 and ω_2, that satisfy this last equation are the natural frequencies of the system with a dynamic vibration absorber. Figures 9–15(a) and (b) show the curves of amplitude $|X(j\omega)|$ versus frequency ω for the systems depicted in Figures 9–14(a) and (b), respectively, when b is negligibly small.

Note that the addition of viscous damping in parallel with the absorber spring k_a relieves excessive vibrations at the two natural frequencies. That is, very large amplitudes at the two resonance frequencies may be reduced to smaller values.

9–6 FREE VIBRATIONS IN MULTI-DEGREES-OF-FREEDOM SYSTEMS

In this section, we shall discuss vibrations that may occur in multi-degrees-of-freedom systems. In particular, we treat free vibrations of a two-degrees-of-freedom system in detail. (Discussions of free vibrations of a three-degrees-of-freedom system are given in **Problem A–9–15**.)

Two-degrees-of-freedom system. A *two-degrees-of-freedom system* requires two independent coordinates to specify the system's configuration. Consider the mechanical system shown in Figure 9–16, which illustrates the two-degrees-of-freedom

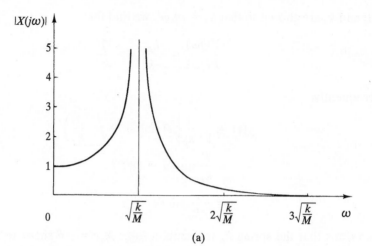

(a)

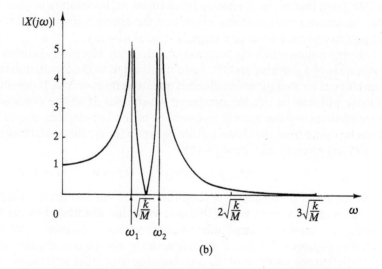

(b)

Figure 9–15 (a) Curve of amplitude versus frequency for the system of Figure 9–14(a); (b) curve of amplitude versus frequency for the system of Figure 9–14(b).

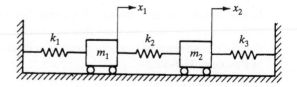

Figure 9–16 Mechanical system with two degrees of freedom.

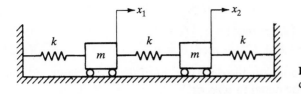

Figure 9–17 Mechanical system with two degrees of freedom.

case. Let us derive a mathematical model of this system. We assume that the masses move without friction. Applying Newton's second law to mass m_1 and mass m_2, we have

$$m_1\ddot{x}_1 = -k_1 x_1 - k_2(x_1 - x_2)$$
$$m_2\ddot{x}_2 = -k_3 x_2 - k_2(x_2 - x_1)$$

Rearranging terms yields

$$m_1\ddot{x}_1 + k_1 x_1 + k_2(x_1 - x_2) = 0 \qquad (9\text{–}19)$$
$$m_2\ddot{x}_2 + k_3 x_2 + k_2(x_2 - x_1) = 0 \qquad (9\text{–}20)$$

These two equations represent a mathematical model of the system. [Free vibrations of the mechanical system described by Equations (9–19) and (9–20) are discussed in **Problem A–9–14**.]

Free vibrations in two-degrees-of-freedom system. Consider the mechanical system shown in Figure 9–17, which is a special case of the system given in Figure 9–16. The equations of motion for the system of Figure 9–17 can be obtained by substituting $m_1 = m_2 = m$ and $k_1 = k_2 = k_3 = k$ into Equations (9–19) and (9–20), yielding

$$m\ddot{x}_1 + 2k x_1 - k x_2 = 0 \qquad (9\text{–}21)$$
$$m\ddot{x}_2 + 2k x_2 - k x_1 = 0 \qquad (9\text{–}22)$$

Let us examine the free vibration of this system. To find the natural frequencies of the free vibration, we assume that the motion is harmonic. That is, we assume that

$$x_1 = A \sin \omega t, \qquad x_2 = B \sin \omega t$$

Then

$$\ddot{x}_1 = -A\omega^2 \sin \omega t, \qquad \ddot{x}_2 = -B\omega^2 \sin \omega t$$

If the preceding expressions are substituted into Equations (9–21) and (9–22), the resulting equations are

$$(-mA\omega^2 + 2kA - kB) \sin \omega t = 0$$
$$(-mB\omega^2 + 2kB - kA) \sin \omega t = 0$$

Since these equations must be satisfied at all times and since $\sin \omega t$ cannot be zero at all times, the quantities in parentheses must be equal to zero. Thus,

$$-mA\omega^2 + 2kA - kB = 0$$
$$-mB\omega^2 + 2kB - kA = 0$$

Rearranging terms, we have

$$(2k - m\omega^2)A - kB = 0 \qquad (9\text{--}23)$$
$$-kA + (2k - m\omega^2)B = 0 \qquad (9\text{--}24)$$

For constants A and B to be nonzero, the determinant of the coefficients of Equations (9–23) and (9–24) must be equal to zero, or

$$\begin{vmatrix} 2k - m\omega^2 & -k \\ -k & 2k - m\omega^2 \end{vmatrix} = 0$$

This determinantal equation determines the natural frequencies of the system and can be rewritten as

$$(2k - m\omega^2)^2 - k^2 = 0$$

or

$$\omega^4 - 4\frac{k}{m}\omega^2 + 3\frac{k^2}{m^2} = 0 \qquad (9\text{--}25)$$

Equation (9–25) can be factored as

$$\left(\omega^2 - \frac{k}{m}\right)\left(\omega^2 - \frac{3k}{m}\right) = 0$$

or

$$\omega^2 = \frac{k}{m}, \qquad \omega^2 = \frac{3k}{m}$$

Consequently, ω^2 has two values, the first representing the first natural frequency ω_1 (first mode) and the second representing the second natural frequency ω_2 (second mode):

$$\omega_1 = \sqrt{\frac{k}{m}}, \qquad \omega_2 = \sqrt{\frac{3k}{m}}$$

It should be remembered that in the one-degree-of-freedom system only one natural frequency exists, whereas the two-degrees-of-freedom system has two natural frequencies.

Note that, from Equation (9–23), we obtain

$$\frac{A}{B} = \frac{k}{2k - m\omega^2} \qquad (9\text{--}26)$$

Also, from Equation (9–24), we have

$$\frac{A}{B} = \frac{2k - m\omega^2}{k} \qquad (9\text{--}27)$$

If we substitute $\omega^2 = k/m$ (first mode) into either Equation (9–26) or (9–27), we obtain, in both cases,

$$\frac{A}{B} = 1$$

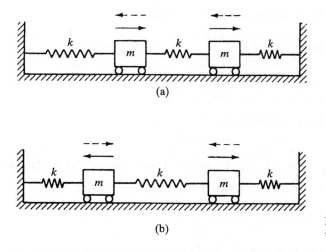

(a)

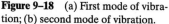

(b)

Figure 9–18   (a) First mode of vibration; (b) second mode of vibration.

If we substitute $\omega^2 = 3k/m$ (second mode) into either Equation (9–26) or (9–27), we have

$$\frac{A}{B} = -1$$

If the system vibrates at either of its two natural frequencies, the two masses must vibrate at the same frequency. From the first equation for the amplitude ratio A/B, at the lowest natural frequency ω_1 the amplitude ratio becomes unity, or $A = B$ (the first mode of vibration), which means that both masses move the same amount in the same direction; that is, the motions are in phase. [See Figure 9–18(a).] At the second natural frequency ω_2, the amplitude ratio becomes -1, or $A = -B$ (the second mode of vibration), so the motions are opposite in phase. [See Figure 9–18(b).] In the present system, the amplitude ratio becomes equal to 1 or -1 when the masses vibrate at a natural frequency. The reason for this is that we assumed that $m_1 = m_2$ and $k_1 = k_2 = k_3$. Without such assumptions, the ratio A/B may not be equal to 1 or -1. (See **Problem A–9–14.**)

Note that it is possible to excite only one of the two modes by properly setting the initial conditions. (See **Problem A–9–16.**) For arbitrary initial conditions, two modes of vibration may occur simultaneously. That is, the vibration of m_1 may consist of the sum of two components: a harmonic motion with amplitude A_1 at the frequency ω_1 and a harmonic motion with amplitude A_2 at the frequency ω_2. In this case, the vibration of m_2 consists of the sum of two harmonic components: one with amplitude B_1 at the frequency ω_1 and one with amplitude B_2 at the frequency ω_2.

Comments.   To find the natural frequencies of the system, we need only the characteristic equation. For example, in the present problem, taking the Laplace transforms of Equations (9–21) and (9–22), we have

$$m[s^2 X_1(s) - sx_1(0) - \dot{x}_1(0)] + 2kX_1(s) - kX_2(s) = 0$$
$$m[s^2 X_2(s) - sx_2(0) - \dot{x}_2(0)] + 2kX_2(s) - kX_1(s) = 0$$

Eliminating $X_2(s)$ from the last two equations, we obtain

$$X_1(s) = \frac{[sx_1(0) + \dot{x}_1(0)]\left(s^2 + \dfrac{2k}{m}\right) + \dfrac{k}{m}[sx_2(0) + \dot{x}_2(0)]}{s^4 + 4\dfrac{k}{m}s^2 + 3\dfrac{k^2}{m^2}}$$

The characteristic equation for the system is

$$s^4 + 4\frac{k}{m}s^2 + 3\frac{k^2}{m^2} = 0$$

If we substitute $s = j\omega$, the characteristic equation can be rewritten as

$$\omega^4 - 4\frac{k}{m}\omega^2 + 3\frac{k^2}{m^2} = 0$$

which is exactly the same as Equation (9–25). The advantage of the method using the assumed harmonic solution is that one can easily visualize the mode of vibration by the sign of the amplitude ratio A/B.

Many-degrees-of-freedom system. Generally, an *n-degrees-of-freedom system* (such as that consisting of n masses and $n + 1$ springs) has n natural frequencies. If free vibration takes place at any one of the system's natural frequencies, all the n masses will vibrate at that frequency, and the amplitude of any mass will bear a fixed value relative to the amplitude of any other mass. The system, however, may vibrate with more than one natural frequency. Then the resultant vibration may appear quite complicated and may seem to be a random vibration, although it is not.

EXAMPLE PROBLEMS AND SOLUTIONS

Problem A–9–1

Assuming that the mechanical system shown in Figure 9–19 is at rest before the excitation force $p(t) = P \sin \omega t$ is applied, derive the complete solution $x(t)$ and the steady-state solution $x_{ss}(t)$. The displacement x is measured from the equilibrium position before the excitation force is applied.

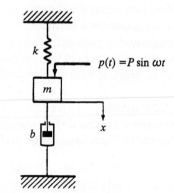

Figure 9–19 Mechanical system.

Solution   The equation of motion for the system is

$$m\ddot{x} + b\dot{x} + kx = P \sin \omega t$$

Noting that $x(0) = 0$ and $\dot{x}(0) = 0$, we find that the Laplace transform of this equation is

$$(ms^2 + bs + k)X(s) = P\frac{\omega}{s^2 + \omega^2}$$

or

$$X(s) = \frac{P\omega}{s^2 + \omega^2} \frac{1}{ms^2 + bs + k}$$

$$= \frac{P\omega}{m} \frac{1}{s^2 + \omega^2} \frac{1}{s^2 + 2\zeta\omega_n s + \omega_n^2}$$

where $\omega_n = \sqrt{k/m}$ and $\zeta = b/(2\sqrt{mk})$. We can expand $X(s)$ as

$$X(s) = \frac{P\omega}{m} \left(\frac{as + c}{s^2 + \omega^2} + \frac{-as + d}{s^2 + 2\zeta\omega_n s + \omega_n^2} \right)$$

By simple calculations, we find that

$$a = \frac{-2\zeta\omega_n}{(\omega_n^2 - \omega^2)^2 + 4\zeta^2\omega_n^2\omega^2}$$

$$c = \frac{\omega_n^2 - \omega^2}{(\omega_n^2 - \omega^2)^2 + 4\zeta^2\omega_n^2\omega^2}$$

$$d = \frac{4\zeta^2\omega_n^2 - (\omega_n^2 - \omega^2)}{(\omega_n^2 - \omega^2)^2 + 4\zeta^2\omega_n^2\omega^2}$$

Hence,

$$X(s) = \frac{P\omega}{m} \frac{1}{(\omega_n^2 - \omega^2)^2 + 4\zeta^2\omega_n^2\omega^2}$$

$$\times \left[\frac{-2\zeta\omega_n s + (\omega_n^2 - \omega^2)}{s^2 + \omega^2} + \frac{2\zeta\omega_n(s + \zeta\omega_n) + 2\zeta^2\omega_n^2 - (\omega_n^2 - \omega^2)}{s^2 + 2\zeta\omega_n s + \omega_n^2} \right]$$

The inverse Laplace transform of $X(s)$ gives

$$x(t) = \frac{P\omega}{m[(\omega_n^2 - \omega^2)^2 + 4\zeta^2\omega_n^2\omega^2]} \left[\frac{(\omega_n^2 - \omega^2)}{\omega} \sin \omega t - 2\zeta\omega_n \cos \omega t \right.$$

$$+ \frac{2\zeta^2\omega_n^2 - (\omega_n^2 - \omega^2)}{\omega_n\sqrt{1 - \zeta^2}} e^{-\zeta\omega_n t} \sin \omega_n \sqrt{1 - \zeta^2}t$$

$$\left. + 2\zeta\omega_n e^{-\zeta\omega_n t} \cos \omega_n \sqrt{1 - \zeta^2}t \right]$$

At steady state $(t \to \infty)$, the terms involving $e^{-\zeta \omega_n t}$ approach zero. Thus, at steady state,

$$
x_{ss}(t) = \frac{P\omega}{m[(\omega_n^2 - \omega^2)^2 + 4\zeta^2 \omega_n^2 \omega^2]} \left(\frac{\omega_n^2 - \omega^2}{\omega} \sin \omega t - 2\zeta \omega_n \cos \omega t \right)
$$

$$
= \frac{P\omega}{(k - m\omega^2)^2 + b^2\omega^2} \left(\frac{k - m\omega^2}{\omega} \sin \omega t - b \cos \omega t \right)
$$

$$
= \frac{P}{\sqrt{(k - m\omega^2)^2 + b^2\omega^2}} \sin \left(\omega t - \tan^{-1} \frac{b\omega}{k - m\omega^2} \right)
$$

Problem A–9–2

Consider the mechanical system shown in Figure 9–20. If $m = 10$ kg, $b = 30$ N-s/m, $k = 500$ N/m, $P = 10$ N, and $\omega = 2$ rad/s, what is the steady-state output $x(t)$? The displacement x is measured from the equilibrium position before the input $p(t)$ is applied.

Solution The equation of motion for the system is

$$
m\ddot{x} + b\dot{x} + 2kx = p(t) = P \sin \omega t
$$

The transfer function of the system is

$$
\frac{X(s)}{P(s)} = \frac{1}{ms^2 + bs + 2k}
$$

where $X(s) = \mathcal{L}[x(t)]$ and $P(s) = \mathcal{L}[p(t)]$. Substituting the given numerical values for m, b, and k into this last equation, we obtain

$$
\frac{X(s)}{P(s)} = G(s) = \frac{1}{10s^2 + 30s + 1000}
$$

Then the sinusoidal transfer function becomes

$$
G(j\omega) = \frac{1}{10(j\omega)^2 + 30j\omega + 1000}
$$

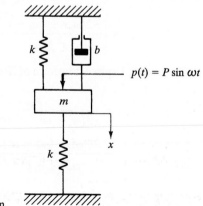

Figure 9–20 Mechanical system.

From Equation (9–6), the steady-state output $x(t)$ is given by

$$x(t) = X\sin(\omega t + \phi)$$
$$= |G(j\omega)|P\sin[\omega t + \underline{/G(j\omega)}]$$

For $\omega = 2$ rad/s, we obtain

$$G(j2) = \frac{1}{(1000 - 40) + j60}$$

Hence,

$$|G(j2)| = \frac{1}{\sqrt{960^2 + 60^2}} = 0.0010396$$

$$\underline{/G(j2)} = -\tan^{-1}\frac{60}{960} = -0.0624 \text{ rad}$$

Thus, for $P = 10$ N, the steady-state output is

$$x(t) = 0.0010396 \times 10 \sin(2t - 0.0624)$$
$$= 0.010396 \sin(2t - 0.0624)$$

Problem A–9–3

Consider the spring–mass system shown in Figure 9–21. The system is initially at rest, or $x(0) = 0$ and $\dot{x}(0) = 0$. At $t = 0$, a force $p(t) = P \sin \omega t$ is applied to the mass. Using the Laplace transform method, determine $x(t)$ for $t \geq 0$. Find the solution $x(t)$ when $m = 1$ kg, $k = 100$ N/m, $P = 50$ N, and $\omega = 5$ rad/s. Assume that the mass m moves without friction.

Solution The equation of motion for the system is

$$m\ddot{x} + kx = P \sin \omega t$$

By defining $\omega_n = \sqrt{k/m}$, this last equation can be written

$$\ddot{x} + \omega_n^2 x = \frac{P}{m} \sin \omega t$$

The Laplace transform of the preceding equation, under the initial conditions $x(0) = 0$ and $\dot{x}(0) = 0$, is

$$(s^2 + \omega_n^2)X(s) = \frac{P}{m}\frac{\omega}{s^2 + \omega^2}$$

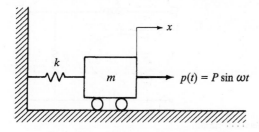

Figure 9–21 Spring–mass system.

Consequently,

$$X(s) = \frac{P\omega}{m} \frac{1}{(s^2 + \omega_n^2)(s^2 + \omega^2)}$$

$$= \frac{P\omega}{m} \left(\frac{1}{\omega^2 - \omega_n^2} \frac{1}{s^2 + \omega_n^2} - \frac{1}{\omega^2 - \omega_n^2} \frac{1}{s^2 + \omega^2} \right)$$

The inverse Laplace transform of this last equation is

$$x(t) = \frac{P}{m} \frac{1}{\omega^2 - \omega_n^2} \left(\frac{\omega}{\omega_n} \sin \omega_n t - \sin \omega t \right)$$

From the given numerical values, we find that $\omega_n = \sqrt{k/m} = \sqrt{100/1} = 10$ rad/s, $P/m = 50$ N/kg, and $\omega/\omega_n = 5/10 = 0.5$. Substituting these numerical values into the equation for $x(t)$, we have

$$x(t) = -\frac{1}{3} \sin 10t + \frac{2}{3} \sin 5t \text{ m}$$

Problem A–9–4

In the electrical circuit of Figure 9–22, assume that voltage e_i is applied to the input terminals and voltage e_o appears at the output terminals. In addition, assume that the input is sinusoidal and is given by

$$e_i(t) = E_i \sin \omega t$$

What is the steady-state current $i(t)$?

Solution   Applying Kirchhoff's voltage law to the circuit yields

$$L\frac{di}{dt} + Ri + \frac{1}{C}\int i \, dt = e_i$$

Then the Laplace transform of this last equation, assuming zero initial conditions, is

$$\left(Ls + R + \frac{1}{Cs}\right)I(s) = E_i(s)$$

Hence, the transfer function between $I(s)$ and $E_i(s)$ becomes

$$\frac{I(s)}{E_i(s)} = \frac{1}{Ls + R + (1/Cs)} = \frac{Cs}{LCs^2 + RCs + 1}$$

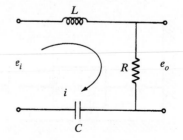

Figure 9–22   Electrical circuit.

The sinusoidal transfer function is

$$\frac{I(j\omega)}{E_i(j\omega)} = G(j\omega) = \frac{Cj\omega}{-LC\omega^2 + RCj\omega + 1}$$

Therefore, the steady-state current $i(t)$ is [see Equation (9–6)]

$$
\begin{aligned}
i(t) &= |G(j\omega)|E_i \sin[\omega t + \underline{/G(j\omega)}] \\
&= \frac{CE_i\omega}{\sqrt{(1 - LC\omega^2)^2 + (RC\omega)^2}} \sin\left(\omega t + 90° - \tan^{-1}\frac{RC\omega}{1 - LC\omega^2}\right) \\
&= \frac{CE_i\omega}{\sqrt{(1 - LC\omega^2)^2 + (RC\omega)^2}} \cos\left(\omega t - \tan^{-1}\frac{RC\omega}{1 - LC\omega^2}\right)
\end{aligned}
$$

Problem A–9–5

Consider the mechanical system shown in Figure 9–23. If the excitation force $p(t) = P \sin \omega t$, where $P = 1$ N and $\omega = 2$ rad/s, is applied, the steady-state amplitude of $x(t)$ is found to be 0.05 m. If the forcing frequency is changed to $\omega = 10$ rad/s, the steady-state amplitude of $x(t)$ is found to be 0.02 m. Determine the values of b and k.

Solution The equation of motion for the system is

$$b\dot{x} + kx = p(t)$$

The transfer function is

$$\frac{X(s)}{P(s)} = \frac{1}{bs + k}$$

Hence, the sinusoidal transfer function is

$$\frac{X(j\omega)}{P(j\omega)} = \frac{1}{bj\omega + k}$$

The amplitude ratio is

$$\left|\frac{X(j\omega)}{P(j\omega)}\right| = \frac{1}{\sqrt{b^2\omega^2 + k^2}}$$

so

$$|X(j\omega)| = \frac{|P(j\omega)|}{\sqrt{b^2\omega^2 + k^2}}$$

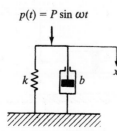

$p(t) = P \sin \omega t$

k $\quad$ b

x

Figure 9–23 Mechanical system.

From the problem statement, if $p(t) = P \sin \omega t = \sin 2t$, the amplitude of $x(t)$ is 0.05 m. Therefore,

$$0.05 = \frac{1}{\sqrt{b^2 \times 2^2 + k^2}}$$

or

$$4b^2 + k^2 = 400 \tag{9-28}$$

If $p(t) = P \sin \omega t = \sin 10t$, then the amplitude of $x(t)$ is 0.02 m. Hence,

$$0.02 = \frac{1}{\sqrt{b^2 \times 10^2 + k^2}}$$

or

$$100b^2 + k^2 = 2500 \tag{9-29}$$

From Equations (9–28) and (9–29), we obtain

$$96b^2 = 2100$$

or

$$b = 4.677 \text{ N-s/m}$$

Also,

$$k^2 = 312.57$$

or

$$k = 17.68 \text{ N/m}$$

Problem A–9–6

In the mechanical system shown in Figure 9–24, assume that the input and output are the displacements p and x, respectively. The displacement x is measured from the equilibrium position. Suppose that $p(t) = P \sin \omega t$. What is the output $x(t)$ at steady state? Assume that the system remains linear throughout the operating period.

Solution   The equation of motion for the system is

$$m\ddot{x} + b(\dot{x} - \dot{p}) + kx = 0$$

or

$$m\ddot{x} + b\dot{x} + kx = b\dot{p}$$

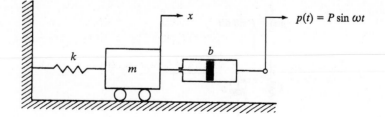

Figure 9–24   Mechanical system.

Hence, the transfer function between $X(s)$ and $P(s)$ is

$$\frac{X(s)}{P(s)} = \frac{bs}{ms^2 + bs + k}$$

Then the sinusoidal transfer function is

$$\frac{X(j\omega)}{P(j\omega)} = \frac{bj\omega}{-m\omega^2 + bj\omega + k}$$

Thus,

$$\left|\frac{X(j\omega)}{P(j\omega)}\right| = \frac{b\omega}{\sqrt{(k - m\omega^2)^2 + b^2\omega^2}}$$

and

$$\phi = \left|\frac{X(j\omega)}{P(j\omega)}\right| = \tan^{-1}\frac{b\omega}{0} - \tan^{-1}\frac{b\omega}{k - m\omega^2}$$

$$= 90° - \tan^{-1}\frac{b\omega}{k - m\omega^2}$$

Noting that $|P(j\omega)| = P$, the output is obtained as

$$x(t) = |X(j\omega)| \sin(\omega t + \phi)$$

$$= \frac{Pb\omega}{\sqrt{(k - m\omega^2)^2 + b^2\omega^2}} \sin\left(\omega t + 90° - \tan^{-1}\frac{b\omega}{k - m\omega^2}\right)$$

The angle $\tan^{-1}[b\omega/(k - m\omega^2)]$ varies from $0°$ to $180°$ as ω increases from zero to infinity. So, for small ω the output leads the input by almost $90°$, and for large ω the output lags the input by almost $90°$.

Problem A–9–7

Find the steady-state displacements $x_1(t)$ and $x_2(t)$ of the mechanical system shown in Figure 9–25. Assume that the viscous damping coefficients b_1 and b_2 are positive, but negligibly small. [This means that, in obtaining equations, we may assume that $b_1 \doteq 0$ and $b_2 \doteq 0$. Since b_1 and b_2 are positive, however small, the system is stable and Equation (9–6) can be used to find the steady-state solution.] The displacements x_1 and x_2 are measured from the respective equilibrium positions in the absence of the excitation force.

Solution The equations of motion for the system are

$$m_1\ddot{x}_1 + b_1\dot{x}_1 + k_1x_1 + b_2(\dot{x}_1 - \dot{x}_2) + k_2(x_1 - x_2) = p(t) = P\sin\omega t$$
$$m_2\ddot{x}_2 + b_2(\dot{x}_2 - \dot{x}_1) + k_2(x_2 - x_1) = 0$$

Since b_1 and b_2 are negligibly small, let us substitute $b_1 = 0$ and $b_2 = 0$ into the equations of motion. Then

$$m_1\ddot{x}_1 + k_1x_1 + k_2(x_1 - x_2) = p(t)$$
$$m_2\ddot{x}_2 + k_2(x_2 - x_1) = 0$$

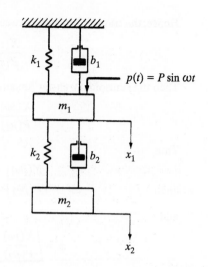

Figure 9-25   Mechanical system.

Taking the Laplace transforms of these two equations, assuming zero initial conditions, we have

$$(m_1 s^2 + k_1 + k_2) X_1(s) - k_2 X_2(s) = P(s)$$
$$(m_2 s^2 + k_2) X_2(s) - k_2 X_1(s) = 0$$

from which it follows that

$$\frac{X_2(s)}{X_1(s)} = \frac{k_2}{m_2 s^2 + k_2}$$

and

$$\frac{X_1(s)}{P(s)} = \frac{m_2 s^2 + k_2}{(m_1 s^2 + k_1 + k_2)(m_2 s^2 + k_2) - k_2^2}$$

Since the system is basically stable, Equation (9–6) can be applied. In so doing, the amplitudes $|X_1(j\omega)|$ and $|X_2(j\omega)|$ are obtained from the sinusoidal transfer functions

$$\frac{X_1(j\omega)}{P(j\omega)} = \frac{k_2 - m_2 \omega^2}{(k_1 + k_2 - m_1 \omega^2)(k_2 - m_2 \omega^2) - k_2^2}$$

and

$$\frac{X_2(j\omega)}{X_1(j\omega)} = \frac{k_2}{k_2 - m_2 \omega^2}$$

Thus, the steady-state solution $x_1(t)$ is

$$x_1(t) = |X_1(j\omega)| \sin\left[\omega t + \left/\frac{X_1(j\omega)}{P(j\omega)}\right.\right]$$

$$= \frac{P(k_2 - m_2 \omega^2)}{(k_1 + k_2 - m_1 \omega^2)(k_2 - m_2 \omega^2) - k_2^2} \sin(\omega t + \theta) \qquad (\theta = 0° \text{ or } 180°)$$

The steady-state solution $x_2(t)$ is

$$x_2(t) = |X_2(j\omega)| \sin\left[\omega t + \left/\frac{X_2(j\omega)}{P(j\omega)}\right.\right]$$

$$= \frac{k_2}{k_2 - m_2\omega^2}|X_1(j\omega)| \sin\left[\omega t + \left/\frac{X_2(j\omega)X_1(j\omega)}{X_1(j\omega)P(j\omega)}\right.\right]$$

$$= \frac{Pk_2}{(k_1 + k_2 - m_1\omega^2)(k_2 - m_2\omega^2) - k_2^2} \sin(\omega t + \theta) \qquad (\theta = 0° \text{ or } 180°)$$

Note that the angles $\underline{/X_1(j\omega)}/P(j\omega)$ and $\underline{/X_2(j\omega)}/P(j\omega)$ are either 0° or 180°. Consequently, the motions of masses m_1 and m_2 are either in phase or 180° out of phase with the excitation. That is, masses m_1 and m_2 move in the same direction if $\omega < \sqrt{k_2/m_2}$ and in the opposite direction if $\omega > \sqrt{k_2/m_2}$. If $\omega = \sqrt{k_2/m_2}$, mass m_1 stays still, but mass m_2 moves sinusoidally.

Problem A–9–8

Find the period of a conical pendulum in which a ball of mass m revolves about a fixed vertical axis at a constant speed, as shown in Figure 9–26.

Solution   As the ball stands out at a constant angle, the vertical component of the tension S in the cord balances with the gravitational force mg, and the horizontal component of S balances with the centrifugal force $m\omega^2 r$ Hence, from geometry,

$$\frac{m\omega^2 r}{mg} = \frac{r}{h}$$

or

$$\omega^2 = \frac{g}{h}$$

Therefore, the period T is

$$T = \frac{2\pi}{\omega} = 2\pi\sqrt{\frac{h}{g}}$$

Note that, since $h = l \cos \theta$, the period is a function of the angle θ.

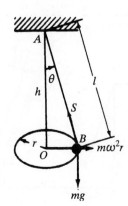

Figure 9–26   Conical pendulum.

Problem A–9–9

A boy riding a bicycle with a constant speed of 800 m/min around a horizontal circular path of radius $r = 50$ m leans inward at angle θ with the vertical, as shown in Figure 9–27. Determine the angle of inclination θ needed to maintain a steady-state circular motion.

Solution   The centripetal force that is necessary for a circular motion is

$$m\omega^2 r = m\frac{v^2}{r}$$

The gravitational force mg can be resolved into two component forces, F and R, as shown in Figure 9–27. The horizontal force $F = mg \tan \theta$ must provide the necessary centripetal force mv^2/r. (Note that F can be supplied by friction if the surface is sufficiently rough. If it is not, the boy must reduce his speed if he is to avoid slipping.) Hence,

$$mg \tan \theta = m\frac{v^2}{r}$$

or

$$\tan \theta = \frac{v^2}{gr}$$

Substituting the given numerical values into this last equation, we find that

$$\tan \theta = \frac{(800/60)^2}{9.807 \times 50} = 0.3626$$

or

$$\theta = 19.93°$$

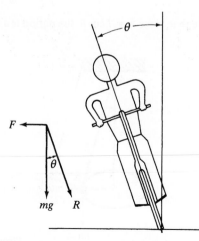

Figure 9–27   Boy riding a bicycle around a circular path.

Problem A–9–10

In rotating systems, if shafts rotate at critical speeds, large vibrations may develop as a result of resonance effects. In Figure 9–28(a), a disk of mass m is mounted on an elastic shaft whose mass is negligible compared with that of the disk, which is placed midway between bearings. Assume that the disk is not perfectly symmetrical and that there is an eccentricity e from the center of the disk. The geometrical center of the disk, the center of mass of the disk, and the center of rotation are denoted by points O, G, and R, respectively. The distance between points R and O is r, and that between points O and G is e. Assume that the equivalent spring constant of the elastic shaft is k, so that the restoring force due to the elastic shaft is kr. What is the critical speed of the system?

Solution   From Figure 9–28(a), the centrifugal force acting on the shaft is $m\omega^2(e + r)$. This force balances with the restoring force of the elastic shaft, kr. So

$$m\omega^2(e + r) = kr \qquad (9\text{–}30)$$

or

$$\omega^2(e + r) = \omega_n^2 r$$

where $\omega_n = \sqrt{k/m}$. Solving for r yields

$$r = \frac{e}{\left(\dfrac{\omega_n}{\omega}\right)^2 - 1}$$

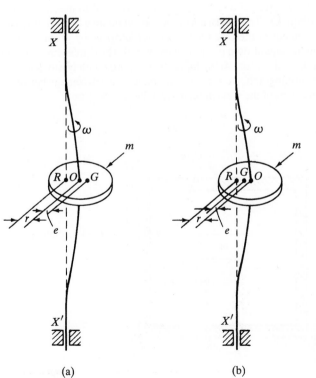

(a)   (b)

Figure 9–28   (a) Rotating system in which the angular speed is lower than the critical speed; (b) rotating system in which the angular speed is higher than the critical speed.

The deflection r tends to increase rapidly as ω approaches ω_n. At $\omega = \omega_n$, resonance occurs. The deflection r increases until Equation (9–30) no longer holds. The critical speed of the shaft is thus

$$\omega_{cr} = \omega_n = \sqrt{\frac{k}{m}}$$

At speeds higher than critical, the center of gravity G will be situated as shown in Figure 9–28(b), and the centrifugal force becomes

$$m\omega^2(r - e)$$

This force balances with the restoring force of the elastic shaft kr; hence,

$$m\omega^2(r - e) = kr$$

Solving for r and noting that $k/m = \omega_n^2$, we have

$$r = \frac{e}{1 - \left(\dfrac{\omega_n}{\omega}\right)^2}$$

For $\omega > \omega_n$, the deflection r decreases and approaches e with increasing ω. For $\omega \gg \omega_n$, the center of gravity of the disk moves toward the line XX', and in this case the disk does not whirl, but the deflected shaft whirls about the center of gravity, G.

Problem A–9–11

Figure 9–29 is a schematic diagram of a seismograph, a device used to measure ground displacement during earthquakes. The displacement of mass m relative to inertial space is denoted by x, the displacement of the case relative to inertial space by y. The displacement x is measured from the equilibrium position when $y = 0$. The displacement y is the input to the system and, in the case of earthquakes, is approximately sinusoidal, or $y(t) = Y \sin \omega t$. In the seismograph, we measure the relative displacement between x and y. Define the displacement of the mass m relative to the case as z, or

$$z = x - y$$

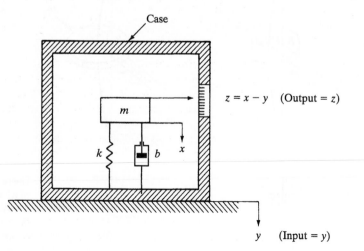

Figure 9–29 Seismograph.

Show that the seismograph measures and records the displacement of its case y accurately if $\omega \gg \omega_n$, where $\omega_n = \sqrt{k/m}$.

Solution   The equation of motion for the seismograph is

$$m\ddot{x} + b(\dot{x} - \dot{y}) + k(x - y) = 0 \tag{9-31}$$

In terms of the relative displacement $z = x - y$, Equation (9-31) becomes

$$m(\ddot{y} + \ddot{z}) + b\dot{z} + kz = 0$$

or

$$m\ddot{z} + b\dot{z} + kz = -m\ddot{y}$$

Taking the Laplace transform of this last equation and assuming zero initial conditions, we find that

$$(ms^2 + bs + k)Z(s) = -ms^2Y(s)$$

Note that the input to the system is the displacement y and that the output is the relative displacement z. The transfer function between $Z(s)$ and $Y(s)$ is

$$\frac{Z(s)}{Y(s)} = \frac{-ms^2}{ms^2 + bs + k}$$

The sinusoidal transfer function is

$$\frac{Z(j\omega)}{Y(j\omega)} = \frac{m\omega^2}{-m\omega^2 + bj\omega + k}$$

Substituting $k/m = \omega_n^2$ and $b/m = 2\zeta\omega_n$ into this last equation gives

$$\frac{Z(j\omega)}{Y(j\omega)} = \frac{\omega^2}{-\omega^2 + 2\zeta\omega_n j\omega + \omega_n^2} = \frac{\beta^2}{1 - \beta^2 + j2\zeta\beta} \tag{9-32}$$

where $\beta = \omega/\omega_n$.

In the seismograph, we want to determine the input displacement $y(t)$ accurately by measuring the relative displacement $z(t)$. A glance at Equation (9-32) tells us that we can do so easily if $\beta \gg 1$, in which case Equation (9-32) reduces to

$$\frac{Z(j\omega)}{Y(j\omega)} \doteq -\frac{\beta^2}{\beta^2} = -1$$

The seismograph measures and records the displacement of its case y accurately if $\beta \gg 1$ or $\omega \gg \omega_n$. In fact, for $\omega \gg \omega_n$, mass m tends to remain fixed in space, and the motion of the case can be seen as a relative motion between the mass and the case.

To produce the condition $\omega \gg \omega_n$, we choose the undamped natural frequency ω_n as low as possible. (Choose a relatively large mass and a spring as soft as the elastic and static deflection limits allow.) Then the seismograph will measure and record the displacements of all frequencies well above the undamped natural frequency ω_n, which is very low.

Problem A-9-12

A schematic diagram of a translational accelerometer is given in Figure 9-30. The system configuration is basically the same as that of the seismograph, but their essential difference lies in the choice of the undamped natural frequency $\omega_n = \sqrt{k/m}$. Let us

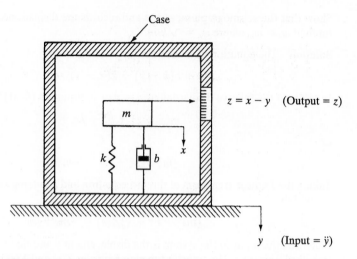

Figure 9–30   Translational accelerometer.

denote the displacement of mass m relative to inertial space by x and that of the case relative to inertial space by y. The displacement x is measured from the equilibrium position when $y = 0$. The input to the translational accelerometer is the acceleration $\ddot{y}$. The output is the displacement of the mass m relative to the case, or $z = x - y$. (We measure and record the relative displacement z, not the absolute displacement x.)

   Show that if the undamped natural frequency ω_n is sufficiently high compared with the frequencies of the input, then the displacement z can be made nearly proportional to $\ddot{y}$.

Solution   The equation of motion for the system is

$$m\ddot{x} + b(\dot{x} - \dot{y}) + k(x - y) = 0$$

In terms of the relative displacement z, this last equation becomes

$$m(\ddot{y} + \ddot{z}) + b\dot{z} + kz = 0$$

or

$$m\ddot{z} + b\dot{z} + kz = -m\ddot{y}$$

The Laplace transform of the preceding equation, assuming zero initial conditions, gives

$$(ms^2 + bs + k)Z(s) = -ms^2Y(s)$$

The transfer function between output $Z(s)$ and input $s^2Y(s)$ [the input is the acceleration $\ddot{y}$, and thus its Laplace transform is $s^2Y(s)$] is

$$\frac{Z(s)}{s^2Y(s)} = \frac{-m}{ms^2 + bs + k} = \frac{-1}{s^2 + 2\zeta\omega_n s + \omega_n^2} \qquad (9\text{–}33)$$

From Equation (9–33), we see that if the undamped natural frequency ω_n is sufficiently high compared with the frequencies of the input, then

$$\frac{Z(s)}{s^2Y(s)} \doteq -\frac{1}{\omega_n^2}$$

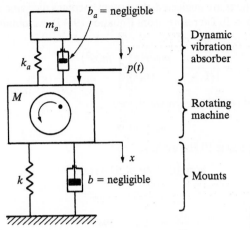

Dynamic
vibration
absorber

Rotating
machine

Mounts

Figure 9–31 Rotating machine with a
dynamic vibration absorber.

Thus, the displacement z is nearly proportional to $\ddot{y}$. Hence, in the accelerometer, we
choose the undamped natural frequency $\omega_n = \sqrt{k/m}$ to be sufficiently high.

Problem A–9–13

A rotating machine with a mass of 100 kg and mounted on an isolator rotates at a constant speed of 10 Hz. An unbalanced mass m located a distance r from the center of the
rotor is exciting vibrations at a frequency ω that is very close to the natural frequency
ω_n of the system, with the result that the machine vibrates violently and a large vibratory force is transmitted to the foundation.

Design a dynamic vibration absorber to reduce the vibration. When the
dynamic vibration absorber is added to the rotating machine as shown in Figure
9–31, the entire system will become a two-degrees-of-freedom system. Determine
the mass m_a and spring constant k_a of the dynamic vibration absorber such that the
lower natural frequency is 20% off the operating frequency. Determine also the
higher natural frequency of the system. Assume that the values of b (the viscous-
friction coefficient of the isolator) and b_a (the viscous-friction coefficient of the
dynamic vibration absorber) are positive, but negligibly small. (Note that, since the
values of b and b_a are positive, however small, the system is stable. Therefore, the
steady-state displacements can be obtained with the use of the sinusoidal transfer
function.)

Solution   The equations of motion for the system are

$$M\ddot{x} + b\dot{x} + kx + b_a(\dot{x} - \dot{y}) + k_a(x - y) = p(t) = m\omega^2 r \sin \omega t$$
$$m_a\ddot{y} + b_a(\dot{y} - \dot{x}) + k_a(y - x) = 0$$

where x and y are the displacements of mass M and mass m_a, respectively, and both x
and y are measured from the respective equilibrium positions. Since $b \doteq 0$ and $b_a \doteq 0$,
the last two equations may be simplified to

$$M\ddot{x} + kx + k_a(x - y) = p(t)$$
$$m_a\ddot{y} + k_a(y - x) = 0$$

When the viscous frictions are neglected, the system becomes the same as that shown in Figure 9–14(b) with $b = 0$. Therefore, from Equation (9–17), we obtain the amplitude $|X(j\omega)|$ of $x(t)$ as:

$$|X(j\omega)| = \left| \frac{m\omega^2 r(k_a - m_a\omega^2)}{(k + k_a - M\omega^2)(k_a - m_a\omega^2) - k_a^2} \right| \tag{9–34}$$

To make this amplitude equal to zero, we choose

$$k_a = m_a\omega^2$$

Since the operating speed is 10 Hz, we have

$$\omega = 10 \times 2\pi = 62.8 \text{ rad/s}$$

so

$$\frac{k_a}{m_a} = 62.8^2 = 3944$$

The two natural frequencies ω_1 and ω_2 (where $\omega_1 < \omega_2$) of the entire system can be found from the characteristic equation. [The denominator of Equation (9–34) is the characteristic polynomial.] We have

$$(k + k_a - M\omega_i^2)(k_a - m_a\omega_i^2) - k_a^2 = 0 \qquad i = 1, 2$$

or

$$\left(1 + \frac{k_a}{k} - \frac{M}{k}\omega_i^2\right)\left(1 - \frac{m_a}{k_a}\omega_i^2\right) - \frac{k_a}{k} = 0 \tag{9–35}$$

Note that in the present system, since the natural frequency $\omega_n = \sqrt{k/M}$ is very close to the operating frequency $\omega = \sqrt{k_a/m_a}$, we can set

$$\sqrt{\frac{k}{M}} = \sqrt{\frac{k_a}{m_a}} = \omega = 62.8$$

Now, the lower natural frequency must be 20% off the operating frequency ω. Since $\omega_1 < \omega$, this means that

$$\omega_1 = 0.8\omega = 0.8 \times 62.8$$

Substituting $\omega_i = \omega_1$ and $k/M = k_a/m_a = \omega^2$ into Equation (9–35), we have

$$\left(1 + \frac{k_a}{k} - \frac{\omega_1^2}{\omega^2}\right)\left(1 - \frac{\omega_1^2}{\omega^2}\right) - \frac{k_a}{k} = 0$$

Substituting $\omega_1/\omega = 0.8$ into this last equation, we obtain

$$\left(1 + \frac{k_a}{k} - 0.8^2\right)(1 - 0.8^2) - \frac{k_a}{k} = 0$$

Solving for k_a/k yields

$$\frac{k_a}{k} = 0.2025$$

It follows that

$$\frac{m_a}{M} = \frac{k_a}{k} = 0.2025$$

Since $M = 100$ kg, we have

$$m_a = 0.2050 \times 100 = 20.25 \text{ kg}$$

Because

$$\frac{k}{M} = \frac{k_a}{m_a} = 62.8^2$$

we obtain

$$k_a = (62.8)^2 m_a = (62.8)^2(20.25) = 79.9 \times 10^3 \text{ N/m}$$

Thus, the mass and spring constant of the dynamic vibration absorber are $m_a = 20.25$ kg and $k_a = 79.9 \times 10^3$ N/m, respectively.

The two natural frequencies ω_1 and ω_2 can be determined by substituting $k_a/k = 0.2025$ into the equation

$$\left(1 + \frac{k_a}{k} - \frac{\omega_i^2}{\omega^2}\right)\left(1 - \frac{\omega_i^2}{\omega^2}\right) - \frac{k_a}{k} = 0 \qquad i = 1,2$$

or

$$\left(1 + 0.2025 - \frac{\omega_i^2}{\omega^2}\right)\left(1 - \frac{\omega_i^2}{\omega^2}\right) - 0.2025 = 0$$

Solving for ω_i/ω, we obtain

$$\frac{\omega_i^2}{\omega^2} = 0.64, \qquad \text{or} \qquad 1.5625$$

Since $\omega_1 < \omega_2$,

$$\frac{\omega_1^2}{\omega^2} = 0.64, \qquad \frac{\omega_2^2}{\omega^2} = 1.5625$$

Therefore,

$$\omega_1 = 0.8\omega = 0.8 \times 62.8 = 50.24 \text{ rad/s} = 8 \text{ Hz}$$

and

$$\omega_2 = 1.25\omega = 1.25 \times 62.8 = 78.5 \text{ rad/s} = 12.5 \text{ Hz}$$

Problem A–9–14

Consider the two-degrees-of-freedom mechanical system shown in Figure 9–32. Obtain the first and second modes of vibration. The displacements x_1 and x_2 are measured from the respective equilibrium positions. Assume that masses m_1 and m_2 move without friction.

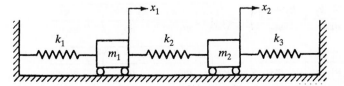

Figure 9–32 Mechanical system.

Solution　The equations of motion for the system are

$$m_1\ddot{x}_1 = -k_1x_1 - k_2(x_1 - x_2)$$
$$m_2\ddot{x}_2 = -k_3x_2 - k_2(x_2 - x_1)$$

which can be rewritten as

$$m_1\ddot{x}_1 + (k_1 + k_2)x_1 - k_2x_2 = 0 \qquad (9\text{–}36)$$
$$m_2\ddot{x}_2 + (k_2 + k_3)x_2 - k_2x_1 = 0 \qquad (9\text{–}37)$$

To find the natural frequencies of the free vibration, assume that the motion is harmonic. That is, assume that

$$x_1 = A\sin\omega t, \qquad x_2 = B\sin\omega t$$

Then

$$\ddot{x}_1 = -A\omega^2\sin\omega t, \qquad \ddot{x}_2 = -B\omega^2\sin\omega t$$

Substituting the harmonic solutions into Equations (9–36) and (9–37), we get

$$[(-m_1\omega^2 + k_1 + k_2)A - k_2B]\sin\omega t = 0$$
$$[-k_2A + (-m_2\omega^2 + k_2 + k_3)B]\sin\omega t = 0$$

Since these two equations must be satisfied at all times, and since $\sin\omega t$ cannot be zero at all times, the quantities in the brackets must be equal to zero. Thus,

$$(-m_1\omega^2 + k_1 + k_2)A - k_2B = 0 \qquad (9\text{–}38)$$
$$-k_2A + (-m_2\omega^2 + k_2 + k_3)B = 0 \qquad (9\text{–}39)$$

For constants A and B to be nonzero, the determinant of the coefficient matrix must be zero, or

$$\begin{vmatrix} -m_1\omega^2 + k_1 + k_2 & -k_2 \\ -k_2 & -m_2\omega^2 + k_2 + k_3 \end{vmatrix} = 0$$

which can be simplified to

$$(-m_1\omega^2 + k_1 + k_2)(-m_2\omega^2 + k_2 + k_3) - k_2^2 = 0$$

or

$$\omega^4 - \left(\frac{k_1 + k_2}{m_1} + \frac{k_2 + k_3}{m_2}\right)\omega^2 + \frac{k_1k_2 + k_2k_3 + k_3k_1}{m_1m_2} = 0$$

Solving this last equation for ω^2, we obtain

$$\omega^2 = \frac{1}{2}\left(\frac{k_1 + k_2}{m_1} + \frac{k_2 + k_3}{m_2}\right) \pm \sqrt{\frac{1}{4}\left(\frac{k_1 + k_2}{m_1} - \frac{k_2 + k_3}{m_2}\right)^2 + \frac{k_2^2}{m_1m_2}}$$

Now, we define

$$\omega_1^2 = \frac{1}{2}\left(\frac{k_1 + k_2}{m_1} + \frac{k_2 + k_3}{m_2}\right) - \sqrt{\frac{1}{4}\left(\frac{k_1 + k_2}{m_1} - \frac{k_2 + k_3}{m_2}\right)^2 + \frac{k_2^2}{m_1m_2}}$$

$$\omega_2^2 = \frac{1}{2}\left(\frac{k_1 + k_2}{m_1} + \frac{k_2 + k_3}{m_2}\right) + \sqrt{\frac{1}{4}\left(\frac{k_1 + k_2}{m_1} - \frac{k_2 + k_3}{m_2}\right)^2 + \frac{k_2^2}{m_1m_2}}$$

The vibration at frequency ω_1 is the first mode of vibration and that at frequency ω_2 is the second mode of vibration. Note that, from Equation (9–38), we obtain

$$\frac{A}{B} = \frac{k_2}{-m_1\omega^2 + k_1 + k_2} \tag{9–40}$$

Also, from Equation (9–39), we get

$$\frac{A}{B} = \frac{-m_2\omega^2 + k_2 + k_3}{k_2} \tag{9–41}$$

Substituting ω_1^2 into Equations (9–40) and (9–41) and writing A/B as A_1/B_1 yields

$$\frac{A_1}{B_1} = \frac{k_2}{-m_1\omega_1^2 + k_1 + k_2} = \frac{-m_2\omega_1^2 + k_2 + k_3}{k_2} = \frac{1}{\lambda_1}$$

Similarly, substituting ω_2^2 into Equations (9–40) and (9–41) and writing A/B as A_2/B_2, we obtain

$$\frac{A_2}{B_2} = \frac{k_2}{-m\omega_2^2 + k_1 + k_2} = \frac{-m_2\omega_2^2 + k_2 + k_3}{k_2} = \frac{1}{\lambda_2}$$

Notice that

$$k_2\lambda_1 = -m_1\omega_1^2 + k_1 + k_2$$

$$= \frac{1}{2}\left[k_1 + k_2 - \frac{m_1}{m_2}(k_2 + k_3)\right]$$

$$+ \sqrt{\frac{1}{4}\left[k_1 + k_2 - \frac{m_1}{m_2}(k_2 + k_3)\right]^2 + \frac{m_1}{m_2}k_2^2} > 0$$

Also,

$$k_2\lambda_2 = -m_1\omega_2^2 + k_1 + k_2$$

$$= \frac{1}{2}\left[k_1 + k_2 - \frac{m_1}{m_2}(k_2 + k_3)\right]$$

$$- \sqrt{\frac{1}{4}\left[k_1 + k_2 - \frac{m_1}{m_2}(k_2 + k_3)\right]^2 + \frac{m_1}{m_2}k_2^2} < 0$$

Thus,

$$\frac{A_1}{B_1} = \frac{1}{\lambda_1} > 0, \qquad \frac{A_2}{B_2} = \frac{1}{\lambda_2} < 0$$

which means that in the first mode of vibration masses m_1 and m_2 move in the same direction, whereas in the second mode of vibration masses m_1 and m_2 move in opposite directions. Figures 9–33(a) and (b) show the first and second modes of vibration, respectively.

Problem A–9–15

Figure 9–34 shows a three-degrees-of-freedom system. To simplify the analysis, we assume that all of the masses are equal and the four springs are identical. We also assume that the masses move without friction. The displacements x_1, x_2, and x_3 are measured from their respective equilibrium positions. Obtain the natural frequencies and the modes of vibration of the system.

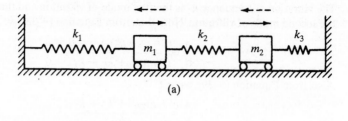

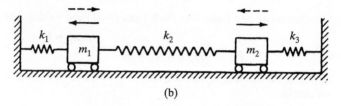

Figure 9–33   (a) First mode of vibration; (b) second mode of vibration.

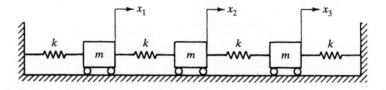

Figure 9–34   Mechanical system with three degrees of freedom.

Solution   The equations of motion for the system are

$$m\ddot{x}_1 + kx_1 + k(x_1 - x_2) = 0$$
$$m\ddot{x}_2 + k(x_2 - x_1) + k(x_2 - x_3) = 0$$
$$m\ddot{x}_3 + k(x_3 - x_2) + kx_3 = 0$$

which can be rewritten as

$$m\ddot{x}_1 + 2kx_1 - kx_2 = 0 \qquad (9\text{–}42)$$
$$m\ddot{x}_2 - kx_1 + 2kx_2 - kx_3 = 0 \qquad (9\text{–}43)$$
$$m\ddot{x}_3 - kx_2 + 2kx_3 = 0 \qquad (9\text{–}44)$$

To obtain the natural frequencies of the system, we assume that the motion is harmonic. That is, we assume that

$$x_1 = A \sin \omega t$$
$$x_2 = B \sin \omega t$$
$$x_3 = C \sin \omega t$$

Then Equations (9–42), (9–43), and (9–44) become, respectively,

$$(-m\omega^2 A + 2kA - kB)\sin \omega t = 0$$
$$(-m\omega^2 B - kA + 2kB - kC)\sin \omega t = 0$$
$$(-m\omega^2 C - kB + 2kC)\sin \omega t = 0$$

Since these three equations must be satisfied at all times, and since $\sin \omega t$ cannot be zero at all times, the quantities in parentheses must be equal to zero. That is,

$$(2k - m\omega^2)A - kB = 0$$
$$-kA + (2k - m\omega^2)B - kC = 0$$
$$-kB + (2k - m\omega^2)C = 0$$

These three equations can be simplified to

$$\left(2 - \frac{m\omega^2}{k}\right)A - B = 0 \tag{9-45}$$

$$-A + \left(2 - \frac{m\omega^2}{k}\right)B - C = 0 \tag{9-46}$$

$$-B + \left(2 - \frac{m\omega^2}{k}\right)C = 0 \tag{9-47}$$

For constants A, B, and C to be nonzero, the determinant of the coefficients of Equations (9–45), (9–46), and (9–47) must be zero, or

$$\begin{vmatrix} 2 - \dfrac{m\omega^2}{k} & -1 & 0 \\ -1 & 2 - \dfrac{m\omega^2}{k} & -1 \\ 0 & -1 & 2 - \dfrac{m\omega^2}{k} \end{vmatrix} = 0$$

Expanding this determinantal equation, we obtain

$$\left(2 - \frac{m\omega^2}{k}\right)^3 - \left(2 - \frac{m\omega^2}{k}\right) - \left(2 - \frac{m\omega^2}{k}\right) = 0$$

or

$$\left(2 - \frac{m\omega^2}{k}\right)\left(2 - \frac{m\omega^2}{k} - \sqrt{2}\right)\left(2 - \frac{m\omega^2}{k} + \sqrt{2}\right) = 0$$

from which we get

$$\frac{m\omega^2}{k} = 2 - \sqrt{2}, \qquad \frac{m\omega^2}{k} = 2, \qquad \frac{m\omega^2}{k} = 2 + \sqrt{2}$$

Thus,

$$\omega = 0.7654\sqrt{\frac{k}{m}}, \qquad \omega = 1.4142\sqrt{\frac{k}{m}}, \qquad \omega = 1.8478\sqrt{\frac{k}{m}}$$

Hence, the first mode of vibration is at $\omega = 0.7654\sqrt{k/m}$, the second mode of vibration is at $\omega = 1.4142\sqrt{k/m}$, and the third mode of vibration is at $\omega = 1.8478\sqrt{k/m}$.

First mode of vibration $(m\omega^2/k = 0.5858)$. From Equations (9–45) and (9–47), we have

$$1.4142A = B$$
$$B = 1.4142C$$

Thus, the amplitude ratio becomes

$$A:B:C = 1:1.4142:1$$

Second mode of vibration $(m\omega^2/k = 2)$**.** From Equations (9–45) and (9–46), we have

$$B = 0$$
$$A = -C$$

Hence, the amplitude ratio becomes

$$A:B:C = 1:0:-1$$

Note that the second mass does not move, because amplitude $B = 0$.

Third mode of vibration $(m\omega^2/k = 3.4142)$**.** From Equations (9–45) and (9–47), we get

$$-1.4142A = B$$
$$B = -1.4142C$$

Thus, the amplitude ratio becomes

$$A:B:C = 1:-1.4142:1$$

Figure 9–35 depicts the three modes of vibration of the system.

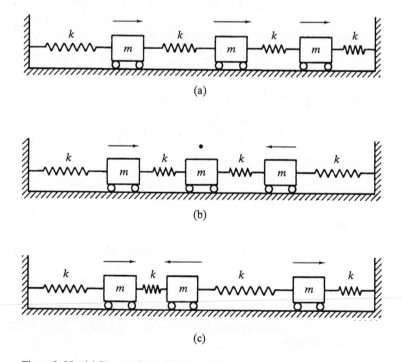

(a)

(b)

(c)

Figure 9–35 (a) First mode of vibration; (b) second mode of vibration; (c) third mode of vibration.

Problem A-9-16

Consider the mechanical system shown in Figure 9-36. Determine the natural frequencies and modes of vibration. In the diagram, the displacements x and y are measured from their respective equilibrium positions. Assume that

$$m = 1 \text{ kg}, \qquad M = 2 \text{ kg}, \qquad k_1 = 10 \text{ N/m}, \qquad k_2 = 40 \text{ N/m}$$

Determine the vibration when the initial conditions are given by

(a) $x(0) = 0.028078 \text{ m}$, $\quad \dot{x}(0) = 0 \text{ m/s}$, $\quad y(0) = 0.1 \text{ m}$, $\quad \dot{y}(0) = 0 \text{ m/s}$

(b) $x(0) = 0.17808 \text{ m}$, $\quad \dot{x}(0) = 0 \text{ m/s}$, $\quad y(0) = -0.1 \text{ m}$, $\quad \dot{y}(0) = 0 \text{ m/s}$

Solution   The equations of motion for the system are

$$M\ddot{x} + k_1(x - y) + k_2 x = 0$$
$$m\ddot{y} + k_1(y - x) = 0$$

Substituting the given numerical values into these two equations, we obtain

$$2\ddot{x} + 10(x - y) + 40x = 0 \tag{9-48}$$
$$\ddot{y} + 10(y - x) = 0 \tag{9-49}$$

To find the natural frequencies of the free vibration, assume that the motion is harmonic. That is, assume that

$$x = A \sin \omega t, \qquad y = B \sin \omega t$$

Then

$$\ddot{x} = -A\omega^2 \sin \omega t, \qquad \ddot{y} = -B\omega^2 \sin \omega t$$

If the preceding expressions are substituted into Equations (9-48) and (9-49), we obtain

$$[-2\omega^2 A + 10(A - B) + 40A] \sin \omega t = 0$$
$$[-\omega^2 B + 10(B - A)] \sin \omega t = 0$$

Since these equations must be satisfied at all times, and since $\sin \omega t$ cannot be zero at all times, the quantities in the brackets must be equal to zero. Thus,

$$-2\omega^2 A + 10(A - B) + 40A = 0$$
$$-\omega^2 B + 10(B - A) = 0$$

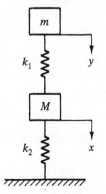

Figure 9-36   Mechanical system with two degrees of freedom.

Rearranging terms yields

$$(50 - 2\omega^2)A - 10B = 0 \tag{9-50}$$
$$-10A + (10 - \omega^2)B = 0 \tag{9-51}$$

For constants A and B to be nonzero, the determinant of the coefficient matrix must be equal to zero, or

$$\begin{vmatrix} 50 - 2\omega^2 & -10 \\ -10 & 10 - \omega^2 \end{vmatrix} = 0$$

which yields

$$\omega^4 - 35\omega^2 + 200 = 0$$

or

$$(\omega^2 - 7.1922)(\omega^2 - 27.808) = 0$$

Hence,

$$\omega_1^2 = 7.1922 \qquad \text{and} \qquad \omega_2^2 = 27.808$$

or

$$\omega_1 = 2.6818 \qquad \text{and} \qquad \omega_2 = 5.2733$$

Note that ω_1 is the frequency of the first mode of vibration and ω_2 is the frequency of the second mode of vibration. Note also that, from Equations (9–50) and (9–51), we obtain

$$\frac{A}{B} = \frac{10}{50 - 2\omega^2}, \qquad \frac{A}{B} = \frac{10 - \omega^2}{10}$$

Substituting $\omega_1^2 = 7.1922$ into A/B, we get

$$\frac{A}{B} = \frac{10}{50 - 2\omega_1^2} = \frac{10 - \omega_1^2}{10} = 0.28078 > 0$$

Similarly, substituting $\omega_2^2 = 27.808$ into A/B, we find that

$$\frac{A}{B} = \frac{10}{50 - 2\omega_2^2} = \frac{10 - \omega_2^2}{10} = -1.7808 < 0$$

Hence, in the first mode of vibration two masses move in the same direction (two motions are in phase), while in the second mode of vibration two masses move in opposite directions (two motions are out of phase). Figure 9–37 shows the first and second modes of vibration.

Next, we shall obtain the vibrations $x(t)$ and $y(t)$, subject to the given initial conditions. Laplace transforming Equations (9–48) and (9–49), we obtain

$$2[s^2 X(s) - sx(0) - \dot{x}(0)] + 10[X(s) - Y(s)] + 40X(s) = 0$$
$$[s^2 Y(s) - sy(0) - \dot{y}(0)] + 10[Y(s) - X(s)] = 0$$

Using the initial conditions that $x(0) \neq 0$, $\dot{x}(0) = 0$, $y(0) \neq 0$, and $\dot{y}(0) = 0$, we can simplify the last two equations as follows:

$$(2s^2 + 50)X(s) = 2sx(0) + 10Y(s) \tag{9-52}$$
$$(s^2 + 10)Y(s) = sy(0) + 10X(s) \tag{9-53}$$

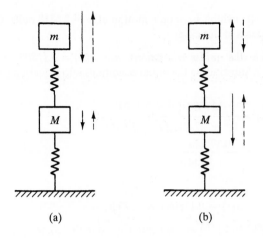

(a)   (b)

Figure 9–37   (a) First mode of vibration;
(b) second mode of vibration.

Eliminating $Y(s)$ from Equations (9–52) and (9–53) and solving for $X(s)$, we obtain

$$X(s) = \frac{(s^2 + 10)sx(0) + 5sy(0)}{s^4 + 35s^2 + 200} \tag{9–54}$$

Case (a), *in which the initial conditions are* $x(0) = 0.028078$, $\dot{x}(0) = 0$, $y(0) = 0.1$, *and* $\dot{y}(0) = 0$: Substituting the initial conditions into Equation (9–54), we get

$$\begin{aligned}
X(s) &= \frac{(s^2 + 10)s \times 0.028078 + 5s \times 0.1}{s^4 + 35s^2 + 200} \\[2mm]
&= \frac{0.028078s(s^2 + 27.808)}{(s^2 + 27.808)(s^2 + 7.1922)} \\[2mm]
&= \frac{0.028078s}{s^2 + 7.1922} \tag{9–55}
\end{aligned}$$

The inverse Laplace transform of $X(s)$ gives

$$x(t) = 0.028078 \cos 2.6818t$$

Substituting Equation (9–55) into Equation (9–53) and solving for $Y(s)$, we obtain

$$Y(s) = \frac{1}{s^2 + 10}\left[sy(0) + \frac{0.28078s}{s^2 + 7.1922} \right]$$

Substituting $y(0) = 0.1$ into this last equation, we get

$$\begin{aligned}
Y(s) &= \frac{0.1}{s^2 + 10}\frac{s^3 + 7.1922s + 2.8078s}{s^2 + 7.1922} \\[2mm]
&= \frac{0.1}{s^2 + 10}\frac{s(s^2 + 10)}{s^2 + 7.1922} \\[2mm]
&= \frac{0.1s}{s^2 + 7.1922}
\end{aligned}$$

Hence,

$$y(t) = 0.1 \cos 2.6818t$$

Notice that both $x(t)$ and $y(t)$ exhibit harmonic motion at $\omega = 2.6818$ rad/s. Only the first mode of vibration appears in this case.

Case (**b**), *in which the initial conditions are* $x(0) = 0.17808$, $\dot{x}(0) = 0$, $y(0) = -0.1$, *and* $\dot{y}(0) = 0$: Substituting the initial conditions into Equation (9–54), we have

$$X(s) = \frac{0.17808s(s^2 + 7.1922)}{(s^2 + 27.808)(s^2 + 7.1922)}$$

$$= \frac{0.17808s}{s^2 + 27.808} \tag{9–56}$$

Hence,

$$x(t) = 0.17808 \cos 5.2733t$$

Substituting Equation (9–56) and $y(0) = -0.1$ into Equation (9–53), we get

$$Y(s) = \frac{-0.1s(s^2 + 10)}{(s^2 + 10)(s^2 + 27.808)}$$

$$= -\frac{0.1s}{s^2 + 27.808}$$

Thus,

$$y(t) = -0.1 \cos 5.2733t$$

In this case, only the second mode of vibration appears.

Note that, for arbitrary initial conditions, both the first and second modes of vibration appear.

PROBLEMS

Problem B–9–1

The spring–mass system shown in Figure 9–38 is initially at rest. If mass m is excited by a sinusoidal force $p(t) = P \sin \omega t$, what is the response $x(t)$? Assume that $m = 1$ kg, $k = 100$ N/m, $P = 5$ N, and $\omega = 2$ rad/s. The displacement $x(t)$ is measured from the equilibrium position before the force $p(t)$ is applied.

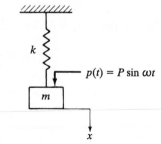

Figure 9–38 Spring–mass system.

Problem B–9–2

Consider the mechanical vibratory system shown in Figure 9–39. Assume that the displacement x is measured from the equilibrium position in the absence of the sinusoidal

Input force
$p(t) = P \sin \omega t$

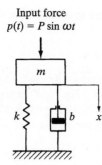

Figure 9–39 Mechanical vibratory system.

excitation force. The initial conditions are $x(0) = 0$ and $\dot{x}(0) = 0$, and the input force $p(t) = P \sin \omega t$ is applied at $t = 0$. Assume that $m = 2$ kg, $b = 24$ N-s/m, $k = 200$ N/m, $P = 5$ N, and $\omega = 6$ rad/s. Obtain the complete solution $x(t)$.

Problem B–9–3

Consider the electrical circuit shown in Figure 9–40. If the input voltage $e_i(t)$ is $E_i \sin \omega t$, what is the output voltage $e_o(t)$ at steady state?

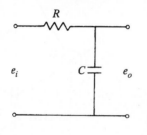

Figure 9–40 Electrical circuit.

Problem B–9–4

Consider the mechanical system shown in Figure 9–41. Obtain the steady-state outputs $x_1(t)$ and $x_2(t)$ when the input $p(t)$ is a sinusoidal force given by

$$p(t) = P \sin \omega t$$

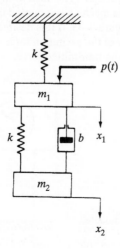

Figure 9–41 Mechanical system.

The output displacements $x_1(t)$ and $x_2(t)$ are measured from the respective equilibrium positions.

Problem B–9–5

Consider a conical pendulum consisting of a stone of mass 0.1 kg attached to the end of 1-m cord and rotated at an angular speed of 1 Hz. Find the tension in the cord. If the maximum allowable tension in the cord is 10 N, what is the maximum angular speed (in hertz) that can be attained without breaking the cord?

Problem B–9–6

In the speed-regulator system of Figure 9–42, what is the frequency ω needed to maintain the configuration shown in the diagram?

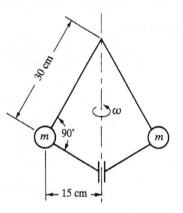

Figure 9–42   Speed-regulator system.

Problem B–9–7

A rotating machine of mass $M = 100$ kg has an unbalanced mass $m = 0.2$ kg a distance $r = 0.5$ m from the center of rotation. (The mass M includes mass m.) The operating speed is 10 Hz. Suppose that the machine is mounted on an isolator consisting of a spring and damper, as shown in Figure 9–43. If it is desired that ζ be 0.2, specify the spring constant k such that only 10% of the excitation force is transmitted to the foundation. Determine the amplitude of the transmitted force.

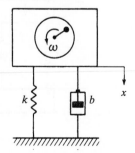

Figure 9–43   Rotating machine mounted on a vibration isolator.

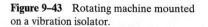

Problem B–9–8

In Figure 9–44, an instrument is attached to a base whose motion is to be measured. The relative motion between mass m and the base recorded by a rotating drum will indicate the motion of the base. Assume that x is the displacement of the mass, y is the displacement of the base, and $z = x - y$ is the motion of the pen relative to the base. If the motion of the base is $y = Y \sin \omega t$, what is the steady-state amplitude ratio of z to y? Show that if $\omega \gg \omega_n$, where $\omega_n = \sqrt{k/m}$, the device can be used for measuring the displacement of the base. Show also that if $\omega \ll \omega_n$, the device can be used for measuring the acceleration of the base.

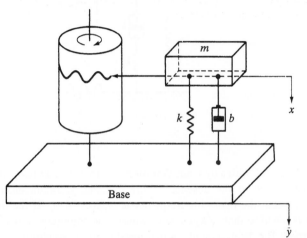

Figure 9–44 Motion- or acceleration-measuring instrument.

Problem B–9–9

Figure 9–45 shows a machine of mass m mounted on a vibration isolator in which spring k_1 is the load-carrying spring and viscous damper b_2 is in series with spring k_2. Determine the force transmissibility when the machine is subjected to a sinusoidal excitation force $p(t) = P \sin \omega t$. Determine also the amplitude of the force transmitted to the foundation. The displacement x is measured from the equilibrium position before the excitation force $p(t)$ is applied.

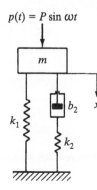

Figure 9–45 Machine mounted on a vibration isolator.

Problem B–9–10

A machine of mass m is mounted on a vibration isolator as shown in Figure 9–46. If the foundation is vibrating according to $p = P \sin \omega t$, where p is the displacement of the foundation, find the vibration amplitude of the machine. Determine the motion transmissibility. The displacement x is measured from the equilibrium position in the absence of the vibration of the foundation.

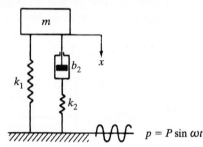

Figure 9–46   Machine mounted on a vibration isolator.

Problem B–9–11

Figure 9–47 shows a machine with a dynamic vibration absorber. The undamped natural frequency of the system in the absence of the dynamic vibration absorber is $\omega_n = \sqrt{k/m}$. Suppose that the operating frequency ω is close to ω_n. If the dynamic vibration absorber is tuned so that $\sqrt{k_a/m_a} = \omega$, what is the amplitude of mass m_a of the vibration absorber? The displacement x is measured from the equilibrium position in the absence of the excitation force $p(t)$.

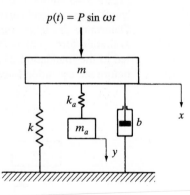

Figure 9–47   Machine with a dynamic vibration absorber.

Problem B–9–12

Consider the pendulum system shown in Figure 9–48. Determine the natural frequencies and modes of vibration.

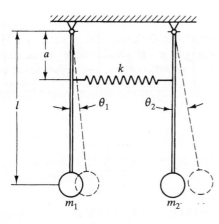

Figure 9–48   Pendulum system.

Problem B–9–13

Consider the pitching motion of an automobile. Figure 9–49(a) shows a simplified model of the auto body and the front and rear springs. Figure 9–49(b) shows the coupling of the translational and rotational vibrations. Determine the natural frequencies and modes of vibration. The vertical displacement x of the center of gravity (point G) is measured from the equilibrium position in the absence of any motions. Assume the following numerical values:

$$m = 2000 \text{ kg}, \qquad l_1 = 1.5 \text{ m}, \qquad l_2 = 2 \text{ m}, \qquad k_1 = 4 \times 10^4 \text{ N/m}$$
$$k_2 = 4 \times 10^4 \text{ N/m}, \qquad J = 2500 \text{ kg-m}^2$$

(J is the moment of inertia of the body about point G, the center of gravity of the body.)

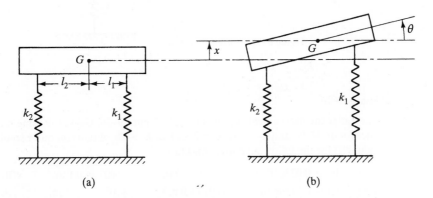

Figure 9–49   (a) Auto body and front and rear springs; (b) coupling of the translational and rotational vibrations.

Problem B–9–14

Consider the mechanical system shown in Figure 9–50. Obtain the first and second modes of vibration. The displacements x_1 and x_2 are measured from their respective equilibrium positions. Assume that the mass elements move without friction.

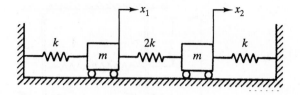

Figure 9–50 Mechanical system.

Problem B–9–15

Consider the mechanical system shown in Figure 9–51. Determine the natural frequencies and modes of vibration. In the diagram, the displacements x and y are measured from their respective equilibrium positions. Assume that

$$m = 1 \text{ kg}, \qquad M = 10 \text{ kg}, \qquad k_1 = 10 \text{ N/m}, \qquad k_2 = 100 \text{ N/m}$$

Determine the vibration when the initial conditions are given by

$$x(0) = 0.05 \text{ m}, \qquad \dot{x}(0) = 0 \text{ m/s}, \qquad y(0) = 0 \text{ m}, \qquad \dot{y}(0) = 0 \text{ m/s}$$

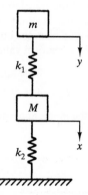

Figure 9–51 Mechanical system.

Problem B–9–16

Consider the mechanical system shown in Figure 9–36. Using the same numerical values for m, M, k_1, and k_2 as given in **Problem A–9–16**, obtain computer solutions for $x(t)$ and $y(t)$ for the following initial conditions:

(a) $x(0) = 0.028078 \text{ m}, \qquad \dot{x}(0) = 0 \text{ m/s}, \qquad y(0) = 0.1 \text{ m}, \qquad \dot{y}(0) = 0 \text{ m/s}$
(b) $x(0) = 0.17808 \text{ m}, \qquad \dot{x}(0) = 0 \text{ m/s}, \qquad y(0) = -0.1 \text{ m}, \qquad \dot{y}(0) = 0 \text{ m/s}$
(c) $x(0) = 0.1 \text{ m}, \qquad \dot{x}(0) = 0 \text{ m/s}, \qquad y(0) = -0.1 \text{ m}, \qquad \dot{y}(0) = 0 \text{ m/s}$

Write a MATLAB program that plots curves $x(t)$ versus t and $y(t)$ versus t for each of the three sets of initial conditions. Plot the resulting curves.

Time-Domain Analysis and Design of Control Systems

10–1 INTRODUCTION

This chapter presents basic information on control systems. Our discussions are limited to time-domain analysis and design based on the transient-response analysis and root-locus analysis.

We begin the chapter by defining some terms that are essential in describing control systems; we then follow with a description of closed-loop and open-loop control systems. Finally, the advantages and disadvantages of closed-loop and open-loop control systems are compared.

Plants.   A *plant* is a piece of equipment—perhaps a set of machine parts functioning together—the purpose of which is to perform a particular operation. In this book, we shall call any physical object that is to be controlled a plant.

Disturbances.   A *disturbance* is a signal that tends to affect the value of the output of a system adversely. If the disturbance is generated within the system, it is called *internal*; an *external* disturbance is generated outside the system and is an input.

Feedback control.   *Feedback control* refers to an operation that, in the presence of disturbances, tends to reduce the difference between the output of a system and the reference input and that does so on the basis of this difference. Here, only unpredictable disturbances are so specified, since predictable or known disturbances can always be compensated for within the system.

Feedback control systems.   A system that maintains a prescribed relationship between the output and the reference input by comparing them and using the difference as a means of control is called a *feedback control system* or simply a *control system.* An example is a room temperature control system. By measuring the actual room temperature and comparing it with the reference temperature (the desired temperature), the thermostat turns the heating or cooling equipment on or off in such a way as to ensure that the temperature of the room remains at a comfortable level, regardless of outside conditions.

Feedback control systems, of course, are not limited to engineering, but can be found in various nonengineering fields as well. The human body, for instance, is a highly advanced feedback control system. Both body temperature and blood pressure are kept constant by means of physiological feedback. In fact, feedback performs the vital function of making the human body relatively insensitive to external disturbances, thus enabling it to function properly in a changing environment.

As another example, consider the control of automobile speed by a human operator. For a given situation, the driver decides on an appropriate speed, which may be the posted speed limit on the road or highway involved. This speed acts as the reference speed. The driver observes the actual speed by looking at the speedometer. If he or she is traveling too slowly, the driver depresses the accelerator and the car speeds up. If the actual speed is too high, the driver releases the pressure on the accelerator and the car slows down. This is a feedback control system with a human operator. The human operator here can easily be replaced by a mechanical, an electrical, or some similar device. Instead of the driver observing the speedometer, an electric generator can be used to produce a voltage that is proportional to the speed. This voltage can be compared with a reference voltage that corresponds to the desired speed. The difference in the voltages can then be used as the error signal to position the throttle to increase or decrease the speed as needed.

Closed-loop control systems.   Feedback control systems are often referred to as *closed-loop control systems.* In practice, the terms *feedback control* and *closed-loop control* are used interchangeably. In a closed-loop control system, the actuating error signal, which is the difference between the input signal and the feedback signal (which may be the output signal itself or a function of the output signal and its derivatives), is fed to the controller so as to reduce the error and bring the output of the system to a desired value. The term *closed-loop control* always implies the use of feedback control action in order to reduce system error.

Open-loop control systems.   Those control systems in which the output has no effect on the control action are called *open-loop control systems.* In other words, in an open-loop control system, the output is neither measured nor fed back for comparison with the input. One practical example is a washing machine. Soaking, washing, and rinsing in the washer operate on a time basis. The machine does not measure the output signal, that is, the cleanliness of the clothes.

In an open-loop control system, the output is not compared with the reference input. Thus, to each reference input, there corresponds a fixed operating condition, and as a result, the accuracy of the system depends on calibration. In the presence of disturbances, an open-loop control system will not perform the desired task. Open-loop control can be used, in practice, only if the relationship between the input and output is known and if there are neither internal nor external disturbances. Clearly, such systems are not feedback control systems. Note that any control system that operates on a time basis is an open-loop control system. For example, traffic control by means of signals operated on a time basis is another example of open-loop control.

Closed-loop versus open-loop control systems.   An advantage of the closed-loop control system is the fact that the use of feedback makes the system response relatively insensitive to external disturbances and internal variations in system parameters. It is thus possible to use relatively inaccurate and inexpensive components to obtain accurate control of a given plant, whereas doing so is impossible in the open-loop case.

From the point of view of stability, the open-loop control system is easier to build, because system stability is not a major problem. By contrast, stability *is* a major problem in the closed-loop control system, which may tend to overcorrect errors that can cause oscillations of constant or changing amplitude.

General requirements of control systems.   A primary requirement of any control system is that it must be stable. In addition to absolute stability, a control system must have a reasonable relative stability; that is, the response must show reasonable damping. Moreover, the speed of response must be reasonably fast. A control system must also be capable of reducing errors to zero or to some small tolerable value.

Because the needs for reasonable relative stability and for steady-state accuracy tend to be incompatible, in designing control systems it is necessary to make the most effective compromise between the two.

Outline of the chapter.   This chapter presents introductory material on control systems analysis and design in the time domain. Specifically, Section 10-1 has presented an introduction to control systems. Section 10-2 deals with block diagrams of control systems. Section 10-3 discusses, first, control actions generally found in industrial automatic controllers and, second, simple electronic controllers. Section 10-4 covers the transient-response analysis of control systems. The responses of first- and second-order control systems to step inputs are examined, and the effects of various control actions on the transient-response characteristics of control systems are discussed. Section 10-5 is concerned with transient-response specifications. Section 10-6 discusses improving transient-response and steady-state characteristics. Velocity feedback to improve the transient response is treated, and then static error constants are defined and used to obtain steady-state errors. Section 10-7 deals with stability analysis and presents Routh's stability criterion. Section 10-8 discusses the root-locus method for analyzing and designing control systems. Basic rules for constructing root-locus plots are presented. Section 10–9 treats the MATLAB approach to plotting root loci. Finally, Section 10–10 examines tuning rules for PID controllers. Ziegler–Nichols rules are detailed and PID-controlled systems are designed, utilizing the root-locus method.

10–2 BLOCK DIAGRAMS AND THEIR SIMPLIFICATION

A system may consist of a number of components. To show the functions performed by each component, block diagrams are frequently used in the analysis and design of control systems. This section first defines the open-loop transfer function, feed-forward transfer function, and closed-loop transfer function. Then the simplification of block diagrams is discussed using general rules for reducing block diagrams. Finally, a MATLAB approach to obtaining transfer functions or state-space representations of series-connected systems, parallel-connected systems, and feedback-connected systems is presented.

Open-loop transfer function and feedforward transfer function.   Figure 10–1 shows the block diagram of a closed-loop system with a feedback element. The ratio of the feedback signal $B(s)$ to the actuating error signal $E(s)$ is called the *open-loop transfer function*. That is,

$$\text{open-loop transfer function} = \frac{B(s)}{E(s)} = G(s)H(s)$$

The ratio of the output $C(s)$ to the actuating error signal $E(s)$ is called the *feedforward transfer function*, so

$$\text{feedforward transfer function} = \frac{C(s)}{E(s)} = G(s)$$

If the feedback transfer function is unity, then the open-loop transfer function and the feedforward transfer function are the same.

Closed-loop transfer function.   For the system shown in Figure 10–1, the output $C(s)$ and input $R(s)$ are related as follows:

$$C(s) = G(s)E(s)$$
$$E(s) = R(s) - B(s) = R(s) - H(s)C(s)$$

Eliminating $E(s)$ from these equations gives

$$C(s) = G(s)[R(s) - H(s)C(s)]$$

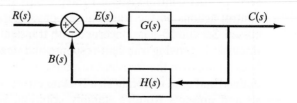

Figure 10–1 Block diagram of a closed-loop system with a feedback element.

or

$$\frac{C(s)}{R(s)} = \frac{G(s)}{1 + G(s)H(s)} \qquad (10\text{--}1)$$

The transfer function relating $C(s)$ to $R(s)$ is called the *closed-loop transfer function*. This transfer function relates the closed-loop system dynamics to the dynamics of the feedforward elements and feedback elements. Since, from Equation (10–1),

$$C(s) = \frac{G(s)}{1 + G(s)H(s)} R(s)$$

the output response of a given closed-loop system clearly depends on both the closed-loop transfer function and the nature of the input.

Block diagram reduction.   Blocks can be connected in series only if the output of one block is not affected by the block that follows. If there are any loading effects between the components, these components must be combined into a single block.

A complicated block diagram involving many feedback loops can be simplified by a step-by-step rearrangement, using rules of block diagram algebra. Some of these important rules are given in Table 10–1. They are obtained by writing the same

TABLE 10–1   Rules of Block Diagram Algebra

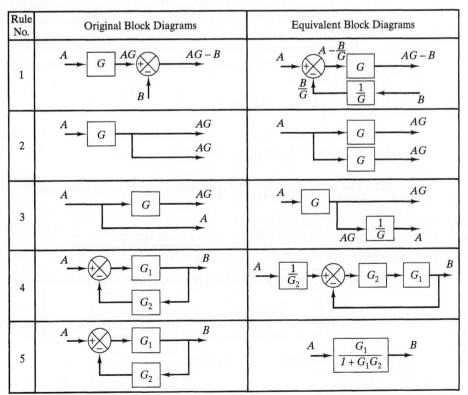

equation in a different way. Simplification of the block diagram by rearrangements and substitutions considerably reduces the labor needed for subsequent mathematical analysis. Note, however, that as the block diagram is simplified, the transfer functions in new blocks become more complex because new poles and new zeros are generated.

In simplifying a block diagram, remember the following:

1. The product of the transfer functions in the feedforward direction must remain the same.

2. The product of the transfer functions around the loop must remain the same.

A general rule for simplifying a block diagram is to move branch points and summing points, interchange summing points, and then eliminate internal feedback loops.

Example 10–1

Consider the system shown in Figure 10–2. Simplify this diagram by eliminating loops.

Moving the summing point of the negative feedback loop containing H_2 outside the positive feedback loop containing H_1, we obtain Figure 10–3(a). Eliminating the positive feedback loop, we have Figure 10–3(b). Then, eliminating the loop containing H_2/G_1 gives Figure 10–3(c). Finally, eliminating the feedback loop results in Figure 10–3(d).

Notice that the numerator of the closed-loop transfer function $C(s)/R(s)$ is the product of the transfer functions of the feedforward path. The denominator of $C(s)/R(s)$ is equal to

$$1 - \sum (\text{product of the transfer functions around each loop})$$

$$= 1 - (G_1 H_1 - G_2 H_2 - G_1 G_2)$$

$$= 1 - G_1 H_1 + G_2 H_2 + G_1 G_2$$

(The positive feedback loop yields a negative term in the denominator.)

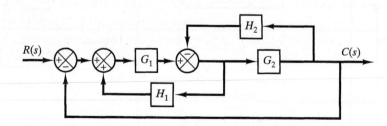

Figure 10–2   System with multiple loops.

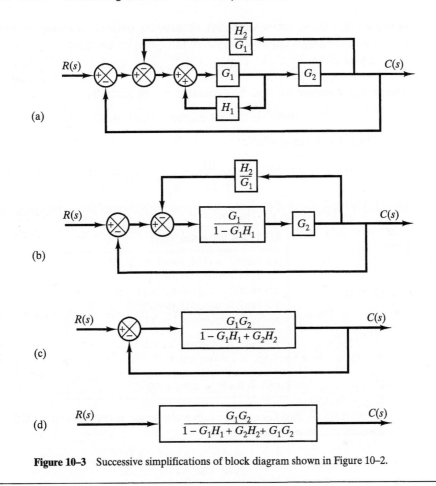

(a)

(b)

(c)

(d)

Figure 10–3   Successive simplifications of block diagram shown in Figure 10–2.

Using MATLAB to obtain transfer functions of series-connected blocks, parallel-connected blocks, and feedback-connected blocks.   A physical system may involve many interconnected blocks. In what follows, we shall consider series-connected blocks, parallel-connected blocks, and feedback-connected blocks. Any linear, time-invariant system may be represented by combinations of series-connected blocks, parallel-connected blocks, and feedback-connected blocks.

Series-connected blocks.   In the system shown in Figure 10–4, G_1 and G_2 are series connected. System G_1 and system G_2 are respectively defined by

$$sys1 = tf(num1,den1)$$
$$sys2 = tf(num2,den2)$$

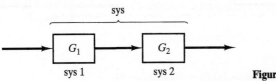

Figure 10–4   Series-connected blocks.

provided that these two systems are themselves defined in terms of transfer functions. Then the series-connected system $G_1 G_2$ can be given by

$$\text{sys} = \text{series(sys1,sys2)}$$

The system's numerator and denominator can be given by

$$[\text{num,den}] = \text{series(num1,den1,num2,den2)}$$

Consider the case where

$$G_1 = \frac{10}{s^2 + 2s + 10}, \qquad G_2 = \frac{5}{s + 5}$$

MATLAB Program 10–1 produces the transfer function of the series-connected system.

MATLAB Program 10–1

```
>> num1 = [10]; den1 = [1   2   10];
>> sys1 = tf(num1,den1);
>> num2 = [5]; den2 = [1   5];
>> sys2 = tf(num2,den2);
>> sys = series(sys1,sys2)

Transfer function:
          50
    ---------------------------------
    s^3 + 7 s^2 + 20 s + 50
```

If systems G_1 and G_2 are given in state-space form, then their MATLAB representations are respectively given by

$$\text{sys1} = \text{ss(A1,B1,C1,D1)}$$
$$\text{sys2} = \text{ss(A2,B2,C2,D2)}$$

The series-connected system $G_1 G_2$ is given by

$$\text{sys} = \text{series(sys1,sys2)}$$

or

$$[\text{A,B,C,D}] = \text{series(A1,B1,C1,D1,A2,B2,C2,D2)}$$

Parallel-connected blocks.   Figures 10–5(a) and (b) show parallel-connected systems. In Figure 10–5(a) two systems G_1 and G_2 are added, while in Figure 10–5(b)

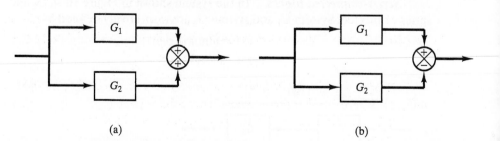

(a)   (b)

Figure 10–5 Parallel-connected systems. (a) $G_1 + G_2$; (b) $G_1 - G_2$.

system G_2 is subtracted from system G_1. If G_1 and G_2 are defined in terms of transfer functions, then

$$sys1 = tf(num1,den1)$$
$$sys2 = tf(num2,den2)$$

and the parallel-connected system $G_1 + G_2$ is given by

$$sys = parallel(sys1,sys2)$$

or

$$[num,den] = parallel(num1,den1,num2,den2)$$

Consider the case where

$$G_1 = \frac{10}{s^2 + 2s + 10}, \qquad G_2 = \frac{5}{s + 5}$$

MATLAB Program 10–2 produces the transfer function of the parallel-connected system.

MATLAB Program 10–2

```
>> num1 = [10]; den1 = [1   2   10];
>> sys1 = tf(num1,den1);
>> num2 = [5]; den2 = [1   5];
>> sys2 = tf(num2,den2);
>> sys = parallel(sys1,sys2)
```

Transfer function:
```
   5 s^2 + 20 s + 100
-------------------------------
s^3 + 7 s^2 + 20 s + 50
```

If G_1 and G_2 are given in state-space form, then

$$sys1 = ss(A1,B1,C1,D1)$$
$$sys2 = ss(A2,B2,C2,D2)$$

and the parallel-connected system $G_1 + G_2$ is given by

$$sys = parallel(sys1,sys2)$$

or

$$[A,B,C,D] = parallel(A1,B1,C1,D1,A2,B2,C2,D2)$$

If the parallel-connected system is $G_1 - G_2$, as shown in Figure 10–5(b), then we define sys1 and sys2 as before, but change sys2 to $-sys2$ in the expression for sys; that is,

$$sys = parallel(sys1,-sys2)$$

Feedback-connected blocks: Figure 10–6(a) shows a negative feedback system, and Figure 10–6(b) shows a positive feedback system.

If G and H are defined in terms of transfer functions, then

$$sysg = [numg,deng]$$
$$sysh = [numh,denh]$$

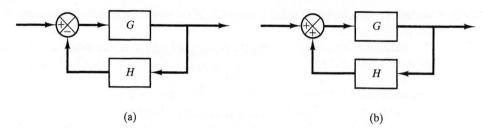

(a)   (b)

Figure 10–6   (a) Negative feedback system; (b) positive feedback system.

and the entire feedback system is given by

$$sys = feedback(sysg, sysh)$$

or

$$[num, den] = feedback(numg, deng, numh, denh)$$

If the system has a unity feedback function, then $H = [1]$ and sys can be given by

$$sys = feedback(sysg, [1])$$

Note that, in treating the feedback system, MATLAB assumes that the feedback is negative. If the system involves a positive feedback, we need to add "+1" in the argument of feedback as follows:

$$sys = feedback(sysg, sysh, +1)$$

Alternatively, we may use "−sysh" in the sys statement; that is,

$$sys = feedback(sysg, -sysh)$$

for the positive feedback system.

Consider the case where

$$G = \frac{5}{s^2 + 2s}, \qquad H = 0.1s + 1$$

MATLAB Program 10–3 produces the transfer function of the feedback-connected system.

MATLAB Program 10–3

```
>> numg = [5]; deng = [1   2   0];
>> sysg = tf(numg,deng);
>> numh = [0.1   1]; denh = [1];
>> sysh = tf(numh,denh);
>> sys = feedback(sysg,sysh)

Transfer function:
          5
---------------------
s^2 + 2.5 s + 5
```

10–3 AUTOMATIC CONTROLLERS

An automatic controller compares the actual value of the plant output with the desired value, determines the deviation, and produces a control signal that will reduce the deviation to zero or a small value. The way in which the automatic controller produces the control signal is called the *control action*.

Here, we describe the fundamental control actions commonly used in industrial automatic controllers. We then briefly discuss an electronic controller.

Control actions.   The control actions normally found in industrial automatic controllers consist of the following: two-position, or on–off; proportional; integral; derivative; and combinations of proportional, integral, and derivative. A good understanding of the basic properties of various control actions is necessary for the engineer to select the one best suited to his or her particular application.

Classifications of industrial automatic controllers.   Industrial automatic controllers can be classified according to their control action as follows:

1. Two-position, or on–off, controllers
2. Proportional controllers
3. Integral controllers
4. Proportional-plus-integral controllers
5. Proportional-plus-derivative controllers
6. Proportional-plus-integral-plus-derivative controllers

Automatic controller, actuator, and sensor (measuring element). Figure 10–7 is a block diagram of an industrial control system consisting of an automatic controller, an actuator, a plant, and a sensor or measuring element. The controller detects the actuating error signal, which is usually at a very low power level,

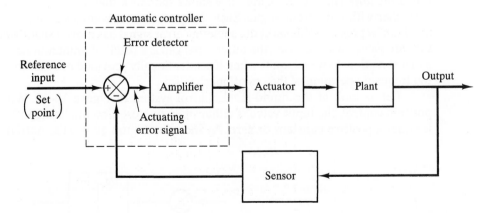

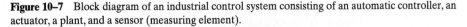

Figure 10–7 Block diagram of an industrial control system consisting of an automatic controller, an actuator, a plant, and a sensor (measuring element).

and amplifies it to a sufficiently high level. (Thus, the automatic controller comprises an error detector and an amplifier.) Quite often, a suitable feedback circuit, together with an amplifier, is used to alter the actuating error signal to produce a better control signal.

The actuator is an element that produces the input to the plant according to the control signal, so that the feedback signal will correspond to the reference input signal.

The sensor or measuring element is a device that converts the output variable into another suitable variable, such as a displacement, pressure, or voltage, which can be used to compare the output with the reference input signal. This element is in the feedback path of the closed-loop system. The set point of the controller must be converted to a reference input of the same units as the feedback signal from the sensor or measuring element.

Two-position, or on–off, control action.   In a two-position control system, the actuating element has only two fixed positions, which are, in many cases, simply on and off. Two-position, or on–off, control is simple and inexpensive and, for this reason, is used extensively in both industrial and household control systems.

To explain the concept, let the output signal from the controller be $m(t)$ and the actuating error signal be $e(t)$. In two-position control, the signal $m(t)$ remains at either a maximum or a minimum value, depending on whether the actuating error signal is positive or negative, so that

$$m(t) = M_1 \qquad e(t) > 0$$
$$ = M_2 \qquad e(t) < 0$$

where M_1 and M_2 are constants. The minimum value M_2 is generally either zero or $-M_1$. As a rule, two-position controllers are electrical devices, and an electric solenoid-operated valve is widely used in such controllers.

Figure 10–8 shows the block diagram of a two-position controller. The range through which the actuating error signal must move before switching occurs is called the *differential gap*. Figure 10–9 shows the block diagram of a two-position controller with a differential gap. Such a gap causes the controller output $m(t)$ to maintain its present value until the actuating error signal has moved slightly beyond the zero value. In some cases, the differential gap is a result of unintentional friction and lost motion; however, quite often it is intentionally provided in order to prevent too frequent operation of the on–off mechanism.

Let us look at the liquid-level control system of Figure 10–10. With two-position control, the input valve is either open or closed, so the liquid inflow rate is either a positive constant or zero. As shown in Figure 10–11, the output signal

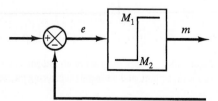

Figure 10–8   Block diagram
of a two-position controller.

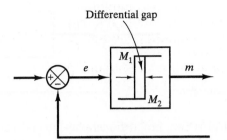

Figure 10–9 Block diagram of a two-position controller with a differential gap.

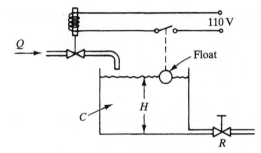

Figure 10–10 Liquid-level control system.

moves continuously between the two limits required, thereby causing the actuating element to shift from one fixed position to the other. Such output oscillation between two limits is a typical response characteristic of a system that is under two-position control.

From Figure 10–11, we see that the amplitude of the output oscillation can be reduced by decreasing the differential gap. This step, however, increases the number of on–off switchings per unit time and reduces the useful life of the component. The magnitude of the differential gap must be determined from such factors as the accuracy required and the life of the component.

Proportional, integral, and derivative control actions.   In addition to two-position or on-off control action, proportional, integral, and derivative control actions are basic control actions found in industrial automatic controllers. For each control action, the output of the controller, $M(s)$, and the actuating error signal $E(s)$ are related by a transfer function of a specific form. In what follows, we illustrate transfer functions

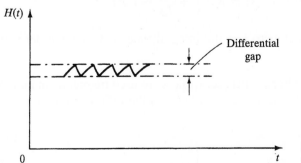

Figure 10–11 Head-versus-time curve of the system shown in Figure 10–10.

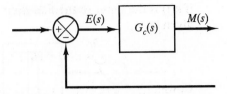

Figure 10–12 Block diagram of a controller.

$M(s)/E(s)$ for proportional control action, proportional-plus-integral control action, proportional-plus-derivative control action, and proportional-plus-integral-plus-derivative control action.

Referring to the controller shown in Figure 10–12, for proportional control action, $M(s)$ and $E(s)$ are related by

$$\frac{M(s)}{E(s)} = G_c(s) = K_p$$

where K_p is termed the *proportional gain*.

For integral control action, the relationship between $M(s)$ and $E(s)$ is

$$\frac{M(s)}{E(s)} = G_c(s) = \frac{K_i}{s}$$

where K_i is called the *integral gain*.

For proportional-plus-integral control action, $M(s)$ and $E(s)$ are related by

$$\frac{M(s)}{E(s)} = G_c(s) = K_p\left(1 + \frac{1}{T_i s}\right)$$

where K_p is the proportional gain and T_i is a constant called the *integral time*.

For proportional-plus-derivative control action, $M(s)$ and $E(s)$ are related by

$$\frac{M(s)}{E(s)} = G_c(s) = K_p(1 + T_d s)$$

where K_p is the proportional gain and T_d is a constant called the *derivative time*.

Finally, for proportional-plus-integral-plus-derivative control action, $M(s)$ and $E(s)$ are related by

$$\frac{M(s)}{E(s)} = G_c(s) = K_p\left(1 + \frac{1}{T_i s} + T_d s\right)$$

where K_p is the proportional gain, T_i is the integral time, and T_d is the derivative time.

Electronic PID controllers.   PID controllers are used frequently in industrial control systems. Since the transfer function $G_c(s)$ of the PID controller is

$$G_c(s) = K_p\left(1 + \frac{1}{T_i s} + T_d s\right) \tag{10–2}$$

if $e(t)$ is the input to the PID controller, then the output $m(t)$ from the controller is given by

$$m(t) = K_p \left[e(t) + \frac{1}{T_i} \int_{-\infty}^{t} e(t)\, dt + T_d \frac{de(t)}{dt} \right]$$

Constants K_p, T_i, and T_d are the controller parameters. Equation (10–2) can also be written as

$$G_c(s) = K_p + \frac{K_i}{s} + K_d s \qquad (10\text{–}3)$$

where

K_p = proportional gain
$K_i = K_p/T_i$ = integral gain
$K_d = K_p T_d$ = derivative gain

In this case, K_p, K_i, and K_d become controller parameters.

In actual PID controllers, instead of adjusting the proportional gain, we adjust the proportional band. The proportional band is proportional to $1/K_p$ and is expressed in percent. (For example, 25% proportional band corresponds to $K_p = 4$.)

Figure 10–13 shows an electronic PID controller that uses operational amplifiers. The transfer function $E(s)/E_i(s)$ is given by

$$\frac{E(s)}{E_i(s)} = -\frac{Z_2}{Z_1}$$

where

$$Z_1 = \frac{R_1}{R_1 C_1 s + 1}, \qquad Z_2 = \frac{R_2 C_2 s + 1}{C_2 s}$$

Thus,

$$\frac{E(s)}{E_i(s)} = -\left(\frac{R_2 C_2 s + 1}{C_2 s} \right)\left(\frac{R_1 C_1 s + 1}{R_1} \right)$$

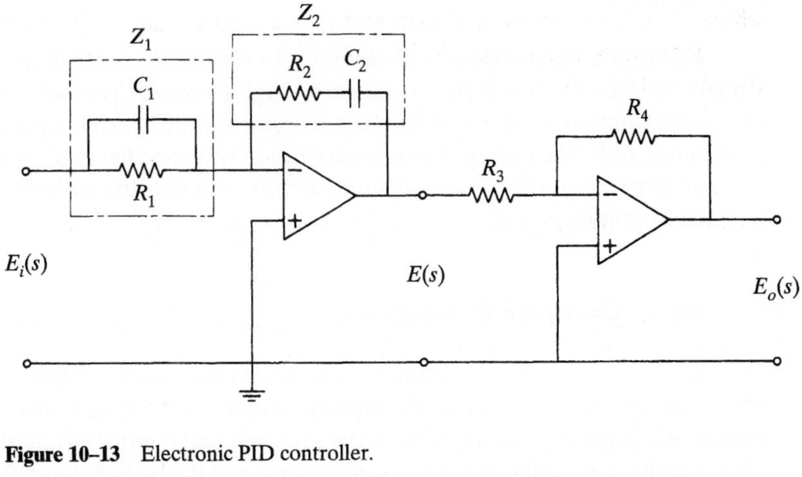

Figure 10–13   Electronic PID controller.

Noting that

$$\frac{E_o(s)}{E(s)} = -\frac{R_4}{R_3}$$

we have

$$\frac{E_o(s)}{E_i(s)} = \frac{E_o(s)}{E(s)} \frac{E(s)}{E_i(s)} = \frac{R_4 R_2}{R_3 R_1} \frac{(R_1 C_1 s + 1)(R_2 C_2 s + 1)}{R_2 C_2 s}$$

$$= \frac{R_4 R_2}{R_3 R_1} \left(\frac{R_1 C_1 + R_2 C_2}{R_2 C_2} + \frac{1}{R_2 C_2 s} + R_1 C_1 s \right)$$

$$= \frac{R_4 (R_1 C_1 + R_2 C_2)}{R_3 R_1 C_2} \left[1 + \frac{1}{(R_1 C_1 + R_2 C_2)s} + \frac{R_1 C_1 R_2 C_2}{R_1 C_1 + R_2 C_2} s \right] \quad (10\text{-}4)$$

Hence,

$$K_p = \frac{R_4 (R_1 C_1 + R_2 C_2)}{R_3 R_1 C_2}$$

$$T_i = R_1 C_1 + R_2 C_2$$

$$T_d = \frac{R_1 C_1 R_2 C_2}{R_1 C_1 + R_2 C_2}$$

In terms of the proportional gain, integral gain, and derivative gain, we have

$$K_p = \frac{R_4 (R_1 C_1 + R_2 C_2)}{R_3 R_1 C_2}$$

$$K_i = \frac{R_4}{R_3 R_1 C_2}$$

$$K_d = \frac{R_4 R_2 C_1}{R_3}$$

Notice that the second operational-amplifier circuit acts as a sign inverter as well as a gain adjuster.

Hydraulic controllers.   In addition to electronic controllers, hydraulic controllers are used extensively in industry. High-pressure hydraulic systems enable very large forces to be derived. Moreover, these systems permit a rapid and accurate positioning of loads. Frequently, a combination of electronic and hydraulic systems is found because of the advantages resulting from a mixture of both electronic control and hydraulic power.

10–4 TRANSIENT-RESPONSE ANALYSIS

In this section, we treat the transient-response analysis of control systems and the effects of integral and derivative control actions on the transient-response performance. We begin with an analysis of the proportional control of a first-order system, after which we describe the effects of integral and derivative control actions on the

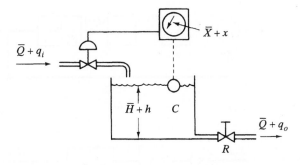

Figure 10–14 Liquid-level control system.

transient performance of the system. Then we present the proportional control of a system with an inertia load and illustrate the fact that adding derivative control action markedly improves the transient performance.

Proportional control of first-order system.   Suppose that the controller in the liquid-level control system of Figure 10–14 is a proportional controller. Suppose also that the reference input to the system is $\overline{X}$. At $t = 0$, a change in the reference input is made from $\overline{X}$ to $\overline{X} + x$. Assume that all the variables shown in the diagram—x, q_i, h, and q_o—are measured from their respective steady-state values $\overline{X}, \overline{Q}, \overline{H}$, and $\overline{Q}$. Assume also that the magnitudes of the variables x, q_i, h, and q_o are sufficiently small, which means that the system can be approximated by a linear mathematical model.

Referring to Section 7–2, the following equation for the liquid-level system can be derived:

$$RC\frac{dh}{dt} + h = Rq_i \tag{10–5}$$

[See Equation (7–4).] So the transfer function between $H(s)$ and $Q_i(s)$ is found to be

$$\frac{H(s)}{Q_i(s)} = \frac{R}{RCs + 1} \tag{10–6}$$

Here, we assume that the gain K_v of the control valve is constant near the steady-state operating condition. Then, since the controller is a proportional one, the change in inflow rate q_i is proportional to the actuating error e (where $e = x - h$), or

$$q_i = K_p K_v e \tag{10–7}$$

where K_p is the gain of the proportional controller. In terms of Laplace-transformed quantities, Equation (10–7) becomes

$$Q_i(s) = K_p K_v E(s)$$

A block diagram of this system appears in Figure 10–15(a). To simplify our analysis, we assume that x and h are the same kind of signal with the same units, so that they can be compared directly. (Otherwise, we must insert a feedback transfer function K_b in the feedback path.) A simplified block diagram is given in Figure 10–15(b), where $K = K_p K_v$.

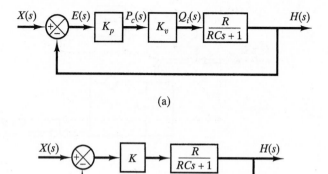

Figure 10–15 (a) Block diagram of the liquid-level control system shown in Figure 10–14; (b) simplified block diagram.

Next, we investigate the response $h(t)$ to a change in the reference input. We shall assume a unit-step change in $x(t)$. The closed-loop transfer function between $H(s)$ and $X(s)$ is given by

$$\frac{H(s)}{X(s)} = \frac{KR}{RCs + 1 + KR} \tag{10–8}$$

Since the Laplace transform of the unit-step function is $1/s$, substituting $X(s) = 1/s$ into Equation (10–8) gives

$$H(s) = \frac{KR}{RCs + 1 + KR}\frac{1}{s}$$

Then the expansion of $H(s)$ into partial fractions results in

$$H(s) = \frac{KR}{1 + KR}\left\{\frac{1}{s} - \frac{1}{s + [(1 + KR)/RC]}\right\} \tag{10–9}$$

Next, by taking the inverse Laplace transforms of both sides of Equation (10–9), we obtain the time solution

$$h(t) = \frac{KR}{1 + KR}(1 - e^{-t/T_1}) \qquad t \geq 0 \tag{10–10}$$

where

$$T_1 = \frac{RC}{1 + KR}$$

Notice that the time constant T_1 of the closed-loop system is different from the time constant RC of the liquid-level system alone.

The response curve $h(t)$ is plotted against t in Figure 10–16. From Equation (10–10), we see that, as t approaches infinity, the value of $h(t)$ approaches $KR/(1 + KR)$, or

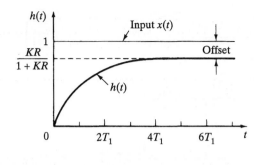

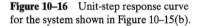

Figure 10–16   Unit-step response curve for the system shown in Figure 10–15(b).

$$h(\infty) = \frac{KR}{1 + KR}$$

Since $x(\infty) = 1$, there is a steady-state error of magnitude $1/(1 + KR)$. Such an error is called *offset*. The value of offset becomes smaller as the gain K becomes larger.

Eliminating offset by the use of integral control.   In the proportional control of a plant whose transfer function does not possess an integrator $1/s$ (and thus, the feed-forward transfer function does not involve an integrator or integrators), there is a steady-state error, or offset, in the response to a step input. Such an offset can be eliminated if integral control action is included in the controller.

Under integral control action, the control signal (the output signal from the controller) at any instant is the area under the actuating-error-signal curve up to that instant. The control signal $m(t)$ can have a nonzero value when the actuating error signal $e(t)$ is zero, as Figure 10–17(a) shows. This situation is impossible in the case of the proportional controller, since a nonzero control signal requires a nonzero actuating error signal. (A nonzero actuating error signal at steady state means that there is an offset.) Figure 10–17(b) shows the curve of $e(t)$ versus t and the corresponding curve of $m(t)$ versus t when the controller is of the proportional type.

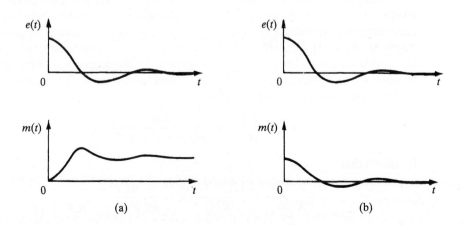

Figure 10–17   (a) Error curve and control signal curve for a system that uses an integral controller; (b) error curve and control signal curve for a system that uses a proportional controller.

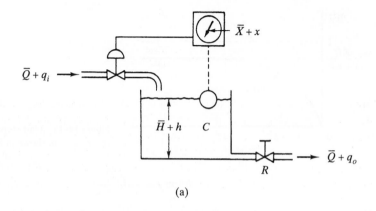

(a)

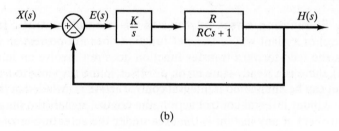

(b)

Figure 10–18   (a) Liquid-level control system; (b) block diagram.

Note that integral control action improves steady-state accuracy by removing offset, or steady-state error. However, it may lead to an oscillatory response of slowly decreasing amplitude or even increasing amplitude, both of which are undesirable.

Integral control of a liquid-level system.   Figure 10–18(a) shows a liquid-level control system. We assume that the controller is an integral one. We also assume that the variables x, q_i, h, and q_o, which are measured from their respective steady-state values $\overline{X}, \overline{Q}, \overline{H}$, and $\overline{Q}$, are small quantities; therefore, the system can be considered linear. Under these assumptions, the block diagram of the system can be obtained as shown in Figure 10–18(b). From this diagram, the closed-loop transfer function between $H(s)$ and $X(s)$ is

$$\frac{H(s)}{X(s)} = \frac{KR}{RCs^2 + s + KR}$$

It follows that

$$\frac{E(s)}{X(s)} = \frac{X(s) - H(s)}{X(s)} = \frac{RCs^2 + s}{RCs^2 + s + KR}$$

Since the system is stable, the steady-state error e_{ss} for the unit-step response can be obtained by applying the final-value theorem:

$$e_{ss} = \lim_{s \to 0} sE(s)$$

$$= \lim_{s \to 0} \frac{s(RCs^2 + s)}{RCs^2 + s + KR} \frac{1}{s}$$

$$= 0$$

Integral control of the liquid-level system thus eliminates the steady-state error in the response to the step input, thereby improving steady-state accuracy. This is an important improvement over proportional control alone, which gives offset.

Note that proportional-plus-integral control action gives just as good a steady-state accuracy as integral control action alone. In fact, the use of proportional-plus-integral control action will enable the transient response to decay faster.

Derivative control action.   Derivative control action, when added to a proportional controller, provides a means of obtaining high sensitivity. An advantage of using derivative control action is that it responds to the rate of change of the actuating error and can produce a significant correction before the magnitude of the actuating error becomes too large. Derivative control thus anticipates the actuating error, initiates an early corrective action, and tends to increase the stability of the system.

Although derivative control does not affect the steady-state error directly, it adds damping to the system and therefore permits the use of a larger value of the system gain, a factor that yields an improvement in steady-state accuracy.

Notice that, because derivative control operates on the rate of change of the actuating error and not on the actuating error itself, this mode is never used alone. It is always used in combination with proportional or proportional-plus-integral control action.

Proportional control of a system with inertia load.   Before considering the effect of derivative control action on a system's performance, let us discuss the proportional control of an inertia load.

In the position control system of Figure 10–19(a), the box with the transfer function K_p represents a proportional controller. Its output is a torque signal T, which is applied to an inertia element (rotor) J. The output of the system is the angular displacement c of the inertia element. For the inertia element, we have

$$J\ddot{c} = T$$

The Laplace transform of this last equation, assuming zero initial conditions, becomes

$$Js^2C(s) = T(s)$$

Hence,

$$\frac{C(s)}{T(s)} = \frac{1}{Js^2}$$

The diagram of Figure 10–19(a) can be redrawn as shown in Figure 10–19(b). From this diagram, the closed-loop transfer function is

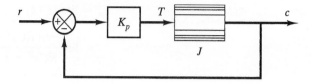

(a)

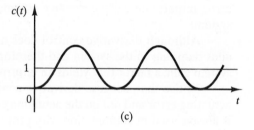

(b)

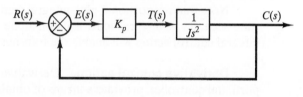

(c)

Figure 10–19   (a) Position control system; (b) block diagram; (c) unit-step response curve.

$$\frac{C(s)}{R(s)} = \frac{K_p}{Js^2 + K_p}$$

Since the roots of the characteristic equation

$$Js^2 + K_p = 0$$

are imaginary, the response to a unit-step input continues to oscillate indefinitely, as shown in Figure 10–19(c).

Control systems exhibiting such sustained oscillations are not acceptable. We shall see that the addition of derivative control will stabilize the system.

Proportional-plus-derivative control of a system with inertia load.   Let us modify the proportional controller to a proportional-plus-derivative controller whose transfer function is $K_p(1 + T_d s)$. The torque developed by the controller is proportional to $K_p(e + T_d \dot{e})$, where e is the actuating error signal. Derivative control action is essentially anticipatory, measuring the instantaneous error velocity and predicting the large overshoot ahead of time in order to produce an appropriate counteraction before too large an overshoot occurs.

For the system shown in Figure 10–20(a), the closed-loop transfer function is given by

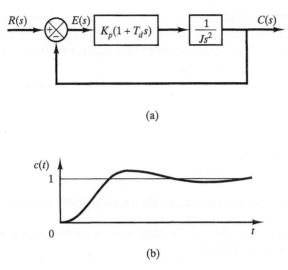

(a)

(b)

Figure 10–20 (a) Block diagram of a position control system that uses a proportional-plus-derivative controller; (b) unit-step response curve.

$$\frac{C(s)}{R(s)} = \frac{K_p(1 + T_d s)}{Js^2 + K_p T_d s + K_p}$$

The characteristic equation

$$Js^2 + K_p T_d s + K_p = 0$$

now has two roots with negative real parts, because the values of J, K_p, and T_d are positive. Thus, derivative control action introduces a damping effect. A typical response curve $c(t)$ to a unit-step input is presented in Figure 10–20(b). Clearly, the response curve is a marked improvement over the original response curve, shown in Figure 10–19(c).

10–5 TRANSIENT-RESPONSE SPECIFICATIONS

Because systems that store energy cannot respond instantaneously, they exhibit a transient response when they are subjected to inputs or disturbances. Consequently, the transient-response characteristics constitute one of the most important factors in system design.

In many practical cases, the desired performance characteristics of control systems can be given in terms of transient-response specifications. Frequently, such performance characteristics are specified in terms of the transient response to a unit-step input, since such an input is easy to generate and is sufficiently drastic. (If the response of a linear system to a step input is known, it is mathematically possible to compute the system's response to any input.)

The transient response of a system to a unit-step input depends on initial conditions. For convenience in comparing the transient responses of various systems, it is common practice to use a standard initial condition: The system is at rest initially, with its output and all time derivatives thereof zero. Then the response characteristics can be easily compared.

Transient-response specifications.   The transient response of a practical control system often exhibits damped oscillations before reaching a steady state. In specifying the transient-response characteristics of a control system to a unit-step input, it is common to name the following:

1. Delay time, t_d
2. Rise time, t_r
3. Peak time, t_p
4. Maximum overshoot, M_p
5. Settling time, t_s

These specifications are defined next and are shown graphically in Figure 10–21.

Delay time.   The delay time t_d is the time needed for the response to reach half the final value the very first time.

Rise time.   The rise time t_r is the time required for the response to rise from 10% to 90%, 5% to 95%, or 0% to 100% of its final value. For underdamped second-order systems, the 0% to 100% rise time is normally used. For overdamped systems, the 10% to 90% rise time is common.

Peak time.   The peak time t_p is the time required for the response to reach the first peak of the overshoot.

Maximum (percent) overshoot.   The maximum overshoot M_p is the maximum peak value of the response curve [the curve of $c(t)$ versus t], measured from $c(\infty)$. If $c(\infty) = 1$, the maximum percent overshoot is $M_p \times 100\%$. If the final

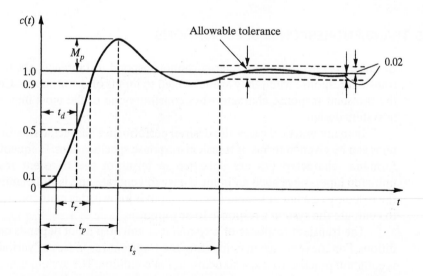

Figure 10–21   Transient-response specifications.

steady-state value $c(\infty)$ of the response differs from unity, then it is common practice to use the following definition of the maximum *percent* overshoot:

$$\text{maximum percent overshoot} = \frac{c(t_p) - c(\infty)}{c(\infty)} \times 100\%$$

The amount of the maximum (percent) overshoot directly indicates the relative stability of the system.

Settling time.   The settling time t_s is the time required for the response curve to reach and stay within 2% of the final value. In some cases, 5%, instead of 2%, is used as the percentage of the final value. Throughout this book, however, we use the 2% criterion. The settling time is related to the largest time constant of the system.

Comments.   If we specify the values of t_d, t_r, t_p, t_s, and M_p, the shape of the response curve is virtually fixed. This fact can be seen clearly from Figure 10–22.

Note that not all these specifications necessarily apply to any given case. For instance, for an overdamped system, the terms *peak time* and *maximum overshoot* do not apply.

Position control system.   The position control system (servo system) shown in Figure 10–23(a) consists of a proportional controller and load elements (inertia and viscous-friction elements). Suppose that we wish to control the output position c in accordance with the input position r.

The equation for the load elements is

$$J\ddot{c} + b\dot{c} = T$$

where T is the torque produced by the proportional controller, whose gain constant is K. Taking Laplace transforms of both sides of this last equation, assuming zero initial conditions, we find that

$$Js^2C(s) + bsC(s) = T(s)$$

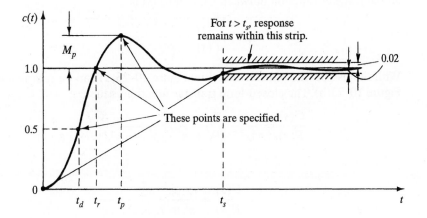

Figure 10–22   Specifications of transient-response curve.

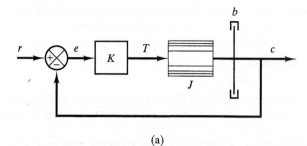

(a)

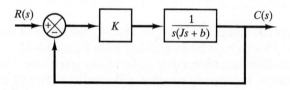

(b)

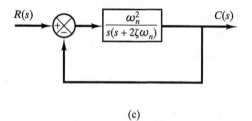

(c)

Figure 10–23   (a) Position control system;
(b) block diagram; (c) block diagram of a
second-order system in standard form.

So the transfer function between $C(s)$ and $T(s)$ is

$$\frac{C(s)}{T(s)} = \frac{1}{s(Js + b)}$$

With the use of this transfer function, Figure 10–23(a) can be redrawn as shown in Figure 10–23(b). The closed-loop transfer function is then

$$\frac{C(s)}{R(s)} = \frac{K}{Js^2 + bs + K} = \frac{K/J}{s^2 + (b/J)s + (K/J)}$$

or

$$\frac{C(s)}{R(s)} = \frac{\omega_n^2}{s^2 + 2\zeta\omega_n s + \omega_n^2} \qquad (10\text{–}11)$$

where

$$\omega_n = \sqrt{\frac{K}{J}} = \text{undamped natural frequency}$$

$$\zeta = \frac{b}{2\sqrt{KJ}} = \text{damping ratio}$$

In terms of ζ and ω_n, the block diagram of Figure 10–23(b) can be redrawn as shown in Figure 10–23(c).

Next, let us consider the unit-step response of this system when $0 < \zeta < 1$. For a unit-step input, we have $R(s) = 1/s$. Then

$$C(s) = \frac{\omega_n^2}{s^2 + 2\zeta\omega_n s + \omega_n^2} \frac{1}{s}$$

$$= \frac{1}{s} - \frac{s + 2\zeta\omega_n}{s^2 + 2\zeta\omega_n s + \omega_n^2}$$

$$= \frac{1}{s} - \frac{\zeta\omega_n}{(s + \zeta\omega_n)^2 + \omega_d^2} - \frac{s + \zeta\omega_n}{(s + \zeta\omega_n)^2 + \omega_d^2} \tag{10–12}$$

where $\omega_d = \omega_n\sqrt{1 - \zeta^2}$. The inverse Laplace transform of Equation (10–12) gives

$$c(t) = 1 - \frac{\zeta}{\sqrt{1 - \zeta^2}} e^{-\zeta\omega_n t} \sin \omega_d t - e^{-\zeta\omega_n t} \cos \omega_d t$$

$$= 1 - e^{-\zeta\omega_n t}\left(\frac{\zeta}{\sqrt{1 - \zeta^2}} \sin \omega_d t + \cos \omega_d t\right) \tag{10–13}$$

or

$$c(t) = 1 - \frac{e^{-\zeta\omega_n t}}{\sqrt{1 - \zeta^2}} \sin\left(\omega_d t + \tan^{-1}\frac{\sqrt{1 - \zeta^2}}{\zeta}\right) \tag{10–14}$$

A family of curves $c(t)$ plotted against t with various values of ζ is shown in Figure 10–24, where the abscissa is the dimensionless variable $\omega_n t$. The curves are functions only of ζ.

A few comments on transient-response specifications.   Except in certain applications in which oscillations cannot be tolerated, it is preferable that the transient response be sufficiently fast as well as reasonably damped. So, in order to get a desirable transient response for a second-order system, the damping ratio ζ may be chosen between 0.4 and 0.8. Small values of ζ ($\zeta < 0.4$) yield excessive overshoot in the transient response, and a system with a large value of ζ ($\zeta > 0.8$) responds sluggishly.

Later we shall see that the maximum overshoot and the rise time conflict with each other. In other words, both the maximum overshoot and the rise time cannot be made smaller simultaneously. If one is made smaller, the other necessarily becomes larger.

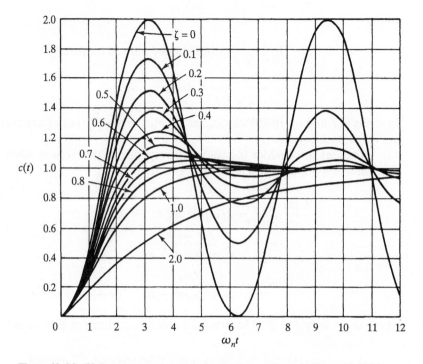

Figure 10–24   Unit-step response curves for the system shown in Figure 10–23(c).

Second-order systems and transient-response specifications.   Let us now obtain the rise time, peak time, maximum overshoot, and settling time of the second-order system given by Equation (10–11). These values will be derived in terms of ζ and ω_n. The system is assumed to be underdamped.

Rise time t_r.   We find the rise time t_r by letting $c(t_r) = 1$ in Equation (10–13), or

$$c(t_r) = 1 = 1 - e^{-\zeta\omega_n t_r}\left(\frac{\zeta}{\sqrt{1 - \zeta^2}} \sin \omega_d t_r + \cos \omega_d t_r\right) \qquad (10\text{--}15)$$

Since $e^{-\zeta\omega_n t_r} \neq 0$, Equation (10–15) yields

$$\frac{\zeta}{\sqrt{1 - \zeta^2}} \sin \omega_d t_r + \cos \omega_d t_r = 0$$

or

$$\tan \omega_d t_r = -\frac{\sqrt{1 - \zeta^2}}{\zeta}$$

Thus, the rise time t_r is

$$t_r = \frac{1}{\omega_d} \tan^{-1}\left(-\frac{\sqrt{1 - \zeta^2}}{\zeta}\right) = \frac{\pi - \beta}{\omega_d} \qquad (10\text{--}16)$$

where β is defined in Figure 10–25. Clearly, to obtain a small value of t_r, we must have a large ω_d.

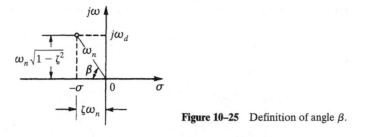

Figure 10–25 Definition of angle β.

Peak time t_p. We obtain the peak time by differentiating $c(t)$ in Equation (10–13) with respect to time t and letting this derivative equal zero. That is,

$$\frac{dc}{dt} = \frac{\omega_n}{\sqrt{1 - \zeta^2}} e^{-\zeta\omega_n t} \sin \omega_d t = 0$$

It follows that

$$\sin \omega_d t = 0$$

or

$$\omega_d t = 0, \pi, 2\pi, 3\pi, \ldots$$

Since the peak time corresponds to the first peak overshoot, we have $\omega_d t_p = \pi$. Then

$$t_p = \frac{\pi}{\omega_d} \tag{10–17}$$

The peak time t_p corresponds to one half-cycle of the frequency of damped oscillation.

Maximum overshoot M_p. The maximum overshoot occurs at the peak time $t_p = \pi/\omega_d$. Thus, from Equation (10–13),

$$M_p = c(t_p) - 1$$

$$= -e^{-\zeta\omega_n(\pi/\omega_d)}\left(\frac{\zeta}{\sqrt{1 - \zeta^2}} \sin \pi + \cos \pi\right)$$

$$= e^{-\zeta\pi/\sqrt{1-\zeta^2}} \tag{10–18}$$

Since $c(\infty) = 1$, the maximum percent overshoot is $e^{-\zeta\pi/\sqrt{1-\zeta^2}} \times 100\%$.

Settling time t_s. For an underdamped second-order system, the transient response for a unit-step input is given by Equation (10–14). Notice that the response curve $c(t)$ always remains within a pair of the envelope curves, as shown in Figure 10–26. [The curves $1 \pm (e^{-\zeta\omega_n t}/\sqrt{1 - \zeta^2})$ are the *envelope curves* of the transient response to a unit-step input.] The time constant of these envelope curves is $1/\zeta\omega_n$. The settling time t_s is four times this time constant, or

$$t_s = \frac{4}{\zeta\omega_n} \tag{10–19}$$

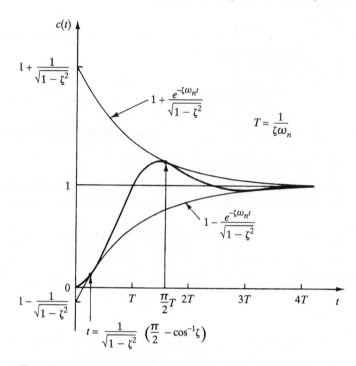

Figure 10–26   Unit-step response curve and its envelope curves.

Note that the settling time is inversely proportional to the undamped natural frequency of the system. Since the value of ζ is usually determined from the requirement of permissible maximum overshoot, the settling time is determined primarily by the undamped natural frequency ω_n. In other words, the duration of the transient period can be varied, without changing the maximum overshoot, by adjusting the undamped natural frequency ω_n.

From the preceding analysis, it is clear that ω_d must be large if we are to have a rapid response. To limit the maximum overshoot M_p and make the settling time small, the damping ratio ζ should not be too small. The relationship between the maximum overshoot and the damping ratio is presented in Figure 10–27. Note that if the damping ratio is between 0.4 and 0.8, then the maximum percent overshoot for a step response is between 25% and 2.5%.

Example 10–2

Determine the rise time, peak time, maximum overshoot, and settling time when the control system shown in Figure 10–28 is subjected to a unit-step input.

Notice that $\omega_n = 1$ rad/s and $\zeta = 0.5$ for this system. So

$$\omega_d = \omega_n\sqrt{1 - \zeta^2} = \sqrt{1 - 0.5^2} = 0.866$$

Rise time t_r: From Equation (10–16), the rise time is

$$t_r = \frac{\pi - \beta}{\omega_d}$$

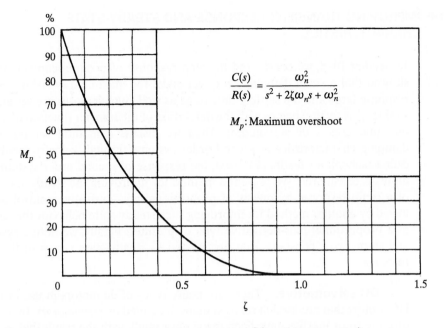

$$\frac{C(s)}{R(s)} = \frac{\omega_n^2}{s^2 + 2\zeta\omega_n s + \omega_n^2}$$

M_p: Maximum overshoot

Figure 10–27   Curve relating maximum overshoot M_p and damping ratio ζ.

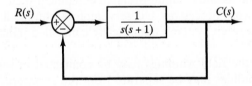

Figure 10–28   Control system.

where $\beta = \sin^{-1} 0.866 = 1.05$ rad. Therefore,

$$t_r = \frac{3.14 - 1.05}{0.866} = 2.41 \text{ s}$$

Peak time t_p: The peak time t_p is given by Equation (10–17):

$$t_p = \frac{\pi}{\omega_d} = \frac{3.14}{0.866} = 3.63 \text{ s}$$

Maximum overshoot M_p: From Equation (10–18), the maximum overshoot is

$$M_p = e^{-\zeta\pi/\sqrt{1-\zeta^2}} = e^{-0.5\times3.14/0.866} = e^{-1.81} = 0.163$$

Settling time t_s: The settling time t_s, defined by Equation (10–19), is

$$t_s = \frac{4}{0.5 \times 1} = 8 \text{ s}$$

10–6 IMPROVING TRANSIENT-RESPONSE AND STEADY-STATE CHARACTERISTICS

In Section 10–5, we considered the step response of position control systems. We showed that a small damping ratio will make the maximum overshoot in the step response large and the settling time large as well. Such features are generally undesirable. This section begins with the derivation of the transfer function of a servo system that uses a dc servomotor. Then we discuss a method for improving the damping characteristics of second-order systems through velocity feedback (also called tachometer feedback). Next, the response of second-order systems to ramp inputs is considered. We present a method for improving the steady-state behavior of a ramp response by means of proportional-plus-derivative control action, followed by another method for improving such steady-state behavior through the use of a proportional-plus-derivative type of prefilter. Finally, we define system types and static error constants that are related to steady-state errors in the transient response.

DC servomotors.   There are many types of dc motors in use in industries. DC motors that are used in servo systems are called *dc servomotors*. In dc servomotors, the rotor inertias have been made very small, with the result that motors with very high torque-to-inertia ratios are commercially available. Some dc servomotors have extremely small time constants. DC servomotors with relatively small power ratings are used in instruments and computer-related equipment such as disk drives, tape drives, printers, and word processors. DC servomotors with medium and large power ratings are used in robot systems, numerically controlled milling machines, and so on.

In dc servomotors, the field windings may be connected in series with the armature, or the field windings may be separate from the armature. (That is, the magnetic field is produced by a separate circuit.) In the latter case, where the field is excited separately, the magnetic flux is independent of the armature current. In some dc servomotors, the magnetic field is produced by a permanent magnet; therefore, the magnetic flux is constant. Such dc servomotors are called *permanent-magnet* dc servomotors. DC servomotors with separately excited fields, as well as permanent-magnet dc servomotors, can be controlled by the armature current. The technique is called *armature control of dc servomotors*.

In the case where the armature current is maintained constant and the speed is controlled by the field voltage, the dc motor is called a *field-controlled* dc motor. (Some speed control systems use field-controlled dc motors.) The requirement of constant armature current, however, is a serious disadvantage. (Providing a constant current source is much more difficult than providing a constant voltage source.) The time constants of the field-controlled dc motor are generally large compared with the time constants of a comparable armature-controlled dc motor.

A dc servomotor may also be driven by an electronic motion controller, frequently called a *servodriver*, as a motor–driver combination. The servodriver controls the motion of the dc servomotor and operates in various modes. Some of the servodriver's features are point-to-point positioning, velocity profiling, and programmable acceleration. Electronic motion controllers that use a pulse-width-modulated driver

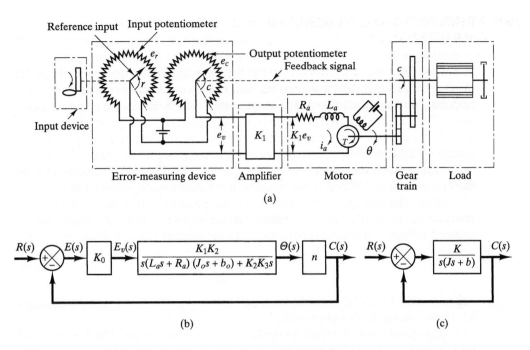

Reference input   Input potentiometer

Input device

Error-measuring device   Amplifier   Motor   Gear train   Load

(a)

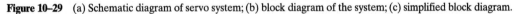

(b)

(c)

Figure 10–29   (a) Schematic diagram of servo system; (b) block diagram of the system; (c) simplified block diagram.

to control a dc servomotor are frequently seen in robot control systems, numerical control systems, and other position or speed control systems.

In what follows we shall discuss a servo system that uses armature control of a dc servomotor.

A servo system.   Consider the servo system shown in Figure 10–29(a). The objective of this system is to control the position of the mechanical load in accordance with the reference position. The operation of the system is as follows: A pair of potentiometers acting as an error-measuring device converts the input and output positions into proportional electric signals. The command input signal determines the angular position r of the wiper arm of the input potentiometer. The angular position r is the reference input to the system, and the electric potential of the arm is proportional to the angular position of the arm. The output shaft position determines the angular position c of the wiper arm of the output potentiometer. The difference between the input angular position r and the output angular position c is the error signal e, or

$$e = r - c$$

The potential difference $e_r - e_c = e_v$ is the error voltage, where e_r is proportional to r and e_c is proportional to c; that is, $e_r = K_0 r$ and $e_c = K_0 c$, where K_0 is a proportionality constant. The error voltage that appears at the potentiometer terminals is amplified by the amplifier, whose gain constant is K_1. The output voltage of this amplifier is applied to the armature circuit of the dc motor. (The amplifier must have very high input impedance, because the potentiometers are essentially high-impedance circuits and do not tolerate current drain. At the same time, the amplifier must have low output impedance, since it feeds into the armature circuit of the motor.) A fixed voltage is applied to the field winding. If an error exists, the motor develops a torque to rotate the output load in such a way as to reduce the error to zero. For constant field current, the torque developed by the motor is

$$T = K_2 i_a$$

where K_2 is the motor torque constant and i_a is the armature current.

Notice that if the sign of the current i_a is reversed, the sign of the torque T will be reversed, which will result in a reversal of the direction of rotor rotation.

When the armature is rotating, a voltage proportional to the product of the flux and the angular velocity is induced in the armature. For a constant flux, the induced voltage e_b is directly proportional to the angular velocity $d\theta/dt$, or

$$e_b = K_3 \frac{d\theta}{dt} \tag{10-20}$$

where e_b is the back emf, K_3 is the back-emf constant of the motor, and θ is the angular displacement of the motor shaft.

The speed of an armature-controlled dc servomotor is controlled by the armature voltage e_a. (The armature voltage $e_a = K_1 e_v$ is the output of the amplifier.) The differential equation for the armature circuit is

$$L_a \frac{di_a}{dt} + R_a i_a + e_b = e_a$$

or

$$L_a \frac{di_a}{dt} + R_a i_a + K_3 \frac{d\theta}{dt} = K_1 e_v \tag{10-21}$$

The equation for torque equilibrium is

$$J_0 \frac{d^2\theta}{dt^2} + b_0 \frac{d\theta}{dt} = T = K_2 i_a \tag{10-22}$$

where J_0 is the inertia of the combination of the motor, load, and gear train, referred to the motor shaft, and b_0 is the viscous-friction coefficient of the combination of the motor, load, and gear train, referred to the motor shaft. The transfer function between the motor shaft displacement and the error voltage is obtained from Equations (10–21) and (10–22) as follows:

$$\frac{\Theta(s)}{E_v(s)} = \frac{K_1 K_2}{s(L_a s + R_a)(J_0 s + b_0) + K_2 K_3 s} \tag{10-23}$$

where $\Theta(s) = \mathcal{L}[\theta(t)]$ and $E_v(s) = \mathcal{L}[e_v(t)]$. We assume that the gear ratio of the gear train is such that the output shaft rotates n times for each revolution of the motor shaft. Thus,

$$C(s) = n\Theta(s) \tag{10-24}$$

where $C(s) = \mathcal{L}[c(t)]$ and $c(t)$ is the angular displacement of the output shaft. The relationship among $E_v(s)$, $R(s)$, and $C(s)$ is

$$E_v(s) = K_0[R(s) - C(s)] = K_0 E(s) \qquad (10\text{–}25)$$

where $R(s) = \mathcal{L}[r(t)]$. The block diagram of this system can be constructed from Equations (10–23), (10–24), and (10–25) as shown in Figure 10–29(b). The transfer function in the feedforward path of this system is

$$G(s) = \frac{C(s)}{\Theta(s)} \frac{\Theta(s)}{E_v(s)} \frac{E_v(s)}{E(s)} = \frac{K_0 K_1 K_2 n}{s[(L_a s + R_a)(J_0 s + b_0) + K_2 K_3]}$$

Since L_a is usually small, it can be neglected, and the transfer function in the feedforward path becomes

$$G(s) = \frac{K_0 K_1 K_2 n}{s[R_a(J_0 s + b_0) + K_2 K_3]}$$

$$= \frac{K_0 K_1 K_2 n / R_a}{J_0 s^2 + \left(b_0 + \dfrac{K_2 K_3}{R_a} \right) s} \qquad (10\text{–}26)$$

The term $[b_0 + (K_2 K_3 / R_a)]s$ indicates that the back emf of the motor effectively increases the viscous friction of the system. The inertia J_0 and the viscous-friction coefficient $b_0 + (K_2 K_3 / R_a)$ are referred to the motor shaft. When J_0 and $b_0 + (K_2 K_3 / R_a)$ are multiplied by $1/n^2$, the inertia and viscous-friction coefficient are expressed in terms of the output shaft. Introducing new parameters defined by

$J = J_0 / n^2 =$ moment of inertia referred to the output shaft
$b = [b_0 + (K_2 K_3 / R_a)]/n^2 =$ viscous-friction coefficient referred to the output shaft
$K = K_0 K_1 K_2 / n R_a$

we can simplify the transfer function $G(s)$ given by Equation (10–26) to

$$G(s) = \frac{K}{J s^2 + b s} \qquad (10\text{–}27)$$

The block diagram of the system shown in Figure 10–29(b) can thus be simplified as shown in Figure 10–29(c).

From Equations (10–26) and (10–27), it can be seen that the transfer functions involve the term $1/s$. Thus, this system possesses an integrating property. In Equation (10–27), the time constant J/b of the motor becomes smaller for a smaller R_a and smaller J_0. With small J_0, as the resistance R_a is reduced, the motor time constant approaches zero, and the motor acts as an ideal integrator.

Tachometers.   A dc tachometer is a generator that produces a voltage proportional to its rotating speed. The device is used as a transducer, converting the velocity of the rotating shaft into a proportional dc voltage. If the input to the tachometer is the shaft position θ and the output is the voltage e, then the transfer function of the dc tachometer is

$$\frac{E(s)}{\Theta(s)} = K_h s$$

where $E(s) = \mathcal{L}[e], \Theta(s) = \mathcal{L}[\theta]$, and K_h is a constant. Including such a tachometer in the feedback path of a position control system will improve the damping characteristics of the system. The use of a tachometer to obtain a velocity feedback signal is referred to as *velocity feedback* or *tachometer feedback.*

Position control systems with velocity feedback. Systems with a small damping ratio exhibit a large maximum overshoot and a long, sustained oscillation in the step response. To increase the effective damping of the system and thus improve the transient-response characteristics, velocity feedback is frequently employed.

Consider the position control system with velocity feedback shown in Figure 10–30(a). Let us assume that the viscous-friction coefficient b is small, so that the damping ratio in the absence of the tachometer is quite small. In the present system, the velocity signal, together with the positional signal, is fed back to the input to produce the actuating error signal. (Note that, in obtaining the velocity signal, it is preferable to use a tachometer rather than physically differentiate the output position signal, because differentiation always accentuates noise signals.)

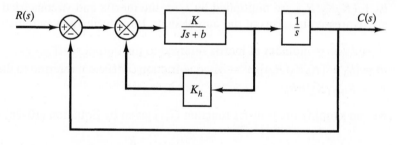

(a)

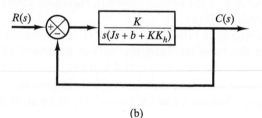

(b)

Figure 10–30 (a) Block diagram of a position control system with velocity feedback; (b) simplified block diagram.

The block diagram of Figure 10–30(a) can be simplified as shown in Figure 10–30(b), giving

$$\frac{C(s)}{R(s)} = \frac{K}{Js^2 + (b + KK_h)s + K}$$

The damping ratio ζ for this system is

$$\zeta = \frac{b + KK_h}{2\sqrt{KJ}} \tag{10–28}$$

and the undamped natural frequency ω_n is

$$\omega_n = \sqrt{\frac{K}{J}}$$

Notice that the undamped natural frequency ω_n is not affected by velocity feedback. Given the values of J and b, the value of K is determined from the requirement on the undamped natural frequency ω_n. The velocity feedback (tachometer feedback) constant K_h is then adjusted so that ζ is between 0.4 and 0.8.

Remember that velocity feedback (tachometer feedback) has the effect of increasing the damping ratio without affecting the undamped natural frequency of a second-order system.

Example 10–3

Assume, for the system shown in Figure 10–30(a), that

$$J = 2 \text{ kg-m}^2$$
$$b = 1 \text{ N-m-s}$$

We wish to determine the values of the gain K and velocity feedback constant K_h so that the maximum overshoot is 0.2 and the peak time is 1 s.

From Equation (10–18), the maximum overshoot is

$$M_p = e^{-\zeta\pi/\sqrt{1-\zeta^2}}$$

This value must be 0.2. Hence,

$$e^{-\zeta\pi/\sqrt{1-\zeta^2}} = 0.2$$

which yields

$$\zeta = 0.456$$

The peak time t_p is specified as 1 s. Therefore, from Equation (10–17),

$$t_p = \frac{\pi}{\omega_d} = 1$$

or

$$\omega_d = 3.14$$

Since ζ is 0.456, ω_n is

$$\omega_n = \frac{\omega_d}{\sqrt{1 - \zeta^2}} = 3.53$$

The undamped natural frequency ω_n is equal to $\sqrt{K/J} = \sqrt{K/2}$, so

$$K = 2\omega_n^2 = 24.92 \text{ N-m}$$

K_h is then obtained from Equation (10–28) as

$$K_h = \frac{2\sqrt{KJ}\zeta - 1}{K} = 0.218 \text{ s}$$

Steady-state error in ramp responses. Position control systems may be subjected to changing inputs that can be approximated by a series of piecewise ramp inputs. In such a ramp response, the steady-state error for a ramp response must be small.

Consider the system shown in Figure 10–31. The transient response of this system when it is subjected to a ramp input can be found by a straightforward method. In the present analysis, we shall examine the steady-state error when the system is subjected to such an input.

From the block diagram, we have

$$\frac{E(s)}{R(s)} = \frac{R(s) - C(s)}{R(s)} = 1 - \frac{C(s)}{R(s)} = \frac{Js^2 + bs}{Js^2 + bs + K}$$

The steady-state error for the unit-ramp response can be obtained as follows: For a unit-ramp input $r(t) = t$, we have $R(s) = 1/s^2$. The steady-state error e_{ss} is then obtained as

$$e_{ss} = \lim_{s \to 0} sE(s) = \lim_{s \to 0} s \frac{Js^2 + bs}{Js^2 + bs + K} \frac{1}{s^2}$$
$$= \lim_{s \to 0} \frac{s^2(Js + b)}{s^2(Js^2 + bs + K)} = \frac{b}{K}$$

To ensure the small steady-state error for the ramp response, the value of K must be large and the value of b small. However, a large value of K and a small value of b will make the damping ratio ζ small and will, in general, result in undesirable transient-response characteristics. Consequently, some means of improving steady-state error for the ramp response without adversely affecting transient-response behavior is necessary. Two such means are discussed next.

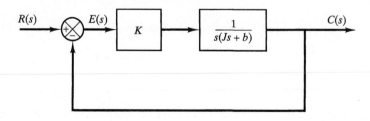

Figure 10–31 Block diagram of a position control system with a proportional controller.

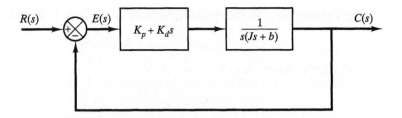

Figure 10–32 Block diagram of a position control system with a proportional-plus-derivative controller.

Proportional-plus-derivative control of second-order systems. A compromise between acceptable transient-response behavior and acceptable steady-state behavior can be achieved through proportional-plus-derivative control action. For the system shown in Figure 10–32, the closed-loop transfer function is

$$\frac{C(s)}{R(s)} = \frac{K_p + K_d s}{Js^2 + (b + K_d)s + K_p} \qquad (10\text{–}29)$$

Therefore,

$$\frac{E(s)}{R(s)} = \frac{R(s) - C(s)}{R(s)} = \frac{Js^2 + bs}{Js^2 + (b + K_d)s + K_p}$$

For a unit-ramp input, $R(s) = 1/s^2$. So it follows that

$$E(s) = \frac{Js^2 + bs}{Js^2 + (b + K_d)s + K_p} \frac{1}{s^2}$$

The steady-state error for a unit-ramp response is

$$e_{ss} = \lim_{s \to 0} sE(s) = \lim_{s \to 0} s\frac{Js^2 + bs}{Js^2 + (b + K_d)s + K_p} \frac{1}{s^2} = \frac{b}{K_p}$$

The characteristic equation is

$$Js^2 + (b + K_d)s + K_p = 0$$

The effective damping of this system is thus $b + K_d$ rather than b. Since the damping ratio ζ of this system is

$$\zeta = \frac{b + K_d}{2\sqrt{K_p J}}$$

it is possible to have both a small steady-state error e_{ss} for a ramp response and a reasonable damping ratio by making b small and K_p large and choosing K_d large enough so that ζ is between 0.4 and 0.8.

Let us examine the unit-step response of this system. If we define

$$\omega_n = \sqrt{\frac{K_p}{J}}, \qquad z = \frac{K_p}{K_d}$$

then Equation (10–29) can be written in terms of ω_n, ζ, and z as

$$\frac{C(s)}{R(s)} = \left(1 + \frac{s}{z}\right)\frac{\omega_n^2}{s^2 + 2\zeta\omega_n s + \omega_n^2}$$

Notice that if a zero $s = -z$ is located near the closed-loop poles, the transient-response behavior differs considerably from that of a second-order system without a zero. Typical step-response curves of this system with $\zeta = 0.5$ and various values of $z/\zeta\omega_n$ are shown in Figure 10–33. From these curves, we see that proportional-plus-derivative control action will make the rise time smaller and the maximum overshoot larger.

Second-order systems with a proportional-plus-derivative type of pre-filter. The steady-state error for the ramp response can be eliminated if the input is introduced to the system through a proportional-plus-derivative type of prefilter, as shown in Figure 10–34, and if the value of k is properly set.

The transfer function $C(s)/R(s)$ for this system is

$$\frac{C(s)}{R(s)} = \frac{(1 + ks)K}{Js^2 + bs + K}$$

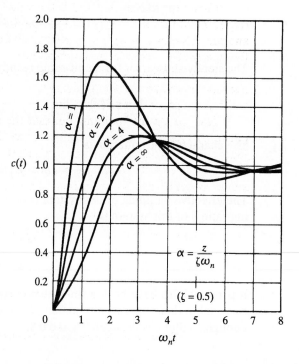

Figure 10–33 Unit-step response curves for the system shown in Figure 10–32 with the damping ratio ζ equal to 0.5.

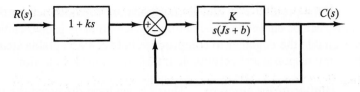

Figure 10–34   Block diagram of a position control system with a proportional-plus-derivative type of prefilter.

Therefore, the difference between $R(s)$ and $C(s)$ is

$$E(s) = R(s) - C(s) = \left[1 - \frac{C(s)}{R(s)}\right]R(s)$$

$$= \frac{Js^2 + (b - Kk)s}{Js^2 + bs + K}R(s)$$

The steady-state error for a ramp response is

$$e_{ss} = \lim_{s \to 0} sE(s) = \lim_{s \to 0} s\frac{Js^2 + (b - Kk)s}{Js^2 + bs + K}\frac{1}{s^2} = \frac{b - Kk}{K}$$

So if we choose

$$k = \frac{b}{K}$$

the steady-state error for the ramp response can be made equal to zero.

Given the values of J and b, the value of K is normally determined from the requirement that $\omega_n = \sqrt{K/J}$. Once the value of K is determined, b/K is a constant and the value of $k = b/K$ becomes constant. The use of such a prefilter eliminates the steady-state error for a ramp response.

Note that the transient response of this system to a unit-step input will exhibit a smaller rise time and a larger maximum overshoot than the corresponding system without the prefilter will show.

It is worthwhile pointing out that the block diagram of a system with a proportional-plus-derivative controller shown in Figure 10–32 can be redrawn as in Figure 10–35. From this diagram, it can be seen that the proportional-plus-derivative controller is, in fact, a combination of a prefilter and velocity feedback in which the values of both k and K_h are chosen to be K_d/K_p.

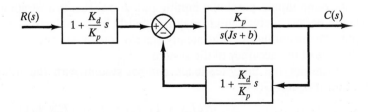

Figure 10–35   Modified block diagram of the system shown in Figure 10–32.

If the prefilter and velocity feedback are provided separately, the values of k and K_h can be chosen independently of each other. A proper choice of these values may enable the engineer to compromise between acceptable steady-state error for the ramp response and acceptable transient-response behavior to the step input.

Higher order systems. Thus far, we have discussed the transient-response analysis of second-order systems. Higher order systems involve more than two closed-loop poles. Since all closed-loop poles more or less contribute to the transient response of such systems, analytical expressions of the output become very complicated, and a computer approach to obtaining the response curve becomes necessary.

A well-designed higher order system may have a pair of complex-conjugate closed-loop poles that are located to the right of all other closed-loop poles, so that the response of the system is dominated by that pair of complex-conjugate closed-loop poles.

Dominant closed-loop poles. The relative dominance of closed-loop poles is determined by the ratio of the real parts of those poles, as well as by the relative magnitudes of the residues evaluated at the closed-loop poles. The magnitudes of the residues depend on both the closed-loop poles and zeros.

If the ratios of the real parts exceed 5 and there are no zeros nearby, then the closed-loop poles nearest the $j\omega$-axis will dominate in the transient-response behavior, because these poles correspond to transient-response terms that decay slowly. Those closed-loop poles which have dominant effects on the transient-response behavior are called *dominant closed-loop* poles. Quite often, the dominant closed-loop poles occur in the form of a complex-conjugate pair. Among all closed-loop poles, the dominant ones are the most important.

The gain of a higher order system is often adjusted so that there will exist a pair of dominant complex-conjugate closed-loop poles. The presence of such poles in a stable system reduces the effect of such nonlinearities as dead zone, backlash, and coulomb friction.

Classification of control systems. System types. Any physical control system inherently suffers steady-state errors in response to certain types of inputs. A system may have no steady-state error to a step input, but may exhibit nonzero steady-state error to a ramp input. (The only way we may be able to eliminate this error is to modify the structure of the system.)

Control systems may be classified according to their ability to follow step inputs, ramp inputs, parabolic inputs, and so on. This is a reasonable classification scheme because actual inputs may frequently be considered combinations of such inputs. The magnitudes of the steady-state errors due to these individual inputs are indicative of the accuracy of the system.

Consider the unity-feedback control system with the following open-loop transfer function $G(s)$:

$$G(s) = \frac{K(T_a s + 1)(T_b s + 1) \cdots (T_m s + 1)}{s^N (T_1 s + 1)(T_2 s + 1) \cdots (T_p s + 1)} \tag{10-30}$$

This transfer function involves the term s^N in the denominator, representing a pole of multiplicity N at the origin. The present classification scheme is based on the number of integrations indicated by the open-loop transfer function. A system is called type 0, type 1, type 2, ..., if $N = 0$, $N = 1$, $N = 2$, ..., respectively. Note that this classification is different from that of the order of a system. As the type number is increased, accuracy is improved; however, increasing the type number aggravates the stability problem. A compromise between steady-state accuracy and relative stability is always necessary. In practice, it is rather exceptional to have type 3 or higher systems, because we find it generally difficult to design stable systems having more than two integrations in the feedforward path.

We shall see later that, if $G(s)$ is written so that each term in the numerator and denominator, except the term s^N, approaches unity as s approaches zero, then the open-loop gain K is directly related to the steady-state error.

Steady-state errors in the transient response.   We have discussed steady-state errors in step and ramp responses both in Section 10–4 and in the current section. Next, we present a systematic discussion of steady-state errors in the transient response.

Consider the system shown in Figure 10–36. The closed-loop transfer function is

$$\frac{C(s)}{R(s)} = \frac{G(s)}{1 + G(s)}$$

The transfer function between the error signal $e(t)$ and the input signal $r(t)$ is

$$\frac{E(s)}{R(s)} = 1 - \frac{C(s)}{R(s)} = \frac{1}{1 + G(s)}$$

where the error $e(t)$ is the difference between the input signal and the output signal.

The final-value theorem provides a convenient way to find the steady-state performance of a stable system. Since

$$E(s) = \frac{1}{1 + G(s)} R(s)$$

if the system is stable, the steady-state error is

$$e_{ss} = \lim_{t \to \infty} e(t) = \lim_{s \to 0} sE(s) = \lim_{s \to 0} \frac{sR(s)}{1 + G(s)}$$

The static error constants defined next are figures of merit of control systems. The higher the constants, the smaller is the steady-state error. In a given system, the

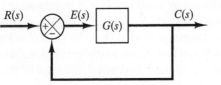

Figure 10–36   Control system.

output may be position, velocity, pressure, temperature, or the like; the physical form of the output is immaterial to the present analysis. Therefore, in what follows, we shall call the output "position," the rate of change of the output "velocity," and so on. This means that, in a temperature control system, "position" represents the output temperature, "velocity" represents the rate of change of the output temperature, and so on.

Static position error constant K_p.   The steady-state error of the system of Figure 10–36 for a unit-step input is

$$e_{ss} = \lim_{s \to 0} \frac{s}{1 + G(s)} \frac{1}{s}$$

$$= \frac{1}{1 + G(0)}$$

The static position error constant K_p is defined by

$$K_p = \lim_{s \to 0} G(s) = G(0)$$

Thus, the steady-state error in terms of the static position error constant K_p is given by

$$e_{ss} = \frac{1}{1 + K_p}$$

For a type 0 system,

$$K_p = \lim_{s \to 0} \frac{K(T_a s + 1)(T_b s + 1) \cdots}{(T_1 s + 1)(T_2 s + 1) \cdots} = K$$

For a type 1 or higher system,

$$K_p = \lim_{s \to 0} \frac{K(T_a s + 1)(T_b s + 1) \cdots}{s^N (T_1 s + 1)(T_2 s + 1) \cdots} = \infty \qquad \text{for } N \geq 1$$

Hence, for a type 0 system, the static position error constant K_p is finite, while for a type 1 or higher system, K_p is infinite.

For a unit-step input, the steady-state error e_{ss} may be summarized as follows:

$$e_{ss} = \frac{1}{1 + K} \qquad \text{for type 0 systems}$$

$$e_{ss} = 0 \qquad \text{for type 1 or higher systems}$$

The foregoing analysis indicates that the response of the feedback control system shown in Figure 10–36 to a step input involves a steady-state error if there is no integration in the feedforward path. (If small errors for step inputs can be tolerated, then a type 0 system is permissible, provided that the gain K is sufficiently large. If the gain K is too large, however, it is difficult to obtain reasonable relative stability.) If zero steady-state error for a step input is desired, the type of the system must be 1 or higher.

Static velocity error constant K_v.   The steady-state error of the system of Figure 10–36 with a unit-ramp input is given by

$$e_{ss} = \lim_{s \to 0} \frac{s}{1 + G(s)} \frac{1}{s^2}$$

$$= \lim_{s \to 0} \frac{1}{sG(s)}$$

The static velocity error constant K_v is defined by

$$K_v = \lim_{s \to 0} sG(s)$$

Thus, the steady-state error in terms of the static velocity error constant K_v is given by

$$e_{ss} = \frac{1}{K_v}$$

The term *velocity error* is used here to express the steady-state error for a ramp input. The dimension of the velocity error is the same as that of the system error. That is, velocity error is an error, not in velocity, but in position due to a ramp input. The values of K_v are obtained as follows:

For a type 0 system,

$$K_v = \lim_{s \to 0} \frac{sK(T_a s + 1)(T_b s + 1) \cdots}{(T_1 s + 1)(T_2 s + 1) \cdots} = 0$$

For a type 1 system,

$$K_v = \lim_{s \to 0} \frac{sK(T_a s + 1)(T_b s + 1) \cdots}{s(T_1 s + 1)(T_2 s + 1) \cdots} = K$$

For a type 2 or higher system,

$$K_v = \lim_{s \to 0} \frac{sK(T_a s + 1)(T_b s + 1) \cdots}{s^N (T_1 s + 1)(T_2 s + 1) \cdots} = \infty \qquad \text{for } N \geq 2$$

The steady-state error e_{ss} for the unit-ramp input can be summarized as follows:

$$e_{ss} = \frac{1}{K_v} = \infty \qquad \text{for type 0 systems}$$

$$e_{ss} = \frac{1}{K_v} = \frac{1}{K} \qquad \text{for type 1 systems}$$

$$e_{ss} = \frac{1}{K_v} = 0 \qquad \text{for type 2 or higher systems}$$

The foregoing analysis indicates that a type 0 system is incapable of following a ramp input in the steady state. The type 1 system with unity feedback can follow the ramp input with a finite error. In steady-state operation, the output velocity is exactly the same as the input velocity, but there is a positional error that is proportional to

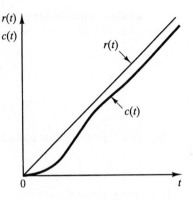

Figure 10–37 Response of a type 1 unity-feedback system to a ramp input.

the velocity of the input and inversely proportional to the gain K. Figure 10–37 is an example of the response of a type 1 system with unity feedback to a ramp input. The type 2 or higher system can follow a ramp input with zero error at steady state.

Static acceleration error constant K_a. The steady-state error of the system of Figure 10–36 with a unit-parabolic input (an acceleration input) defined by

$$r(t) = \frac{t^2}{2} \qquad \text{for } t \geq 0$$
$$= 0 \qquad \text{for } t < 0$$

is given by

$$e_{ss} = \lim_{s \to 0} \frac{s}{1 + G(s)} \frac{1}{s^3}$$
$$= \frac{1}{\lim_{s \to 0} s^2 G(s)}$$

The static acceleration error constant K_a is defined by the equation

$$K_a = \lim_{s \to 0} s^2 G(s)$$

The steady-state error is then

$$e_{ss} = \frac{1}{K_a}$$

Note that the acceleration error—the steady-state error due to a parabolic input—is an error in position.

The values of K_a are obtained as follows:
For a type 0 system,

$$K_a = \lim_{s \to 0} \frac{s^2 K(T_a s + 1)(T_b s + 1) \cdots}{(T_1 s + 1)(T_2 s + 1) \cdots} = 0$$

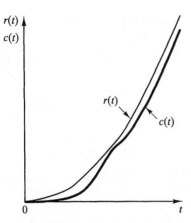

Figure 10–38   Response of a type 2 unity-feedback system to a parabolic input.

For a type 1 system,

$$K_a = \lim_{s \to 0} \frac{s^2 K (T_a s + 1)(T_b s + 1) \cdots}{s(T_1 s + 1)(T_2 s + 1) \cdots} = 0$$

For a type 2 system,

$$K_a = \lim_{s \to 0} \frac{s^2 K (T_a s + 1)(T_b s + 1) \cdots}{s^2 (T_1 s + 1)(T_2 s + 1) \cdots} = K$$

For a type 3 or higher system,

$$K_a = \lim_{s \to 0} \frac{s^2 K (T_a s + 1)(T_b s + 1) \cdots}{s^N (T_1 s + 1)(T_2 s + 1) \cdots} = \infty \qquad \text{for } N \geq 3$$

Thus, the steady-state error for the unit parabolic input is

$$e_{ss} = \infty \qquad \text{for type 0 and type 1 systems}$$

$$e_{ss} = \frac{1}{K} \qquad \text{for type 2 systems}$$

$$e_{ss} = 0 \qquad \text{for type 3 or higher systems}$$

Note that both type 0 and type 1 systems are incapable of following a parabolic input in the steady state. The type 2 system with unity feedback can follow a parabolic input with a finite error signal. Figure 10–38 shows an example of the response of a type 2 system with unity feedback to a parabolic input. The type 3 or higher system with unity feedback follows a parabolic input with zero error at steady state.

Summary.   Table 10–2 summarizes the steady-state errors for type 0, type 1, and type 2 unity-feedback systems subjected to various inputs. The finite values for steady-state errors appear on the diagonal line. Above the diagonal, the steady-state errors are infinity; below the diagonal, they are zero.

Remember that the terms *position error*, *velocity error*, and *acceleration error* indicate steady-state deviations in the output position. A finite velocity error implies that, after transients have died out, the input and output move at the same velocity, but have a finite position difference.

TABLE 10–2 Steady-State Errors of Unity-Feedback Systems in Terms of Gain K, where K is defined in Equation (10–30)

	Step Input $r(t) = 1$	Ramp Input $r(t) = t$	Acceleration Input $r(t) = \frac{1}{2}t^2$
Type 0 system	$\dfrac{1}{1 + K}$	∞	∞
Type 1 system	0	$\dfrac{1}{K}$	∞
Type 2 system	0	0	$\dfrac{1}{K}$

The error constants K_p, K_v, and K_a describe the ability of a unity-feedback system to reduce or eliminate steady-state error. Therefore, these constants are indicative of the steady-state performance of the system. It is generally desirable to increase the error constants, while maintaining the transient response within an acceptable range. To improve the steady-state performance, we can increase the type of the system by adding an integrator or integrators to the feedforward path. This, however, introduces an additional stability problem. The design of a satisfactory system with more than two integrators in series in the feedforward path is generally difficult.

Example 10–4

Obtain the steady-state error in the unit-ramp response of the system shown in Figure 10–31, using the static velocity error constant K_v.

For the system of Figure 10–31,

$$K_v = \lim_{s \to 0} s \frac{K}{s(Js + b)} = \frac{K}{b}$$

Hence,

$$e_{ss} = \frac{1}{K_v} = \frac{b}{K}$$

which is the same result as obtained earlier.

10–7 STABILITY ANALYSIS

The most important question about the closed-loop control system is concerned with stability. For any practical purpose, the system must be stable. Thus, stability analysis is most important in control systems analysis.

Stability analysis in the complex plane.   The stability of a linear closed-loop system can be determined from the location of the closed-loop poles in the s-plane. If any of these poles lie in the right-half s-plane, then, with increasing time, they give rise to the dominant mode, and the transient response increases monoton-ically or oscillates with increasing amplitude. Either of these motions represents an unstable system. For such a system, as soon as the power is turned on, the output may increase with time. If no saturation takes place in the system and no mechanical stop is provided, then the system may eventually be damaged and fail, since the re-sponse of a real physical system cannot increase indefinitely. Therefore, closed-loop poles in the right-half s-plane are not permissible in the usual linear control system. If all closed-loop poles lie to the left of the $j\omega$-axis, any transient response eventual-ly reaches equilibrium. This represents a stable system.

Whether a linear system is stable or unstable is a property of the system itself and does not depend on the input, or driving function, of the system. The poles of the input, or driving function, do not affect the stability of the system, but contribute only to steady-state response terms in the solution. Thus, the problem of absolute stability can be solved readily by choosing no closed-loop poles in the right-half s-plane, including the $j\omega$-axis. (Mathematically, closed-loop poles on the $j\omega$-axis will yield oscillations, the amplitude of which neither decays nor grows with time. In practical cases, where noise is present, however, the amplitude of oscillations may increase at a rate determined by the noise power level. Therefore, a control system should not have closed-loop poles on the $j\omega$-axis.) The absolute stability of higher order systems can be examined easily with the use of Routh's stability criterion.

Routh's stability criterion.   Most linear closed-loop control systems have closed-loop transfer functions of the form

$$\frac{C(s)}{R(s)} = \frac{b_0 s^m + b_1 s^{m-1} + \cdots + b_{m-1} s + b_m}{a_0 s^n + a_1 s^{n-1} + \cdots + a_{n-1} s + a_n} = \frac{B(s)}{A(s)}$$

where the a's and b's are constants and $m \le n$. The locations of the roots of the characteristic equation (the denominator of the preceding equation) determine the stability of the closed-loop system. A simple criterion, known as *Routh's stability cri-terion*, enables us to determine the number of closed-loop poles that lie in the right-half s-plane without having to factor the polynomial. This criterion applies only to polynomials with a finite number of terms. When Routh's stability criterion is applied to a control system, information about the absolute stability of the system can be obtained directly from the coefficients of the characteristic equation.

The procedure used in Routh's stability criterion is as follows:

1. Write the polynomial in s in the following form:

$$a_0 s^n + a_1 s^{n-1} + \cdots + a_{n-1} s + a_n = 0 \qquad (10\text{–}31)$$

where the coefficients are real quantities. Assume that $a_n \ne 0$; that is, any zero root has been removed.

2. If any of the coefficients are zero or negative in the presence of at least one positive coefficient, there is a root or roots that are imaginary or that have pos-itive real parts. In such a case, the system is not stable. If we are interested in only the absolute stability of the system, there is no need to follow the proce-dure further. Note that all the coefficients must be positive. This is a necessary

condition, as may be seen from the following argument: A polynomial in s having real coefficients can always be factored into linear and quadratic factors, such as $(s + a)$ and $(s^2 + bs + c)$, where $a, b,$ and c are real. The linear factors yield real roots, and the quadratic factors yield complex roots of the polynomial. The factor $(s^2 + bs + c)$ yields roots having negative real parts only if b and c are both positive. For all roots to have negative real parts, the constants $a,$ $b, c,$ and so on, in all factors must be positive. The product of any number of linear and quadratic factors containing only positive coefficients always yields a polynomial with positive coefficients. It is important to note that the condition that all the coefficients be positive is not sufficient to assure stability. The necessary, but not sufficient, condition for stability is that the coefficients of Equation (10–31) all be present and have a positive sign. (If all a's are negative, they can be made positive by multiplying both sides of the equation by -1.)

3. If all of the coefficients of the polynomial are positive, arrange the coefficients of the polynomial in rows and columns according to the following pattern:

$$
\begin{array}{cccccc}
s^n & a_0 & a_2 & a_4 & a_6 & \cdots \\
s^{n-1} & a_1 & a_3 & a_5 & a_7 & \cdots \\
s^{n-2} & b_1 & b_2 & b_3 & b_4 & \cdots \\
s^{n-3} & c_1 & c_2 & c_3 & c_4 & \cdots \\
s^{n-4} & d_1 & d_2 & d_3 & d_4 & \cdots \\
\vdots & \vdots & \vdots & & & \\
s^2 & e_1 & e_2 & & & \\
s^1 & f_1 & & & & \\
s^0 & g_1 & & & &
\end{array}
$$

The coefficients $b_1, b_2, b_3,$ and so on, are evaluated as follows:

$$b_1 = \frac{a_1 a_2 - a_0 a_3}{a_1}$$

$$b_2 = \frac{a_1 a_4 - a_0 a_5}{a_1}$$

$$b_3 = \frac{a_1 a_6 - a_0 a_7}{a_1}$$

$$\vdots$$

The evaluation of the b's is continued until the remaining ones are all zero. The same pattern of cross multiplying the coefficients of the two previous rows is followed in evaluating the c's, d's, e's, and so on. That is,

$$c_1 = \frac{b_1 a_3 - a_1 b_2}{b_1}$$

$$c_2 = \frac{b_1 a_5 - a_1 b_3}{b_1}$$

$$c_3 = \frac{b_1 a_7 - a_1 b_4}{b_1}$$

$$\vdots$$

and

$$d_1 = \frac{c_1 b_2 - b_1 c_2}{c_1}$$

$$d_2 = \frac{c_1 b_3 - b_1 c_3}{c_1}$$

$$\vdots$$

The process continues until the nth row has been completed. The finished array of coefficients is triangular. Note that, in developing the array, an entire row may be divided or multiplied by a positive number in order to simplify the subsequent numerical calculation without altering the stability conclusion.

Routh's stability criterion states that the number of roots of Equation (10–31) with positive real parts is equal to the number of changes in sign of the coefficients of the first column of the array. (The column consisting of s^n, s^{n-1}, ..., s^0 on the far left side of the table is used for identification purposes only. In counting the column number, this column is not included. The first column of the array means the first numerical column.) Note that the exact values of the terms in the first column need not be known; only the signs are needed. The necessary and sufficient condition that all roots of Equation (10–31) lie in the left-half s-plane is that all the coefficients of Equation (10–31) be positive and all terms in the first column of the array have positive signs.

Example 10–5

Let us apply Routh's stability criterion to the third-order polynomial

$$a_0 s^3 + a_1 s^2 + a_2 s + a_3 = 0$$

where all the coefficients are positive numbers. The array of coefficients becomes

s^3	a_0	a_2
s^2	a_1	a_3
s^1	$\dfrac{a_1 a_2 - a_0 a_3}{a_1}$	
s^0	a_3	

The condition that all roots have negative real parts is given by

$$a_1 a_2 > a_0 a_3$$

Example 10–6

Consider the polynomial

$$s^4 + 2s^3 + 3s^2 + 4s + 5 = 0$$

Let us follow the procedure just presented and construct the array of coefficients. (The first two rows can be obtained directly from the given polynomial. The remaining terms are obtained from these rows. If any coefficients are missing, they may be replaced by zeros in the array.) The completed array is as follows:

$$
\begin{array}{ccc|ccccl}
s^4 & 1 & 3 & 5 & s^4 & 1 & 3 & 5 \\
s^3 & 2 & 4 & 0 & s^3 & 2 & 4 & 0 \\
 & & & & & 1 & 2 & 0 & \text{The second row is divided by 2.} \\
s^2 & 1 & 5 & & s^2 & 1 & 5 \\
s^1 & -6 & & & s^1 & -3 \\
s^0 & 5 & & & s^0 & 5
\end{array}
$$

In this example, the number of changes in sign of the coefficients in the first column of the array is 2. This means that there are two roots with positive real parts. Note that the result is unchanged when the coefficients of any row are multiplied or divided by a positive number in order to simplify the computation.

Special cases. If a term in the first column of the array in any row is zero, but the remaining terms are not zero or there is no remaining term, then the zero term is replaced by a very small positive number ϵ and the rest of the array is evaluated. For example, consider the following equation:

$$s^3 + 2s^2 + s + 2 = 0 \tag{10–32}$$

The array of coefficients is

$$
\begin{array}{ccc}
s^3 & 1 & 1 \\
s^2 & 2 & 2 \\
s^1 & 0 \approx \epsilon \\
s^0 & 2
\end{array}
$$

If the sign of the coefficient above the zero (ϵ) is the same as that below it, it indicates that there is a pair of imaginary roots. Actually, Equation (10–32) has two roots at $s = \pm j$.

If, however, the sign of the coefficient above the zero (ϵ) is opposite that below it, then there is one sign change. For example, for the equation

$$s^3 - 3s + 2 = (s - 1)^2(s + 2) = 0$$

the array of coefficients is as follows:

One sign change:
$$
\begin{array}{ccc}
s^3 & 1 & -3 \\
s^2 & 0 \approx \epsilon & 2 \\
s^1 & -3 - \dfrac{2}{\epsilon} \\
s^0 & 2
\end{array}
$$
One sign change:

There are two sign changes of the coefficients in the first column of the array. This agrees with the correct result indicated by the factored form of the polynomial equation.

If all of the coefficients in any derived row are zero, then there are roots of equal magnitude lying radially opposite each other in the s-plane; that is, there are two real roots of equal magnitude and opposite sign and/or two conjugate imaginary roots. In such a case, the evaluation of the rest of the array can be continued by forming an auxiliary polynomial with the coefficients of the last row and by using the coefficients of the derivative of this polynomial in the next row. Such roots of equal magnitude and lying radially opposite each other in the s-plane can be found by solving the auxiliary polynomial, which is always even. For a $2n$-degree auxiliary polynomial, there are n pairs of equal and opposite roots. For example, consider the following equation:

$$s^5 + 2s^4 + 24s^3 + 48s^2 - 25s - 50 = 0$$

The array of coefficients is

$$
\begin{array}{llll}
s^5 & 1 & 24 & -25 \\
s^4 & 2 & 48 & -50 & \leftarrow \text{Auxiliary polynomial } P(s) \\
s^3 & 0 & 0
\end{array}
$$

The terms in the s^3 row are all zero. The auxiliary polynomial is then formed from the coefficients of the s^4 row. The auxiliary polynomial $P(s)$ is

$$P(s) = 2s^4 + 48s^2 - 50$$

which indicates that there are two pairs of roots of equal magnitude and opposite sign. These pairs are obtained by solving the auxiliary polynomial equation $P(s) = 0$. The derivative of $P(s)$ with respect to s is

$$\frac{dP(s)}{ds} = 8s^3 + 96s$$

The terms in the s^3 row are replaced by the coefficients of the last equation, that is, 8 and 96. The array of coefficients then becomes

$$
\begin{array}{llll}
s^5 & 1 & 24 & -25 \\
s^4 & 2 & 48 & -50 \\
s^3 & 8 & 96 & \qquad \leftarrow \text{Coefficients of } dP(s)/ds \\
s^2 & 24 & -50 \\
s^1 & 112.7 & 0 \\
s^0 & -50
\end{array}
$$

We see that there is one change in sign in the first column of the new array. Thus, the original equation has one root with a positive real part. By solving for roots of the auxiliary polynomial equation,

$$2s^4 + 48s^2 - 50 = 0$$

we obtain

$$s^2 = 1, \qquad s^2 = -25$$

or

$$s = \pm 1, \qquad s = \pm j5$$

These two pairs of roots are a part of the roots of the original equation. As a matter of fact, the original equation can be written in factored form as follows:

$$(s + 1)(s - 1)(s + j5)(s - j5)(s + 2) = 0$$

Relative stability analysis. Routh's stability criterion provides the answer to the question of absolute stability. In many practical cases, however, this answer is not sufficient; we usually require information about the relative stability of the system. A useful approach to examining relative stability is to shift the s-plane axis and apply Routh's stability criterion. That is, we substitute

$$s = \hat{s} - \sigma \qquad (\sigma = \text{positive constant})$$

into the characteristic equation of the system, write the polynomial in terms of $\hat{s}$, and apply Routh's stability criterion to the new polynomial in $\hat{s}$. The number of changes of sign in the first column of the array developed for the polynomial in $\hat{s}$ is equal to the number of roots that are located to the right of the vertical line $s = -\sigma$. Thus, this test reveals the number of roots that lie to the right of the vertical line $s = -\sigma$.

Application of Routh's stability criterion to control systems analysis. Routh's stability criterion is of limited usefulness in linear control systems analysis, mainly because it does not suggest how to improve relative stability or how to stabilize an unstable system. It is possible, however, to determine the effects of changing one or two parameters of a system by examining the values that cause instability. We close this section with a brief consideration of the problem of determining the stability range of a parameter value.

Consider the system shown in Figure 10–39. Let us determine the range of K for stability. The closed-loop transfer function is

$$\frac{C(s)}{R(s)} = \frac{K}{s(s + 1)(s + 2) + K}$$

The characteristic equation is

$$s^3 + 3s^2 + 2s + K = 0$$

The array of coefficients becomes

$$
\begin{array}{c c c}
s^3 & 1 & 2 \\
s^2 & 3 & K \\
s^1 & \dfrac{6 - K}{3} & 0 \\
s^0 & K &
\end{array}
$$

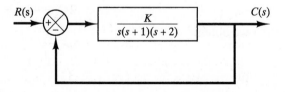

Figure 10–39   Control system.

For stability, K must be positive, and all coefficients in the first column of the array must be positive. Therefore,

$$6 > K > 0$$

When $K = 6$, the system becomes oscillatory, and, mathematically, the oscillation is sustained at constant amplitude.

10–8 ROOT-LOCUS ANALYSIS

The basic characteristic of the transient response of a closed-loop system is closely related to the location of the closed-loop poles. If the system has a variable loop gain, then the location of the closed-loop poles depends on the value of the loop gain chosen. It is important, therefore, that the designer know how the closed-loop poles move in the s-plane as the loop gain is varied.

From the design viewpoint, in some systems simple gain adjustment may move the closed-loop poles to desired locations. Then the design problem may become merely the selection of an appropriate gain value. If, however, the gain adjustment alone does not yield a desired result, the addition of a compensator to the system will become necessary.

A simple method for finding the roots of the characteristic equation has been developed by W. R. Evans and is used extensively in control engineering. In this method, called the *root-locus method*, the roots of the characteristic equation are plotted for all values of a system parameter. The roots corresponding to a particular value of this parameter can then be located on the resulting graph. Note that the parameter is usually the gain, but any other variable of the open-loop transfer function may be used. Unless otherwise stated, we shall assume that the gain of the open-loop transfer function is the parameter to be varied through all values, from zero to infinity.

By using the root-locus method, the designer can predict the effects on the location of the closed-loop poles of varying the gain value or adding open-loop poles and/or open-loop zeros. Therefore, it is desired that the designer have a good understanding of the method for generating the root loci of the closed-loop system, both by hand and with the use of computer software like MATLAB.

Root-locus method.   The basic idea behind the root-locus method is that the values of s that make the transfer function around the loop equal to -1 must satisfy the characteristic equation of the system.

The locus of roots of the characteristic equation of the closed-loop system as the gain is varied from zero to infinity gives the method its name. Such a plot clearly shows the contributions of each open-loop pole or zero to the locations of the closed-loop poles.

In designing a linear control system, we find that the root-locus method proves quite useful, since it indicates the manner in which the open-loop poles and zeros should be modified so that the response meets system performance specifications. The method is particularly suited to obtaining approximate results very quickly.

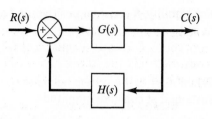

Figure 10–40   Control system.

Angle and magnitude conditions.   Consider the system shown in Figure 10–40. The closed-loop transfer function is

$$\frac{C(s)}{R(s)} = \frac{G(s)}{1 + G(s)H(s)} \tag{10–33}$$

The characteristic equation for this closed-loop system is obtained by setting the denominator of the right-hand side of Equation (10–33) equal to zero. That is,

$$1 + G(s)H(s) = 0$$

or

$$G(s)H(s) = -1 \tag{10–34}$$

Here, we assume that $G(s)H(s)$ is a ratio of polynomials in s. Since $G(s)H(s)$ is a complex quantity, Equation (10–34) can be split into two equations by equating the angles and magnitudes of both sides, respectively, to obtain the following:

Angle condition:

$$\underline{/G(s)H(s)} = \pm180°(2k + 1) \qquad k = 0, 1, 2, \ldots \tag{10–35}$$

Magnitude condition:

$$|G(s)H(s)| = 1 \tag{10–36}$$

The values of s that fulfill both the angle and magnitude conditions are the roots of the characteristic equation, or the closed-loop poles. A plot of the points in the complex plane satisfying the angle condition alone is the root locus. The roots of the characteristic equation (the closed-loop poles) corresponding to a given value of the gain can be determined from the magnitude condition. The details of applying the angle and magnitude conditions to obtain the closed-loop poles are presented later in this section.

In many cases, $G(s)H(s)$ involves a gain parameter K, and the characteristic equation may be written as

$$1 + \frac{K(s + z_1)(s + z_2) \cdots (s + z_m)}{(s + p_1)(s + p_2) \cdots (s + p_n)} = 0 \tag{10–37}$$

Then the root loci for the system are the loci of the closed-loop poles as the gain K is varied from zero to infinity.

Note that, to begin sketching the root loci of a system by the root-locus method, we must know the locations of the poles and zeros of $G(s)H(s)$. Remember that the angles of the complex quantities originating from the open-loop poles and open-loop zeros to the test point s are measured in the counterclockwise direction.

Note that, because the open-loop complex-conjugate poles and complex-conjugate zeros, if any, are always located symmetrically about the real axis, the root loci are always symmetrical with respect to that axis. Therefore, we need to construct only the upper half of the root loci and draw the mirror image of the upper half in the lower half *s*-plane.

Illustrative example.   Although computer approaches to the construction of the root loci are easily available, let us use graphical computation, combined with inspection, to determine and plot the root loci upon which the roots of the characteristic equation of the closed-loop system must lie. Such a graphical approach enhances one's understanding of how the closed-loop poles move in the complex plane as the open-loop poles and zeros are moved.

The first step in the procedure for constructing a root-locus plot is to use the angle condition to seek out the loci of possible roots. Then, if necessary, the loci can be scaled, or graduated, in gain with the use of the magnitude condition.

Because graphical measurements of angles and magnitudes are involved in the analysis, we find it necessary to use the same divisions on the abscissa as on the ordinate in sketching the root locus on graph paper.

Example 10–7

Consider the system shown in Figure 10–41. For this system,

$$G(s) = \frac{K}{s(s + 1)(s + 2)}, \qquad H(s) = 1$$

We assume that the value of the gain K is nonnegative. Let us sketch the root-locus plot and determine the value of K such that the damping ratio ζ of a pair of dominant complex-conjugate closed-loop poles is 0.5.

For the given system, the angle condition becomes

$$\angle G(s) = \frac{K}{\angle s(s + 1)(s + 2)}$$
$$= -\angle s - \angle s + 1 - \angle s + 2$$
$$= \pm 180°(2k + 1) \qquad k = 0, 1, 2, \ldots$$

The magnitude condition is

$$|G(s)| = \left| \frac{K}{s(s + 1)(s + 2)} \right| = 1$$

A typical procedure for sketching the root-locus plot is as follows:

1. *Determine the root loci on the real axis.* The first step in constructing a root-locus plot is to locate the open-loop poles $s = 0$, $s = -1$, and $s = -2$ in the complex

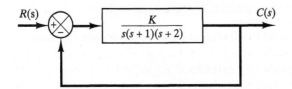

R(*s*) $\dfrac{K}{s(s+1)(s+2)}$ *C*(*s*)

Figure 10–41   Control system.

plane. (There are no open-loop zeros in this system.) The locations of the open-loop poles are indicated by crosses. (The locations of the open-loop zeros in this book will be indicated by small circles.) Note that the starting points of the root loci (the points corresponding to $K = 0$) are open-loop poles. The number of individual root loci for the given system is three, which is the same as the number of open-loop poles.

To determine the root loci on the real axis, we select a test point s. If the test point is on the positive real axis, then

$$\underline{/s} = \underline{/s + 1} = \underline{/s + 2} = 0°$$

This shows that the angle condition cannot be satisfied. Hence, there is no root locus on the positive real axis. Next, select a test point on the negative real axis between 0 and -1. Then

$$\underline{/s} = 180°, \qquad \underline{/s + 1} = \underline{/s + 2} = 0°$$

Thus,

$$-\underline{/s} - \underline{/s + 1} - \underline{/s + 2} = -180°$$

and the angle condition is satisfied. Therefore, the portion of the negative real axis between 0 and -1 forms a portion of the root locus. If a test point is selected between -1 and -2, then

$$\underline{/s} = \underline{/s + 1} = 180°, \qquad \underline{/s + 2} = 0°$$

and

$$-\underline{/s} - \underline{/s + 1} - \underline{/s + 2} = -360°$$

It can be seen that the angle condition is not satisfied. Therefore, the negative real axis from -1 to -2 is not a part of the root locus. Similarly, if a test point is located on the negative real axis from -2 to $-\infty$, the angle condition is satisfied. Thus, root loci exist on the negative real axis between 0 and -1 and between -2 and $-\infty$.

2. *Determine the asymptotes of the root loci.* The asymptotes of the root loci as s approaches infinity can be determined as follows: If a test point s is selected very far from the origin, then angles $\underline{/s}$, $\underline{/s + 1}$, and $\underline{/s + 2}$ may be considered the same. Therefore, the root loci for very large values of s must be asymptotic to straight lines whose angles are given by

$$\text{angles of asymptotes} = \frac{\pm 180°(2k + 1)}{3} \qquad k = 0, 1, 2, \ldots$$

Since the angle repeats itself as k is varied, the distinct angles for the asymptotes are determined to be 60°, $-60°$, and 180°. Thus, there are three asymptotes. The one having an angle of 180° is the negative real axis.

Before we can draw these asymptotes in the complex plane, we must find the point where they intersect the real axis. Since

$$G(s) = \frac{K}{s(s + 1)(s + 2)} \tag{10–38}$$

if a test point is located very far from the origin, then $G(s)$ may be written as

$$G(s) = \frac{K}{s^3 + 3s^2 + \cdots} \tag{10–39}$$

The characteristic equation is

$$G(s) = -1$$

so, from Equation (10–39), the characteristic equation may be written as

$$s^3 + 3s^2 + \cdots = -K$$

For a large value of s, this last equation may be approximated by

$$(s + 1)^3 = 0$$

If the abscissa of the intersection of the asymptotes and the real axis is denoted by $s = \sigma_a$, then

$$\sigma_a = -1$$

and the point of origin of the asymptotes is $(-1, 0)$. The asymptotes are almost part of the root loci in regions very far from the origin.

3. *Determine the breakaway point.* To plot root loci accurately, we must find the breakaway point, where the root-locus branches originating from the poles at 0 and -1 break away (as K is increased) from the real axis and move into the complex plane. The breakaway point corresponds to a point in the s-plane where multiple roots of the characteristic equation occur.

A simple method for finding the breakaway point is available. The method is as follows: Let us write the characteristic equation as

$$f(s) = B(s) + KA(s) = 0 \tag{10–40}$$

where $A(s)$ and $B(s)$ do not contain K. Note that $f(s) = 0$ has multiple roots at points where

$$\frac{df(s)}{ds} = 0$$

This can be seen from the following reasoning: Suppose that $f(s)$ has multiple roots of order r. Then $f(s)$ may be written as

$$f(s) = (s - s_1)^r (s - s_2) \cdots (s - s_n)$$

If we differentiate this equation with respect to s and set $s = s_1$, then we get

$$\frac{df(s)}{ds}\bigg|_{s=s_1} = 0 \tag{10–41}$$

This means that multiple roots of $f(s)$ will satisfy Equation (10–41). From Equation (10–40), we obtain

$$\frac{df(s)}{ds} = B'(s) + KA'(s) = 0 \tag{10–42}$$

where

$$A'(s) = \frac{dA(s)}{ds}, \qquad B'(s) = \frac{dB(s)}{ds}$$

From Equation (10–42), the particular value of K that will yield multiple roots of the characteristic equation is

$$K = -\frac{B'(s)}{A'(s)}$$

If we substitute this value of K into Equation (10–40), we get

$$f(s) = B(s) - \frac{B'(s)}{A'(s)} A(s) = 0$$

or

$$B(s)A'(s) - B'(s)A(s) = 0 \qquad (10\text{–}43)$$

If Equation (10–43) is solved for s, the points where multiple roots occur can be obtained. On the other hand, from Equation (10–40), we have

$$K = -\frac{B(s)}{A(s)}$$

and

$$\frac{dK}{ds} = -\frac{B'(s)A(s) - B(s)A'(s)}{A^2(s)}$$

If dK/ds is set equal to zero, we get Equation (10–43). Therefore, the breakaway points can be determined simply from the roots of

$$\frac{dK}{ds} = 0$$

Note that not all the solutions of Equation (10–43) or of $dK/ds = 0$ correspond to actual breakaway points. If a point at which $df(s)/ds = 0$ is on a root locus, it is an actual breakaway point. Stated differently, if, at a point at which $df(s)/ds = 0$, the value of K takes a real positive value, then that point is an actual breakaway point.

For the present example, the characteristic equation $G(s) + 1 = 0$ is given by

$$\frac{K}{s(s + 1)(s + 2)} + 1 = 0$$

or

$$K = -(s^3 + 3s^2 + 2s)$$

Setting $dK/ds = 0$, we obtain

$$\frac{dK}{ds} = -(3s^2 + 6s + 2) = 0$$

or

$$s = -0.4226, \qquad s = -1.5774$$

Since the breakaway point must lie on a root locus between 0 and -1, it is clear that $s = -0.4226$ corresponds to the actual breakaway point. The point $s = -1.5774$ is not on the root locus; hence, this point is not an actual breakaway point. In fact, evaluating the values of K corresponding to $s = -0.4226$ and $s = -1.5774$ yields

$$K = 0.3849 \qquad \text{for } s = -0.4226$$
$$K = -0.3849 \qquad \text{for } s = -1.5774$$

Note that when the two branches enter the complex region from the breakaway point, they leave the real axis at angles of $\pm 90°$.

4. *Determine the points where the root loci cross the imaginary axis.* These points can be found easily by substituting $s = j\omega$ into the characteristic equation, equating both the real and imaginary parts, to zero, and then solving for ω and K. For the present system, the characteristic equation is

$$s^3 + 3s^2 + 2s + K = 0$$

Substituting $s = j\omega$ into the characteristic equation, we obtain

$$(j\omega)^3 + 3(j\omega)^2 + 2(j\omega) + K = 0$$

or

$$(K - 3\omega^2) + j(2\omega - \omega^3) = 0$$

Equating both the real and imaginary parts of this last equation to zero yields

$$K - 3\omega^2 = 0, \qquad 2\omega - \omega^3 = 0$$

from which it follows that

$$\omega = \pm\sqrt{2}, \qquad K = 6 \qquad \text{or} \qquad \omega = 0, \qquad K = 0$$

Thus, root loci cross the imaginary axis at $\omega = \pm\sqrt{2}$, and the value of K at the crossing points is 6. Also, a root-locus branch on the real axis touches the imaginary axis at $\omega = 0$.

5. *Choose a test point in the broad neighborhood of the $j\omega$-axis and the origin*, as shown in Figure 10–42, and apply the angle condition. If a test point is on the root loci, then the sum of the three angles, $\theta_1 + \theta_2 + \theta_3$, must be 180°. If the test point does not satisfy the angle condition, select another test point until it satisfies the condition. (The sum of the angles at the test point will indicate in which direction the test point should be moved.) Continue this process and locate a sufficient number of points satisfying the angle condition.

6. *Draw the root loci,* based on the information obtained in the foregoing steps, as shown in Figure 10–43.

7. *Determine a pair of dominant complex-conjugate closed-loop poles such that the damping ratio ζ is 0.5.* Closed-loop poles with $\zeta = 0.5$ lie on lines passing through the origin and making the angles $\pm\cos^{-1}\zeta = \pm\cos^{-1}0.5 = \pm 60°$ with

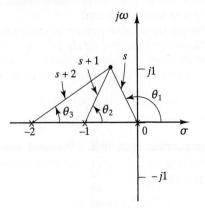

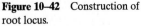

Figure 10–42   Construction of root locus.

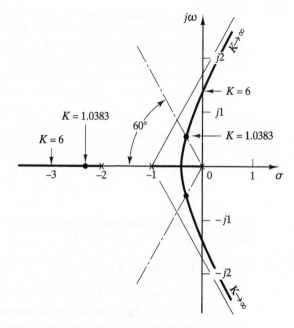

Figure 10–43   Root-locus plot.

the negative real axis. From Figure 10–43, such closed-loop poles having $\zeta = 0.5$ are obtained as follows:

$$s_1 = -0.3337 + j0.5780, \qquad s_2 = -0.3337 - j0.5780$$

The value of K that yields these poles is found from the magnitude condition as follows:

$$K = |s(s + 1)(s + 2)|_{s=-0.3337+j0.5780}$$
$$= 1.0383$$

Using this value of K, we find the third pole at $s = -2.3326$.

Note that, from step 4, it can be seen that for $K = 6$ the dominant closed-loop poles lie on the imaginary axis at $s = \pm j\sqrt{2}$. With this value of K, the system will exhibit sustained oscillations. For $K > 6$, the dominant closed-loop poles lie in the right-half s-plane, resulting in an unstable system.

Finally, note that, if necessary, the root loci can be easily graduated in terms of K with the use of the magnitude condition. We simply pick out a point on a root locus, measure the magnitudes of the three complex quantities $s, s + 1$, and $s + 2$, and multiply these magnitudes; the product is equal to the gain value K at that point, or

$$|s| \cdot |s + 1| \cdot |s + 2| = K$$

General rules for constructing root loci.   We next summarize the general rules and procedure for constructing the root loci of the system shown in Figure 10–44. First, obtain the characteristic equation

$$1 + G(s)H(s) = 0$$

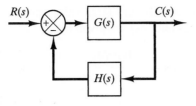

Figure 10-44   Control system.

Then rearrange this equation into the form

$$1 + \frac{K(s + z_1)(s + z_2)\cdots(s + z_m)}{(s + p_1)(s + p_2)\cdots(s + p_n)} = 0 \qquad (10\text{-}44)$$

so that the parameter of interest appears as the multiplicative factor. In the present discussion, we assume that the parameter of interest is the gain K, where $K > 0$.

1. *Locate the poles and zeros of $G(s)H(s)$ on the s-plane. The root-locus branches start from open-loop poles and terminate at zeros (finite zeros or zeros at infinity).*

   Note that the root loci are symmetrical about the real axis of the s-plane, because the complex poles and complex zeros occur only in conjugate pairs.

   Find the starting points and terminating points of the root loci, and find also the number of separate root loci. The points on the root loci corresponding to $K = 0$ are open-loop poles. Each root locus thus originates at a pole of the open-loop transfer function $G(s)H(s)$. As K is increased to infinity, each root-locus approaches either a zero of the open-loop transfer function or infinity in the complex plane.

   A root-locus plot will have just as many branches as there are roots of the characteristic equation. Since the number of open-loop poles generally exceeds that of zeros, the number of branches usually equals that of poles. If the number of closed-loop poles is the same as the number of open-loop poles, then the number of individual root-locus branches terminating at finite open-loop zeros is equal to the number m of open-loop zeros. The remaining $n - m$ branches terminate at infinity ($n - m$ implicit zeros at infinity) along asymptotes.

2. *Determine the root loci on the real axis.* Root loci on the real axis are determined by open-loop poles and zeros lying on the axis. The complex-conjugate poles and zeros of the open-loop transfer function have no effect on the location of the root loci on the real axis, because the angle contribution of a pair of complex-conjugate poles or zeros is 360° on that axis. If the total number of real poles and real zeros to the right of a test point is odd, then that point lies on a root locus. The root locus and its complement form alternate segments along the real axis.

3. *Determine the asymptotes of root loci.* If the test point s is located far from the origin, then the angle of each complex quantity may be considered the same. One open-loop zero then cancels the effects of one open-loop pole and vice versa.

Therefore, the root loci for very large values of s must be asymptotic to straight lines whose angles (slopes) are given by

$$\text{angles of asymptotes} = \frac{\pm 180°(2k + 1)}{n - m} \qquad k = 0, 1, 2, \ldots$$

where

$$n = \text{number of finite poles of } G(s)H(s)$$
$$m = \text{number of finite zeros of } G(s)H(s)$$

Here, $k = 0$ corresponds to the asymptotes making the smallest angle with the real axis. Although k assumes an infinite number of values, as k is increased, the angle repeats itself, and the number of distinct asymptotes is $n - m$. Note that if the number of asymptotes is odd, then one of the asymptotes is the negative real axis.

All the asymptotes intersect on the real axis. The point at which they do so is obtained as follows: If both the numerator and denominator of the open-loop transfer function are expanded, the result is

$$G(s)H(s) = \frac{K[s^m + (z_1 + z_2 + \cdots + z_m)s^{m-1} + \cdots + z_1 z_2 \cdots z_m]}{s^n + (p_1 + p_2 + \cdots + p_n)s^{n-1} + \cdots + p_1 p_2 \cdots p_n}$$

If a test point is located very far from the origin, then dividing the denominator by the numerator yields

$$G(s)H(s) = \frac{K}{s^{n-m} + [(p_1 + p_2 + \cdots + p_n) - (z_1 + z_2 + \cdots + z_m)]s^{n-m-1} + \cdots}$$

Since the characteristic equation is

$$G(s)H(s) = -1$$

it may be written

$$s^{n-m} + [(p_1 + p_2 + \cdots + p_n) - (z_1 + z_2 + \cdots + z_m)] s^{n-m-1} + \cdots = -K \tag{10–45}$$

For a large value of s, Equation (10–45) may be approximated by

$$\left[s + \frac{(p_1 + p_2 + \cdots + p_n) - (z_1 + z_2 + \cdots + z_m)}{n - m} \right]^{n-m} = 0$$

If the abscissa of the intersection of the asymptotes and the real axis is denoted by $s = \sigma_a$, then

$$\sigma_a = -\frac{(p_1 + p_2 + \cdots + p_n) - (z_1 + z_2 + \cdots + z_m)}{n - m} \tag{10–46}$$

Because all the complex poles and zeros occur in conjugate pairs, σ_a is always a real quantity. Once the intersection of the asymptotes and the real axis is found, the asymptotes can be readily drawn in the complex plane.

It is important to note that the asymptotes show the behavior of the root loci for $|s| \gg 1$. A root locus branch may lie on one side of the corresponding asymptote or may cross the corresponding asymptote from one side to the other side.

4. *Find the breakaway and break-in points.* [A break-in point is a point where a root locus in the complex plane enters the real axis (or other root locus) as the gain K is increased. (A break-in point, like a breakaway point, corresponds to a point in the s-plane where multiple roots of the characteristic equation occur.)] Because of the conjugate symmetry of the root loci, the breakaway points and break-in points either lie on the real axis or occur in complex-conjugate pairs.

   If a root locus lies between two adjacent open-loop poles on the real axis, then there exists at least one breakaway point between the two poles. Similarly, if the root locus lies between two adjacent zeros (one zero may be located at $-\infty$) on the real axis, then there always exists at least one break-in point between the two zeros. If the root locus lies between an open-loop pole and a zero (finite or infinite) on the real axis, then there may exist no breakaway or break-in points, or there may exist both breakaway and break-in points.

   Suppose that the characteristic equation is given by

   $$B(s) + KA(s) = 0$$

   The breakaway points and break-in points correspond to multiple roots of the characteristic equation. As shown earlier, the breakaway and break-in points can be determined from the roots of

   $$\frac{dK}{ds} = -\frac{B'(s)A(s) - B(s)A'(s)}{A^2(s)} = 0 \qquad (10\text{–}47)$$

   where the prime indicates differentiation with respect to s. It is important to note that the breakaway points and break-in points must be roots of Equation (10–47), but not all roots of Equation (10–47) are breakaway or break-in points. If a real root of Equation (10–47) lies on the root-locus portion of the real axis, then it is an actual breakaway or break-in point. If a real root of Equation (10–47) is not on the root-locus portion of the real axis, then this root corresponds to neither a breakaway point nor a break-in point. If two roots $s = s_1$ and $s = -s_1$ of Equation (10–47) are a complex-conjugate pair, and if it is not certain whether they are on root loci, then it is necessary to check the corresponding K value. If the value of K corresponding to a root $s = s_1$ of $dK/ds = 0$ is positive, point $s = s_1$ is an actual breakaway or break-in point. (Since K is assumed to be nonnegative, if the value of K thus obtained is negative, then point $s = s_1$ is neither a breakaway nor a break-in point.)

5. *Determine the angle of departure (angle of arrival) of the root locus from a complex pole (at a complex zero).* To sketch the root loci with reasonable accuracy, we must find the directions of the root loci near the complex poles and zeros. If a test point is chosen and moved in the very vicinity of a complex

pole (or a complex zero), the sum of the angular contributions from all other poles and zeros can be considered to remain the same. Therefore, the angle of departure (or angle of arrival) of the root locus from a complex pole (or at a complex zero) can be found by subtracting from 180° the sum of all the angles of vectors from all other poles and zeros to the complex pole (or complex zero) in question, with appropriate signs included. That is,

Angle of departure from a complex pole = 180°
− (sum of the angles of vectors to a complex pole in question from other poles)
+ (sum of the angles of vectors to a complex pole in question from zeros)
Angle of arrival at a complex zero = 180°
− (sum of the angles of vectors to a complex zero in question from other zeros)
+ (sum of the angles of vectors to a complex zero in question from poles)

The angle of departure is shown in Figure 10–45.

6. *Find the points where the root loci may cross the imaginary axis.* The points where the root loci intersect the $j\omega$-axis can be found by letting $s = j\omega$ in the characteristic equation, equating both the real part and the imaginary part to zero, and solving for ω and K. The values of ω thus found give the frequencies at which root loci cross the imaginary axis. The K value corresponding to each crossing frequency gives the gain at the crossing point.

7. *Taking a series of test points in the broad neighborhood of the origin of the s-plane, sketch the root loci.* Determine the root loci in the broad neighborhood of the $j\omega$-axis and the origin. The most important part of the root loci is on neither the real axis nor the asymptotes, but is in the broad neighborhood of the $j\omega$-axis and the origin. The shape of the root loci in this important region in the s-plane must be obtained with sufficient accuracy.

8. *Determine the closed-loop poles.* A particular point on each root-locus branch is a closed-loop pole if the value of K at that point satisfies the magnitude condition. Conversely, the magnitude condition enables us to determine the value

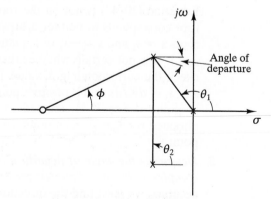

Figure 10–45 Construction of the root locus.
[Angle of departure = $180° − (\theta_1 + \theta_2) + \phi$.]

of the gain K at any specific root location on the locus. (If necessary, the root loci may be graduated in terms of K. The root loci are continuous with K.)

The value of K corresponding to any point s on a root locus can be obtained using the magnitude condition, or

$$K = \frac{\text{product of lengths between point } s \text{ to poles}}{\text{product of lengths between point } s \text{ to zeros}}$$

This value can be evaluated either graphically or analytically.

If the gain K of the open-loop transfer function is given in the problem, then, applying the magnitude condition, we can find the correct locations of the closed-loop poles for a given K on each branch of the root loci by a trial-and-error approach or with the use of MATLAB (see Section 10–9).

Constructing root loci when a variable parameter does not appear as a multiplicative factor.   In some cases, the variable parameter K may not appear as a multiplicative factor of $G(s)H(s)$. In such cases, it may be possible to rewrite the characteristic equation so that K does appear as a multiplicative factor of $G(s)H(s)$. Example 10–8 illustrates how to proceed.

Example 10–8

Consider the system shown in Figure 10–46. Draw a root-locus diagram, and then determine the value of k such that the damping ratio of the dominant closed-loop poles is 0.4.

Here, the system involves velocity feedback. The open-loop transfer function is

$$\text{open loop transfer function} = \frac{20}{s(s+1)(s+4) + 20ks}$$

Notice that the adjustable variable k does not appear as a multiplicative factor. The characteristic equation for the system is

$$s^3 + 5s^2 + 4s + 20 + 20ks = 0 \qquad (10\text{–}48)$$

If we define

$$20k = K$$

then Equation (10–48) becomes

$$s^3 + 5s^2 + 4s + Ks + 20 = 0 \qquad (10\text{–}49)$$

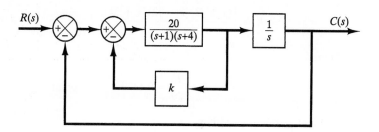

Figure 10–46   Control system.

Dividing both sides of Equation (10–49) by the sum of the terms that do not contain K, we get

$$1 + \frac{Ks}{s^3 + 5s^2 + 4s + 20} = 0$$

or

$$1 + \frac{Ks}{(s + j2)(s - j2)(s + 5)} = 0 \tag{10–50}$$

Equation (10–50) is now of the form of Equation (10–37).

We shall now sketch the root loci of the system given by Equation (10–50). Notice that the open-loop poles are located at $s = j2$, $s = -j2$, and $s = -5$, and the open-loop zero is located at $s = 0$. The root locus exists on the real axis between 0 and -5. If the test point s is located far from the origin, then the angles $\underline{/s}$, $\underline{/s + j2}$, $\underline{/s - j2}$, and $\underline{/s + 5}$ may be considered the same. Since one open-loop zero cancels the effects of one open-loop pole and vice versa, the root loci for very large values of s must be asymptotic to straight lines whose angles are given by

$$\text{angles of asymptotes} = \frac{\pm 180°(2k + 1)}{3 - 1} = \pm 90°$$

The intersection of the asymptotes with the real axis can be found from

$$\lim_{s \to \infty} \frac{Ks}{s^3 + 5s^2 + 4s + 20} = \lim_{s \to \infty} \frac{K}{s^2 + 5s + \cdots} = \lim_{s \to \infty} \frac{K}{(s + 2.5)^2}$$

as

$$\sigma_a = -2.5$$

The angle of departure (angle θ) from the pole at $s = j2$ is obtained as follows:

$$\theta = 180° - 90° - 21.8° + 90° = 158.2°$$

Thus, the angle of departure from the pole $s = j2$ is 158.2°. Figure 10–47 shows a root-locus plot of the system.

Note that the closed-loop poles with $\zeta = 0.4$ must lie on straight lines passing through the origin and making the angles $\pm 66.42°$ with the negative real axis. In the present case, there are two intersections of the root-locus branch in the upper-half s-plane with the straight line of angle 66.42°. Thus, two values of K will give the damping ratio ζ of the closed-loop poles equal to 0.4. At point P, the value of K is

$$K = \left| \frac{(s + j2)(s - j2)(s + 5)}{s} \right|_{s = -1.0490 + j2.4065} = 8.9801$$

Hence,

$$k = \frac{K}{20} = 0.4490 \qquad \text{at point } P$$

At point Q, the value of K is

$$K = \left| \frac{(s + j2)(s - j2)(s + 5)}{s} \right|_{s = -2.1589 + j4.9652} = 28.260$$

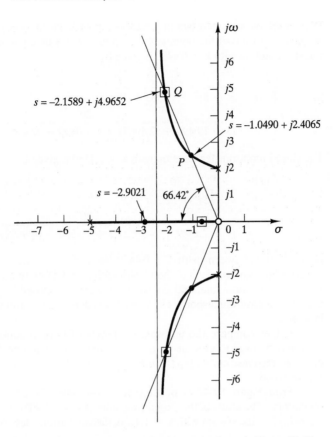

Figure 10–47   Root-locus plot for the system shown in Figure 10–46.

Thus,

$$k = \frac{K}{20} = 1.4130 \qquad \text{at point } Q$$

Consequently, we have two solutions for this problem. For $k = 0.4490$, the three closed-loop poles are located at

$$s = -1.0490 + j2.4065, \qquad s = -1.0490 - j2.4065, \qquad s = -2.9021$$

For $k = 1.4130$, the three closed-loop poles are located at

$$s = -2.1589 + j4.9652, \qquad s = -2.1589 - j4.9652, \qquad s = -0.6823$$

It is important to point out that the zero at the origin is the open-loop zero, not the closed-loop zero. This is evident because the original system shown in Figure 10–46 does not have a closed-loop zero, since

$$\frac{C(s)}{R(s)} = \frac{20}{s(s + 1)(s + 4) + 20(1 + ks)}$$

The open-loop zero at $s = 0$ was introduced in the process of modifying the characteristic equation such that the adjustable variable $K = 20k$ was to appear as a multiplicative factor.

We have obtained two different values of k which satisfy the requirement that the damping ratio of the dominant closed-loop poles be equal to 0.4. The closed-loop transfer function with $k = 0.4490$ is given by

$$\frac{C(s)}{R(s)} = \frac{20}{s^3 + 5s^2 + 12.98s + 20}$$

$$= \frac{20}{(s + 1.0490 + j2.4065)(s + 1.0490 - j2.4065)(s + 2.9021)}$$

The closed-loop transfer function with $k = 1.4130$ is given by

$$\frac{C(s)}{R(s)} = \frac{20}{s^3 + 5s^2 + 32.26s + 20}$$

$$= \frac{20}{(s + 2.1589 + j4.9652)(s + 2.1589 - j4.9652)(s + 0.6823)}$$

Notice that the system with $k = 0.4490$ has a pair of dominant complex-conjugate closed-loop poles, whereas in the system with $k = 1.4130$, the real closed-loop pole at $s = -0.6823$ is dominant and the complex-conjugate closed-loop poles are not dominant. In this case, the response characteristic is determined primarily by the real closed-loop pole.

Let us compare the unit-step responses of both systems. MATLAB Program 10–4 may be used to plot the unit-step response curves in one diagram. The resulting unit-step response curves [$c_1(t)$ for $k = 0.4490$ and $c_2(t)$ for $k = 1.4130$] are shown in Figure 10–48.

From Figure 10–48 we notice that the response of the system with $k = 0.4490$ is oscillatory. (The effect of the closed-loop pole at $s = -2.9021$ on the unit-step response is small.) For the system with $k = 1.4130$, the oscillations due to the closed-loop poles

---

MATLAB Program 10–4

```
>> % Enter numerators and denominators of systems
>> % with k = 0.4490 and k = 1.4130, respectively.
>>
>> num1 = [0   0   0   20];
>> den1 = [1   5   12.98   20];
>> num2 = [0   0   0   20];
>> den2 = [1   5   32.26   20];
>> t = 0:0.1:10;
>> [c1,x1,t] = step(num1,den1,t);
>> [c2,x2,t] = step(num2,den2,t);
>> plot(t,c1,t,c2)
>> text(3.1,1.1,'k = 0.4490')
>> text(4.8,0.86,'k = 1.4130')
>> grid
>> title('Unit-Step Responses of Two Systems')
>> xlabel('t (sec)')
>> ylabel('Outputs c_1 and c_2')
```

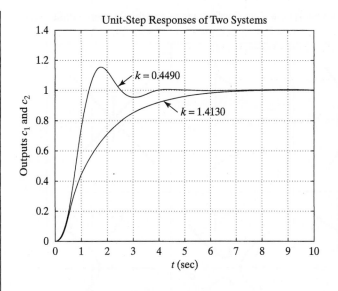

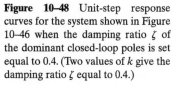

Figure 10–48 Unit-step response curves for the system shown in Figure 10–46 when the damping ratio ζ of the dominant closed-loop poles is set equal to 0.4. (Two values of k give the damping ratio ζ equal to 0.4.)

at $s = -2.1589 \pm j4.9652$ damp out much faster than the purely exponential response due to the closed-loop pole at $s = -0.6823$.

The system with $k = 0.4490$ (which exhibits a faster response with relatively small overshoot) has a much better response characteristic than the system with $k = 1.4130$ (which exhibits a slow overdamped response). Therefore, we should choose $k = 0.4490$ for the present system.

Typical pole–zero configurations and corresponding root loci.   In concluding this section, we present several open-loop pole–zero configurations and their corresponding root loci in Table 10–3. The pattern of the root loci depends only on the relative separation of the open-loop poles and zeros. If the number of open-loop poles exceeds the number of finite zeros by three or more, there is a value of the gain K beyond which root loci enter the right-half s-plane; thus, the system can become unstable. A stable system must have all its closed-loop poles in the left-half s-plane.

Note that once we have some experience with the method, we can easily evaluate the changes in the root loci due to the changes in the number and location of the open-loop poles and zeros by visualizing the root-locus plots resulting from various pole–zero configurations.

Summary.   From the preceding discussions, it should be clear that it is possible to sketch a reasonably accurate root-locus diagram for a given system by following simple rules. (The reader is urged to study various root-locus diagrams shown in the solved problems at the end of the chapter.) At preliminary design stages, we may not need the precise locations of the closed-loop poles; often, their approximate locations are all that is needed to make an estimate of a system's performance. In that case, it is important that the designer have the capability of quickly sketching the root loci for a given system.

TABLE 10–3   Open-Loop Pole–Zero Configurations and the Corresponding Root Loci

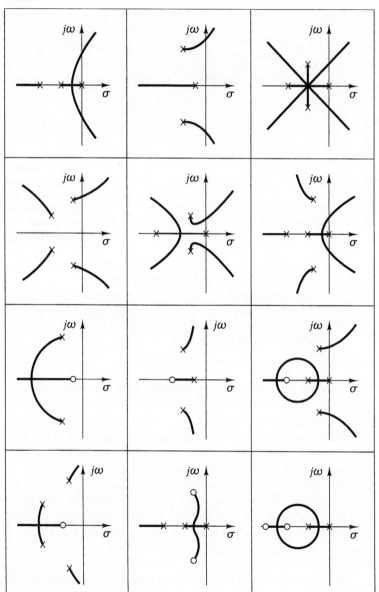

10–9 ROOT-LOCUS PLOTS WITH MATLAB

In this section, we present the MATLAB approach to generating root-locus plots.

   Plotting root loci with MATLAB.   In plotting root loci with MATLAB, we deal with the system equation given in the form of Equation (10–37), which may be written as

$$1 + K\frac{\text{num}}{\text{den}} = 0$$

where num is the numerator polynomial and den is the denominator polynomial. Note that both vectors num and den must be written in descending powers of s.

A MATLAB command commonly used for plotting root loci is

$$\text{rlocus(num,den)}$$

With this command, the root-locus plot is drawn on the screen. The gain vector K is automatically determined.

Note that command

$$\text{rlocus(num,den,K)}$$

utilizes the user-supplied gain vector K. (The vector K contains all the gain values for which the closed-loop poles are to be computed.)

If the preceding two commands are invoked with left-hand arguments, that is,

$$[r,K] = \text{rlocus(num,den)}$$
$$[r,K] = \text{rlocus(num,den,K)}$$

the screen will show the matrix r and gain vector K. (r has length K rows and length den $-$ 1 columns containing the locations of complex roots. Each row of the matrix corresponds to a gain from vector K.) The plot command

$$\text{plot(r,'-')}$$

plots the root loci.

If it is desired to plot the root loci with marks 'o' and 'x', it is necessary to use the following command:

$$r = \text{rlocus(num,den)}$$
$$\text{plot(r,'o')} \quad \text{or} \quad \text{plot(r,'x')}$$

Plotting root loci with marks 'o' or 'x' is instructive, since each calculated closed-loop pole is shown graphically; in some portion of the root loci those marks are densely situated, and in another portion of the root loci they are sparsely situated. Through an internal adaptive step-size routine, MATLAB supplies its own set of gain values that are used to calculate a root-locus plot. Also, MATLAB uses the automatic axis-scaling feature of the *plot* command.

Finally, note that, since the gain vector is determined automatically, root-locus plots of

$$G(s)H(s) = \frac{K(s+1)}{s(s+2)(s+3)}$$

$$G(s)H(s) = \frac{10K(s+1)}{s(s+2)(s+3)}$$

$$G(s)H(s) = \frac{200K(s+1)}{s(s+2)(s+3)}$$

are all the same. The num and den set is the same for all three systems. The num and den are

$$\text{num} = [0 \quad 0 \quad 1 \quad 1]$$
$$\text{den} = [1 \quad 5 \quad 6 \quad 0]$$

Example 10–9

Consider a system whose open-loop transfer function $G(s)H(s)$ is

$$G(s)H(s) = \frac{K}{s(s + 0.5)(s^2 + 0.6s + 10)}$$

$$= \frac{K}{s^4 + 1.1s^3 + 10.3s^2 + 5s}$$

There are no open-loop zeros. Open-loop poles are located at $s = -0.3 + j3.1480$, $s = -0.3 - j3.1480$, $s = -0.5$, and $s = 0$.

To set the plot region on the screen to be square, enter the command axis ('equal'). With this command, a line with unity slope is at a true 45° angle and is not skewed by the irregular shape of the screen. Entering MATLAB Program 10–5 into the computer, we obtain the root-locus plot shown in Figure 10–49.

MATLAB Program 10–5

```
>> num = [1];
>> den = [1    1.1    10.3    5    0];
>> [r, K] = rlocus(num,den);
>> plot(r,'-')
>> axis('equal'); v = [-4    4    -4    4]; axis(v)
>> grid
>> title('Root-Locus Plot of G(s) = K/[s(s+0.5) (s^2+0.6s+10)]')
>> xlabel ('Real Axis'); ylabel ('Imaginary Axis')
```

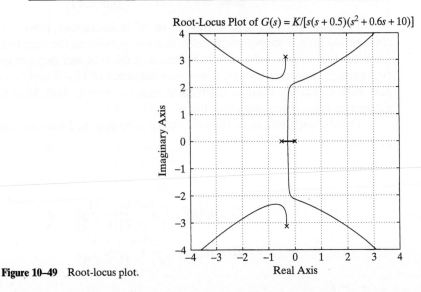

Figure 10–49   Root-locus plot.

Example 10–10

Consider the system shown in Figure 10–50. Plot the root loci with a square aspect ratio so that a line with unity slope is a true 45° line.

MATLAB Program 10–6 produces a root-locus plot in a square region. The resulting plot is shown in Figure 10–51. Note that this system is stable only for a limited range of gain K. (For the exact range of K, see **Problem A–10–10**.)

MATLAB Program 10–6

```
>> num = [1   1];
>> den = [1   3   12   −16   0];
>> K1 = [0:0.01:5]; K2 = [5:0.02:100]; K3 = [100:0.5:500];
>> K = [K1   K2   K3];
>> [r, K] = rlocus(num,den,K);
>> plot(r,'-')
>> axis('equal'); v = [−6   6   −6   6]; axis (v)
>> grid
>> title ('Root-Locus Plot of G(s) = K(s+1)/[s(s−1) (s^2+4s+16)]')
>> xlabel('Real Axis'); ylabel('Imaginary Axis')
```

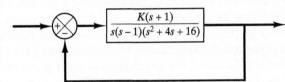

$$\dfrac{K(s+1)}{s(s-1)(s^2+4s+16)}$$

Figure 10–50   Control system.

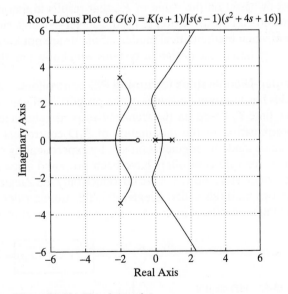

Root-Locus Plot of $G(s) = K(s+1)/[s(s-1)(s^2+4s+16)]$

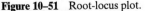

Figure 10–51   Root-locus plot.

10–10 TUNING RULES FOR PID CONTROLLERS

It is interesting to note that more than half of the industrial controllers in use today utilize PID or modified PID control schemes. Analog PID controllers are mostly hydraulic, pneumatic, electric, and electronic types or their combinations. Currently, many of these are transformed into digital forms through the use of microprocessors.

Because most PID controllers are adjusted on-site, many different types of tuning rules have been proposed in the literature. With these tuning rules, PID controllers can be delicately and finely tuned. Also, automatic tuning methods have been developed, and some PID controllers may possess on-line automatic tuning capabilities. Many practical methods for bumpless switching (from manual operation to automatic operation) and gain scheduling are commercially available.

The usefulness of PID controls lies in their general applicability to most control systems. In the field of process control systems, it is a well-known fact that both basic and modified PID control schemes have proved their usefulness in providing satisfactory control, although they may not provide optimal control in many given situations.

PID control of plants.   Figure 10–52 shows the PID control of a plant. If a mathematical model of the plant can be derived, then it is possible to apply various design techniques for determining the parameters of the controller that will meet the transient and steady-state specifications of the closed-loop system. However, if the plant is so complicated that its mathematical model cannot be easily obtained, then an analytical approach to the design of a PID controller is not possible. In that case, we must resort to experimental approaches to the tuning of PID controllers.

The process of selecting the controller parameters to meet given performance specifications is known as *controller tuning*. Ziegler and Nichols suggested rules for tuning PID controllers (to set values K_p, T_i, and T_d) based on experimental step responses or based on the value of K_p that results in marginal stability when only the proportional control action is used. Ziegler–Nichols rules, presented next, are convenient when mathematical models of plants are not known. (These rules can, of course, be applied to the design of systems *with* known mathematical models.)

Ziegler–Nichols rules for tuning PID controllers.   Ziegler and Nichols proposed rules for determining values of the proportional gain K_p, integral time T_i, and derivative time T_d based on the transient-response characteristics of a given plant. Such determination of the parameters of PID controllers or tuning of PID controllers can be made by engineers on-site by experimenting on the plant. (Numerous tuning rules for PID controllers have been proposed since Ziegler and Nichols offered their rules. Here, however, we introduce only the Ziegler–Nichols tuning rules.)

The two methods called Ziegler–Nichols tuning rules are aimed at obtaining 25% maximum overshoot in the step response. (See Figure 10–53.)

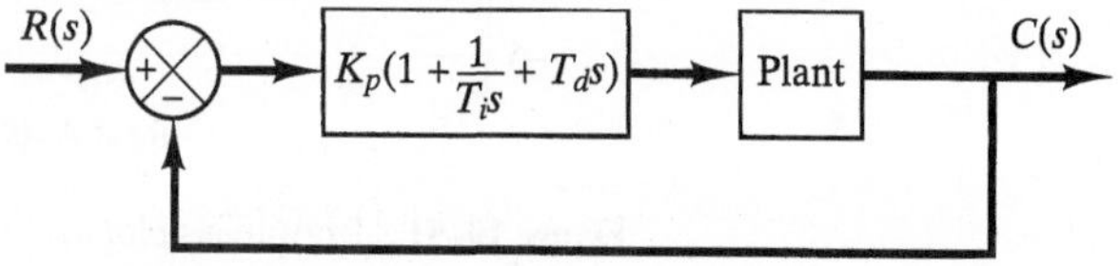

Figure 10–52   PID control of a plant.

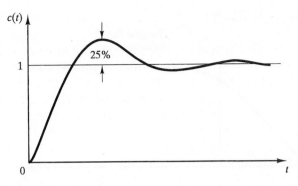

Figure 10–53 Unit-step response curve showing 25% maximum overshoot.

First method. In the first Ziegler–Nichols method, we obtain the response of the plant to a unit-step input experimentally, as shown in Figure 10–54. If the plant involves neither integrators nor dominant complex-conjugate poles, then such a unit-step response curve may look like an S-shaped curve, as shown in Figure 10–55. (If the response does not exhibit an S-shaped curve, this method does not apply.) Step-response curves of this nature may be generated experimentally or from a dynamic simulation of the plant.

The S-shaped curve may be characterized by two constants—the delay time L and a time constant T—determined by drawing a line tangent to the S-shaped curve at the inflection point. These constants are determined by the intersections of the tangent line with the time axis and the line $c(t) = K$, as shown in Figure 10–55. The transfer function $C(s)/U(s)$ may then be approximated by a first-order system with a transport lag as follows:

$$\frac{C(s)}{U(s)} = \frac{Ke^{-Ls}}{Ts + 1}$$

Ziegler and Nichols suggested setting the values of K_p, T_i, and T_d according to the formula shown in Table 10–4.

Notice that the PID controller tuned by the first Ziegler–Nichols method gives

$$G_c(s) = K_p\left(1 + \frac{1}{T_i s} + T_d s\right)$$

$$= 1.2\frac{T}{L}\left(1 + \frac{1}{2Ls} + 0.5Ls\right)$$

$$= 0.6T\frac{\left(s + \frac{1}{L}\right)^2}{s}$$

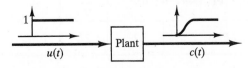

Figure 10–54 Unit-step response of a plant.

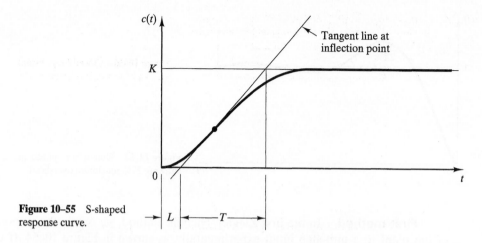

Figure 10–55 S-shaped response curve.

Thus, the PID controller has a pole at the origin and double zeros at $s = -1/L$.

Second method. In the second Ziegler–Nichols method, we first set $T_i = \infty$ and $T_d = 0$. Using the proportional control action only (see Figure 10–56), we increase K_p from 0 to a critical value K_{cr} at which the output first exhibits sustained oscillations. (If the output does not exhibit sustained oscillations for whatever value K_p may take, then this method does not apply.) Thus the critical gain K_{cr} and the corresponding period P_{cr} are determined experimentally. (See Figure 10–57.) Ziegler and Nichols suggested that we set the values of the parameters K_p, T_i, and T_d according to the formula shown in Table 10–5.

Notice that the PID controller tuned by the second Ziegler–Nichols method gives

$$G_c(s) = K_p\left(1 + \frac{1}{T_i s} + T_d s\right)$$

$$= 0.6K_{cr}\left(1 + \frac{1}{0.5P_{cr}s} + 0.125P_{cr}s\right)$$

$$= 0.075K_{cr}P_{cr}\frac{\left(s + \dfrac{4}{P_{cr}}\right)^2}{s}$$

TABLE 10–4 Ziegler–Nichols Tuning Rule Based on Step Response of Plant (First Method)

Type of Controller	K_p	T_i	T_d
P	$\dfrac{T}{L}$	∞	0
PI	$0.9\dfrac{T}{L}$	$\dfrac{L}{0.3}$	0
PID	$1.2\dfrac{T}{L}$	$2L$	$0.5L$

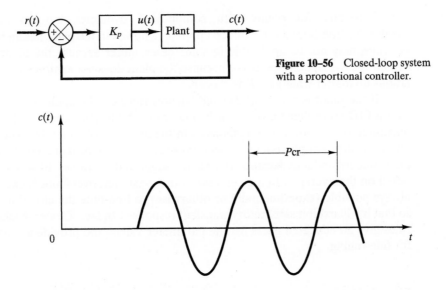

Figure 10–56   Closed-loop system with a proportional controller.

Figure 10–57   Sustained oscillation with period P_{cr}.

Thus, the PID controller has a pole at the origin and double zeros at $s = -4/P_{cr}$.

Comments.   Ziegler–Nichols tuning rules (and other tuning rules presented in the literature) have been widely used to tune PID controllers in process control systems for which the plant dynamics are not precisely known. Over many years, such tuning rules have proved to be highly useful. Ziegler–Nichols tuning rules can, of course, be applied to plants whose dynamics are known. (If plant dynamics are known, many analytical and graphical approaches to the design of PID controllers are available in addition to Ziegler–Nichols tuning rules.)

If the transfer function of the plant is known, a unit-step response may be calculated or the critical gain K_{cr} and critical period P_{cr} may be calculated. Then, with those calculated values, it is possible to determine the parameters K_p, T_i, and T_d from Table 10–4 or Table 10–5. However, the real usefulness of Ziegler–Nichols tuning rules (and other tuning rules) becomes apparent when the plant dynamics are *not* known, so that no analytical or graphical approaches to the design of controllers are available.

TABLE 10–5   Ziegler–Nichols Tuning Rule Based on Critical Gain K_{cr} and Critical Period P_{cr} (Second Method)

Type of Controller	K_p	T_i	T_d
P	$0.5K_{cr}$	∞	0
PI	$0.45K_{cr}$	$\dfrac{1}{1.2}P_{cr}$	0
PID	$0.6K_{cr}$	$0.5P_{cr}$	$0.125P_{cr}$

Generally, for plants with complicated dynamics, but no integrators, Ziegler–Nichols tuning rules can be applied. If, however, the plant has an integrator, the rules may not be applicable in some cases, either because the plant does not exhibit the S-shaped response or because the plant does not exhibit sustained oscillations no matter what gain K is chosen.

If the plant is such that Ziegler–Nichols rules can be applied, then the plant with a PID controller tuned by such rules will exhibit approximately 10% to 60% maximum overshoot in step response. On the average (obtained by experimenting on many different plants), the maximum overshoot is approximately 25%. (This is quite understandable, because the values suggested in Tables 10–4 and 10–5 are based on the average.) In a given case, if the maximum overshoot is excessive, it is always possible (experimentally or otherwise) to fine-tune the closed-loop system so that it will exhibit satisfactory transient responses. In fact, Ziegler–Nichols tuning rules give an educated guess for the parameter values and provide a starting point for fine-tuning.

Example 10–11

Consider the control system shown in Figure 10–58, in which a PID controller is used to control the system. The PID controller has the transfer function

$$G_c(s) = K_p\left(1 + \frac{1}{T_i s} + T_d s\right)$$

Although many analytical methods are available for the design of a PID controller for this system, let us apply a Ziegler–Nichols tuning rule for the determination of the values of parameters K_p, T_i, and T_d. Then we will obtain a unit-step response curve and check whether the designed system exhibits approximately 25% maximum overshoot. If the maximum overshoot is excessive (40% or more), fine-tune the system and reduce the amount of the maximum overshoot to approximately 25%.

Since the plant has an integrator, we use the second Ziegler–Nichols method. Setting $T_i = \infty$ and $T_d = 0$, we obtain the closed-loop transfer function

$$\frac{C(s)}{R(s)} = \frac{K_p}{s(s + 1)(s + 5) + K_p}$$

The value of K_p that makes the system marginally stable so that sustained oscillations occur can be obtained by the use of Routh's stability criterion. Since the characteristic equation for the closed-loop system is

$$s^3 + 6s^2 + 5s + K_p = 0$$

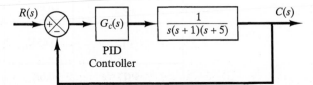

Figure 10–58   PID-controlled system.

the Routh array becomes

$$
\begin{array}{c|cc}
s^3 & 1 & 5 \\
s^2 & 6 & K_p \\
s^1 & \dfrac{30 - K_p}{6} & \\
s^0 & K_p &
\end{array}
$$

Examining the coefficients of the first column of the Routh array, we find that sustained oscillation will occur if $K_p = 30$. Thus, the critical gain K_{cr} is

$$K_{cr} = 30$$

With the gain K_p set equal to $K_{cr}(= 30)$, the characteristic equation becomes

$$s^3 + 6s^2 + 5s + 30 = 0$$

To find the frequency of the sustained oscillations, we substitute $s = j\omega$ into this characteristic equation and obtain

$$(j\omega)^3 + 6(j\omega)^2 + 5(j\omega) + 30 = 0$$

or

$$6(5 - \omega^2) + j\omega(5 - \omega^2) = 0$$

from which we find the frequency of the sustained oscillation to be $\omega^2 = 5$ or $\omega = \sqrt{5}$. Hence, the period of sustained oscillation is

$$P_{cr} = \frac{2\pi}{\omega} = \frac{2\pi}{\sqrt{5}} = 2.8099$$

From Table 10–5, we determine

$$K_p = 0.6K_{cr} = 18$$
$$T_i = 0.5P_{cr} = 1.405$$
$$T_d = 0.125P_{cr} = 0.35124$$

The transfer function of the PID controller is thus

$$G_c(s) = K_p\left(1 + \frac{1}{T_i s} + T_d s\right)$$
$$= 18\left(1 + \frac{1}{1.405s} + 0.35124s\right)$$
$$= \frac{6.3223(s + 1.4235)^2}{s}$$

The PID controller has a pole at the origin and double zero at $s = -1.4235$. A block diagram of the control system with the designed PID controller is shown in Figure 10–59.

Next, let us examine the unit-step response of the system. The closed-loop transfer function $C(s)/R(s)$ is given by

$$\frac{C(s)}{R(s)} = \frac{6.3223s^2 + 18s + 12.811}{s^4 + 6s^3 + 11.3223s^2 + 18s + 12.811}$$

The unit-step response of this system can be obtained easily with MATLAB. (See MATLAB Program 10–7.) The resulting unit-step response curve is shown in Figure 10–60. The maximum overshoot in the unit-step response is approximately 62%, is

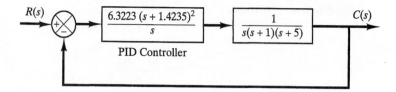

Figure 10–59 Block diagram of the system with PID controller designed by the use of the second Ziegler–Nichols tuning method.

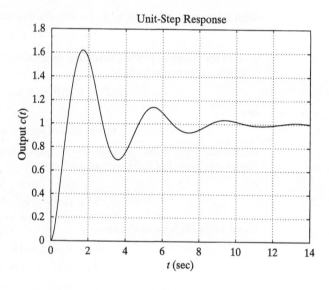

Figure 10–60 Unit-step response curve of PID-controlled system designed by the use of the second Ziegler–Nichols tuning method.

excessive. It can be reduced by fine-tuning the controller parameters, possibly on the computer. We find that, by keeping $K_p = 18$ and by moving the double zero of the PID controller to $s = -0.65$, that is, using the PID controller

$$G_c(s) = 18\left(1 + \frac{1}{3.077s} + 0.7692s\right) = 13.846\frac{(s + 0.65)^2}{s} \qquad (10\text{–}51)$$

MATLAB Program 10–7

```
>> t = 0:0.01:14;
>> num = [0   0   6.3223   18   12.811];
>> den = [1   6   11.3223   18   12.811];
>> step(num,den,t)
>> grid
>> title('Unit-Step Response')
>> xlabel('t'); ylabel('Output c(t)')
```

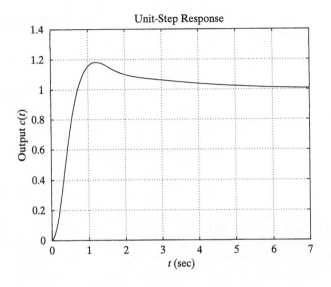

Figure 10–61   Unit-step response of the system shown in Figure 10–58 with PID controller with parameters $K_p = 18, T_i = 3.077$, and $T_d = 0.7692$, or $G_c(s) = 13.846(s + 0.65)^2/s$.

the maximum overshoot in the unit-step response can be reduced to approximately 18%. (See Figure 10–61.) If the proportional gain K_p is increased to 39.42, without changing the location of the double zero ($s = -0.65$), that is, if the PID controller

$$G_c(s) = 39.42\left(1 + \frac{1}{3.077s} + 0.7692s\right) = 30.322\,\frac{(s + 0.65)^2}{s} \qquad (10\text{–}52)$$

is used, then the speed of response is increased, but the maximum overshoot is also increased, to approximately 28%, as shown in Figure 10–62. Since the maximum overshoot in this case is fairly close to 25% and the response is faster than the system with $G_c(s)$ given by Equation (10–51), we may consider $G_c(s)$, as given by Equation (10–52), to be acceptable. Then the tuned values of K_p, T_i, and T_d become

$$K_p = 39.42, \qquad T_i = 3.077, \qquad T_d = 0.7692$$

It is interesting to observe that these values are approximately twice the values suggested by the second Ziegler–Nichols tuning method. The important thing to note here is that the Ziegler–Nichols rule has provided a starting point for fine-tuning.

It is instructive to note that, in the case where the double zero is located at $s = -1.4235$, increasing the value of K_p increases the speed of response, but as far as the percentage maximum overshoot is concerned, varying gain K_p has very little effect. The reason for this may be seen from the root-locus analysis. Figure 10–63 shows the root-locus diagram for the system designed with the second Ziegler–Nichols tuning method. Since the dominant branches of root loci are along the $\zeta = 0.3$ lines for a considerable range of K, varying the value of K (from 6 to 30) will not change the damping ratio of the dominant closed-loop poles very much. However, varying the location of the double zero

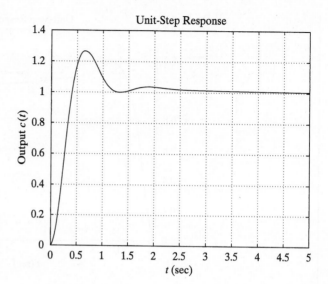

Figure 10–62 Unit-step response of the system shown in Figure 10–58 with PID controller with parameters $K_p = 39.42$, $T_i = 3.077$, and $T_d = 0.7692$, or $G_c(s) = 30.322(s + 0.65)^2/s$.

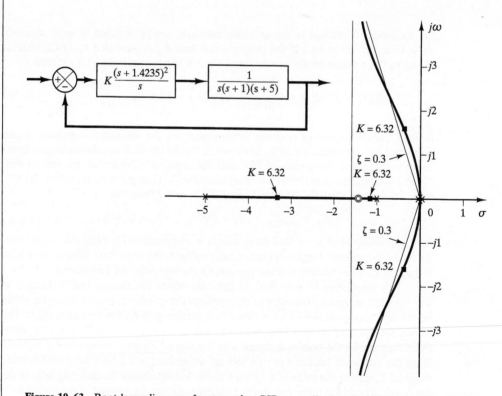

Figure 10–63 Root-locus diagram of system when PID controller has double zero at $s = -1.4235$.

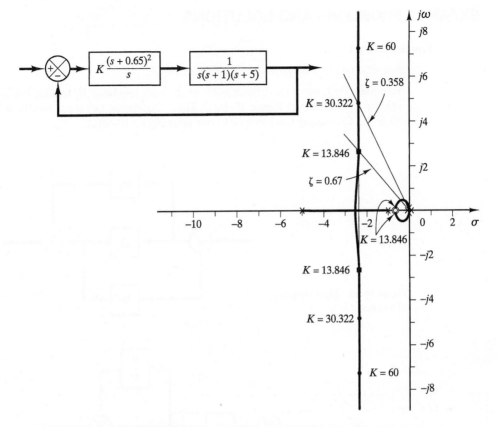

Figure 10–64 Root-locus diagram of system when PID controller has double zero at $s = -0.65$. $K = 13.846$ corresponds to $G_c(s)$ given by Equation (10–51) and $K = 30.322$ corresponds to $G_c(s)$ given by Equation (10–52).

has a considerable effect on the maximum overshoot, because the damping ratio of the dominant closed-loop poles can be changed significantly. This can also be seen from the root-locus analysis. Figure 10–64 shows the root-locus diagram for the system with PID controller with a double zero at $s = -0.65$. Notice the change of the root-locus configuration, making it possible to change the damping ratio of the dominant closed-loop poles.

In the figure, notice that, in the case where the system has gain $K = 30.322$, the closed-loop poles at $s = -2.35 \pm j4.82$ act as dominant poles. Two additional closed-loop poles are very near the double zero at $s = -0.65$, with the result that these closed-loop poles and the double zero almost cancel each other. The dominant pair of closed-loop poles indeed determines the nature of the response. By contrast, when the system has $K = 13.846$, the closed-loop poles at $s = -2.35 \pm j2.62$ are not quite dominant, because the two other closed-loop poles near the double zero at $s = -0.65$ have a considerable effect on the response. The maximum overshoot in the step response in this case (18%) is much larger than in the case of a second-order system having only dominant closed-loop poles. (In the latter case, the maximum overshoot in the step response would be approximately 6%.)

EXAMPLE PROBLEMS AND SOLUTIONS

Problem A–10–1

Simplify the block diagram shown in Figure 10–65.

Solution   First, move the branch point of the path involving H_1 outside the loop involving H_2 as shown in Figure 10–66(a). Then, eliminating two loops results in Figure 10–66(b). Combining two blocks into one gives Figure 10–66(c).

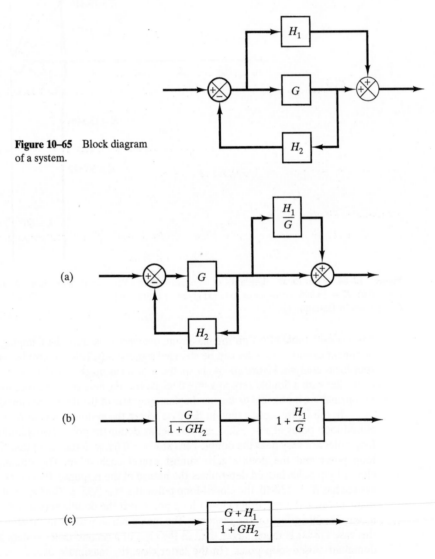

Figure 10–65   Block diagram of a system.

Figure 10–66   Block diagrams showing steps for simplifying the system of Figure 10–65.

Problem A–10–2

For the block diagram shown in Figure 10–67, derive the transfer function relating $C(s)$ and $R(s)$.

Solution The signal $X(s)$ is the sum of two signals $G_1 R(s)$ and $R(s)$:

$$X(s) = G_1 R(s) + R(s)$$

The output signal $C(s)$ is the sum of $G_2 X(s)$ and $R(s)$:

$$C(s) = G_2 X(s) + R(s) = G_2[G_1 R(s) + R(s)] + R(s)$$

We thus have

$$\frac{C(s)}{R(s)} = G_1 G_2 + G_2 + 1$$

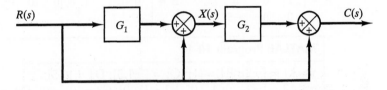

Figure 10–67 Block diagram of a system.

Problem A–10–3

Consider the system shown in Figure 10–68. A state-space representation of G_1 is

$$\dot{x} = -5x + u$$
$$y = -1.5x + 0.5u$$

or

$$A1 = [-5], \qquad B1 = [1], \qquad C1 = [-1.5], \qquad D1 = [0.5]$$

A state space representation of G_2 is

$$\begin{bmatrix} \dot{x}_1 \\ \dot{x}_2 \end{bmatrix} = \begin{bmatrix} -2 & 0 \\ 1 & 0 \end{bmatrix}\begin{bmatrix} x_1 \\ x_2 \end{bmatrix} + \begin{bmatrix} 1 \\ 0 \end{bmatrix} y$$

$$z = \begin{bmatrix} 0 & 10 \end{bmatrix}\begin{bmatrix} x_1 \\ x_2 \end{bmatrix} + [0]y$$

or

$$A2 = \begin{bmatrix} -2 & 0 \\ 1 & 0 \end{bmatrix}, \qquad B2 = \begin{bmatrix} 1 \\ 0 \end{bmatrix}, \qquad C2 = \begin{bmatrix} 0 & 10 \end{bmatrix}, \qquad D2 = [0]$$

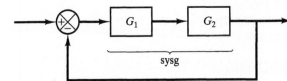

sysg

Figure 10–68 Closed-loop system.

Define

$$sys1 = ss(A1,B1,C1,D1)$$
$$sys2 = ss(A2,B2,C2,D2)$$

Obtain a state-space expression of the series connection of G_1 and G_2 with MATLAB. Use the command

$$sysg = series(sys1,sys2)$$

Then plot the unit-step response of the closed-loop system shown in Figure 10–68 by using the following commands:

$$sys = feedback(sysg,[1]); \qquad step(sys)$$

Solution   MATLAB Program 10–8 produces sysg characterized by

$$A = \begin{bmatrix} -2 & 0 & -1.5 \\ 1 & 0 & 0 \\ 0 & 0 & -5 \end{bmatrix}, \qquad B = \begin{bmatrix} 0.5 \\ 0 \\ 1 \end{bmatrix}, \qquad C = [0 \quad 10 \quad 0], \qquad D = [0]$$

MATLAB Program 10–8

```
>> A1 = [−5]; B1 = [1]; C1 = [−1.5]; D1 = [0.5];
>> A2 = [−2   0; 1   0]; B2 = [1; 0]; C2 = [0   10]; D2 = [0];
>> sys1 = ss(A1,B1,C1,D1);
>> sys2 = ss(A2,B2,C2,D2);
>> sysg = series(sys1,sys2)

a =
          x1    x2    x3
    x1    −2     0   −1.5
    x2     1     0     0
    x3     0     0    −5

b =
          u1
    x1    0.5
    x2     0
    x3     1

c =
          x1    x2    x3
    y1     0    10     0

d =
          u1
    y1     0

Continuous-time model.
>> sys = feedback(sysg,[1]);
>> step(sys)
>> grid
>> title('Unit-Step Response')
>> xlabel('t')
>> ylabel('Output z(t)')
```

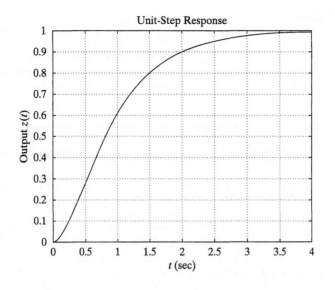

Figure 10–69 Unit-step response curve.

MATLAB Program 10–8 also produces the unit-step response of the closed-loop system, as shown in Figure 10–69.

Problem A–10–4

Consider the control system shown in Figure 10–70, in which a proportional-plus-integral controller is used to control the load element consisting of a moment of inertia and viscous friction. Show that if the system is stable this controller eliminates steady-state error in the response to a step reference input and a step disturbance input. Show also that if the controller is replaced by an integral controller, as shown in Figure 10–71, the system becomes unstable.

Solution For the control system shown in Figure 10–70, the closed-loop transfer function between $C(s)$ and $R(s)$ in the absence of the disturbance input $D(s)$ is

$$\frac{C(s)}{R(s)} = \frac{K_p + \dfrac{K_p}{T_i s}}{Js^2 + bs + K_p + \dfrac{K_p}{T_i s}}$$

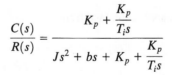

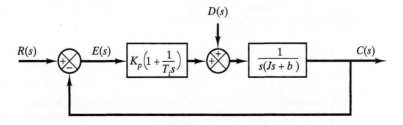

Figure 10–70 Proportional-plus-integral control of a load element consisting of moment of inertia and viscous friction.

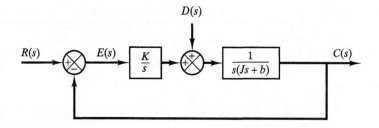

Figure 10–71 Integral control of a load element consisting of moment of inertia and viscous friction.

The error signal can be obtained from

$$\frac{E(s)}{R(s)} = \frac{R(s) - C(s)}{R(s)} = \frac{Js^2 + bs}{Js^2 + bs + K_p + \dfrac{K_p}{T_i s}}$$

Hence, for the unit-step input $R(s) = 1/s$,

$$E(s) = \frac{Js^2 + bs}{Js^2 + bs + K_p + \dfrac{K_p}{T_i s}} \frac{1}{s}$$

If the system is stable, the steady-state error e_{ss} can be obtained as follows:

$$e_{ss} = \lim_{s \to 0} sE(s) = \lim_{s \to 0} \frac{Js^2 + bs}{Js^2 + bs + K_p + \dfrac{K_p}{T_i s}} = 0$$

For the system shown in Figure 10–70, the closed-loop transfer function between $C(s)$ and $D(s)$ in the absence of the reference input $R(s)$ is

$$\frac{C(s)}{D(s)} = \frac{s}{Js^3 + bs^2 + K_p s + \dfrac{K_p}{T_i}}$$

The error signal is obtained from

$$E(s) = -C(s) = -\frac{s}{Js^3 + bs^2 + K_p s + \dfrac{K_p}{T_i}} D(s)$$

If the system is stable, the steady-state error e_{ss} in the response to a step disturbance torque of magnitude $D\,[D(s) = D/s]$ is

$$e_{ss} = \lim_{s \to 0} sE(s) = \lim_{s \to 0} \frac{-s^2}{Js^3 + bs^2 + K_p s + \dfrac{K_p}{T_i}} \frac{D}{s} = 0$$

Thus, the steady-state error to the step disturbance torque is zero.

If the controller were an integral controller, as in Figure 10–71, then the system would always become unstable, because the characteristic equation

$$Js^3 + bs^2 + K = 0$$

will have at least one root with a positive real part. Such an unstable system cannot be used in practice.

Note that in the system of Figure 10–70, the proportional control action tends to stabilize the system, while the integral control action tends to eliminate or reduce steady-state error in response to various inputs.

Problem A–10–5

Determine the values of K and k such that the closed-loop system shown in Figure 10–72 has a damping ratio ζ of 0.7 and an undamped natural frequency ω_n of 4 rad/s.

Solution   The closed-loop transfer function is

$$\frac{C(s)}{R(s)} = \frac{K}{s^2 + (2 + Kk)s + K}$$

Noting that

$$\omega_n = \sqrt{K}, \qquad 2\zeta\omega_n = 2 + Kk$$

we obtain

$$K = \omega_n^2 = 4^2 = 16$$

and

$$2 + Kk = 2\zeta\omega_n = 2 \times 0.7 \times 4 = 5.6$$

Thus,

$$Kk = 3.6$$

or

$$k = \frac{3.6}{16} = 0.225$$

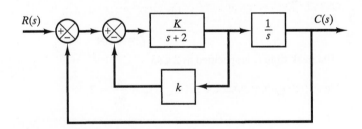

Figure 10–72   Closed-loop system.

Problem A–10–6

Determine the values of K and k of the closed-loop system shown in Figure 10–73 so that the maximum overshoot in unit-step response is 25% and the peak time is 2 s. Assume that $J = 1$ kg-m^2.

Solution   The closed-loop transfer function is

$$\frac{C(s)}{R(s)} = \frac{K}{Js^2 + Kks + K}$$

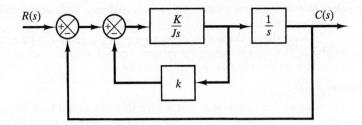

Figure 10–73   Closed-loop system.

Substituting $J = 1$ kg-m^2 into this last equation, we have

$$\frac{C(s)}{R(s)} = \frac{K}{s^2 + Kks + K}$$

Note that

$$\omega_n = \sqrt{K}, \qquad 2\zeta\omega_n = Kk$$

The maximum overshoot M_p is

$$M_p = e^{-\zeta\pi/\sqrt{1-\zeta^2}}$$

which is specified as 25%. Hence,

$$e^{-\zeta\pi/\sqrt{1-\zeta^2}} = 0.25$$

from which it follows that

$$\frac{\zeta\pi}{\sqrt{1 - \zeta^2}} = 1.386$$

or

$$\zeta = 0.404$$

The peak time t_p is specified as 2 s, so

$$t_p = \frac{\pi}{\omega_d} = 2$$

or

$$\omega_d = 1.57$$

The undamped natural frequency ω_n is then

$$\omega_n = \frac{\omega_d}{\sqrt{1 - \zeta^2}} = \frac{1.57}{\sqrt{1 - 0.404^2}} = 1.72$$

Therefore, we obtain

$$K = \omega_n^2 = 1.72^2 = 2.95 \text{ N-m}$$

$$k = \frac{2\zeta\omega_n}{K} = \frac{2 \times 0.404 \times 1.72}{2.95} = 0.471 \text{ s}$$

Problem A–10–7

When the closed-loop system shown in Figure 10–74(a) is subjected to a unit-step input, the system output responds as shown in Figure 10–74(b). Determine the values of K and T from the response curve.

Solution The maximum overshoot of 25.4% corresponds to $\zeta = 0.4$. From the response curve, we have

$$t_p = 3$$

Consequently,

$$t_p = \frac{\pi}{\omega_d} = \frac{\pi}{\omega_n\sqrt{1-\zeta^2}} = \frac{\pi}{\omega_n\sqrt{1-0.4^2}} = 3$$

It follows that

$$\omega_n = 1.143$$

From the block diagram, we have

$$\frac{C(s)}{R(s)} = \frac{K}{Ts^2 + s + K}$$

from which we obtain

$$\omega_n = \sqrt{\frac{K}{T}}, \qquad 2\zeta\omega_n = \frac{1}{T}$$

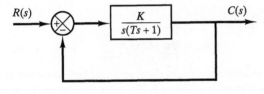

$R(s)$ $\dfrac{K}{s(Ts+1)}$ $C(s)$

(a)

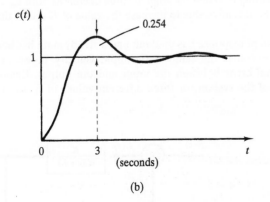

(b)

Figure 10–74 (a) Closed-loop system; (b) unit-step response curve.

Therefore,

$$T = \frac{1}{2\zeta\omega_n} = \frac{1}{2 \times 0.4 \times 1.143} = 1.094$$

$$K = \omega_n^2 T = 1.143^2 \times 1.094 = 1.429$$

Problem A–10–8

For the closed-loop system shown in Figure 10–75, discuss the effects that varying the values of K and b have on the steady-state error in a unit-ramp response. Sketch typical unit-ramp response curves for a small value, a medium value, and a large value of K.

Solution The closed-loop transfer function is

$$\frac{C(s)}{R(s)} = \frac{K}{Js^2 + bs + K}$$

Therefore,

$$\frac{E(s)}{R(s)} = \frac{R(s) - C(s)}{R(s)} = \frac{Js^2 + bs}{Js^2 + bs + K}$$

For a unit-ramp input, $R(s) = 1/s^2$. Thus,

$$E(s) = \frac{Js^2 + bs}{Js^2 + bs + K} \frac{1}{s^2}$$

The steady-state error is

$$e_{ss} = \lim_{s \to 0} sE(s) = \frac{b}{K}$$

We see that we can reduce the steady-state error e_{ss} by increasing the gain K or decreasing the viscous-friction coefficient b. However, increasing the gain or decreasing the viscous-friction coefficient causes the damping ratio to decrease, with the result that the transient response of the system becomes more oscillatory. On the one hand, doubling K decreases e_{ss} to half its original value, whereas ζ is decreased to 0.707 of its original value, since ζ is inversely proportional to the square root of K. On the other hand, decreasing b to half its original value decreases both e_{ss} and ζ to half their original values. So it is advisable to increase the value of K rather than decrease the value of b.

After the transient response has died out and a steady state has been reached, the output velocity becomes the same as the input velocity. However, there is a steady-state positional error between the input and the output. Examples of the unit-ramp response of the system for three different values of K are illustrated in Figure 10–76.

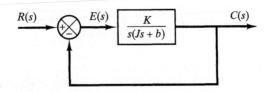

Figure 10–75   Closed-loop system.

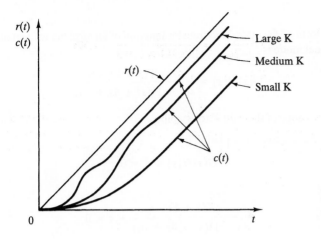

Figure 10–76 Unit-ramp response curves for the system shown in Figure 10–75.

Problem A–10–9

Consider the following characteristic equation:

$$s^4 + Ks^3 + s^2 + s + 1 = 0$$

Determine the range of K for stability.

Solution   The Routh array of coefficients is

s^4	1	1	1
s^3	K	1	0
s^2	$\dfrac{K-1}{K}$	1	0
s^1	$1 - \dfrac{K^2}{K-1}$	0	
s^0	1		

For stability, we require that

$$K > 0$$

$$\frac{K-1}{K} > 0$$

$$1 - \frac{K^2}{K-1} > 0$$

From the first and second conditions, K must be greater than unity. For $K > 1$, notice that the term $1 - [K^2/(K-1)]$ is always negative, since

$$\frac{K-1-K^2}{K-1} = \frac{-1 + K(1-K)}{K-1} < 0$$

Thus, the three conditions cannot be fulfilled simultaneously. Therefore, there is no value of K that allows stability of the system.

Problem A–10–10

A simplified form of the open-loop transfer function of an airplane with an autopilot in the longitudinal mode is

$$G(s)H(s) = \frac{K(s+1)}{s(s-1)(s^2+4s+16)}$$

Determine the range of the gain K for stability.

Solution   The characteristic equation is

$$G(s)H(s) + 1 = 0$$

or

$$\frac{K(s+1)}{s(s-1)(s^2+4s+16)} + 1 = 0$$

which can be rewritten as

$$s(s-1)(s^2+4s+16) + K(s+1) = 0$$

or

$$s^4 + 3s^3 + 12s^2 + (K-16)s + K = 0$$

The Routh array for this characteristic equation is

s^4	1	12	K
s^3	3	$K-16$	0
s^2	$\dfrac{52-K}{3}$	K	0
s^1	$\dfrac{-K^2+59K-832}{52-K}$	0	
s^0	K		

The values of K that make the s^1 term in the first column of the array equal to zero are $K = 35.68$ and $K = 23.32$. Hence, the fourth term in the first column of the array becomes

$$\frac{-(K-35.68)(K-23.32)}{52-K}$$

The condition for stability is that all terms in the first column of the array be positive. Thus, we require that

$$52 > K$$
$$35.68 > K > 23.32$$
$$K > 0$$

from which we conclude that K must be greater than 23.32, but smaller than 35.68, or

$$35.68 > K > 23.32$$

A root-locus plot for this system is shown in Figure 10–77. From the plot, it is clear that the system is stable for $35.68 > K > 23.32$.

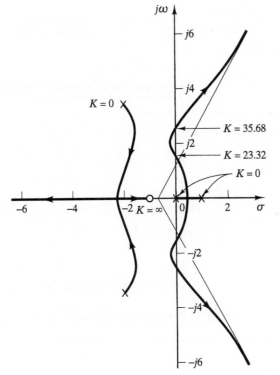

Figure 10–77 Root-locus plot.

Problem A–10–11

A control system with

$$G(s) = \frac{K}{s^2(s + 1)}, \qquad H(s) = 1$$

is unstable for all positive values of the gain K.

Plot the root loci of the system. Using the plot, show that the system can be stabilized by adding a zero on the negative real axis or by modifying $G(s)$ to $G_1(s)$, where

$$G_1(s) = \frac{K(s + a)}{s^2(s + 1)} \qquad (0 \le a < 1)$$

Solution A root-locus plot for the system with

$$G(s) = \frac{K}{s^2(s + 1)}, \qquad H(s) = 1$$

is shown in Figure 10–78(a). Since two branches lie in the right half-plane, the system is unstable for any value of $K > 0$.

The addition of a zero to the transfer function $G(s)$ bends the right half-plane branches to the left and brings all root-locus branches to the left half-plane, as shown in the root-locus plot of Figure 10–78(b). Thus, the system with

$$G_1 = \frac{K(s + a)}{s^2(s + 1)}, \qquad H(s) = 1 \qquad (0 \le a < 1)$$

is stable for all $K > 0$.

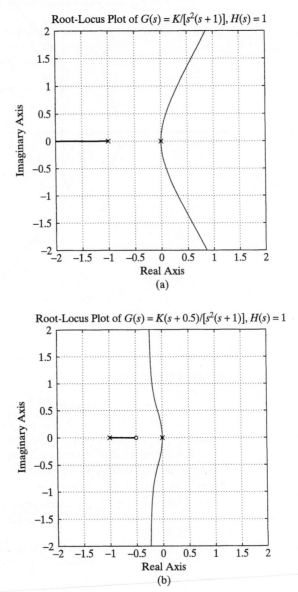

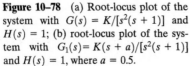

Figure 10–78 (a) Root-locus plot of the system with $G(s) = K/[s^2(s + 1)]$ and $H(s) = 1$; (b) root-locus plot of the system with $G_1(s) = K(s + a)/[s^2(s + 1)]$ and $H(s) = 1$, where $a = 0.5$.

Problem A–10–12

Consider the control system shown in Figure 10–79. Plot the root loci for the system. Then determine the value of the gain K such that the damping ratio ζ of the dominant closed-loop poles is 0.5. Using MATLAB, determine all closed-loop poles. Finally, plot the unit-step response curve with MATLAB.

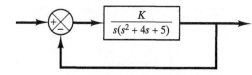

Figure 10–79   Control system.

Solution   The open-loop transfer function is

$$G(s) = \frac{K}{s(s^2 + 4s + 5)}$$

Hence, for this system,

$$num = [1]$$
$$den = [1 \quad 4 \quad 5 \quad 0]$$

MATLAB Program 10–9 produces a root-locus plot for the system, as well as the $\zeta = 0.5$ line (a line radiating from the origin and having an angle of 60° from the negative real axis if the axes are square). The root-locus plot is shown in Figure 10–80.

MATLAB Program 10–9

```
>> num = [1];
>> den = [1   4   5   0];
>> K1 = [0:0.001:2.5]; K2 = [2.5:0.01:100]; K3 = [100:0.5:1000];
>> K = [K1   K2   K3];
>> [r, K] = rlocus(num,den,K);
>> plot(r,'-')
>> axis('equal'); v = [−4   4   −4   4]; axis(v)
>> hold
Current plot held
>> x = [0   −2]; y = [0   3.464]; line(x,y)
>> grid
>> title ('Root-Locus Plot')
>> xlabel('Real Axis'); ylabel('Imaginary Axis')
>> text(−3.5,2.5,'\zeta = 0.5')
>> hold
Current plot released
```

Next, we shall determine the value of the gain K such that the dominant closed-loop poles have a damping ratio ζ of 0.5. We may write the dominant closed-loop poles as

$$s = x \pm j\sqrt{3}x$$

where x is an unknown constant to be determined. Since the characteristic equation for the system is

$$s^3 + 4s^2 + 5s + K = 0$$

substituting $s = x + j\sqrt{3}x$ into this last equation, we obtain

$$(x + j\sqrt{3}x)^3 + 4(x + j\sqrt{3}x)^2 + 5(x + j\sqrt{3}x) + K = 0$$

or

$$-8x^3 - 8x^2 + 5x + K + 2\sqrt{3}j(4x^2 + 2.5x) = 0$$

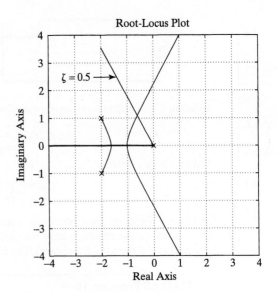

Figure 10–80   Root-locus plot.

Equating the real part and imaginary part to zero, respectively, we get

$$-8x^3 - 8x^2 + 5x + K = 0 \tag{10–53}$$

$$4x^2 + 2.5x = 0 \tag{10–54}$$

Noting that $x \neq 0$, we obtain, from Equation (10–54),

$$4x + 2.5 = 0$$

or

$$x = -0.625$$

Substituting $x = -0.625$ into Equation (10–53), we get

$$
\begin{aligned}
K &= 8x^3 + 8x^2 - 5x \\
&= 8(-0.625)^3 + 8(-0.625)^2 - 5(-0.625) \\
&= 4.296875
\end{aligned}
$$

Thus, we determine that K equals 4.296875.

To determine all closed-loop poles, we may enter MATLAB Program 10–10 into the computer.

MATLAB Program 10–10

```
p = [1   4   5   4.296875];
roots(p)
ans =

-2.7500
-0.6250 + 1.0825i
-0.6250 - 1.0825i
```

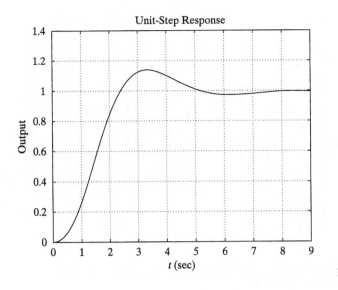

Figure 10–81   Unit-step response.

Thus, the closed-loop poles are located at $s = -0.625 \pm j1.0825$ and $s = -2.75$.

The unit-step response curve can be plotted by entering MATLAB Program 10–11 into the computer. The resulting unit-step response curve is shown in Figure 10–81.

MATLAB Program 10–11

```
>> num = [4.2969];
>> den = [1   4   5   4.2969];
>> step(num,den)
>> grid
>> title('Unit-Step Response')
>> xlabel('t')
>> ylabel('Output')
```

Problem A–10–13

Consider the system shown in Figure 10–82. Design a compensator such that the dominant closed-loop poles are located at $s = -2 \pm j2\sqrt{3}$. Plot the unit-step response curve of the designed system with MATLAB.

Solution From Figure 10–83(a), if the closed-loop pole is to be located at $s = -2 + j2\sqrt{3}$, the sum of the angle contributions of the open-loop poles (at $s = 0$ and $s = -2$) is $-120° - 90° = -210°$. This means that, to have the closed-loop pole at

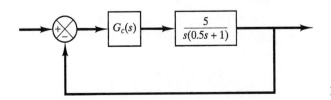

Figure 10–82   Control system.

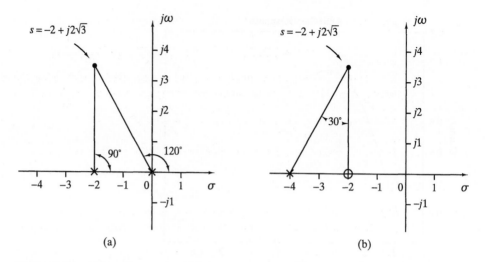

Figure 10–83 (a) Open-loop poles and a desired closed-loop pole; (b) compensator pole–zero configuration to contribute phase lead angle of 30°.

$s = -2 + j2\sqrt{3}$, we must add 30° to the open-loop transfer function. Stated differently, the angle deficiency of the given open-loop transfer function at the desired closed-loop pole $s = -2 + j2\sqrt{3}$ is

$$180° - 120° - 90° = -30°$$

The compensator must contribute 30°. A compensator that contributes a positive angle is called a *lead compensator*, a possible form of which is given by

$$G_c(s) = K\frac{s + a}{s + b}$$

Let us choose the zero of the lead compensator at $s = -2$. Then the pole of the compensator must be located at $s = -4$ in order to have a phase lead angle of 30°. [See Figure 10–83(b).] thus,

$$G_c(s) = K\frac{s + 2}{s + 4}$$

The gain K is determined from the magnitude condition

$$\left| K\frac{s + 2}{s + 4}\frac{5}{s(0.5s + 1)} \right|_{s=-2+j2\sqrt{3}} = 1$$

or

$$K = \left| \frac{s(s + 4)}{10} \right|_{s=-2+j2\sqrt{3}} = 1.6$$

Hence,

$$G_c(s) = 1.6\frac{s + 2}{s + 4}$$

The open-loop transfer function of the compensated system is

$$G_c(s)\,\frac{5}{s(0.5s + 1)} = 1.6\frac{s + 2}{s + 4}\frac{10}{s(s + 2)} = \frac{16}{s(s + 4)}$$

Next, we shall obtain unit-step responses of the original system and the compensated system. The original system has the following closed-loop transfer function:

$$\frac{C(s)}{R(s)} = \frac{10}{s^2 + 2s + 10}$$

The compensated system has the following closed-loop transfer function:

$$\frac{C(s)}{R(s)} = \frac{16}{s^2 + 4s + 16}$$

MATLAB Program 10–12 plots the unit-step response curves of the original and compensated systems. The resulting unit-step response curves are shown in Figure 10–84.

MATLAB Program 10–12

```
>> num = [10]; den = [1   2   10];
>> numc = [16]; denc = [1   4   16];
>> t = 0:0.01:5;
>> c1 = step(num,den,t);
>> c2 = step(numc,denc,t);
>> plot(t,c1,'.',t,c2,'-')
>> grid
>> title('Unit-Step Responses of Original System and Compensated System')
>> xlabel('t (sec)'); ylabel('Outputs')
>> text(1.75,1.27,'Original system')
>> text(1.75,1.12,'Compensated system')
```

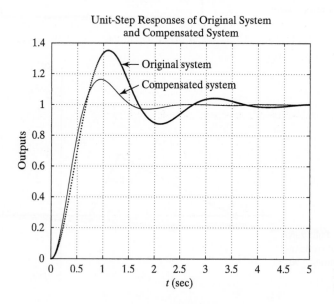

Figure 10–84 Unit-step response curves.

Problem A–10–14

Consider the system shown in Figure 10–85. Design a compensator such that the dominant closed-loop poles are located at $s = -1 \pm j1$. Then obtain the unit-step and unit-ramp responses of the uncompensated and compensated systems.

Solution For a desired closed-loop pole at $s = -1 + j1$, the angle contribution of the two open-loop poles at the origin is $-135° - 135° = -270°$. Hence, the angle deficiency is

$$180° - 135° - 135° = -90°$$

The compensator must contribute 90°.

Let us use a lead compensator of the form

$$G_c(s) = K\frac{s + a}{s + b}$$

and let us choose the zero of the lead compensator at $s = -0.5$. (Note that the choice of the zero at $s = -0.5$ is, in a sense, arbitrary. The zero should not be too close to the origin and should be located somewhere between -0.4 and -0.8, so that the lead compensator can provide a 90° phase lead.) To obtain a phase lead angle of 90°, the pole of the compensator must be located at $s = -3$. (See Figure 10–86.) Thus,

$$G_c(s) = K\frac{s + 0.5}{s + 3}$$

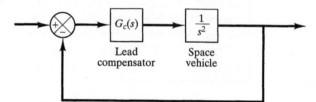

Figure 10–85   Control system.

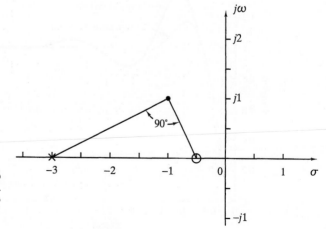

Figure 10–86   Pole–zero location of lead compensator contributing 90° phase lead.

where the gain K must be determined from the magnitude condition

$$\left| K \frac{s + 0.5}{s + 3} \frac{1}{s^2} \right|_{s = -1 + j1} = 1$$

or

$$K = \left| \frac{(s + 3)s^2}{s + 0.5} \right|_{s = -1 + j1} = 4$$

Hence, the lead compensator becomes

$$G_c(s) = 4 \frac{s + 0.5}{s + 3}$$

Then the feedforward transfer function becomes

$$G_c(s) \frac{1}{s^2} = \frac{4s + 2}{s^3 + 3s^2}$$

A root-locus plot of the system is shown in Figure 10–87.

Note that the closed-loop transfer function is

$$\frac{C(s)}{R(s)} = \frac{4s + 2}{s^3 + 3s^2 + 4s + 2}$$

The closed-loop poles are located at $s = -1 \pm j1$ and $s = -1$.

Next, we shall obtain the unit-step and unit-ramp responses of the uncompensated and compensated systems. MATLAB Program 10–13 obtains the unit-step response curves, which are shown in Figure 10–88.

MATLAB Program 10–14 obtains the unit-ramp response curves, which are shown in Figure 10–89.

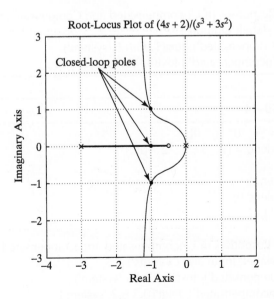

Figure 10–87   Root-locus plot.

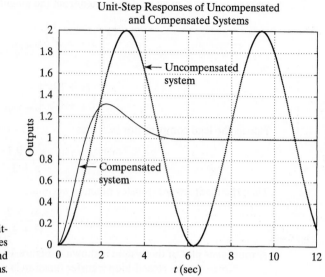

Figure 10–88 Unit-step response curves of uncompensated and compensated systems.

MATLAB Program 10–13

```
>> num = [1]; den = [1   0   1];
>> numc = [4   2]; denc = [1   3   4   2];
>> t = 0:0.02:12;
>> c1 = step(num,den,t);
>> c2 = step(numc,denc,t);
>> plot(t,c1,'.',t,c2,'-')
>> grid
>> title('Unit-Step Responses of Uncompensated and Compensated Systems')
>> xlabel('t (sec)'); ylabel('Outputs')
>> text(1.95,0.73,'Compensated'); text(1.95,0.6,'system')
>> text(4.9,1.65,'Uncompensated'); text(4.9,1.53,'system')
```

MATLAB Program 10–14

```
>> num = [1]; den = [1   0   1   0];
>> numc = [4   2]; denc = [1   3   4   2   0];
>> t = 0:0.02:15;
>> c1 = step(num,den,t);
>> c2 = step(numc,denc,t);
>> plot(t,t,'.',t,c1,'-.',t,c2,'-')
>> grid
>> title('Unit-Ramp Responses of Uncompensated and Compensated Systems')
>> xlabel('t (sec)'); ylabel('Input and Outputs')
>> text(6.3,1.85,'Compensated'); text(6.3,0.85,'system')
>> text(10.3,7.2,'Uncompensated'); text(10.3,6.2,'system')
>> text(12.8,10.9,'Input')
```

Unit-Ramp Responses of Uncompensated
and Compensated Systems

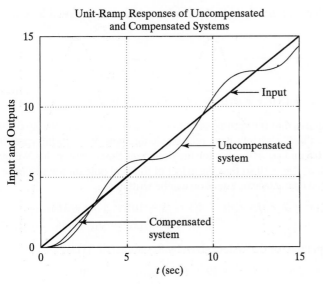

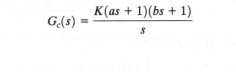

Figure 10–89 Unit-ramp response curves of uncompensated and compensated systems.

Problem A–10–15

Consider the system shown in Figure 10–90, which represents PID control of a second-order plant $G(s)$. Assume that disturbances $D(s)$ enter the system as shown in the diagram. It is assumed that the reference input $R(s)$ is normally held constant, and the response characteristics to disturbances are a very important consideration in this system.

Design a control system such that the response to any step disturbance will be damped out quickly (in 2 to 3 s in terms of the 2% settling time). Choose the configuration of the closed-loop poles such that there is a pair of dominant closed-loop poles. Then obtain the response to the unit-step disturbance input. Obtain also the response to the unit-step reference input.

Solution The PID controller has the transfer function

$$G_c(s) = \frac{K(as + 1)(bs + 1)}{s}$$

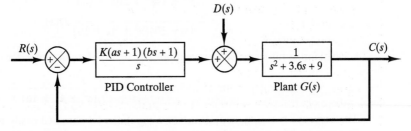

Figure 10–90 PID-controlled system.

For the disturbance input in the absence of the reference input, the closed-loop transfer function becomes

$$\frac{C_d(s)}{D(s)} = \frac{s}{s(s^2 + 3.6s + 9) + K(as + 1)(bs + 1)}$$

$$= \frac{s}{s^3 + (3.6 + Kab)s^2 + (9 + Ka + Kb)s + K} \tag{10–55}$$

The specification requires that the settling time of the response to the unit-step disturbance be 2 to 3 s and that the system have reasonable damping. We may interpret the specification as $\zeta = 0.5$ and $\omega_n = 4$ rad/s for the dominant closed-loop poles. We may choose the third pole at $s = -10$ so that the effect of this real pole on the response is small. Then the desired characteristic equation can be written as

$$(s + 10)(s^2 + 2 \times 0.5 \times 4s + 4^2) = (s + 10)(s^2 + 4s + 16)$$
$$= s^3 + 14s^2 + 56s + 160$$

The characteristic equation for the system given by Equation (10–55) is

$$s^3 + (3.6 + Kab)s^2 + (9 + Ka + Kb)s + K = 0$$

Hence, we require

$$3.6 + Kab = 14$$
$$9 + Ka + Kb = 56$$
$$K = 160$$

which yields

$$ab = 0.065, \qquad a + b = 0.29375$$

The transfer function of the PID controller now becomes

$$G_c(s) = \frac{K[abs^2 + (a + b)s + 1]}{s}$$

$$= \frac{160(0.065s^2 + 0.29375s + 1)}{s}$$

$$= \frac{10.4(s^2 + 4.5192s + 15.385)}{s}$$

With this PID controller, the response to the disturbance is given by

$$C_d(s) = \frac{s}{s^3 + 14s^2 + 56s + 160} D(s)$$

$$= \frac{s}{(s + 10)(s^2 + 4s + 16)} D(s)$$

Clearly, for a unit-step disturbance input, the steady-state output is zero, since

$$\lim_{t \to \infty} c_d(t) = \lim_{s \to 0} sC_d(s) = \lim_{s \to 0} \frac{s^2}{(s + 10)(s^2 + 4s + 16)} \frac{1}{s} = 0$$

The response to a unit-step disturbance input can be obtained easily with MATLAB Program 10–15, which produces the response curve shown in Figure 10–91(a). From the response curve, we see that the settling time is approximately 2.7 s. Thus, the response damps out quickly and the system designed here is acceptable.

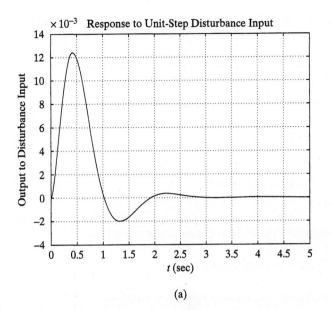

(a)

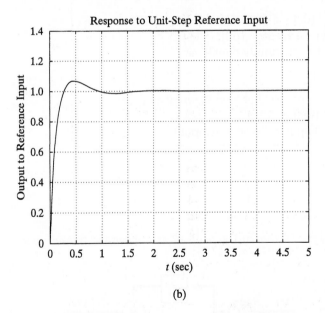

(b)

Figure 10–91 (a) Response to unit-step disturbance input; (b) response to unit-step reference input.

For the reference input $r(t)$, the closed-loop transfer function is

$$\frac{C_r(s)}{R(s)} = \frac{10.4(s^2 + 4.5192s + 15.385)}{s^3 + 14s^2 + 56s + 160}$$

$$= \frac{10.4s^2 + 47s + 160}{s^3 + 14s^2 + 56s + 160}$$

The response to a unit-step reference input can also be obtained with the use of MATLAB Program 10–15. The resulting response curve is shown in Figure 10–91(b).

The response curve shows that the maximum overshoot is 7.3% and the settling time is 1.7 s. The system has quite acceptable response characteristics.

MATLAB Program 10–15

```
>> % ----- Response to unit-step disturbance input -----
>>
>> numd = [1   0];
>> dend = [1   14   56   160];
>> t = 0:0.01:5;
>> [c1,x1,t] = step(numd,dend,t);
>> plot(t,c1)
>> grid
>> title('Response to Unit-Step Disturbance Input')
>> xlabel('t (sec)')
>> ylabel('Output to Disturbance Input')
>>
>> % ----- Response to unit-step reference input -----
>>
>> numr = [10.4   47   160];
>> denr = [1   14   56   160];
>> [c2, x2, t] = step(numr,denr,t);
>> plot(t,c2)
>> grid
>> title('Response to Unit-Step Reference Input')
>> xlabel('t (sec)')
>> ylabel('Output to Reference Input')
```

PROBLEMS

Problem B–10–1

Simplify the block diagram shown in Figure 10–92 and obtain the transfer function $C(s)/R(s)$.

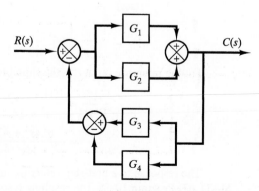

Figure 10–92 Block diagram of a system.

Problem B–10–2

Simplify the block diagram shown in Figure 10–93 and obtain the transfer function $C(s)/R(s)$.

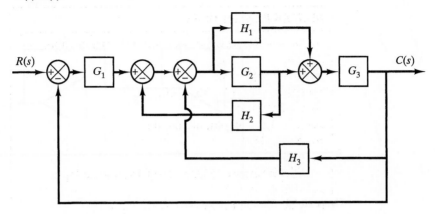

Figure 10–93   Block diagram of a system.

Problem B–10–3

Use the following series and feedback commands of MATLAB to obtain the closed-loop transfer function $C(s)/R(s)$ of the system shown in Figure 10–94:

$$\text{sysg} = \text{series(sys1,sys2)}$$
$$\text{sys} = \text{feedback(sysg,[1])}$$

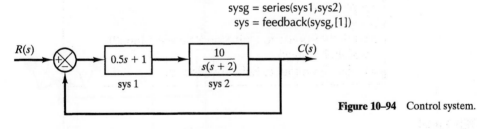

Figure 10–94   Control system.

Problem B–10–4

Use the parallel, series, and feedback commands of MATLAB to obtain the closed-loop transfer function $C(s)/R(s)$ of the system shown in Figure 10–95.

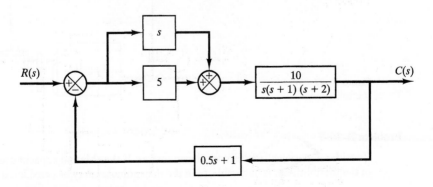

Figure 10–95   Control system.

Problem B–10–5

Derive the transfer function $E_o(s)/E_i(s)$ of the electronic controller shown in Figure 10–96. What is the control action of this controller?

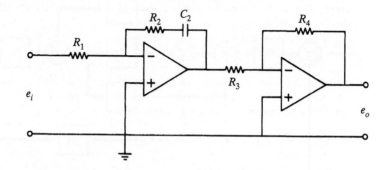

Figure 10–96   Electronic controller.

Problem B–10–6

Explain the operation of the speed-control system shown in Figure 10–97.

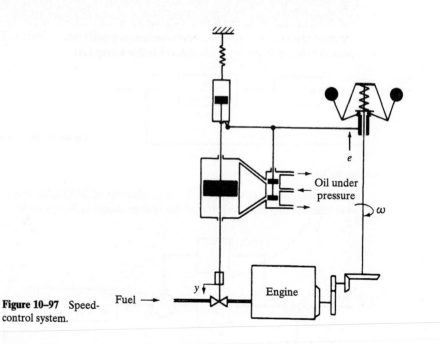

Figure 10–97   Speed-control system.

Problem B–10–7

Consider a glass-walled mercury thermometer. If the thermal capacitance of the glass of the thermometer is negligible, then the thermometer may be considered a first-order system, and its transfer function may be given by

$$\frac{\Theta(s)}{\Theta_b(s)} = \frac{1}{Ts + 1}$$

where $\Theta(s)$ is the Laplace transform of the thermometer temperature θ and $\Theta_b(s)$ is the Laplace transform of the bath temperature θ_b, both temperatures measured from the ambient temperature.

Assume that a glass-walled mercury thermometer is used to measure the temperature of a bath and that the thermal capacitance of the glass is negligible. Assume also that the time constant of the thermometer is not known, so it is experimentally determined by lowering the device into a pail of water held at 10°C. Figure 10–98 shows the temperature response observed during the test. [$\theta(0) = 30$°C.] Find the time constant. If the thermometer is placed in a bath, the temperature of which is increasing linearly at a rate of 10°C/min, how much steady-state error does the thermometer show?

If the thermal capacitance of the glass of a mercury thermometer is not negligible, the thermometer may be considered a second-order system and the transfer function may be modified to

$$\frac{1}{(T_1s + 1)(T_2s + 1)}$$

where T_1 and T_2 are time constants. Sketch a typical temperature response curve (θ versus t) when such a thermometer with two time constants is placed in a bath held at a constant temperature θ_b, where both the thermometer temperature θ and the bath temperature θ_b are measured from the ambient temperature.

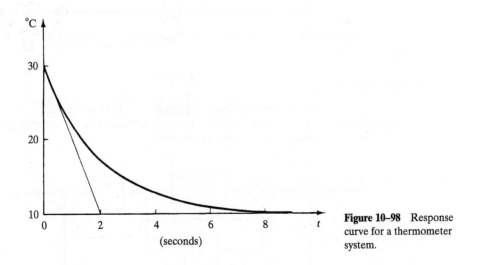

Figure 10–98 Response curve for a thermometer system.

Problem B–10–8

Obtain the unit-step response of the control system shown in Figure 10–99.

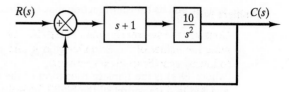

Figure 10–99 Control system.

Problem B–10–9

Consider a system defined by

$$\frac{C(s)}{R(s)} = \frac{\omega_n^2}{s^2 + 2\zeta\omega_n s + \omega_n^2}$$

Determine the values of ζ and ω_n so that the system responds to a step input with approximately 5% overshoot and with a settling time of 2 s.

Problem B–10–10

Figure 10–100 shows a position control system with velocity feedback. What is the response $c(t)$ to the unit step input?

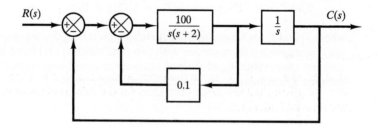

Figure 10–100   Block diagram of a position control system with velocity feedback.

Problem B–10–11

Consider the system shown in Figure 10–101. Determine the value of k such that the damping ratio ζ is 0.5. Then obtain the rise time t_r, peak time t_p, maximum overshoot M_p, and settling time t_s in the unit-step response.

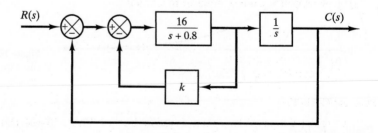

Figure 10–101   Control system.

Problem B–10–12

Consider the system shown in Figure 10–102, which involves velocity feedback. Determine the values of the amplifier gain K and the velocity feedback gain K_h so that the following specifications are satisfied:

1. Damping ratio of the closed-loop poles is 0.5
2. Settling time ≤ 2 s
3. Static velocity error constant $K_v \geq 50$ s^{-1}
4. $0 < K_h < 1$

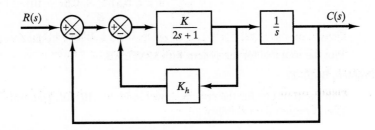

Figure 10–102   Control system.

Problem B–10–13

Find the response $c(t)$ of the system shown in Figure 10–103 when the input $r(t)$ is a unit ramp. Also, find the steady-state error for the unit-ramp response. Assume that the system is underdamped.

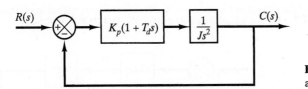

Figure 10–103   Block diagram of a system.

Problem B–10–14

Determine the range of the gain K required for the stability of a unity-feedback control system whose open-loop transfer function is

$$G(s) = \frac{K}{s(s + 1)(s + 5)}$$

Use Routh's stability criterion.

Problem B–10–15

Consider the closed-loop control system shown in Figure 10–104. Determine the range of the gain K required for stability. Plot a root-locus diagram for the system.

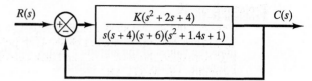

$R(s)$ $\dfrac{K(s^2 + 2s + 4)}{s(s + 4)(s + 6)(s^2 + 1.4s + 1)}$ $C(s)$

Figure 10–104 Control system.

Problem B–10–16

Consider the system whose open-loop transfer function is

$$G(s)H(s) = \frac{K}{s(s + 0.5)(s^2 + 0.6s + 10)}$$

$$= \frac{K}{s^4 + 1.1s^3 + 10.3s^2 + 5s}$$

Plot the root loci for the system with MATLAB.

Problem B–10–17

Plot the root loci for the system shown in Figure 10–105. Determine the range of the gain K required for stability.

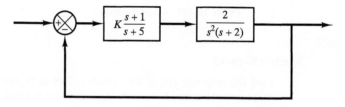

$K\dfrac{s + 1}{s + 5}$ $\dfrac{2}{s^2(s + 2)}$

Figure 10–105 Control system.

Problem B–10–18

Consider the system shown in Figure 10–106. Design a compensator such that the dominant closed-loop poles are located at $s = -1 \pm j\sqrt{3}$.

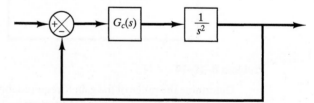

$G_c(s)$ $\dfrac{1}{s^2}$

Figure 10–106 Control system.

Problem B–10–19

Consider the system shown in Figure 10–107. Plot the root loci for the system. Determine the value of K such that the damping ratio ζ of the dominant closed-loop poles is 0.6. Then determine all closed-loop poles. Plot the unit-step response curve with MATLAB.

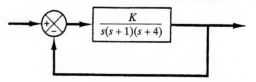

$\dfrac{K}{s(s + 1)(s + 4)}$

Figure 10–107 Control system.

Problem B–10–20

Consider the system shown in Figure 10–108. It is desired to design a PID controller $G_c(s)$ such that the dominant closed-loop poles are located at $s = -1 \pm j1$. For the PID controller, choose $a = 0.5$, and then determine the values of K and b. Draw a root-locus plot with MATLAB.

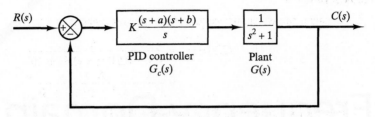

$R(s)$ $K\dfrac{(s+a)(s+b)}{s}$ $\dfrac{1}{s^2+1}$ $C(s)$

PID controller   Plant
$G_c(s)$ $G(s)$

Figure 10–108 PID-controlled system.

Frequency-Domain Analysis and Design of Control Systems

11–1 INTRODUCTION

This chapter deals with the frequency-response approach to the analysis and design of control systems. By the term *frequency response*, we mean the steady-state response of a system to a sinusoidal input.

An advantage of the frequency-response approach is that frequency-response tests are, in general, simple and can be made accurately by use of readily available sinusoidal signal generators and precise measurement equipment. Often, the transfer functions of complicated components can be determined experimentally by frequency-response tests.

Frequency-response analysis and design of linear control systems is based on the Nyquist stability criterion, which enables us to investigate both the absolute and relative stabilities of linear closed-loop systems from a knowledge of their open-loop frequency-response characteristics.

In this chapter, we first present Bode diagrams (logarithmic plots). We then discuss the Nyquist stability criterion, after which the concept of the phase margin and gain margin is introduced. Finally, the frequency-response approach to the design of control systems is treated. MATLAB approaches to obtain Bode diagrams and Nyquist plots are included in this chapter.

It is noted that although the frequency response of a control system presents a qualitative picture of the transient response, the correlation between the frequency and transient responses is indirect, except in the case of second-order systems. In designing a closed-loop system, we adjust the frequency-response characteristic of the open-loop transfer function by using several design criteria in order to obtain acceptable transient-response characteristics for the system.

Outline of the chapter.   Section 11-1 has given introductory remarks. Section 11-2 presents Bode diagrams of transfer-function systems. In particular, first-order systems and second-order systems are examined in detail. The determination of static error constants from Bode diagrams is also discussed. Section 11-3 treats plotting Bode diagrams with MATLAB. Section 11-4 deals with Nyquist plots and the Nyquist stability criterion. The concept of phase margin and gain margin is introduced. Section 11-5 discusses plotting Nyquist diagrams with MATLAB. Finally, Section 11-6 presents the Bode diagram approach to the design of control systems. Specifically, we discuss the design of the lead compensator, lag compensator, and lag–lead compensator.

11-2 BODE DIAGRAM REPRESENTATION OF THE FREQUENCY RESPONSE

A useful way to represent frequency-response characteristics of dynamic systems is the *Bode diagram*. (Bode diagrams are also called *logarithmic plots of frequency response*.) In this section we treat basic materials associated with Bode diagrams, using first- and second-order systems as examples. We then discuss the problem of identifying the transfer function of a system from the Bode diagram.

Bode diagrams.   A sinusoidal transfer function may be represented by two separate plots, one giving the magnitude versus frequency of the function, the other the phase angle (in degrees) versus frequency. A Bode diagram consists of two graphs: a curve of the logarithm of the magnitude of a sinusoidal transfer function and a curve of the phase angle; both curves are plotted against the frequency on a logarithmic scale.

The standard representation of the logarithmic magnitude of $G(j\omega)$ is $20 \log |G(j\omega)|$, where the base of the logarithm is 10. The unit used in this representation is the decibel, usually abbreviated dB. In the logarithmic representation, the curves are drawn on semilog paper, using a logarithmic scale for frequency and a linear scale for either magnitude (but in decibels) or phase angle (in degrees). (The frequency range of interest determines the number of logarithmic cycles required on the abscissa.)

The main advantage of using a Bode diagram is that multiplication of magnitudes can be converted into addition. Furthermore, a simple method for sketching an approximate log-magnitude curve is available. The method, based on asymptotic approximation by straight-line asymptotes, is sufficient if only rough information on the frequency-response characteristics is needed. Should exact curves be desired, corrections can be made easily. The phase-angle curves are readily drawn if a template for the phase-angle curve of $1 + j\omega$ is available.

Note that the experimental determination of a transfer function can be made simple if frequency-response data are presented in the form of a Bode diagram.

The logarithmic representation is useful in that it shows both the low- and high-frequency characteristics of the transfer function in one diagram. Expanding the low-frequency range by means of a logarithmic scale for the frequency is highly advantageous, since characteristics at low frequencies are most important in practical systems. Although it is not possible to plot the curves right down to zero frequency (because $\log 0 = -\infty$), this does not create a serious problem.

Number–decibel conversion line.   A number–decibel conversion line is shown in Figure 11–1. The decibel value of any number can be obtained from this line. As a number increases by a factor of 10, the corresponding decibel value increases by a factor of 20. This relationship may be seen from the formula

$$20 \log (K \times 10^n) = 20 \log K + 20n$$

Note that, when expressed in decibels, the reciprocal of a number differs from its value only in sign; that is, for the number K,

$$20 \log K = -20 \log \frac{1}{K}$$

Bode diagram of gain K.   A number greater than unity has a positive value in decibels, while a number smaller than unity has a negative value. The log-magnitude curve of a constant gain K is a horizontal straight line at the magnitude of $20 \log K$ decibels. The phase angle of the gain K is zero. Varying K in the transfer function

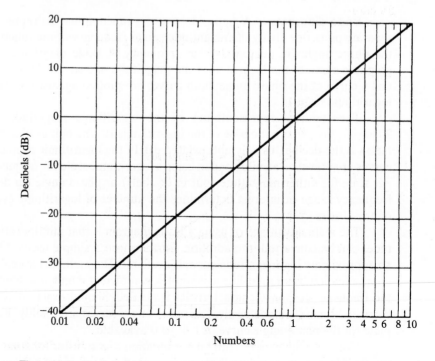

Figure 11–1   Number–decibel conversion line.

raises or lowers the log-magnitude curve of the transfer function by the corresponding constant amount, but does not affect the phase angle.

Bode diagrams of integral and derivative factors.   The log magnitude of $1/(j\omega)$ in decibels is

$$20 \log \left| \frac{1}{j\omega} \right| = -20 \log \omega \ \text{dB}$$

The phase angle of $1/j\omega$ is constant and equal to $-90°$.

If the log magnitude $-20 \log \omega \, \text{dB}$ is plotted against ω on a logarithmic scale, the resulting curve is a straight line. Since

$$(-20 \log 10\omega) \ \text{dB} = (-20 \log \omega - 20) \ \text{dB}$$

the slope of the line is -20 dB/decade.

Similarly, the log magnitude of $j\omega$ in decibels is

$$20 \log |j\omega| = 20 \log \omega \ \text{dB}$$

The phase angle of $j\omega$ is constant and equal to $90°$. The log-magnitude curve is a straight line with a slope of 20 dB/decade. Figures 11–2 and 11–3 show Bode diagrams of $1/j\omega$ and $j\omega$, respectively.

Bode diagram of first-order system.   Consider the sinusoidal transfer function

$$\frac{X(j\omega)}{P(j\omega)} = G(j\omega) = \frac{1}{Tj\omega + 1}$$

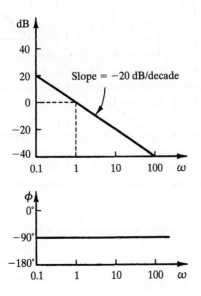

Figure 11–2   Bode diagram of $G(j\omega) = 1/(j\omega)$.

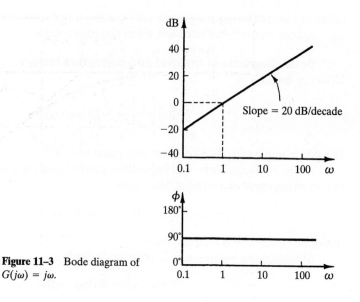

Figure 11–3 Bode diagram of $G(j\omega) = j\omega$.

The log magnitude of this first-order sinusoidal transfer function, in decibels, is

$$20 \log \left| \frac{1}{Tj\omega + 1} \right| = -20 \log \sqrt{\omega^2 T^2 + 1} \text{ dB}$$

For low frequencies such that $\omega \ll 1/T$, the log magnitude may be approximated by

$$-20 \log \sqrt{\omega^2 T^2 + 1} \doteq -20 \log 1 = 0 \text{ dB}$$

Thus, the log-magnitude curve at low frequencies is the constant 0-dB line. For high frequencies such that $\omega \gg 1/T$,

$$-20 \log \sqrt{\omega^2 T^2 + 1} \doteq -20 \log \omega T \text{ dB}$$

This is an approximate expression for the high-frequency range. At $\omega = 1/T$, the log magnitude equals 0 dB; at $\omega = 10/T$, the log magnitude is -20 dB. Hence, the value of $-20 \log \omega T$ dB decreases by 20 dB for every decade of ω. For $\omega \gg 1/T$, the log-magnitude curve is therefore a straight line with a slope of -20 dB/decade (or -6 dB/octave).

The preceding analysis shows that the logarithmic representation of the frequency-response curve of the factor $1/(Tj\omega + 1)$ can be approximated by two straight-line asymptotes: a straight line at 0 dB for the frequency range $0 < \omega < 1/T$ and a straight line with slope -20 dB/decade (or -6 dB/octave) for the frequency range $1/T < \omega < \infty$. The exact log-magnitude curve, the asymptotes, and the exact phase-angle curve are shown in Figure 11–4.

The frequency at which the two asymptotes meet is called the *corner* frequency or *break* frequency. For the factor $1/(Tj\omega + 1)$, the frequency $\omega = 1/T$ is the corner frequency, since, at $\omega = 1/T$, the two asymptotes have the same value. (The low-frequency asymptotic expression at $\omega = 1/T$ is 20 log 1 dB = 0 dB, and the

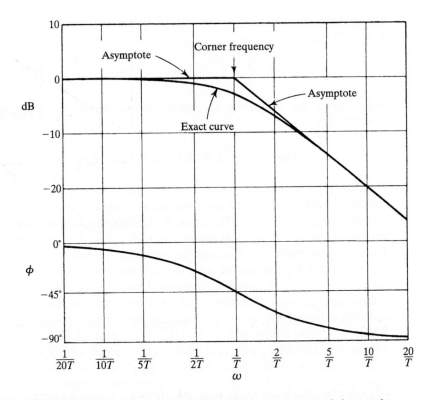

Figure 11–4 Log-magnitude curve together with the asymptotes and phase-angle curve of $1/(j\omega T + 1)$.

high-frequency asymptotic expression at $\omega = 1/T$ is also 20 log 1 dB = 0 dB.) The corner frequency divides the frequency-response curve into two regions: a curve for the low-frequency region and a curve for the high-frequency region. The corner frequency is very important in sketching logarithmic frequency-response curves.

The exact phase angle ϕ of the factor $1/(Tj\omega + 1)$ is

$$\phi = -\tan^{-1} \omega T$$

At zero frequency, the phase angle is 0°. At the corner frequency, the phase angle is

$$\phi = -\tan^{-1}\frac{T}{T} = -\tan^{-1} 1 = -45°$$

At infinity, the phase angle becomes −90°. Because it is given by an inverse-tangent function, the phase angle is skew symmetric about the inflection point at $\phi = -45°$.

The error in the magnitude curve caused by the use of asymptotes can be calculated. The maximum error occurs at the corner frequency and is approximately equal to −3 dB, since

$$-20 \log \sqrt{1 + 1} + 20 \log 1 = -10 \log 2 = -3.03 \text{ dB}$$

The error at the frequency one octave below the corner frequency, that is, at $\omega = 1/(2T)$, is

$$-20 \log \sqrt{\frac{1}{4} + 1} + 20 \log 1 = -20 \log \frac{\sqrt{5}}{2} = -0.97 \text{ dB}$$

The error at the frequency one octave above the corner frequency, that is, at $\omega = 2/T$, is

$$-20 \log \sqrt{2^2 + 1} + 20 \log 2 = -20 \log \frac{\sqrt{5}}{2} = -0.97 \text{ dB}$$

Thus, the error one octave below or above the corner frequency is approximately equal to -1 dB. Similarly, the error one decade below or above the corner frequency is approximately -0.04 dB. The error, in decibels, involved in using the asymptotic expression for the frequency response curve of $1/(Tj\omega + 1)$ is shown in Figure 11–5. The error is symmetric with respect to the corner frequency.

Since the asymptotes are easy to draw and are sufficiently close to the exact curve, the use of such approximations in drawing Bode diagrams is convenient in establishing the general nature of the frequency-response characteristics quickly and with a minimum amount of calculation. Any straight-line asymptotes must have slopes of $\pm 20n$ dB/decade $(n = 0, 1, 2, \dots)$; that is, their slopes must be 0 dB/decade, ± 20 dB/decade, ± 40 dB/decade, and so on. If accurate frequency-response curves are desired, corrections may easily be made by referring to the curve given in Figure 11–5. In practice, an accurate frequency-response curve can be drawn by introducing a correction of 3 dB at the corner frequency and a correction of 1 dB at points one octave below and above the corner frequency and then connecting these points by a smooth curve.

Note that varying the time constant T shifts the corner frequency to the left or to the right, but the shapes of the log-magnitude and the phase-angle curves remain the same.

The transfer function $1/(Tj\omega + 1)$ has the characteristics of a low-pass filter. For frequencies above $\omega = 1/T$, the log magnitude falls off rapidly toward $-\infty$,

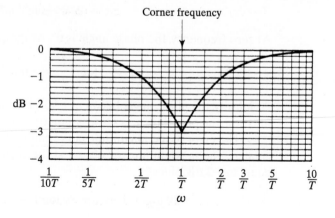

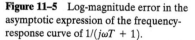

Figure 11–5 Log-magnitude error in the asymptotic expression of the frequency-response curve of $1/(j\omega T + 1)$.

essentially because of the presence of the time constant. In the low-pass filter, the output can follow a sinusoidal input faithfully at low frequencies, but as the input frequency is increased, the output cannot follow the input because a certain amount of time is required for the system to build up in magnitude. Therefore, at high frequencies, the amplitude of the output approaches zero and the phase angle of the output approaches $-90°$. Therefore, if the input function contains many harmonics, then the low-frequency components are reproduced faithfully at the output, while the high-frequency components are attenuated in amplitude and shifted in phase. Thus, a first-order element yields exact, or almost exact, duplication only for constant or slowly varying phenomena.

The shapes of phase-angle curves are the same for any factor of the form $(Tj\omega + 1)^{\pm 1}$. Hence, it is convenient to have a template for the phase-angle curve on cardboard, to be used repeatedly for constructing phase-angle curves for any function of the form $(Tj\omega + 1)^{\pm 1}$. If no such template is available, we have to locate several points on the curve. The phase angles of $(Tj\omega + 1)^{\pm 1}$ are

$$\pm 45° \quad \text{at} \quad \omega = \frac{1}{T}$$

$$\pm 26.6° \quad \text{at} \quad \omega = \frac{1}{2T}$$

$$\pm 5.7° \quad \text{at} \quad \omega = \frac{1}{10T}$$

$$\pm 63.4° \quad \text{at} \quad \omega = \frac{2}{T}$$

$$\pm 84.3° \quad \text{at} \quad \omega = \frac{10}{T}$$

Bode diagram of second-order system.   Next, we shall consider a second-order system in the standard form

$$G(s) = \frac{\omega_n^2}{s^2 + 2\zeta\omega_n s + \omega_n^2}$$

The sinusoidal transfer function $G(j\omega)$ is

$$G(j\omega) = \frac{\omega_n^2}{(j\omega)^2 + 2\zeta\omega_n(j\omega) + \omega_n^2}$$

or

$$G(j\omega) = \frac{1}{\left(j\dfrac{\omega}{\omega_n}\right)^2 + 2\zeta\left(j\dfrac{\omega}{\omega_n}\right) + 1} \tag{11-1}$$

If $\zeta > 1$, $G(j\omega)$ can be expressed as a product of two first-order terms with real poles. If $0 < \zeta < 1$, $G(j\omega)$ is a product of two complex-conjugate terms. Asymptotic approximations to the frequency-response curves are not accurate for this $G(j\omega)$

with low values of ζ, because the magnitude and phase of $G(j\omega)$ depend on both the corner frequency and the damping ratio ζ.

Noting that

$$|G(j\omega)| = \left| \frac{1}{\left(j\dfrac{\omega}{\omega_n}\right)^2 + 2\zeta\left(j\dfrac{\omega}{\omega_n}\right) + 1} \right|$$

or

$$|G(j\omega)| = \frac{1}{\sqrt{\left(1 - \dfrac{\omega^2}{\omega_n^2}\right)^2 + \left(2\zeta\dfrac{\omega}{\omega_n}\right)^2}} \tag{11-2}$$

we may obtain the asymptotic frequency-response curve as follows: Since

$$20 \log \left| \frac{1}{\left(j\dfrac{\omega}{\omega_n}\right)^2 + 2\zeta\left(j\dfrac{\omega}{\omega_n}\right) + 1} \right| = -20 \log \sqrt{\left(1 - \dfrac{\omega^2}{\omega_n^2}\right)^2 + \left(2\zeta\dfrac{\omega}{\omega_n}\right)^2}$$

for low frequencies such that $\omega \ll \omega_n$, the log magnitude becomes

$$-20 \log 1 = 0 \text{ dB}$$

The low-frequency asymptote is thus a horizontal line at 0 dB. For high frequencies $\omega \gg \omega_n$, the log magnitude becomes

$$-20 \log \frac{\omega^2}{\omega_n^2} = -40 \log \frac{\omega}{\omega_n} \text{ dB}$$

The equation for the high-frequency asymptote is a straight line having the slope -40 dB/decade, since

$$-40 \log \frac{10\,\omega}{\omega_n} = -40 - 40 \log \frac{\omega}{\omega_n}$$

The high-frequency asymptote intersects the low-frequency one at $\omega = \omega_n$ since at this frequency

$$-40 \log \frac{\omega_n}{\omega_n} = -40 \log 1 = 0 \text{ dB}$$

This frequency, ω_n, is the corner frequency for the quadratic function considered.

The two asymptotes just derived are independent of the value of ζ. Near the frequency $\omega = \omega_n$, a resonant peak occurs, as may be expected from Equation (11–1). The damping ratio ζ determines the magnitude of this resonant peak. Errors obviously exist in the approximation by straight-line asymptotes. The magnitude of the error is large for small values of ζ. Figure 11–6 shows the exact log-magnitude curves, together with the straight-line asymptotes and the exact phase-angle curves for the quadratic function given by Equation (11–1) with several values of ζ. If corrections

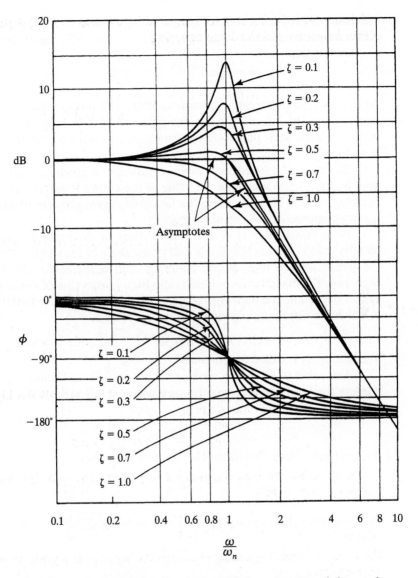

Figure 11–6 Log-magnitude curves together with the asymptotes and phase-angle curves of the quadratic sinusoidal transfer function given by Equation (11–1).

are desired in the asymptotic curves, the necessary amounts of correction at a sufficient number of frequency points may be obtained from Figure 11–6.

The phase angle of the quadratic function given by Equation (11–1) is

$$\phi = \left/\frac{1}{\left(j\dfrac{\omega}{\omega_n}\right)^2 + 2\zeta\left(j\dfrac{\omega}{\omega_n}\right) + 1}\right. = -\tan^{-1}\frac{2\zeta\dfrac{\omega}{\omega_n}}{1 - \left(\dfrac{\omega}{\omega_n}\right)^2} \qquad (11\text{–}3)$$

The phase angle is a function of both ω and ζ. At $\omega = 0$, the phase angle equals $0°$. At the corner frequency $\omega = \omega_n$, the phase angle is $-90°$, regardless of ζ, since

$$\phi = -\tan^{-1}\left(\frac{2\zeta}{0}\right) = -\tan^{-1}\infty = -90°$$

At $\omega = \infty$, the phase angle becomes $-180°$. The phase-angle curve is skew symmetric about the inflection point, the point where $\phi = -90°$. There are no simple ways to sketch such phase curves; one needs to refer to the phase-angle curves shown in Figure 11–6.

To obtain the frequency-response curves of a given quadratic transfer function, we must first determine the value of the corner frequency ω_n and that of the damping ratio ζ. Then, by using the family of curves given in Figure 11–6, the frequency-response curves can be plotted.

Note that Figure 11–6 shows the effects of the input frequency ω and the damping ratio ζ on the amplitude and phase angle of the steady-state output. From the figure, we see that, as the damping ratio is increased, the amplitude ratio decreases. The maximum amplitude ratio for a given value of ζ occurs at a frequency that is less than the undamped natural frequency ω_n. Notice that the frequency ω_r at which the amplitude ratio is a maximum occurs at

$$\omega_r = \omega_n\sqrt{1 - 2\zeta^2}$$

This frequency is called the *resonant frequency*.

The value of ω_r can be obtained as follows: From Equation (11–2), since the numerator of $|G(j\omega)|$ is constant, a peak value of $|G(j\omega)|$ will occur when

$$g(\omega) = \left(1 - \frac{\omega^2}{\omega_n^2}\right)^2 + \left(2\zeta\frac{\omega}{\omega_n}\right)^2 \tag{11–4}$$

is a minimum. Since Equation (11–4) can be written as

$$g(\omega) = \left[\frac{\omega^2 - \omega_n^2(1 - 2\zeta^2)}{\omega_n^2}\right]^2 + 4\zeta^2(1 - \zeta^2)$$

the minimum value of $g(\omega)$ occurs at $\omega = \omega_n\sqrt{1 - 2\zeta^2}$. Thus the resonant frequency ω_r is

$$\omega_r = \omega_n\sqrt{1 - 2\zeta^2} \qquad 0 \le \zeta \le 0.707 \tag{11–5}$$

As the damping ratio ζ approaches zero, the resonant frequency approaches ω_n. For $0 < \zeta \le 0.707$, the resonant frequency ω_r is less than the damped natural frequency $\omega_d = \omega_n\sqrt{1 - \zeta^2}$, which is exhibited in the transient response. From Equation (11–5), it can be seen that, for $\zeta > 0.707$, there is no resonant peak. The magnitude $|G(j\omega)|$ decreases monotonically with increasing frequency ω. (The magnitude is less than 0 dB for all values of $\omega > 0$; recall that, for $0.7 < \zeta \le 1$, the step response is oscillatory, but the oscillations are well damped and are hardly perceptible.)

The magnitude of the resonant peak M_r can be found by substituting Equation (11–5) into Equation (11–2). For $0 \le \zeta \le 0.707$,

$$M_r = |G(j\omega)|_{\text{max}} = |G(j\omega_r)| = \frac{1}{2\zeta\sqrt{1 - \zeta^2}} \tag{11–6}$$

As ζ approaches zero, M_r approaches infinity. This means that, if the undamped system is excited at its natural frequency, the magnitude of $G(j\omega)$ becomes infinite. For $\zeta > 0.707$,

$$M_r = 1 \qquad (11\text{–}7)$$

The relationship between M_r and ζ given by Equations (11–6) and (11–7) is shown in Figure 11–7.

The phase angle of $G(j\omega)$ at the frequency where the resonant peak occurs can be obtained by substituting Equation (11–5) into Equation (11–3). Thus, at the resonant frequency ω_r,

$$\angle G(j\omega_r) = -\tan^{-1}\frac{\sqrt{1 - 2\zeta^2}}{\zeta} = -90° + \sin^{-1}\frac{\zeta}{\sqrt{1 - \zeta^2}}$$

Minimum-phase systems and nonminimum-phase systems.   Transfer functions having neither poles nor zeros in the right-half s-plane are called *minimum-phase transfer functions*, whereas those having poles and/or zeros in the right-half s-plane are called *nonminimum-phase transfer functions*. Systems with minimum-phase transfer functions are called *minimum-phase systems*; those with nonminimum-phase transfer functions are called *nonminimum-phase systems*.

For systems with the same magnitude characteristic, the range in phase angle of the minimum-phase transfer function is minimum for all such systems, while the range in phase angle of any nonminimum-phase transfer function is greater than this minimum.

Consider as an example the two systems whose sinusoidal transfer functions are, respectively,

$$G_1(j\omega) = \frac{1 + j\omega T}{1 + j\omega T_1} \qquad \text{and} \qquad G_2(j\omega) = \frac{1 - j\omega T}{1 + j\omega T_1}, \qquad 0 < T < T_1$$

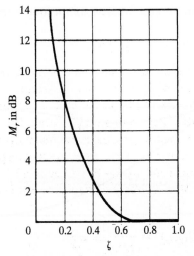

Figure 11–7   Curve of M_r versus ζ for the second-order system $1/[(j\omega/\omega_n)^2 + 2\zeta(j\omega/\omega_n) + 1]$.

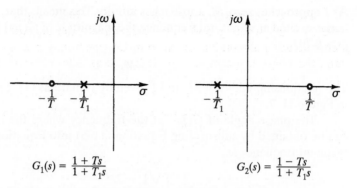

$$G_1(s) = \frac{1 + Ts}{1 + T_1 s} \qquad\qquad G_2(s) = \frac{1 - Ts}{1 + T_1 s}$$

Figure 11–8 Pole–zero configurations of minimum-phase and nonmini-mum-phase systems (G_1: minimum phase, G_2: nonminimum phase).

The pole–zero configurations of these systems are shown in Figure 11–8. The two sinusoidal transfer functions have the same magnitude characteristics, but they have different phase-angle characteristics, as shown in Figure 11–9. The two systems differ from each other by the factor

$$G(j\omega) = \frac{1 - j\omega T}{1 + j\omega T}$$

The magnitude of the factor $(1 - j\omega T)/(1 + j\omega T)$ is always unity. But the phase angle equals $-2\tan^{-1} \omega T$ and varies from 0 to $-180°$ as ω is increased from zero to infinity.

For a minimum-phase system, the magnitude and phase-angle characteristics are directly related. That is, if the magnitude curve of a system is specified over the entire frequency range from zero to infinity, then the phase-angle curve is uniquely determined, and vice versa. This relationship, however, does not hold for a nonmini-mum-phase system.

Nonminimum-phase situations may arise (1) when a system includes a nonmin-imum-phase element or elements and (2) in the case where a minor loop is unstable.

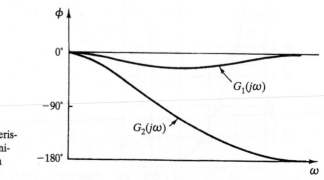

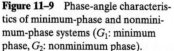

Figure 11–9 Phase-angle characteris-tics of minimum-phase and nonmini-mum-phase systems (G_1: minimum phase, G_2: nonminimum phase).

For a minimum-phase system, the phase angle at $\omega = \infty$ becomes $-90°(q - p)$, where p and q are, respectively, the degrees of the numerator and denominator polynomials of the transfer function. For a nonminimum-phase system, the phase angle at $\omega = \infty$ differs from $-90°(q - p)$. In either system, the slope of the log-magnitude curve at $\omega = \infty$ is equal to $-20(q - p)$ dB/decade. It is therefore possible to detect whether the system is minimum phase by examining both the slope of the high-frequency asymptote of the log-magnitude curve and the phase angle at $\omega = \infty$. If the slope of the log-magnitude curve as ω approaches infinity is $-20(q - p)$ dB/decade and the phase angle at $\omega = \infty$ is equal to $-90°(q - p)$, the system is minimum phase.

Nonminimum-phase systems are slow in response because of their faulty behavior at the start of the response. In most practical control systems, excessive phase lag should be carefully avoided. In designing a system, if a fast response is of primary importance, nonminimum-phase components should not be used.

Example 11–1

Consider the mechanical system shown in Figure 11–10. An experimental Bode diagram for this system is shown in Figure 11–11. The ordinate of the magnitude curve is the amplitude ratio of the output to the input, measured in decibels—that is, $|X(j\omega)/P(j\omega)|$ in dB. The units of $|X(j\omega)/P(j\omega)|$ are m/N. The phase angle is $\angle X(j\omega)/P(j\omega)$ in degrees. The input is a sinusoidal force of the form

$$p(t) = P \sin \omega t$$

where P is the amplitude of the sinusoidal input force and the input frequency is varied from 0.01 to 100 rad/s. The displacement x is measured from the equilibrium position before the sinusoidal force is applied. Note that the amplitude ratio $|X(j\omega)/P(j\omega)|$ does not depend on the absolute value of P. (This is because, if the input amplitude is doubled, the output amplitude is also doubled. Therefore, we can choose any convenient amplitude P.) Determine the numerical values of m, b, and k from the Bode diagram.

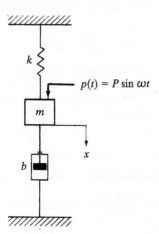

Figure 11–10 Mechanical system.

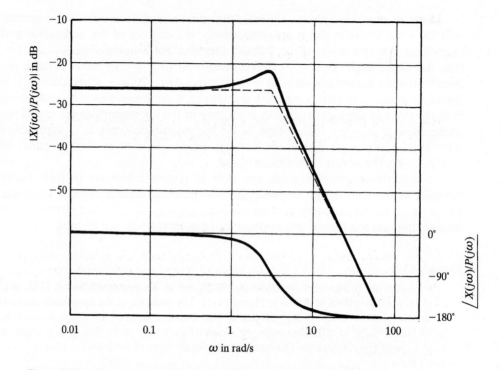

Figure 11–11   Experimental Bode diagram for the system shown in Figure 11–10.

First, we need to determine the transfer function of the system. The system equation is

$$m\ddot{x} + b\dot{x} + kx = p(t) = P \sin \omega t$$

The Laplace transform of this last equation, with zero initial condition, gives

$$(ms^2 + bs + k)X(s) = P(s)$$

where $P(s) = \mathscr{L}[p(t)]$. The transfer function for the system is

$$\frac{X(s)}{P(s)} = \frac{1}{ms^2 + bs + k}$$

This mechanical system possesses poles only in the left-half s-plane, so it is a *minimum-phase system*. For a minimum-phase system, the transfer function can be uniquely determined solely from the magnitude curve of the Bode diagram.

The sinusoidal transfer function is

$$\frac{X(j\omega)}{P(j\omega)} = \frac{1}{m(j\omega)^2 + bj\omega + k} \tag{11–8}$$

Now, from the Bode diagram, we find that

$$\frac{X(j0+)}{P(j0+)} = -26 \text{ dB}$$

Hence,

$$\frac{X(j0+)}{P(j0+)} = \frac{1}{k} = -26 \text{ dB} = 0.0501$$

or

$$k = 19.96 \text{ N/m}$$

Also from the Bode diagram, the corner frequency ω_n is seen to be 3.2 rad/s. Since the corner frequency of the system given by Equation (11–8) is $\sqrt{k/m}$, it follows that

$$\omega_n = \sqrt{\frac{k}{m}} = 3.2$$

Thus,

$$m = \frac{k}{(3.2)^2} = \frac{19.96}{10.24} = 1.949 \text{ kg}$$

Next, we need to estimate the value of the damping ratio ζ. Comparing the Bode diagram of Figure 11–11 with the Bode diagram of the standard second-order system shown in Figure 11–6, we find the damping ratio ζ to be approximately 0.32, or $\zeta = 0.32$. Then, noting that

$$\frac{b}{m} = 2\zeta\omega_n$$

we obtain

$$b = 2\zeta\omega_n m = 2 \times 0.32 \times 3.2 \times 1.949 = 3.992 \text{ N-s/m}$$

We have thus determined the values of m, b, and k to be as follows:

$$m = 1.949 \text{ kg}, \qquad b = 3.992 \text{ N-s/m}, \qquad k = 19.96 \text{ N/m}$$

Relationship between system type and log-magnitude curve.   Consider the unity-feedback control system. The static position, velocity, and acceleration error constants describe the low-frequency behavior of type 0, type 1, and type 2 systems, respectively. For a given system, only one of the static error constants is finite and significant. (The larger the value of the finite static error constant, the higher the loop gain is as ω approaches zero.)

The type of the system determines the slope of the log-magnitude curve at low frequencies. Thus, information concerning the existence and magnitude of the steady-state error in the response of a control system to a given input can be determined by observing the low-frequency region of the log-magnitude curve of the system.

Determination of static position error constants.   Consider the unity-feedback control system shown in Figure 11–12. Assume that the open-loop transfer function is given by

$$G(s) = \frac{K(T_a s + 1)(T_b s + 1)\cdots(T_m s + 1)}{s^N(T_1 s + 1)(T_2 s + 1)\cdots(T_p s + 1)}$$

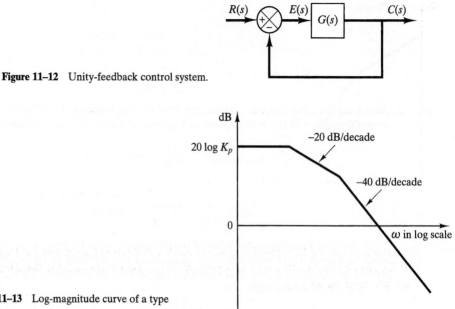

Figure 11–12   Unity-feedback control system.

Figure 11–13   Log-magnitude curve of a type 0 system.

or

$$G(j\omega) = \frac{K(T_a j\omega + 1)(T_b j\omega + 1)\cdots(T_m j\omega + 1)}{(j\omega)^N(T_1 j\omega + 1)(T_2 j\omega + 1)\cdots(T_p j\omega + 1)}$$

Figure 11–13 is an example of the log-magnitude plot of a type 0 system. In such a system, the magnitude of $G(j\omega)$ equals K_p at low frequencies, or

$$\lim_{\omega \to 0} G(j\omega) = K_p$$

It follows that the low-frequency asymptote is a horizontal line at $20 \log K_p$ dB.

Determination of static velocity error constants.   Consider again the unity-feedback control system shown in Figure 11–12. Figure 11–14 is an example of the log-magnitude plot of a type 1 system. The intersection of the initial -20-dB/decade segment (or its extension) with the line $\omega = 1$ has the magnitude $20 \log K_v$. This may be seen as follows: For a type 1 system,

$$G(j\omega) = \frac{K_v}{j\omega} \qquad \text{for } \omega \ll 1$$

Thus,

$$20 \log \left| \frac{K_v}{j\omega} \right|_{\omega=1} = 20 \log K_v$$

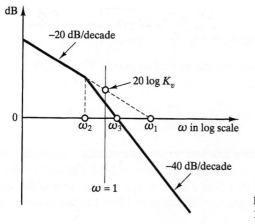

Figure 11–14 Log-magnitude curve of a type 1 system.

The intersection of the initial -20-dB/decade segment (or its extension) with the 0-dB line has a frequency numerically equal to K_v. To see this, define the frequency at this intersection to be ω_1; then

$$\left| \frac{K_v}{j\omega_1} \right| = 1$$

or

$$K_v = \omega_1$$

As an example, consider the type 1 system with unity feedback whose open-loop transfer function is

$$G(s) = \frac{K}{s(Js + F)}$$

If we define the corner frequency to be ω_2 and the frequency at the intersection of the -40-dB/decade segment (or its extension) with the 0-dB line to be ω_3, then

$$\omega_2 = \frac{F}{J}, \qquad \omega_3^2 = \frac{K}{J}$$

Since

$$\omega_1 = K_v = \frac{K}{F}$$

it follows that

$$\omega_1\omega_2 = \omega_3^2$$

or

$$\frac{\omega_1}{\omega_3} = \frac{\omega_3}{\omega_2}$$

On the Bode diagram,

$$\log \omega_1 - \log \omega_3 = \log \omega_3 - \log \omega_2$$

Thus, the ω_3 point is just midway between the ω_2 and ω_1 points. The damping ratio ζ of the system is then

$$\zeta = \frac{F}{2\sqrt{KJ}} = \frac{\omega_2}{2\omega_3}$$

Determination of static acceleration error constants. Consider once more the unity-feedback control system shown in Figure 11–12. Figure 11–15 is an example of the log-magnitude plot of a type 2 system. The intersection of the initial -40-dB/decade segment (or its extension) with the $\omega = 1$ line has the magnitude of $20 \log K_a$. Since, at low frequencies,

$$G(j\omega) = \frac{K_a}{(j\omega)^2} \qquad \text{for } \omega \ll 1$$

it follows that

$$20 \log \left| \frac{K_a}{(j\omega)^2} \right|_{\omega=1} = 20 \log K_a$$

The frequency ω_a at the intersection of the initial -40-dB/decade segment (or its extension) with the 0-dB line gives the square root of K_a numerically. This can be seen from the following:

$$20 \log \left| \frac{K_a}{(j\omega_a)^2} \right| = 20 \log 1 = 0$$

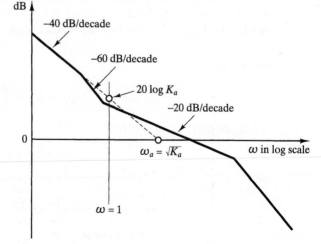

Figure 11–15   Log-magnitude curve of a type 2 system.

which yields

$$\omega_a = \sqrt{K_a}$$

Cutoff frequency and bandwidth. Referring to Figure 11–16, the frequency ω_b at which the magnitude of the closed-loop frequency response is 3 dB below its zero-frequency value is called the *cutoff frequency*. Thus,

$$\left|\frac{C(j\omega)}{R(j\omega)}\right| < \left|\frac{C(j0)}{R(j0)}\right| - 3 \text{ dB} \qquad \text{for } \omega > \omega_b$$

For systems in which $|C(j0)/R(j0)| = 0$ dB,

$$\left|\frac{C(j\omega)}{R(j\omega)}\right| < -3 \text{ dB} \qquad \text{for } \omega > \omega_b$$

The closed-loop system filters out the signal components whose frequencies are greater than the cutoff frequency and transmits those signal components with frequencies lower than the cutoff frequency.

The frequency range $0 \le \omega \le \omega_b$ in which the magnitude of the closed loop does not drop -3 dB is called the *bandwidth* of the system. The bandwidth indicates the frequency where the gain starts to fall off from its low-frequency value. Thus, the bandwidth indicates how well the system will track an input sinusoid. Note that, for a given ω_n, the rise time increases with increasing damping ratio ζ. On the other hand, the bandwidth decreases with increasing ζ. Therefore, the rise time and the bandwidth are inversely proportional to each other.

The specification of the bandwidth may be determined by the following factors:

1. The ability to reproduce the input signal. A large bandwidth corresponds to a small rise time, or a fast response. Roughly speaking, we can say that the bandwidth is proportional to the speed of the response.

2. The necessary filtering characteristics for high-frequency noise.

For the system to follow arbitrary inputs accurately, it is necessary that it have a large bandwidth. From the viewpoint of noise, however, the bandwidth should not be too large. Thus, there are conflicting requirements on the bandwidth, and a

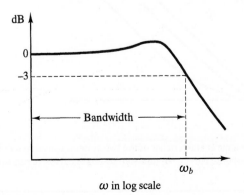

Figure 11–16   Logarithmic plot showing cutoff frequency ω_b and bandwidth.

compromise is usually necessary for good design. Note that a system with a large bandwidth requires high-performance components, so the cost of components usually increases with the bandwidth.

Cutoff rate. The cutoff rate is the slope of the log-magnitude curve near the cutoff frequency. The cutoff rate indicates the ability of a system to distinguish a signal from noise.

Note that a closed-loop frequency response curve with a steep cutoff characteristic may have a large resonant peak magnitude, which implies that the system has a relatively small stability margin.

Example 11–2

Consider the following two systems:

$$\text{System I: } \frac{C(s)}{R(s)} = \frac{1}{s + 1}, \qquad \text{System II: } \frac{C(s)}{R(s)} = \frac{1}{3s + 1}$$

Compare the bandwidths of these systems. Show that the system with the larger bandwidth has a faster speed of response and can follow the input much better than the system with a smaller bandwidth.

Figure 11–17(a) shows the closed-loop frequency-response curves for the two systems. (Asymptotic curves are represented by dashed lines.) We find that the bandwidth of system I is $0 \leq \omega \leq 1$ rad/s and that of system II is $0 \leq \omega \leq 0.33$ rad/s. Figures 11–17(b) and (c) show, respectively, the unit-step response and unit-ramp response curves for the two systems. Clearly, system I, whose bandwidth is three times wider than that of system II, has a faster speed of response and can follow the input much better.

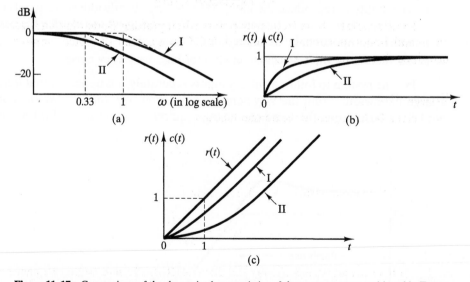

Figure 11–17 Comparison of the dynamic characteristics of the two systems considered in Example 11–2: (a) closed-loop frequency-response curves; (b) unit-step response curves; (c) unit-ramp response curves.

11–3 PLOTTING BODE DIAGRAMS WITH MATLAB

In MATLAB, the command 'bode' computes magnitudes and phase angles of the frequency response of continuous-time, linear, time-invariant systems.

When the command 'bode' (without left-hand arguments) is entered in the computer, MATLAB produces a Bode plot on the screen. When invoked with left-hand arguments, as in

$$[\text{mag,phase,w}] = \text{bode(num,den,w)}$$

'bode' returns the frequency response of the system in matrices mag, phase, and w. No plot is drawn on the screen. The matrices mag and phase contain the magnitudes and phase angles respectively, of the frequency response of the system, evaluated at user-specified frequency points. The phase angle is returned in degrees. The magnitude can be converted to decibels with the statement

$$\text{magdB} = 20^* \log 10(\text{mag})$$

To specify the frequency range, use the command logspace(d1,d2) or logspace(d1,d2,n). logspace(d1,d2) generates a vector of 50 points logarithmically equally spaced between decades 10^{d1} and 10^{d2}. That is, to generate 50 points between 0.1 rad/s and 100 rad/s, enter the command

$$w = \text{logspace}(-1,2)$$

logspace(d1,d2,n) generates n points logarithmically equally spaced between decades 10^{d1} and 10^{d2}. For example, to generate 100 points between 1 rad/s and 1000 rad/s, enter the following command:

$$w = \text{logspace}(0,3,100)$$

To incorporate these frequency points when plotting Bode diagrams, use the command bode(num,den,w) or bode(A,B,C,D,iu,w), each of which employs the user-specified frequency vector w.

Example 11–3

Plot a Bode diagram of the transfer function

$$G(s) = \frac{25}{s^2 + 4s + 25}$$

When the system is defined in the form

$$G(s) = \frac{\text{num}(s)}{\text{den}(s)}$$

use the command bode(num,den) to draw the Bode diagram. [When the numerator and denominator contain the polynomial coefficients in descending powers of s, bode(num,den) draws the Bode diagram.] MATLAB Program 11–1 plots the Bode diagram for this system. The resulting Bode diagram is shown in Figure 11–18.

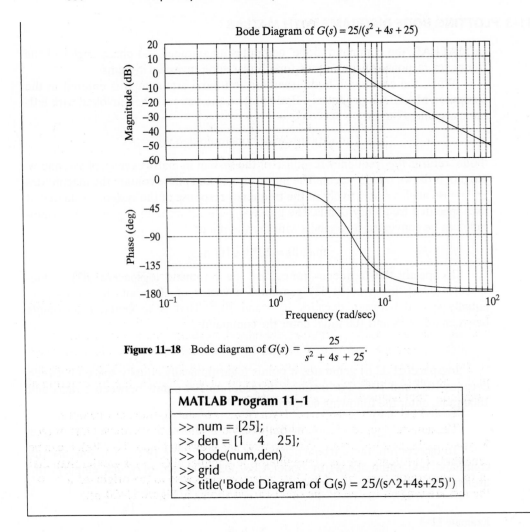

Figure 11–18   Bode diagram of $G(s) = \dfrac{25}{s^2 + 4s + 25}$.

MATLAB Program 11–1
>> num = [25]; >> den = [1   4   25]; >> bode(num,den) >> grid >> title('Bode Diagram of G(s) = 25/(s^2+4s+25)')

11–4 NYQUIST PLOTS AND THE NYQUIST STABILITY CRITERION

In this section, we first discuss Nyquist plots and then the Nyquist stability criterion. We then define the phase margin and gain margin, which are frequently used for determining the relative stability of a control system. Finally, we discuss conditionally stable systems.

Nyquist plots.   The *Nyquist plot* of a sinusoidal transfer function $G(j\omega)$ is a plot of the magnitude of $G(j\omega)$ versus the phase angle of $G(j\omega)$ in polar coordinates as ω is varied from zero to infinity. Thus, the polar plot is the locus of vectors $|G(j\omega)|\underline{/G(j\omega)}$ as ω is varied from zero to infinity. Note that, in polar plots, a positive (negative) phase angle is measured counterclockwise (clockwise) from the positive real axis. The Nyquist plot is often called the *polar plot*. An example of such a

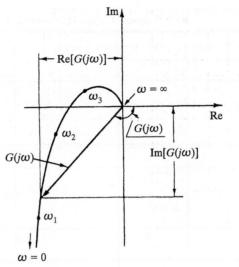

Figure 11–19   Nyquist plot.

plot is shown in Figure 11–19. Each point on the polar plot of $G(j\omega)$ represents the terminal point of a vector at a particular value of ω. The projections of $G(j\omega)$ on the real and imaginary axes are the real and imaginary components of the function.

An advantage in using a Nyquist plot is that it depicts the frequency-response characteristics of a system over the entire frequency range in a single plot. One disadvantage is that the plot does not clearly indicate the contribution of each individual factor of the open-loop transfer function.

Table 11–1 shows examples of Nyquist plots of simple transfer functions.

The general shapes of the low-frequency portions of the Nyquist plots of type 0, type 1, and type 2 minimum-phase systems are shown in Figure 11–20(a). It can be seen that, if the degree of the denominator polynomial of $G(j\omega)$ is greater than that of the numerator, then the $G(j\omega)$ loci converge clockwise to the origin. At $\omega = \infty$, the loci are tangent to one or the other axis, as shown in Figure 11–20(b).

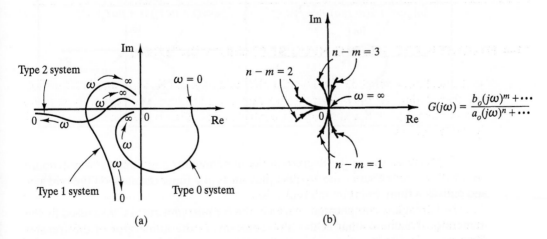

(a)   (b)

Figure 11–20   (a) Nyquist plots of type 0, type 1, and type 2 systems; (b) Nyquist plots in the high-frequency range.

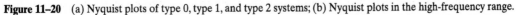

TABLE 11–1   Nyquist Plots of Simple Transfer Functions

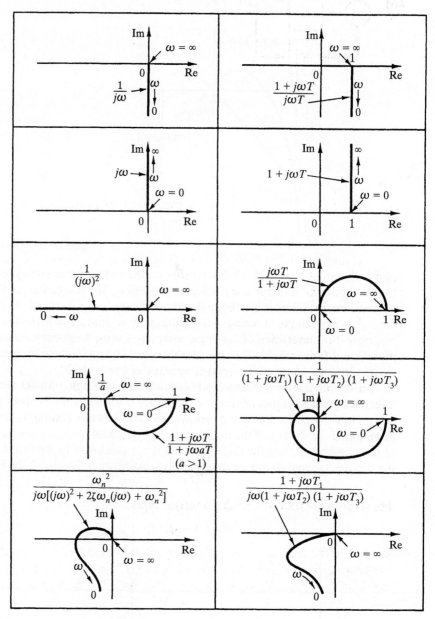

For the case where the degrees of the denominator and numerator polynomials of $G(j\omega)$ are the same, the Nyquist plot starts at a finite distance on the real axis and ends at a finite point on the real axis.

Note that any complicated shapes in the Nyquist plot curves are caused by the numerator dynamics—that is, the time constants in the numerator of the transfer function.

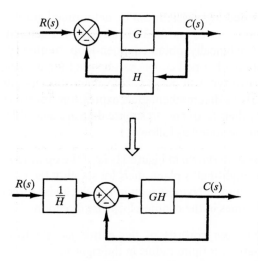

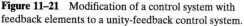

Figure 11–21   Modification of a control system with feedback elements to a unity-feedback control system.

Nyquist stability criterion.   In designing a control system, we require that the system be stable. Furthermore, it is necessary that the system have adequate relative stability. In what follows, we shall show that the Nyquist plot indicates not only whether a system is stable, but also the *degree* of stability of a stable system. The Nyquist plot also gives information as to how stability may be improved if that is necessary.

In the discussion that follows, we shall assume that the systems considered have unity feedback. Note that it is always possible to reduce a system with feedback elements to a unity-feedback system, as shown in Figure 11–21. Hence, the extension of relative stability analysis for the unity-feedback system to nonunity-feedback systems is possible.

Now consider the system shown in Figure 11–22. The closed-loop transfer function is

$$\frac{C(s)}{R(s)} = \frac{G(s)}{1 + G(s)}$$

For stability, all roots of the characteristic equation

$$1 + G(s) = 0$$

must lie in the left-half s-plane. The Nyquist stability criterion relates the open-loop frequency response $G(j\omega)$ to the number of zeros and poles of $1 + G(s)$ that lie in the right-half s-plane. This criterion, due to H. Nyquist, is useful in control engineering

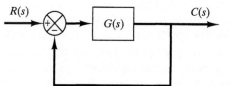

Figure 11–22   Unity-feedback control system.

because the absolute stability of the closed-loop system can be determined graphically from open-loop frequency-response curves and there is no need for actually determining the closed-loop poles. Analytically obtained open-loop frequency-response curves, as well as experimentally obtained curves, can be used for the stability analysis. This confluence of the two types of curve is convenient because, in designing a control system, it often happens that mathematical expressions for some of the components are not known; only their frequency-response data are available.

The Nyquist stability criterion can be stated as follows:

Nyquist stability criterion: In the system shown in Figure 11–22, if the open-loop transfer function $G(s)$ has P poles in the right-half s-plane, then, for stability, the $G(s)$ locus as a representative point s traces out the Nyquist path in the clockwise direction must encircle the $-1 + j0$ point P times in the counterclockwise direction.

The Nyquist path is a closed contour that consists of the entire $j\omega$-axis from $\omega = -\infty$ to $+\infty$ and a semicircular path of infinite radius in the right-half s-plane. Thus, the Nyquist path encloses the entire right-half s-plane. The direction of the path is clockwise.

Remarks on the Nyquist stability criterion

1. The Nyquist stability criterion can be expressed as

$$Z = N + P \tag{11–9}$$

where

Z = number of zeros of $1 + G(s)$ in the right-half s-plane.
N = number of clockwise encirclements of the $-1 + j0$ point
P = number of poles of $G(s)$ in the right-half s-plane

If P is not zero, then, for a stable control system, we must have $Z = 0$, or $N = -P$, which means that we must have P counterclockwise encirclements of the $-1 + j0$ point.

If $G(s)$ does not have any poles in the right-half s-plane, then, from Equation (11–9), we must have $Z = N$ for stability. For example, consider the system with the following open-loop transfer function:

$$G(s) = \frac{K}{s(T_1 s + 1)(T_2 s + 1)}$$

Figure 11–23 shows the Nyquist path and $G(s)$ loci for a small and large value of the gain K. Since the number of poles of $G(s)$ in the right-half s-plane is zero, for this system to be stable, it is necessary that $N = Z = 0$, or that the $G(s)$ locus not encircle the $-1 + j0$ point.

For small values of K, there is no encirclement of the $-1 + j0$ point; hence, the system is stable for small values of K. For large values of K, the locus of $G(s)$ encircles the $-1 + j0$ point twice in the clockwise direction, indicating two closed-loop poles in the right-half s-plane, and the system is unstable. For good accuracy, K should be large. From the stability viewpoint,

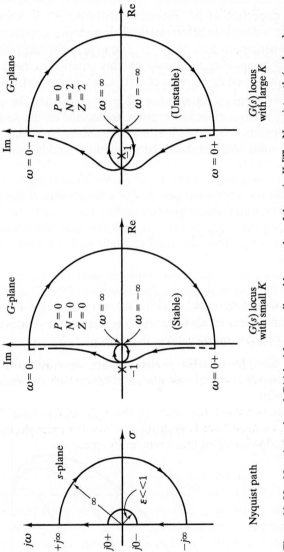

Figure 11–23 Nyquist path and $G(s)$ loci for a small and large value of the gain K. [The Nyquist path (a closed contour) must not pass through a pole or zero. Because the given $G(s)$ has a pole at the origin, the contour must be modified by use of a semicircle with an infinitesimal radius ε as shown in the figure. From $s = j0-$ to $s = j0+$ the representative point moves along the semicircle of radius ε. The area that the modified closed contour avoids is very small and approaches zero as the radius ε approaches zero. Thus, the Nyquist path encloses the entire right-half s plane.]

however, a large value of K causes poor stability or even instability. To compromise between accuracy and stability, it is necessary to insert a compensator into the system.

2. We must be careful when testing the stability of multiple-loop systems, since they may include poles in the right-half s-plane. (Note that, although an inner loop may be unstable, the entire closed-loop system can be made stable by proper design.) Simple inspection of the encirclements of the $-1 + j0$ point by the $G(j\omega)$ locus is not sufficient to detect instability in multiple-loop systems. In such cases, however, whether any pole of $1 + G(s)$ is in the right-half s-plane may be determined easily by applying the Routh stability criterion to the denominator of $G(s)$ or by actually finding the poles of $G(s)$ with the use of MATLAB.

3. If the locus of $G(j\omega)$ passes through the $-1 + j0$ point, then the zeros of the characteristic equation, or closed-loop poles, are located on the $j\omega$-axis. This is not desirable for practical control systems. For a well-designed closed-loop system, none of the roots of the characteristic equation should lie on the $j\omega$-axis.

Phase and gain margins.   Figure 11–24 shows Nyquist plots of $G(j\omega)$ for three different values of the open-loop gain K. For a large value of the gain K, the system is unstable. As the gain is decreased to a certain value, the $G(j\omega)$ locus passes through the $-1 + j0$ point. This means that, with this gain, the system is on the verge of instability and will exhibit sustained oscillations. For a small value of the gain K, the system is stable.

In general, the closer the $G(j\omega)$ locus comes to encircling the $-1 + j0$ point, the more oscillatory is the system response. The closeness of the $G(j\omega)$ locus to the $-1 + j0$ point can be used as a measure of the margin of stability. (This does not apply, however, to conditionally stable systems.) It is common practice to represent the closeness in terms of phase margin and gain margin.

Phase margin.   The *phase margin* is that amount of additional phase lag at the gain crossover frequency required to bring the system to the verge of instability.

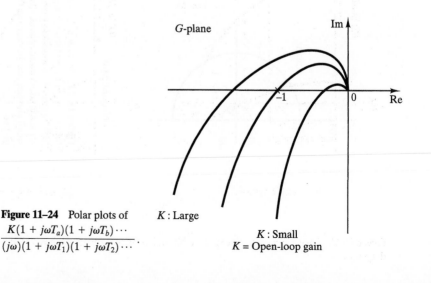

Figure 11–24   Polar plots of
$$\frac{K(1 + j\omega T_a)(1 + j\omega T_b)\cdots}{(j\omega)(1 + j\omega T_1)(1 + j\omega T_2)\cdots}.$$

K : Large

K : Small
K = Open-loop gain

The gain crossover frequency is the frequency at which $|G(j\omega)|$, the magnitude of the open-loop transfer function, is unity. The phase margin γ is 180° plus the phase angle ϕ of the open-loop transfer function at the gain crossover frequency, or

$$\gamma = 180° + \phi$$

On the Nyquist plot, a line may be drawn from the origin to the point at which the unit circle crosses the $G(j\omega)$ locus. The angle from the negative real axis to this line is the phase margin, which is positive for $\gamma > 0$ and negative for $\gamma < 0$. For a minimum-phase system to be stable, the phase margin must be positive.

Figures 11–25(a) and (b) illustrate the phase margins of a stable system and an unstable system in Nyquist plots and Bode diagrams. In the Bode diagrams, the critical point in the complex plane corresponds to the 0-dB line and −180° line.

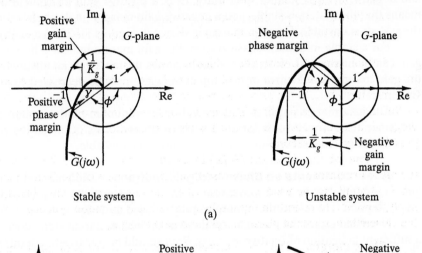

Stable system   Unstable system

(a)

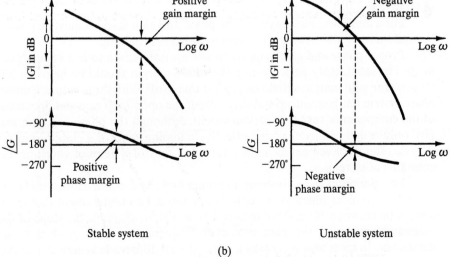

Stable system   Unstable system

(b)

Figure 11–25   Phase and gain margins of stable and unstable systems: (a) Nyquist plots; (b) Bode diagrams.

Gain margin. The *gain margin* is the reciprocal of the magnitude $|G(j\omega)|$ at the frequency at which the phase angle is $-180°$. Defining the phase crossover frequency ω_1 to be the frequency at which the phase angle of the open-loop transfer function equals $-180°$ gives the gain margin K_g:

$$K_g = \frac{1}{|G(j\omega_1)|}$$

In terms of decibels,

$$K_g \text{ dB} = 20 \log K_g = -20 \log |G(j\omega_1)|$$

Expressed in decibels, the gain margin is positive if K_g is greater than unity and negative if K_g is smaller than unity. Thus, a positive gain margin (in decibels) means that the system is stable, and a negative gain margin (in decibels) means that the system is unstable. The gain margin is shown in Figures 11–25(a) and (b).

For a stable minimum-phase system, the gain margin indicates how much the gain can be increased before the system becomes unstable. For an unstable system, the gain margin is indicative of how much the gain must be decreased to make the system stable.

The gain margins of first- and second-order minimum-phase systems are infinite, since the Nyquist plots for such systems do not cross the negative real axis. Thus, such first- and second-order systems cannot be unstable.

A few comments on phase and gain margins. The phase and gain margins of a control system are a measure of the closeness of the Nyquist plot to the $-1 + j0$ point. Therefore, these margins may be used as design criteria.

Note that either the gain margin alone or the phase margin alone does not give a sufficient indication of relative stability. *Both* should be given in the determination of relative stability.

For a minimum-phase system, the phase and gain margins must be positive for the system to be stable. Negative margins indicate instability.

Proper phase and gain margins ensure against variations in a system's components. For satisfactory performance, the phase margin should be between 30° and 60°, and the gain margin should be greater than 6 dB. With these values, a minimum-phase system has guaranteed stability, even if the open-loop gain and time constants of the components vary to a certain extent. Although the phase and gain margins give only rough estimates of the effective damping ratio of a closed-loop system, they do offer a convenient means for designing control systems or adjusting the gain constants of systems.

For minimum-phase systems, the magnitude and phase characteristics of the open-loop transfer function are definitely related. The requirement that the phase margin be between 30° and 60° means that, in a Bode diagram, the slope of the log-magnitude curve at the gain crossover frequency is more gradual than -40 dB/decade. In most practical cases, a slope of -20 dB/decade is desirable at the gain crossover frequency for stability. If the slope is -40 dB/decade, the system could be either stable or unstable. (Even if the system is stable, however, the phase margin is

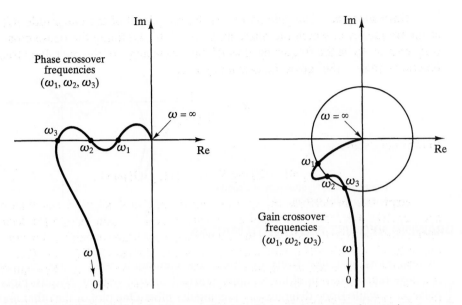

Figure 11-26 Nyquist plots showing more than two phase or gain crossover frequencies.

small.) If the slope at the gain crossover frequency is -60 dB/decade or steeper, the system will be unstable.

For a nonminimum-phase system with unstable open loop, the stability condition will not be satisfied unless the $G(j\omega)$ plot encircles the $-1 + j0$ point. Hence, such a stable nonminimum-phase system will have negative phase and gain margins.

It is also important to point out that conditionally stable systems will have two or more phase crossover frequencies, and some higher order systems with complicated numerator dynamics may also have two or more gain crossover frequencies, as shown in Figure 11-26. For stable systems having two or more gain crossover frequencies, the phase margin is measured at the highest gain crossover frequency.

Conditionally stable systems. Figure 11-27 is an example of a $G(j\omega)$ locus for which the closed-loop system can be made stable or unstable by varying the open-loop gain. If the open-loop gain is increased sufficiently, the $G(j\omega)$ locus encloses the $-1 + j0$ point twice, and the system becomes unstable. If the open-loop gain is decreased sufficiently, again the $G(j\omega)$ locus encloses the $-1 + j0$ point twice. The system is stable only for the limited range of values of the open-loop gain for which the $-1 + j0$ point is completely outside the $G(j\omega)$ locus. Such a system is *conditionally stable.*

Such a conditionally stable system becomes unstable when large input signals are applied, since a large signal may cause saturation, which in turn reduces the open-loop gain of the system.

For stable operation of the conditionally stable system considered here, the critical point $-1 + j0$ must not be located in the regions between *OA* and *BC* shown in Figure 11-27.

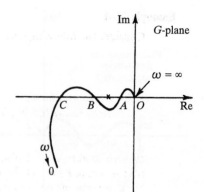

Figure 11–27   Nyquist plot of a conditionally stable system.

11–5 DRAWING NYQUIST PLOTS WITH MATLAB

Nyquist plots, just like Bode diagrams, are commonly used in the frequency-response representation of linear, time-invariant control systems. Nyquist plots are polar plots, while Bode diagrams are rectangular plots. One plot or the other may be more convenient for a particular operation, but a given operation can always be carried out in either plot.

The command 'nyquist' computes the frequency response for continuous-time, linear, time-invariant systems. When invoked without left-hand arguments, 'nyquist' produces a Nyquist plot on the screen. That is, the command

$$nyquist(num,den)$$

draws the Nyquist plot of the transfer function

$$G(s) = \frac{num(s)}{den(s)}$$

where num and den contain the polynomial coefficients in descending powers of s.

The command

$$nyquist(num,den,w)$$

employs the user-specified frequency vector w, which gives the frequency points in radians per second at which the frequency response will be calculated.

When invoked with left-hand arguments, as in

$$[re,im,w] = nyquist(num,den)$$

or

$$[re,im,w] = nyquist(num,den,w)$$

MATLAB returns the frequency response of the system in the matrices re, im, and w. No plot is drawn on the screen. The matrices re and im contain the real and imaginary parts of the frequency response of the system, evaluated at the frequency points specified in the vector w. Note that re and im have as many columns as outputs and one row for each element in w.

Example 11–4

Consider the following open-loop transfer function:

$$G(s) = \frac{1}{s^2 + 0.8s + 1}$$

Draw a Nyquist plot with MATLAB.

Since the system is given in the form of the transfer function, the command

nyquist(num,den)

may be used to draw a Nyquist plot. MATLAB Program 11–2 produces the Nyquist plot shown in Figure 11–28. In this plot, the ranges for the real axis and imaginary axis are automatically determined.

If we wish to draw the Nyquist plot using manually determined ranges—for example, from −2 to 2 on the real axis and from −2 to 2 on the imaginary axis—we enter the following command into the computer:

v=[−2   2   −2   2];
axis(v);

Alternatively, we may combine these two lines into one as follows:

axis([−2   2   −2   2]);

MATLAB Program 11–2

```
>> num = [1];
>> den = [1   0.8   1];
>> nyquist(num,den)
>> title('Nyquist Plot of G(s) = 1/(s^2+0.8s+1)')
```

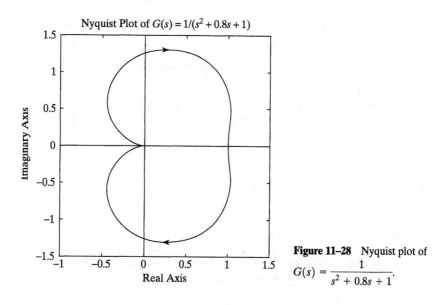

Nyquist Plot of $G(s) = 1/(s^2 + 0.8s + 1)$

Figure 11–28 Nyquist plot of $G(s) = \dfrac{1}{s^2 + 0.8s + 1}$.

Example 11–5

Draw a Nyquist plot for

$$G(s) = \frac{1}{s(s + 1)}$$

MATLAB Program 11–3 will produce a correct Nyquist plot on the computer, even though a warning message "Divide by zero" may appear on the screen. The resulting Nyquist plot is shown in Figure 11–29. Notice that the plot includes the loci for both $\omega > 0$ and $\omega < 0$. If we wish to draw the Nyquist plot for only the positive frequency region ($\omega > 0$), then we need to use the commands

[re,im,w] = nyquist(num,den,w)

plot(re,im)

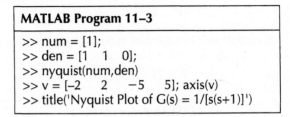

MATLAB Program 11–3

```
>> num = [1];
>> den = [1   1   0];
>> nyquist(num,den)
>> v = [–2    2    –5    5]; axis(v)
>> title('Nyquist Plot of G(s) = 1/[s(s+1)]')
```

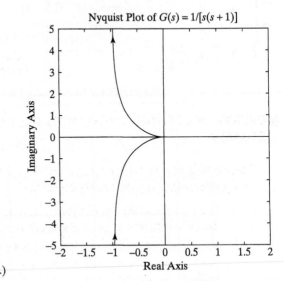

Figure 11–29 Nyquist plot of
$G(s) = 1/s(s + 1)$. (The plot shows
Nyquist loci for both $\omega > 0$ and $\omega < 0$.)

MATLAB Program 11–4 uses these two lines of commands. The resulting Nyquist plot is presented in Figure 11–30.

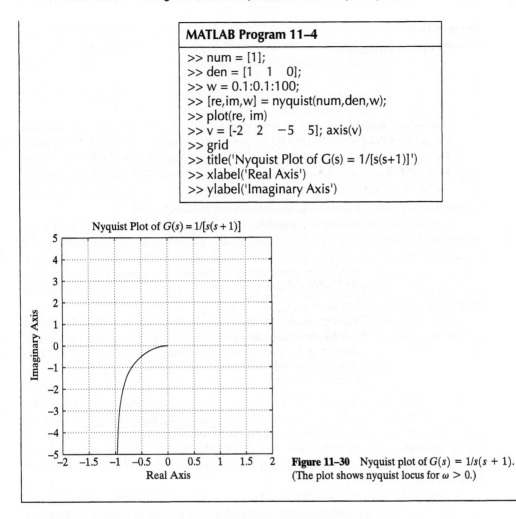

MATLAB Program 11–4

```
>> num = [1];
>> den = [1   1   0];
>> w = 0.1:0.1:100;
>> [re,im,w] = nyquist(num,den,w);
>> plot(re, im)
>> v = [-2   2   -5   5]; axis(v)
>> grid
>> title('Nyquist Plot of G(s) = 1/[s(s+1)]')
>> xlabel('Real Axis')
>> ylabel('Imaginary Axis')
```

Nyquist Plot of $G(s) = 1/[s(s+1)]$

Figure 11–30 Nyquist plot of $G(s) = 1/s(s + 1)$. (The plot shows nyquist locus for $\omega > 0$.)

11–6 DESIGN OF CONTROL SYSTEMS IN THE FREQUENCY DOMAIN

This section discusses control systems design based on the Bode diagram approach, an approach that is particularly useful for the following reasons:

1. In the Bode diagram, the low-frequency asymptote of the magnitude curve is indicative of one of the static error constants K_p, K_v, or K_a.

2. Specifications of the transient response can be translated into those of the frequency response in terms of the phase margin, gain margin, bandwidth, and so forth. These specifications can be easily handled in the Bode diagram. In particular, the phase and gain margins can be read directly from the Bode diagram.

3. The design of a compensator or controller to satisfy the given specifications (in terms of the phase margin and gain margin) can be carried out in the Bode diagram in a simple and straightforward manner.

In this section we present the lead, lag, and lag–lead compensation techniques. Before we begin design problems, we shall briefly explain each of these compensations.

Lead compensation is commonly used for improving stability margins. Lead compensation increases the system bandwidth; thus, the system has a faster response. However, a system using lead compensation may be subjected to high-frequency noise problems due to its increased high-frequency gains.

Lag compensation reduces the system gain at higher frequencies without reducing the system gain at lower frequencies. The system bandwidth is reduced, and thus the system has a slower speed of response. Because of the reduced high-frequency gain, the total system gain can be increased, and thereby low-frequency gain can be increased and the steady-state accuracy can be improved. Also, any high-frequency noises involved in the system can be attenuated.

Lag-lead compensation is a combination of lag compensation and lead compensation. A compensator that has characterstics of both a lag compensator and a lead compensator is known as a *lag–lead compensator*. With the use of a lag–lead compensator, the low-frequency gain can be increased (and hence the steady-state accuracy can be improved), while at the same time the system bandwidth and stability margins can be increased.

The PID controller is a special case of a lag–lead controller. The PD control action, which affects the high-frequency region, increases the phase-lead angle and improves the system stability, as well as increasing the system bandwidth (and thus increasing the speed of response). That is, the PD controller behaves in much the same way as a lead compensator. The PI control action affects the low-frequency portion and, in fact, increases the low-frequency gain and improves steady-state accuracy. Therefore, the PI controller acts as a lag compensator. The PID control action is a combination of the PI and PD control actions. The design techniques for PID controllers basically follow those of lag–lead compensators. (In industrial control systems, however, each of the PID control actions in the PID controller may be adjusted experimentally.)

In what follows, we first discuss the design of a lead compensator. Then we treat the design of a lag compensator, followed by the design of a lag–lead compensator. We use MATLAB to obtain step and ramp responses of the designed systems to verify their transient-response performance.

Frequency characteristics of lead, lag, and lag–lead compensators. Before we discuss design problems, we shall examine the frequency characteristics of the lead compensator, lag compensator, and lag–lead compensator.

Characteristics of a lead compensator. Consider a lead compensator having the following transfer function:

$$K_c \alpha \frac{Ts + 1}{\alpha Ts + 1} = K_c \frac{s + \dfrac{1}{T}}{s + \dfrac{1}{\alpha T}} \qquad 0 < \alpha < 1$$

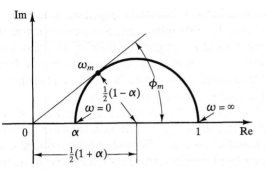

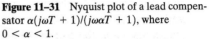

Figure 11–31   Nyquist plot of a lead compensator $\alpha(j\omega T + 1)/(j\omega\alpha T + 1)$, where $0 < \alpha < 1$.

The lead compensator has a zero at $s = -1/T$ and a pole at $s = -1/(\alpha T)$. Since $0 < \alpha < 1$, the zero is always located to the right of the pole in the complex plane. Note that, for a small value of α, the pole is located far to the left. The minimum value of α is limited by the physical construction of the lead compensator and is usually taken to be about 0.05. (This means that the maximum phase lead that may be produced by a lead compensator is about 65°.)

Figure 11–31 shows the Nyquist plot of

$$K_c\alpha\frac{j\omega T + 1}{j\omega\alpha T + 1} \qquad 0 < \alpha < 1$$

with $K_c = 1$. For a given value of α, the angle between the positive real axis and the tangent line drawn from the origin to the semicircle gives the maximum phase lead angle ϕ_m. We shall call the frequency at the tangent point ω_m. From Figure 11–31 the phase angle at $\omega = \omega_m$ is ϕ_m, where

$$\sin\phi_m = \frac{\dfrac{1-\alpha}{2}}{\dfrac{1+\alpha}{2}} = \frac{1-\alpha}{1+\alpha} \tag{11-10}$$

Equation (11–10) relates the maximum phase-lead angle to the value of α.

Figure 11–32 shows the Bode diagram of a lead compensator when $K_c = 1$ and $\alpha = 0.1$. The corner frequencies for the lead compensator are $\omega = 1/T$ and $\omega = 1/(\alpha T) = 10/T$. Examining Figure 11–32, we see that ω_m is the geometric mean of the two corner frequencies, or

$$\log\omega_m = \frac{1}{2}\left(\log\frac{1}{T} + \log\frac{1}{\alpha T}\right)$$

Hence,

$$\omega_m = \frac{1}{\sqrt{a}\,T} \tag{11-11}$$

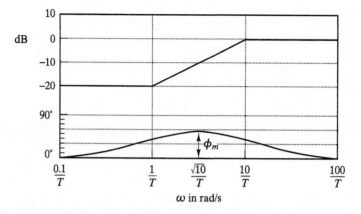

Figure 11–32 Bode diagram of a lead compensator $\alpha(j\omega T + 1)/(j\omega\alpha T + 1)$, where $\alpha = 0.1$.

As the figure indicates, the lead compensator is basically a high-pass filter. (High frequencies are passed, but low frequencies are attenuated.)

Characteristics of a lag compensator. Consider a lag compensator that has the following transfer function:

$$G_c(s) = K_c\beta\frac{Ts + 1}{\beta Ts + 1} = K_c\frac{s + \dfrac{1}{T}}{s + \dfrac{1}{\beta T}} \qquad \beta > 1$$

In the complex plane, the lag compensator has a zero at $s = -1/T$ and a pole at $s = -1/(\beta T)$. The pole is located to the right of the zero.

Figure 11–33 shows a Bode diagram of the given lag compensator, where $K_c = 1$ and $\beta = 10$. The corner frequencies are at $\omega = 1/T$ and $\omega = 1/(\beta T)$. As seen from Figure 11–33, where the values of K_c and β are set equal to 1 and 10, respectively, the magnitude of the lag compensator becomes 10 (or 20 dB) at low frequencies and unity (or 0 dB) at high frequencies. Thus, the lag compensator is essentially a low-pass filter.

Characteristics of a lag–lead compensator. Consider the lag–lead compensator given by

$$G_c(s) = K_c\left(\frac{s + \dfrac{1}{T_1}}{s + \dfrac{\gamma}{T_1}}\right)\left(\frac{s + \dfrac{1}{T_2}}{s + \dfrac{1}{\beta T_2}}\right) \qquad (11\text{–}12)$$

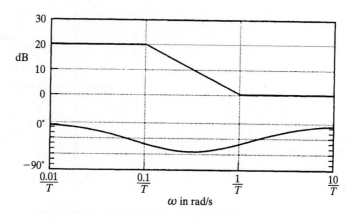

Figure 11–33   Bode diagram of a lag compensator
$\beta(j\omega T + 1)/(j\omega\beta T + 1)$, where $\beta = 10$.

where $\gamma > 1$ and $\beta > 1$. The term

$$\frac{s + \dfrac{1}{T_1}}{s + \dfrac{\gamma}{T_1}} = \frac{1}{\gamma}\left(\frac{T_1 s + 1}{\dfrac{T_1}{\gamma}s + 1}\right) \qquad \gamma > 1$$

produces the effect of the lead network, and the term

$$\frac{s + \dfrac{1}{T_2}}{s + \dfrac{1}{\beta T_2}} = \beta\left(\frac{T_2 s + 1}{\beta T_2 s + 1}\right) \qquad \beta > 1$$

produces the effect of the lag network.

In designing a lag–lead compensator, we frequently choose $\gamma = \beta$. (This is not necessary. We can, of course, chose $\gamma \neq \beta$.) In what follows, we shall consider the case where $\gamma = \beta$. The Nyquist plot of the lag–lead compensator with $K_c = 1$ and $\gamma = \beta$ is shown in Figure 11–34. It can be seen that, for $0 < \omega < \omega_1$, the compensator acts as a lag compensator, while for $\omega_1 < \omega < \infty$, it acts as a lead compensator. The frequency ω_1 is the frequency at which the phase angle is zero. It is given by

$$\omega_1 = \frac{1}{\sqrt{T_1 T_2}}$$

(For a derivation of this equation, see **Problem A–11–14**.)

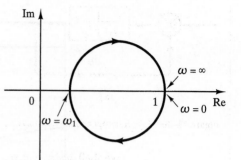

Figure 11–34 Nyquist plot of the lag–lead compensator given by Equation (11–12) with $K_c = 1$ and $\gamma = \beta$.

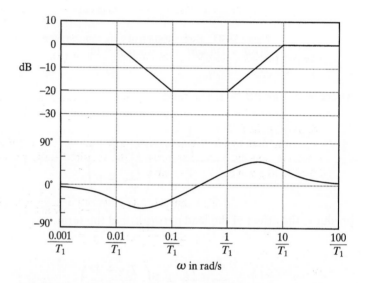

Figure 11–35 Bode diagram of the lag–lead compensator given by Equation (11–12) with $K_c = 1$, $\gamma = \beta = 10$ and $T_2 = 10T_1$.

Figure 11–35 shows the Bode diagram of a lag–lead compensator when $K_c = 1$, $\gamma = \beta = 10$, and $T_2 = 10T_1$. Notice that the magnitude curve has the value 0 dB in the low- and high-frequency regions.

Example 11–6 Design of a Lead Compensator

Consider the system shown in Figure 11–36(a). We wish to design a compensator such that the closed-loop system will satisfy the following requirements:

$$\text{static velocity error constant} = K_v = 20 \text{ s}^{-1}$$
$$\text{phase margin} = 50°$$
$$\text{gain margin} \geq 10 \text{ dB}$$

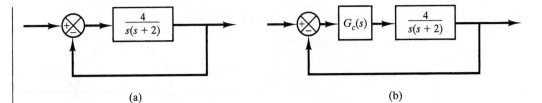

Figure 11–36   (a) Uncompensated system; (b) compensated system.

We shall design a lead compensator $G_c(s)$ of the form

$$G_c(s) = K_c \alpha \frac{Ts + 1}{\alpha Ts + 1} = K_c \frac{s + \dfrac{1}{T}}{s + \dfrac{1}{\alpha T}} \qquad 0 < \alpha < 1$$

The compensated system is shown in Figure 11–36 (b).
    We define

$$G_1(s) = KG(s) = \frac{4K}{s(s + 2)}$$

where $K = K_c \alpha$.
    The first step in the design is to adjust the gain K to meet the steady-state performance specification or to provide the required static velocity error constant K_v. Since K_v is given as 20 s^{-1}, we have

$$K_v = \lim_{s \to 0} sG_c(s)G(s) = \lim_{s \to 0} s \frac{Ts + 1}{\alpha Ts + 1} G_1(s)$$

$$= \lim_{s \to 0} \frac{s4K}{s(s + 2)} = 2K = 20$$

or

$$K = 10$$

With $K = 10$, the compensated system will satisfy the steady-state requirement.
    We shall next plot the Bode diagram of

$$G_1(s) = \frac{40}{s(s + 2)}$$

MATLAB Program 11–5 produces the Bode diagram shown in Figure 11–37. From this plot, the phase margin is found to be 17°. The gain margin is $+\infty$ dB.
    Since the specification calls for a phase margin of 50°, the additional phase lead necessary to satisfy the phase-margin requirement is 33°. A lead compensator can contribute this amount.
    Noting that the addition of a lead compensator modifies the magnitude curve in the Bode diagram, we realize that the gain crossover frequency will be shifted to the right. Accordingly, we must offset the increased phase lag of $G_1(j\omega)$ due to this increase in the gain crossover frequency. Taking the shift of the gain crossover frequency into consideration, we may assume that ϕ_m, the maximum phase lead required, is approximately 38°. (This means that approximately 5° has been added to compensate for the shift in

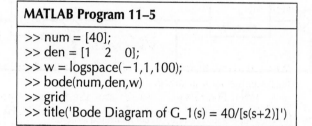

MATLAB Program 11–5

```
>> num = [40];
>> den = [1   2   0];
>> w = logspace(−1,1,100);
>> bode(num,den,w)
>> grid
>> title('Bode Diagram of G_1(s) = 40/[s(s+2)]')
```

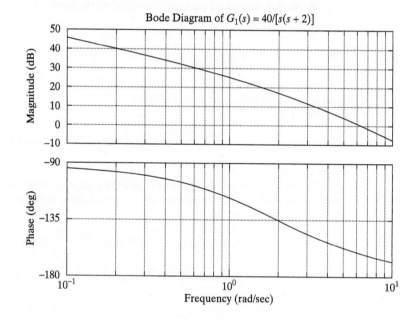

Figure 11–37   Bode diagram of $G_1(s) = 40/[s(s + 2)]$.

the gain crossover frequency.) Since

$$\sin \phi_m = \frac{1 - \alpha}{1 + \alpha}$$

$\phi_m = 38°$ corresponds to $\alpha = 0.2379$. Note that $\alpha = 0.24$ corresponds to $\phi_m = 37.8°$. Whether we choose $\phi_m = 38°$ or $\phi_m = 37.8°$ does not make much difference in the final solution. Hence, let us choose $\alpha = 0.24$.

Once the attenuation factor α has been determined on the basis of the required phase-lead angle, the next step is to determine the corner frequencies $\omega = 1/T$ and $\omega = 1/(\alpha T)$ of the lead compensator. Notice that the maximum phase-lead angle ϕ_m occurs at the geometric mean of the two corner frequencies, or $\omega = 1/(\sqrt{\alpha}T)$.

The amount of the modification in the magnitude curve at $\omega = 1/(\sqrt{\alpha}T)$ due to the inclusion of the term $(Ts + 1)/(\alpha Ts + 1)$ is

$$\left| \frac{1 + j\omega T}{1 + j\omega \alpha T} \right|_{\omega = \frac{1}{\sqrt{\alpha}T}} = \left| \frac{1 + j\dfrac{1}{\sqrt{\alpha}}}{1 + j\alpha \dfrac{1}{\sqrt{\alpha}}} \right| = \frac{1}{\sqrt{\alpha}}$$

Note that

$$\frac{1}{\sqrt{\alpha}} = \frac{1}{\sqrt{0.24}} = 0.2041 = 6.2 \text{ dB}$$

We need to find the frequency point where the total magnitude becomes 0 dB when the lead compensator is added.

From Figure 11–37, we see that the frequency point where the magnitude of $G_1(j\omega)$ is −6.2 dB occurs between $\omega = 1$ and 10 rad/s. Hence, we plot a new Bode diagram of $G_1(j\omega)$ in the frequency range between $\omega = 1$ and 10 to locate the exact point where $G_1(j\omega) = -6.2$ dB. MATLAB Program 11–6 produces the Bode diagram in this frequency range, which is shown in Figure 11–38. From the diagram, we find that the frequency point where $|G_1(j\omega)| = -6.2$ dB is $\omega = 9$ rad/s. Let us select this frequency to be the new gain crossover frequency, or $\omega_c = 9$ rad/s. Noting that this frequency corresponds to $1/(\sqrt{\alpha}T)$, or

$$\omega_c = \frac{1}{\sqrt{\alpha}T}$$

MATLAB Program 11–6

```
>> num = [40];
>> den = [1   2   0];
>> w = logspace(0,1,100);
>> bode(num,den,w)
>> grid
>> title('Bode Diagram of G(s) = 40/[s(s+2)]')
```

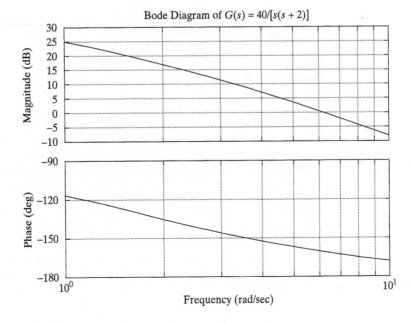

Figure 11–38 Bode diagram of $G_1(s) = 40/[s(s + 2)]$.

we obtain

$$\frac{1}{T} = \omega_c\sqrt{\alpha} = 9\sqrt{0.24} = 4.409$$

and

$$\frac{1}{\alpha T} = \frac{\omega_c}{\sqrt{\alpha}} = \frac{9}{\sqrt{0.24}} = 18.371$$

The lead compensator thus determined is

$$G_c(s) = K_c\frac{s + 4.409}{s + 18.371} = K_c\alpha\frac{0.227s + 1}{0.0544s + 1}$$

where K_c is determined as

$$K_c = \frac{K}{\alpha} = \frac{10}{0.24} = 41.667$$

Thus, the transfer function of the compensator becomes

$$G_c(s) = 41.667\frac{s + 4.409}{s + 18.371} = 10\frac{0.227s + 1}{0.0544s + 1}$$

MATLAB Program 11–7 produces the Bode diagram of this lead compensator, which is shown in Figure 11–39. Note that

$$\frac{G_c(s)}{K}G_1(s) = \frac{G_c(s)}{10}10G(s) = G_c(s)G(s)$$

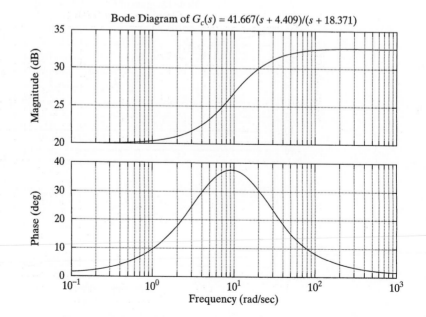

Bode Diagram of $G_c(s) = 41.667(s + 4.409)/(s + 18.371)$

Figure 11–39   Bode diagram of the compensator.

MATLAB Program 11–7

```
>> numc = [41.667      183.71];
>> denc = [1      18.371];
>> w = logspace(-1,3,100);
>> bode(numc,denc,w)
>> grid
>> title('Bode Diagram of G_c(s) = 41.667 (s + 4.409)/(s+18.371)')
```

The open-loop transfer function of the designed system is

$$G_c(s)G(s) = 41.667 \frac{s + 4.409}{s + 18.371} \frac{4}{s(s + 2)}$$

$$= \frac{166.668s + 734.839}{s^3 + 20.371s^2 + 36.742s}$$

MATLAB Program 11–8 will produce the Bode diagram of $G_c(s)G(s)$, which is shown in Figure 11–40. From the figure, notice that the phase margin is approximately 50° and

---

MATLAB Program 11–8

```
>> num = [166.668      734.839];
>> den = [1      20.371      36.742      0];
>> w = logspace(-1,3,100);
>> bode(num,den,w)
>> grid
>> title('Bode Diagram of G_c(s)G(s)')
```

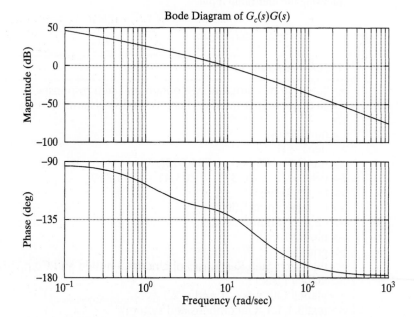

Figure 11–40   Bode diagram of $G_c(s)G(s)$.

the gain margin is $+\infty$ dB. Since the static velocity error constant is $20\,\mathrm{s}^{-1}$ ($734.839/36.742 = 20$), all requirements are satisfied. Hence, the designed system is satisfactory.

Unit-step response   We shall check the unit-step response of the designed system. We plot both the unit-step response of the designed system and that of the original, un-compensated system.

The closed-loop transfer function of the original, uncompensated system is

$$\frac{C_1(s)}{R_1(s)} = \frac{4}{s^2 + 2s + 4}$$

The closed-loop transfer function of the compensated system is

$$\frac{C_2(s)}{R_2(s)} = \frac{41.667(s + 4.409) \times 4}{(s + 18.371)s(s + 2) + 41.667(s + 4.409) \times 4}$$

$$= \frac{166.668s + 734.839}{s^3 + 20.371s^2 + 203.41s + 734.839}$$

MATLAB Program 11–9 produces the unit-step responses of the uncompensated and compensated systems. The resulting response curves are shown in Figure 11–41.

Unit-ramp response   It is worthwhile to check the unit-ramp response of the compen-sated system. Since $K_v = 20\,\mathrm{s}^{-1}$, the steady-state error following the unit-ramp input will be $1/K_v = 0.05$. The static velocity error constant of the uncompensated system is $2\,\mathrm{s}^{-1}$. Hence, the original uncompensated system will have a large steady-state error following the unit-ramp input.

MATLAB Program 11–9

```
>> % —— In this program, we obtain unit-step responses
>> % of uncompensated and compensated systems ——
>>
>> num1 = [4];
>> den1 = [1   2   4];
>> num2 = [166.668    734.839];
>> den2 = [1    20.371    203.41    734.839];
>> t = 0:0.01:6;
>> c1 = step(num1,den1,t);
>> c2 = step(num2,den2,t);
>> plot(t,c1,t,c2)
>> grid
>> title('Unit-Step Responses of Uncompensated and Compensated Systems')
>> xlabel('t (sec)')
>> ylabel('Outputs c_1 and c_2')
>> text(3.1,1.1,'Uncompensated system')
>> text(0.8,1.24,'Compensated system')
```

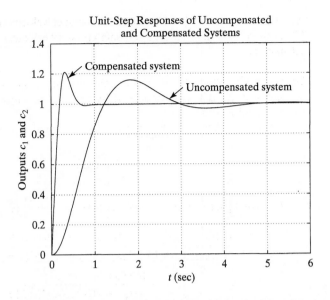

Figure 11–41 Unit-step responses of uncompensated and compensated systems.

MATLAB Program 11–10 produces the unit-ramp response curves. [Note that the unit-ramp response is obtained as the unit-step response of $C_i(s)/sR(s)$, where $i = 1, 2$ and $R(s)$ is a unit-step input.] The resulting curves are shown in Figure 11–42. The compensated system has a steady-state error equal to one-tenth that of the original, uncompensated system.

MATLAB Program 11–10

```
>> % —— In this program, we obtain unit-ramp responses
>> % of uncompensated and compensated systems ——
>>
>> num1 = [4];
>> den1 = [1   2   4   0];
>> num2 = [166.668      734.839];
>> den2 = [1      20.371     203.41      734.839    0];
>> t = 0:0.01:4;
>> c1 = step(num1,den1,t);
>> c2 = step(num2,den2,t);
>> plot(t,c1,t,c2,t,t,'-')
>> grid
>> title('Unit-Ramp Responses of Uncompensated and Compensated Systems')
>> xlabel('t (sec)')
>> ylabel('Input and Outputs c_1 and c_2')
>> text(1.85,3.35,'Input')
>> text(2.7,1.1,'Uncompensated')
>> text(2.7,0.85,'system')
>> text(0.4,2.35,'Compensated')
>> text(0.4,2.1'system')
```

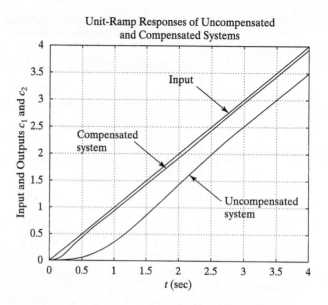

Figure 11–42 Unit-ramp responses of uncompensated and compensated systems.

Example 11–7   Design of a Lag Compensator

Consider the system shown in Figure 11–43. The open-loop transfer function is given by

$$G(s) = \frac{1}{s(s + 1)(0.5s + 1)}$$

It is desired to compensate the system so that the static velocity error constant K_v is $5\ s^{-1}$, the phase margin is at least $40°$, and the gain margin is at least 10 dB.

We shall use a lag compensator of the form

$$G_c(s) = K_c\beta\frac{Ts + 1}{\beta Ts + 1} = K_c\frac{s + \dfrac{1}{T}}{s + \dfrac{1}{\beta T}} \qquad \beta > 1$$

Define

$$K_c\beta = K$$

and

$$G_1(s) = KG(s) = \frac{K}{s(s + 1)(0.5s + 1)}$$

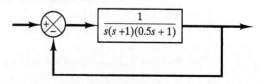

Figure 11–43 Control system.

The first step in the design is to adjust the gain K to meet the required static velocity error constant. Thus,

$$K_v = \lim_{s \to 0} sG_c(s)G(s) = \lim_{s \to 0} s\frac{Ts + 1}{\beta Ts + 1}G_1(s) = \lim_{s \to 0} sG_1(s)$$

$$= \lim_{s \to 0} \frac{sK}{s(s + 1)(0.5s + 1)} = K = 5$$

or

$$K = 5$$

With $K = 5$, the compensated system satisfies the steady-state performance requirement.

We shall next plot the Bode diagram of

$$G_1(j\omega) = \frac{5}{j\omega(j\omega + 1)(0.5j\omega + 1)}$$

The magnitude curve and phase-angle curve of $G_1(j\omega)$ are shown in Figure 11–44. From this plot, the phase margin is found to be $-20°$, which means that the gain adjusted but uncompensated system is unstable.

Noting that the addition of a lag compensator modifies the phase curve of the Bode diagram, we must allow 5° to 12° to the specified phase margin to compensate for the modification of the phase curve. Since the frequency corresponding to a phase margin of 40° is 0.7 rad/s, the new gain crossover frequency (of the compensated system) must be chosen to be near this value. To avoid overly large time constants for the lag compensator, we shall choose the corner frequency $\omega = 1/T$ (which corresponds to the zero of the lag compensator) to be 0.1 rad/s. Since this corner frequency is not too far below the new gain crossover frequency, the modification in the phase curve may not be

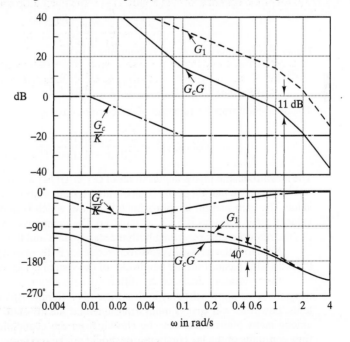

Figure 11–44   Bode diagrams for $G_1 = KG$ (gain-adjusted, but uncompensated system), G_c/K (gain-adjusted compensator), and G_cG (compensated system).

small. Hence, we add about 12° to the given phase margin as an allowance to account for the lag angle introduced by the lag compensator. The required phase margin is now 52°. The phase angle of the uncompensated open-loop transfer function is −128° at about $\omega = 0.5$ rad/s, so we choose the new gain crossover frequency to be 0.5 rad/s. To bring the magnitude curve down to 0 dB at this new gain crossover frequency, the lag compensator must give the necessary attenuation, which in this case is −20 dB. Hence,

$$20 \log \frac{1}{\beta} = -20$$

or

$$\beta = 10$$

The other corner frequency $\omega = 1/(\beta T)$ which corresponds to the pole of the lag compensator, is then determined as

$$\frac{1}{\beta T} = 0.01 \text{ rad/s}$$

Thus, the transfer function of the lag compensator is

$$G_c(s) = K_c(10) \frac{10s + 1}{100s + 1} = K_c \frac{s + \dfrac{1}{10}}{s + \dfrac{1}{100}}$$

Since the gain K was determined to be 5 and β was determined to be 10, we have

$$K_c = \frac{K}{\beta} = \frac{5}{10} = 0.5$$

Hence, the compensator $G_c(s)$ is determined to be

$$G_c(s) = 5 \frac{10s + 1}{100s + 1}$$

The open-loop transfer function of the compensated system is thus

$$G_c(s)G(s) = \frac{5(10s + 1)}{s(100s + 1)(s + 1)(0.5s + 1)}$$

The magnitude and phase-angle curves of $G_c(j\omega)G(j\omega)$ are also shown in Figure 11–44.

The phase margin of the compensated system is about 40°, which is the required value. The gain margin is approximately 11 dB, which is quite acceptable. The static velocity error constant is 5 s⁻¹, as required. The compensated system, therefore, satisfies the requirements regarding both the steady state and the relative stability.

Note that the new gain crossover frequency is decreased from approximately 2 to 0.5 rad/s. This means that the bandwidth of the system is reduced.

Compensators designed by different methods or by different designers (even using the same approach) may look sufficiently different. Any of the well-designed systems, however, will give similar transient and steady-state performance. The best among many alternatives may be chosen from the economic consideration that the time constants of the lag compensator should not be too large.

Finally, we shall examine the unit-step response and unit-ramp response of the compensated system and the original, uncompensated system. The closed-loop transfer functions of the compensated and uncompensated systems are

$$\frac{C(s)}{R(s)} = \frac{50s + 5}{50s^4 + 150.5s^3 + 101.5s^2 + 51s + 5}$$

and

$$\frac{C(s)}{R(s)} = \frac{1}{0.5s^3 + 1.5s^2 + s + 1}$$

respectively. MATLAB Program 11–11 will produce the unit-step and unit-ramp responses of the compensated and uncompensated systems. The resulting unit-step response curves and unit-ramp response curves are shown in Figures 11–45 and 11–46,

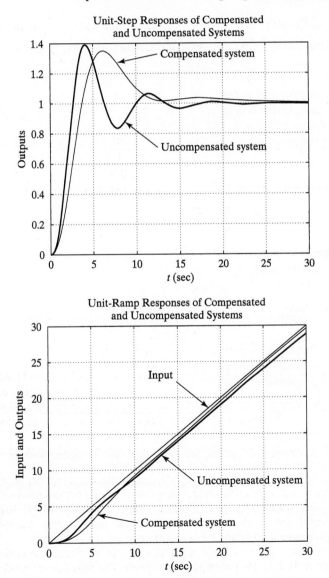

Figure 11–45   Unit-step response curves for the compensated and uncompensated systems.

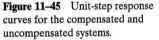

Figure 11–46   Unit-ramp response curves for the compensated and uncompensated systems.

MATLAB Program 11–11

```
>> % —— Unit-step response ——
>>
>> num = [1];
>> den = [0.5   1.5   1   1];
>> numc = [50   5];
>> denc = [50   150.5   101.5   51   5];
>> t = 0:0.01:30;
>> [c1,x1,t] = step(num,den,t);
>> [c2,x2,t] = step(numc,denc,t);
>> plot(t,c1,'.',t,c2,'-')
>> grid
>> title('Unit-Step Responses of Compensated and Uncompensated Systems')
>> xlabel('t (sec)')
>> ylabel('Outputs')
>> text(12.6,1.32,'Compensated system')
>> text(12.6,0.7,'Uncompensated system')
>>
>> % —— Unit-ramp response ——
>>
>> num1 = [1];
>> den1 = [0.5   1.5   1   1   0];
>> num1c = [50   5];
>> den1c = [50   150.5   101.5   51   5   0];
>> t = 0:0.01:30;
>> [y1, z1, t] = step(num1,den1,t);
>> [y2, z2, t] = step(num1c,den1c,t);
>> plot (t,y1,'.',t,y2,'-',t,t,'--')
>> grid
>> title('Unit-Ramp Responses of Compensated and Uncompensated Systems')
>> xlabel('t (sec)')
>> ylabel('Input and Outputs')
>> text(10.7,2.5,'Compensated system')
>> text(17.2,8.3,'Uncompensated system')
>> text(12, 23,'Input')
```

respectively. From the response curves, we find that the designed system is satisfactory.

Note that the zero and the poles of the designed closed-loop systems are as follows:

$$\text{zero at } s = -0.1$$

$$\text{poles at } s = -0.2859 \pm j0.5196, \qquad s = -0.1228, \qquad s = -2.3155$$

The dominant closed-loop poles are very close to the $j\omega$-axis, with the result that the response is slow. Also, a pair consisting of the closed-loop pole at $s = -0.1228$ and the zero at $s = -0.1$ produces a slowly decreasing tail of small amplitude.

A few comments on lag compensation

1. Lag compensators are essentially low-pass filters. Therefore, lag compensation permits a high gain at low frequencies (which improves the steady-state performance of a system) and reduces gain in the higher critical range of frequencies so as to improve the phase margin. Note that in lag compensation we utilize the attenuation characteristic of the lag compensator at high frequencies, rather than the phase-lag characteristic. (The phase-lag characteristic is of no use for compensation purposes.)

2. Suppose that the zero and pole of a lag compensator are located at $s = -z$ and $s = -p$, respectively. Then the exact location of the zero and pole is not critical, provided that they are close to the origin and the ratio z/p is equal to the required multiplication factor of the static velocity error constant.

   Note, however, that the zero and pole of the lag compensator should not be located unnecessarily close to the origin, because the lag compensator will create an additional closed-loop pole in the same region as the zero and pole of the lag compensator.

   The closed-loop pole located near the origin gives a very slowly decaying transient response, although its magnitude will become very small because the zero of the lag compensator will almost cancel the effect of this pole. However, the transient response (decay) due to this pole is so slow that the settling time will be adversely affected.

   It is also noted that in the system compensated by a lag compensator the transfer function between the plant disturbance and the system error may not involve a zero that is near the closed-loop pole that is located close to the origin. Therefore, the transient response to the disturbance input may last very long.

3. The attenuation due to the lag compensator will shift the gain crossover frequency to a lower frequency at which the phase margin is acceptable. Thus, the lag compensator will reduce the bandwidth of the system and will result in a slower transient response. [The phase-angle curve of $G_c(j\omega)G(j\omega)$ is relatively unchanged near and above the new gain crossover frequency.]

4. Since the lag compensator tends to integrate the input signal, it acts approximately as a proportional-plus-integral controller. Because of this, a lag-compensated system tends to become less stable. To avoid this undesirable feature, the time constant T should be made sufficiently larger than the largest time constant of the system.

5. Conditional stability may occur when a system having saturation or limiting is compensated by the use of a lag compensator. When saturation or limiting takes place in the system, it reduces the effective loop gain. Then the system becomes less stable, and may even become unstable, as shown in Figure 11–47. To avoid this possibility, the system must be designed so that the effect of lag

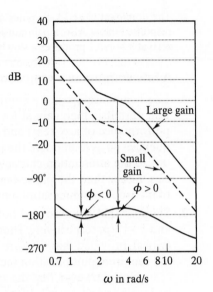

Figure 11–47 Bode diagram of a conditionally stable system.

compensation becomes significant only when the amplitude of the input to
the saturating element is small. (This can be done by means of minor feed-
back-loop compensation.)

Example 11–8   Design of a Lag–Lead Compensator

Consider the unity-feedback system whose open-loop transfer function is

$$G(s) = \frac{K}{s(s + 1)(s + 4)}$$

Design a compensator $G_c(s)$ such that the static velocity error constant is 10 s^{-1}, the
phase margin is 50°, and the gain margin is 10 dB or more.

We shall design a lag–lead compensator of the form

$$G_c(s) = K_c \frac{\left(s + \dfrac{1}{T_1}\right)\left(s + \dfrac{1}{T_2}\right)}{\left(s + \dfrac{\beta}{T_1}\right)\left(s + \dfrac{1}{\beta T_2}\right)}$$

Then the open-loop transfer function of the compensated system is $G_c(s)G(s)$. Since
the gain K of the plant is adjustable, let us assume that $K_c = 1$. Then $\lim\limits_{s \to 0} G_c(s) = 1$.
From the requirement on the static velocity error constant, we obtain

$$K_v = \lim_{s \to 0} sG_c(s)G(s) = \lim_{s \to 0} sG_c(s)\frac{K}{s(s + 1)(s + 4)}$$

$$= \frac{K}{4} = 10$$

Hence,

$$K = 40$$

We shall first plot a Bode diagram of the uncompensated system with $K = 40$. MATLAB Program 11–12 may be used to plot this Bode diagram, which is shown in Figure 11–48.

From Figure 11–48 the phase margin of the uncompensated system is found to be $-16°$, which indicates that the gain adjusted but uncompensated system is unstable. The next step in the design of a lag–lead compensator is to choose a new gain crossover frequency. From the phase-angle curve for $G(j\omega)$, we notice that the phase crossover frequency is $\omega = 2$ rad/s. We may choose the new gain crossover frequency to be 2 rad/s so that the phase-lead angle required at $\omega = 2$ rad/s is about 50°. A single lag–lead compensator can provide this amount of phase-lead angle quite easily.

Once we choose the gain crossover frequency to be 2 rad/s, we can determine the corner frequencies of the phase-lag portion of the lag–lead compensator. Let us choose the corner frequency $\omega = 1/T_2$ (which corresponds to the zero of the phase-lag portion of the compensator) to be 1 decade below the new gain crossover frequency; that is, $\omega = 0.2$ rad/s. For another corner frequency $\omega = 1/(\beta T_2)$, we need the value of β, which can be determined from a consideration of the lead portion of the compensator.

MATLAB Program 11–12
``` >> num = [40]; >> den = [1   5   4   0]; >> w = logspace(-1,1,100); >> bode(num,den,w) >> grid >> title('Bode Diagram of G(s) = 40/[s(s+1)(s+4)]') ```

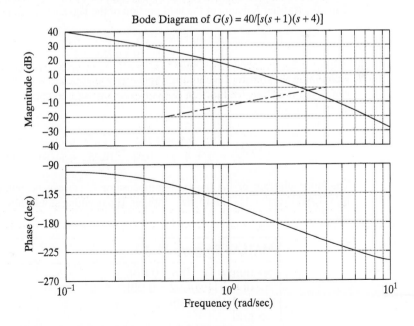

Figure 11–48   Bode diagram of $G(s) = 40/[s(s + 1)(s + 4)]$.

For the lead compensator, the maximum phase-lead angle ϕ_m is given by Equation (11–10). By substituting $\alpha = 1/\beta$ into that equation, we have

$$\sin \phi_m = \frac{\beta - 1}{\beta + 1}$$

Notice that $\beta = 10$ corresponds to $\phi_m = 54.9°$. Since we need a 50° phase margin, we may choose $\beta = 10$. (Note that we will be using several degrees less than the maximum angle, 54.9°.) Thus,

$$\beta = 10$$

Then the corner frequency $\omega = 1/(\beta T_2)$ (which corresponds to the pole of the phase-lag portion of the compensator) becomes

$$\omega = 0.02$$

The transfer function of the phase-lag portion of the lag–lead compensator becomes

$$\frac{s + 0.2}{s + 0.02} = 10\left(\frac{5s + 1}{50s + 1}\right)$$

The phase-lead portion can be determined as follows: Since the new gain crossover frequency is $\omega = 2$ rad/s, from Figure 11–48, $|G(j2)|$ is found to be 6 dB. Hence, if the lag–lead compensator contributes -6 dB at $\omega = 2$ rad/s, then the new gain crossover frequency is as desired. From this requirement, it is possible to draw a straight line of slope 20 dB/decade passing through the point (-6 dB, 2 rad/s). (Such a line has been drawn manually in Figure 11–48.) The intersections of this line and the 0-dB line and -20-dB line determine the corner frequencies. From this consideration, for the lead portion, the corner frequencies can be determined as $\omega = 0.4$ rad/s and $\omega = 4$ rad/s. Thus, the transfer function of the lead portion of the lag–lead compensator becomes

$$\frac{s + 0.4}{s + 4} = \frac{1}{10}\left(\frac{2.5s + 1}{0.25s + 1}\right)$$

Combining the transfer functions of the lag and lead portions of the compensator, we can obtain the transfer function $G_c(s)$ of the lag–lead compensator. Since we chose $K_c = 1$, we have

$$G_c(s) = \frac{s + 0.4}{s + 4} \frac{s + 0.2}{s + 0.02} = \frac{(2.5s + 1)(5s + 1)}{(0.25s + 1)(50s + 1)}$$

The Bode diagram of the lag–lead compensator $G_c(s)$ can be obtained by entering MATLAB Program 11–13 into the computer. The resulting plot is shown in Figure 11–49.

MATLAB Program 11–13
``` >> num = [1   0.6   0.08]; >> den = [1   4.02   0.08]; >> bode(num,den) >> grid >> title('Bode Diagram of Lag-Lead Compensator') ```

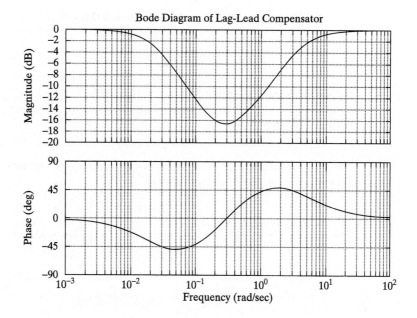

Figure 11–49   Bode diagram of the designed lag–lead compensator.

The open-loop transfer function of the compensated system is

$$G_c(s)G(s) = \frac{(s + 0.4)(s + 0.2)}{(s + 4)(s + 0.02)} \frac{40}{s(s + 1)(s + 4)}$$

$$= \frac{40s^2 + 24s + 3.2}{s^5 + 9.02s^4 + 24.18s^3 + 16.48s^2 + 0.32s}$$

The magnitude and phase-angle curves of the designed open-loop transfer function $G_c(s)G(s)$ are shown in the Bode diagram of Figure 11–50, obtained from MATLAB Program 11–14. From the diagram, we see that the requirements on the phase margin, gain margin, and static velocity error constant are all satisfied.

We next investigate the transient-response characteristics of the designed system.

Unit-step response   Noting that

$$G_c(s)G(s) = \frac{40(s + 0.4)(s + 0.2)}{(s + 4)(s + 0.02)s(s + 1)(s + 4)}$$

MATLAB Program 11–14

```
>> num = [40   24   3.2];
>> den = [1   9.02   24.18   16.48   0.32   0];
>> bode(num,den)
>> grid
>> title('Bode Diagram of G_c(s)G(s)')
```

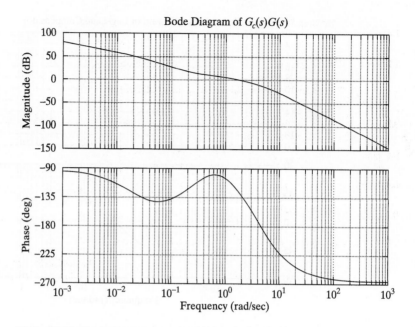

Figure 11–50   Bode diagram of $G_c(s)G(s)$.

we have

$$\frac{C(s)}{R(s)} = \frac{G_c(s)G(s)}{1 + G_c(s)G(s)}$$

$$= \frac{40(s + 0.4)(s + 0.2)}{(s + 4)(s + 0.02)s(s + 1)(s + 4) + 40(s + 0.4)(s + 0.2)}$$

$$= \frac{40s^2 + 24s + 3.2}{s^5 + 9.02s^4 + 24.18s^3 + 56.48s^2 + 24.32s + 3.2}$$

To obtain the unit-step response, we may use MATLAB Program 11–15, which produces the unit-step response curve shown in Figure 11–51. (Note that the gain adjusted but uncompensated system is unstable.)

MATLAB Program 11–15

```
>> % —— Unit-step response ——
>>
>> num = [40   24   3.2];
>> den = [1   9.02   24.18   56.48   24.32   3.2];
>> t = 0:0.05:40;
>> step(num,den,t)
>> grid
>> title('Unit-Step Response of Designed System')
>> xlabel('t'); ylabel('Output')
```

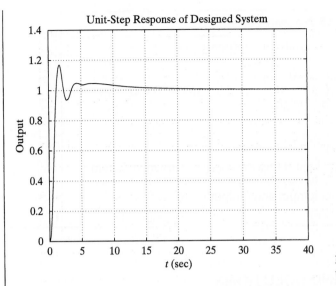

Figure 11–51   Unit-step response of the designed system.

Unit-ramp response   The unit-ramp response of this system may be obtained by entering MATLAB Program 11–16 into the computer. Here, we converted the unit-ramp response of $G_cG/(1 + G_cG)$ into the unit-step response of $G_cG/[s(1 + G_cG)]$. The unit-ramp response curve obtained from the program is shown in Figure 11–52.

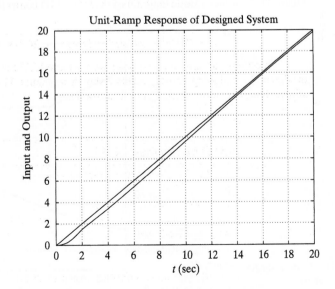

Figure 11–52   Unit-ramp response of the designed system.

```
MATLAB Program 11–16

>> % —— Unit-ramp response ——
>>
>> num = [40   24   3.2];
>> den = [1   9.02   24.18   56.48   24.32   3.2   0];
>> t = 0:0.05:20;
>> c = step(num,den,t);
>> plot(t,c,t,t,'–.')
>> grid
>> title('Unit-Ramp Response of Designed System')
>> xlabel('t (sec)')
>> ylabel('Input and Output')
```

EXAMPLE PROBLEMS AND SOLUTIONS

Problem A–11–1

Plot a Bode diagram of a PID controller given by

$$G_c(s) = 2.2 + \frac{2}{s} + 0.2s$$

Solution The controller transfer function $G_c(s)$ can be written as

$$G_c(s) = 2\frac{(0.1s + 1)(s + 1)}{s}$$

Figure 11–53 shows a Bode diagram of the given PID controller.

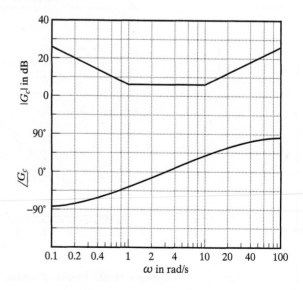

Figure 11–53 Bode diagram of PID controller given by
$G_c(s) = 2(0.1s + 1)(s + 1)/s.$

Problem A–11–2

Consider the mechanical system shown in Figure 11–54. Assume that $x(0) = 0$ and $p(0) = 0$. The numerical values for b_1, b_2, k_1, and k_2 are given as follows:

$$b_1 = 1 \text{ N-s/m}, \qquad b_2 = 2.85 \text{ N-s/m}, \qquad k_1 = 4 \text{ N/m}, \qquad k_2 = 57 \text{ N/m}$$

Assuming the displacement p as the input and the displacement x as the output, obtain the transfer function $X(s)/P(s)$. Then plot a Bode diagram for the system.

Solution   The equations of motion for the system are

$$b_1(\dot{p} - \dot{x}) + k_1(p - x) = b_2(\dot{x} - \dot{y})$$
$$b_2(\dot{x} - \dot{y}) = k_2 y$$

Taking the Laplace transforms of these two equations, substituting zero initial conditions, and eliminating $Y(s)$, we find that

$$\frac{X(s)}{P(s)} = \frac{(b_1 s + k_1)(b_2 s + k_2)}{(b_1 s + k_1)(b_2 s + k_2) + b_2 k_2 s}$$

Substituting the given numerical values into this last equation, we obtain the transfer function $X(s)/P(s)$ as follows:

$$\frac{X(s)}{P(s)} = \frac{(s + 4)(2.85s + 57)}{(s + 4)(2.85s + 57) + 2.85 \times 57s}$$

$$= \frac{(s + 4)(s + 20)}{(s + 1)(s + 80)}$$

The sinusoidal transfer function is

$$\frac{X(j\omega)}{P(j\omega)} = \frac{(j\omega + 4)(j\omega + 20)}{(j\omega + 1)(j\omega + 80)}$$

$$= \frac{(1 + 0.25j\omega)(1 + 0.05j\omega)}{(1 + j\omega)(1 + 0.0125j\omega)}$$

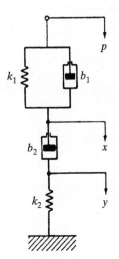

Figure 11–54   Mechanical system.

The corner frequencies are $\omega = 1$ rad/s, $\omega = 4$ rad/s, $\omega = 20$ rad/s, and $\omega = 80$ rad/s. Figure 11–55 shows a Bode diagram for this system. (Both the accurate magnitude curve and the approximate curve by asymptotes are shown.)

Notice that the system acts as a band-stop filter. That is, for $1 < \omega < 80$, the output is attenuated, and for $0 < \omega < 1$ and $80 < \omega$, the output can follow the input faithfully.

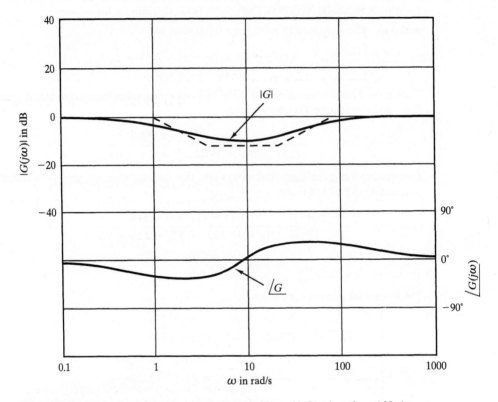

Figure 11–55   Bode diagram for the system shown in Figure 11–54, where $b_1 = 1$ N-s/m, $b_2 = 2.85$ N-s/m, $k_1 = 4$ N/m, and $k_2 = 57$ N/m.

Problem A–11–3

Draw a Bode diagram of the following nonminimum-phase system:

$$\frac{C(s)}{R(s)} = 1 - Ts$$

Obtain the unit-ramp response of the system and plot $c(t)$ versus t.

Solution   The Bode diagram of the system is shown in Figure 11–56. For a unit-ramp input, $R(s) = 1/s^2$, we have

$$C(s) = \frac{1 - Ts}{s^2} = \frac{1}{s^2} - \frac{T}{s}$$

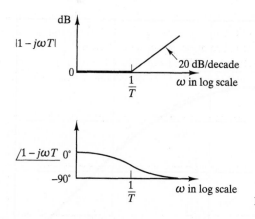

Figure 11–56   Bode diagram of $1 - j\omega T$.

The inverse Laplace transform of $C(s)$ gives

$$c(t) = t - T \qquad \text{for } t \geq 0$$

Figure 11–57 shows the response curve $c(t)$ versus t. (Note the faulty behavior at the start of the response.) A characteristic property of such a nonminimum-phase system is that the transient response starts out in a direction opposite that of the input, but eventually comes back in the same direction as the input.

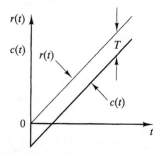

Figure 11–57   Unit-ramp response of the system considered in Problem A–11–3.

Problem A–11–4

A Bode diagram of a dynamic system is given in Figure 11–58. Determine the transfer function of the system from the diagram.

Solution   We first draw straight-line asymptotes to the magnitude curve, as shown in Figure 11–58. [The asymptotes must have slopes of $\pm 20n$ dB/decade $(n = 0, 1, 2, \dots)$.] The intersections of these asymptotes are corner frequencies. Notice that there are two corner frequencies, $\omega = 1$ rad/s and $\omega = 5$ rad/s.

To determine the transfer function of the system, we need to examine the low-frequency region. We have

$$G(j0+) = 14 \text{ dB}$$

Thus,

$$G(j0+) = 5.01$$

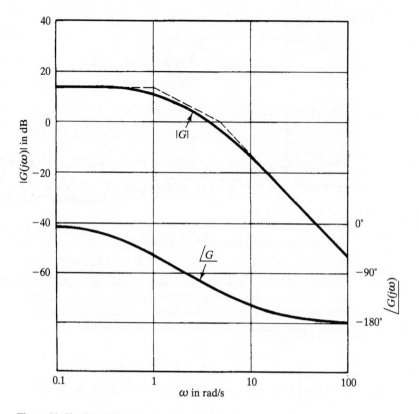

Figure 11–58   Bode diagram of a dynamic system.

Since, from $\omega = 1$ rad/s to $\omega = 5$ rad/s, the asymptote has the slope of -20 dB/decade, the transfer function must have a term $1/(1 + j\omega)$. From $\omega = 5$ rad/s to $\omega = \infty$, the asymptote has a slope of -40 dB/decade. [This means that an additional -20-dB/decade slope has been added to the slope from $\omega = 5$ rad/s to $\omega = \infty$. Hence, the transfer function must involve a term $1/(1 + j0.2\omega)$.]

Now, combining all the terms together, we find the transfer function to be

$$G(j\omega) = \frac{5.01}{(1 + j\omega)(1 + 0.2j\omega)}$$

Notice that the given phase-angle curve starts from $0°$ and approaches $-180°$. The sinusoidal transfer function $G(j\omega)$ determined here is of second order, and the phase angle of the transfer function agrees with the given phase-angle curve. Consequently, the transfer function determined here is

$$G(s) = \frac{5.01}{(s + 1)(0.2s + 1)}$$

Problem A–11–5

For the standard second-order system

$$\frac{C(s)}{R(s)} = \frac{\omega_n^2}{s^2 + 2\zeta\omega_n s + \omega_n^2} \qquad (11\text{--}13)$$

show that the bandwidth ω_b is given by

$$\omega_b = \omega_n(1 - 2\zeta^2 + \sqrt{4\zeta^4 - 4\zeta^2 + 2})^{1/2}$$

Note that ω_b/ω_n is a function only of ζ. Plot a curve of ω_b/ω_n versus ζ.

Solution The bandwidth ω_b is determined from $|C(j\omega_b)/R(j\omega_b)| = -3$ dB. Quite often, instead of -3 dB, we use -3.01 dB, which is equal to 0.707. Thus,

$$\left|\frac{C(j\omega_b)}{R(j\omega_b)}\right| = \left|\frac{\omega_n^2}{(j\omega_b)^2 + 2\zeta\omega_n(j\omega_b) + \omega_n^2}\right| = 0.707$$

Then

$$\frac{\omega_n^2}{\sqrt{(\omega_n^2 - \omega_b^2)^2 + (2\zeta\omega_n\omega_b)^2}} = 0.707$$

from which we get

$$\omega_n^4 = 0.5[(\omega_n^2 - \omega_b^2)^2 + 4\zeta^2\omega_n^2\omega_b^2]$$

Dividing both sides of this last equation by ω_n^4, we obtain

$$1 = 0.5\left\{\left[1 - \left(\frac{\omega_b}{\omega_n}\right)^2\right]^2 + 4\zeta^2\left(\frac{\omega_b}{\omega_n}\right)^2\right\}$$

Solving this last equation for $(\omega_b/\omega_n)^2$ yields

$$\left(\frac{\omega_b}{\omega_n}\right)^2 = -2\zeta^2 + 1 \pm \sqrt{4\zeta^4 - 4\zeta^2 + 2}$$

Since $(\omega_b/\omega_n)^2 > 0$, we take the plus sign in this last equation. Then

$$\omega_b^2 = \omega_n^2(1 - 2\zeta^2 + \sqrt{4\zeta^4 - 4\zeta^2 + 2})$$

or

$$\omega_b = \omega_n(1 - 2\zeta^2 + \sqrt{4\zeta^4 - 4\zeta^2 + 2})^{1/2}$$

Figure 11–59 shows a curve relating ω_b/ω_n to ζ.

Problem A–11–6

Transport lag, or dead time, is a feature of nonminimum-phase behavior and has an excessive phase lag with no attenuation at high frequencies. Transport lags normally exist in thermal, hydraulic, and pneumatic systems.

If $x(t)$ and $y(t)$ are the input and output, respectively, of a transport lag, then

$$y(t) = x(t - L)$$

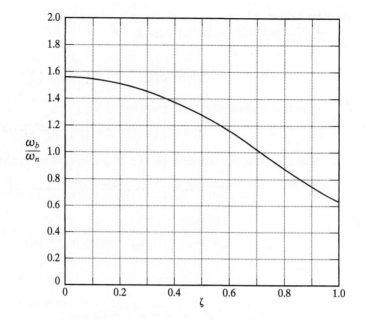

Figure 11–59 Curve of ω_b/ω_n versus ζ (where ω_b is the bandwidth) for the standard second-order system defined by Equation (11–13).

where L is the dead time. The transfer function of the transport lag is

$$\text{transfer function of transport lag} = \left.\frac{\mathscr{L}[x(t-L)1(t-L)]}{\mathscr{L}[x(t)1(t)]}\right|_{\text{zero initial condition}}$$

$$= \frac{X(s)e^{-Ls}}{X(s)} = e^{-Ls}$$

Consider the transport lag given by

$$G(j\omega) = e^{-j\omega L}$$

The magnitude of the transport lag is always equal to unity, since

$$|G(j\omega)| = |\cos \omega L - j \sin \omega L| = 1$$

Therefore, the log magnitude of the transport lag $e^{-j\omega L}$ is equal to 0 dB.

Draw the phase angle curve of the transport lag on the Bode diagram.

Solution   For

$$G(j\omega) = e^{-j\omega L}$$

we have

$$\underline{/G(j\omega)} = -\omega L \qquad \text{(radians)}$$
$$= -57.3\omega L \qquad \text{(degrees)}$$

The phase angle thus varies linearly with the frequency ω. The phase-angle characteristic of transport lag is shown in Figure 11–60.

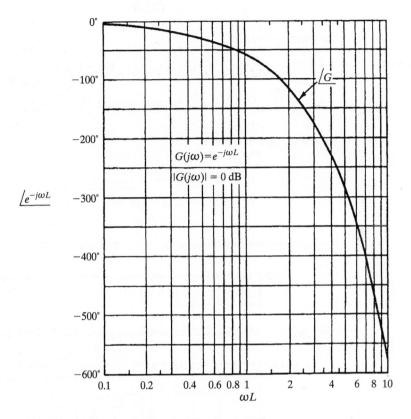

Figure 11–60 Phase-angle characteristic of transport lag.

Problem A–11–7

Consider the thermal system shown in Figure 11–61. The plant involves a dead time of L seconds. The controller is a proportional one. If $L = 0$, such a system is stable for all values of the gain K ($0 < K < \infty$). However, if the plant involves dead time, the closed-loop system can become unstable if K exceeds a certain critical value. Determine the critical value of the gain K when $T = 1$ s and $L = 0.5$ s.

Solution We first plot a Bode diagram when $K = 1$. We then determine the value of K that will cause the system to be critically stable.

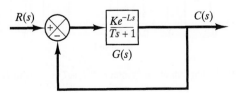

Figure 11–61 Thermal system.

The sinusoidal transfer function $G(j\omega)$ with $K = 1, T = 1$, and $L = 0.5$ is

$$G(j\omega) = \frac{e^{-0.5j\omega}}{j\omega + 1}$$

The log magnitude is

$$20 \log |G(j\omega)| = 20 \log |e^{-0.5j\omega}| + 20 \log \left| \frac{1}{j\omega + 1} \right|$$

$$= 0 + 20 \log \left| \frac{1}{j\omega + 1} \right|$$

The phase angle of $G(j\omega)$ is

$$\underline{/G(j\omega)} = \underline{/e^{-0.5j\omega}} + \underline{\left/ \frac{1}{j\omega + 1} \right.}$$

$$= -0.5\omega - \tan^{-1}\omega \qquad \text{(radians)}$$

$$= 57.3(-0.5\omega - \tan^{-1}\omega) \qquad \text{(degrees)}$$

The log-magnitude and phase-angle curves are shown in Figure 11–62.

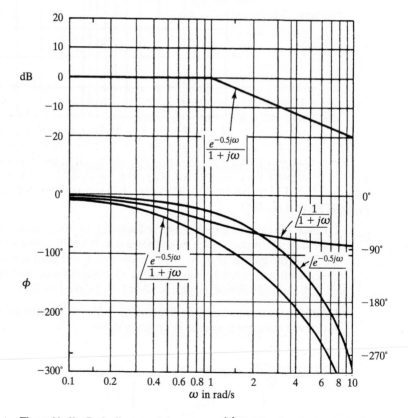

Figure 11–62   Bode diagram of the system $e^{-j\omega L}/(j\omega T + 1)$ with $L = 0.5$ and $T = 1$.

The phase crossover frequency is approximately $\omega = 3.7$ rad/s. (The exact value is $\omega = 3.6732$ rad/s.) The gain value at this frequency is approximately -11 dB. (The exact value is -10.99 dB.) This means that if the gain K is increased by 10.99 dB, the system will become critically stable. Thus, the critical value of the gain K is

$$K = 10.99 \text{ dB} = 3.544$$

Problem A–11–8

Consider the closed-loop system shown in Figure 11–63. Determine the critical value of K for stability by use of the Nyquist stability criterion.

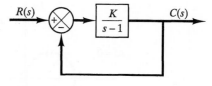

Figure 11–63 Closed-loop system.

Solution   The Nyquist plot of

$$G(j\omega) = \frac{K}{j\omega - 1}$$

is a circle with center at $-K/2$ on the negative real axis and radius $K/2$, as shown in Figure 11–64(a). As ω is increased from $-\infty$ to ∞, the $G(j\omega)$ locus makes a counterclockwise rotation. In this system, $P = 1$, because there is one pole of $G(s)$ in the right-half s-plane. For the closed-loop system to be stable, Z must be equal to zero. Therefore, $N = Z - P$ must be equal to -1, or there must be one counterclockwise encirclement of the $-1 + j0$ point for stability. (If there is no encirclement of the $-1 + j0$ point, the system is unstable.) Thus, for stability, K must be greater than unity, and $K = 1$ gives the stability limit. Figure 11–64(b) shows both stable and unstable cases of $G(j\omega)$ plots.

Problem A–11–9

Draw a Nyquist plot of

$$G(s) = \frac{20(s^2 + s + 0.5)}{s(s + 1)(s + 10)}$$

On the plot, locate the frequency points where $\omega = 0.1, 0.2, 0.4, 0.6, 1.0, 2.0, 4.0, 6.0, 10.0, 20.0,$ and 40.0 rad/s.

Solution   Noting that

$$G(j\omega) = \frac{2(-\omega^2 + j\omega + 0.5)}{j\omega(j\omega + 1)(0.1j\omega + 1)}$$

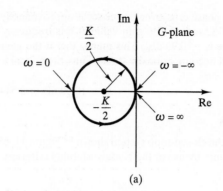

(a)

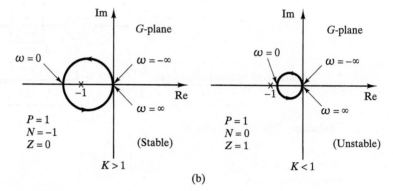

(b)

Figure 11–64   (a) Nyquist plot of $K/(j\omega - 1)$; (b) Nyquist plots of $K/(j\omega - 1)$ for stable and unstable cases.

we have

$$|G(j\omega)| = \frac{2\sqrt{(0.5 - \omega^2)^2 + \omega^2}}{\omega\sqrt{1 + \omega^2}\sqrt{1 + 0.01\,\omega^2}}$$

$$\angle G(j\omega) = \tan^{-1}\left(\frac{\omega}{0.5 - \omega^2}\right) - 90° - \tan^{-1}\omega - \tan^{-1}(0.1\,\omega)$$

The magnitude and phase angle may be computed as shown in Table 11–2. Figure 11–65 shows the Nyquist plot. Notice the existence of a loop in the Nyquist plot.

Problem A–11–10

Draw a Nyquist plot of the following system with MATLAB:

$$G(s) = \frac{20(s^2 + s + 0.5)}{s(s + 1)(s + 10)}$$

Draw the plot for only the positive frequency region.

Solution   To draw the Nyquist plot for only the positive frequency region, we use the following commands:

```
[re,im,w] = nyquist(num,den,w)
        plot (re, im)
```

TABLE 11–2 Magnitude and Phase of $G(j\omega)$ Considered in Problem A–11–9

ω	$|G(j\omega)|$	$\angle G(j\omega)$
0.1	9.952	−84.75°
0.2	4.918	−78.96°
0.4	2.435	−64.46°
0.6	1.758	−47.53°
1.0	1.573	−24.15°
2.0	1.768	−14.49°
4.0	1.801	−22.24°
6.0	1.692	−31.10°
10.0	1.407	−45.03°
20.0	0.893	−63.44°
40.0	0.485	−75.96°

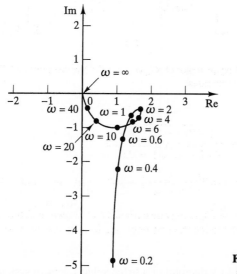

Figure 11–65 Nyquist plot of $G(s)$ $= 20(s^2 + s + 0.5)/[s(s + 1)(s + 10)]$.

The frequency region may be divided into several subregions by using different increments. For example, the frequency region of interest may be divided into three subregions as follows:

$$w1 = 0.1{:}0.1{:}10;$$
$$w2 = 10{:}2{:}100;$$
$$w3 = 100{:}10{:}500;$$
$$w = [w1 \quad w2 \quad w3]$$

MATLAB Program 11–17 uses this frequency region. Using this program, we obtain the Nyquist plot shown in Figure 11–66.

MATLAB Program 11–17

```
>> num = [20   20   10];
>> den = [1   11   10   0];
>> w1 = 0.1:0.1:10; w2 = 10:2:100; w3 = 100:10:500;
>> w = [w1   w2   w3];
>> [re,im,w] = nyquist(num,den,w);
>> plot(re,im)
>> v = [-3   3   -5   1]; axis (v)
>> grid
>> title('Nyquist Plot of G(s) = 20(s^2+s+0.5)/[s(s+1)(s+10)]')
>> xlabel('Real Axis'), ylabel ('Imaginary Axis')
```

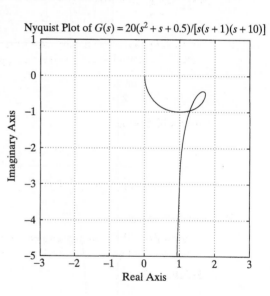

Nyquist Plot of $G(s) = 20(s^2 + s + 0.5)/[s(s + 1)(s + 10)]$

Figure 11–66 Nyquist plot of $G(s) = 20(s^2 + s + 0.5)/[s(s + 1)(s + 10)]$.

Problem A–11–11

Figure 11–67 shows a block diagram of a space vehicle control system. Determine the gain K such that the phase margin is 50°. What is the gain margin in this case?

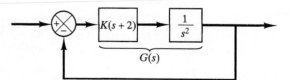

$G(s)$

Figure 11–67 Space vehicle control system.

Solution Since

$$G(j\omega) = \frac{K(j\omega + 2)}{(j\omega)^2}$$

we have

$$\underline{/G(j\omega)} = \underline{/j\omega + 2} - 2\underline{/j\omega} = \tan^{-1}\frac{\omega}{2} - 180°$$

The requirement that the phase margin be 50° means that $\underline{/G(j\omega_c)}$ must be equal to $-130°$, where ω_c is the gain crossover frequency, or

$$\underline{/G(j\omega_c)} = -130°$$

Hence, we set

$$\tan^{-1}\frac{\omega_c}{2} = 50°$$

from which we obtain

$$\omega_c = 2.3835 \text{ rad/s}$$

Since the phase curve never crosses the $-180°$ line, the gain margin is $+\infty$ dB. Noting that the magnitude of $G(j\omega)$ must be equal to 0 dB at $\omega = 2.3835$, we have

$$\left|\frac{K(j\omega + 2)}{(j\omega)^2}\right|_{\omega=2.3835} = 1$$

from which we get

$$K = \frac{2.3835^2}{\sqrt{2^2 + 2.3835^2}} = 1.8259$$

This K value will give the phase margin of 50°.

Problem A–11–12

Draw a Bode diagram of the open-loop transfer function $G(s)$ of the closed-loop system shown in Figure 11–68. Determine the phase margin and gain margin.

Solution Note that

$$G(j\omega) = \frac{20(j\omega + 1)}{j\omega(j\omega + 5)[(j\omega)^2 + 2j\omega + 10]}$$

$$= \frac{0.4(j\omega + 1)}{j\omega(0.2j\omega + 1)\left[\left(\dfrac{j\omega}{\sqrt{10}}\right)^2 + \dfrac{2}{10}j\omega + 1\right]}$$

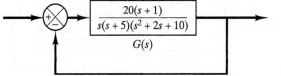

$$\frac{20(s + 1)}{s(s + 5)(s^2 + 2s + 10)}$$

$G(s)$

Figure 11–68 Closed-loop system.

The quadratic term in the denominator has the corner frequency of $\sqrt{10}$ rad/s and the damping ratio ζ of 0.3162, or

$$\omega_n = \sqrt{10}, \qquad \zeta = 0.3162$$

The Bode diagram of $G(j\omega)$ is shown in Figure 11–69. From this diagram, we find the phase margin to be 100° and the gain margin to be +13.3 dB.

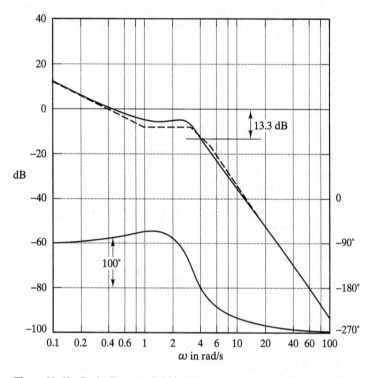

Figure 11–69   Bode diagram of $G(s)$ of the system shown in Figure 11–68.

Problem A–11–13

Consider the control system shown in Figure 11–70. Determine the value of the gain K such that the phase margin is 60°.

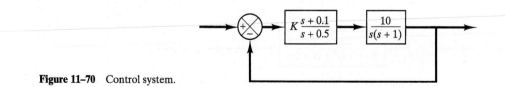

Figure 11–70   Control system.

Solution   The open-loop transfer function is

$$G(s) = K\frac{s + 0.1}{s + 0.5}\frac{10}{s(s + 1)}$$

$$= \frac{K(10s + 1)}{s^3 + 1.5s^2 + 0.5s}$$

Let us plot the Bode diagram of $G(s)$ when $K = 1$. MATLAB Program 11–18 may be used for this purpose. Figure 11–71 shows the Bode diagram produced by the program. From the diagram, the required phase margin of 60° occurs at the frequency $\omega = 1.15$ rad/s. The magnitude of $G(j\omega)$ at this frequency is found to be 14.5 dB. Then the gain K must satisfy the following equation:

$$20 \log K = -14.5 \text{ dB}$$

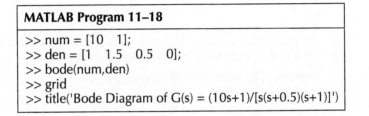

MATLAB Program 11–18

```
>> num = [10   1];
>> den = [1   1.5   0.5   0];
>> bode(num,den)
>> grid
>> title('Bode Diagram of G(s) = (10s+1)/[s(s+0.5)(s+1)]')
```

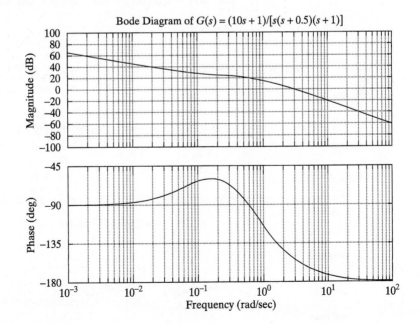

Figure 11–71   Bode diagram of $G(s) = (10s + 1)/[s(s + 0.5)(s + 1)]$.

or

$$K = 0.188$$

Thus, we have determined the value of the gain K.

Problem A–11–14

Consider a lag–lead compensator $G_c(s)$ defined by

$$G_c(s) = K_c \frac{\left(s + \dfrac{1}{T_1}\right)\left(s + \dfrac{1}{T_2}\right)}{\left(s + \dfrac{\beta}{T_1}\right)\left(s + \dfrac{1}{\beta T_2}\right)}$$

Show that at frequency

$$\omega_1 = \frac{1}{\sqrt{T_1 T_2}}$$

the phase angle of $G_c(j\omega)$ becomes zero. (This compensator acts as a lag compensator for $0 < \omega < \omega_1$ and as a lead compensator for $\omega_1 < \omega < \infty$.)

Solution   The angle of $G_c(j\omega)$ is given by

$$\angle G_c(j\omega) = \left\lfloor j\omega + \frac{1}{T_1} + \left\lfloor j\omega + \frac{1}{T_2} - \left\lfloor j\omega + \frac{\beta}{T_1} - \left\lfloor j\omega + \frac{1}{\beta T_2}\right.\right.\right.\right.$$

$$= \tan^{-1}\omega T_1 + \tan^{-1}\omega T_2 - \tan^{-1}\omega T_1/\beta - \tan^{-1}\omega T_2\beta$$

At $\omega = \omega_1 = 1/\sqrt{T_1 T_2}$, we have

$$\angle G_c(j\omega_1) = \tan^{-1}\sqrt{\frac{T_1}{T_2}} + \tan^{-1}\sqrt{\frac{T_2}{T_1}} - \tan^{-1}\frac{1}{\beta}\sqrt{\frac{T_1}{T_2}} - \tan^{-1}\beta\sqrt{\frac{T_2}{T_1}}$$

Since

$$\tan\left(\tan^{-1}\sqrt{\frac{T_1}{T_2}} + \tan^{-1}\sqrt{\frac{T_2}{T_1}}\right) = \frac{\sqrt{\dfrac{T_1}{T_2}} + \sqrt{\dfrac{T_2}{T_1}}}{1 - \sqrt{\dfrac{T_1}{T_2}}\sqrt{\dfrac{T_2}{T_1}}} = \infty$$

or

$$\tan^{-1}\sqrt{\frac{T_1}{T_2}} + \tan^{-1}\sqrt{\frac{T_2}{T_1}} = 90°$$

and also

$$\tan^{-1}\frac{1}{\beta}\sqrt{\frac{T_1}{T_2}} + \tan^{-1}\beta\sqrt{\frac{T_2}{T_1}} = 90°$$

we have

$$\angle G_c(j\omega_1) = 90° - 90° = 0°$$

Thus, the angle of $G_c(j\omega_1)$ becomes $0°$ at $\omega = \omega_1 = 1/\sqrt{T_1 T_2}$.

Problem A–11–15

Consider the system shown in Figure 11–72. Design a lead compensator such that the closed-loop system will have the phase margin of 50° and the gain margin of not less than 10 dB. Assume that

$$G_c(s) = K_c\alpha\left(\frac{Ts + 1}{\alpha Ts + 1}\right) \qquad 0 < \alpha < 1$$

It is desired that the bandwidth of the closed-loop system be $1 \sim 2$ rad/s.

Solution   Notice that

$$G_c(j\omega)G(j\omega) = K_c\alpha\left(\frac{Tj\omega + 1}{\alpha Tj\omega + 1}\right)\frac{0.2}{(j\omega)^2(0.2j\omega + 1)}$$

Since the bandwidth of the closed-loop system is close to the gain crossover frequency, we choose the gain crossover frequency to be 1 rad/s. At $\omega = 1$, the phase angle of $G(j\omega)$ is –191.31°. Hence, the lead network needs to supply $50° + 11.31° = 61.31°$ at $\omega = 1$. Accordingly, α can be determined from

$$\sin \phi_m = \sin 61.31° = \frac{1 - \alpha}{1 + \alpha} = 0.8772$$

The result is

$$\alpha = 0.06541$$

Noting that the maximum phase lead angle ϕ_m occurs at the geometric mean of the two corner frequencies, we have

$$\omega_m = \sqrt{\frac{1}{T}\frac{1}{\alpha T}} = \frac{1}{\sqrt{\alpha}\,T} = \frac{1}{\sqrt{0.06541}\,T} = \frac{3.910}{T} = 1$$

Thus,

$$\frac{1}{T} = \frac{1}{3.910} = 0.2558$$

and

$$\frac{1}{\alpha T} = \frac{0.2558}{0.06541} = 3.910$$

Hence,

$$G_c(j\omega)G(j\omega) = 0.06541K_c\frac{3.910j\omega + 1}{0.2558j\omega + 1}\frac{0.2}{(j\omega)^2(0.2j\omega + 1)}$$

or

$$\frac{G_c(j\omega)G(j\omega)}{0.06541K_c} = \frac{3.910j\omega + 1}{0.2558j\omega + 1}\frac{0.2}{(j\omega)^2(0.2j\omega + 1)}$$

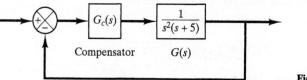

Compensator $G(s)$

Figure 11–72   Closed-loop system.

A Bode diagram for $G_c(j\omega)G(j\omega)/(0.06541K_c)$ is shown in Figure 11–73. By simple calculations (or from the Bode diagram), we find that the magnitude curve must be raised by 2.306 dB so that the magnitude equals 0 dB at $\omega = 1$ rad/s. Hence, we set

$$20 \log 0.06541K_c = 2.306$$

or

$$0.06541K_c = 1.3041$$

which yields

$$K_c = 19.94$$

The magnitude and phase curves of the compensated system show that the system has the phase margin of 50° and the gain margin of 16 dB. Consequently, the design specifications are satisfied.

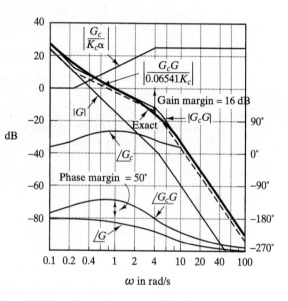

Figure 11–73   Bode diagram of the system shown in Figure 11–72.

Problem A–11–16

Consider the unity-feedback system whose open-loop transfer function is

$$G(s) = \frac{K}{s(s + 1)(s + 2)}$$

Design a compensator such that the system will have the static velocity error constant of 10 s^{-1}, the phase margin of 50°, and the gain margin of 10 dB or more.

Solution   Let us use a lag-lead compensator of the form

$$G_c(s) = K_c \frac{\left(s + \dfrac{1}{T_1}\right)\left(s + \dfrac{1}{T_2}\right)}{\left(s + \dfrac{\beta}{T_1}\right)\left(s + \dfrac{1}{\beta T_2}\right)}$$

Then the open-loop transfer function of the compensated system is $G_c(s)G(s)$. Since the gain K of the plant is adjustable, let us assume that $K_c = 1$. Then $\lim_{s \to 0} G_c(s) = 1$. From the requirement on the static velocity error constant, we obtain

$$K_v = \lim_{s \to 0} sG_c(s)G(s) = \lim_{s \to 0} sG_c(s) \frac{K}{s(s + 1)(s + 2)}$$

$$= \frac{K}{2} = 10$$

Hence,

$$K = 20$$

We shall next draw the Bode diagram of the uncompensated system with $K = 20$, as shown in Figure 11–74. The phase margin of the uncompensated system is found to be $-32°$, which indicates that the uncompensated system with $K = 20$ is unstable.

The next step in the design of a lag–lead compensator is to choose a new gain crossover frequency. From the phase-angle curve for $G(j\omega)$, we notice that $\underline{/G(j\omega)} = -180°$ at $\omega = 1.5$ rad/s. It is convenient to choose the new gain crossover

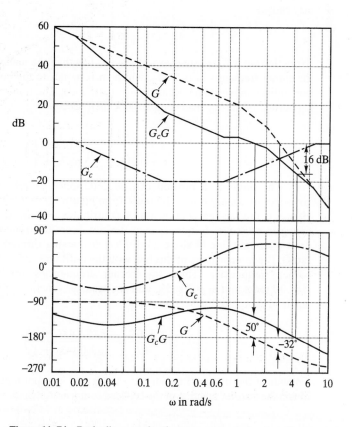

Figure 11–74 Bode diagrams for the uncompensated system, the compensator, and the compensated system. (G, uncompensated system; G_c, compensator; G_cG, compensated system.)

frequency to be 1.5 rad/s so that the phase-lead angle required at $\omega = 1.5$ rad/s is about 50°, which is quite possible with the use of a single lag–lead network.

Once we choose the gain crossover frequency to be 1.5 rad/s, we can determine one of the corner frequencies of the phase-lag portion of the lag–lead compensator. Let us choose the corner frequency $\omega = 1/T_2$ (which corresponds to the zero of the phase-lag portion of the compensator) to be 1 decade below the new gain crossover frequency, or $\omega = 0.15$ rad/s.

For the lead compensator, the maximum phase-lead angle ϕ_m is given by Equation (11–10), with $\alpha = 1/\beta$ in the present case. Substituting $\alpha = 1/\beta$ in Equation (11–10), we obtain

$$\sin \phi_m = \frac{1 - \dfrac{1}{\beta}}{1 + \dfrac{1}{\beta}} = \frac{\beta - 1}{\beta + 1}$$

Notice that $\beta = 10$ corresponds to $\phi_m = 54.9°$. Since we need a 50° phase margin, we may choose

$$\beta = 10$$

(Note that we will be using several degrees less than the maximum angle, 54.9°.) Then the corner frequency $\omega = 1/\beta T_2$ (which corresponds to the pole of the phase-lag portion of the compensator) becomes $\omega = 0.015$ rad/s. The transfer function of the phase-lag portion of the lag–lead compensator then becomes

$$\frac{s + 0.15}{s + 0.015} = 10\left(\frac{6.67s + 1}{66.7s + 1}\right)$$

The phase-lead portion can be determined as follows: Since the new gain crossover frequency is $\omega = 1.5$ rad/s, from Figure 11–74, $G(j1.5)$ is found to be 13 dB. Hence, if the lag–lead compensator contributes -13 dB at $\omega = 1.5$ rad/s, then the new gain crossover frequency is as desired. From this requirement, it is possible to draw a straight line of slope 20 dB/decade, passing through the point (-13 dB, 1.5 rad/s). The intersections of this line with the 0-dB line and -20-dB line determine the corner frequencies. Thus, the corner frequencies for the lead portion are $\omega = 0.7$ rad/s and $\omega = 7$ rad/s, and the transfer function of the lead portion of the lag–lead compensator becomes

$$\frac{s + 0.7}{s + 7} = \frac{1}{10}\left(\frac{1.43s + 1}{0.143s + 1}\right)$$

Combining the transfer functions of the lag and lead portions of the compensator, we obtain the transfer function of the lag–lead compensator. Since we chose $K_c = 1$, we have

$$G_c(s) = \left(\frac{s + 0.7}{s + 7}\right)\left(\frac{s + 0.15}{s + 0.015}\right) = \left(\frac{1.43s + 1}{0.143s + 1}\right)\left(\frac{6.67s + 1}{66.7s + 1}\right)$$

The magnitude and phase-angle curves of the lag–lead compensator just designed are shown in Figure 11–74. The open-loop transfer function of the compensated system is

$$G_c(s)G(s) = \frac{(s + 0.7)(s + 0.15)20}{(s + 7)(s + 0.015)s(s + 1)(s + 2)}$$

$$= \frac{10(1.43s + 1)(6.67s + 1)}{s(0.143s + 1)(66.7s + 1)(s + 1)(0.5s + 1)} \qquad (11\text{--}14)$$

The magnitude and phase-angle curves of the system of Equation (11–14) are also shown in Figure 11–74. The phase margin of the compensated system is 50°, the gain margin is 16 dB, and the static velocity error constant is 10 s^{-1}. All the requirements are therefore met, and the design has been completed.

Let us now examine the transient-response characteristics of the compensated system. (The uncompensated system with $K = 20$ is unstable.) The closed-loop transfer function of the compensated system is

$$\frac{C(s)}{R(s)} = \frac{95.381s^2 + 81s + 10}{4.7691s^5 + 47.7287s^4 + 110.3026s^3 + 163.724s^2 + 82s + 10}$$

The unit-step and unit-ramp response curves obtained with MATLAB are shown in Figures 11–75 and 11–76, respectively.

Note that the designed closed-loop control system has the following closed-loop zeros and poles:

$$\text{zeros at } s = -0.1499, \qquad s = -0.6993$$
$$\text{poles at } s = -0.8973 \pm j1.4439$$
$$s = -0.1785, \qquad s = -0.5425, \qquad s = -7.4923$$

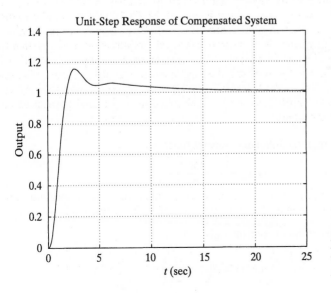

Unit-Step Response of Compensated System

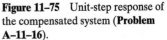

Figure 11–75 Unit-step response of the compensated system (**Problem A–11–16**).

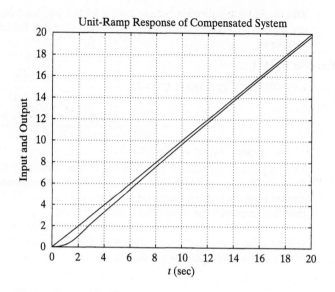

Figure 11–76   Unit-ramp response of the compensated system
(**Problem A–11–16**).

The pole at $s = -0.1785$ and the zero at $s = -0.1499$ are located very close to each other. Such a pair of pole and zero produces a long tail of small amplitude in the step response, as seen in Figure 11-75. Also, the pole at $s = -0.5425$ and the zero at $s = -0.6993$ are located fairly close to each other. This pair adds an amplitude to the long tail.

PROBLEMS

Problem B–11–1

Draw a Bode diagram of

$$G(s) = \frac{10s}{s + 10}$$

Problem B–11–2

Draw a Bode diagram of the following transfer function:

$$G(s) = \frac{s + 5}{s^2 + 2s + 10}$$

Problem B–11–3

Draw Bode diagrams of the PI controller given by

$$G_c(s) = 5\left(1 + \frac{1}{2s}\right)$$

and the PD controller given by

$$G_c(s) = 5(1 + 0.5s)$$

Problem B-11-4

Consider a PID controller given by

$$G_c(s) = 30.3215\frac{(s + 0.65)^2}{s}$$

Draw a Bode diagram of the controller.

Problem B-11-5

Draw Bode diagrams of the lead network and lag network shown in Figures 11–77(a) and (b), respectively.

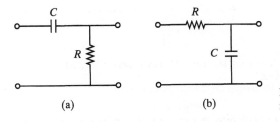

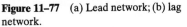

Figure 11–77 (a) Lead network; (b) lag network.

(a)   (b)

Problem B-11-6

In the mechanical system of Figure 11–78, $x(t)$ is the input displacement and $\theta(t)$ is the output angular displacement. Assume that the masses involved are negligibly small and that all motions are restricted to be small; therefore, the system can be considered linear. The initial conditions for x and θ are zeros, or $x(0-) = 0$ and $\theta(0-) = 0$.

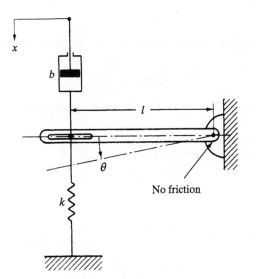

Figure 11–78 Mechanical system.

Consider the case where the input displacement $x(t)$ is given by

$$x(t) = X \sin \omega t$$

What is the steady-state output $\theta(t)$? Also, draw a Bode diagram of

$$G(j\omega) = \frac{\Theta(j\omega)}{X(j\omega)}$$

when $l = 0.1$ m, $k = 2$ N/m, and $b = 0.2$ N-s/m.

Problem B–11–7

Given

$$G(s) = \frac{\omega_n^2}{s^2 + 2\zeta\omega_n s + \omega_n^2}$$

show that

$$|G(j\omega_n)| = \frac{1}{2\zeta}$$

Problem B–11–8

Plot a Bode diagram of the following $G(s)$ with MATLAB.

$$G(s) = \frac{320(s + 2)}{s(s + 1)(s^2 + 8s + 64)}$$

Problem B–11–9

Plot a Bode diagram of the following $G(s)$ with MATLAB.

$$G(s) = \frac{20(s^2 + s + 0.5)}{s(s + 1)(s + 10)}$$

Problem B–11–10

Consider a unity-feedback system with the following feedforward-transfer function:

$$G(s) = \frac{1}{s(s^2 + 0.8s + 1)}$$

Draw a Nyquist plot of $G(s)$ with MATLAB. Is this system stable?

Problem B–11–11

Draw a Nyquist plot of the following $G(s)$ with MATLAB.

$$G(s) = \frac{20(s + 1)}{s(s + 5)(s^2 + 2s + 10)}$$

Problem B–11–12

Consider a unity-feedback control system with the following open-loop transfer function:

$$G(s) = \frac{s^2 + 2s + 1}{s^3 + 0.2s^2 + s + 1}$$

Draw a Nyquist plot of $G(s)$ and examine the stability of the closed-loop system.

Problem B–11–13

A system with the open-loop transfer function

$$G(s)H(s) = \frac{K}{s^2(T_1s + 1)}$$

is inherently unstable. The system can be stabilized by adding derivative control. Sketch the Nyquist plots for the open-loop transfer function with and without derivative control.

Problem B–11–14

Figure 11–79 shows a block diagram of a process control system. Determine the range of the gain K for stability.

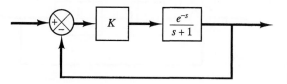

Figure 11–79 Process control system.

Problem B–11–15

Consider a unity-feedback control system with the open-loop transfer function

$$G(s) = \frac{K}{s(s^2 + s + 4)}$$

Determine the value of the gain K such that the phase margin is 50°. What is the gain margin with this gain K?

Problem B–11–16

Consider the system shown in Figure 11–80. Draw a Bode diagram of the open-loop transfer function and determine the value of the gain K such that the phase margin is 50°. What is the gain margin of the system with this gain K?

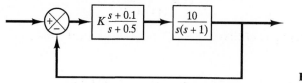

Figure 11–80 Control system.

Problem B–11–17

Referring to the control system shown in Figure 11–81, design a lead compensator $G_c(s)$ such that the phase margin is 45°, the gain margin is not less than 8 dB, and the static velocity error constant K_v is 4.0 s^{-1}. Plot unit-step and unit-ramp response curves of the compensated system with MATLAB.

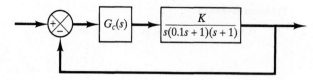

Figure 11–81   Control system.

Problem B–11–18

Consider the system shown in Figure 11–82. Design a compensator such that the static velocity error constant K_v is 50 s^{-1}, the phase margin is 50°, and the gain margin is not less than 8 dB. Plot unit-step and unit-ramp response curves of the compensated and uncompensated systems with MATLAB.

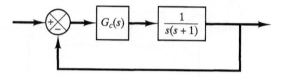

Figure 11–82   Control system.

Appendix A
Systems of Units

In this appendix, we review systems of units commonly used in engineering fields, such as the cgs system of units, mks system of units, metric engineering system of units, British engineering system of units, and International System of units (SI).

Units. A physical quantity can be measured only by comparison with a like quantity. A distinct amount of a physical quantity is called a *unit*. (To be useful, the unit should be a convenient, practical size.) Any physical quantity of the same kind can be compared with it, and its value can be stated in terms of a ratio and the unit used.

Basic units and derived units. The general unit of a physical quantity is defined as its dimension. A system of units can be developed by choosing, for each basic dimension of the system, a specific unit (e.g., the meter for length, the kilogram for mass, and the second for time). Such a unit is called a *basic unit*. The corresponding physical quantity is called a *basic quantity*. All units that are not basic are called *derived units*.

Systematic units. *Systematic units* are systematically derived units within a unit system. They can be obtained by replacing the general units (dimensions) by the basic units of the system.

If we define the dimensions of length, mass, and time as $[L]$, $[M]$, and $[T]$, respectively, then physical quantities may be expressed as $[L]^x[M]^y[T]^z$. For instance, the dimension of acceleration is $[L][T]^{-2}$ and that of force is $[L][M][T]^{-2}$. In the mks system of units, the systematic unit of acceleration is therefore 1 m/s^2 and that of force is 1 kg-m/s^2.

Absolute systems of units and gravitational systems of units. Systems of units in which mass is taken as a basic unit are called *absolute systems of units*, whereas those in which force rather than mass is taken as a basic unit are called *gravitational systems of units*.

The cgs system of units. The *cgs system of units* is an absolute system of units based on the centimeter, gram mass, and second. This system has been widely used in science. Among its disadvantages are the facts that the derived units for force and energy are too small for practical purposes and that the system does not combine with the practical electrical units to form a comprehensive system of units.

The mks system of units. The *mks system of units* is an absolute system of units based on the meter, kilogram mass, and second. In this system, the derived units for force and energy are a convenient size in an engineering sense, and all the practical electrical units fit in as natural units to form a comprehensive system of units.

Metric engineering system of units. The *metric engineering system of units* is a gravitational system of units based on the meter, kilogram force, and second. (Since the standard of force is defined as the weight of the prototype standard mass of the kilogram, the basic unit of force is variable, but this factor is not a serious disadvantage.)

British engineering system of units. The *British engineering system of units* is a gravitational system of units based on the foot, pound force, and second. This system is the one that has been used in the United States. The derived unit of mass is lb$_f$-s^2/ft and is called a slug (1 slug = 1 lb$_f$-s^2/ft).

International System of units (SI). The *International System of units* (Système International d'Unités) is the internationally agreed-upon system of units for expressing the values of physical quantities. (See Table A–1.) In this system, four more basic units are added to the customary three (meter, kilogram, second) of the mks absolute system of units. The four added basic units are the ampere as the unit of electric current, the kelvin as the unit of thermodynamic temperature, the candela as the unit of luminous intensity, and the mole as the unit of amount of substance. Thus, in SI units, the meter, kilogram, second, ampere, kelvin, candela, and mole constitute the seven basic units. Two auxiliary SI units are the radian, which is the unit of a plane angle, and the steradian, which is the unit of a solid angle. Table A–1 lists the seven basic

TABLE A-1   International System of Units (SI)

Type of Units	Quantity	Unit	Symbol	Dimension
Basic units	Length	meter	m	
	Mass	kilogram	kg	
	Time	second	s	
	Electric current	ampere	A	
	Temperature	kelvin	K	
	Luminous intensity	candela	cd	
	Amount of substance	mole	mol	
Auxiliary units	Plane angle	radian	rad	
	Solid angle	steradian	sr	
Derived units	Acceleration	meter per second squared	m/s^2	
	Activity (of radioactive source)	1 per second	s^{-1}	
	Angular acceleration	radian per second squared	rad/s^2	
	Angular velocity	radian per second	rad/s	
	Area	square meter	m^2	
	Density	kilogram per cubic meter	kg/m^3	
	Dynamic viscosity	newton second per square meter	$N\text{-}s/m^2$	$m^{-1}\,kg\,s^{-1}$
	Electric capacitance	farad	F	$m^{-2}\,kg^{-1}\,s^4\,A^2$
	Electric charge	coulomb	C	As
	Electric field strength	volt per meter	V/m	$m\,kg\,s^{-3}\,A^{-1}$
	Electric resistance	ohm	Ω	$m^2\,kg\,s^{-3}\,A^{-2}$
	Entropy	joule per kelvin	J/K	$m^2\,kg\,s^{-2}\,K^{-1}$
	Force	newton	N	$m\,kg\,s^{-2}$
	Frequency	hertz	Hz	s^{-1}
	Illumination	lux	lx	$m^{-2}\,cd\,sr$
	Inductance	henry	H	$m^2\,kg\,s^{-2}\,A^{-2}$
	Kinematic viscosity	square meter per second	m^2/s	
	Luminance	candela per square meter	cd/m^2	
	Luminous flux	lumen	lm	cd sr
	Magnetic field strength	ampere per meter	A/m	
	Magnetic flux	weber	Wb	$m^2\,kg\,s^{-2}\,A^{-1}$
	Magnetic flux density	tesla	T	$kg\,s^{-2}\,A^{-1}$
	Magnetomotive force	ampere turn	A	A
	Power	watt	W	$m^2\,kg\,s^{-3}$
	Pressure	pascal (newton per square meter)	Pa (N/m^2)	$m^{-1}\,kg\,s^{-2}$

TABLE A–1 (*continued*)

Type of Units	Quantity	Unit	Symbol	Dimension
Derived units (cont.)	Radiant intensity	watt per steradian	W/sr	$m^2\,kg\,s^{-3}\,sr^{-1}$
	Specific heat	joule per kilogram kelvin	J/kg-K	$m^2\,s^{-2}\,K^{-1}$
	Thermal conductivity	watt per meter kelvin	W/m-K	$m\,kg\,s^{-3}\,K^{-1}$
	Velocity	meter per second	m/s	
	Voltage	volt	V	$m^2\,kg\,s^{-3}\,A^{-1}$
	Volume	cubic meter	m^3	
	Wave number	1 per meter	m^{-1}	
	Work, energy, quantity of heat	joule	J	$m^2\,kg\,s^{-2}$

units, the two auxiliary units, and some of the derived units of the International System of units. (Multiples and submultiples of the units are indicated by a series of 16 prefixes for various powers of 10; see Table A–2.)

TABLE A–2 Prefixes and Abbreviated Prefixes for Multiples and Submultiples in Powers of 10

	Prefix	*Abbreviated Prefix*
10^{18}	exa	E
10^{15}	peta	P
10^{12}	tera	T
10^{9}	giga	G
10^{6}	mega	M
10^{3}	kilo	k
10^{2}	hecto	h
10	deka	da
10^{-1}	deci	d
10^{-2}	centi	c
10^{-3}	milli	m
10^{-6}	micro	μ
10^{-9}	nano	n
10^{-12}	pico	p
10^{-15}	femto	f
10^{-18}	atto	a

The seven basic SI units are defined in the following way:

Meter.   The meter is the length equal to 1 650 763.73 wavelengths of radiation in vacuum corresponding to the unperturbed transition between levels $2P_{10}$ and $5d_5$ of the atom of krypton 86, the orange–red line.

Kilogram.   The kilogram is the mass of a particular cylinder (of diameter 39 mm and height 39 mm) of platinum–iridium alloy, called the International Prototype Kilogram, which is preserved in a vault at Sèvres, France, by the International Bureau of Weights and Measures.

Second.   The second is the duration of 9 192 631 770 periods of the radiation corresponding to the transition between the two hyperfine levels of the fundamental state of the atom of cesium 133.

Ampere.   The ampere is a constant current that, if maintained in two straight, parallel conductors of infinite length, of negligible circular cross sections, and placed 1 meter apart in a vacuum, will produce between these conductors a force equal to 2×10^{-7} newton per meter of length.

Kelvin.   The kelvin is the fraction 1/273.16 of the thermodynamic temperature of the triple point of water. (Note that the triple point of water is 0.01° C.)

Candela.   The candela is the luminous intensity, in the direction of the normal, of a blackbody surface 1/600 000 square meter in area at the temperature of solidification of platinum under a pressure of 101 325 newtons per square meter.

Mole.   The mole is the amount of substance of a system that contains as many elementary entities as there are atoms in 0.012 kilogram of carbon 12.

The two auxiliary units of SI (radian and steradian) are defined as follows:

Radian.   The radian is a unit of plane angular measurement equal to the angle at the center of a circle subtended by an arc equal in length to the radius. (The dimension of the radian is zero, since it is a ratio of quantities of the same dimension.)

Steradian.   The steradian is a unit of measure of solid angles that is expressed as the solid angle subtended at the center of a sphere by a portion of the surface whose area is equal to the square of the radius of the sphere. (The dimension of the steradian is also zero, since it is a ratio of quantities of the same dimension.)

Appendix B
Conversion Tables

Conversion tables for mass, length, area, volume, energy, power, pressure, and temperature are presented in Tables B–1 through B–9.

TABLE B–1 Conversion Table for Mass

g	kg	lb	oz	grain	slug
1	10^{-3}	2.205×10^{-3}	3.527×10^{-2}	15.432	6.852×10^{-5}
10^3	1	2.205	35.27	15.432×10^3	6.852×10^{-2}
453.6	0.4536	1	16	7000	3.108×10^{-2}
28.35	2.835×10^{-2}	0.0625	1	437.5	1.943×10^{-3}
6.480×10^{-2}	6.480×10^{-5}	1.429×10^{-4}	2.286×10^{-3}	1	4.440×10^{-6}
1.459×10^4	14.59	32.17	514.78	2.252×10^5	1

TABLE B–2 Conversion Tables for Length

cm	m	in.	ft	yd
1	0.01	0.3937	0.03281	0.01094
100	1	39.37	3.281	1.0936
2.54	0.0254	1	0.08333	0.02778
30.48	0.3048	12	1	0.3333
91.44	0.9144	36	3	1

TABLE B–2 (*continued*)

km	mile	nautical mile	ft
1	0.6214	0.5400	3280.84
1.6093	1	0.8690	5280
1.852	1.151	1	6076

TABLE B–3 Conversion Table for Length (from in. to mm)

in.	mm	in.	mm	in.	mm	in.	mm
1/32	0.794	9/32	7.144	17/32	13.494	25/32	19.844
1/16	1.587	5/16	7.937	9/16	14.287	13/16	20.638
3/32	2.381	11/32	8.731	19/32	15.081	27/32	21.431
1/8	3.175	3/8	9.525	5/8	15.875	7/8	22.225
5/32	3.969	13/32	10.319	21/32	16.669	29/32	23.019
3/16	4.762	7/16	11.112	11/16	17.462	15/16	23.812
7/32	5.556	15/32	11.906	23/32	18.256	31/32	24.606
1/4	6.350	1/2	12.700	3/4	19.050	1	25.400

TABLE B–4 Conversion Tables for Area

cm^2	m^2	$in.^2$	ft^2	yd^2
1	10^{-4}	0.155	1.0764×10^{-3}	1.196×10^{-4}
10^4	1	1550	10.764	1.196
6.452	6.452×10^{-4}	1	6.944×10^{-3}	7.716×10^{-4}
929.0	0.09290	144	1	0.1111
8361	0.8361	1296	9	1

km^2	$mile^2$
1	0.3861
2.590	1

TABLE B–5 Conversion Tables for Volume

mm^3	cm^3	in.^{3}
1	10^{-3}	6.102×10^{-5}
10^3	1	6.102×10^{-2}
1.639×10^4	16.39	1

m^3	ft^3	yd^3
1	35.315	1.308
2.832×10^{-2}	1	3.704×10^{-2}
0.7646	27	1

U.S. gallon	liter	barrel
1	3.785	2.381×10^{-2}
0.2642	1	0.6290×10^{-2}
42	159	1

TABLE B–6 Conversion Table for Energy

J	kg$_f$-m	ft-lb$_f$	kWh	kcal	Btu
1	0.10197	0.7376	2.778×10^{-7}	2.389×10^{-4}	9.480×10^{-4}
9.807	1	7.233	2.724×10^{-6}	2.343×10^{-3}	9.297×10^{-3}
1.356	0.1383	1	3.766×10^{-7}	3.239×10^{-4}	1.285×10^{-3}
3.600×10^6	3.671×10^5	2.655×10^6	1	860	3413
4186	426.9	3087	1.163×10^{-3}	1	3.968
1055	107.6	778	2.930×10^{-4}	0.2520	1

TABLE B–7 Conversion Table for Power

kW	kg$_f$-m/s	ft-lb$_f$/s	British horsepower hp	kcal/s	Btu/s
1	101.97	737.6	1.341	0.2389	0.9480
9.807×10^{-3}	1	7.233	1.315×10^{-2}	2.343×10^{-3}	9.297×10^{-3}
1.356×10^{-3}	0.1383	1	1.818×10^{-3}	3.239×10^{-4}	1.285×10^{-3}
0.7457	76.04	550	1	0.1782	0.7069
4.186	426.9	3087	5.613	1	3.968
1.055	107.6	778.0	1.414	0.2520	1

TABLE B-8 Conversion Table for Pressure

Pa or N/m²	bar (10⁵ N/m²)	kg_f/cm²	lb_f/in.²	atm (standard atmospheric pressure)	mm Hg	in. Hg	m H₂O
1	1×10^{-5}	1.0197×10^{-5}	1.450×10^{-4}	9.869×10^{-6}	7.501×10^{-3}	2.953×10^{-4}	1.0197×10^{-4}
1×10^{5}	1	1.0197	14.50	0.9869	750.1	29.53	10.197
9.807×10^{4}	0.9807	1	14.22	0.9678	735.6	28.96	10.000
6.895×10^{3}	0.06895	0.07031	1	0.06805	51.71	2.036	0.7031
1.0133×10^{5}	1.0133	1.0332	14.70	1	760	29.92	10.33
1.3332×10^{2}	1.3332×10^{-3}	1.3595×10^{-3}	19.34×10^{-3}	1.3158×10^{-3}	1	3.937×10^{-2}	1.360×10^{-2}
3.386×10^{3}	0.03386	0.03453	0.4912	0.03342	25.4	1	0.3453
9.807×10^{3}	0.09807	0.10000	1.422	0.09678	73.55	2.896	1

TABLE B–9   Conversion Table for Temperature

°C	°F	°C	°F	°C	°F
−50	−58	16	60.8	44	111.2
−40	−40	18	64.4	46	114.8
−30	−22	20	68.0	48	118.4
−20	−4	22	71.6	50	122.0
−10	14	24	75.2	55	131.0
−5	23	26	78.8	60	140.0
0	32	28	82.4	65	149.0
2	35.6	30	86.0	70	158.0
4	39.2	32	89.6	75	167.0
6	42.8	34	93.2	80	176.0
8	46.4	36	96.8	85	185.0
10	50.0	38	100.4	90	194.0
12	53.6	40	104.0	95	203.0
14	57.2	42	107.6	100	212.0

To convert from Fahrenheit to Celsius, subtract 32 and multiply by 5/9:

$$t_C = \frac{5}{9}(t_F - 32)$$

To convert from Celsius to Fahrenheit, multiply by 9/5 and add 32:

$$t_F = \frac{9}{5}t_C + 32$$

Absolute zero temperature occurs at −273.15° on the Celsius scale and at −459.67° on the Fahrenheit scale. Absolute temperatures on the two scales are $t_C + 273.15$ and $t_F + 459.67$. Note that in most calculations the constants used are 273 and 460. Note also that

$$t_C \text{ degrees Celsius} = (t_C + 273.15) \text{ kelvin}$$

Appendix C
Vector–Matrix Algebra

C–1 INTRODUCTION

In deriving mathematical models of modern dynamic systems, one finds that the differential equations involved may become very complicated due to the multiplicity of inputs and outputs. To simplify the mathematical expressions of the system equations, it is advantageous to use vector–matrix notation, such as that used in the state-space representation of dynamic systems. For theoretical work, the notational simplicity gained by using vector–matrix operations is most convenient and is, in fact, essential for the analysis and design of modern dynamic systems. With vector–matrix notation, one can handle large, complex problems with ease by following the systematic format of representing the system equations and dealing with them mathematically by computer.

The principal objective of this appendix is to present definitions associated with matrices and the basic matrix algebra necessary for the analysis of dynamic systems.

C–2 DEFINITIONS ASSOCIATED WITH MATRICES

Matrix.   A matrix is defined as a rectangular array of elements that may be real numbers, complex numbers, functions, or operators. The number of columns, in general, is not necessarily the same as the number of rows. Consider the matrix

$$\mathbf{A} = \begin{bmatrix} a_{11} & a_{12} & \cdots & a_{1m} \\ a_{21} & a_{22} & \cdots & a_{2m} \\ \vdots & \vdots & & \vdots \\ a_{n1} & a_{n2} & \cdots & a_{nm} \end{bmatrix}$$

where a_{ij} denotes the (i, j)th element of $\mathbf{A}$. This matrix has n rows and m columns and is called an $n \times m$ matrix. The first index represents the row number, the second index the column number. The matrix $\mathbf{A}$ is sometimes written (a_{ij}).

Equality of two matrices. Two matrices are said to be equal if and only if their corresponding elements are equal. Note that equal matrices must have the same number of rows and the same number of columns.

Vector. A matrix having only one column, such as

$$\begin{bmatrix} x_1 \\ x_2 \\ \vdots \\ x_n \end{bmatrix}$$

is called a *column* vector. A column vector having n elements is called an n-vector or n-dimensional vector.

A matrix having only one row, such as

$$\begin{bmatrix} x_1 & x_2 & \cdots & x_n \end{bmatrix}$$

is called a *row* vector.

Square matrix. A square matrix is a matrix in which the number of rows is equal to the number of columns. A square matrix is sometimes called a *matrix of order n*, where n is the number of rows (or columns).

Diagonal matrix. If all the elements other than the main diagonal elements of a square matrix $\mathbf{A}$ are zero, $\mathbf{A}$ is called a *diagonal* matrix and is written as

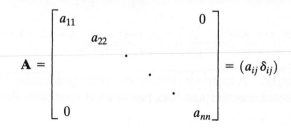

where the δ_{ij} are the Kronecker deltas, defined by

$$\delta_{ij} = 1 \qquad \text{if } i = j$$
$$= 0 \qquad \text{if } i \neq j$$

Note that all of the elements that are not explicitly written in the foregoing matrix are zero. The diagonal matrix is sometimes written

$$\text{diag}(a_{11}, a_{22}, \ldots, a_{nn})$$

Identity matrix or unity matrix. The identity matrix or unity matrix $\mathbf{I}$ is a matrix whose elements on the main diagonal are equal to unity and whose other elements are equal to zero; that is,

$$\mathbf{I} = \begin{bmatrix} 1 & 0 & \cdots & 0 \\ 0 & 1 & \cdots & 0 \\ \vdots & \vdots & & \vdots \\ 0 & 0 & \cdots & 1 \end{bmatrix} = \text{diag}(1, 1, \ldots, 1)$$

Zero matrix. A zero matrix is a matrix whose elements are all zero.

Determinant of a matrix. For each square matrix, there exists a determinant. The determinant has the following properties:

1. If any two consecutive rows or columns are interchanged, the determinant changes its sign.
2. If any row or any column consists only of zeros, then the value of the determinant is zero.
3. If the elements of any row (or any column) are exactly k times those of another row (or another column), then the value of the determinant is zero.
4. If, to any row (or any column), any constant times another row (or column) is added, the value of the determinant remains unchanged.
5. If a determinant is multiplied by a constant, then only one row (or one column) is multiplied by that constant. Note, however, that the determinant of k times an $n \times n$ matrix $\mathbf{A}$ is k^n times the determinant of $\mathbf{A}$, or

$$|k\mathbf{A}| = k^n|\mathbf{A}|$$

6. The determinant of the product of two square matrices $\mathbf{A}$ and $\mathbf{B}$ is the product of determinants, or

$$|\mathbf{AB}| = |\mathbf{A}||\mathbf{B}|$$

Singular matrix. A square matrix is called *singular* if the associated determinant is zero. In a singular matrix, not all the rows (or not all the columns) are independent of each other.

Nonsingular matrix. A square matrix is called *nonsingular* if the associated determinant is nonzero.

Transpose. If the rows and columns of an $n \times m$ matrix $\mathbf{A}$ are interchanged, the resulting $m \times n$ matrix is called the *transpose* of $\mathbf{A}$. The transpose of the matrix $\mathbf{A}$ is denoted by $\mathbf{A}'$. That is, if

$$\mathbf{A} = \begin{bmatrix} a_{11} & a_{12} & \cdots & a_{1m} \\ a_{21} & a_{22} & \cdots & a_{2m} \\ \vdots & \vdots & & \vdots \\ a_{n1} & a_{n2} & \cdots & a_{nm} \end{bmatrix}$$

then

$$\mathbf{A}' = \begin{bmatrix} a_{11} & a_{21} & \cdots & a_{n1} \\ a_{12} & a_{22} & \cdots & a_{n2} \\ \vdots & \vdots & & \vdots \\ a_{1m} & a_{2m} & \cdots & a_{nm} \end{bmatrix}$$

Note that $(\mathbf{A}')' = \mathbf{A}$.

Symmetric matrix. If a square matrix $\mathbf{A}$ is equal to its transpose, or

$$\mathbf{A} = \mathbf{A}'$$

then the matrix $\mathbf{A}$ is called a *symmetric* matrix.

Skew-symmetric matrix. If a square matrix $\mathbf{A}$ is equal to the negative of its transpose, or

$$\mathbf{A} = -\mathbf{A}'$$

then the matrix $\mathbf{A}$ is called a *skew-symmetric* matrix.

Conjugate matrix. If the complex elements of a matrix $\mathbf{A}$ are replaced by their respective conjugates, then the resulting matrix is called the *conjugate* of $\mathbf{A}$ and is denoted by $\overline{\mathbf{A}} = (\overline{a}_{ij})$, where $\overline{a}_{ij}$ is the complex conjugate of a_{ij}. For example, if

$$\mathbf{A} = \begin{bmatrix} 0 & 1 & 0 \\ -1 + j & -3 - j3 & -1 + j4 \\ -1 + j & -1 & -2 + j3 \end{bmatrix}$$

then

$$\overline{\mathbf{A}} = \begin{bmatrix} 0 & 1 & 0 \\ -1 - j & -3 + j3 & -1 - j4 \\ -1 - j & -1 & -2 - j3 \end{bmatrix}$$

Conjugate transpose. The conjugate transpose is the conjugate of the transpose of a matrix. Given a matrix $\mathbf{A}$, the conjugate transpose is denoted by $\overline{\mathbf{A}}'$ or $\mathbf{A}^*$; that is,

$$\overline{\mathbf{A}}' = \mathbf{A}^* = (\overline{a}_{ji})$$

For example, if

$$\mathbf{A} = \begin{bmatrix} 1 & j2 & 1 + j5 \\ 2 + j & j & 3 - j \\ 3 & 1 & 1 + j3 \end{bmatrix}$$

then

$$\overline{\mathbf{A}}' = \mathbf{A}^* = \begin{bmatrix} 1 & 2 - j & 3 \\ -j2 & -j & 1 \\ 1 - j5 & 3 + j & 1 - j3 \end{bmatrix}$$

Note that

$$(\mathbf{A}^*)^* = \mathbf{A}$$

If $\mathbf{A}$ is a real matrix (i.e., a matrix whose elements are real), the conjugate transpose $\mathbf{A}^*$ is the same as the transpose $\mathbf{A}'$.

Hermitian matrix. A matrix whose elements are complex quantities is called a *complex* matrix. If a complex matrix $\mathbf{A}$ satisfies the relationship

$$\mathbf{A} = \mathbf{A}^* \qquad \text{or} \qquad a_{ij} = \overline{a}_{ji}$$

where $\overline{a}_{ji}$ is the complex conjugate of a_{ji}, then $\mathbf{A}$ is called a *Hermitian* matrix. An example is

$$\mathbf{A} = \begin{bmatrix} 1 & 4 + j3 \\ 4 - j3 & 2 \end{bmatrix}$$

If a Hermitian matrix $\mathbf{A}$ is written as $\mathbf{A} = \mathbf{B} + j\mathbf{C}$, where $\mathbf{B}$ and $\mathbf{C}$ are real matrices, then

$$\mathbf{B} = \mathbf{B}' \qquad \text{and} \qquad \mathbf{C} = -\mathbf{C}'$$

In the preceding example,

$$\mathbf{A} = \mathbf{B} + j\mathbf{C} = \begin{bmatrix} 1 & 4 \\ 4 & 2 \end{bmatrix} + j\begin{bmatrix} 0 & 3 \\ -3 & 0 \end{bmatrix}$$

Skew-Hermitian matrix. If a matrix $\mathbf{A}$ satisfies the relationship

$$\mathbf{A} = -\mathbf{A}^*$$

then $\mathbf{A}$ is called a *skew-Hermitian* matrix. An example is

$$\mathbf{A} = \begin{bmatrix} j5 & -2 + j3 \\ 2 + j3 & j \end{bmatrix}$$

If a skew-Hermitian matrix $\mathbf{A}$ is written as $\mathbf{A} = \mathbf{B} + j\mathbf{C}$, where $\mathbf{B}$ and $\mathbf{C}$ are real matrices, then

$$\mathbf{B} = -\mathbf{B}' \qquad \text{and} \qquad \mathbf{C} = \mathbf{C}'$$

In the present example,

$$\mathbf{A} = \mathbf{B} + j\mathbf{C} = \begin{bmatrix} 0 & -2 \\ 2 & 0 \end{bmatrix} + j\begin{bmatrix} 5 & 3 \\ 3 & 1 \end{bmatrix}$$

C–3 MATRIX ALGEBRA

This section presents the essentials of matrix algebra, as well as additional definitions. It is important to remember that some matrix operations obey the same rules as those in ordinary algebra, but others do not.

Addition and subtraction of matrices. Two matrices $\mathbf{A}$ and $\mathbf{B}$ can be added if they have the same number of rows and the same number of columns. If $\mathbf{A} = (a_{ij})$ and $\mathbf{B} = (b_{ij})$, then

$$\mathbf{A} + \mathbf{B} = (a_{ij} + b_{ij})$$

Thus, each element of $\mathbf{A}$ is added to the corresponding element of $\mathbf{B}$. Similarly, subtraction of matrices is defined as

$$\mathbf{A} - \mathbf{B} = (a_{ij} - b_{ij})$$

As an example, consider

$$\mathbf{A} = \begin{bmatrix} 1 & 2 & 3 \\ 4 & 5 & 6 \end{bmatrix} \quad \text{and} \quad \mathbf{B} = \begin{bmatrix} 5 & 2 & 3 \\ 1 & 4 & 1 \end{bmatrix}$$

Then

$$\mathbf{A} + \mathbf{B} = \begin{bmatrix} 6 & 4 & 6 \\ 5 & 9 & 7 \end{bmatrix} \quad \text{and} \quad \mathbf{A} - \mathbf{B} = \begin{bmatrix} -4 & 0 & 0 \\ 3 & 1 & 5 \end{bmatrix}$$

Multiplication of a matrix by a scalar. The product of a matrix and a scalar is a matrix in which each element is multiplied by the scalar; that is, for a matrix $\mathbf{A}$ and a scalar k,

$$k\mathbf{A} = \begin{bmatrix} ka_{11} & ka_{12} & \cdots & ka_{1m} \\ ka_{21} & ka_{22} & \cdots & ka_{2m} \\ \vdots & \vdots & & \vdots \\ ka_{n1} & ka_{n2} & \cdots & ka_{nm} \end{bmatrix}$$

Multiplication of a matrix by a matrix. Multiplication of a matrix by a matrix is possible between conformable matrices (matrices such that the number of columns of the first matrix equals the number of rows of the second). Otherwise, multiplication of two matrices is not defined.

Let $\mathbf{A}$ be an $n \times m$ matrix and $\mathbf{B}$ be an $m \times p$ matrix. Then the product $\mathbf{AB}$, which we read "$\mathbf{A}$ postmultiplied by $\mathbf{B}$" or "$\mathbf{B}$ premultiplied by $\mathbf{A}$," is defined as follows:

$$\mathbf{AB} = \mathbf{C} = (c_{ij}) = \left(\sum_{k=1}^{m} a_{ik}b_{kj} \right) \qquad i = 1, 2, \ldots, n; j = 1, 2, \ldots, p$$

The product matrix $\mathbf{C}$ has the same number of rows as $\mathbf{A}$ and the same number of columns as $\mathbf{B}$. Thus, the matrix $\mathbf{C}$ is an $n \times p$ matrix.

Note that even if $\mathbf{A}$ and $\mathbf{B}$ are conformable for $\mathbf{AB}$, they may not be conformable for $\mathbf{BA}$, in which case $\mathbf{BA}$ is not defined.

The associative and distributive laws hold for matrix multiplication; that is,

$$(\mathbf{AB})\mathbf{C} = \mathbf{A}(\mathbf{BC})$$
$$(\mathbf{A} + \mathbf{B})\mathbf{C} = \mathbf{AC} + \mathbf{BC}$$
$$\mathbf{C}(\mathbf{A} + \mathbf{B}) = \mathbf{CA} + \mathbf{CB}$$

If $\mathbf{AB} = \mathbf{BA}$, then $\mathbf{A}$ and $\mathbf{B}$ are said to commute. Note that, in general, $\mathbf{AB} \neq \mathbf{BA}$. To show this, let

$$\mathbf{A} = \begin{bmatrix} 1 & 0 \\ 2 & 1 \\ 3 & 4 \end{bmatrix} \qquad \text{and} \qquad \mathbf{B} = \begin{bmatrix} 2 & 1 & 5 \\ 1 & 0 & 0 \end{bmatrix}$$

Then

$$\mathbf{AB} = \begin{bmatrix} 2 & 1 & 5 \\ 5 & 2 & 10 \\ 10 & 3 & 15 \end{bmatrix} \qquad \text{and} \qquad \mathbf{BA} = \begin{bmatrix} 19 & 21 \\ 1 & 0 \end{bmatrix}$$

Clearly, $\mathbf{AB} \neq \mathbf{BA}$. As another example, let

$$\mathbf{A} = \begin{bmatrix} 1 & 2 \\ 2 & 4 \end{bmatrix} \qquad \text{and} \qquad \mathbf{B} = \begin{bmatrix} 1 & 2 \\ 2 & 2 \end{bmatrix}$$

Then

$$\mathbf{AB} = \begin{bmatrix} 5 & 6 \\ 10 & 12 \end{bmatrix} \qquad \text{and} \qquad \mathbf{BA} = \begin{bmatrix} 5 & 10 \\ 6 & 12 \end{bmatrix}$$

Again, $\mathbf{AB} \neq \mathbf{BA}$.

Because matrix multiplication is, in general, not commutative, we must preserve the order of the matrices when we multiply one matrix by another. (This is the reason why we often use the terms "premultiplication" or "postmultiplication," to indicate whether the matrix is multiplied from the left or the right.)

An example of the case where $\mathbf{AB} = \mathbf{BA}$ is given next.

$$\mathbf{A} = \begin{bmatrix} 1 & 0 \\ 0 & 3 \end{bmatrix}, \qquad \mathbf{B} = \begin{bmatrix} 2 & 0 \\ 0 & 5 \end{bmatrix}$$

AB and **BA** are given by

$$\mathbf{AB} = \mathbf{BA} = \begin{bmatrix} 2 & 0 \\ 0 & 15 \end{bmatrix}$$

Clearly, **A** and **B** commute in this case.

Power of a matrix.   The kth power of a square matrix **A** is defined to be

$$\mathbf{A}^k = \underbrace{\mathbf{AA}\cdots\mathbf{A}}_{k}$$

Note that, for a diagonal matrix $\mathbf{A} = \text{diag}(a_{11}, a_{22}, \ldots, a_{nn})$,

$$\mathbf{A}^k = \begin{bmatrix} a_{11}^k & & & & 0 \\ & a_{22}^k & & & \\ & & \cdot & & \\ & & & \cdot & \\ & & & & \cdot \\ 0 & & & & a_{nn}^k \end{bmatrix}$$
$$= \text{diag}(a_{11}^k, a_{22}^k, \ldots, a_{nn}^k)$$

Further properties of matrices.   The transposes of $\mathbf{A} + \mathbf{B}$ and **AB** are respectively given by

$$(\mathbf{A} + \mathbf{B})' = \mathbf{A}' + \mathbf{B}'$$
$$(\mathbf{AB})' = \mathbf{B}'\mathbf{A}'$$

To prove the last relationship, note that the (i, j)th element of **AB** is

$$\sum_{k=1}^{m} a_{ik}b_{kj} = c_{ij}$$

The (i, j)th element of $\mathbf{B}'\mathbf{A}'$ is

$$\sum_{k=1}^{m} b_{ki}a_{jk} = \sum_{k=1}^{m} a_{jk}b_{ki} = c_{ji}$$

which is equal to the (j, i)th element of **AB** or the (i, j)th element of $(\mathbf{AB})'$. Hence, $(\mathbf{AB})' = \mathbf{B}'\mathbf{A}'$. As an example, consider

$$\mathbf{A} = \begin{bmatrix} 1 & 4 \\ 2 & 3 \end{bmatrix} \quad \text{and} \quad \mathbf{B} = \begin{bmatrix} 4 & 2 \\ 5 & 6 \end{bmatrix}$$

Then

$$\mathbf{AB} = \begin{bmatrix} 24 & 26 \\ 23 & 22 \end{bmatrix}$$

$$\mathbf{B}'\mathbf{A}' = \begin{bmatrix} 4 & 5 \\ 2 & 6 \end{bmatrix}\begin{bmatrix} 1 & 2 \\ 4 & 3 \end{bmatrix} = \begin{bmatrix} 24 & 23 \\ 26 & 22 \end{bmatrix}$$

Clearly, $(\mathbf{AB})' = \mathbf{B}'\mathbf{A}'$.

In a similar way, we obtain for the conjugate transposes of $\mathbf{A} + \mathbf{B}$ and $\mathbf{AB}$,

$$(\mathbf{A} + \mathbf{B})^* = \mathbf{A}^* + \mathbf{B}^*$$
$$(\mathbf{AB})^* = \mathbf{B}^*\mathbf{A}^*$$

Rank of matrix. A matrix $\mathbf{A}$ is said to have rank m if there exists an $m \times m$ submatrix $\mathbf{M}$ of $\mathbf{A}$ such that the determinant of $\mathbf{M}$ is nonzero and the determinant of every $r \times r$ submatrix (where $r \geq m + 1$) of $\mathbf{A}$ is zero.

As an example, consider the following matrix:

$$\mathbf{A} = \begin{bmatrix} 1 & 2 & 3 & 4 \\ 0 & 1 & -1 & 0 \\ 1 & 0 & 1 & 2 \\ 1 & 1 & 0 & 2 \end{bmatrix}$$

Note that $|\mathbf{A}| = 0$. One of a number of largest submatrices whose determinant is not equal to zero is

$$\begin{bmatrix} 1 & 2 & 3 \\ 0 & 1 & -1 \\ 1 & 0 & 1 \end{bmatrix}$$

Hence, the rank of the matrix $\mathbf{A}$ is 3.

C–4 MATRIX INVERSION

Minor M_{ij}. If the ith row and jth column are deleted from an $n \times n$ matrix $\mathbf{A}$, the resulting matrix is an $(n - 1) \times (n - 1)$ matrix. The determinant of this $(n - 1) \times (n - 1)$ matrix is called the minor M_{ij} of the matrix $\mathbf{A}$.

Cofactor A_{ij}. The cofactor A_{ij} of the element a_{ij} of the $n \times n$ matrix $\mathbf{A}$ is defined by the equation

$$A_{ij} = (-1)^{i+j} M_{ij}$$

That is, the cofactor A_{ij} of the element a_{ij} is $(-1)^{i+j}$ times the determinant of the matrix formed by deleting the ith row and the jth column from $\mathbf{A}$. Note that the cofactor A_{ij} of the element a_{ij} is the coefficient of the term a_{ij} in the expansion of the determinant $|\mathbf{A}|$, since it can be shown that

$$a_{i1}A_{i1} + a_{i2}A_{i2} + \cdots + a_{in}A_{in} = |\mathbf{A}|$$

If $a_{i1}, a_{i2}, \ldots, a_{in}$ are replaced by $a_{j1}, a_{j2}, \ldots, a_{jn}$, then

$$a_{j1}A_{i1} + a_{j2}A_{i2} + \cdots + a_{jn}A_{in} = 0 \qquad i \neq j$$

because the determinant of $\mathbf{A}$ in this case possesses two identical rows. Hence, we obtain

$$\sum_{k=1}^{n} a_{jk}A_{ik} = \delta_{ji}|\mathbf{A}|$$

Similarly,

$$\sum_{k=1}^{n} a_{ki} A_{kj} = \delta_{ij} |\mathbf{A}|$$

Adjoint matrix.   The matrix $\mathbf{B}$ whose element in the ith row and jth column equals A_{ji} is called the adjoint of $\mathbf{A}$ and is denoted by adj $\mathbf{A}$, or

$$\mathbf{B} = (b_{ij}) = (A_{ji}) = \text{adj } \mathbf{A}$$

That is, the adjoint of $\mathbf{A}$ is the transpose of the matrix whose elements are the cofactors of $\mathbf{A}$, or

$$\text{adj } \mathbf{A} = \begin{bmatrix} A_{11} & A_{21} & \cdots & A_{n1} \\ A_{12} & A_{22} & \cdots & A_{n2} \\ \vdots & \vdots & & \vdots \\ A_{1n} & A_{2n} & \cdots & A_{nn} \end{bmatrix}$$

Note that the element of the jth row and ith column of the product $\mathbf{A}(\text{adj } \mathbf{A})$ is

$$\sum_{k=1}^{n} a_{jk} b_{ki} = \sum_{k=1}^{n} a_{jk} A_{ik} = \delta_{ji} |\mathbf{A}|$$

Hence, $\mathbf{A}(\text{adj } \mathbf{A})$ is a diagonal matrix with diagonal elements equal to $|\mathbf{A}|$, or

$$\mathbf{A}(\text{adj } \mathbf{A}) = |\mathbf{A}| \mathbf{I}$$

Similarly, the element in the jth row and ith column of the product $(\text{adj } \mathbf{A})\mathbf{A}$ is

$$\sum_{k=1}^{n} b_{jk} a_{ki} = \sum_{k=1}^{n} A_{kj} a_{ki} = \delta_{ij} |\mathbf{A}|$$

Hence, we have the relationship

$$\mathbf{A}(\text{adj } \mathbf{A}) = (\text{adj } \mathbf{A})\mathbf{A} = |\mathbf{A}| \mathbf{I} \tag{C-1}$$

For example, given the matrix

$$\mathbf{A} = \begin{bmatrix} 1 & 2 & 0 \\ 3 & -1 & -2 \\ 1 & 0 & -3 \end{bmatrix}$$

we find that the determinant of $\mathbf{A}$ is 17 and that

$$\text{adj } \mathbf{A} = \begin{bmatrix} \begin{vmatrix} -1 & -2 \\ 0 & -3 \end{vmatrix} & -\begin{vmatrix} 2 & 0 \\ 0 & -3 \end{vmatrix} & \begin{vmatrix} 2 & 0 \\ -1 & -2 \end{vmatrix} \\ -\begin{vmatrix} 3 & -2 \\ 1 & -3 \end{vmatrix} & \begin{vmatrix} 1 & 0 \\ 1 & -3 \end{vmatrix} & -\begin{vmatrix} 1 & 0 \\ 3 & -2 \end{vmatrix} \\ \begin{vmatrix} 3 & -1 \\ 1 & 0 \end{vmatrix} & -\begin{vmatrix} 1 & 2 \\ 1 & 0 \end{vmatrix} & \begin{vmatrix} 1 & 2 \\ 3 & -1 \end{vmatrix} \end{bmatrix}$$

$$= \begin{bmatrix} 3 & 6 & -4 \\ 7 & -3 & 2 \\ 1 & 2 & -7 \end{bmatrix}$$

Thus,

$$\mathbf{A}(\text{adj } \mathbf{A}) = \begin{bmatrix} 1 & 2 & 0 \\ 3 & -1 & -2 \\ 1 & 0 & -3 \end{bmatrix} \begin{bmatrix} 3 & 6 & -4 \\ 7 & -3 & 2 \\ 1 & 2 & -7 \end{bmatrix}$$

$$= 17 \begin{bmatrix} 1 & 0 & 0 \\ 0 & 1 & 0 \\ 0 & 0 & 1 \end{bmatrix}$$

$$= |\mathbf{A}| \, \mathbf{I}$$

Inverse of a matrix. If, for a square matrix $\mathbf{A}$, a matrix $\mathbf{B}$ exists such that $\mathbf{BA} = \mathbf{AB} = \mathbf{I}$, then $\mathbf{B}$ is denoted by $\mathbf{A}^{-1}$ and is called the *inverse* of $\mathbf{A}$. The inverse of a matrix $\mathbf{A}$ exists if the determinant of $\mathbf{A}$ is nonzero or $\mathbf{A}$ is nonsingular.

By definition, the inverse matrix $\mathbf{A}^{-1}$ has the property that

$$\mathbf{AA}^{-1} = \mathbf{A}^{-1}\mathbf{A} = \mathbf{I}$$

where $\mathbf{I}$ is the identity matrix. If $\mathbf{A}$ is nonsingular and $\mathbf{AB} = \mathbf{C}$, then $\mathbf{B} = \mathbf{A}^{-1}\mathbf{C}$. This can be seen from the equation

$$\mathbf{A}^{-1}\mathbf{AB} = \mathbf{IB} = \mathbf{B} = \mathbf{A}^{-1}\mathbf{C}$$

If $\mathbf{A}$ and $\mathbf{B}$ are nonsingular matrices, then the product $\mathbf{AB}$ is a nonsingular matrix. Moreover,

$$(\mathbf{AB})^{-1} = \mathbf{B}^{-1}\mathbf{A}^{-1}$$

The preceding equation may be proved as follows:

$$(\mathbf{B}^{-1}\mathbf{A}^{-1})\mathbf{AB} = \mathbf{B}^{-1}(\mathbf{A}^{-1}\mathbf{A})\mathbf{B} = \mathbf{B}^{-1}\mathbf{IB} = \mathbf{B}^{-1}\mathbf{B} = \mathbf{I}$$

Similarly,

$$(\mathbf{AB})(\mathbf{B}^{-1}\mathbf{A}^{-1}) = \mathbf{I}$$

Note that

$$(\mathbf{A}^{-1})^{-1} = \mathbf{A}$$
$$(\mathbf{A}^{-1})' = (\mathbf{A}')^{-1}$$
$$(\mathbf{A}^{-1})^{*} = (\mathbf{A}^{*})^{-1}$$

From Equation (C–1) and the definition of the inverse matrix, we have

$$\mathbf{A}^{-1} = \frac{\text{adj } \mathbf{A}}{|\mathbf{A}|}$$

Hence, the inverse of a matrix is the transpose of the matrix of its cofactors, divided by the determinant of the matrix. That is, if

$$\mathbf{A} = \begin{bmatrix} a_{11} & a_{12} & \cdots & a_{1n} \\ a_{21} & a_{22} & \cdots & a_{2n} \\ \vdots & \vdots & & \vdots \\ a_{n1} & a_{n2} & \cdots & a_{nn} \end{bmatrix}$$

then

$$\mathbf{A}^{-1} = \frac{\text{adj } \mathbf{A}}{|\mathbf{A}|} = \begin{bmatrix} \dfrac{A_{11}}{|\mathbf{A}|} & \dfrac{A_{21}}{|\mathbf{A}|} & \cdots & \dfrac{A_{n1}}{|\mathbf{A}|} \\ \dfrac{A_{12}}{|\mathbf{A}|} & \dfrac{A_{22}}{|\mathbf{A}|} & \cdots & \dfrac{A_{n2}}{|\mathbf{A}|} \\ \vdots & \vdots & & \vdots \\ \dfrac{A_{1n}}{|\mathbf{A}|} & \dfrac{A_{2n}}{|\mathbf{A}|} & \cdots & \dfrac{A_{nn}}{|\mathbf{A}|} \end{bmatrix}$$

where A_{ij} is the cofactor of a_{ij} of the matrix $\mathbf{A}$. Thus, the terms in the ith column of $\mathbf{A}^{-1}$ are $1/|\mathbf{A}|$ times the cofactors of the ith row of the original matrix $\mathbf{A}$. For example, if

$$\mathbf{A} = \begin{bmatrix} 1 & 2 & 0 \\ 3 & -1 & -2 \\ 1 & 0 & -3 \end{bmatrix}$$

then the adjoint of $\mathbf{A}$ and the determinant $|\mathbf{A}|$ are respectively found to be

$$\text{adj } \mathbf{A} = \begin{bmatrix} 3 & 6 & -4 \\ 7 & -3 & 2 \\ 1 & 2 & -7 \end{bmatrix} \qquad \text{and} \qquad |\mathbf{A}| = 17$$

Hence, the inverse of $\mathbf{A}$ is

$$\mathbf{A}^{-1} = \frac{\text{adj } \mathbf{A}}{|\mathbf{A}|} = \begin{bmatrix} \frac{3}{17} & \frac{6}{17} & -\frac{4}{17} \\ \frac{7}{17} & -\frac{3}{17} & \frac{2}{17} \\ \frac{1}{17} & \frac{2}{17} & -\frac{7}{17} \end{bmatrix}$$

In what follows, we give formulas for finding inverse matrices for the 2×2 matrix and the 3×3 matrix. For the 2×2 matrix

$$\mathbf{A} = \begin{bmatrix} a & b \\ c & d \end{bmatrix} \qquad \text{where } ad - bc \neq 0$$

the inverse matrix is given by

$$\mathbf{A}^{-1} = \frac{1}{ad - bc} \begin{bmatrix} d & -b \\ -c & a \end{bmatrix}$$

For the 3×3 matrix

$$\mathbf{A} = \begin{bmatrix} a & b & c \\ d & e & f \\ g & h & i \end{bmatrix} \qquad \text{where } |\mathbf{A}| \neq 0$$

the inverse matrix is given by

$$
\mathbf{A}^{-1} = \frac{1}{|\mathbf{A}|}
\begin{bmatrix}
\begin{vmatrix} e & f \\ h & i \end{vmatrix} & -\begin{vmatrix} b & c \\ h & i \end{vmatrix} & \begin{vmatrix} b & c \\ e & f \end{vmatrix} \\[3mm]
-\begin{vmatrix} d & f \\ g & i \end{vmatrix} & \begin{vmatrix} a & c \\ g & i \end{vmatrix} & -\begin{vmatrix} a & c \\ d & f \end{vmatrix} \\[3mm]
\begin{vmatrix} d & e \\ g & h \end{vmatrix} & -\begin{vmatrix} a & b \\ g & h \end{vmatrix} & \begin{vmatrix} a & b \\ d & e \end{vmatrix}
\end{bmatrix}
$$

Remarks on cancellation of matrices. Cancellation of matrices is not valid in matrix algebra. Consider, for example, the product of the two singular matrices

$$
\mathbf{A} = \begin{bmatrix} 2 & 1 \\ 6 & 3 \end{bmatrix} \neq \mathbf{0} \qquad \text{and} \qquad \mathbf{B} = \begin{bmatrix} 1 & -2 \\ -2 & 4 \end{bmatrix} \neq \mathbf{0}
$$

Then

$$
\mathbf{AB} = \begin{bmatrix} 2 & 1 \\ 6 & 3 \end{bmatrix}\begin{bmatrix} 1 & -2 \\ -2 & 4 \end{bmatrix} = \begin{bmatrix} 0 & 0 \\ 0 & 0 \end{bmatrix} = \mathbf{0}
$$

Clearly, $\mathbf{AB} = \mathbf{0}$ implies neither that $\mathbf{A} = \mathbf{0}$ nor that $\mathbf{B} = \mathbf{0}$. In fact, $\mathbf{AB} = \mathbf{0}$ implies one of the following three statements:

1. $\mathbf{A} = \mathbf{0}$
2. $\mathbf{B} = \mathbf{0}$
3. Both $\mathbf{A}$ and $\mathbf{B}$ are singular.

We can easily prove that, if both $\mathbf{A}$ and $\mathbf{B}$ are nonzero matrices and $\mathbf{AB} = \mathbf{0}$, then both $\mathbf{A}$ and $\mathbf{B}$ are singular: Assume that $\mathbf{A}$ and $\mathbf{B}$ are not singular. Then a matrix $\mathbf{A}^{-1}$ exists with the property that

$$
\mathbf{A}^{-1}\mathbf{AB} = \mathbf{B} = \mathbf{0}
$$

which contradicts the assumption that $\mathbf{B}$ is a nonzero matrix. Thus, we conclude that both $\mathbf{A}$ and $\mathbf{B}$ must be singular if $\mathbf{A} \neq \mathbf{0}$ and $\mathbf{B} \neq \mathbf{0}$.

Similarly, notice that if $\mathbf{A}$ is singular, then neither $\mathbf{AB} = \mathbf{AC}$ nor $\mathbf{BA} = \mathbf{CA}$ implies that $\mathbf{B} = \mathbf{C}$. If, however, $\mathbf{A}$ is a nonsingular matrix, then $\mathbf{AB} = \mathbf{AC}$ implies that $\mathbf{B} = \mathbf{C}$ and $\mathbf{BA} = \mathbf{CA}$ also implies that $\mathbf{B} = \mathbf{C}$.

C–5 DIFFERENTIATION AND INTEGRATION OF MATRICES

The derivative of an $n \times m$ matrix $\mathbf{A}(t)$ is defined to be the $n \times m$ matrix, each element of which is the derivative of the corresponding element of the original matrix, provided that all the elements $a_{ij}(t)$ have derivatives with respect to t. That is,

$$\frac{d}{dt}\mathbf{A}(t) = \left(\frac{d}{dt}a_{ij}(t)\right) = \begin{bmatrix} \frac{d}{dt}a_{11}(t) & \frac{d}{dt}a_{12}(t) & \cdots & \frac{d}{dt}a_{1m}(t) \\ \frac{d}{dt}a_{21}(t) & \frac{d}{dt}a_{22}(t) & \cdots & \frac{d}{dt}a_{2m}(t) \\ \vdots & \vdots & & \vdots \\ \frac{d}{dt}a_{n1}(t) & \frac{d}{dt}a_{n2}(t) & \cdots & \frac{d}{dt}a_{nm}(t) \end{bmatrix}$$

Similarly, the integral of an $n \times m$ matrix $\mathbf{A}(t)$ is defined to be

$$\int \mathbf{A}(t)\, dt = \left(\int a_{ij}(t)\, dt\right) = \begin{bmatrix} \int a_{11}(t)\, dt & \int a_{12}(t)\, dt & \cdots & \int a_{1m}(t)\, dt \\ \int a_{21}(t)\, dt & \int a_{22}(t)\, dt & \cdots & \int a_{2m}(t)\, dt \\ \vdots & \vdots & & \vdots \\ \int a_{n1}(t)\, dt & \int a_{2n}(t)\, dt & \cdots & \int a_{nm}(t)\, dt \end{bmatrix}$$

Differentiation of the product of two matrices. If the matrices $\mathbf{A}(t)$ and $\mathbf{B}(t)$ can be differentiated with respect to t, then

$$\frac{d}{dt}[\mathbf{A}(t)\mathbf{B}(t)] = \frac{d\mathbf{A}(t)}{dt}\mathbf{B}(t) + \mathbf{A}(t)\frac{d\mathbf{B}(t)}{dt}$$

Here again the multiplication of $\mathbf{A}(t)$ and $d\mathbf{B}(t)/dt$ [or $d\mathbf{A}(t)/dt$ and $\mathbf{B}(t)$] is, in general, not commutative.

Differentiation of $\mathbf{A}^{-1}(t)$. If a matrix $\mathbf{A}(t)$ and its inverse $\mathbf{A}^{-1}(t)$ are differentiable with respect to t, then the derivative of $\mathbf{A}^{-1}(t)$ is given by

$$\frac{d\mathbf{A}^{-1}(t)}{dt} = -\mathbf{A}^{-1}(t)\frac{d\mathbf{A}(t)}{dt}\mathbf{A}^{-1}(t)$$

The derivative may be obtained by differentiating $\mathbf{A}(t)\mathbf{A}^{-1}(t)$ with respect to t. Since

$$\frac{d}{dt}[\mathbf{A}(t)\mathbf{A}^{-1}(t)] = \frac{d\mathbf{A}(t)}{dt}\mathbf{A}^{-1}(t) + \mathbf{A}(t)\frac{d\mathbf{A}^{-1}(t)}{dt}$$

and

$$\frac{d}{dt}\mathbf{A}(t)\mathbf{A}^{-1}(t) = \frac{d}{dt}\mathbf{I} = 0$$

we obtain

$$\mathbf{A}(t)\frac{d\mathbf{A}^{-1}(t)}{dt} = -\frac{d\mathbf{A}(t)}{dt}\mathbf{A}^{-1}(t)$$

or

$$\frac{d\mathbf{A}^{-1}(t)}{dt} = -\mathbf{A}^{-1}(t)\frac{d\mathbf{A}(t)}{dt}\mathbf{A}^{-1}(t)$$

Appendix D
Introduction to MATLAB

D–1 INTRODUCTION

MATLAB is a matrix-based system for performing mathematical and engineering calculations. We may think of MATLAB as a language of technical computing. All variables handled in MATLAB are matrices. That is, MATLAB has only one data type: a matrix, or rectangular array, of numbers. MATLAB has an extensive set of routines for obtaining graphical outputs.

This section presents background materials necessary for the effective use of MATLAB in solving control engineering problems. First, we introduce MATLAB commands and mathematical functions. Then we present matrix operators, relational and logical operators, and special characters used in MATLAB. Finally, we introduce the semicolon operator, MATLAB ways to enter vectors and matrices into the computer, the colon operator, and other important materials that we must become familiar with before writing MATLAB programs to solve system dynamics and control engineering problems.

MATLAB is used with a variety of toolboxes. (A toolbox is a collection of special files called M-files.) For control systems analysis and design, MATLAB is used with the control system toolbox. When we refer to MATLAB in this book, we include the base programs of MATLAB and the control system toolbox. (Student editions of MATLAB include the control system toolbox.)

MATLAB is basically command driven. Therefore, the user must know various commands that are used in solving computational problems. Table D–1 lists various types of MATLAB commands and predefined functions that are frequently used in solving system dynamics and control engineering problems.

TABLE D–1   MATLAB Commands and Matrix Functions

Commands and Matrix Functions Commonly Used in Solving Control Engineering Problems	Explanations of What Commands Do and Matrix Functions Mean
abs	Absolute value, complex magnitude.
acker	Compute a state-feedback gain matrix for pole placement, using Ackermann's formula.
angle	Phase angle.
ans	Answer when expression is not assigned.
atan	Arctangent.
axis	Manual axis scaling.
bode	Plot Bode diagram.
clear	Clear workspace.
clf	Clear current figure.
computer	Type of computer.
conj	Complex conjugate.
connect	Derive state-space model for block diagram interconnection.
conv	Convolution, multiplication.
corrcoef	Correlation coefficients.
cos	Cosine.
cosh	Hyperbolic cosine.
cov	Covariance.
ctrb	Compute the controllability matrix.
c2d	Conversion of continuous-time models to discrete-time models.
deconv	Deconvolution, division.
det	Determinant.
diag	Diagonal matrix.
eig	Eigenvalues and eigenvectors.
end	Terminate scope of for, while, switch, try, and if statements.
exit	Terminate program.
exp	Exponential base e.
expm	Matrix exponential.
eye	Identity matrix.
feedback	Feedback connection of two LTI models.
filter	Direct filter implementation.
for	Repeat statements a specified number of times.
format long	Fifteen-digit scaled fixed point. (Example: 1.33333333333333)

TABLE D–1 (*continued*)

Commands and Matrix Functions Commonly Used in Solving Control Engineering Problems	Explanations of What Commands Do and Matrix Functions Mean
format long e	Fifteen-digit floating point. (Example: 1.33333333333333e + 000)
format short	Five-digit scaled fixed point. (Example: 1.3333)
format short e	Five-digit floating point. (Example: 1.3333e + 000)
freqs	Laplace transform frequency response.
freqz	z-Transform frequency response.
gram	Controllability and observability gramians.
grid	Toggles the major lines of the current axes.
grid off	Removes major and minor grid lines from the current axes.
grid on	Adds major grid lines to the current axes.
help	Lists all primary help topics.
hold	Toggles the hold state.
hold off	Returns to the default mode whereby plot commands erase the previous plots and reset all axis properties before drawing new plots.
hold on	Holds the current plot and all axis properties so that subsequent graphing commands add to the existing graph.
i	$\sqrt{-1}$
imag	Imaginary part.
impulse	Impulse response of LTI models.
inf	Infinity (∞)
inv	Inverse
j	$\sqrt{-1}$
legend	Graph legend.
length	Length of vector.
linspace	Linearly spaced vector.
load	Load workspace variables from disk.
log	Natural logarithm.
loglog	Loglog x–y plot.
logm	Matrix logarithm.
logspace	Logarithmically spaced vector.
log10	Log base 10.
lqe	Linear quadratic estimator design.
lqr	Linear quadratic regulator design.
lsim	Simulate time response of LTI models to arbitrary inputs.
lyap	Solve continuous-time Lyapunov equations.
margin	Gain and phase margins and crossover frequencies.
max	Maximum value.

TABLE D–1 (continued)

Commands and Matrix Functions Commonly Used in Solving Control Engineering Problems	Explanations of What Commands Do and Matrix Functions Mean
mean	Mean value.
median	Median value.
mesh	Three-dimensional mesh surface.
meshgrid	X and Y arrays for three-dimensional plots.
min	Minimum value.
minreal	Minimal realization and pole–zero cancellation.
NaN	Not a number.
ngrid	Generate grid lines for a Nichols plot.
nichols	Draw the Nichols plot of the LTI model.
num2str	Convert number to string.
nyquist	Plot Nyquist frequency response.
obsv	Compute the observability matrix.
ode45	Solve nonstiff differential equations, medium-order method.
ode23	Solve nonstiff differential equations, low order-method.
ode113	Solve nonstiff differential equations, variable-order method.
ones	Constant.
ord2	Generate continuous-time second-order system.
pade	Pade approximation of time delays.
parallel	Parallel interconnection of two LTI models.
pi	Pi(π)
place	Compute a state-feedback gain matrix for pole placement.
plot	Linear x–y plot.
polar	Polar plot.
pole	Compute the poles of LTI models.
poly	Convert roots to polynomial.
polyfit	Polynomial curve fitting.
polyval	Polynomial evaluation.
polyvalm	Matrix polynomial evaluation.
printsys	Print system in pretty format.
prod	Product of elements.
pzmap	Pole–zero map of LTI models.
quit	Terminate program
rand	Generate random numbers and matrices.
rank	Calculate the rank of a matrix.
real	Real part.
rem	Remainder or modulus.
residue	Partial-fraction expansion.

TABLE D–1 (*continued*)

Commands and Matrix Functions Commonly Used in Solving Control Engineering Problems	Explanations of What Commands Do and Matrix Functions Mean
rlocfind	Find root-locus gains for a given set of roots.
rlocus	Plot root loci.
rmodel	Generate random stable continuous-time nth-order test models.
roots	Polynomial roots.
semilogx	Semilog x–y plot (x-axis logarithmic).
semilogy	Semilog x–y plot (y-axis logarithmic).
series	Interconnect two LTI models in series.
shg	Show graph window.
sign	Signum function.
sin	Sine.
sinh	Hyperbolic sine.
size	Size of matrix.
sqrt	Square root.
sqrtm	Matrix square root.
ss	Create state-space model or convert LTI model to state-space model.
ss2tf	Convert state-space model to transfer-function model.
std	Standard deviation.
step	Plot unit-step response.
subplot	Create axes in tiled positions.
sum	Sum of elements.
switch	Switch among several cases based on expression.
tan	Tangent.
tanh	Hyperbolic tangent.
text	Arbitrarily positioned text.
tf	Create transfer-function model or convert LTI model to transfer-function model.
tf2ss	Convert transfer-function model to state-space model.
tf2zp	Convert transfer-function model to zero–pole model.
title	Plot title.
trace	Trace of a matrix.
who	Lists all the variables currently in memory.
whos	List all the variables in the current workspace, together with information about their size, bytes, class, etc.
xlabel	x-axis label.
ylabel	y-axis label.
zero	Transmission zeros of LTI systems.
zeros	Zeros array.
zpk	Create zero–pole-gain models or convert to zero–pole-gain format.
zp2tf	Convert zero–pole model to transfer-function model.

Accessing and exiting MATLAB. On most systems, once MATLAB has been installed, to invoke MATLAB, execute the command MATLAB. To exit MATLAB, execute the command exit or quit.

MATLAB has an on-line help facility that may be invoked whenever the need arises. The command help will display a list of predefined functions and operators for which on-line help is available. The command

help 'function name'

will give information on the purpose and use of the specific function named. The command

help help

will give information on how to use the on-line help.

Matrix operators. The following notation is used in matrix operations:

+	Addition
−	Subtraction
*	Multiplication
^	Power
'	Conjugate transpose

(If multiple operations are involved, the order of the arithmetic operations can be altered with the use of parentheses.)

Relational and logical operators. The following relational operators are used in MATLAB:

<	Less than
<=	Less than or equal
>	Greater than
>=	Greater than or equal
==	Equal
~=	Not equal

The first four operators compare the real parts only. The last two ($==$, $\sim=$) compare both the real and imaginary parts. Note that "$=$" is used in an assignment statement while "$==$" is used in a relation.

The logical operators are

&	AND
|	OR
~	NOT

Special characters.   The following special characters are used in MATLAB:

[ ]	Used to form vectors and matrices
( )	Arithmetic expression precedence
,	Separate subscripts and function arguments
;	End rows, suppress printing
:	Subscripting, vector generation
!	Execute operating system command
%	Comment

Use of semicolon.   The semicolon is used to suppress printing. If the last character of a statement is a semicolon, printing is suppressed; the command is still executed, but the result is not displayed. This is a useful feature, since one may not need to print intermediate results. Also, in entering a matrix, a semicolon is used to indicate the end of a row, except the last row.

Use of colon.   The colon plays an important role in MATLAB, being involved in creating vectors, subscripting matrices, and specifying iterations. For example, the statement

$$t = 1{:}5$$

generates a row vector containing the numbers from 1 to 5 with unit increment—that is,

$$t =$$
$$1 \quad 2 \quad 3 \quad 4 \quad 5$$

An increment of other than unity can be used. For example,

$$t = 1{:}0.5{:}3$$

will result in

$$t =$$
$$1.0000 \quad 1.5000 \quad 2.0000 \quad 2.5000 \quad 3.0000$$

Negative increments may be used. For example, the statement

$$t = 5{:}-1{:}2$$

gives

$$t =$$
$$5 \quad 4 \quad 3 \quad 2$$

Other MATLAB commands that generate sequential data, such as linspace and logspace, are presented later in this section.

The colon is frequently used to subscript matrices. $A(:,j)$ is the jth column of A and $A(i,:)$ is the ith row of **A**. For example, if matrix **A** is given by

$$A = \begin{bmatrix} 1 & 2 & 3 \\ 4 & 5 & 6 \\ 7 & 8 & 9 \end{bmatrix}$$

then A(:,3) gives the third element in all of the rows (i.e., the third column), as follows:

$$3$$
$$6$$
$$9$$

A(2,:) gives the second row of **A**, namely,

$$4 \quad 5 \quad 6$$

A(:) returns a long column vector consisting elements of the first column, second column, and third column:

$$1$$
$$4$$
$$7$$
$$2$$
$$5$$
$$8$$
$$3$$
$$6$$
$$9$$

Note that A(i,j) denotes the entry in the ith row, jth column of matrix **A**. For example, A(2,3) is 6.

An individual vector can be referenced with indexes inside parentheses. For example, if a vector **x** is given by

$$x = [2 \quad 4 \quad 6 \quad 8 \quad 10]$$

then x(3) is the third element of **x** and x([1 2 3]) gives the first three elements of **x** (that is, 2, 4, 6).

Entering vectors in MATLAB programs.   In entering vectors and matrices in MATLAB programs, no dimension statements or type statements are needed. Vectors, which are $1 \times n$ or $n \times 1$ matrices, are used to hold ordinary one-dimensional sampled data signals, or sequences. One way to introduce a sequence into MATLAB is to enter it as an explicit list of elements separated by blank spaces or commas, as in

$$x = [1 \quad 2 \quad 3 \quad -4 \quad -5]$$

or

$$x = [1,2,3,-4,-5]$$

For readability, it is better to provide spaces between the elements. The values must always be entered within square brackets.

The statement

$$x = [1 \quad 2 \quad 3 \quad -4 \quad -5]$$

creates a simple five-element real sequence in a row vector. The sequence can be turned into a column vector by transposition. That is,

$$y = x'$$

results in

$$y =$$
$$1$$
$$2$$
$$3$$
$$-4$$
$$-5$$

How to enter matrices in MATLAB programs.   A matrix

$$\mathbf{A} = \begin{bmatrix} 1.2 & 10 & 15 \\ 3 & 5.5 & 2 \\ 4 & 6.8 & 7 \end{bmatrix}$$

may be entered in MATLAB programs by a row vector as follows:

$$A = [1.2 \quad 10 \quad 15; 3 \quad 5.5 \quad 2; 4 \quad 6.8 \quad 7]$$

Again, the values must be entered within square brackets. As with vectors, the elements of any row must be separated by blanks (or by commas). The end of each row, except the last, is indicated by a semicolon.

Note that the elements of the matrix **A** are automatically displayed after the statement is executed following the carriage return:

$$A =$$

1.2000	10.0000	15.0000
3.0000	5.5000	2.0000
4.0000	6.8000	7.0000

If we add a semicolon at the end of matrix statement such that

$$A = [1.2 \quad 10 \quad 15; 3 \quad 5.5 \quad 2; 4 \quad 6.8 \quad 7];$$

the output is suppressed and no output will be seen on the screen.

A large matrix may be spread across several input lines. For example, consider the matrix

$$\mathbf{B} = \begin{bmatrix} 1.5630 & 2.4572 & 3.1113 & 4.1051 \\ 3.2211 & 1.0000 & 2.5000 & 3.2501 \\ 1.0000 & 2.0000 & 0.6667 & 0.0555 \\ 0.2345 & 0.9090 & 1.0000 & 0.3333 \end{bmatrix}$$

This matrix may be spread across four input lines, as follows:

$$B = [1.5630 \quad 2.4572 \quad 3.1113 \quad 4.1051$$
$$3.2211 \quad 1.0000 \quad 2.5000 \quad 3.2501$$
$$1.0000 \quad 2.0000 \quad 0.6667 \quad 0.0555$$
$$0.2345 \quad 0.9090 \quad 1.0000 \quad 0.3333]$$

Note that carriage returns replace the semicolons.

As another example, a matrix

$$C = \begin{bmatrix} 1 & e^{-0.02} \\ \sqrt{2} & 3 \end{bmatrix}$$

may be entered as follows:

$$C = [1 \quad \exp(-0.02); \text{sqrt}(2) \quad 3]$$

After the carriage return, the following matrix will be seen on the screen:

```
C =
   1.0000   0.9802
   1.4142   3.0000
```

Generating vectors. Besides the colon operator, the linspace and logspace commands generate sequential data:

$$x = \text{linspace}(n1,n2,n)$$
$$w = \text{logspace}(d1,d2,n)$$

The linspace command generates a vector from n1 to n2 with n points (including both endpoints). See the following example:

```
>> x = linspace (–10, 10, 5)
x =
   –10   –5   0   5   10
```

The logspace command generates a logarithmically spaced vector from 10^{d1} to 10^{d2} with n points (again, including both endpoints). Frequently, n is chosen to be 50 or 100, but it can be any number. For example,

$$w = \text{logspace} (-1,1,10)$$

generates 10 points from 0.1 to 10, as shown below. (Note that 10 points include both endpoints.)

```
>> w = logspace (–1,1,10)
w =
  Columns 1 through 8
   0.1000   0.1668   0.2783   0.4642   0.7743   1.2915   2.1544   3.5938
  Columns 9 through 10
   5.9948   10.0000
```

Transpose and conjugate transpose. The apostrophe or prime denotes the conjugate transpose of a matrix. If the matrix is real, the conjugate transpose is simply the transpose. An entry such as

$$A = [1 \quad 2 \quad 3;4 \quad 5 \quad 6;7 \quad 8 \quad 9]$$

will produce the following matrix on the screen:

$$A =$$

$$\begin{array}{ccc} 1 & 2 & 3 \\ 4 & 5 & 6 \\ 7 & 8 & 9 \end{array}$$

Also, if

$$B = A'$$

is entered, then

$$B =$$

$$\begin{array}{ccc} 1 & 4 & 7 \\ 2 & 5 & 8 \\ 3 & 6 & 9 \end{array}$$

appears on the screen.

Entering complex numbers.   Complex numbers may be entered using the function i or j. For example, a number $1 + j\sqrt{3}$ may be entered as

$$x = 1+\text{sqrt}(3)*i$$

or

$$x = 1+\text{sqrt}(3)*j$$

This complex number, $1 + j\sqrt{3} = 2 \exp[(\pi/3)j]$, may also be entered as

$$x = 2*\exp((\text{pi}/3)*j)$$

It is important to note that, when complex numbers are entered as matrix elements within brackets, we avoid blank spaces. For example, $1 + j5$ should be entered as

$$x = 1+5*j$$

If spaces are provided around the $+$ sign, as in

$$x = 1 + 5*j$$

two separate numbers are represented.

If i and j are used as variables, a new complex unit may be generated as follows:

$$ii = \text{sqrt}(-1)$$

or

$$jj = \text{sqrt}(-1)$$

Then $-1 + j\sqrt{3}$ may be entered as

$$x = -1+\text{sqrt}(3)*ii$$

or

$$x = -1+\text{sqrt}(3)*jj$$

If we defined ii $= \sqrt{-1}$ and want to change ii to the predefined i $= \sqrt{-1}$, enter clear ii in the computer. Then the predefined variable i can be reset.

Entering complex matrices.   For the complex matrix

$$\mathbf{X} = \begin{bmatrix} 1 & j \\ -j5 & 2 \end{bmatrix}$$

an entry like

$$X = [1 \quad j; -j*5 \quad 2]$$

will produce the following matrix on the screen:

X =
        1.0000   0 + 1.0000i
             0 − 5.0000i   2.0000

Note that

$$Y = X'$$

will yield

Y =
        1.0000   0 + 5.0000i
             0 − 1.0000i   2.0000

which is

$$\mathbf{Y} = \begin{bmatrix} 1 & j5 \\ -j & 2 \end{bmatrix}$$

Since the prime gives the complex conjugate transpose, for an unconjugated transpose use one of the following two entries:

$$Y.' \qquad \text{or} \qquad conj(Y')$$

If we enter

$$Y.'$$

then the screen shows

ans =
        1.0000   0 − 1.0000i
             0 + 5.0000i   2.0000

Also, if

$$conj(Y')$$

is entered, then the screen shows

ans =
        1.0000   0 − 1.0000i
             0 + 5.0000i   2.0000

Entering a long statement that will not fit on one line. A statement is normally terminated with a carriage return or the enter key. If the statement being entered is too long for one line, an ellipsis consisting of three or more periods (...), followed by the carriage return, can be used to indicate that the statement continues on the next line. An example is

$$x = 1.234 + 2.345 + 3.456 + 4.567 + 5.678 + 6.789...$$
$$+ 7.890 + 8.901 - 9.012;$$

Note that the blank spaces around the $=$, $+$, and $-$ signs are optional. Such spaces are often provided to improve readability.

Entering several statements on one line. Several statements can be placed on one line if they are separated by commas or semicolons. Examples are

$$x1 = [1 \quad 2 \quad 3], x2 = [4 \quad 5 \quad 6], x3 = [7 \quad 8 \quad 9]$$

and

$$x1 = [1 \quad 2 \quad 3]; x2 = [4 \quad 5 \quad 6]; x3 = [7 \quad 8 \quad 9]$$

Selecting the output format. All computations in MATLAB are performed in double precision. However, the displayed output may have a fixed point with four decimal places. For example, for the vector

$$\mathbf{x} = [1/3 \quad 0.00002]$$

MATLAB exhibits the following output:

```
x =
    0.3333    0.0000
```

If at least one element of a matrix is not an exact integer, there are four possible output formats. The displayed output can be controlled with the use of the following commands:

```
format short
format long
format short e
format long e
```

Once invoked, the chosen format remains in effect until changed.

For systems analysis, format short and format long are commonly used. Whenever MATLAB is invoked and no format command is entered, MATLAB shows the numerical results in format short, as follows:

```
x = [1/3   0.00002];

x

x =
     0.3333      0.0000
```

For x = [1/3   0.00002], the commands format short; x and format long; x yield the following output:

```
format short; x
x =
      0.3333    0.0000
format long; x
x =
      0.33333333333333      0.00002000000000
```

If all elements of a matrix or vector are exact integers, then format short and format long yield the same result as shown below.

```
y = [2   5   40];
y
y =
      2   5   40
format short; y
y =
      2   5   40
format long; y
y =
      2   5   40
```

Utility matrices.   In MATLAB, the functions

```
ones (n)
ones (m,n)
zeros (n)
zeros (m,n)
```

generate special matrices. The function ones(n) produces an $n \times n$ matrix of ones, while ones(m,n) produces an $m \times n$ matrix of ones. Similarly, zeros(n) produces an $n \times n$ matrix of zeros, while zeros(m,n) produces an $m \times n$ matrix of zeros.

Identity matrix.   We often need to enter an identity matrix **I** in MATLAB programs. The statement eye(n) gives an $n \times n$ identity matrix. That is,

```
eye(5)
ans =
      1   0   0   0   0
      0   1   0   0   0
      0   0   1   0   0
      0   0   0   1   0
      0   0   0   0   1
```

The statement eye(A) returns an identity matrix the same size as the matrix **A**.

Diagonal matrix. If x is a vector, the statement diag(x) produces a diagonal matrix with x on the diagonal line. For example, for a vector

$$x = [ones(1,n)]$$

diag([ones(1,n)]) gives the following $n \times n$ identity matrix:

```
diag([ones(1,5)])
ans =
     1    0    0    0    0
     0    1    0    0    0
     0    0    1    0    0
     0    0    0    1    0
     0    0    0    0    1
```

If **A** is a square matrix, then diag(A) is a vector consisting of the diagonal of **A**, and diag(diag(A)) is a diagonal matrix with elements of diag(A) appearing on the diagonal line. See the following MATLAB output.

```
A = [1   2   3;4   5   6;7   8   9];
diag(A)
ans =
     1
     5
     9
diag(diag(A))
ans =
     1    0    0
     0    5    0
     0    0    9
```

Note that diag(1:5) gives

```
diag(1:5)
ans =
     1    0    0    0    0
     0    2    0    0    0
     0    0    3    0    0
     0    0    0    4    0
     0    0    0    0    5
```

and diag(0:4) gives

```
diag(0:4)
ans =
     0   0   0   0   0
     0   1   0   0   0
     0   0   2   0   0
     0   0   0   3   0
     0   0   0   0   4
```

Hence, diag(1:5) – diag(0:4) is an identity matrix.

It is important to note that diag(0,n) is quite different from diag(0:n).diag(0,n) is an $(n + 1) \times (n + 1)$ matrix consisting of all zero elements. See the following MAT-LAB output.

```
diag(0,4)
ans =
     0   0   0   0   0
     0   0   0   0   0
     0   0   0   0   0
     0   0   0   0   0
     0   0   0   0   0
```

Variables in MATLAB.   A convenient feature of MATLAB is that variables need not be dimensioned before they are used. This is because a variable's dimensions are generated automatically upon the first use of the variable. (The dimensions of the variables can be altered later if necessary.) Such variables (and their dimensions) remain in memory until the command exit or quit is entered.

Suppose that we enter the following statements in MATLAB workspace:

>> A = [1   2   3;4   5   6;7   8   9];
>> 15/31;
>>x = [3   4   5];

To obtain a list of the variables in the workspace, simply type the command who. Then all of the variables currently in the workspace appear on the screen as shown below.

```
>> who
Your variables are:

A    ans    x
```

If, instead of entering the command who, we enter the command whos, the screen will show a list of all the variables in the current workspace, together with information about their size, number of bytes, and class.

The following MATLAB output is illustrative:

```
>> whos
    Name        Size            Bytes    Class
    A           3×3                72    double array
    ans         1×1                 8    double array
    x           1×3                24    double array
Grand total is 13 elements using 104 bytes
```

The command clear will clear all nonpermanent variables from the workspace. If it is desired to clear only a particular variable, say, 'x', from the workspace, enter the command clear x.

How to save variables when exiting from MATLAB. When 'exit' or 'quit' is entered, all variables in MATLAB are lost. If the command save is entered before exiting, then all variables can be kept in a disk file named matlab.mat. When we later reenter MATLAB, the command load will restore the workspace to its former state.

If you need to know the time and date. Statement clock gives the year, month, day, hour, minute, and second. That is, clock returns a six-element row vector containing the current time and date in decimal form:

clock
ans=

[year   month   day   hour   minute   second]

or

```
>> clock
ans =
   1.0e + 003 *
    2.0030     0.0010     0.0200     0.0170     0.0210     0.0259
```

Statement date gives the current date:

```
> date
ans =
20-Jan-2003
```

Correcting mistyped characters. Use the arrow keys on the keypad to edit mistyped commands or to recall previous command lines. For example, if we typed

x = (1   1   2]

then the parenthesis must be corrected. Instead of retyping the entire line, hit the *up-arrow* key. The incorrect line will be displayed again. Using the *left-arrow* key, move the cursor over the parenthesis, type [, and then hit the *delete* key.

How MATLAB is used.   MATLAB is usually used in a command-driven mode. When single-line commands are entered, MATLAB processes them immediately and displays the results. MATLAB is also capable of executing sequences of commands that are stored in files.

The commands that are typed may be accessed later by using the *up-arrow* key. It is possible to scroll through some of the latest commands that are entered and recall a particular command line.

How to enter comments in a MATLAB program.   If you desire to enter comments that are not to be executed, use the % symbol at the start of the line. That is, the % symbol indicates that the rest of the line is a comment and should be ignored.

D–2 ADDITION, SUBTRACTION, MULTIPLICATION, AND DIVISION WITH MATLAB

Addition and subtraction.   Matrices of the same dimension may be added or subtracted. Consider the following matrices:

$$\mathbf{A} = \begin{bmatrix} 2 & 3 \\ 4 & 5 \\ 6 & 7 \end{bmatrix}, \qquad \mathbf{B} = \begin{bmatrix} 1 & 0 \\ 2 & 3 \\ 0 & 4 \end{bmatrix}$$

If we enter

$$A = [2 \quad 3;4 \quad 5;6 \quad 7]$$

then the screen shows

$$A =$$

$$\begin{matrix} 2 & 3 \\ 4 & 5 \\ 6 & 7 \end{matrix}$$

If matrix **B** is entered as

$$B = [1 \quad 0;2 \quad 3;0 \quad 4]$$

then the screen shows

$$B =$$
$$\begin{matrix} 1 & 0 \\ 2 & 3 \\ 0 & 4 \end{matrix}$$

For the addition of two matrices, such as A + B, we enter

$$C = A + B$$

Then matrix **C** appears on the screen as

$$C =$$

$$\begin{matrix} 3 & 3 \\ 6 & 8 \\ 6 & 11 \end{matrix}$$

If a vector **x** is given by

$$\mathbf{x} = \begin{bmatrix} 5 \\ 4 \\ 6 \end{bmatrix}$$

then we enter this vector as

$$x = [5;4;6]$$

The screen shows the column vector as follows:

$$x =$$

$$\begin{matrix} 5 \\ 4 \\ 6 \end{matrix}$$

The following entry will subtract 1 from each element of vector **x**:

$$y = x - 1$$

The screen will show

$$y =$$

$$\begin{matrix} 4 \\ 3 \\ 5 \end{matrix}$$

Matrix multiplication.   Consider the matrices

$$\mathbf{x} = \begin{bmatrix} 1 \\ 2 \\ 3 \end{bmatrix}, \qquad \mathbf{y} = \begin{bmatrix} 4 \\ 5 \\ 6 \end{bmatrix}, \qquad \mathbf{A} = \begin{bmatrix} 1 & 1 & 2 \\ 3 & 4 & 0 \\ 1 & 2 & 5 \end{bmatrix}$$

or, in MATLAB,

$$x = [1;2;3]; \qquad y = [4;5;6]; \qquad A = [1 \quad 1 \quad 2;3 \quad 4 \quad 0;1 \quad 2 \quad 5]$$

If the expression

$$z = x'*y$$

is entered into the computer, the result is

$$z = 32$$

(Multiplication of matrices is denoted by *.) Note that when the variable name and "=" are not included in the expression, as in

$$x'*y$$

the result is assigned to the generic variable ans:

```
>> x'*y
ans =
        32
```

Also, the entry

$$x*y'$$

will yield a 3 × 3 matrix, as follows:

```
>> x*y'
ans =
        4       5       6
        8      10      12
       12      15      18
```

Similarly, if we enter

$$y*x'$$

then the screen shows

```
ans =
        4     8    12
        5    10    15
        6    12    18
```

Matrix–vector products are a special case of general matrix–matrix products. For example, an entry like

$$b = A * x$$

will produce

```
b =
        9
       11
       20
```

Note that a scalar can multiply, or be multiplied by, any matrix. For example, entering

$$5*A$$

gives

```
ans =
        5     5    10
       15    20     0
        5    10    25
```

and an entry such as

$$A * 5$$

will also give

ans =

5	5	10
15	20	0
5	10	25

Magnitude and phase angle of a complex number. The magnitude and phase angle of a complex number $z = x + iy = re^{i\theta}$ are respectively given by

$$r = abs(z)$$
$$theta = angle(z)$$

and the statement

$$z = r*exp(i*theta)$$

converts them back to the original complex number z.

Array multiplication. Array, or element-by-element, multiplication is denoted by '.*'. If **x** and **y** have the same dimension, then

$$x.*y$$

denotes the array whose elements are simply the products of the individual elements of x and y. For example, if

$$\mathbf{x} = [1 \quad 2 \quad 3], \qquad \mathbf{y} = [4 \quad 5 \quad 6]$$

then

$$z = x.*y$$

results in

$$\mathbf{z} = [4 \quad 10 \quad 18]$$

Obtaining squares of entries of vector x. For a vector **x**, x.^2 gives the vector of the square of each element. For example, for

$$x = [1 \quad 2 \quad 3]$$

x.^2 is given as shown in the following MATLAB output:

```
x = [1   2   3];
x.^2

ans =

        1       4       9
```

Also, for the vector

$$\mathbf{y} = [2 + 5j \quad 3 + 4j \quad 1 - j]$$

y.^2 is given as follows:

```
y = [2+5*i    3+4*i    1−i];
y.^2

ans =

 −21.0000 + 20.0000i    − 7.0000 + 24.0000i        0 − 2.0000i
```

Similarly, if matrices **A** and **B** have the same dimensions, then **A.*B** denotes the array whose elements are simply the products of the corresponding elements of **A** and **B**. For example, if

$$\mathbf{A} = \begin{bmatrix} 1 & 2 & 3 \\ 0 & 9 & 8 \end{bmatrix}, \qquad \mathbf{B} = \begin{bmatrix} 4 & 5 & 6 \\ 7 & 6 & 5 \end{bmatrix}$$

then

$$C = A.*B$$

results in

$$\mathbf{C} = \begin{bmatrix} 4 & 10 & 18 \\ 0 & 54 & 40 \end{bmatrix}$$

Obtaining squares of entries of matrix A.   For a matrix **A**, A.^2 gives the matrix consisting of the square of each element. For example, for matrices

$$\mathbf{A} = \begin{bmatrix} 1 & 2 \\ 3 & 4 \end{bmatrix}, \qquad \mathbf{B} = \begin{bmatrix} 1+j & 2-2j \\ 3+4j & 5-j \end{bmatrix}$$

A.^2 and B.^2 are given as follows:

```
A = [1    2;3    4];
A.^2

ans =

      1     4
      9    16

B = [1+ i    2−2*i;3+4*i    5 − i];
B.^2

ans =

            0 +  2.0000i                0 −   8.0000i
      − 7.0000 + 24.0000i         24.0000 − 10.0000i
```

Absolute values.   The command abs(A) gives the matrix consisting of the absolute value of each element of **A**. If **A** is complex, abs(A) returns the complex modulus (magnitude):

$$abs(A) = sqrt(real(A).^2 + imag(A).^2)$$

The command angle(A) returns the phase angles, in radians, of the elements of the complex matrix **A**. The angles lie between $-\pi$ and π. The following example is illustrative:

```
A = [2+2*i   1+3*i;4+5*i   6-i];
abs(A)

ans =

        2.8284      3.1623
        6.4031      6.0828

angle(A)

ans =

        0.7854       1.2490
        0.8961      -0.1651
```

Array division.   The expressions x./y, x.\y, A./B, and A.\B give the quotients of the individual elements. Thus, for

$$\mathbf{x} = [1 \quad 2 \quad 3], \qquad \mathbf{y} = [4 \quad 5 \quad 6]$$

the statement

$$u = x./y$$

gives

$$\mathbf{u} = [0.25 \quad 0.4 \quad 0.5]$$

and the statement

$$v = x.\backslash y$$

results in

$$\mathbf{v} = [4 \quad 2.5 \quad 2]$$

Similarly, for matrices **A** and **B**, where

$$\mathbf{A} = \begin{bmatrix} 1 & 2 & 3 \\ 1 & 9 & 8 \end{bmatrix}, \qquad \mathbf{B} = \begin{bmatrix} 4 & 5 & 6 \\ 7 & 6 & 5 \end{bmatrix}$$

the statement

$$C = A./B$$

gives

$$\mathbf{C} = \begin{bmatrix} 0.2500 & 0.4000 & 0.5000 \\ 0.1429 & 1.5000 & 1.6000 \end{bmatrix}$$

and the statement

$$D = A.\backslash B$$

yields

$$\mathbf{D} = \begin{bmatrix} 4.0000 & 2.5000 & 2.0000 \\ 7.0000 & 0.6667 & 0.6250 \end{bmatrix}$$

Note that whenever a division of a number by zero occurs, MATLAB gives a warning, as in the following outputs:

```
>> 5/0
Warning: Divide by zero.

ans =

    Inf
```

```
>> 0/0
Warning: Divide by zero.

ans =

    NaN
```

("Inf" denotes "infinity" and NaN means "not a number.")

D–3 COMPUTING MATRIX FUNCTIONS

In this section, we discuss computations of norms, eigenvalues, eigenvectors, and polynomial evaluation, among other topics.

Norms.   The norm of a matrix is a scalar that gives some measure of the size of the matrix. Several different definitions are commonly used. One is

$$norm(\mathbf{A}) = \text{largest singular value of } \mathbf{A}$$

Similarly, several definitions are available for the norm of a vector. One commonly used definition is

$$norm(x) = sum(abs(x).^2)^0.5$$

The following example illustrates the use of the norm command:

```
>> x = [2    3    6];
>> norm(x)

ans =

     7
```

Characteristic equation.   The roots of the characteristic equation are the same as the eigenvalues of a matrix $\mathbf{A}$. The characteristic equation of $\mathbf{A}$ is computed with

$$p = poly(A)$$

For example, if

$$\mathbf{A} = \begin{bmatrix} 0 & 1 & 0 \\ 0 & 0 & 1 \\ -6 & -11 & -6 \end{bmatrix}$$

then the command poly(A) will yield

p = poly(A)
p =
1.0000   6.0000   11.0000   6.0000

This is the MATLAB representation of the characteristic equation

$$s^3 + 6s^2 + 11s + 6 = 0$$

Note that polynomials are represented as row vectors containing the polynomial coefficients in descending order; that is, in the present example,

$$p = [1 \quad 6 \quad 11 \quad 6]$$

The roots of the characteristic equation $p = 0$ can be obtained by entering the command $r = \text{roots}(p)$:

r = roots(p)
r =
− 3.0000
− 2.0000
− 1.0000

The roots are $s = -3$, $s = -2$, and $s = -1$. Note that the commands poly and roots can be combined into a single expression, such as

$$\text{roots}(\text{poly}(A))$$

The roots of the characteristic equation may be reassembled back into the original polynomial with the command q = poly(r). For r = [−3   −2   −1], poly(r) will produce the polynomial equation

$$s^3 + 6s^2 + 11s + 6 = 0$$

The following MATLAB output is illustrative:

>> r = [−3   −2   −1];
>> q = poly(r)
q =
1   6   11   6

Addition or subtraction of polynomials.   If the two polynomials are of the same order, add the arrays that describe their coefficients. If the polynomials are of different order (n and m, where $m < n$), then add $n - m$ zeros to the left-hand side of the coefficient array of the lower order polynomial. The following MATLAB output is illustrative:

```
>> a = [3    10    25    36    50];
>> b = [0    0    1    2    10];
>> a+b

ans =

        3      10      26      38      60
```

For the subtraction of b from a, consider the subtraction as an addition of a and $-b$.

Eigenvalues and eigenvectors.   If **A** is an $n \times n$ matrix, then the n numbers λ that satisfy

$$\mathbf{Ax} = \lambda\mathbf{x}$$

are the eigenvalues of **A**. They are obtained with the command

$$\mathrm{eig(A)}$$

which returns the eigenvalues in a column vector.

If **A** is real and symmetric, the eigenvalues will be real. But if **A** is not symmetric, the eigenvalues will frequently be complex numbers.

For example, with

$$\mathbf{A} = \begin{bmatrix} 0 & 1 & 0 \\ -1 & 0 & 2 \\ 3 & 0 & 5 \end{bmatrix}$$

the command eig(A) produces the eigenvalues shown in the following output:

```
>>A = [0    1    0;-1    0    2;3    0    5];
>> eig(A)

ans =

    5.2130
    -0.1065 + 1.4487i
    -0.1065 - 1.4487i
```

MATLAB functions may have single- or multiple-output arguments. For example, as seen previously, eig(A) produces a column vector consisting of the eigenvalues of **A**, while the double-assignment statement

$$[X,D] = eig(A)$$

produces eigenvalues and eigenvectors. The diagonal elements of the diagonal matrix **D** are the eigenvalues, and the columns of **X** are the corresponding eigenvectors such that

$$\mathbf{AX} = \mathbf{XD}$$

For example, if

$$\mathbf{A} = \begin{bmatrix} 0 & 1 & 0 \\ 0 & 0 & 1 \\ -6 & -11 & -6 \end{bmatrix}$$

then the statement

$$[X,D] = eig(A)$$

gives the following result:

```
>> A = [0    1    0;0    0    1;-6    -11    -6];
>> [X, D] = eig(A)

X =

    -0.5774          0.2182         -0.1048
     0.5774         -0.4364          0.3145
    -0.5774          0.8729         -0.9435

D =

    -1.0000               0               0
          0         -2.0000               0
          0               0         -3.0000
```

The eigenvectors are scaled so that the norm of each is unity.

If the eigenvalues of a matrix are distinct, the eigenvectors are always independent and the eigenvector matrix **X** will diagonalize the original matrix **A** if **X** is applied as a similarity transformation matrix. However, if a matrix has repeated eigenvalues (multiple eigenvalues), it is not diagonalizable unless it has a full (independent) set of eigenvectors. If the eigenvectors are not independent, the original matrix is said to be defective. Even if a matrix is defective, the solution from eig satisfies the relationship $\mathbf{AX} = \mathbf{XD}$.

Convolution (product of polynomials).   Consider the polynomials

$$a(s) = 3s^4 + 10s^3 + 25s^2 + 36s + 50$$
$$b(s) = s^2 + 2s + 10$$

The product of the polynomials is the convolution of the coefficients. The product of polynomials $a(s)$ and $b(s)$ can be obtained by entering the command c = conv(a,b).

```
>> a = [3     10     25     36     50];
>> b = [1     2      10];
>> % Define the product of a and b as c.
>> c = conv(a,b)

c =
        3     16     75     186     372     460     500
```

The foregoing is the MATLAB representation of the polynomial

$$c(s) = 3s^6 + 16s^5 + 75s^4 + 186s^3 + 372s^2 + 460s + 500$$

Deconvolution (division of polynomials).   To divide the polynomial $a(s)$ by $b(s)$, use the deconvolution command

$$[q,r] = deconv(a,b)$$

The following MATLAB output is illustrative:

```
>> a = [3     10     25     36     50];
>> b = [1     2      10];
>> % Define the quotient and remainder of a/b as q and r, respectively.
>> [q, r] = deconv (a,b)

q =
        3     4    -13

r =
        0     0     0     22     180
```

This MATLAB output means that

$$3s^4 + 10s^3 + 25s^2 + 36s + 50$$
$$= (s^2 + 2s + 10)(3s^2 + 4s - 13) + 22s + 180$$

Polynomial evaluation.   If p is a vector whose elements are the coefficients of a polynomial in descending powers, then polyval(p,s) is the value of the polynomial, evaluated at s. For example, to evaluate the polynomial

$$p(s) = 3s^2 + 2s + 1$$

at $s = 5$, enter the command

$$p = [3\ \ 2\ \ 1];$$
$$polyval(p,5)$$

Then we get

$$ans =$$
$$86$$

The command polyvalm(J) evaluates the polynomial p in a matrix sense. For example, consider the matrix

$$\mathbf{J} = \begin{bmatrix} -2 + j2\sqrt{3} & 0 & 0 \\ 0 & -2 - j2\sqrt{3} & 0 \\ 0 & 0 & -10 \end{bmatrix}$$

The command poly(J) gives the characteristic polynomial for $\mathbf{J}$:

```
>> J = [-2+i*2*sqrt (3)    0    0;0    -2-i*2*sqrt(3)    0;0    0    -10];
>> p = poly (J)

p =

    1.0000   14.0000   56.0000   160.0000
```

This is the MATLAB expression for the characteristic polynomial for $\mathbf{J}$:

$$\text{poly}(\mathbf{J}) = \phi(\mathbf{J}) = \mathbf{J}^3 + 14\mathbf{J}^2 + 56\mathbf{J} + 160\mathbf{I}$$

Here, $\mathbf{I}$ is the identity matrix. For the matrix

$$\mathbf{A} = \begin{bmatrix} 0 & 1 & 0 \\ 0 & 0 & 1 \\ -6 & -11 & -6 \end{bmatrix}$$

the command polyvalm(poly(J),A) evaluates the following $\phi(\mathbf{A})$:

$$\phi(\mathbf{A}) = \mathbf{A}^3 + 14\mathbf{A}^2 + 56\mathbf{A} + 160\mathbf{I} = \begin{bmatrix} 154 & 45 & 8 \\ -48 & 66 & -3 \\ 18 & -15 & 84 \end{bmatrix}$$

See the following MATLAB output:

```
>> J = [-2+i*2*sqrt (3)    0    0;0    -2-i*2*sqrt(3)    0;0    0    -10];
>> A = [0    1    0;0    0    1;-6    -11    -6];
>> polyvalm (poly (J), A)

ans =

    154.0000       45.0000       8.0000
    -48.0000       66.0000      -3.0000
     18.0000      -15.0000      84.0000
```

Matrix exponential. The command expm(A) gives the matrix exponential of an $n \times n$ matrix **A**. That is,

$$\text{expm}(\mathbf{A}) = \mathbf{I} + \mathbf{A} + \frac{\mathbf{A}^2}{2!} + \frac{\mathbf{A}^3}{3!} + \cdots$$

Note that a transcendental function is interpreted as a matrix function if an "m" is appended to the function name, as in expm(A) or sqrtm(A).

As an example, consider the matrix

$$\mathbf{A} = \begin{bmatrix} 0 & 1 & 0 \\ 0 & 0 & 1 \\ -6 & -11 & -6 \end{bmatrix}$$

then the matrix exponential $e^{\mathbf{A}}$ can be obtained as follows:

```
>> A = [0   1   0;0   0   1;-6   -11   -6];
>> expm(A)

ans =

    0.7474      0.4530      0.0735
   -0.4410     -0.0611      0.0121
   -0.0723     -0.5735     -0.1334
```

The following MATLAB output affords another example:

```
>> expm(eye (3))

ans =

    2.7183          0          0
         0     2.7183          0
         0          0     2.7183
```

D–4 PLOTTING RESPONSE CURVES

MATLAB has an extensive set of routines for obtaining graphical output. The plot command creates linear x–y plots. (Logarithmic or polar plots are created by substituting the word loglog, semilogx, semilogy, or polar for plot.) All such commands are used the same way: They only affect how the axis is scaled and how the data are displayed.

x–y plot. If **x** and **y** are vectors of the same length, the command

$$\text{plot}(x,y)$$

plots the values in y against the values in x.

Plotting multiple curves.   To plot multiple curves on a single graph, use the plot command with multiple arguments:

$$\text{plot(X1, Y1, X2, Y2, } \ldots \text{, Xn, Yn)}$$

The variables X1, Y1, X2, Y2, and so on, are pairs of vectors. Each x–y pair is graphed, generating multiple curves on the plot. Multiple arguments have the benefit of allowing vectors of different lengths to be displayed on the same graph. Each pair uses a different type of line.

Plotting more than one curve on a single graph may also be accomplished by using the command hold on (or the command hold), either of which freezes the current plot and inhibits erasure and rescaling. Hence, subsequent curves will be plotted over the original curve. Entering the command hold off (or the command hold again) releases the current plot.

Adding grid lines, title of the graph, *x*-axis label, and *y*-axis label. Once a graph is on the screen, grid lines may be drawn, the graph may be titled, and *x*- and *y*-axes may be labeled. MATLAB commands to produce a grid, title, *x*-axis label, and *y*-axis label are as follows:

> grid   (grid lines)
> title   (graph title)
> xlabel (*x*-axis label)
> ylabel (*y*-axis label)

Note that, once the command display has been brought back, grid lines, the graph title, and *x* and *y* labels can be put on the plot by successively entering the preceding commands.

Writing text on the graph.   To write a text on the plot, use the command text. The command text (X, Y, 'string') adds the text in quotes to location (X,Y) on the current plot, where (X, Y) is expressed in units of the current plot. For example, the statement

$$\text{text(3,0.45,'sin t')}$$

will write sin *t* horizontally, beginning at point (3,0.45).

Similarly, the command text (X, Y, Z, 'string') adds text to a three-dimensional plot.

Another command used frequently to add text to a two-dimensional plot is gtext. The command gtext('string') displays the graph window, puts up a crosshair, and waits for a mouse button or keyboard key to be pressed. The crosshair can be positioned with the mouse. Pressing a mouse button or any key writes the text string onto the graph at the selected location.

Imaginary and complex data.   If **z** ia a complex vector, then plot(z) is equivalent to plot(real(z), imag(z)). That is, plot(z) will plot the imaginary part of **z** against the real part of **z**.

Polar plots. The command polar (theta, rho) will give a plot of the angle theta (in radians) versus the radius rho, in polar coordinates. Subsequent use of the grid command draws polar grid lines.

Logarithmic plots. The following commands produce the indicated plots:

log log: a plot using $\log_{10}$–$\log_{10}$ scales
semilogx: a plot using semilog scales; the x axis is $\log_{10}$, while the
             y axis is linear
semilogy: a plot using semilog scales; the y axis is $\log_{10}$, while the
             x axis is linear

Automatic plotting algorithms. In MATLAB, a plot is automatically scaled. If another plot is requested, the old plot is erased and the axis is automatically rescaled. The automatic plotting algorithms for transient-response curves, root loci, Bode diagrams, Nyquist plots, and the like are designed to work with a wide range of systems, but are not always perfect. Thus, in certain situations, it may become desirable to override the automatic axis scaling feature of the plot command and to select the plotting limits manually.

Manual axis scaling. If it is desired to plot a curve in a region specified by

$$v = [\text{x-min} \quad \text{x-max} \quad \text{y-min} \quad \text{y-max}]$$

enter the command axis(v), where v is a four-element vector. This command sets the axis scaling to the prescribed limits. For logarithmic plots, the elements of v are $\log_{10}$ of the minima and maxima.

Executing axis(v) freezes the current axis scaling for subsequent plots. Entering the plot command and typing axis again resumes automatic scaling.

The command axis('equal') sets the plot region on the screen to be square. With an equal aspect ratio, a line with unity slope is at a true 45°, not skewed by the irregular shape of the screen. The command axis('normal') sets the aspect ratio back to its normal appearance.

Plot types. The command

$$\text{plot(x,y,'o')}$$

draws a point plot using o-mark symbols. Note that the o-marks will not be connected with lines or curves. To connect them with solid lines or curves, plot the data twice by entering the command

$$\text{plot(x,y,x,y,'o')}$$

The command

$$\text{plot(X1,Y1,':',X2,Y2,'+')}$$

uses a dotted line for the first curve and the plus symbol (+) for the second curve. Available line and point types are as follows:

Line types		Point types	
solid	-	point	•
dashed	--	plus	+
dotted	:	star	*
dash-dot	-•	circle	O
		x-mark	×

Available colors on the screen. The statements

$$plot(X,Y,'r')$$
$$plot(X,Y,'+g')$$

indicate the use of a red line on the first graph and green + marks on the second. The available colors are as follows:

red	r
green	g
blue	b
white	w
yellow	y
magenta	m
cyan	c
black	black

(Other colors can be generated using a color map.)

Plotting and printing curves. Let us enter MATLAB Program D–1 into the computer and print the resulting plot.

MATLAB Program D–1

```
>> t = 0 : 0. 01*pi:2*pi;
>> alpha = 3;
>> y = sin(alpha*t);
>> plot(t,y)
>> grid
>> title('Plot of sin(\alphat)    (\alpha = 3)')
>> xlabel('t (sec)')
>> ylabel('sin (\alphat)')
```

Note that the vector t is a partition of the domain $0 \le t \le 2\pi$ with mesh size 0.01π, while y is a vector giving the values of the sine at the nodes of the partition.

Note that, in graphics screen mode, pressing any key will cause MATLAB to show the command screen. By using the up-arrow key, enter any one of the last several commands (plot, grid, title, xlabel, or ylabel). MATLAB will then show the current graphics screen. MATLAB will also show the current graphics screen, if the command shg (show graph) is entered.

Figure D–1 shows a plot of the output $y = \sin(\alpha t)$ versus t, reduced to 50% of the actual print size. The letter size of the title, xlabel, and ylabel may be adequate for the plot as seen on the screen. However, if the plot size is reduced to half or less, the letters become too small.

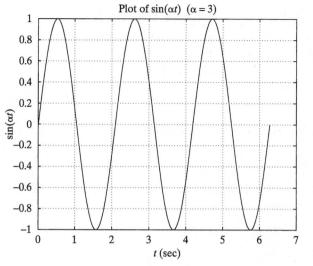

Plot of $\sin(\alpha t)$ $(\alpha = 3)$

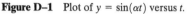

Figure D–1   Plot of $y = \sin(\alpha t)$ versus t.

To enlarge the letter size, specify a larger fontsize in the title, xlabel, and ylabel. (See MATLAB Program D–2.) The size of the plot in Figure D–2 is half that of the actual print from the computer. Notice that in Figure D–2 the letter size of the title, xlabel, and ylabel appears adequate.

MATLAB Program D–2

```
>> t = 0 : 0.01*pi:2*pi;
>> alpha = 3;
>> y = sin(alpha*t);
>> plot(t,y)
>> grid
>> title('Plot of sin(\alphat) (\alpha = 3)', 'Fontsize', 20)
>> xlabel('t (sec)','Fontsize', 20)
>> ylabel('sin(\alphat)','Fontsize', 20)
```

To generate greek letters, such as α, β, γ, δ, ω, or ζ, use '\character'. See below.

α	\alpha	θ	\theta
β	\beta	ζ	\zeta
γ	\gamma	Δ	\Delta
δ	\delta	Θ	\Theta
ω	\omega	Σ	\Sigma
σ	\sigma	Ω	\Omega
			etc.

In addition to setting the font size with commands such as

'Fontsize', 15
'Fontsize', 20

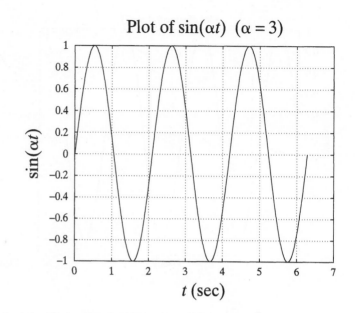

Figure D–2 Plot of sin(αt) versus t.

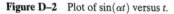

it is possible to set the font angle and font name, using commands such as

'Fontangle', 'italic'
'Fontname', 'Times New Roman'

Subscripts can be obtained by using "_". For example,

y_1 generates y_1
y_2 generates y_2

Superscripts can be generated by using "^". For example,

x^2 generates x^2
x^3 generates x^3

As another example, let us plot the graph of

$$y = x^2$$

over the interval $0 \le x \le 3$ with increments of 0.1. MATLAB Program D–3 plots this graph.

MATLAB Program D–3

```
>> x = 0:0.1:3;
>> y = x.^2;
>> plot(x,y)
>> grid
>> title('Plot of y = x^2','Fontsize',20,'Fontangle','italic')
>> xlabel('x','Fontsize',20,'Fontangle','italic')
>> ylabel('y','Fontsize',20,'Fontangle','italic')
```

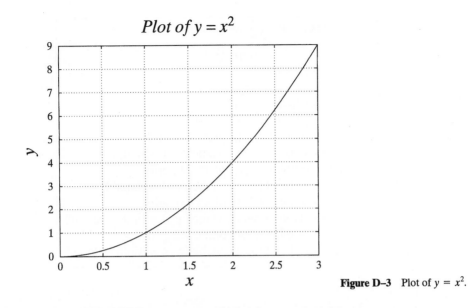

Figure D–3   Plot of $y = x^2$.

Note that it is necessary that '^2' be preceded by a period to ensure that it operates entrywise. Figure D–3 shows the resulting plot.

Use of subplot command.   Multiple curves on one screen may be split into multiple windows with the use of the command

$$\text{subplot}(m,n,p)$$

The graph display is subdivided into m × n smaller subwindows numbered left to right and top to bottom. The integer p specifies the window. For example, subplot(235) or subplot(2,3,5) splits the graph display into six subwindows numbered from left to right and top to bottom. The integer p = 5 means the fifth window (that is, the window located in the second row and second column).

Let us next plot the four curves

$$y_1 = \sin t$$
$$y_2 = \sin 2t$$
$$y_3 = \sin t + \sin 2t$$
$$y_4 = (\sin t)(\sin 2t)$$

for $0 \le t \le 2\pi$. We shall plot the curves in four subwindows, one curve to a window. MATLAB Program D–4 produces the plots shown in Figure D–4.

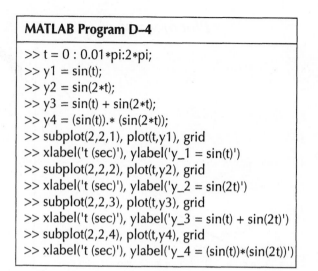

```
MATLAB Program D–4

>> t = 0 : 0.01*pi:2*pi;
>> y1 = sin(t);
>> y2 = sin(2*t);
>> y3 = sin(t) + sin(2*t);
>> y4 = (sin(t)).* (sin(2*t));
>> subplot(2,2,1), plot(t,y1), grid
>> xlabel('t (sec)'), ylabel('y_1 = sin(t)')
>> subplot(2,2,2), plot(t,y2), grid
>> xlabel('t (sec)'), ylabel('y_2 = sin(2t)')
>> subplot(2,2,3), plot(t,y3), grid
>> xlabel('t (sec)'), ylabel('y_3 = sin(t) + sin(2t)')
>> subplot(2,2,4), plot(t,y4), grid
>> xlabel('t (sec)'), ylabel('y_4 = (sin(t))*(sin(2t))')
```

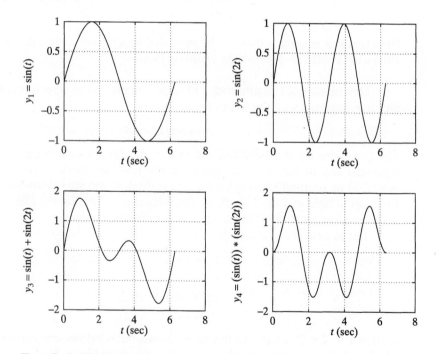

Figure D–4   Plots of $\sin t$, $\sin 2t$, $\sin t + \sin 2t$, and $(\sin t)(\sin 2t)$.

References

B–1 Barnet, S., "Matrices, Polynomials, and Linear Time-Invariant Systems," *IEEE Trans. Automatic Control*, **AC-18** (1973), pp. 1–10.

B–2 Bellman, R., *Introduction to Matrix Analysis*. New York: McGraw-Hill Book Company, 1960.

B–3 Bode, H. W., *Network Analysis and Feedback Design*. New York: Van Nostrand Reinhold, 1945.

C–1 Cannon, R., *Dynamics of Physical Systems*. New York: McGraw-Hill Book Company, 1967.

C–2 Churchill, R. V., *Operational Mathematics*, 3rd ed. New York: McGraw-Hill Book Company, 1972.

C–3 Coddington, E. A., and N. Levinson, *Theory of Ordinary Differential Equations*. New York: McGraw-Hill Book Company, 1955.

D–1 Dorf, R. C., and R. H. Bishop, *Modern Control Systems*, 9th ed. Upper Saddle River, NJ: Prentice Hall, 2001.

E–1 Evans, W. R., "Graphical Analysis of Control Systems," *AIEE Trans. Part II*, **67** (1948), pp. 547–51.

E–2 Evans, W. R., "Control System Synthesis by Root Locus Method," *AIEE Trans. Part II*, **69** (1950), pp. 66–9.

E–3 Evans, W. R., "The Use of Zeros and Poles for Frequency Response or Transient Response," *ASME Trans.*, **76** (1954), pp. 1135–44.

E–4 Evans, W. R., *Control System Dynamics*. New York: McGraw-Hill Book Company, 1954.

F–1 Franklin, G. F., J. D. Powell, and A. Emami-Naeini, *Feedback Control of Dynamic Systems*, 4th ed. Upper Saddle River, NJ: Prentice Hall, 2002.

G–1 Gantmacher, F. R., *Theory of Matrices*, Vols. I and II. New York: Chelsea Publishing Company, Inc., 1959.

G–2 Gardner, M. F., and J. L. Barnes, *Transients in Linear Systems*. New York: John Wiley & Sons, Inc., 1942.

H–1 Halmos, P. R., *Finite Dimensional Vector Spaces*. New York: Van Nostrand Reinhold, 1958.

I–1 Irwin, J. D., *Basic Engineering Circuit Analysis*. New York: Macmillan, Inc., 1984.

K–1 Kuo, B. C., *Automatic Control Systems*, 6th ed. Englewood Cliffs, NJ: Prentice Hall, 1991.

L–1 Levin, W. S., *The Control Handbook*. Boca Raton, FL: CRC Press, 1996.

L–2 Levin, W. S., *Control System Fundamentals*. Boca Raton, FL: CRC Press, 2000.

M–1 MathWorks, Inc., MATLAB, version 6, Natick, MA: MathWorks, Inc., 2000.

N–1 Noble, B., and J. Daniel, *Applied Linear Algebra*, 2nd ed. Englewood Cliffs, NJ: Prentice Hall, 1977.

N–2 Nyquist, H., "Regeneration Theory," *Bell System Tech. J.*, **11** (1932), pp. 126–47.

O–1 Ogata, K., *State Space Analysis of Control Systems*. Englewood Cliffs, NJ: Prentice Hall, 1967.

O–2 Ogata, K., *Solving Control Engineering Problems with MATLAB*. Englewood Cliffs, NJ: Prentice Hall, 1994.

O–3 Ogata, K., *Designing Linear Control Systems with MATLAB*. Englewood Cliffs, NJ: Prentice Hall, 1994.

O–4 Ogata, K., *Modern Control Engineering*, 4th ed. Upper Saddle River, NJ: Prentice Hall, 2002.

P–1 Phillips, C. L., and R. D. Harbor, *Feedback Control Systems*. Englewood Cliffs, NJ: Prentice Hall, 1988.

R–1 Rowell, G., and D. Wormley, *System Dynamics*. Upper Saddle River, NJ: Prentice Hall, 1997.

S–1 Smith, R. J., *Electronics: Circuits and Devices*, 2nd ed. New York: John Wiley & Sons, Inc., 1980.

U–1 Umez-Eronini, E., *System Dynamics and Control*. Pacific Grove, CA: Brooks/Cole Publishing Company, 1999.

W–1 Webster, J. G., *Wiley Encyclopedia of Electrical and Electronics Engineering*, vol. 4. New York: John Wiley & Sons, Inc., 1999.

Z–1 Ziegler, J. G., and N. B. Nichols, "Optimum Settings for Automatic Controllers," *ASME Trans.* **64** (1942), pp. 759–68.

Z–2 Ziegler, J. G., and N. B. Nichols, "Process Lags in Automatic Control Circuits," *ASME Trans.* **65** (1943), pp. 433–44.

Index